T0318744

MYCORRHIZAL MEDIATION OF SOIL

MYCORRHIZAL MEDIATION OF SOIL

FERTILITY, STRUCTURE, AND CARBON STORAGE

Edited By

NANCY COLLINS JOHNSON

CATHERINE GEHRING

JAN JANSA

ELSEVIER

AMSTERDAM • BOSTON • HEIDELBERG • LONDON
NEW YORK • OXFORD • PARIS • SAN DIEGO
SAN FRANCISCO • SINGAPORE • SYDNEY • TOKYO

Elsevier
Radarweg 29, PO Box 211, 1000 AE Amsterdam, Netherlands
The Boulevard, Langford Lane, Kidlington, Oxford OX5 1GB, United Kingdom
50 Hampshire Street, 5th Floor, Cambridge, MA 02139, United States

Library of Congress Cataloging-in-Publication Data
A catalog record for this book is available from the Library of Congress

British Library Cataloguing-in-Publication Data
A catalogue record for this book is available from the British Library

ISBN: 978-0-12-804312-7

Publisher: Cathleen Sether
Acquisition Editor: Candice Janco
Editorial Project Manager: Emily Thomson
Production Project Manager: Mohanapriyan Rajendran
Cover Designer: Victoria Pearson

Typeset by TNQ Books and Journals

Contents

II

MYCORRHIZAL MEDIATION OF SOIL FERTILITY

NANCY COLLINS JOHNSON

11. Integrating Ectomycorrhizas Into Sustainable Management of Temperate Forests

M.D. JONES

12. Mycorrhizal Mediation of Soil Fertility Amidst Nitrogen Eutrophication and Climate Change

M.F. ALLEN AND E.B. ALLEN

III

MYCORRHIZAL MEDIATION OF SOIL STRUCTURE AND SOIL-PLANT WATER RELATIONS

CATHERINE GEHRING

13. Introduction: Mycorrhizas and Soil Structure, Moisture, and Salinity

C.A. GEHRING

14. Mycorrhizas and Soil Aggregation

A. LEHMANN, E.F. LEIFHEIT AND M.C. RILLIG

15. Arbuscular Mycorrhizal Fungi and Soil Salinity

M. MIRANSARI

16. Mycorrhizas, Drought, and Host-Plant Mortality

C.A. GEHRING, R.L. SWATY AND R.J. DECKERT

21. Magnitude, Dynamics, and Control of the Carbon Flow to Mycorrhizas

K.J. FIELD, S.J. DAVIDSON, S.A. ALGHAMDI AND D.D. CAMERON

22. Trading Carbon Between Arbuscular Mycorrhizal Fungi and Their Hyphae-Associated Microbes

B. DRIGO AND S. DONN

23. Immobilization of Carbon in Mycorrhizal Mycelial Biomass and Secretions

R.D. FINLAY AND K.E. CLEMMENSEN

24. Mycorrhizal Interactions With Saprotrophs and Impact on Soil Carbon Storage

E. VERBRUGGEN, R. PENA, C.W. FERNANDEZ AND J.L. SOONG

25. Biochar—Arbuscular Mycorrhiza Interaction in Temperate Soils

R.T. KOIDE

26. Integrating Mycorrhizas Into Global Scale Models: A Journey Toward Relevance in the Earth's Climate System

E.R. BRZOSTEK, K.T. REBEL, K.R. SMITH AND R.P. PHILLIPS

List of Contributors

L.K. Abbott University of Western Australia, Perth, WA, Australia

S.A. Alghamdi University of Sheffield, Sheffield, United Kingdom; King Abudul Aziz University, Jeddah, Kingdom of Saudi Arabia

E.B. Allen University of California, Riverside, CA, United States

M.F. Allen University of California, Riverside, CA, United States

P.M. Antunes Algoma University, Sault Ste. Marie, ON, Canada

P. Bonfante Università degli Studi di Torino, Torino, Italy

E.R. Brzostek West Virginia University, Morgantown, WV, United States

D.D. Cameron University of Sheffield, Sheffield, United Kingdom

K.E. Clemmensen Swedish University of Agricultural Sciences, Uppsala, Sweden

S.J. Davidson University of Sheffield, Sheffield, United Kingdom

R.J. Deckert Northern Arizona University, Flagstaff, AZ, United States

I.A. Dickie Lincoln University, Lincoln, New Zealand

S. Donn Western Sydney University, Penrith, NSW, Australia

B. Drigo Western Sydney University, Penrith, NSW, Australia; University of South Australia, Mawson Lakes, SA, Australia

C.W. Fernandez University of Minnesota, St. Paul, MN, United States

K.J. Field University of Leeds, Leeds, United Kingdom

R.D. Finlay Swedish University of Agricultural Sciences, Uppsala, Sweden

C.A. Gehring Northern Arizona University, Flagstaff, AZ, United States

M. Giovannetti Università degli Studi di Torino, Torino, Italy; Gregor Mendel Institute (GMI), Austrian Academy of Sciences, Vienna Biocenter (VBC), Vienna, Austria

C. Hamel Agriculture and Agri-Food Canada, Quebec, QC, Canada

A. Hodge University of York, York, United Kingdom

J. Jansa Academy of Sciences of the Czech Republic, Prague, Czech Republic

N.C. Johnson Northern Arizona University, Flagstaff, AZ, United States

M.D. Jones University of British Columbia Okanagan Campus, Kelowna, BC, Canada

R.T. Koide Brigham Young University, Provo, UT, United States

A. Koyama Algoma University, Sault Ste. Marie, ON, Canada

T.W. Kuyper Wageningen University, Wageningen, The Netherlands

J.R. Leake The University of Sheffield, Sheffield, United Kingdom

A. Lehmann Freie Universität Berlin, Berlin, Germany

E.F. Leifheit Freie Universität Berlin, Berlin, Germany

R.M. Miller Argonne National Laboratory, Lemont, IL, United States

M. Miransari AbtinBerkeh Scientific Ltd. Company, Isfahan, Iran

R. Pena University of Goettingen, Goettingen, Germany

R.P. Phillips Indiana University, Bloomington, IN, United States

B.J. Pickles University of Reading, Reading, England, United Kingdom

C. Plenchette Consultant in Agronomy and Soil Biology and Microbiology, Quetigny, France

J.I. Querejeta Spanish Research Council (CEBAS-CSIC), Murcia, Spain

D.J. Read The University of Sheffield, Sheffield, United Kingdom

K.T. Rebel University of Utrecht, Utrecht, Netherlands

M.C. Rillig Freie Universität Berlin, Berlin, Germany

A. Salvioli Università degli Studi di Torino, Torino, Italy

S.W. Simard University of British Columbia, Vancouver, BC, Canada

K.R. Smith West Virginia University, Morgantown, WV, United States

M.M. Smits Hasselt University, Diepenbeek, Belgium

J.L. Soong University of Antwerp, Wilrijk, Belgium

R.L. Swaty The Nature Conservancy's LANDFIRE Team, Evanston, IL, United States

F.P. Teste Grupo de Estudios Ambientales, IMASL-CONICET, Universidad Nacional de San Luis, San Luis, Argentina; The University of Western Australia, Crawley (Perth), WA, Australia

K.K. Treseder University of California, Irvine, CA, United States

E. Verbruggen University of Antwerp, Wilrijk, Belgium

V. Volpe Università degli Studi di Torino, Torino, Italy

H. Wallander Lund University, Lund, Sweden

G.W.T. Wilson Oklahoma State University, Stillwater, OK, United States

Preface

Mycorrhizas have long been recognized for their key roles in plant health and production of food and fiber, but increasingly they are recognized for moving matter and energy through ecosystems. These widespread plant–fungus partnerships function at the interface between living roots and soil. In this regard, they integrate the biotic and abiotic components of ecosystems. The purpose of this book is to review and synthesize information about mycorrhizas and provide new insights about their roles in soil fertility, structure, and carbon storage. Current global carbon models vary widely in their predictions of the dynamics of the terrestrial carbon pool, ranging from a large sink to a large source. A better understanding of mycorrhizal mediation of soil could inform earth system models and may improve the accuracy of their predictions.

The idea for this book arose while planning the Eighth International Conference on Mycorrhizas that was hosted in Flagstaff Arizona in August 2015. The goal of the conference was to unite mycorrhizal science across disciplines and scales. A well-timed query from Marisa LaFleur at Elsevier sparked the idea to integrate the knowledge and perspectives of a broad range of mycorrhizal researchers into a single edited volume designed to explore mycorrhizal mediation of soil processes. After the conference, chapter authors were selected based on their expertise in particular areas so that a broad spectrum of knowledge and ideas is represented in the book.

This has truly been a team effort, with the contributions of more than 50 scientists included in the 26 chapters of the book. We express sincere gratitude to the chapter authors for contributing their expertise, time, and hard work toward synthesizing and communicating complex information and ideas. Without these contributions, this book would not be possible. We also wish to thank our publisher Candice Janco and project manager Emily Thompson who worked patiently and tirelessly throughout the process of creating this book. Finally, we thank the many peer reviewers who contributed their expertise by evaluating each of the chapters. The scientific merit of the final product is much improved through this critical and constructive evaluation.

Nancy Collins Johnson, Catherine Gehring, and Jan Jansa
August 2016

COVER PHOTOS

San Francisco Peaks from Flagstaff Arizona, USA, courtesy of Rick Johnson Photography. Micrographs of *Gigaspora gigantea* spore, ectomycorrhizal roots, and *Rhizophagus* sp. hyphal network courtesy of Jan Jansa and Jan Borovička.

LIST OF SCIENTIFIC REVIEWERS

Lyn Abbott	University of Western Australia
Ricardo Aroca	Spanish National Research Council
Heike Bücking	South Dakota State University
Frank Graf	Swiss Federal Institute for Forest Snow and Landscape Research
Tom Horton	State University of New York
Iver Jakobsen	Technical University of Denmark
Christina Kaiser	University of Vienna
Peter Kennedy	University of Minnesota
Roger Koide	Brigham Young University
Robert Koller	Institute of Bio- and Geosciences: Plant Sciences, Forschungszentrum Jülich
Adam Langley	Villanova University
Johannes Lehmann	Cornell University
Björn Lindahl	Swedish University of Agricultural sciences
Lyla Taylor	University of Sheffield
Mark Tibbett	University of Reading
Kathleen Treseder	University of California, Irvine

CHAPTER

1

Mycorrhizas: At the Interface of Biological, Soil, and Earth Sciences

N.C. Johnson[1], J. Jansa[2]

[1]Northern Arizona University, Flagstaff, AZ, United States; [2]Academy of Sciences of the Czech Republic, Prague, Czech Republic

1.1 SUCCESSFUL COEXISTENCE OF PLANTS AND FUNGI

Mycorrhizas are symbioses between plants and fungi. These nutritional partnerships have evolved independently hundreds of times in multiple lineages in the plant and fungal kingdoms (Bidartondo et al., 2011; Tedersoo and Smith, 2013). Estimates suggest that approximately 50,000 species of fungi form mycorrhizal associations with approximately 250,000 species of plants (van der Heijden et al., 2015). Photosynthetic plants require minerals and water to synthesize organic compounds, but these essential resources are often in limited supply. Mycorrhizas greatly expand the capacity of plants to acquire nutrients and water because fungi have several traits that make them particularly well suited for mining minerals from organic and inorganic substrates and improving the hydraulic properties of soil. Fungal mycelium is composed of fine, threadlike hyphae that can access pores too small for roots to enter; consequently by associating with mycorrhizal fungi, plants substantially increase their contact with mineral particles and organic residues in the soil. Also, fungi have the capacity to synthesize organic acids and other compounds that may depolymerize organic compounds and solubilize mineral nutrients (Smith and Read, 2008). From the perspective of biological markets, plants can often increase their fitness by investing photosynthate in fungal partnerships (Werner et al., 2014). The prevalence of mycorrhizas in all types of vegetation throughout the Earth's history suggests that this symbiosis is a mechanism for innovation in a constantly changing environment.

Although mycorrhizas have evolved independently many times, four general types are often recognized based upon the identities of the plant and fungal partners (Table 1.1). Arbuscular mycorrhizas, the most ancient and widespread type of mycorrhiza, form between fungi in the phylum Glomeromycota and the majority of plant species, ranging from nonvascular plants to angiosperms (Parniske, 2008; Brundrett, 2009). Ectomycorrhizas form between thousands of Basidiomycota or Ascomycota and many important trees and

Mycorrhizal Mediation of Soil
http://dx.doi.org/10.1016/B978-0-12-804312-7.00001-2

TABLE 1.1 Four Major Types of Mycorrhizas Distinguished by the Taxa of Host Plants, Fungal Symbionts, and the Biomes Where the Mycorrhizas Are Most Common

Mycorrhizal Type	Host Plants	Main Fungal Symbionts	Predominant Biomes
Arbuscular mycorrhizal (AM)	~200,000 Species of angiosperms, gymnosperms, bryophytes, and pteridophytes	~300–1600 species of Glomeromycota	Tropical and temperate forests, grasslands, savannas, shrublands, deserts, and most agricultural crops including fruit trees
Ectomycorrhizal (EcM)	~6000 Species of angiosperms and gymnosperms	~20,000 species of Basidiomycota and Ascomycota	Boreal (taiga), temperate, and tropical forests; tundra; and agroforestry
Ericoid mycorrhizal (ErM)	Members of the Ericaceae, Epacridaceae, and Empetraceae families, and some bryophytes	>150 species of Ascomycota (primarily) and some Basidiomycota	Tundra, boreal, and temperate forests
Orchid mycorrhizal	All Orchidaceae	~25,000 species of Basidiomycota	Tropical and temperate biomes

Information derived from Smith, S.E., Read, D.J., 2008. Mycorrhizal Symbiosis. Academic Press, New York and van der Heijden, M.G.A., Martin, F.M., Selosse, M.-A., Sanders, I.R., 2015. Mycorrhizal ecology and evolution: the past, the present, and the future. New Phytologist 205, 1406–1423.

woody species that are often stand dominants in temperate and boreal forests (Tedersoo et al., 2010; Tedersoo and Smith, 2013). Ericoid mycorrhizas are associations between highly specialized Ascomycota (and a few Basidomycota) and several large families within the order Ericales and also bryophytes that occur in vast areas of tundra, boreal forests, and heathlands (Straker, 1996; van der Heijden et al., 2015). Orchid mycorrhizas are associations between Basidiomycota and all members of the Orchidaceae, a remarkably diverse family encompassing 9% of all angiosperm species (Brundrett, 2009). Orchid mycorrhizas are particularly interesting because orchids have a prolonged seedling stage, during which they are unable to photosynthesize and are entirely dependent upon mycorrhizal symbioses to provide an exogenous supply of carbohydrate (Smith and Read, 2008). In this regard, orchid mycorrhizas represent an unconventional carbon dynamic in which a heterotrophic fungus provides carbohydrate to its plant host. Although orchid mycorrhizas are fascinating, only Chapter 21 will consider orchid mycorrhizas. The other chapters will mainly focus on the arbuscular mycorrhizal (AM), ectomycorrhizal (EcM), and ericoid mycorrhizal (ErM) symbioses because of their key roles as mediators of soil fertility, structure, and carbon storage.

1.2 MYCORRHIZAL RESEARCH: PAST, PRESENT, AND FUTURE

It has long been recognized that fungi inhabit plant roots. Frank coined the term *mycorrhiza* in 1885; but, many published descriptions of fungal associations with plant roots predate this classic paper (Frank, 1885; Trappe and Berch, 1985). In a subsequent paper, Frank suggested that EcM fungi aided nitrogen uptake from humus and thus increased the growth of host plants (Frank, 1894). This insight initiated a period of experimentation on the nutritional

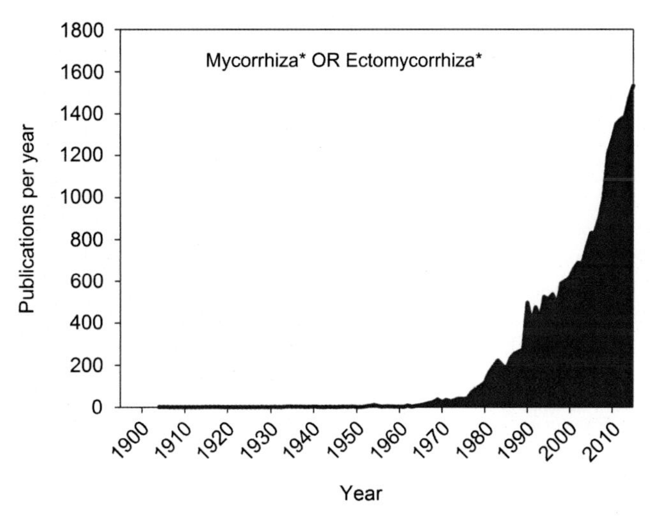

FIGURE 1.1 Web of Science analysis of the number of publications per year with: "mycorrhiza* OR ectomycorrhiza*" as the sole topic search terms. This analysis was performed on July 26, 2016.

role of EcM symbioses (Harley, 1985). Although AM fungi were described inside plant roots several decades before Frank's publication, the identity of the fungal partner remained elusive because Glomeromycota are obligate biotrophs that cannot be cultured using traditional mycological methods (Koide and Mosse, 2004). Consequently most of the early studies of mycorrhizas focused on orchid, EcM, and ErM symbioses. The first serious work on the importance of AM symbioses for plant nutrition was published much later by Mosse (1957).

Frank recognized the nutritional benefits of mycorrhizas for plants more than 130 years ago, but recognition of their link to soil fertility, structure, and carbon is more recent. A Web of Science analysis of published works since 1900 shows that the number of publications on mycorrhizas remained relatively low until the late 1960s and then steadily increased, with more than 1500 publications on mycorrhizas in 2015 alone (Fig. 1.1). Stunning as this might seem, the growth of total publications about mycorrhizas has closely tracked the exponential growth of all scientific publications, which have been increasing at a rate of approximately 3% annually (Bornmann and Mutz, 2015). More advanced analyses show that the first publications related to both mycorrhizas and fertility occurred in 1976 (Fig. 1.2A), and the first publications related to mycorrhizas and soil structure or soil carbon occurred in the mid-1980s (Fig. 1.2B and C). It is notable that the growth of these three groups of publications is considerably higher (9%, 15%, and 17% for fertility, structure, and carbon, respectively) than the background growth of scientific publications as a whole. This coarse analysis of literature suggests that there is a growing interest in mycorrhizal mediation of soil properties and processes, and mycorrhizas are becoming integrated into the interdisciplinary fields of soil science and earth science.

During the past century, mycorrhizal literature has progressed from descriptions of botanical curiosities to reports that mycorrhizas may be primary controllers of ecosystem responses and feedbacks to climate change (Terrer et al., 2016). The burgeoning literature about mycorrhizas challenges even the most conscientious students and researchers. This book addresses the need for a review and synthesis of recent scientific discoveries and ideas about mycorrhizas.

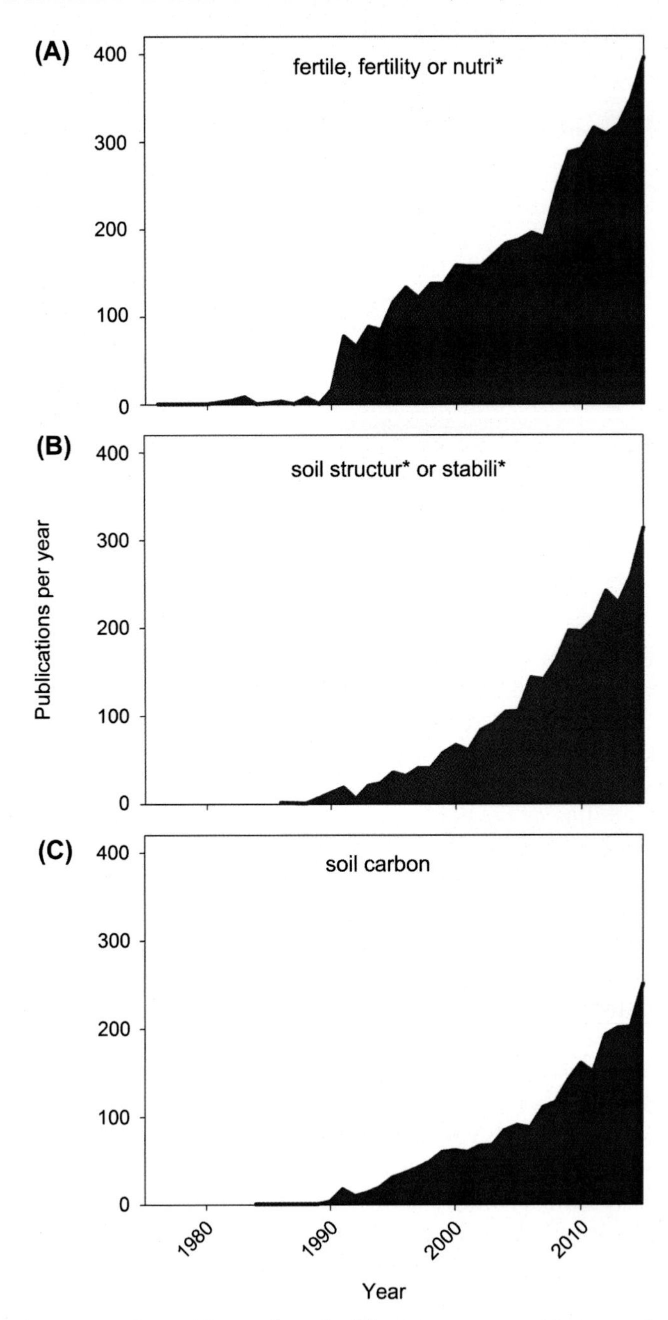

FIGURE 1.2 Web of Science analysis of the number of publications per year with the search terms: (A) mycorrhiza* OR ectomycorrhiza* AND fertile OR fertility OR nutri*; (B) mycorrhiza* OR ectomycorrhiza* AND soil OR soils AND structur* or stabili*, (C) mycorrhiza* OR ectomycorrhiza* AND soil OR soils AND carbon. This analysis was performed on July 26, 2016. *Thanks to Joseph Sweet and Mary DeJong for their assistance.*

1.3 GOALS AND OBJECTIVES

The purpose of this book is to explore the ways that mycorrhizas facilitate the formation and functioning of soils. Mycorrhizas integrate the biosphere with the atmosphere, lithosphere, and even the hydrosphere (Chapter 2). This book examines mycorrhizas at the interface of biological, soil, and earth sciences. The perspectives of more than 50 experts, including some of the early pioneers in mycorrhizal research, are complied in this volume. Each chapter summarizes the state of the art, and often articulates more questions than answers, largely because relatively few studies have examined mycorrhizas from a soil science or earth science perspective. This perspective is unique and complementary to excellent books that are available on mycorrhizal biology (Smith and Read, 2008), ecology (Allen, 1991; van der Heijden and Sanders, 2002; Horton, 2015), and management for sustainable agriculture(Solaiman et al., 2014).

The book is organized in four sections that explore mycorrhizal mediation of: (I) soil development, (II) fertility, (III) soil structure and plant–water relations, and (IV) carbon flux and soil carbon storage. Section I examines the role of mycorrhizas in soil-forming processes across a range of temporal and spatial scales. In Chapter 2, Leake and Read describe the tandem evolution of plants, mycorrhizal fungi, and soil, and provide evidence that through increased weathering the early mycorrhizal symbioses influenced climate as well as soil development. In Chapter 3, Smits and Wallander delve deeper into the mechanism of mycorrhizal weathering of parent material in modern-day ecosystems. Then in Chapter 4, Johnson, Miller, and Wilson examine how climate, parent material, and topography influence the formation of mycorrhizas and in turn how mycorrhizas influence these three physicochemical soil-forming factors. Finally, Teste and Dickie consider temporal changes in mycorrhizas during the course of ecosystem development in Chapter 5. The goals of Sections II, III, and IV are reviewed in Chapters 6, 13, and 19, respectively.

References

Allen, M.F., 1991. The Ecology of Mycorrhizae. Cambridge University Press, New York.

Bidartondo, M.I., Read, D.J., Trappe, J.M., Merckx, V., Ligrone, R., Duckett, G., 2011. The dawn of symbiosis between plants and fungi. Biology Letters 7, 574–577.

Bornmann, L., Mutz, R., 2015. Growth rates of modern science: a bibliometric analysis based on the number of publications and cited references. Journal of the Association for Information Science and Technology 66, 2215–2222.

Brundrett, M.C., 2009. Mycorrhizal associations and other means of nutrition of vascular plants: understanding the global diversity of host plants by resolving conflicting information and developing reliable means of diagnosis. Plant and Soil 320, 37–77.

Frank, A.B., 1885. Uber die auf wurzelsymbiose beruhende ernahrung gewisser Baume durch unterirdische pilze (on the nutritional dependence of certain trees on root symbiosis with belowground fungi). Berichte der Deutschen Botanischen Gesellschaft 3, 128–145.

Frank, A.B., 1894. Die bedeutung der mykorrhiza-pilze fur die gemeine Kiefer. Forstwissenschaftliches Centralblatt 16, 183–190.

Harley, J.L., 1985. Mycorhiza: the first 65 years, from the time of Frank till 1950. In: Molina, R. (Ed.), 6th North American Conference on Mycorrhizae. Forest Research Laboratory, Bend, Oregon, pp. 26–33.

Horton, T.R. (Ed.), 2015. Mycorrhizal Networks. Ecological Studies, vol. 244. Springer.

Koide, R.T., Mosse, B., 2004. A history of research on arbuscular mycorrhiza. Mycorrhiza 14, 145–163.

Mosse, B., 1957. Growth and chemical composition of mycorrhizal and non-mycorrhizal apples. Nature 179, 923–924.

Parniske, M., 2008. Arbuscular mycorrhiza: the mother of plant root endosymbioses. Nature Reviews Microbiology 6, 763–775.
Smith, S.E., Read, D.J., 2008. Mycorrhizal Symbiosis. Academic Press, New York.
Solaiman, Z., Abbott, L.K., Ajit, V. (Eds.), 2014. Mycorrhizal Fungi: Use in Sustainable Agriculture and Land Restoration. Springer.
Straker, C.J., 1996. Ericoid mycorrhiza: ecological and host specificity. Mycorrhiza 6, 215–225.
Tedersoo, L., May, T.W., Smilth, M.E., 2010. Ectomycorrhizal lifestyle in fungi: global diversity, distribution, and evolution of phylogenetic lineages. Mycorrhia 20, 217–263.
Tedersoo, L., Smith, M.E., 2013. Lineages of ectomycorrhizal fungi revisited: foraging strategies and novel lineages revealed by sequences from belowground. Fungal Biology Reviews 27, 83–99.
Terrer, C., Vicca, S., Hungate, B.A., Phillips, R.P., Prentice, I.C., 2016. Mycorrhizal association as a primary control of the CO_2 fertilization effect. Science 353, 72–74.
Trappe, J.M., Berch, S.M., 1985. The prehistory of mycorrhizae: A.B. Frank's predecessors. In: Molina, R. (Ed.), 6th North American Conference on Mycorrhizae. Forest Research Laboratory, Bend, Oregon, pp. 2–11.
van der Heijden, M.G.A., Martin, F.M., Selosse, M.-A., Sanders, I.R., 2015. Mycorrhizal ecology and evolution: the past, the present, and the future. New Phytologist 205, 1406–1423.
van der Heijden, M.G.A., Sanders, I.R. (Eds.), 2002. Mycorhizal Ecology. Springer, Berlin, Heidelberg, New York.
Werner, G.D.A., Strassmann, J.E., Ivens, A.B.F., Engelmoer, D.J.P., Verbruggen, E., Queller, D.C., Noe, R., Johnson, N.C., Hammerstein, P., Kiers, E.T., 2014. The evolution of microbial markets. Proceedings National Academy of Science United States of America 16, 1237–1244.

SECTION I

MYCORRHIZAL MEDIATION OF SOIL DEVELOPMENT

CHAPTER

2

Mycorrhizal Symbioses and Pedogenesis Throughout Earth's History

J.R. Leake, D.J. Read

The University of Sheffield, Sheffield, United Kingdom

2.1 THE IMPORTANCE OF RECIPROCAL EFFECTS OF PLANT–MYCORRHIZA–SOIL INTERACTIONS IN THE EVOLUTION AND ASSEMBLY OF TERRESTRIAL ECOSYSTEMS

Soil provides the foundation of terrestrial biomes, being both the product of biosphere–geosphere–atmosphere interactions, and a pivotal factor in controlling the assembly of terrestrial ecosystems and their roles in carbon (C) and nutrient cycles (Jenny, 1980; Bardgett, 2005). The biological, chemical, and physical properties of soils and the functions they provide underpin or directly deliver most of the ecosystem services upon which humans depend, including food, fiber, biofuels, fresh water, and regulating services such as hydrological functioning affecting flooding incidence, greenhouse gas emissions, and climate (MEA, 2005). The importance of the biotic component in soil development was made explicit in Jenny's model (1941, 1980), which defined the five factors that drive pedogenesis. This model sees soil (S) as a function (f) of climate (c), organisms (or), relief (topography) (r), parent material (p) and time (t), together with stochastic factors (…) such as fires and floods, so that $S = f(cl, or, r, p, t, \ldots)$. In this model, organisms are recognized both as drivers of soil development, and responsive to the environmental context in which they occur and the soils they have helped to create, so that $Or = f(s, cl, r, p, t, \ldots)$. Jenny (1941) emphasized that soil is an open system in which substances may be added or removed, and that soil and environment form coupled systems that can be changed when the functions that have created them change. Together these insights provide a conceptual model of pedogenesis with far-reaching consequences both for considering the effects of terrestrial organisms and their evolution on processes of soil development, and

Mycorrhizal Mediation of Soil
http://dx.doi.org/10.1016/B978-0-12-804312-7.00002-4

their long-term effects and reciprocal interactions with the wider environment. Soil plays an especially important role in global C cycling, storing about 75% of the organic C stock in contemporary terrestrial ecosystems (Hiederer and Köchy, 2011; Scharlemann et al., 2014). This is approximately three times the preindustrial atmospheric carbon dioxide (CO_2) C stock of the late Holocene epoch. This reflects the accumulation of plant photosynthate–C stabilized in soil organic matter pools through plant–microbial interactions.

Whereas the importance of plants as agents of soil formation were well established after Jenny (1941), the potential role of coevolutionary advances in the almost universal symbiosis between plants and mycorrhizal fungi in impacting soil forming processes and global biogeochemical cycles over the past 400-500 Ma, has not been considered in depth previously. Nonetheless, reciprocal effects of plants on soil development and of soil properties on plants in different climate zones underpins our understanding of the present day global distribution of soil types, biomes, dominant types of mycorrhizal plant communities (Read, 1991; Read et al., 2004; Soudzilovskaia et al., 2015), and their specific effects on soil C and nitrogen (N) cycling processes (Clemmensen et al., 2013; Augusto et al., 2015). Contemporary global patterns of plant-root colonization intensity by arbuscular mycorrhizal (AM) fungi, and the independently and more recently evolved ectomycorrhizal (EcM) fungi, are similarly explained by their climate and soil chemistry type preferences (Soudzilovskaia et al., 2015). However, only recently have the potential long-term reciprocal effects of evolutionary advances in land plants and mycorrhizal symbionts on biogeochemical cycles, the atmosphere, and climate been recognized (Taylor et al., 2009), and the role of evolution in processes of pedogenesis started to be investigated (Morris et al., 2015). In turn, changes caused by advances in plants and their symbionts will have fed back to influence both the evolution of terrestrial ecosystems and their roles in pedogenesis and biogeochemical cycles.

Two important questions arise from reciprocal plant–soil interactions involving mycorrhizas. First, how has the coevolution of plants and mycorrhizal fungi changed the rates and processes of soil formation and altered global biogeochemical cycles through the period of Earth's history during which land plants and their symbionts have evolved and diversified? Second, how have changes in the Earth's system and climates over the history of land–plant evolution, in part caused by effects of pedogenesis on global biogeochemistry, influenced the trajectory of evolutionary advances in plant and mycorrhiza functional traits?

To address these questions, we integrate advances in palynology, paleontology, paleopedology, and phylogenetics, with experimental studies of plant and mycorrhizal fungal below-ground C allocation and mineral weathering under past and present atmospheric CO_2 concentrations, in taxa selected across key nodes of evolutionary advancement. This synthesis seeks to better understand the evolutionary history of mycorrhizas and how their role in pedogenesis has changed with evolutionary advancement in both plant and fungal partners. We consider how these organisms and their interactions with soils have bioengineered the planet through feedbacks on biogeochemical cycles of calcium (Ca), phosphorus (P), C, and silicon (Si). These regulate Earth's atmospheric CO_2 concentration and climate over geological timescales via the geochemical C cycle, in turn affecting rates of terrestrial and marine productivity (Berner, 2006; Algeo and Scheckler, 1998; Lenton, 2001; Taylor et al., 2009).

2.2 PLANTS AND MYCORRHIZAS AS AGENTS OF PEDOGENESIS: COUPLING PLANT PHOTOSYNTHATE ENERGY TO THE ACTIONS OF FUNGAL MYCELIAL NETWORKS

Central to the role of plants and associated mycorrhizal fungi in pedogenesis is their transformative effects on soil chemistry and physical properties through their inputs of organic matter, selective uptake of elements and water, and capacities to directly and indirectly accelerate chemical and physical weathering processes that liberate essential nutrient elements from rocks and minerals (Haselwandter, 2008; Leake et al., 2008; Taylor et al., 2009; Brantley et al., 2011; Chapters 3 and 4). Mycorrhizas can affect these actions directly and indirectly through their effects in improving plant nutrition, biomass production, and plant health. Their effects are intrinsically scale-dependent in that the evolution of larger plants with deeper roots (Algeo and Scheckler, 1998; Morris et al., 2015) and more extensive mycorrhizal hyphal networks (Quirk et al., 2012) tend to have larger effects on mineral weathering, in assisting the formation of pedogenic minerals such as clays, in accumulating soil organic matter, and in developing soil structure such as soil aggregates and pores. In turn, these factors control the core soil functions including C, water, and nutrient storage capacity and some of the major pathways by which particles, fluids, and gases pass into and out of soil. These soil functions will have fed back to influence the assembly of terrestrial ecosystems as plants and mycorrhizal fungi have evolved, and selected for traits that increase their fitness through the mutualistic interactions between the plant and fungal partners.

The driving force behind the effects of the rise of terrestrial ecosystems on soil and biogeochemistry is conversion of atmospheric CO_2 into organic C by photosynthesis, empowering biomass accumulation, proton-pumping, and secretion of organic acids and enzymes such as reductases that change the solubility of elements and together enable the selective uptake of essential nutrients from soils via mycorrhizal fungi and roots (Taylor et al., 2009). About 85% of extant vascular plant species form mycorrhizal associations (Brundrett, 2009), typically investing between 5% and 15% of their photosynthate in supporting mycorrhizal fungal partners to facilitate nutrient and water uptake from soil. The "C energy hypothesis" links below-ground allocation of photosynthate to mineral weathering and element mass-transfers by mycorrhizas (Leake et al., 2008; Quirk et al., 2014). This underpins C-for-nutrient exchanges between autotrophic plants and mutualistic soil fungi, and defines the bioenergetics of mycorrhizal mycelial "networks of power and influence" that are typically two to three orders of magnitude longer than roots (Leake et al., 2004). Globally, we estimate mycorrhizal hyphae in the top 10 cm of soils extend over distances of approximately 4.5×10^{17} km, which is half the width of our galaxy (9.5×10^{17} km), and the surface area of these hyphae, assuming a mean diameter of only 4 μm, is nearly 2.5 times the area of the continental land masses[1]. Most of this mycelium is relatively short-lived and replaced several times per year.

[1] This estimate is based on multiplying the typical lengths of mycorrhizal mycelium in different ecosystems dominated by plants associated with different mycorrhizal types and different land use types reported by Leake et al. (2004) by the areal extent of different major biomes and land uses in which these plants are representatives, assuming a soil depth of only 10 cm, because most hyphal length data are only for topsoil, and hyphal lengths tend to decrease with depth, in parallel with root densities.

This vast mycorrhizal hyphosphere interacts with soil minerals, organic matter, and other soil microorganisms in ways that inevitably affect the chemical, physical, and biological properties of soils. C flow through mycorrhizal hyphae into the mycorrhizosphere supports other microorganisms, some of which operate synergistically in enhancing nutrient uptake (Chapters 9 and 22). AM fungal mycelium supplies C that supports phosphate-solubilizing and plant-growth promoting rhizobacteria such as *Pseudomonas* and *Burkholderia* that ultimately impact soil processes and soil functions, such as mineralization of organic P (Zhang et al., 2014, 2016), and weathering of mineral P, such as the Ca phosphate apatite via local acidification (Taktek et al., 2015). The effects of mycorrhizal plants on pedogenesis were therefore conceptualized by Brantley et al. (2011) as "solar to-chemical conversion of energy by plants regulates flows of carbon, water, and nutrients through plant-microbe soil networks, thereby controlling the location and extent of biological weathering." This is consistent with the suggestion that photosynthate allocation affects landscape evolution, the potential energy of present day net primary production being three to seven orders of magnitude greater than the kinetic energy generated by tectonic uplift and the exogenic forces of denudation that have traditionally been considered to control changes in topography over time (Phillips, 2009). If only a small percentage of plant photosynthate allocated to roots and mycorrhizas is directly involved in processes of geomorphological significance, such as physical or chemical weathering of rocks and minerals, the effects of mycorrhizas on soils and landscapes would be extremely important over geological time.

Key to these long-term pedogenic and global biogeochemical effects of the coevolution of plants and mycorrhizal fungi is their enhancement of weathering of continental silicate rocks. Release of Ca and magnesium (Mg) into the oceans from weathering of terrestrial silicate rocks such as basalt ultimately removes CO_2 from the atmosphere for millions of years in the geochemical C cycle (Fig. 2.1), which precipitates marine carbonates such as limestone and chalk until these rocks are either uplifted onto land or subducted and CO_2 is released by volcanic degassing (Berner, 2006). In addition, the dissolution of continental silicates and the fluvial export of dissolved and particulate Si into the oceans supports the productivity and C sequestration by marine organisms that build Si skeletons, such as sponges that date back to at least 600 Ma (Yin et al., 2015), Radiolaria that first appeared over 540 Ma (Braun et al., 2007) ago, and diatoms that first evolved no earlier than 250 Ma ago (Medlin, 2011).

According to Conley and Carey (2015), the rise and expansion of grassy biomes that now dominate the terrestrial Si cycle led to increased rates of dissolved Si transport to the oceans, which in turn supported the coincident rise, over the past 65 Ma, of the diatoms that contribute about 50% of marine net C fixation. Intriguingly, AM fungi may themselves be involved in the global Si cycle because strong linear relationships have been found between P and Si contents of AM fungal vesicles (Olssen et al., 2011). Although it remains to be proven whether these fungi are directly involved in Si uptake by plants, evidence is emerging of their involvement in processes of silicate weathering. Dissolution and trenching of primary silicate minerals by AM fungi growing in association with liverworts and trees has been demonstrated by Quirk et al. (2014, 2015). These observations suggest roles for plant–mycorrhiza associations in the global biogeochemical cycling of Si, as well as of P and Ca. These activities would be expected to have influenced terrestrial and marine productivity, as well as atmospheric CO_2 concentrations, with effects that are likely to have become increasingly important from as early as the Ordovician period when liverwort-like plants first evolved and likely developed symbiotic associations with soil fungi (Section 2.3).

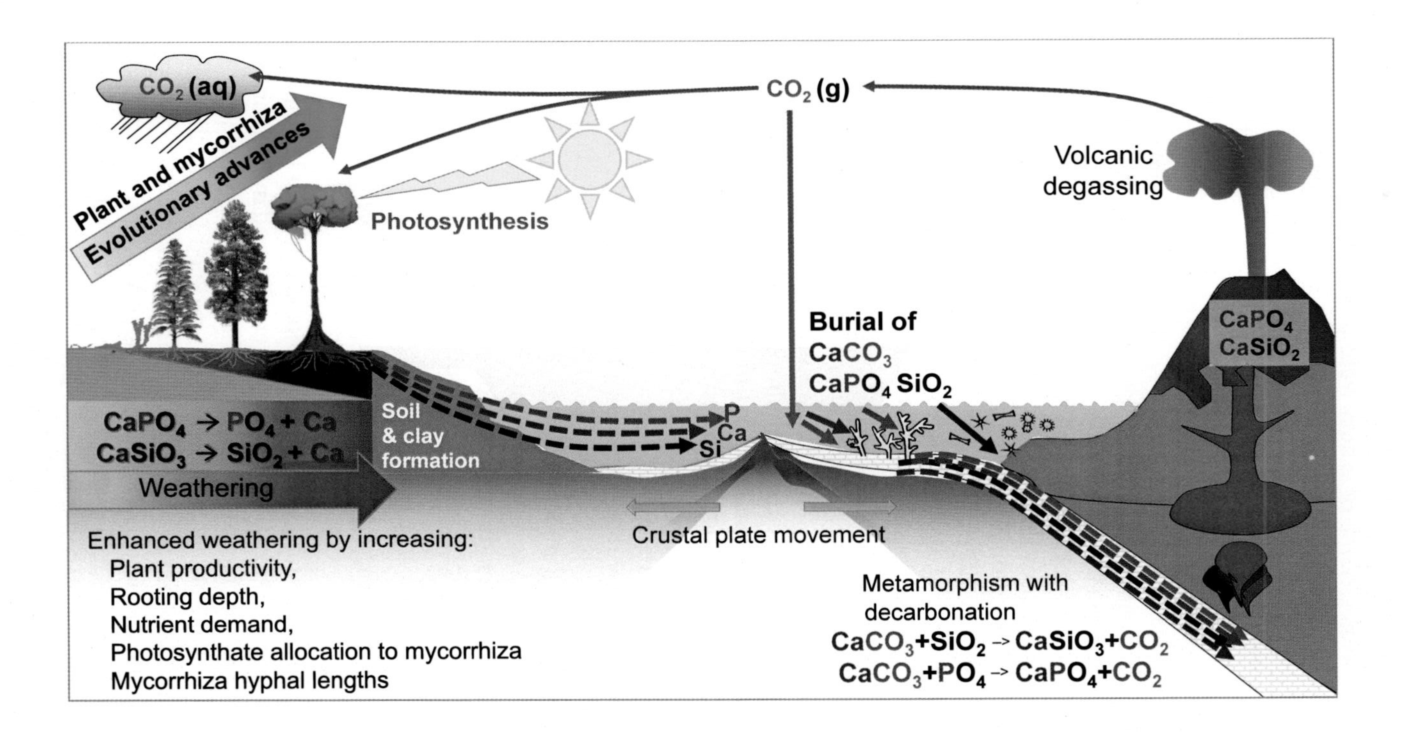

FIGURE 2.1 The effects of evolutionary advancement in plants and mycorrhizas in the geochemical carbon cycle, increasing the weathering of calcium (Ca)-, phosphorus (P)-, and silicon (Si)-bearing minerals and generating clays. Plants and their mycorrhizal fungi have increased the rates of dissolution of continental silicates especially calcium silicate ($CaSiO_3$), and apatite (Ca phosphate-$CaPO_4$), but a portion of the Ca, P, and Si released from rocks is washed into the oceans where these elements increase productivity. Some of the Ca and P end up in limestone and chalk deposits produced by marine organisms such as corals and foraminifera, thereby sequestering carbon dioxide (CO_2) that was dissolved in the oceans into calcium carbonate ($CaCO_3$) rock for millions of years. Dissolved Si is used in sponges, radiolarians, and diatoms that can accumulate on the sea floor. The ocean sediments are recycled by subduction or uplift by tectonic forces, with volcanic degassing and eruptions of base-rich igneous rocks such as basalt returning Ca, P, Si, and other elements back to the continents, thereby reinvigorating ecosystems with new nutrient supplies through weathering. Note for simplicity that magnesium is not shown in the figure, but follows parallel pathways to Ca and is co-involved in sequestering CO_2 into dolomitic limestones. *aq*, Liquid state; *g*, gaseous state.

In considering the broader effects of plant evolution on biogeochemical cycling of silica and ocean productivity, Trembath-Reichert et al. (2015) indicate that many extant members of early diverging land plant lineages, some of which were dominant treelike components of Devonian and Carboniferous terrestrial floras, are involved in amorphous silica (SiO_2) biomineralization and cycling. These include lycophytes and monilophytes in the Marratiales, Equisetales, and Osmundales, some, but not all, of which may have formed mycorrhizal partnerships with AM fungi.

Plants in partnership with mycorrhizal fungi contribute to a number of other processes affecting weathering and pedogenesis (Taylor et al., 2009). Roots and mycorrhizal hyphae directly interact with soil minerals through exudation of protons associated with cation uptake and low–molecular-weight organic chelators, both of which facilitate mineral dissolution, ion exchange, and element leaching through soil profiles. Roots and mycorrhizal fungal exudates provide C sources that support specialist mycorrhizosphere bacteria such as species of *Pseudomonas* involved in mineral dissolution (Zhang et al., 2014, 2016) and plant growth promotion (Chapter 22). The anchoring and binding of particles by roots and mycorrhizal hyphae stabilizes soil, thereby intensifying chemical alteration by preventing erosion and reburial in less chemically and biologically reactive environments. Mycorrhizal fungi in the Glomeromycota are also strongly implicated in contributing directly to soil organic matter (Verbruggen et al., 2016) through hyphal enmeshment, generating water-stable macroaggregates that store organic C (Wilson et al., 2009), improve soil drainage, and reduce surface run-off and erosion. Studies that have attributed C stabilization to an AM fungal hydrophobic protein glomalin, which is a hologue of heat shock protein 60 and mainly bound to the hyphae (Bendini et al., 2009; Hammer and Rillig, 2011), remain contentious because resolving the relative importance of hyphae versus their products on soil properties is difficult. In parallel, EcM fungi, through selective uptake of the most labile forms of organic N and P enhance the recalcitrance of soil organic matter (Clemmensen et al., 2013) and secrete siderophores and organic acids that facilitate mineral dissolution (Haselwandter, 2008). Inputs of EcM fungal necromass that is often rich in polyphenolic acids such as tannins and melanin may also be important in C sequestration. In addition, the extent to which mycorrhizas increase plant growth, productivity, biomass, and resultant nutrient and hydrological fluxes through evapotranspiration will contribute incrementally to the overall effect of plants on pedogenesis (Taylor et al., 2009).

2.3 EVOLUTIONARY ORIGINS OF PLANTS AND MYCORRHIZAS

Important developments in our understanding of the evolution of early land plants and fungal symbionts in soil have come from the fossil record, phylogenetics, and molecular genetics, giving increasingly strong evidence of coevolution of plant–fungal symbioses in the Middle Ordovician era, 470 Ma. The first land plants (embryophytes) evolved from a charophycean algal ancestor that was preadapted with signaling pathways involved in initiating symbiosis with fungi, including strigolactones (Delaux et al., 2015). Phylogenetic and molecular genetic information supports the view that bryophytes were the earliest land plants, liverworts being the basal group, with hornworts probably being the sister group to vascular plants (Qiu et al., 2006). Extant basal liverworts ubiquitously host symbiotic fungi (Bidartondo et al., 2011) in mutualistic mycorrhiza-like partnerships in which plant photosynthate is exchanged for fungal-acquired soil nutrients including P and N (Field et al., 2014, 2015). The earliest fossil evidence of land plants comes from spores, and the nature of early terrestrial plant community components, including cyanobacteria (Strother and Wellman, 2016), putative lichens, cryptophytes, and basal tracheophytes, have been reviewed (Steemans et al., 2010; Edwards et al., 2014, 2015; Gerrienne et al., 2016). The earliest putative fossil of the obligately biotrophic and asexual Glomeromycota fungi is provided by spores that date to the same epoch 460-455 Ma (Redecker et al., 2000). In slightly more recent rocks at 450-440 Ma, trilete spores that are associated with vascular land plants have been found (Steemans et al., 2009). However, the earliest fossil of a liverwort thallus dates only to the Early Devonian age (411-407 Ma) (Guo et al., 2012) with a second, better preserved and different species found in the Middle Devonian era, which shows both a few rhizoids and some fungi, but mutualistic associations with fungi have not been reported (Hernick et al., 2008).

The first vascular plants (tracheophytes) date to about 430 Ma in the Silurian period with the appearance of the diminutive sporophytes of *Cooksonia*, which were only a few centimeters tall. The basal portions of these plants, which may be thalloidlike gametophytes (Gerrienne et al., 2006), have been found in a more recent deposit dated to approximately 415 Ma, and indicate very shallow depth of interaction with soil, but the preservation has been too poor to resolve whether they formed symbioses with fungi. By the late Silurian age, more advanced early trachaeophytes, *Rhynia* and *Zosterophyllum*, which had better developed vascular tissues, lacked roots but had prostrate rhizomes with rhizoids, and lacked leaves but had stomata and much more robust shoots than *Cooksonia*, appeared (Edwards et al., 2015). In the same community, one of the earliest lycophytes, *Drepanophycus*, was found. Like its sister genus *Baragwanathia*, which first appeared about 425 Ma (Rickards, 2000), *Drepanophycus* has both microphylls and adventitious rootlike appendages arising directly from the substantial aerial shoots that attained lengths of up to over 1 m (Edwards et al., 2015). Despite these anatomical advances, the pre-Devonian mycorrhizal status of all these plants remains unknown.

The earliest unequivocal fossil evidence for mycorrhiza-like symbiotic fungal associations with land plants (Remy et al., 1994) comes from the early Devonian Rhynie Chert (411.5 Ma), at a time when, with the exception of the rootlike appendages of the sporophytes of lycophytes, plants had not evolved roots (Strullu-Derrien et al., 2015). The exceptionally well-preserved early land-plant flora at this site has enabled detailed studies of plant–fungal interactions at the cellular scale, including studies of fungal vesicles and highly branched fungal structures in cells that appear to be arbuscules, which in modern plants provide the major interfaces for nutrient-for-C exchange in the symbiosis, together with spores that are definitively members of the Glomeromycota (Remy et al., 1994; Dotzler et al., 2006, 2009; Strullu-Derrien et al., 2014). Spores of the derived Glomeromycota genera *Acaulospora* and *Scutellospora* in the Rhynie Chert reveal remarkably little change to these fungal structures over more than 400 million years (Dotzler et al., 2006, 2009) and establish that major diversification of AM fungi had occurred by the early Devonian age. This corroborates that these plant–fungal symbioses coevolved with earlier land plants through the epochs from which we have no fossil record of plant–fungal interactions.

The mycorrhizal fungal partnerships in the Lycopodiophyta are especially critical in understanding the evolutionary significance of these symbioses in land-plant evolution and Earth's history. This is the earliest group of vascular plants for which we have records of root-like structures, dating back to the Silurian age (Kenrick and Strullu-Derrien, 2014), together with substantial shoots with the earliest microphyll leaves. The earliest fossil record of fungal symbiosis with lycopods comes from rhizomes supporting roots of *Asteroxylon mackiei* in the Rhynie Chert, but the aseptate fungal partner found in the prostrate rhizomes is enigmatic, chytrid-like, possibly zoosporic, and appears to be neither of Mucoromycotina nor Glomeromycota affinities (Strullu-Derrien et al., 2015). The functional role of the fungal partner is unclear but on the basis of the available evidence cannot be assumed to be a mutualistic symbiont. The absence of evidence of AM symbionts in *Asteroxylon* is notable, and surprising, in view of the host plant typically being of substantial biomass and nutrient demand, having shoots up to 12 mm wide and up to at least 45 cm long in an environment in which co-occurring early vascular plants like *Horneophyton* were forming mutualistic associations with both groups of endosymbiotic fungi: *Aglaophyton major* associated with AM fungi and contained arbuscules, and *Nothia aphylla* hosted fungal partners (Krings et al., 2007) that have been interpreted as likely Mucoromycotina (Selosse and Strullu-Derrien, 2015). From the late Silurian age, lycopsids advanced in biomass, rooting depth, and structural complexity, and in the early Devonian period, 411 Ma, *Drepanophycus* had developed rooting axes with vascular tissues that extended decimeters down into soil, and from which finer bifurcating rootlets of 0.4–0.7 mm wide and 5–6 mm long emerged laterally (Matsunaga and Tomescu, 2016). The rootlets either lacked root hairs or these have not been preserved. By the Middle Devonian era (398-385 Ma), lycophytes had already achieved trunks more than 4 m long and 11 cm in diameter (Stein et al., 2012). They were minor components of the earliest forest ecosystems that were dominated by more than 8-m tall cladoxylopsid tree-fern–like plants and aneurophytalean progymnosperms (Stein et al., 2012). By the Carboniferous era, over 300 Ma, arborescent lycopsids had become dominant components of extensive swamp forest ecosystems, rising to over 30 m tall with stems of 1 m in diameter (Krings et al., 2011). They were then major players in soil C sequestration, including the formation of coal, and clearly were important in pedogenesis. Well-preserved fossils of the rootlike appendages of an arborescent lycopsid in the Carboniferous period (313-304 Ma), has shown evidence of AM symbiosis with arbuscules, vesicles, and spores apparently of the Glomeromycota type (Krings et al., 2011). In contemporary wetland forests in the southern United States, the dominant trees in the permanently waterlogged sites form AM associations, confirming that waterlogging is not incompatible with this symbiosis (Jurgensen et al., 1997).

The arborescent lycophytes have since become extinct. Cytological studies of fungal symbionts in extant lycophytes have noted unusual features that are distinct from Glomeromycota type AM associations and have been followed with the molecular studies by Rimington et al. (2015). They investigated sporophytes of 20 lycopod species, and AM fungi were found in seven of these, four of which hosted Mucoromycotina and three that hosted Glomeromycota. The incidence of fungal colonization of these plants was surprisingly low, with only 17 out of 101 plant samples apparently supporting symbiotic fungi. Because the ability to form mycorrhiza-like associations with soil fungi first occurred with the gametophytes of nonvascular plants, the role of mycorrhiza in the gametophytes of early lycopsids is of particular interest but not apparently preserved in the fossil record. Most extant lycophytes have subterranean achlorophyllous mycoheterotrophic gametophytes that are obligately dependent on their mycorrhizal fungal partners for growth, sexual reproduction, and establishment of their autotrophic sporophyte generations (Winther and Friedman, 2007), but this likely evolved in response to the increasing competition from closed canopies developed under faster-growing modern plants.

In contrast to the lycopods, the later diverging Monilophytes (ferns and fern allies) show a much higher occurrence of AM symbiosis in extant taxa, with the exception of the basal Equisetopsida, which diverged in the Devonian age, and in which the single extant lineage *Equisetum* shows inconsistent and generally low colonization by AM fungi (Hodson et al., 2009). The Equisetopsida rose to global importance in the Carboniferous period both as forest understory and in the form of giant horsetails like *Calamites*, and are likely to have played a significant role in the silica cycle (Trembath-Reichert et al., 2015); however, on the basis of current evidence, mycorrhizas are unlikely to have played a role in this. Within extant fern species, 33 out of 58 samples formed arbuscular mycorrhizas, with a very clear specificity to Glomeromycota and only the derived genus *Anagramma* out of 18 fern species containing Mucoromycotina, and in this case as a dual partnership with Glomeromycota (Rimington et al., 2015). The extent of compatibility of more advanced vascular plant lineages with Mucoromycotina remains to be thoroughly investigated, but there is emerging evidence that this symbiosis may be widespread (M. Bidartondo pers. comm).

The earliest forest ecosystems in the Middle Devonian age (398-385 Ma) were dominated by more than 8-m tall cladoxylopsid tree-fern–like plants and aneurophytalean progymnosperms (Stein et al., 2012) that share intermediate features between ferns and gymnosperms (Stein et al., 2007). The rise of the first forests with plants that produced woody tissues (lignophytes) and substantial roots that penetrated over 1 m into the soil heralded a period of intensified silicate mineral weathering with organic and inorganic C sequestration, drawing down over 90% of atmospheric CO_2, mainly into marine carbonate sediments (Figs. 2.1 and 2.2), and nearly doubling atmospheric oxygen concentrations from 400-350 Ma (Royer et al., 2004; Berner, 2006; Algeo and Scheckler, 1998; Lenton, 2001; Morris et al., 2015). Whereas there is no fossil evidence for mycorrhizas in the first forests, the presence of AM fungi in both phylogenetically ancestral and later diverging groups to cladoxylopsids and aneurophytes, and the constancy of mycorrhization in present day trees, strongly implicates the symbionts in facilitating fulfillment of the nutrient demands of these increasingly large plants as they evolved. The rise of the Middle Devonian forests has been linked to the development of new major soil orders and soil processes, including organic rich histosols and clay-rich alfisols (Morris et al., 2015). These innovations reflect both the increasing quantities of organic matter sequestered in soils as a result of the large biomass, litter inputs, and likely the C allocation to mycorrhizal partners from trees causing intensification of weathering processes and stabilization of soil, leading to neoformed expandable mixed-layer clays of the smectite group (Morris et al., 2015) (Figs. 2.1 and 2.2). Although smectite clays have also been formed on the planet Mars, apparently in the absence of life (Mahaffy et al., 2015), this probably took billions of years of chemical weathering, whereas on Earth these clays are most strongly developed and formed quickly under forests that act as "clay mineral factories" (Kennedy et al., 2006). This is consistent with the mineralogical shift in paleosols from mica, illite and chlorite toward increasing dominance by smectite and kaolinite from the Middle Devonian age onwards, when forest ecosystems started to become locally dominant (Morris et al., 2015).

The phylogeny and divergence times of the next major land plant group, the gymnosperms, has been evaluated using nuclear genes (Lu et al., 2014) indicating the crown age of the group is about 350 Ma, with all five extant lineages (cycads, ginkgos, cupressophytes, Pinaceae, and gnetophytes) originating before 300 Ma. However, the crown ages of all families except Ginkgoaceae and Sciadopityaceae are younger than 200 Ma, suggestive of major extinction events possibly caused by extreme cooling in the Carboniferous age (Lu et al., 2014)

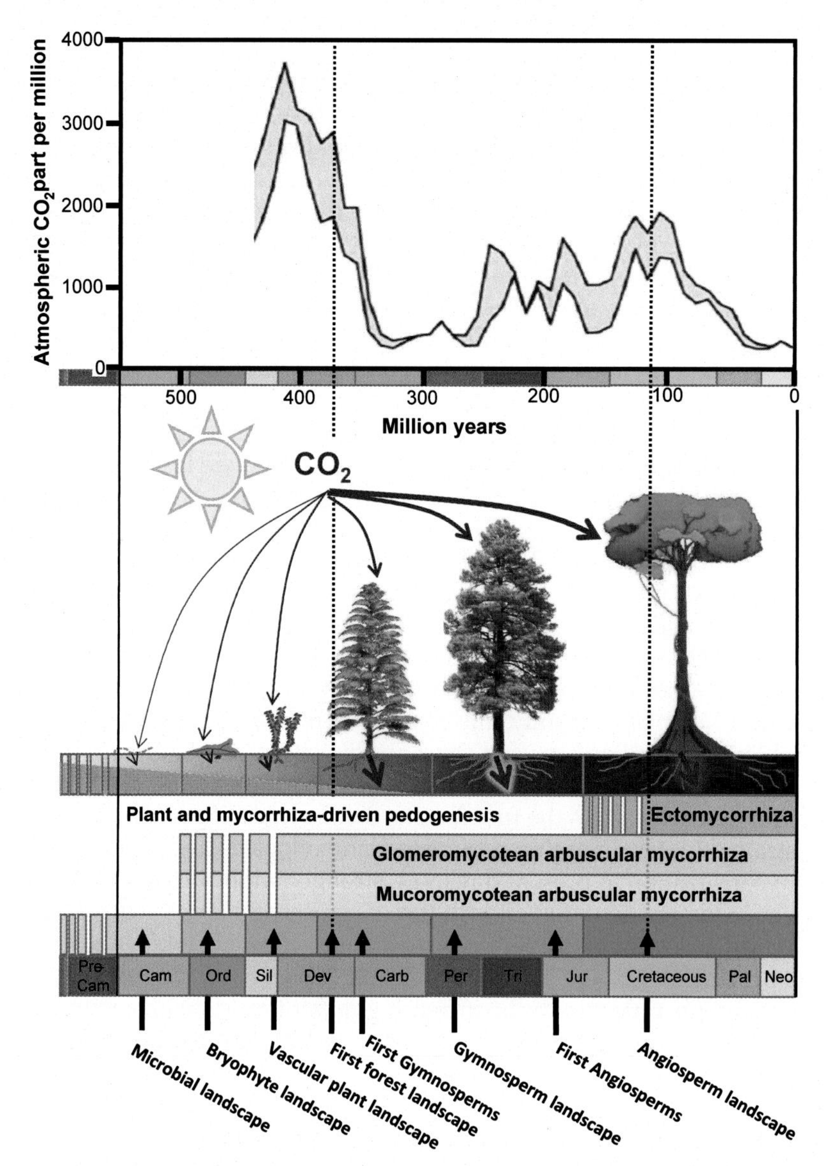

FIGURE 2.2 Key stages in the evolution of plants and mycorrhizas and changes in atmospheric carbon dioxide (CO_2) (based on modeled estimate ranges from Berner, 2006; in Field et al., 2012) linked to changes in rates of mineral weathering and episodic volcanic degassing of CO_2 over the past 450 million years. Only the most important types of plants are shown with respect to key evolutionary advances in structure and mycorrhizal associations. The greening of the continents with the evolutionary advance in plants from microbial mats of algae to liverworts with rhizoids and mycorrhiza-like partnerships with Mucoromycetes and Glomeromycota, to vascular plants with roots and mycorrhizas and the development of arborescence in the first forests led to progressive intensification of mineral weathering (see also Fig. 2.1). The rise of forest ecosystems initially dominated by aneurophytalean progymnosperms in which roots (and mycorrhizas) could interact with soil to depths of 1 m or more coincides with a rapid fall in atmospheric CO_2, in the Devonian (*dotted line*), and the sequestering of carbon (C) in limestone and coal in the Carboniferous era. The rise of angiosperms and ectomycorrhizal associations with both angiosperms and gymnosperms has further intensified weathering since the Cretaceous (*dotted line*). *Cam*, Cambrian; *Carb*, Carboniferous; *Dev*, Devonian; *Jur*, Jurassic; *Neo*, Neogene; *Ord*, Ordovician; *Pal*, Paleogene; *Per*, Permian; *Pre-Cam*, Precambrian; *Sil*, Silurian. *The key stages in greening of the earth by different dominant landscapes are adapted from Gerrienne et al. (2016).*

after the draw-down of atmospheric CO_2 into carbonate rocks. The cycads (Cycadaceae and Zamiaceae) that exclusively form AM symbioses (Fisher and Vovides, 2004) are confirmed as basal groups, but most extant species diversified mainly over the past 150 Ma and were followed by the AM Ginkoaceae. The next diverging group was the Pinaceace, which have developed mycorrhizas with the independently evolved EcM fungi from a wide diversity of basidiomycete and ascomycete fungi and have risen to dominance in the Boreal forest region in the Northern Hemisphere (Fig. 2.2). However, in its sister group, the Gnetales, the basal *Ephedra* together with *Welwitschia* form arbuscular mycorrhizas, whereas *Gnetum* is EcM, sometimes specialized on a clade of hypogeous basidiomycetes in the Boletales that have been shown to facilitate P release from apatite (Bechem and Alexander, 2012). All of the later diverging groups retain associations with AM fungi (Araucariaceae and its sister Podocarpaceae, the Sciadopidaceae, and the Cephalotaxaceae-Taxaceae that are sister to the Cupressaceae) (Lu et al., 2014).

The most recent, successful, and diverse group of land plants, the Angiosperms that have risen from about 167 to 199 Ma and rapidly diversified from the Cretaceous period (Bell et al., 2010) to dominate most of the world's biomes today (Fig. 2.2), are strongly mycorrhizal, with over 85% of a sample of over 10,000 species forming the symbiosis (Brundrett, 2009). AM fungi associate with 72% of plant species, including many temperate and tropical trees as well as forbs and grasses. EcM symbiosis involving both basidiomycete and ascomycete fungi is found in only 2% of plant species, although most of these are comprised of trees and woody shrubs, so this is the second most important type of mycorrhiza in terms of global plant biomass and extent of pedogenic processes at biome scales (Read, 1991). Because EcM fine roots are enveloped by a fungal sheath, the establishment of this kind of mycorrhiza was the most profound alteration in root functioning to occur in plant evolutionary history with the plant interface with soil being entirely mediated by the fungal partners and has far-reaching implications for soil biogeochemistry and weathering processes. Of the other kinds of angiosperm symbioses, ericoid mycorrhizas, which are found in 1.4% of angiosperm species, are the most important for facilitating organic C sequestration in soils and pedogenic processes such as podsolization and metal solubilization via siderophores (Haselwandter, 2008). However, they have low rates of photosynthesis and low productivity as a consequence of growing in nutrient-poor and climatologically harsh environments such as montane and tundra environments that constrain their rates of pedogenesis. Orchids, most of which are rare, and a high proportion of which are epiphytes, comprise 9% of angiosperm species and are obligately mycorrhizal for seedling establishment but are not significant in pedogenesis.

The long-presumed ancestral position of Glomeromycota in the evolution of plant symbioses with soil fungi (Nicolson, 1967; Pirozynski and Malloch, 1975; Redecker et al., 2000; Parniske, 2008; Tisserant et al., 2013) has been called into question since the discovery that the earliest diverging extant liverworts in the Haplomitriosida exclusively form mutualistic symbiosis with partially saprotrophic fungi in the Mucoromycotina, which are basal or sister to the Glomeromycota (Bidartondo et al., 2011). The coevolution of both types of fungal associations in early land plants (Fig. 2.2) is further supported by fossil evidence from the leafless early vascular plant *Horneophyton lignieri* plants in Rhynie Chert, presented by Strullu-Derrien et al. (2014). In its rootless corms that have tufts of rhizoids, there was colonization that shared features of Mucoromycotina fungi in extant land plants, and in the aerial axis Glomeromycota-type associations forming arbuscule-like structures, vesicles, and spores were found. In modern samples of the fungi, both groups of endosymbiotic fungi have been found to host related but phylogenetically distinct Mollicutes-related endobacteria (Desirò et al., 2015). The Mucoromycotina is apparently exclusively associated with a later diverging and narrower clade of these bacteria compared with the Glomeromycota, but widespread occurrence of these bacteria in both types of mycorrhizal fungi supports the idea that a tripartite symbiosis between plants, fungi, and endosymbiotic bacteria may have been instrumental in facilitating the colonization of the land by plants (Desirò et al., 2015), although the functional roles of the endosymbiotic bacteria have yet to be determined. Importantly, for consideration of plant–mycorrhiza–soil interactions, these close relationships with bacteria emphasize the potential importance of mycorrhizal fungi in influencing soil microbial communities and resulting soil processes, as seen in relation to mycorrhizal hyphosphere bacteria like *Pseudomonas* and *Burkholderia* (Zhang et al., 2014, 2016).

Symbiosis genes controlling mycorrhiza formation have been found to be distributed and highly conserved across major groups of land plants from basal Haplomitriopsida liverworts to angiosperms, suggesting vertical inheritance throughout the land-plant phylogeny (Wang et al., 2010). Intriguingly, mycorrhiza-formation genes from Haplomitriopsida, which only hosts Mucoromycotina fungi, can recover functional Glomeromycota mycorrhiza with vesicles and arbuscles in a mutant of the angiosperm *Medicago truncatula* (Wang et al., 2010). This does not resolve whether the earliest AM-like fungal associations were formed with Mucoromycotina but is consistent with the evidence that algal ancestors of land plants were preadapted with signaling pathways involved in initiating symbiosis with fungi (Delaux et al., 2015). Before Bidartondo et al. (2011), the use of targeted Glomeromycota-specific primers for amplification and detection of symbiotic fungal partners in living plants with AM associations had excluded detection of the Mucoromycotina. Now the application of DNA-based methods that detect both fungal groups, together with advances in cytological characterization of arbuscular mycorrhizas formed by them, has revealed that in addition to the Haplomitriopsida, other basal extant plants enter into symbioses with both groups of fungi, sometimes simultaneously (Desirò et al., 2013; Rimington et al., 2015). These include thalloid liverworts, hornworts (Desirò et al., 2013), and one of the most basal tracheophyte groups, the lycopods (Rimington et al., 2015).

Genomic studies and molecular-clock analyses of basidiomycete and ascomycete EcM fungi and their host-plant phylogenies suggest that this type of symbiosis first developed in the Pinaceae between 180 and 154 Ma in the Jurassic age, and in angiosperms in the Rosids from 110 to 88 Ma in the Cretaceous age (Kohler et al., 2015). The fungal partners evolved the capacity to form symbioses from a diverse group of saprotrophs including specialist wood-decay fungi, and this polyphyletic evolution of the EcM lifestyle was marked by convergent losses of some components of their ancestral saprotrophic apparatus. However, most have retained abilities to compete strongly with saprotrophs for organic forms of nutrients, thereby affecting decomposition processes, nutrient cycling, and pedogenesis (Chapters 20 and 21). Many of these fungi have sophisticated multicellular mycelial cord structures that allow them to translocate photosynthate-derived C long distances to exploratory and absorptive mycelia that are actively foraging for nutrients and proliferating around mineral grains such as apatite, which contains weatherable essential elements that, along with water, are transported back to the host plant (Smits et al., 2012).

2.4 COEVOLUTION OF PLANTS, MYCORRHIZAS, AND PHOTOSYNTHATE-DRIVEN WEATHERING AND PEDOGENESIS

The evolution, diversification, and greening of the continents by land plants of increasing biomass, structural complexity, shoot height, rooting depth, nutrient demand, and capacity to fix C, transport water, and regulate its loss through the evolution of stomata and leaves (Figs. 2.1 and 2.2) have been recognized as pivotal steps that transformed Earth-surface processes over the past 500 Ma (Beerling, 2007). One of the key drivers of establishment of mycorrhiza-like symbioses involving allocation of photosynthate C to fungal partnerships is the low bioavailability of soil P to plants. P readily forms insoluble precipitates with elements including Ca and iron, which are 27 times more abundant in the Earth's crust, and with aluminum (Al) and Si that are 90 and 293 times more abundant than P (Brantley et al., 2011). In soil solution, the concentration of phosphate ions is typically 100–1000 times less than other biologically essential elements such as N, potassium, Ca, and Mg (Marschner, 1995), and plant demand rapidly causes depletion zones to develop around plant rhizoids, roots, and root hairs. Mycorrhizal hyphae are able to access some of the smallest pores in soils, in which nutrients are often most abundant. For example, whereas fine roots typically have diameters of 100–500 μm, and root hairs are typically 10–15 μm, AM arterial hyphae are much finer than roots at 20–30 μm, and their fine distal absorptive hyphae are only about 2–7 μm in diameter (Leake et al., 2004).

The primary source of P in early stages of soil formation is provided by the weathering of the mineral apatite, composed mainly of Ca and P, which accounts for over 95% of all P in the Earth's crust (Jahnke, 1992). Apatite occurs in small amounts in most igneous rocks, both as inclusions in other minerals such as silicates and as larger crystals (Deer et al., 2001). The weathering of apatite, particularly in soils, is an important driver of long-term biogeochemical cycles that impact the Earth system. On land surfaces that are not actively eroding or being uplifted, over periods of hundreds to millions of years, apatite stocks become progressively depleted by weatherg, causing terrestrial ecosystems to become progressively more P-limited (Peltzer et al., 2010). The P is transformed mainly by biological and chemical actions linked to the activities of roots, mycorrhizas, and mycorrhizosphere microorganisms into secondary iron and Al phosphates, organic P in soil organic matter, and ecosystem biomass (Walker and Syers, 1976). In a model 120,000-year chronosequence in New Zealand, microbial biomass, which includes mycorrhizal fungi, rose rapidly to hold 70–80% of the combined plant plus microbial biomass P in the first 1000 years, and remained the dominant biotic pool throughout the rest of the sequence (Turner et al., 2013). This confirms the critical role of plant-microbe–soil interactions in P cycling processes, especially through transformations of organic P. The increasing P limitation to ecosystems over time has been shown to be reflected in a shift in symbionts in experimental bioassays of plants grown on soil on another 120,000 year chronosequence in Australia (Albornoz et al., 2016). At 100 years the soil was N limited, at 1000 years N and P, co-limited, and by 120,000 years was P limited. Nodulation declined with increasing P limitation and there was a marked shift from AM to EcM symbioses, possibly reflecting increased requirements to access organic P (Albornoz et al., 2016) or forms of mineral-bound P requiring intensive weathering, for example via secretion of organic acids to release Al-bound P.

Because in long-term chronosequences the loss of P in runoff and leaching of soluble organic P is not balanced by weathering inputs once the apatite fund in the active soil layers are depleted, most of the original fund of P is lost to the oceans, where it provides the primary source of nutrient, affecting marine primary production (Filippelli, 2008) and the draw-down of atmospheric CO_2 into marine sediments containing organic and inorganic C (Figs. 2.1 and 2.2). P is returned to the continental land surface through plate tectonics, causing uplift of buried rocks containing apatite, and through volcanic extrusions, resulting from subduction zones at the plate margins. Of the volcanic rocks, large basaltic provinces are especially important in providing large volumes of fast-weathering Ca silicate minerals in which apatite occurs.

Experimental studies, with extant simple thalloid liverworts in the Marchantiopsida that diverged about 370 Ma, suggest that mycorrhiza-like symbioses will have been extremely beneficial for P nutrition, establishment, and success of liverwort-like early land plants under the high atmospheric CO_2 concentrations that occurred in the Ordovician age (Humphreys et al., 2010) (Fig. 2.2). These plants, which lack leaves, stomata, and roots, interact with a shallow depth of soil via simple rhizoids, but in symbioses with Glomeromycotan AM fungi support hyphal lengths of 100–400 m per plant, increasing uptake of the growth-limiting nutrient P and thereby increasing their biomass production, photosynthetic output, and rate of reproduction. Simulated high CO_2 atmospheres representative of the Ordovician and Silurian eras, during which land plants first evolved and started to diversify, amplified the benefits of the symbiosis and suggests the environmental conditions will have strongly favored the establishment of mycorrhiza-like partnerships in the gametophyte-dominant phase of liverwort early land plants (Humphreys et al., 2010).

Subsequent experiments of a similar kind have quantified the C-for-P efficiency of the mycorrhiza-like symbioses in basal Haplomitriopsida liverworts *Haplomitrium* and *Treubia* that exclusively associate with Mucoromycotina; in *Marchantia* and *Preissia* species that only host Glomeromycotan fungi; and two species, *Neohodgsonia* and *Allinsonia*, that host both groups of fungi simultaneously (Field et al., 2015). These studies support the view that evolutionary advancement in plant–fungal symbioses has resulted in selection for increased C allocation to the fungal partners. The switch from exclusive associations with Mucoromycotina in the basal liverworts to Glomeromycota fungal partnerships in later diverging thalloid liverworts was found to significantly increase the amount of photosynthate C allocated through hyphal networks into a rhizoid-excluding soil compartment, with the dual associations having additive C demands that are greater than those of the single partners. Simulated high atmospheric CO_2 only marginally increased photosynthate allocation to the Mucoromycotina fungi, but greatly increased it from the Glomeromycota and dual association hosts (Field et al., 2015). The evolutionary advance from forming AM symbioses with Glomeromycota fungi to the independently evolved associations with EcM-forming basidiomycetes is also associated with a significant increase in the proportion of photosynthate allocated to the fungal partners (Quirk et al., 2014). The lower branching intensity and wider average diameter of roots of AM compared with EcM trees (Comas et al., 2014) is likely to be as a result of trade-offs in photosynthate allocation between root tissue and mycorrhizal fungal biomass. Greater C allocation to EcM fungi is reflected in the length of mycorrhizal mycelium associated with weathering mineral grains of granite and basalt buried in root-excluding mesh bags, being more than double under established ectomycorrhizas compared with AM trees in an arboretum (Quirk et al., 2012), resulting in Ca silicate weathering rates under EcM trees that were more than double those under AM trees. This effect has also been seen in much higher rates of carbonate mineral weathering under EcM trees compared with AM trees in a second arboretum (Thorley et al., 2015), confirming important differences in pedogenic processes caused by these two major types of mycorrhizas (Chapter 4).

Direct measurements of photosynthate C allocation via mycorrhizal hyphae into mesh cores containing basalt grains in mesocosms using $^{14}CO_2$ tracer supplied to tree shoots demonstrated a clear interdependence between the more than threefold higher ^{14}C allocation to Basidiomycota EcM fungal mycelia and the threefold higher rate of weathering of Ca silicate (Quirk et al., 2014). One reason for the high weathering rates achieved by some EcM fungi is their active exudation of low–molecular-weight organic acids that are particularly effective at chelating metal ions such as Ca, and their tolerance of very acidic pH conditions. The exudation of organic acids is dependent on photosynthate C supplied to the fungi, and appears to be tightly regulated, with oxalate secretion by EcM *Paxillus involutus* found to be both mineral-specific and linearly related to rates of Ca weathering from minerals (Schmalenberger et al., 2015). Highly targeted allocation of labeled photosynthate received by *P. involutus* has been shown with 17 times more ^{14}C transferred to weather apatite grains compared with grains of quartz, resulting in a nearly threefold increase in apatite weathering (Smits et al., 2012). Both AM and EcM fungi showed strong preferential growth on basalt grains compared with granite or quartz, an effect also seen in AM fungi in pot experiments in which the greatest effects of the symbiosis on element mobilization was on basalt and rhyolite, compared with granite and schist, with positive correlations between AM infection rates in roots and total uptake of P, Ca, and Mg across all mineral substrates (Burghelea et al., 2015). Interestingly, Tedersoo et al. (2014) in their analysis of global diversity and geography of soil fungi, including mycorrhizal fungi, found that although distance from the equator and climate were the strongest drivers of fungal diversity, the edaphic factors of Ca, P, and pH have strong effects on the distributions of many fungal groups. This is consistent with the central roles that the weathering of Ca phosphate minerals plays in long-term ecosystem dynamics (Peltzer et al., 2010; Turner et al., 2013) and the evidence of mineral-specific growth and secretory responses of mycorrhizal fungi in response to Ca-bearing minerals (Schmalenberger et al., 2015). The centrality of Ca and P interactions in biogeochemical cycling of both elements is likely to have been important in influencing the evolution of mycorrhizal fungal adaptations (including associations with bacteria) to mobilize these elements from different rocks.

The role of increasing plant biomass, rooting depth, and stature on mycorrhiza-driven weathering has been quantified by Quirk et al. (2015). In incubation experiments lasting for up to 12 months with freshly cleaved mineral coupons, Quirk et al. (2012, 2014, 2015) provided evidence of mycorrhizal fungal trenching and physical deformation of the important rock-forming phyllosilicates biotite, muscovite, and phlogopite. This included interlayer growth and spore production between exfoliating layers apparently by glomeralean AM fungi in symbiosis with thalloid liverworts, gymnosperm, and angiosperm trees. Mineral weathering by thalloid liverworts partnering AM fungi increased P weathering from basalt grains 9–13 times relative to plant-free controls, compared with 5–7 times amplification by liverworts lacking fungal symbionts, and this was associated with a threefold to sevenfold amplification of Ca weathering by liverworts with AM-like associations (Quirk et al., 2015). The etching and trenching of silicates by the AM fungi increased with the length of hyphae produced around basalt grains and with higher atmospheric CO_2 concentrations. Paleosol studies support this direct involvement of AM fungi in silicate weathering. Microscopic and mineralogical studies of a 16-Ma Miocene calcrete paleosol in which there is excellent preservation of rhizoliths, AM spores, and putative AM mycelium, has revealed hyphal attachment to mineral grains that caused etching and preferential hyphal colonization of exfoliated biotite that is most deeply weathered in proximity to these fungi (Sanz-Montero and Rodríguez-Aranda, 2012). Similarly, Koele et al. (2014) reported the formation of dissolution "tunnels" in feldspars under New Zealand forests that have only ever hosted AM fungal partners, but direct involvement of the fungi in this process was not proven in this case. Studies of EcM fungal weathering at the nanometer scale have revealed alteration of phyllosilicate and selective element depletion extending through and changing the structure of the crystal lattices through multiple layers underneath fungal hyphae (Bonneville et al., 2009). This effect has also been seen with saprotrophic fungi (Li et al., 2016) and is not detected by conventional methods for studying mineral weathering, but accounted for 40–50% of biological weathering in this case.

Comparisons between the weathering effects of AM-associated liverworts and AM- and EcM-associated trees grown under parallel conditions in the same controlled environment chambers, taking into account the depths of liverwort rhizoids and AM mycelia (0.1 m), and tree roots and mycelia (0.75 m), indicate early land plants lacking roots would appear to be at least 10-fold less effective at enhancing the total weathering flux than later-evolving trees (Quirk et al., 2015). Dissolution trenches in phyllosilicates buried under EcM trees were two to three times wider and two to four times deeper than those beneath AM trees (Quirk et al., 2012, 2014). Furthermore, the evolutionary advances that have led to angiosperm dominance over gymnosperms in the global flora are reflected in AM and EcM fungal lengths in mesh bags of minerals buried under mature angiosperm trees, being double those found under nearby gymnosperms in the same arboretum (Quirk et al., 2012). Similarly, Ca dissolution rates from basalt were significantly higher under EcM-angiosperm trees compared with AM-gymnosperm trees (Quirk et al., 2012). This evolution-linked intensification of weathering was confirmed by Thorley et al. (2015), who found weathering rates of calcite increased 1.3 times in AM angiosperm trees versus gymnosperm trees and 2.3 times in EcM angiosperm trees versus gymnosperm trees.

These findings suggest that the early terrestrial ecosystems comprising algae, bacteria, cyanobacteria, fungi, probable lichens, and other cryptophytes through the Ordovician era (Fig. 2.2) will have had very limited effects on biogeochemical cycles compared with later-evolving rooted plants and especially forests, as Edwards et al. (2015) concluded. They identified that the primary constraint on terrestrial weathering by plants at this stage in Earth's history was the limited depth of interaction with soils achieved by rootless plants, the modest biomass production, and the much lower recalcitrance and persistence of organic matter production compared with later-evolving plants with woody tissues (Figs. 2.1 and 2.2). This contradicted the bold assertions of Lenton et al. (2012) based on extrapolations of enhanced weathering rates generated by a moss grown on 15-mm beds of granite or andesite grains under highly controlled laboratory experiments, to global weathering rate models through the Ordovician age, which postulated major early impacts of land plants bioengineering the planet via impacts on global biogeochemical cycles. Bryophytes in the moss clade have evolved sophisticated multicellular rhizoids that may function in a manner similar to mycorrhizal fungi, and probably as a result, mosses do not form symbioses with soil fungi. However, in effectively assuming that biotic and abiotic weathering of the continents only extend to 15 mm deep, Lenton et al. (2012) vastly exaggerated the global impacts of mosses (as a proxy for Ordovician land plants) compared with the weathering rates generated by the evolution of roots and mycorrhizal fungi through increasing the depth of weathering to over 1-m depth in later-evolving forests. The evolutionary history of land plants and mycorrhizal fungi points to selection processes having increased below-ground photosynthate allocation and intensification of mineral weathering in response to major innovations such as the development of shoots, roots, stomata, leaves, and arborescence, together with changes in mycorrhizal fungal partnerships involving more intensive interactions with minerals (Quirk et al., 2012). Nonetheless, this does not preclude plant symbioses with mycorrhizal fungi playing an important role in the early stages of the evolution of terrestrial ecosystems through facilitating plant nutrition, water relations, and anchorage to the substrate, but also in the initiation of the formation of soils, particularly through organic matter inputs and effects on soil structure such as aggregation.

2.5 FEEDBACK BETWEEN PLANT-DRIVEN PEDOGENESIS, GLOBAL BIOGEOCHEMICAL CYCLES, AND THE EVOLUTION OF PLANTS AND MYCORRHIZAL FUNCTIONING

Evolutionary advances in plants and mycorrhizal fungi have increased photosynthate fluxes into soils, intensified mineral weathering, and pedogenesis, strengthening biosphere–geosphere–ocean–atmosphere interactions through effects on the biogeochemical cycles of C, Ca, P, and Si, affecting the composition of the atmosphere and global climate over more than 400 Ma. The draw-down of over 90% of atmospheric CO_2 from the end of the Ordovician age to the end of the Devonian age with the rise of terrestrial ecosystems, in which the steepest decline coincided with appearance of the first forests from the Middle Devonian period, led to a major global mass extinction and a switch from greenhouse to icehouse conditions, with much drier climates and fluctuating glaciations (Royer et al., 2004). These atmospheric and climatological changes have been implicated in driving the evolution of leaves and plant-shoot adaptations to balance the competing trade-offs between C fixation, water loss, and thermal regulation (Beerling, 2007). The first widespread appearance of megaphyll leaves occurs in the Progymnosperms such as the earliest tree *Archaeopteris* in the Middle Devonian age, in which stomata were eight times more common per surface area than those in the axial shoots of Early Devonian plants, reflecting the falling CO_2 conditions. Megaphylls evolved independently in progymnosperms and pteridosperms and showed a parallel 25-fold increase in leaf area from 380 to 340 Ma coincident with falling CO_2 (Osborne et al., 2004), and this would have helped to maintain photosynthate production against the falling CO_2 supply. Atmospheric and climate changes were probably instrumental in the development of plant adaptations to drought, including increasingly effective roots and vascular systems for uptake and transport of water. Mineral weathering rates decline with falling global temperatures as a result of reduced hydrological flushing by rain and slower chemical reaction rates. However, as CO_2 concentrations fall below 400 ppm, there is a steepening decline in plant-driven mycorrhiza weathering as allocation of photosynthate to fungal partners sharply falls and biomass production and nutrient demand decrease (Quirk et al., 2014). Young trees grown with either 1500 or 200 ppm CO_2, reflecting the atmospheric changes from the Devonian-Carboniferous and from the Cretaceous-Quaternary periods, demonstrated a threefold drop in tree-driven mycorrhizal fungal weathering fluxes of Ca and Mg from silicate rock grains (Quirk et al., 2014). This acts as a "C starvation" brake on depletion of atmospheric CO_2 and may have helped to stabilize Earth's atmospheric CO_2 minimum, which appears not to have fallen below 180–200 ppm throughout land plant history.

Conversely, after episodes of volcanic degassing, which returns CO_2 to the atmosphere and can quickly re-establish greenhouse conditions, such as during the late Jurassic-Middle Cretaceous age (160-100 Ma), the high CO_2 conditions increase rates of plant-driven mycorrhizal weathering of silicates (Quirk et al., 2014, 2015), helping prevent runaway greenhouse heating.

The evolution of woody plants and lignin, before microorganisms capable of readily degrading these recalcitrant polymers, led to their accumulation in the carboniferous coal swamps and contributed to the doubling of atmospheric oxygen concentrations from the Middle Devonian age to the end of the Carboniferous age (Beerling, 2007) that will have strongly selected for plants to develop adaptations to fire. The evolution of woody plants eventually led to the evolution of specialist basidiomycete and ascomycete fungi with the peroxidase enzymes that can degrade lignin or lignocellulose, reducing the global sequestration of C in dead wood (Floudas et al., 2012). The first fungal ligninolytic manganese peroxidase arose about 295 Ma, which is slightly older than the first definitive fossils of the kind of wood decay fungus with this enzyme dated to about 260 Ma (Floudas et al., 2012). Many of these specialist decay fungi then went on to independently evolve the capacity to form ectomycorrhizas with woody gymnosperm and angiosperm plants (Floudas et al., 2012) from 180 Ma (Kohler et al., 2015). Taking into account the higher rates of mineral weathering achieved by EcM versus AM fungi and by angiosperms versus gymnosperms, Taylor et al. (2012) have incorporated these functions into a global process-based model that couples the Hadley Center general circulation model, the Sheffield dynamic global vegetation model, and a mineral weathering model that incorporates effects of root exudates and mycorrhizal fungi, to estimate the contribution of plants, mycorrhizas, atmospheric CO_2, and climate interactions on Ca and Mg silicate weathering over the past 216 Ma. The models suggest that the importance of mycorrhizas in mineral weathering was almost as large as that of plants alone from the last 67 million years, in which angiosperm trees and EcM forests rose to increasing dominance. These simulations build on previous work that modeled the effect of the almost simultaneous rise of angiosperms and ectomycorrhizas on atmospheric CO_2 through their effects on mineral weathering and the geochemical C cycle (Taylor et al., 2011), taking into account the vertical distribution of roots and associated mycorrhizal hyphae in typical soils across major bioclimatic regions of the Earth. These simulations, linked to a process-based weathering model, indicated that the intensification of weathering attributed to the rise of angiosperms in the GEOCARB family of models that estimate atmospheric CO_2 concentrations for the past 500 Ma (Berner, 2006) should be attributed jointly to the rise of EcM symbiosis, as well as the rise of angiosperms. Furthermore, the importance of fungal hyphal lengths and exudation of low–molecular-weight organic chelates, such as oxalic acid on silicate weathering and global atmospheric CO_2, was evaluated in sensitivity analyses that indicated that these factors that depend on deployment of plant photosynthate will have contributed significantly to draw-down of atmospheric CO_2 over the past 100 Ma (Taylor et al., 2011), and have intensified with the evolutionary advancement in plants and mycorrhizal fungi.

2.6 CONCLUSIONS

The increasing body of evidence from evolutionary biology, paleontology, geochemistry, and earth system sciences reviewed here suggests that the coevolution of plants and mycorrhizal fungi have led to progressive intensification of pedogenesis and biogeochemical cycles, and through this have bioengineered the Earth's surface, ocean chemistry, atmospheric composition, and global climate. This bioengineering probably became significant from the late Silurian period and is largely attributable to the increased C energy inputs to soils achieved by the evolution of plant roots, the production of lignin, and deployment of photosynthate C to mycorrhizal fungi interacting with minerals particularly to release P. Efficient nutrient mobilization and uptake by mycorrhizal fungi has facilitated increased photosynthesis and biomass production by plants. Although the extent to which mycorrhizas were important in early land plants directly and indirectly impacted the Earth's system remains uncertain, particularly for the period 470-411 Ma for which we have evidence of land plants but no unequivocal evidence of mycorrhiza-like associations, the ubiquity of the symbiosis in most land plant groups is consistent with it being functionally important throughout land-plant history. Together with hyphosphere-associated bacteria, mycorrhizal fungi have contributed to mineral weathering by trenching, tunneling, and delamination of silicate minerals in ways that are distinct from, and probably pre-date the evolution of roots as shown in mycorrhiza-like symbioses with thalloid liverworts.

Mineral weathering rates have intensified with evolutionary advancement in plant photosynthetic capacity and have resulted in neoformed clay minerals as plants developed greater biomass and rooting depth, and with an increasing proportion of their photosynthate being allocated to mycorrhizal fungi. Increased photosynthate supports longer mycelial networks, greater exudation of organic chelates, and more mineral-dissolving rhizobacteria that together accelerate base cation release from silicates and generate pedogenic clays. Because the actions of mycorrhizal fungi are dependent on host photosynthate, as the concentrations of atmospheric CO_2 decrease as a result of weathering of Ca silicates and sequestration of CO_2 in ocean carbonates, the rate of weathering falls. Conversely, when atmospheric CO_2 concentrations increase, for example, through volcanic degassing or current anthropogenic burning of fossil fuels, the rates of photosynthesis and weathering will increase over millions of years helping to stabilize the lower and upper ranges of atmospheric CO_2 concentrations over geological time. The lengths of mycorrhizal mycelia in soils today are so large and turn over so fast that over a million years the cumulative total length produced in soils would be in excess of 4.8×10^{10} light years, which is larger than the diameter of the known universe of 9.1×10^{9} light years. It is perhaps not surprising that effects of these mycelia scale from the removal of atoms from silicates, through effects on pedogenesis and soil C stocks, to important global impacts on the Earth's system over geological time.

Acknowledgments

JRL acknowledges collaborators Prof. D.J. Beerling FRS, Dr. Joe Quirk, Dr. Jenny Morris, Dr. Katie Field, Rachel Thorley, Dr. Lyla Taylor and Prof. Duncan Cameron, and financial support from the NERC in the following projects: Evolutionary rise of deep-rooting forests and enhanced chemical weathering: quantitative investigations into the current paradigm (NE/J007471/1), *Origin and co-evolution of land plant-fungal symbioses during the "greening of the Earth"* (NE/I024089/1), *Functional and evolutionary significance of symbiotic fungal associations in lower land plants* (NE/F019033), and from the Leverhulme Trust *Ecosystem CO_2 starvation and Earth's minimum atmospheric CO_2 concentration: an experimental assessment*.

References

Albornoz, F.E., Lambers, H., Turner, B.L., Teste, F.P., Laliberté, E., 2016. Shifts in symbiotic associations in plants capable of forming root symbioses across a long-term soil chronosequence. Ecology and Evolution 6, 2368–2377.

Algeo, T.J., Scheckler, S.E., 1998. Terrestrial-marine teleconnections in the Devonian: links between the evolution of land plants, weathering processes, and marine anoxic events. Philosophical Transactions of the Royal Society B-Biological Sciences 353, 113–128.

Augusto, L., De Schrijver, A., Vesterdal, L., Smolander, A., Prescott, C., Ranger, J., 2015. Influences of evergreen gymnosperm and deciduous angiosperm tree species on the functioning of temperate and boreal forests. Biological Reviews 90, 444–466.

Bardgett, R., 2005. The Biology of Soil. A Community and Ecosystems Approach. Oxford University Press. 242 pages.

Bechem, E.E.T., Alexander, I.J., 2012. Phosphorus nutrition of ectomycorrhizal *Gnetum africanum* plantlets from Cameroon. Plant and Soil 353, 379–393.

Bendini, S., Pellegrino, E., Avio, L., Pellegrini, S., Bazzoffi, P., Argese, E., Giovannetti, M., 2009. Changes in soil aggregation and glomalin-related soil protein content as affected by the arbuscular mycorrhizal fungal species *Glomus mosseae* and *Glomus intraradices*. Soil Biology & Biochemistry 41, 1491–1496.

Bell, C.D., Soltis, D.E., Soltis, P.S., 2010. The age and diversification of angiosperms re-visited. American Journal of Botany 97, 1296–1303.

Beerling, D., 2007. The Emerald Planet. How Plants Changed Earth's History. Oxford University Press. 288 pages.

Berner, R.A., 2006. GEOCARBSULF: a combined model for Phanerozoic atmospheric O_2 and CO_2. Geochimemica et Cosmochimica Acta 70, 5653–5664.

Bidartondo, M.I., Read, D.J., Trappe, J.M., Merckx, V., Ligrone, R., Duckett, J.G., 2011. The dawn of symbiosis between plants and fungi. Biology Letters 7, 574–577.

Bonneville, S., Smits, M.M., Brown, A., Harrington, J., Leake, J.R., Brydson, R., Benning, L.G., 2009. Plant driven fungal weathering: early stages of mineral alteration at the nanometer scale. Geology 37, 615–618.

Brantley, S.L., Megonigal, J.P., Scatena, F.N., Balogh-Brunstad, Z., Barnes, R.T., Bruns, M.A., Van Cappellen, P., Dontsova, K., Hartnett, H.E., Hartshorn, A.S., Heimsath, A., Herndon, E., Jin, L., Keller, C.K., Leake, J.R., McDowell, W.H., Meinzer, F.C., Mozdzer, T.J., Petsch, S., Pett-Ridge, J., Pregitzer, K.S., Raymond, P.A., Riebe, C.S., Shumaker, K., Sutton-Grier, A., Walter, R., Yoo, K., 2011. Twelve testable hypotheses on the geobiology of weathering. Geobiology 9, 140–165.

Braun, A., Chen, J., Waloszek, D., Maas, A., 2007. First early Cambrian Radiolaria. In: Vickers-Rich, P., Komarower, P. (Eds.), The Rise and Fall of the Ediacaran Biota. Geological Society London Special Publications, vol. 286, pp. 143–149.

Brundrett, M.C., 2009. Mycorrhizal associations and other means of nutrition of vascular plants: understanding the global diversity of host plants by resolving conflicting information and developing reliable means of diagnosis. Plant and Soil 320, 37–77.

Burghelea, C., Zaharescu, D.G., Dontsova, K., Maier, R., Huxman, T., Chorover, J., 2015. Mineral nutrient mobilization by plants from rock: influence of rock type and arbuscular mycorrhiza. Biogeochemistry. http://dx.doi.org/10.1007/s10533-015-0092-5.

Clemmensen, K.E., Bahr, A., Ovaskainen, O., Dahlberg, A., Ekblad, A., Wallander, H., Stenlid, J., Finlay, R.D., Wardle, D.A., Lindahl, B.D., 2013. Roots and associated fungi drive long-term carbon sequestration in boreal forest. Science 339, 1615–1618.

Comas, L.H., Callahan, H.S., Midford, P.E., 2014. Patterns in root traits of woody species hosting arbuscular and ectomycorrhizas: implications for the evolution of belowground strategies. Ecology and Evolution 4, 2979–2990.

Conley, D.J., Carey, J.C., 2015. Silica cycling over geologic time. Nature Geoscience 8, 431–432.

Deer, W.A., Howie, R.A., Zussman, J., 2001. Rock forming minerals. Framework Silicates, vol. 4a. Feldspars Geological Society Publishing House, Bath, p. 984.

Delaux, P.M., Radhakrishnan, G.V., Jayaraman, D., Cheema, J., Malbreil, M., Volkening, J.D., Sekimoto, H., Nishiyama, T., Melkonian, M., Pokorny, L., Rothfels, C.J., Sederoff, H.W., Stevenson, D.W., Surek, B., Zhang, Y., Sussman, M.R., Dunand, C., Morris, R.J., Roux, C., Wong, G.K.-S., Oldroyd, G.E.D., Ané, J.-M., 2015. Algal ancestor of land plants was preadapted for symbiosis. PNAS 112, 13390–13395.

Desirò, A., Duckett, J.G., Pressel, S., Villarreal, J.C., Bidartondo, M.I., 2013. Fungal symbioses in hornworts: a chequered history. Proceedings of the Royal Society B 280, 20130207.

Desirò, A., Faccio, A., Kaech, A., Bidartondo, M.I., Bonfante, P., 2015. Endogone, one of the oldest plant-associated fungi, host unique Mollicutes-related endobacteria. New Phytologist 205, 1464–1472.

Dotzler, N., Krings, M., Taylor, T.N., Agerer, R., 2006. Germination shields in Scutellospora (Glomeromycota: Diversisporales, Gigasporaceae) from the 400 million-year-old Rhynie chert. Mycological Progress 5, 178–184.

Dotzler, N., Walker, C., Krings, M., Hass, H., Kerp, H., Taylor, T.N., Agerer, R., 2009. Acaulosporoid glomeromycotan spores with a germination shield from the 400-million-year-old Rhynie chert. Mycological Progress 8, 9–18.

Edwards, D., Morris, J.L., Richardson, J.B., Kenrick, P., 2014. Cryptospores and cryptophytes reveal hidden diversity in early land floras. New Phytologist 202, 50–78.

Edwards, D., Cherns, L., Raven, J.A., 2015. Could land-based early photosynthesizing ecosystems have bioengineered the planet in mid-palaeozoic times? Palaeontology 58, 803–837.

Field, K.J., Cameron, D.D., Leake, J.R., Tille, S., Bidartondo, M.I., Beerling, D.J., 2012. Contrasting arbuscular mycorrhizal responses of vascular and non-vascular plants to a simulated Palaeozoic CO_2 decline. Nature Communications. http://dx.doi.org/10.1038/ncomms1831.

Field, K.J., Rimington, W.R., Bidartondo, M.I., Allinson, K.E., Beerling, D.J., Cameron, D.D., Duckett, J.G., Leake, J.R., Pressel, S., 2014. First evidence of mutualism between ancient plant lineages (Haplomitriopsida liverworts) and Mucoromycotina fungi and its response to simulated Palaeozoic changes in atmospheric CO_2. New Phytologist 205, 743–756. http://dx.doi.org/10.1111/nph.13024.

Field, K.J., Rimington, W.R., Bidartondo, M.I., Allinson, K.E., Beerling, D.J., Cameron, D.D., Duckett, J.G., Leake, J.R., Pressel, S., 2015. Functional analysis of liverworts in dual symbiosis with Glomeromycota and Mucoromycotina fungi under a simulated Palaeozoic CO_2 decline. ISME Journal 2015, 1–13.

Filippelli, G., 2008. The global phosphorus cycle: past, present, and future. Elements 4, 89–95.

Fisher, J.B., Vovides, A.P., 2004. Mycorrhizae are present in cycad roots. The Botanical Review 70, 16–23.

Floudas, D., Binder, M., Riley, R., Barry, K., Blanchette, R.A., Henrissat, B., Martínez, A.T., Otillar, R., Spatafora, J.W., Yadav, J.S., Aerts, A., Benoit, I., Boyd, A., Carlson, A., Copeland, A., Coutinho, P.M., de Vries, R.P., Ferreira, P., Findley, K., Foster, B., Gaskell, J., Glotzer, D., Górecki, P., Heitman, J., Hesse, C., Hori, C., Igarashi, K., Jurgens, J.A., Kallen, N., Kersten, P., Kohler, A., Kües, U., Arun Kumar, T.K.A., Kuo, A., LaButti, K., Larrondo, L.F., Lindquist, E., Ling, A., Lombard, V., Lucas, S., Lundell, T., Martin, R., McLaughlin, D.J., Morgenstern, I., Morin, E., Murat, C., Nagy, L.G., Nolan, M., Ohm, R.A., Patyshakuliyeva, A., Rokas, A., Ruiz-Dueñas, F.J., Sabat, G., Salamov, A., Samejima, M., Schmutz, J., Slot, J.C., St John, F., Stenlid, J., Sun, H., Sun, S., Syed, K., Tsang, A., Wiebenga, A., Young, D., Pisabarro, A., Eastwood, D.C., Martin, F., Cullen, D., Grigoriev, I.V., Hibbett, D.S., 2012. The Paleozoic origin of enzymatic lignin decomposition reconstructed from 31 fungal genomes. Science 336, 1715–1719.

Gerrienne, P., Dilcher, D.L., Bergamaschi, S., Milagres, I., Pereira, E., Rodrigues, M.A.C., 2006. An exceptional specimen of the early land plant *Cooksonia paranensis*, and a hypothesis on the life cycle of the earliest eutracheophytes. Review of Palaeobotany and Palynology 142, 123–130.

Gerrienne, P., Servais, T., Vecoli, M., 2016. Plant evolution and terrestrialization during Palaeozoic times- the phylogenetic context. Review of Palaeobotany and Palynology 227, 4–18.

Guo, C.-Q., Edwards, D., Wu, P.-C., Duckett, J.G., Hueber, F.M., Li, C.-S., 2012. *Riccardiothallus devonicus* gen. et sp. nov., the earliest simple thalloid liverwort from the Lower Devonian of Yunnan, China. Review of Palaeobotany and Palynology 176–177, 35–40.

Haselwandter, K., 2008. Structure and function of siderophores produced by mycorrhizal fungi. Mineralogical Magazine 72, 61–64.

Hammer, E.C., Rillig, M.C., 2011. The influence of different stresses on glomalin levels in an arbuscular mycorrhizal fungus—salinity increases glomalin content. PLoS One 6, e28426. http://dx.doi.org/10.1371/journal.pone.0028426.

Hernick, L.V., Landing, E., Bartowski, K.E., 2008. Earth's oldest liverworts- *Metzgeriothallus sharonae* sp. Nov. from the Middle Devonian (Givetian) of eastern New York, USA. Review of Palaeobotany and Palynology 148, 154–162.

Hiederer, R., Köchy, M., 2011. Global Soil Organic Carbon Estimates and the Harmonized World Soil Database. EUR 25225 EN. Publications Office of the EU, Luxembourg.

Hodson, E., Shahid, F., Basinger, J., Kaminskyj, S., 2009. Fungal endorhizal associates of Equisetum species from Western and Arctic Canada. Mycological Progress 8, 19–27.

Humphreys, C.P., Franks, P.J., Rees, M., Bidartondo, M.I., Leake, J.R., Beerling, D.J., 2010. Mutualistic mycorrhiza-like symbiosis in the most ancient group of land plants. Nature Communications 1, 7.

Jahnke, R.A., 1992. The phosphorus cycle. In: Butcher, S.S., Charlson, R.J., Orians, G.H., Wolfe, G.V. (Eds.), Global Biogeochemical Cycles. Academic Press, London, pp. 301–315.

Jenny, H., 1941. Factors of Soil Formation. A System of Quantitative Pedology. McGraw-Hill, New York. 281 pages.

Jenny, H., 1980. The Soil Resource: Origin and Behaviour. Ecological Studies, vol. 37. Springer-Verlag. 377 pages.

Jurgensen, M.F., Richter, D.L., Davis, M.M., McKevlin, M.R., Craft, M.H., 1997. Mycorrhizal relationships in bottomland hardwood forests of the southern United States. Wetlands Ecology and Management 4, 223–233.

Kennedy, M., Droser, M., Mayer, L.M., Pevear, D., Mrofka, D., 2006. Late Precambrian oxygenation; inception of the clay mineral factory. Science 311, 1446–1449.

Kenrick, P., Strullu-Derrien, C., 2014. The origin and early evolution of roots. Plant Physiology 166, 570–580.

Koele, N., Dickie, I.A., Blum, J.D., Gleason, J.D., de Graaf, L., 2014. Ecological significance of mineral weathering in ectomycorrhizal and arbuscular mycorrhizal ecosystems from a field-based comparison. Soil Biology & Biochemistry 69, 63–70.

Kohler, A., Kuo, A., Nagy, L.G., Morin, E., Barry, K.W., Buscot, F., Canbäck, B., Choi, C., Cichocki, N., Clum, A., Colpaert, J., Copeland, A., Costa, M.D., Doré, J., Floudas, D., Gay, G., Girlanda, M., Henrissat, B., Herrmann, S., Hess, J., Högberg, N., Johansson, T., Khouja, H.R., LaButti, K., Lahrmann, U., Levasseur, A., Lindquist, E.A., Lipzen, A., Marmeisse, R., Martino, E., Murat, C., Ngan, C.Y., Nehls, U., Plett, J.M., Pringle, A., Ohm, R.A., Perotto, S., Peter, M., Riley, R., Rineau, F., Ruytinx, J., Salamov, A., Shah, F., Sun, H., Tarkka, M., Tritt, A., Veneault-Fourrey, C., Zuccaro, A., Mycorrhizal Genomics Initiative Consortium, Tunlid, A., Grigoriev, I.V., Hibbett, D.S., Martin, F., 2015. Convergent losses of decay mechanisms and rapid turnover of symbiosis genes in mycorrhizal mutualists. Nature Genetics 47, 410–415.

Krings, M., Taylor, T.N., Hass, H., Kerp, H., Dotzler, N., Hermsen, E.J., 2007. Fungal endophytes in a 400-million-yr-old land plant: infection pathways, spatial distribution, and host responses. New Phytologist 174, 648–657.

Krings, M., Taylor, T.N., Taylor, E.L., Dotzler, N., Walker, C., 2011. Arbuscular-mycorrhiza-like fungi in Carboniferous arborescent lycopsids. New Phytologist 191, 311–314.

Leake, J.R., Johnson, D., Donnelly, D., Muckle, G.E., Boddy, L., Read, D.J., 2004. Networks of power and influence: the role of mycorrhizal mycelium in controlling plant communities and agro-ecosystem functioning. Canadian Journal of Botany 82, 1016–1045.

Leake, J.R., Duran, A.L., Hardy, K.E., Johnson, I., Beerling, D.J., Banwart, S.A., Smits, M.M., 2008. Biological weathering in soil: the role of symbiotic root-associated fungi biosensing minerals and directing photosynthate-energy into grain-scale mineral weathering. Mineralogical Magazine 72, 85–89.

Lenton, T.M., 2001. The role of land plants, phosphorus weathering and fire in the rise and regulation of atmospheric oxygen. Global Change Biology 7, 613–629.

Lenton, T.M., Crouch, M., Johnson, M., Pires, N., Dolan, L., 2012. First plants cooled the Ordovician. Nature Geosciences 5, 86–89.

Li, Z., Liu, L., Chen, J., Teng, H.H., 2016. Cellular dissolution at hypha-and spore-mineral interfaces revealing unrecognized mechanisms and scales of fungal weathering. Geology. http://dx.doi.org/10.1130/G37561.1.

Lu, Y., Ran, J.-H., Guo, D.-M., Yang, Z.U., Wang, X.Q., 2014. Phylogeny and divergence times of gymnosperms inferred from single-copy nuclear genes. PLoS One 9, e107679.

Mahaffy, P.R., Webster, C.R., Stern, J.C., Brunner, A.E., Atreya, S.K., Conrad, P.G., Domagal-Goldman, S., Eigenbrode, J.L., Flesch, G.J., Christensen, L.E., Franz, H.B., Freissinet, C., Glavin, D.P., Grotzinger, J.P., Jones, J.H., Leshin, L.A., Malespin, C., McAdam, A.C., Ming, D.W., Navarro-Gonzalez, R., Niles, P.B., Owen, T., Pavlov, A.A., Steele, A., Trainer, M.G., Williford, K.H., Wray, J.J., 2015. The imprint of atmospheric evolution in the D/H of Hesperian clay minerals on Mars. Science 347, 412–414.

Marschner, H., 1995. Mineral Nutrition of Higher Plants, second ed. Academic Press London, p. 889.

Matsunaga, K.K.S., Tomescu, A.M.F., 2016. Root evolution at the base of the lycophyte clade – insights from an Early Devonian lycophyte. Annals of Botany. http://dx.doi.org/10.1093/aob/mcw006.

Medlin, L.K., 2011. A review of the evolution of the diatoms from the origin of the lineage to their populations. In: Seckbach, J., Kociolek, J. (Eds.), The Diatom World. Springer Science, Netherlands, pp. 93–118.

MEA, 2005. Millennium Ecosystem Assessment. Ecosystems and Human Well-being: Synthesis. Island Press, Washington, DC. ISBN: 1-59726-040-1.

Morris, J.L., Leake, J.R., Stein, W.E., Berry, C.M., Marshall, J.E.A., Wellman, C.H., Milton, A., Hillier, S., Mannolini, F., Quirk, J., Beerling, D.J., 2015. Investigating Devonian trees as geoengineers of past climates: linking palaeosols to palaeobotany and experimental geobiology. Palaeontology 2015, 1–15.

Nicolson, T.H., 1967. Vesicular-arbuscular mycorrhiza: a universal plant symbiosis. Science Progress 55, 561–581.

Olssen, P.A., Hammer, E.C., Pallon, J., Van Aarle, I.M., 2011. Elemental composition in vesicles of an arbuscular mycorrhizal fungus, as revealed by PIXE analysis. Fungal Biology 115, 643–648.

Osborne, C.P., Beerling, D.J., Lomax, B.H., Chaloner, W.G., 2004. Biophysical constraints on the origin of leaves inferred from the fossil record. Proceedings of the National Academy of Sciences of the United States of America 101, 10360–10362.

Parniske, M., 2008. Arbuscular mycorrhiza: the mother of plant root endosymbioses. Nature Reviews Microbiology 10, 763–775.

Peltzer, D.A., Wardle, D.A., Allison, V.J., Baisden, W.T., Bardgett, R.D., Chadwick, O.A., Condron, L.M., Parfitt, R.L., Porder, S., Richardson, S.J., Turner, B.L., Vitousek, P.M., Walker, J., Walker, L.R., 2010. Understanding ecosystem retrogression. Ecological Monographs 80, 509–529.

Phillips, J.D., 2009. Biological energy in landscape evolution. American Journal of Science 309, 271–290.

Pirozynski, K.A., Malloch, D.W., 1975. The origin of land plants: a matter of mycotrophism. Biosystems 6, 153–164.

Qiu, Y.-L., Li, L., Wang, B., Chen, Z., Knoop, V., Groth-Malonek, M., Dombrovska, O., Lee, J., Kent, L., Rest, J., Estabrook, G.F., Hendry, T.A., Taylor, D.W., Testa, C.M., Ambros, M., Crandall-Stotler, B., Duff, R.J., Stech, M., Frey, W., Quandt, D., Davis, C.C., 2006. The deepest divergences in land plants inferred from phylogenomic evidence. Proceedings of the National Academy of Sciences of the United States of America 103, 15511–15516.

Quirk, J., Beerling, D.J., Banwart, S.A., Leake, J.R., 2012. Evolution of trees and mycorrhizal fungi intensifies silicate mineral weathering. Biology Letters 8, 1006–1011.

Quirk, J., Leake, J.R., Banwart, S.A., Taylor, L.L., Beerling, D.J., 2014. Weathering by tree-root-associating fungi diminishes under simulated Cenozoic atmospheric CO_2 decline. Biogeosciences 11, 321–331.

Quirk, J., Leake, J.R., Johnson, D.A., Taylor, L.L., Saccone, L., Beerling, D.J., 2015. Constraining the role of early land plants in Palaeozoic weathering and global cooling. Proceedings of the Royal Society B. 282, 20151115. http://dx.doi.org/10.1098/rspb.2015.1115.

Read, D.J., 1991. Mycorrhizas in ecosystems. Experientia 47, 376–391.

Read, D.J., Leake, J.R., Perez-Moreno, J., 2004. Mycorrhizal fungi as drivers of ecosystem processes in heathland and boreal forest biomes. Canadian Journal of Botany 82, 1243–1263.

Redecker, D., Kodner, R., Graham, L.E., 2000. Glomalean fungi from the Ordovician. Science 289, 1920–1921.

Remy, W., Taylor, T.N., Hass, H., Kerp, H., 1994. Four-hundred-million-year-old vesicular-arbuscular mycorrhizae. Proceedings of the National Academy of Sciences of the United States of America 91, 11841–11843.

Rickards, R.B., 2000. The age of the earliest club mosses: the Silurian Baragwanathia flora in Victoria, Australia. Geological Magazine 137, 207–209.

Rimington, W.R., Pressel, S., Duckett, J.G., Bidartondo, M.I., 2015. Fungal associations of basal vascular plants: reopening a closed book? New Phytologist 205, 1394–1398.

Royer, D.L., Berner, R.A., Montanez, I.P., Tabor, N.J., Beerling, D.J., 2004. CO_2 as a primary driver of Phanerozoic climate. GSA Today 14, 4–10.

Sanz-Montero, M.E., Rodríguez-Aranda, J.P., 2012. Endomycorrhizae in Miocene paleosols: implications in biotite weathering and accumulation of dolomite in plant roots (SW Madrid Basin, Spain). Palaeogeography, Palaeoclimatology, Palaeoecology 333–334, 121–130.

Scharlemann, J.P.W., Tanner, E.V.J., Hiederer, R., Kapos, V., 2014. Global soil carbon: understanding and managing the largest terrestrial carbon pool. Carbon Management 5, 81–91.

Schmalenberger, A., Duran, A.L., Bray, A.W., Bridge, J., Bonneville, S., Benning, L.G., Romero-Gonzalez, M.E., Leake, J.R., Banwart, S.A., 2015. Oxalate release by ectomycorrhizal *Paxillus involutus* is mineral-specific and controls calcium weathering from minerals. Scientific Reports 5, 12187.

Selosse, M.A., Strullu-Derrien, C., 2015. Origins of the terrestrial flora: a symbiosis with fungi? BIO Web of Conferences 4, 00009. http://dx.doi.org/10.1051/bioconf/20150400009.

Smits, M.M., Bonneville, S., Benning, L.G., Banwart, S.A., Leake, J.R., 2012. Plant-driven weathering of apatite – the role of an ectomycorrhizal fungus. Geobiology 10, 445–456.

Soudzilovskaia, N.A., Douma, J.C., Akhmetzhanova, A.A., Bodegom, P.M., Cornwell, W.K., Moens, E.J., Treseder, K.K., Tibbett, M., Wang, Y.-P., Cornelissen, J.H.C., 2015. Global patterns of plant root colonization intensity by mycorrhizal fungi explained by climate and soil chemistry. Global Ecology and Biogeography 24, 371–382.

Steemans, P., Hérissé, A.L., Melvin, J., Miller, M.A., Paris, F., Verniers, J., Wellman, C.H., 2009. Origin and radiation of the earliest vascular land plants. Science 324, 353.

Steemans, P., Wellman, C.H., Gerrienne, P., 2010. Palaeogeographic and palaeclimate considerations based on Ordovician to Lochkovian vegetation. In: Vecoli, M., Clément, G., Meyer-Berthaud, B. (Eds.), The Terrestrialization Process: Modelling Complex Interactions at the Biosphere-geosphere Interface. Geological Society London. Special Publications, vol. 339, pp. 49–58.

Stein, W.E., Berry, C.M., Hernick, L.V., Mannolini, F., 2012. Surprisingly complex community discovered in the mid-Devonian fossil forest at Gilboa. Nature 483, 78–81.

Stein, W.E., Mannolini, F., Hernick, L.V., Landing, E., Berry, C.M., 2007. Giant cladoxylopsid trees resolve the enigma of the Earth's earliest forest stumps at Gilboa. Nature 446, 904–907.

Strother, P.K., Wellman, C.H., 2016. Palaeoecology of a billion-year-old non-marine cyanobacterium from the Torridon group and Nonesuch formation. Palaeontology 59, 89–108.

Strullu-Derrien, C., Kenrick, P., Pressel, S., Duckett, J.G., Rioult, J.-P., Strullu, D.-G., 2014. Fungal associations in *Horneophyton ligneri* from the Rhynie Chert (c. 407 million year old) closely resemble those in extant lower land plants: novel insights into ancestral plant-fungus symbioses. New Phytologist 203, 964–979.

Strullu-Derrien, C., Wawrzyniak, Z., Goral, T., Kenrick, P., 2015. Fungal colonization of the rooting system of the early land plant Asteroxylon mackiei from the 407-Myr-old Rhynie Chert (Scotland, UK). Botanical Journal of the Linnean Society 179, 201–213.

Taktek, S., Martin Trépanier, M., Servin, P.M., St-Arnaud, M., Piché, Y., Fortin, J.A., Antoun, H., 2015. Trapping of phosphate solubilizing bacteria on hyphae of the arbuscular mycorrhizal fungus *Rhizophagus irregularis* DAOM 197198. Soil Biology & Biochemistry 90, 1–9.

Taylor, L.L., Leake, J.R., Quirk, J., Hardy, K., Banwart, S.A., Beerling, D.J., 2009. Biological weathering and the long-term carbon cycle: integrating mycorrhizal evolution and function into the current paradigm. Geobiology 7, 171–191.

Taylor, L.L., Banwart, S.A., Leake, J.R., Beerling, D.J., 2011. Modelling the evolutionary rise of ectomycorrhiza on sub-surface weathering environments and the geochemical carbon cycle. American Journal of Science 311, 369–403.

Taylor, L.L., Banwart, S.A., Valdes, P.J., Leake, J.R., Beerling, D.J., 2012. Evaluating the effects of the co-evolution of terrestrial ecosystems, climate and CO_2 on continental weathering over geological time: a global-scale process-based approach. Philosophical Transactions of the Royal Society, B 367, 565–582.

Tedersoo, L., Bahram, M., Põlme, S., Kõljalg, U., Yorou, N.S., Wijesundera, R., Villarreal Ruiz, L., Vasco-Palacios, A.M., Thu, P.Q., Suija, A., Smith, M.E., Sharp, C., Saluveer, E., Saitta, A., Rosas, M., Riit, T., Ratkowsky, D., Pritsch, K., Põldmaa, K., Piepenbring, M., Phosri, C., Peterson, M., Parts, K., Pärtel, K., Otsing, E., Nouhra, E., Njouonkou, A.L., Nilsson, R.H., Morgado, L.N., Mayor, J., May, T.W., Majuakim, L., Lodge, D.J., Lee, S.S., Larsson, K.H., Kohout, P., Hosaka, K., Hiiesalu, I., Henkel, T.W., Harend, H., Guo, L.D., Greslebin, A., Grelet, G., Geml, J., Gates, G., Dunstan, W., Dunk, C., Drenkhan, R., Dearnaley, J., De Kesel, A., Dang, T., Chen, X., Buegger, F., Brearley, F.Q., Bonito, G., Anslan, S., Abell, S., Abarenkov, K., 2014. Global diversity and geography of soil fungi. Science. 346, 1256688. http://dx.doi.org/10.1126/science.1256688. 25430773.

Thorley, R.M.S., Taylor, L.L., Banwart, S.A., Leake, J.R., Beerling, D.J., 2015. The role of forest trees and their mycorrhizal fungi in carbonate rock weathering and its significance for global carbon cycling. Plant, Cell and Environment 38, 1947–1961.

Tisserant, E., Malbreil, M., Kuo, A., Kohler, A., Symeonidi, A., Balestrini, R., Charrong, P., Duensingh, N., dit Frey, N.F., Gianinazzi-Pearsoni, V., Gilbert, L.B., Handa, Y., Herr, J.R., Hijri, M., Koul, R., Kawaguchi, M., Krajinski, F., Lammers, P.J., Masclaux, F.G., Murat, C., Morin, E., Ndikumana, S., Pagni, M., Petitpierre, D., Requena, N., Rosikiewicz, P., Riley, R., Saitop, K., San Clemente, H., Shapiro, H., van Tuinen, D., Bécard, G., Bonfante, P., Paszkowski, U., Shachar-Hill, Y.Y., Tuskan, G.A., Young, J.P.W., Sanders, I.R., Henrissat, B., Rensing, S.A., Grigoriev, I.V., Corradi, N., Roux, C., Martin, F., 2013. Genome of an arbuscular mycorrhizal fungus provides insight into the oldest plant symbiosis. Proceedings of the National Academy of Sciences of the United States of America 110, 20117–20122. http://dx.doi.org/10.1073/pnas.1313452110.

Trembath-Reichert, E., Wilson, J.P., McGlynn, S.E., Fischer, W.W., 2015. Four hundred million years of silica biomineralization in land plants. Proceedings of the National Academy of Sciences of the United States of America 112, 5449–5454.

Turner, B.L., Lambers, H., Condron, L.M., Cramer, M.D., Leake, J.R., Richardson, A.E., Smith, S.E., 2013. Soil microbial biomass and the fate of phosphorus during long-term ecosystem development. Plant and Soil 367, 225–234.

Verbruggen, E., Jansa, J., Hammer, E.C., Rillig, M.C., 2016. Do arbuscular mycorrhizal fungi stabilize litter-derived carbon in soil? Journal of Ecology 104, 261–269.

Walker, T.W., Syers, J.K., 1976. The fate of phosphorus during pedogenesis. Geoderma 15, 1–19.

Wang, B., Yeun, L.H., Xue, J.-Y., Liu, Y., Ané, J.M., Qiu, Y.L., 2010. Presence of three mycorrhizal genes in the common ancestor of land plants suggests a key role of mycorrhizas in the colonization of land by plants. New Phytologist 186, 514–525.

Wilson, G.T., Rice, C.W., Rillig, M.C., Springer, A., Hartnett, D.C., 2009. Soil aggregation and carbon sequestration are tightly correlated with the abundance of arbuscular mycorrhizal fungi: results from long-term field experiments. Ecology Letters 12, 452–461.

Winther, J.L., Friedman, W.E., 2007. Arbuscular mycorrhizal associations in Lycopodiaceae. New Phytologist 177, 790–801.

Yin, Z., Zhu, M., Davidson, E.H., Bottjer, D.J., Zhao, F., Tafforeau, P., 2015. Sponge grade body fossil with cellular resolution dating 60 Myr before the Cambrian. Proceedings of the National Academy of Sciences of the United States of America E1453–E1460. http://dx.doi.org/10.1073/pnas.1414577112.

Zhang, L., Fan, J., Ding, X., He, X., Zhang, F., Feng, G., 2014. Hyphosphere interactions between an arbuscular mycorrhizal fungus and a phosphate solubilizing bacterium promote phytate mineralization in soil. Soil Biology & Biochemistry 74, 177–183.

Zhang, L., Xu, M., Liu, Y., Zhang, F., Hodge, A., Feng, G., 2016. Carbon and phosphorus exchange may enable cooperation between an arbuscular mycorrhizal fungus and a phosphate-solubilizing bacterium. New Phytologist 210, 1022–1032.

CHAPTER

3

Role of Mycorrhizal Symbiosis in Mineral Weathering and Nutrient Mining from Soil Parent Material

M.M. Smits[1], *H. Wallander*[2]

[1]Hasselt University, Diepenbeek, Belgium; [2]Lund University, Lund, Sweden

3.1 INTRODUCTION

Rocks are the primary source of all plant nutrients, except nitrogen. These nutrients are bound into various crystalline structures (minerals). Minerals are either formed during rock formation from magma (primary minerals) or formed during soil formation (secondary minerals). Secondary minerals are formed when the local soil solution is saturated in respect to that mineral. In contrast to secondary minerals, primary minerals are formed in the Earth mantle at high temperature and pressure. At the Earth surface these minerals may be thermodynamically unstable and will eventually completely dissolve. This dissolution process is extremely slow for most minerals. It has been estimated that it takes more than 30 million years to dissolve a 1-mm diameter quartz grain under natural soil conditions (Lasaga, 1984). Nonetheless, soil mineral weathering provides an essential input of plant nutrients into ecosystems, avoiding or delaying nutrient limitations (Chadwick et al., 1999).

In addition, mineral weathering produces cations that counteract soil acidification, thereby improving the availability of most plant nutrients (van Breemen et al., 1983). In addition, clays are formed as a weathering product of feldspars and micas (Oades, 1988). Clay particles contribute, with their negatively charged surfaces, to the cation exchange capacity (CEC) of the soil, reducing the leaching of positively charged nutrients such as K^+ and NH_4^+. Clay content correlates positively with water holding capacity and soil organic matter (SOM) content (Sollins et al., 1996).

Moreover, weathering of Ca- and Mg-silicate minerals plays a central role in the global carbon cycle because large amounts of Ca and Mg, released by the weathering process, will be locked up as carbonates in marine sediments (Sundquist, 1985). In the long term, atmospheric CO_2 is regulated by the weathering rates of these minerals, which is influenced by climate and mountain uplift (Berner, 2003; Raymo and Ruddiman, 1992).

Mycorrhizal Mediation of Soil
http://dx.doi.org/10.1016/B978-0-12-804312-7.00003-6

Nearly 100 years ago the vast amounts of nutrients locked in soil minerals triggered the question of whether or not plants can actively tap into this potential nutrient source (Haley, 1923; Turk, 1919). Five decades later, studies appear on the role of microorganisms, including mycorrhizal fungi, in mineral weathering (Webley et al., 1963; Duff et al., 1963; Sperber, 1958; Boyle et al., 1967; Boyle and Voigt, 1973). More recently, a publication with the provocative title "Rock Eating Fungi" appeared in the journal *Nature* (Jongmans et al., 1997). This publication presented evidence of, presumably mycorrhizal, fungal hyphae drilling their way (chemically and/or physically) into feldspar grains. This paper initiated renewed interest in the topic. A series of reviews have been published since then covering the research up to 2009 (Finlay et al., 2009; Hoffland et al., 2004; Landeweert et al., 2001).

Since 2009 more evidence of mycorrhizal weathering has been published based on in vitro and microcosm research. A new perspective is the influence of the emergence of different types of mycorrhizal fungi during the evolution of land plants on mineral weathering rates and thus the global carbon cycle (Taylor et al., 2011). The gap between laboratory-based studies and the real world has been bridged by several field-based studies and mathematical modeling (Sverdrup, 2009; Taylor et al., 2011). An interesting approach has been developed by Quirk et al. (2012, 2014). They created weathering arenas in soil-based microcosms with natural mycorrhizal inoculum. Although the authors claimed to demonstrate a substantial increase in Ca-silicate dissolution rates (2–3 × compared with nonplant controls), we do not agree. Their sequential dissolution approach does not quantify the total amount of Ca-silicates (see data supplement of Quirk et al., 2012). It only dissolves Ca-silicates present on the surface of the basalt grains (<5% of total present). The actual sequential digestion will not oxidize organic matter, and any organic layer present on the basalt grains will block the Ca-silicate dissolution (reducing the estimated Ca-silicates and consequently inducing the calculated weathered Ca-silicate). Thus far, in our opinion, evidence of a substantial role of mycorrhizal fungi on soil mineral weathering has been missing from an experimental and modeling perspective. What is clear from both field and laboratory experiments is that trees have a major effect on soil weathering. Any positive effect of mycorrhizal fungi on tree growth rates will have an indirect positive effect on soil weathering.

In this chapter we briefly introduce the basic mechanisms of physical and chemical weathering because insight in these mechanisms is of vital importance for the interpretation of results from laboratory-based experiments and field and modeling studies. Next, we give an overview of the recent literature on this topic and set them in perspective with the current knowledge on mineral dissolution kinetics. At the end of this chapter we propose future research directions.

3.2 MECHANISMS OF MINERAL WEATHERING

The most visible outcome of weathering is the breakup of rocks and minerals into smaller fragments. This so-called physical weathering acts on all scales, from the erosion of complete mountaintops to micrometer-scale cracks in mineral crystals. Well-known mechanisms of physical weathering are thermal stress and mechanical force caused by freezing water and penetrating tree roots. Fungal hyphae, colonizing cracks and voids in mineral grains, can also produce substantial mechanical force. They can build up high osmotic pressure in their tissues (up to 20 μN/μm). This is enough pressure to penetrate bulletproof material (Howard et al., 1991) and to widen existing cracks in mineral grains and rock fragments. The outcome of physical weathering is an increase in mineral surface area exposed to the soil solution.

Less visible is the chemical alteration or dissolution of minerals. Although in principle most primary minerals dissolve in soil solution, certain compounds accelerate the process. The most common, and by far quantitatively most important weathering agents, are protons. Protons, and hydroxide under alkaline conditions, attack the ion bindings in the mineral crystal lattice. This process is called hydrolysis (or carbonation when carbonic acid is the main proton donor). Biotic processes have a strong influence on the soil solution pH via the exudation of protons in exchange of positively charged nutrients such as NH_4^+ and K^+, the exudation of organic acids, and the release of CO_2 into the soil solution.

Organic acids, such as oxalic acid and citric acid, not only contribute to proton-driven weathering, but their deprotonated anions (in this case oxalate and citrate) also interact in a similar way as protons and hydroxide with the mineral crystal lattice. In fact, many of the deprotonated anions of organic acids are stronger weathering agents than protons and hydroxide. They behave as strong complexants with metals, including Al^{3+}, a central element in most mineral crystal lattices.

Siderophores are another set of organic compounds with metal-complexing properties. These molecules bind strongly with Fe^{3+} and play a key role in the release and uptake of Fe into bacteria, fungi, and plants (Kraemer et al., 2014; Ahmed and Holmström, 2014). Primary minerals containing substantial amounts of Fe, such as hornblende and biotite, show enhanced dissolution rates in the presence of microbial or fungal siderophores (Kalinowski et al., 2000; Sokolova et al., 2010).

To understand the effect of mycorrhizal fungi on weathering, we first need to determine the limiting step in the dissolution process. After decades of research it is well established that under normal, far from equilibrium conditions, the rate-limiting step is the formation of so-called activated surface complexes. That is the complexation of weathering agents such as protons and organic ligands with metals in the mineral crystal lattice (Furrer and Stumm, 1986; Wieland et al., 1988). Transition state theory can describe the kinetics of this step (Lasaga, 1984). For a single weathering agent, its effect on dissolution rate can be described with a simple equation:

$$R = A \cdot k \cdot (\text{agent})^n \tag{3.1}$$

where R is the dissolution rate, A is the mineral surface area, k is the specific rate coefficient, (agent) is the activity of the weathering agent, and n is the reaction order. It is extremely important to notice that, to our knowledge, for all tested weathering agents on all tested primary minerals, the reaction order is less than 1 (between 0.5 and 0.8). This has major, and counterintuitive, consequences in understanding the effect of soil solution heterogeneity on soil-scale weathering rates (see Smits, 2009, and Section 3.4 of this chapter).

3.3 FUNGAL WEATHERING IN THE LABORATORY

Fueled by carbohydrate supplied by the host, many ectomycorrhizal (EcM) fungi have the capacity to exude organic acids and acidify their surrounding substrate when growing in axenic cultures and in symbiosis with plants under laboratory conditions (Rosling, 2009;

Hoffland et al., 2004; Schmalenberger et al., 2015). Using flow-through systems, Calvaruso et al. (2013) estimated weathering rates to be 10 times higher when EcM pine seedlings were present compared with unplanted systems and attributed this to exudation of organic acids and the acidifying effects by the EcM fungi. However, in our opinion there is a large step to transfer these results to natural systems.

Other microorganisms in the soil can perform similar activities. For instance, many soil bacteria have a strong capacity to acidify their surroundings when sufficient carbon resources are available. Trees foster specific communities of bacteria in the rhizosphere, or mycorrhizasphere (Calvaruso et al., 2013; Collignon et al., 2011), and these organisms may be the active partners of weathering reactions in the soil (for review see Uroz et al., 2009). In addition, wood-decomposing fungi produce large amounts of oxalic acids when degrading wood (Dutton and Evans, 1996), which may have secondary effects on P release from the soil (Fransson et al., 2004).

It is not clear if organic acid release and acidification is a primary mechanism to release nutrients from soil minerals or a secondary effect related to the way that microorganisms are growing in the soil. Protons are exuded in response to uptake of positively charged ions regardless of whether minerals are present or not, and exudation of organic acids can occur in response to many different factors such as pathogenesis and detoxification of heavy metals (Dutton and Evans, 1996). Van Schöll (2006a) studied organic acid production in relation to nutrient availability for several EcM fungi growing in vitro or in symbiosis with a host tree. There was no clear trend in responses among the different species growing under different nutrient levels.

Furthermore, concentrations of organic acids in soil solution (van Hees et al., 2000) seldom reach levels that have significant effects on mineral dissolution. On the other hand, fungal hyphae attach to surfaces to be able to grow and proliferate through the soil, which may physically and chemically affect the surface of the minerals (McMaster et al., 2008). When attached to the surfaces, fungi form a layer of extracellular polymeric substances (EPS) around the hyphae. Balogh-Brunstad et al. (2008) suggested that sufficient concentrations of organic acids may accumulate under biolayers of EPS such that mineral dissolution is influenced, and they called for more experiments to confirm this possible effect.

Nanoscale observations using atomic force microscopy on fungal colonization of mineral surfaces demonstrated that a 40- to 80-nm thick EPS formed around hyphal tips of the EcM fungus *Paxillus involutus* (Gazzè et al., 2013). This material fused to form a biofilm that covered most of the mineral surface where the fungus was growing. Saccone et al. (2011) found similar biolayers formed by *P. involutus* growing on hornblende, chlorite, and biotite surfaces. Furthermore, the hornblende surface was less resistant to mechanical forcing under the biolayer compared with freshly cleaved surfaces, suggesting that enhanced weathering had occurred. In addition, several studies have demonstrated dissolution channels on minerals where EcM fungal hyphae have grown. Sometimes these tracks can be 50 nm deep (Gazzè et al., 2012).

In some laboratory experiments with axenic cultures of microorganisms it has been found that attachment to biotite surfaces yielded stronger dissolution compared with when a membrane separated them. This demonstrates that not only the chemicals produced but also the physical attachment is important for mineral dissolution, at least in experiments with bacteria (Holmström, 2015). However, in other experiments no additional effect on weathering was found when hyphae of EcM fungi were attached to minerals in axenic growth in solution cultures (Balogh-Brunstad et al., 2008).

One problem with axenic solution experiments is that no sink (i.e., plant) is available to consume the released elements. Nutrient removal from the soil solution will change the saturation state of minerals containing these nutrients. This process will increase the dissolution rate, although its effect is only substantial under close to saturated conditions. In one study with axenic microcosms with pine seedlings in symbiosis with *P. involutus*, Bonneville et al. (2011) found high levels of K, Mg, and Fe removal under hyphae attached to biotite, which suggests that both hyphal attachment and a sink for element removal (the plant) are important for mineral dissolution to occur. Furthermore, it could be demonstrated that biotite surfaces were strongly acidified under the hyphae, which suggests that specific chemical conditions occur under biolayers formed by the fungal mycelium. Schmalenberger et al. (2015) demonstrated mineral-specific exudation of oxalate by *P. involutus* using labeled $^{14}CO_2$ given to the host plant. Oxalate was exuded in response to minerals in the following sequence: Gabbro > limestone, olivine and basalt > granite and quartz.

Active weathering by EcM fungi presumes that elements provided by the minerals are in short supply and that fungal weathering activity will increase if these essential nutrients become growth limiting. Unfortunately, very few studies have measured weathering capacity by EcM fungi under different nutrient conditions, but results by Rosling et al. (2007) do not suggest that oxalic acid exudation by soil fungi is enhanced by P limitation in a study with no EcM fungi present. On the other hand, work by van Schöll et al. (2006a) demonstrated that limitation of nutrients (P, Mg, K) affected the composition of organic acids exuded by EcM fungi (more oxalate) but not the total amounts. It is interesting to note that they found different results when EcM fungi were grown in pure cultures versus when grown in symbiosis with plant seedlings, which again highlights the importance of having in a nutrient sink (the plant) for accurate interpretations of the effects of EcM fungi on weathering. Smits et al. (2012) demonstrated significant weathering of apatite by *P. involutus* when P was in limiting supply whereas this effect diminished when sufficient P was supplied. In addition, van Schöll et al. (2006b) demonstrated significant weathering of muscovite by *P. involutus* when K was in low supply whereas no effect on hornblende was found under Mg deficiency. Other fungal species that were tested had no effect on weathering of muscovite under K limitation. Carefully designed experiments under controlled conditions are necessary to increase our understanding of fungal weathering under varying nutrient conditions (e.g., K- or P- vs. N-limited conditions). Such approaches could be to use trenched plots with windows of meshes with different sizes in a similar way as was used by Heinemeyer et al. (2011) to estimate the EcM contribution to soil respiration. Such experiments are necessary to predict how forests will respond to future changes in climate and nitrogen deposition.

3.4 FROM LABORATORY TO FIELD

Although numerous experiments clearly show that EcM fungi affect mineral dissolution (see previous section), the significance of mycorrhizal weathering on soil- and global-scale weathering rates remains controversial. This ongoing controversy is caused by the challenge of quantifying the mycorrhizal contribution in the field and insufficient consideration of other soil processes interacting with mineral weathering. Most field research has been focused on established ecosystems in temperate and boreal climates with substantial amounts of SOM accumulation. This is important because recycling nutrients within SOM contributes a major part of plant nutritional requirements (Chapters 4 and 8). However, on fresh surfaces no organic matter has been built up and mineral weathering is presumably the major source of plant nutrients. Modeling is a useful tool to test the scope of mycorrhizal weathering concepts. In this section we will review the available field data and modeling work on mycorrhizal weathering in this area of research.

Without doubt vegetation has a substantial positive effect on soil mineral weathering (Berner, 1992); however, uncertainty remains about the degree to which associated mycorrhizal fungi contribute to this effect. A direct estimation of the contribution of mycorrhizal fungi on the weathering process is challenging because of the slow kinetics of soil mineral weathering and the complex soil matrix. Three different approaches have been adopted to address the effect of mycorrhizal fungi on weathering: (1) historical weathering markers, (2) stable isotopes to trace the source of tree nutrients, and (3) quantification of incubated minerals in contrasting soils. In addition, modeling can be a powerful tool to scale up proposed mycorrhizal weathering mechanisms to soil- or even global-scale weathering processes.

3.4.1 Historical Weathering Markers

Tunnels, as described in Jongmans et al. (1997), are the only quantifiable fungal markers of weathering that remain visible over geological time. Unfortunately, fungal tunneling either reflects only a small portion of the total effect of fungi on the weathering process or the fungal effect is negligible because tunneling contributes less than 0.5% of total mineral weathering (Smits et al., 2005). In a recent paper, Koele et al. (2014) showed that mineral tunneling is not exclusively found under EcM vegetation, but also in forest soils, never exposed to EcM vegetation. This suggests that tunnels can be formed by other means, such as arbuscular mycorrhizal (AM) or saptrophic fungi. Work by Quirk and others indeed shows that AM fungi create weathering trenches on mineral surfaces (Quirk et al., 2012, 2014). It demonstrates that both types of mycorrhizal fungi can interact with mineral surfaces in a similar way. However, it remains unclear how much these trenches contribute to total dissolution rates.

3.4.2 Isotope Tracers

Stable isotopes, especially of Ca and Sr, have been used extensively to source the origin of Ca in drainage water; when applied to plant tissues, they can be used to trace plant nutrients back to their primary source. Isotope tracing has been mostly used to study apatite weathering. Apatite is a calcium–phosphate mineral, and because P has no stable isotopes, the uptake dynamics can only be studied via the Ca ion (or potentially the $^{18}O/^{16}O$ in the phosphate group). In most rocks and soils apatite is the sole primary P source. However, its contribution to the soil solution Ca pool is minor compared with other minerals. If the Ca isotope ratio in the plant is more similar to the signature in apatite than to the signature in the soil solution, then it indicates that the plant directly acquires Ca from apatite. Blum et al. (2002) applied this technique to a temperate mixed forest. However, because in their study area the different mineral sources had similar Ca isotope ratios, they instead used the ratio between Ca and Sr to calculate the contribution of the different Ca sources in plant uptake. In 1926 Fay (1926) already warned for the use of the Ca/Sr ratio to trace sources of Ca because of fractionation during basic chemical transformations (e.g., Ca/Sr-oxalate precipitation/dissolution). If not taken into account, then the strong fractionation of Ca and Sr potentially leads to wrong conclusions. Most of the Ca taken up by trees comes from litter recycling. In a comparable mixed forest, also in the northeastern United States, the annual Ca import from weathering in the rooting zone is less than 0.3% of the annual Ca uptake, which was a 4 times smaller flux than the annual atmospheric deposition (Dijkstra and Smits, 2002). A closer look at the data presented in Blum et al. (2002) clearly separates EcM trees with a high Ca/Sr ratio (the two evergreen species) and EcM trees with a low Ca/Sr ratio in their leaves (the two deciduous species). The EcM deciduous trees group together with the AM deciduous trees. Although in principle this difference could be explained by host-specific mycorrhizal communities, with the two evergreen species hosting EcM fungi with stronger capability to weather apatite than EcM fungi hosting the deciduous species, a more obvious explanation is that Ca/Sr fractionation is different during throughfall and litter recycling between these evergreen and deciduous trees (hence the similar Ca/Sr ratios of EcM and AM deciduous trees). Up to now, isotope techniques have not provided convincing evidence of a major mycorrhizal contribution to mineral weathering.

3.4.3 Mineral Incubations

Long-term soil incubation of minerals in mesh bags is an in situ approach to study the weathering process. Gobran et al. (2005) noted that a strong advantage of mesh-bag incubation studies, compared with microcosm experiments, is that weathering rates are measured under real soil conditions. The drawback of this approach is that it is impossible to distinguish mycorrhizal weathering actions from other weathering actions. What is possible is the creation of root-exclusion zones. A study by Turpault et al. (2009) shows that dissolution of Labradorite (a Ca-rich feldspar) in root-exclusion zones occurs at half of the rate observed where roots are present. Furthermore, this effect diminished in plots that were previously fertilized with Ca. Apatite showed more complex dynamics. In the topsoil, the apatite dissolution rate is higher in root-exclusion zones, whereas at 20-cm depth it is reversed. In contrast to Labradorite, liming did not have a strong effect on the apatite dissolution dynamics.

It is interesting to note that the same dynamics of apatite dissolution rate appeared in a very different setting. Smits et al. (2014) studied apatite loss from the soil profile in a vegetation gradient in Norway. Overall, apatite dissolution rate shows a clear correlation with pH, with 74% of all variation in dissolution rate explained by pH. The remaining variation showed in the topsoil a negative correlation and at 30- to 40-cm depth a positive relation with fungal biomass. A possible explanation, given in Smits et al. (2014), is that in the topsoil mineral dissolution is inhibited by the adsorption of fulvic and humic acids (produced by rhizosphere-induced SOM degradation). This illustrates the strong interrelationship of mineral dissolution with other soil processes (Fig. 3.1).

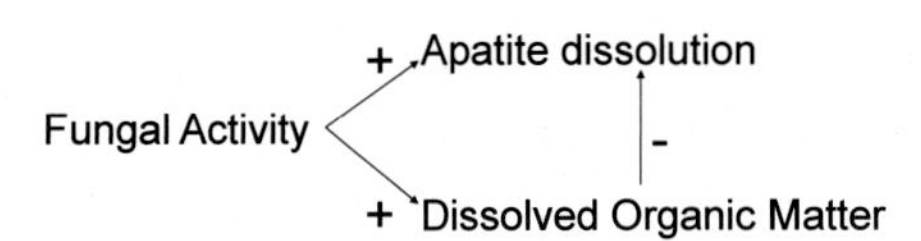

FIGURE 3.1 The direct effects of mycorrhizal fungi in topsoil may increase apatite dissolution, but fungal production of dissolved organic compound may indirectly reduce apatite dissolution, as described in Smits et al. (2014).

3.4.4 Modeling

Modeling is a powerful tool to integrate different processes into one theoretical framework. Models of mineral dissolution rates must account for key factors and processes as well as the scales at which these factors and processes function. Soils are extremely heterogeneous environments, and most processes occur at the micrometer to millimeter scale. Because of the limitation of computational power, simulation models are restricted in their spatial scale. How these small-scale mechanisms are upscaled can have major consequences on the model behavior at the whole-soil level.

The most well-known weathering model, PROFILE (Sverdrup and Warfvinge, 1993), is based on a series of chemical dissolution reactions simulated at the scale of user-defined soil layers, based on the expected average soil solution in each layer. The parameterization of this model is based on batch experiments with organic and inorganic weathering agents. On the basis of PROFILE and its application in many boreal forest soils, Sverdrup (2009) concluded that protons are the major weathering agent in these soils and that organic chelators such as oxalate only play a minor role. The main critique of the PROFILE model from a fungal point of view is that simulating weathering on a soil-layer scale ignores the potential importance of local (fungal scale) high concentrations of fungal weathering agents (Finlay et al., 2009). However, as illustrated in Smits (2009), because of the specific dissolution kinetics of the main dissolution reactions, local concentrations of weathering agents do not automatically lead to higher dissolution rates (see Eq. (3.1)). Following from the kinetics (based on the transition state theory; see previous sections), the PROFILE modeling approach only underestimates the action of certain weathering agents if they are exuded to and stay within a micrometer range of specific mineral surfaces (Smits, 2009). On the other hand, if fungal action takes place specifically at surface sites with weaker crystal structure (e.g. crystal steps, fresh cracks), then the fungal effect on overall dissolution rate will be higher. Indeed, circumstantial evidence based on scanning electron microscopy studies of soil minerals suggests a preferential fungal colonization along crystal steps. It is tempting to believe that these hyphae sense the more easily available nutrients along these steps ("biosensing," see Leake et al., 2008); however, an alternative explanation could be that hyphal growth is determined by the topographic structure of the mineral surface.

In contrast to the PROFILE model, the weathering module developed by Taylor et al. (2011) incorporates exudation of fungal weathering agents on the mineral grain scale. The model explicitly assumes that in systems with EcM trees all tree–soil interactions (uptake and exudation) take place via the EcM fungi acting close to nutrient-bearing minerals rather than close to quartz, which is the most common mineral in topsoils. In this case, the presence of EcM fungi has a major effect on mineral weathering. It is interesting to note that, similar to in the PROFILE model, protons are the major weathering agents (Lyla Taylor, *personal communication*). Although this model neatly explains the increased silicate weathering during the increase of EcM fungi over the past 120 million years, we have major issues with the reality of the assumption made that all EcM action takes place only at nutrient-bearing minerals. The main part of protons exuded by plant roots and mycorrhizal fungi is in exchange for nutrients (cations and NH_4^+). If most of these nutrients come from organic matter breakdown, then most protons are exuded where these nutrients are taken up. That is mostly at CEC sites (clays and organic matter) or close to the degrading organic matter itself and not specifically close to nutrient-bearing minerals. Fungal length measurements on individual grains from the topsoil in a boreal pine forest show a preferential colonization of feldspars over quartz, but because quartz was the dominant mineral, most fungal length was still found on the quartz grains (Smits, 2009). In addition, mineral mesh-bag incubation studies do not show a clear effect of incubated mineral type on fungal colonization rates (Rosenstock et al., 2016). These observations undermine the validity of the assumption that all EcM exchange processes take place only at nutrient-bearing minerals. Without this specific assumption the Taylor model will not show different weathering rates for EcM and non-EcM trees if tree-scale uptake and exudation processes are the same. If fungi–mineral contact does not play an important role in the tree-mycorrhizal effect on mineral dissolution, then any difference among tree species in specific mineral dissolution would be attributed to differences in soil acidification and SOM dynamics and not directly to mycorrhizal type.

3.5 CONCLUSIONS AND FUTURE RESEARCH DIRECTIONS

Despite the high number of laboratory experiments demonstrating fungal-mineral interactions, there is no clear evidence that these laboratory-scale observed processes play a significant role on soil-scale mineral dissolution rates. Field and modeling studies indicate that protons are the dominant chemical weathering agents. Organic weathering agents such as oxalate and citrate only play a minor role. These organic weathering agents could potentially play a more important role if they stay close to mineral surfaces or are preferentially exuded near more reactive parts of the mineral surfaces (e.g. crystal steps and cracks). The observations of "trenches" and "canals" in the shape of fungal hyphae (e.g., Saccone et al., 2011) suggest distinct local weathering environments, although it cannot be excluded that these shapes are artifacts of the hyphal removal procedure. To better estimate the potential role of these local weathering environments on total dissolution rates, the local liquid chemistry should be studied on the micrometer scale and upscaled to soil-scale dissolution rates. To achieve this, detailed information about fungal distribution over different minerals and hyphosphere chemistry is needed. Future research on mycorrhizal weathering should ideally include whole ecosystem dynamics, including SOM behavior. We recommend including early-stage primary succession ecosystems (see also Chapter 5) on low reactive surfaces (e.g., fresh granites) because nutrient inputs from mineral weathering are not minor compared with nutrient recycling. Mesocosm systems as presented by Quirk et al. (2012, 2014) are ideal systems to bridge laboratory and field studies, but care must be taken to properly quantify total dissolution fluxes. The use of stable isotopes (by choosing minerals and soils with distinct isotope ratios) improves the sensitivity over total mass balance approaches, but care must be taken in respect to fractionation.

References

Ahmed, E., Holmström, S.J.M., 2014. Siderophores in environmental research: roles and applications. Microbial Biotechnology 7, 196–208.

Balogh-Brunstad, Z., Keller, C.K., Dickinson, T.J., Stevens, F., Li, C., Bormann, B.T., 2008. Biotite weathering and nutrient uptake by ectomycorrhizal fungus *Suillus tomentosus*, in liquid-culture experiments. Geochimica et Cosmochimica Acta 72, 2601–2618.

Berner, R.A., 1992. Weathering plants, and the long-term carbon cycle. Geochimica et Cosmochimica Acta 56, 3225–3231.

Berner, R.A., 2003. The long-term carbon cycle fossil fuels and atmospheric composition. Nature 426, 323–326.

Blum, J.D., Klaue, A., Nezat, C.A., Driscoll, C.T., Johnson, C.E., Siccama, T.G., et al., 2002. Mycorrhizal weathering of apatite as an important calcium source in base-poor forest ecosystems. Nature 417, 729–731.

Bonneville, S., Morgan, D.J., Schmalenberger, A., Bray, A., Brown, A., Banwart, S.A., et al., 2011. Tree-mycorrhiza symbiosis accelerate mineral weathering: Evidences from nanometer-scale elemental fluxes at the hypha–mineral interface. Geochimica et Cosmochimica Acta 75, 6988–7005.

Boyle, J.R., Voigt, G.K., 1973. Biological weathering of silicate minerals. Plant and Soil 38, 191–201.

Boyle, J.R., Voigt, G.K., Sawhney, B.L., 1967. Biotite flakes: alteration by chemical and biological treatment. Science 155, 193–195.

Calvaruso, C., Turpault, M.P., Frey-Klett, P., Uroz, S., Pierret, M.C., Tosheva, Z., et al., 2013. Increase of apatite dissolution rate by Scots pine roots associated or not with *Burkholderia glathei* PML1(12)Rp in open-system flow microcosms. Geochimica et Cosmochimica Acta 106, 287–306.

Chadwick, O., Derry, L., Vitousek, P., Huebert, B., Hedin, L., 1999. Changing sources of nutrients during four million years of ecosystem development. Nature 397, 491–497.

Collignon, C., Calvaruso, C., Turpault, M.P., 2011. Temporal dynamics of exchangeable K Ca and Mg in acidic bulk soil and rhizosphere under Norway spruce (*Picea abies* Karst.) and beech (*Fagus sylvatica* L.) stands. Plant and Soil 349, 355–366.

Dijkstra, F.A., Smits, M.M., 2002. Tree species effects on calcium cycling: the role of calcium uptake in deep soils. Ecosystems 5, 385–398.

Duff, R.B., Webley, D.M., Scott, R.O., 1963. Solubilization of minerals and related materials by 2-ketogluconic acid-producing bacteria. Soil Science 95, 105–114.

Dutton, M.V., Evans, C.S., 1996. Oxalate production by fungi: its role in pathogenicity and ecology in the soil environment. Canadian Journal of Microbiology 42, 881–895.

Fay, M., 1926. Strontium as a source of error in blood calcium determinations. American Journal of Physiology 77, 73–75.

Finlay, R.D., Wallander, H., Smits, M.M., Holmström, S., van Hees, P.A.W., Lian, B., et al., 2009. The role of fungi in biogenic weathering in boreal forest soils. Fungal Biology Reviews 23, 101–106.

Fransson, A.M., Valeur, I., Wallander, H., 2004. The wood-decaying fungus *Hygrophoropsis aurantiaca* increases P availability in acid forest humus soil while N addition hampers this effect. Soil Biology and Biochemistry 36, 1699–1705.

Furrer, G., Stumm, W., 1986. The coordination chemistry of weathering: I. Dissolution kinetics of delta-Al2O3 and BeO. Geochimica et Cosmochimica Acta 50, 1847–1860.

Gazzè, S.A., Saccone, L., Ragnarsdottir, K.V., Smits, M.M., Duran, A.L., Leake, J.R., et al., 2012. Nanoscale channels on ectomycorrhizal-colonized chlorite: evidence for plant-driven fungal dissolution. Journal of Geophysical Research: Biogeosciences 117, G00N09.

Gazzè, S.A., Saccone, L., Smits, M.M., Duran, A.L., Leake, J.R., Banwart, S.A., et al., 2013. Nanoscale observations of extracellular polymeric substances deposition on phyllosilicates by an ectomycorrhizal fungus. Geomicrobiology Journal 30, 721–730.

Gobran, G.R., Turpault, M.P., Courchesne, F., 2005. Contribution of rhizospheric processes to mineral weathering in forest soils. In: Huang, P.M., Gobran, G.R. (Eds.), Biogeochemistry of Trace Elements in the Rhizosphere. Elsevier, Amsterdam, The Netherlands.

Haley, D.E., 1923. Availability of potassium in orthoclase for plant nutrition. Soil Science 15, 167–180.

Heinemeyer, A., Di Bene, C, Lloyd, A.R., Tortorella, D., Baxter, R., Huntley, B.Gelsomino, A., Ineson, P., et al., 2011. Soil respiration: implications of the plant–soil continuum and respiration chamber collar–insertion depth on measurement and modelling of soil CO_2 efflux rates in three ecosystems. European Journal of Soil Science 62(1), 82–94.

Hoffland, E., Kuyper, T.W., Wallander, H., Plassard, C., Gorbushina, A.A., Haselwandter, K., et al., 2004. The role of fungi in weathering. Frontiers in Ecology and the Environment 2, 258–264.

Holmström, S.J., 2015. Microbe–mineral interactions: the impact of surface attachment on mineral weathering and element selectivity by microorganisms. Chemical Geology 403, 13–23.

Howard, R.J., Ferrari, M.A., Roach, D.H., Money, N.P., 1991. Penetration of hard substrates by a fungus employing enormous turgor pressures. Proceedings of the National Academy of Sciences 88, 11281–11284.

Jongmans, A.G., van Breemen, N., Lundström, U., van Hees, P.A.W., Finlay, R.D., Srinivasan, M., et al., 1997. Rock-eating fungi. Nature 389, 682–683.

Kalinowski, B., Liermann, L., Brantley, S., Barnes, A., Pantano, C., 2000. X-ray photoelectron evidence for bacteria-enhanced dissolution of hornblende. Geochimica et Cosmochimica Acta 64, 1331–1343.

Koele, N., Dickie, I.A., Blum, J.D., Gleason, J.D., de Graaf, L., 2014. Ecological significance of mineral weathering in ectomycorrhizal and arbuscular mycorrhizal ecosystems from a field-based comparison. Soil Biology and Biochemistry 69, 63–70.

Kraemer, S.M., Duckworth, O.W., Harrington, J.M., Schenkeveld, W.D.C., 2014. Metallophores and trace metal biogeochemistry. Aquatic Geochemistry 21, 159–195.

Landeweert, R., Hoffland, E., Finlay, R.D., Kuyper, T.W., van Breemen, N., 2001. Linking plants to rocks: ectomycorrhizal fungi mobilize nutrients from minerals. Trends in Ecology and Evolution 16, 248–254.

Lasaga, A.C., 1984. Chemical kinetics of water-rock interactions. J. Geophys. Res 89, 4009–4025.

Leake, J.R., Duran, A.L., Hardy, K.E., Johnson, I., Beerling, D.J., Banwart, S.A., Smits, M.M., 2008. Biological weathering in soil: the role of symbiotic root-associated fungi biosensing minerals and directing photosynthate-energy into grain-scale mineral weathering. Mineralogical Magazine 72(1), 85–89.

McMaster, T.J., Smits, M.M., Haward, S.J., Leake, J.R., Banwart, S., Ragnarsdottir, K.V., 2008. High-resolution imaging of biotite dissolution and measurement of activation energy. Mineralogical Magazine 72, 115–120.

Oades, J.M., 1988. The retention of organic matter in soils. Biogeochemistry 5, 35–70.

Quirk, J., Beerling, D.J., Banwart, S.A., Kakonyi, G., Romero-Gonzalez, M.E., Leake, J.R., 2012. Evolution of trees and mycorrhizal fungi intensifies silicate mineral weathering. Biology Letters 8, 1006–1011.

Quirk, J., Leake, J.R., Banwart, S.A., Taylor, L.L., Beerling, D.J., 2014. Weathering by tree-root-associating fungi diminishes under simulated Cenozoic atmospheric CO_2 decline. Biogeosciences 11, 321–331.

Raymo, M.E., Ruddiman, W.F., 1992. Tectonic forcing of late Cenozoic climate. Nature 359, 117–122.

Rosenstock, N.P., Berner, C., Smits, M.M., Krám, P., Wallander, H., 2016. The role of phosphorus magnesium and potassium availability in soil fungal exploration of mineral nutrient sources in Norway spruce forests. New Phytologist 211, 542–553.

Rosling, A., 2009. Trees mycorrhiza and minerals—field relevance of in vitro experiments. Geomicrobiology Journal 26, 389–401.

Rosling, A., Suttle, K.B., Johansson, E., van Hees, P.A.W., Banfield, J.F., 2007. Phosphorous availability influences the dissolution of apatite by soil fungi. Geobiology 5, 265–280.

Saccone, L., Gazzè, S.A., Duran, A.L., Leake, J.R., Banwart, S.A., Ragnarsdóttir, K.V., et al., 2011. High resolution characterization of ectomycorrhizal fungal-mineral interactions in axenic microcosm experiments. Biogeochemistry 111, 411–425.

Schmalenberger, A., Duran, A.L., Bray, A.W., Bridge, J., Bonneville, S., Benning, L.G., et al., 2015. Oxalate secretion by ectomycorrhizal *Paxillus involutus* is mineral-specific and controls calcium weathering from minerals. Scientific Reports 5, 12187.

Smits, M.M., 2009. Scale matters? Exploring the effect of scale on fungal–mineral interactions. Fungal Biology Reviews 23, 132–137.

Smits, M.M., Bonneville, S., Benning, L.G., Banwart, S.A., Leake, J.R., 2012. Plant-driven weathering of apatite—the role of an ectomycorrhizal fungus. Geobiology 10, 445–456.

Smits, M.M., Hoffland, E., Jongmans, A.G., van Breemen, N., 2005. Contribution of mineral tunneling to total feldspar weathering. Geoderma 125, 59–69.

Smits, M.M., Johansson, L., Wallander, H., 2014. Soil fungi appear to have a retarding rather than a stimulating role on soil apatite weathering. Plant and Soil 385, 217–228.

Sverdrup, H., 2009. Chemical weathering of soil minerals and the role of biological processes. Fungal Biology Reviews 23, 94–100.

Sverdrup, H., Warfvinge, P., 1993. Calculating field weathering rates using a mechanistic geochemical model PROFILE. Applied Geochemistry 8, 273–283.

Sokolova, T.A., Tolpeshta, I.I., Topunova, I.V., 2010. Biotite weathering in podzolic soil under conditions of a model field experiment. Eurasian Soil Science 43, 1150–1158.

Sollins, P., Homann, P., Caldwell, B.A., 1996. Stabilization and destabilization of soil organic matter: mechanisms and controls. Geoderma 74, 65–105.

Sperber, J., 1958. Solution of apatite by soil microorganisms producing organic acids. Australian Journal of Agricultural Research 9, 782.

Sundquist, E.T., 1985. Geological perspectives on carbon dioxide and the carbon cycle. In: Sundquist, E.T., Broecker, W.S. (Eds.), The Carbon Cycle and Atmospheric CO_2: Natural Variations Archean to Present. Wiley-Blackwell, Hoboken (NJ), USA.

Taylor, L., Banwart, S.A., Leake, J.R., Beerling, D.J., 2011. Modeling the evolutionary rise of ectomycorrhiza on sub-surface weathering environments and the geochemical carbon cycle. American Journal of Science 311, 369–403.

Turk, E.D., 1919. Potassium-bearing minerals as a source of potassium for plant growth. Soil Science 8, 269–302.

Turpault, M.P., Nys, C., Calvaruso, V., 2009. Rhizosphere impact on the dissolution of test minerals in a forest ecosystem. Geoderma 153, 147–154.

Uroz, S., Calvaruso, C., Turpault, M.P., Frey-Klett, P., 2009. Mineral weathering by bacteria: ecology actors and mechanisms. Trends in Microbiology 17, 378–387.

van Breemen, N., Mulder, J., Driscoll, C.T., 1983. Acidification and alkalinization of soils. Plant and Soil 75, 283–308.

van Hees, P.A.W., Lundström, U., Giesler, R., 2000. Low molecular weight organic acids and their Al-complexes in soil solution—composition distribution and seasonal variation in three podzolized soils. Geoderma 94, 173–200.

van Schöll, L., Hoffland, E., van Breemen, N., 2006a. Organic anion exudation by ectomycorrhizal fungi and *Pinus sylvestris* in response to nutrient deficiencies. New Phytologist 170, 153–163.

van Schöll, L., Smits, M.M., Hoffland, E., 2006b. Ectomycorrhizal weathering of the soil minerals muscovite and hornblende. New Phytologist 171, 805–814.

Webley, D.M., Henderson, M.E.K., Taylor, I.F., 1963. The microbiology of rocks and weathered stones. Journal of Soil Science 14, 102–112.

Wieland, E., Wehrli, B., Stumm, W., 1988. The coordination chemistry of weathering: III. A generalization on the dissolution rates of minerals. Geochimica et Cosmochimica Acta 52, 1969–1981.

CHAPTER

4

Mycorrhizal Interactions With Climate, Soil Parent Material, and Topography

N.C. Johnson[1], *R.M. Miller*[2], *G.W.T. Wilson*[3]

[1]Northern Arizona University, Flagstaff, AZ, Unites States; [2]Argonne National Laboratory, Lemont, IL, United States; [3]Oklahoma State University, Stillwater, OK, United States

4.1 INTRODUCTION

Soils are generated by interactions among living and nonliving factors and are shaped by physical and chemical processes and natural selection. Soil characteristics arise as emergent properties of these complex interactions. The well-known "five-factor model" considers soils as complex adaptive systems shaped by the interactions of climate, parent material, topography, time, and living organisms (Dokuchaev, 1883; Jenny, 1941, 1980). Mycorrhizas respond to all five soil-forming factors and they also influence each factor, and in this regard they are integral to soil development and ecosystem structure (Chapter 2). The spatial distribution of many soil orders can be largely predicted from the geographic patterns in climate, soil parent material, and topography. This chapter examines how mycorrhizas interact with these three state factors with the goal of identifying testable hypotheses to help integrate mycorrhizas into soil and earth system research.

Mycorrhizal symbioses encompass a vast diversity of fungus-plant partnerships and they occur in almost all terrestrial and even some aquatic ecosystems (Chapter 1; Table 1.1). The multiple evolutionary events that generated mycorrhizal symbioses between autotrophic plants and heterotrophic fungi suggest that these nutritional partnerships represent a highly stable evolutionary strategy. Our analysis of mycorrhizas in the five-factor model focuses on the three physicochemical factors: climate, soil parent material, and topography; the "organism factor" is narrowly defined as symbioses among host plants and mycorrhizal fungi (Fig. 4.1). Interactions of mycorrhizas with other soil biotia are explored in Chapters 9 and 22. The importance of time to mycorrhizal development and temporal aspects of mycorrhizal mediation of soil processes and properties are considered in Chapter 5 in the context of succession.

Mycorrhizal Mediation of Soil
http://dx.doi.org/10.1016/B978-0-12-804312-7.00004-8

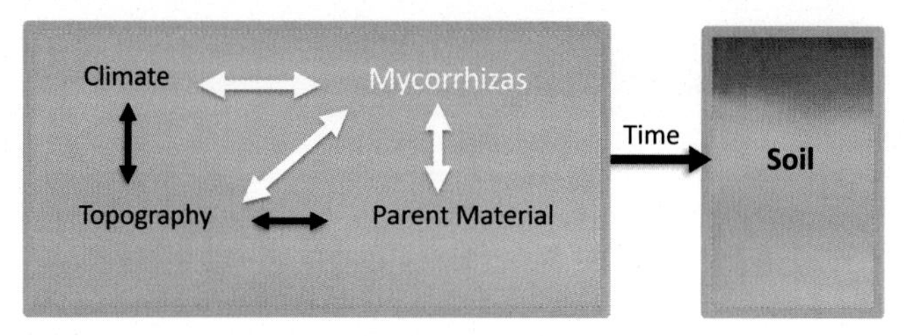

FIGURE 4.1 This chapter explores mycorrhizal interactions with climate, parent material, and topography—three factors in the five-factor model of pedogenesis.

Plants and their associated mycorrhizal fungi are inextricably connected to each other and their local environmental conditions. Geographic variation in climate, parent material, and topography define the abiotic environment to which all living organisms adapt. This chapter examines the ways that climate, parent material, and topography may act similar to environmental filters (i.e., selection pressures) that structure the evolution and ecology of mycorrhizas. Identifying key environmental filters that constrain net primary production and plant and fungal adaptations to these filters will help guide future research and provide insights for incorporating mycorrhizas into predictive models.

4.2 MYCORRHIZAL INTERACTIONS WITH CLIMATE

Climate, defined as long-term precipitation and temperature patterns, is a powerful force in the formation of soils. As the universal solvent, water is a major driver of chemical transformations in the soil profile. The quantity and timing of precipitation within a region determines the total soil water potential and ultimately the amount of physical and chemical "work" that is possible for processes such as leaching, weathering, and erosion. Rates of chemical reactions increase with increasing soil temperature because the frequency of collisions among reaction molecules increases as soils get warmer. The interaction between water availability and temperature influences chemical transformations in the soil and controls evaporation and transpiration, which are the major forces that move water through the soil profile and link soils and plants with the atmosphere.

To a large extent, climate controls the distribution of living organisms. Minimal, optimal, and lethal temperatures are key parameters that delineate the niches of plants and their associated mycorrhizal fungi (e.g., Koske, 1987; Lekberg et al., 2011; Liu et al., 2015). Per-capita growth rates increase with warming from the minimal temperature for survival up to a lethal temperature, which cause populations to crash and give way to a different set of organisms with higher temperature limits. Water availability also controls biotic potential, and the quantity and timing of precipitation has a defining influence on the primary production and composition of plant and fungal communities. Mycorrhizal symbioses are essential plant adaptations for nutrient acquisition in a remarkably wide range of climatic conditions.

Mycorrhizas not only respond to climate, they also influence climate. As discussed in Chapter 2, the evolution of deep-rooted plants and their mycorrhizas in the Phanerozoic can be linked to enhanced mineral weathering and the decarbonation of the oceans (Fig. 2.1). This process has been linked to a 90% reduction in atmospheric CO_2 and a dramatic cooling of the climate during the Devonian period (Fig. 2.2). There is increasing interest in managing mycorrhizas as carbon sinks, with the ultimate goal of mitigating future climate change (Section IV). Achieving this goal will require a better understanding of the factors that govern the allocation of photosynthate to root and fungal growth. Similarly, future studies are needed to gain a more mechanistic understanding of the factors that control the residence time of mycorrhizal biomass in soils and the differences in the decomposition dynamics of arbuscular mycorrhizal (AM), ectomycorrhizal (EcM), and ericoid mycorrhizal (ErM) fungi.

4.2.1 Environmental Predictors of Mycorrhizas

The dominant types of mycorrhizas in a region are strongly tied to climate. It has been suggested that AM symbioses predominate in hot or seasonally warm environments with a wide precipitation range, EcM symbioses generally predominate in moist locations with a wide temperature range, and ErM symbioses predominate in cool and wet environments (Chapter 1, Table 1.1; Read, 1991; van der Heijden and Sanders, 2002). A recent literature analysis showed that global-scale climate and soil patterns could explain approximately 50% of the variance in root colonization intensity of AM and EcM fungi (Soudzilovskaia et al., 2015). This study compiled data on climatic and soil factors with AM colonization data from 4887 plant species in 233 sites and EcM colonization data from 125 plant species in 92 sites. These authors found that AM colonization intensity was best predicted by (1) the mean temperature of the warmest month (19.5°C was optimal), (2) seasonality of temperatures (a 60-day or longer frost-free period was optimal), and (3) a relatively low soil carbon to nitrogen (C/N) ratio (11.8 was optimal). Intensity of EcM colonization was best predicted by (1) precipitation seasonality (low precipitation variation is optimal), (2) soil pH (acidic soil optimal), and (3) the interaction of soil pH and C/N ratio (highest colonization in acid soils when C/N > 11.5). The study by Soudzilovskaia et al. (2015) corroborates previous observations and provides quantitative evidence that the geographic distributions of AM symbioses are more strongly structured by variation in temperature and EcM symbioses are more strongly structured by moisture. The minimal overlap in the optimal soil C/N ratio of AM and EcM colonization is intriguing and will be revisited in the context of soil mineralization processes and contrasting foraging strategies of host plants (see Section 4.2.3).

Temperature affects the rate of chemical reactions and has a profound influence on the growth of living cells. All organisms have a range of minimal, optimal, and maximal (lethal) temperatures. Although AM symbioses occur in seasonally cold climates (Cripps and Eddington, 2005), their optimal and lethal temperatures are relatively high. It has been shown that AM fungi easily survive subzero temperatures when they are dormant, but active growth of AM fungi requires relatively warm temperatures (Tibbett and Cairney, 2007). Some growth of AM fungi inside plant roots has been demonstrated at temperatures below 10°C, but warmer temperatures are necessary for active growth of extraradical hyphae into the soil (Gavito et al., 2003; Hawkes et al., 2008). Even if provided ample photosynthate, AM fungi appear to be incapable of extraradical growth or transfer of phosphorus (P) to their host at temperatures below 10°C (Gavito et al., 2003). At the other extreme, AM fungi seem to be remarkably heat tolerant. Studies from thermal soils indicate that AM fungal hyphae are active in soils up to 35°C whereas root growth was evident up to only 30°C (Bunn et al., 2009). Glomeromycota evolved when the Earth was warmer than today, and they seem to have maintained a preference for warm temperatures.

Unlike AM fungi, which are all members of the phylum Glomeromycota, EcM fungi are polyphyletic, and this diverse evolutionary history may be expected to provide a wider range of optimal temperatures. It has been estimated that EcM symbiosis has evolved in 78–82 different fungal lineages that comprise 251–256 genera (Tedersoo and Smith, 2013). Although EcM symbioses are increasingly being discovered in the tropics, they have primarily been studied in higher latitudes. These studies suggest that EcM are cold tolerant rather than cold loving; growth rates of many different taxa of EcM fungi in culture have been shown to decrease at temperatures below 12°C, and the responses of individual fungal isolates to temperatures below 6°C are variable (Tibbett and Cairney, 2007). The mycelia of many EcM fungi have the capacity to tolerate temperatures below −40°C, with no mortality at −8°C (Lehto et al., 2008). The most likely mechanism to tolerate freezing and associated intracellular dehydration is to increase the concentration of polyols and trehalose (Tibbett and Cairney, 2007). Tropical mycorrhizas and their fungi are understudied (Tedersoo and Smith, 2013); future work is likely to reveal additional EcM taxa in warm climates and the physiological adaptations of these fungi to high temperatures.

4.2.2 Distribution of Soil Orders and Mycorrhizas Corresponds to Climate

The interaction of temperature and moisture on the physicochemical and biological properties of soils generates predictable spatial patterns in the occurrences of many of the major US Department of Agriculture designated soil orders and biomes (Fig. 4.2; NRCS, 1999). Histosols form in waterlogged environments at any temperature, Spodosols form in cold-wet environments, and Oxisols and Ultisols form in warm-wet environments. Alfisols,

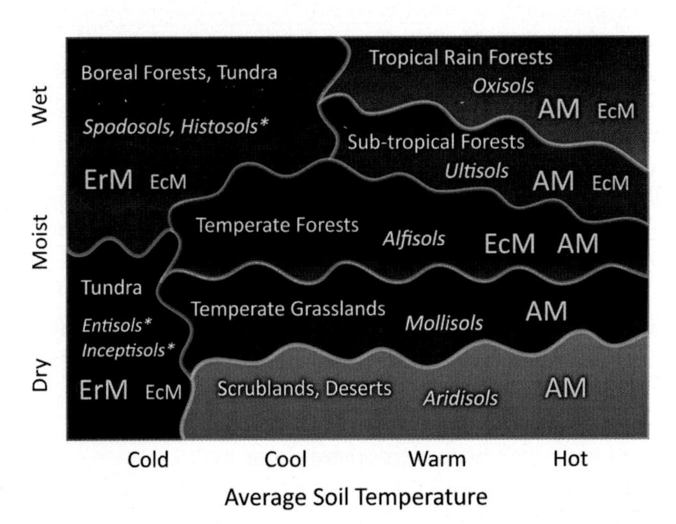

FIGURE 4.2 Climatic conditions under which several US Department of Agriculture soil orders, biome types, and arbuscular mycorrhizal (AM), ectomycorrhizal (EcM), and ericoid mycorrhizal (ErM) symbioses predominate. *, Histosols can be found in any of the temperature conditions where soils remain waterlogged for long periods of time, and Entisols and Inceptisols are very young soils that can occur under any combination of climatic factors. *Modified from Brady, N.C., 1984. The Nature and Properties of Soils. Macmillan Publishing Company, New York. Figure created by Kara Gibson.*

Mollisols, and Aridisols can form across a wide range of temperatures, but they segregate across a moisture gradient that corresponds to vegetation: Alfisols generally form in forests, Mollisols generally form in grasslands, and Aridisols generally form in deserts and shrublands. The distribution of Entisols and Inceptisols are controlled by time, soil parent material, and topography and can occur at any climate. Different types of mycorrhizas vary predictably across temperature and moisture gradients, and these patterns correspond to different environmental pressures faced by plant and fungal partners (Fig. 4.2).

Ericaceous plants and their symbiotic Ascomycota (and occasional Basidiomycota) thrive in the Spodosols and Histosols that form in boreal forests, tundra, and heathlands. EcM shrubs and trees with their symbiotic Basidiomycota and Ascomycota also thrive in these cold, wet environments because they can successfully acquire essential nutrients from the acidic organic-rich soils that develop due to very slow decomposition rates. To a large extent, nitrogen (N) and other essential nutrients are sequestered in recalcitrant organic matter in arctic, subarctic, and alpine soils. EcM and ErM fungi have the enzymatic capability to directly acquire nutrients from complex organic compounds and circumvent the mineralization process (Wu, 2011). Nutrients in these environments are tightly conserved within living mycelial networks (Lindahl et al., 2002).

Arbuscular mycorrhizas predominate in the Oxisols and Ultisols that form in warm, wet environments. Very little soil organic matter accumulates in these soil orders because decomposition is extremely rapid in most tropical and subtropical environments (Aerts, 1997). Weathering and intense leaching of these soils have removed most of the silica and silicate minerals, leaving acidic, clay-rich soils dominated by hydrous oxides of iron and aluminum. Phosphates bind tightly to these oxides and clays, resulting in low soil solution P concentration (see Section 4.3.1). Glomeromycota and associated bacteria can effectively acquire P and other immobile nutrients from these highly weathered tropical and subtropical soils, and AM symbioses are generally extremely important for plant nutrition (e.g., Jemo et al., 2014).

Most tropical rainforest trees form AM symbioses (Janos, 1980), but there are a few exceptions to this rule. All members of the Dipterocarpaceae and some members of the Caesalpiniaceae, Euphorbiaceae, Fagaceae, and Mytaceae form EcM symbioses in tropical forests (Torti et al., 1997; Wang and Qiu, 2006; Tedersoo et al., 2007). Although tree diversity in lowland tropical rainforests is generally extremely high, often exceeding 100 species per hectare (Wright, 2002), a disproportionate number of EcM tree species occur in low-diversity monodominant stands (Connell and Lowman, 1989). There is evidence that EcM trees maintain monodominance by altering N cycling through the formation of recalcitrant litter that is accessible to EcM fungi, but not AM fungi (McGuire et al., 2010; Corrales et al., 2016). Thus mycorrhizas may mediate plant community composition in tropical rainforests by influencing the outcome of plant competition for N.

Alfisols generally form in temperate forests and are mineral soils with rather well-developed horizons with high organic matter content in the surface and translocation of clays to deeper horizons. Temperate forest trees associate with either EcM or AM fungi. Phillips et al. (2013) found compelling support for the hypothesis that forest stands dominated by AM and EcM symbioses have distinct biogeochemical syndromes, and they proposed the "Mycorrhizal-Associated Nutrient Economy" framework to summarize the contrasting traits of AM and EcM stands (Table 4.1). According to this framework, AM-dominated stands have an inorganic nutrient economy whereas EcM-dominated stands have an organic nutrient economy.

TABLE 4.1 Biogeochemical Syndromes of Soil in Temperate Forest Stands Dominated by AM or EcM Tree Species

Trait	AM Dominated	EcM Dominated
Litter decomposition rate	Faster	Slower
Dissolved organic carbon	Lower	Higher
Organic N:inorganic N ratio	Lower	Higher
Nitrification rate	Faster	Slower
Phosphatase enzyme activity	Lower	Higher
Soil pH	Higher	Lower
Tree response to nutrient hot spots	Proliferate fine roots	Proliferate fungal hyphae

Modified from Phillips, R.P., Brzostek, E., Midgley, M.G., 2013. The mycorrhizal-associated nutrient economy: a new framework for predicting carbon-nutrient couplings in temperate forests. New Phytologist 199, 41–51. Foraging strategies of AM and EcM trees is from Chen, W., Koide, R.T., Adams, T.S., DeForest, J.L., Cheng, L., Eissenstat, D.M., 2016. Root morphology and mycorrhizal symbioses together shape nutrient foraging strategies of temperate trees. Proceedings National Academy of Science USA 113, 8741–8746.

Nutrients in the soil are often distributed in distinct patches or "hot spots," and the responses of AM and EcM trees to this heterogeneity have been shown to vary in a predictable way (Chen et al., 2016). Exposure to nutrient hot spots induced production of more fine roots by AM trees, but in EcM trees, it induced production of more fungal hyphae (Table 4.1). These differences in mineralization processes and foraging strategies may have important ramifications for the influence of AM and EcM trees on soil formation as well as their responses to anthropogenic enrichment of N and CO_2 (Terrer et al., 2016).

Grassland soils called Mollisols form in drier environments with insufficient water to support forest vegetation or in moist regions that are frequently burned. Most perennial grassland plants form AM symbioses, and the association is particularly important to warm-season C_4 grasses (Wilson and Hartnett, 1998). Cool-season C_3 grasses may depend less on AM symbioses because they are active when low soil temperatures restrict the metabolism of AM fungi. Root morphology and plasticity indicate that C_3 and C_4 grasses have very different mycorrhizal strategies (Miller et al., 2012). In general, cool-season C_3 grasses form very fine roots that can easily forage for nutrients in small pore spaces in the soil whereas warm-season C_4 grasses form coarse roots that are highly colonized by AM fungi. In general, C_4 grasses rely heavily on AM fungi to acquire essential nutrients and C_3 grasses do not (Johnson et al., 2008). In contrast, AM fungi appear to improve the drought tolerance of C_3 grasses more than C_4 grasses (Worchel et al., 2013).

The extensive root systems of grassland plants and associated AM fungi contribute to the formation of some of the most productive agricultural soils on Earth. Mollisols are extremely fertile, largely because their high organic matter content provides cation exchange sites and maintains favorable soil moisture. Glomeromycotan fungi are a major carbon sink and account for 20–30% of the microbial biomass carbon in grassland soils (Miller et al., 1995; Olsson et al., 1999). Hyphae of AM fungi are largely composed of chitin, a relatively recalcitrant carbohydrate, and rapid production and turnover of AM hyphal networks could contribute substantial amounts of chitinous residues to the soil matrix (Staddon et al., 2003; Miller et al.,

2012). It has been estimated that the amount of organic carbon derived directly from AM fungi ranges from 54 to 900 kg/ha (to a depth of 30 cm; Zhu and Miller, 2003). Misguided carbon sequestration attempts may inadvertently lose this important soil carbon sink if EcM trees are planted in formerly grassland soils. In the páramo grasslands of Ecuador, up to 30% of soil carbon was lost within 20 years of conversion to a pine plantation because EcM fungi can access nutrients via decomposition of soil organic compounds that are inaccessible to AM fungi (Chapela et al., 2001).

As their name implies, Aridisols form in dry regions where evapotranspiration far exceeds precipitation. These soils generally have very little organic matter, and because there is little leaching, they have a high content of salts and base cations. Many of the plants inhabiting arid and semiarid regions depend heavily on AM symbioses for mineral nutrition and drought tolerance (Chapter 16; Bethlenfalvay et al., 1984; Cui and Nobel, 1992). Both AM fungi and biological soil crust communities have been shown to be important stabilizers of erosion-prone soil in arid lands (Chaudhary et al., 2009). In addition, AM symbioses help plants tolerate soil salinity, which is a common stressor in arid and semiarid regions (Chapter 15).

4.2.3 Climatic and Mycorrhizal Mediation of Decomposition

The relationship between mycorrhizas and decomposition is complex. Although this topic is explored in great detail in Section IV of this book, it is briefly discussed here in the context of the importance of climate to mycorrhizal mediation of soil development. Decomposition is often defined as the process by which diverse assemblages of saprotrophic bacteria, fungi, and soil invertebrates breakdown complex organic compounds into simpler compounds as they use litter as an energy source. This simple definition overlooks the importance of litter as a source of essential nutrients as well as energy. The availability of N, P, and sulfur is strongly linked to decomposition because these elements are integral parts of organic matter. According to Perry (1994), one of three things can happen to the nutrients in litter during decomposition: (1) immobilization, (2) mineralization, or (3) direct transfer to a plant if the decomposer is a mycorrhizal fungus. In the last case, the breakdown of complex organic compounds is driven by fungal acquisition of a limiting nutrient rather than saprotrophic use of organic carbon as an energy source. A better understanding of the influence of climate on mycorrhizal symbioses can help predict the fate of nutrients in litter during decomposition.

The shift in dominant types of mycorrhizas observed across temperature and precipitation gradients illustrated in Fig. 4.2 is strongly linked to the process of decomposition (Read, 1991). Metabolic rates of microbial decomposers are highest in warm, moist environments and slowest in cold environments and in places with either too much or too little water. Organic matter accumulates in areas with slow decomposition rates, and in these areas N and other essential plant nutrients are generally tied up in complex organic compounds or immobilized in living organisms. Successful plant taxa have adapted to these soils by partnering with mycorrhizal fungi that can short-circuit the decomposition process and directly access nutrients from litter and dying vegetation (Wu, 2011). The Ascomycota in ErM symbioses utilize powerful catabolic enzymes to liberate N from complex organic molecules (Read and Perez-Moreno, 2003; Mitchell and Gibson, 2006). The diverse Basidiomycota and Ascomycota involved in EcM symbioses exhibit a wide range of saprotrophic abilities, and their relative catabolic capabilities have been linked to the relative abundance of organic versus mineral nutrients in their ambient soil (Koide et al., 2008). It should not be assumed that AM fungi have no role in decomposition simply because they do not produce extracellular enzymes that decompose complex organic compounds. Although the Glomeromycota appear to have no saprotrophic capability to utilize organic carbon as an energy source, it is well known that they are capable of utilizing organic forms of N and P, and in so doing they can be important forces in the decomposition process (Chapter 8; Tu et al., 2006; Atul-Nayyar et al., 2009; Whiteside et al., 2012).

Talbot et al. (2008) have described mycorrhizas "decomposers in disguise" and present three distinct models of how mycorrhizas directly and indirectly contribute to decomposition. The Plan B Hypothesis describes the situation in which EcM and ErM fungi breakdown carbon compounds in litter when their host plants are not allocating sufficient amounts of photosynthate to support the symbiosis. In other words, symbiotic carbon is "plan A" and saprotrophic carbon is "plan B." This scenario is expected to be particularly common during the winter when host plants are dormant (Buee et al., 2007) or when forests are thinned, which suddenly leaves many mycorrhizal fungi without their photobionts (Mosca et al., 2007). More recent analyses suggest that there is little support for this hypothesis (Chapter 20; Lindahl and Tunlid, 2015).

The Coincidental Decomposer Hypothesis describes when mycorrhizal fungi decompose soil organic matter in the process of acquiring organic forms of N or P (Talbot et al., 2008; Lindahl and Tunlid, 2015). This process is most likely to occur in systems where N is extremely limited. The strong selection pressure for strategies to acquire organic N in a temperate forest soil is exemplified by the evolution of a strain of *Laccaria bicolor* that appears to actively kill collembola and transport the animal-derived N to its host *Pinus strobus* (Klironomos and Hart, 2001). In this fascinating case, the "coincidental fungal decomposer" is technically a predator rather than a saprotroph because *L. bicolor* appears to actively kill collembola and rapidly harvest the organic N within the carcasses. Coincidental decomposition is likely to be most prevalent in tundra and cold forest ecosystems where most soil nutrients occur in organic compounds, but all types of mycorrhizas can be coincidental decomposers, even AM symbioses, because Glomeromycota utilize organic forms of N and P (Chapter 8).

The Priming Effect Hypothesis describes the scenario when high inputs of plant photosynthate "prime" the decomposition activities of mycorrhizal fungi by enhancing their ability to access soil carbon that is spatially or energetically unavailable to traditional saprotrophs (Talbot et al., 2008). Mycorrhizal fungi symbiotically receive sugars from their host plants whereas saprotrophs generally must expend more energy to scavenge for energy-rich carbon compounds. Consequently, mycorrhizal fungi can outcompete saprotrophic microbes for carbon substrates in carbon-limited environments. Indeed, this appears to be the case in boreal forests and EcM tropical forest stands, where saprotrophic fungi primarily occur in the surface litter layer that contains high concentrations of labile carbon whereas EcM and ErM fungi mostly occur deeper in the organic horizon where carbon is bound in more recalcitrant compounds (McGuire et al., 2013).

Originally, EcM and ErM symbioses were the focus of the Priming Effect Hypothesis (Talbot et al., 2008), but the principle can be expanded to include AM symbioses (Chapter 20). Glomeromycota produce copious amounts of fine hyphae in the soils around plant roots, which influence the abundance and composition of the surrounding microbial communities through the delivery of labile carbon (Chapter 22; Toljander et al., 2008; Atul-Nayyar et al., 2009; Herman et al., 2012). Under elevated levels of atmospheric CO_2, it has been shown that AM fungi cause a priming effect by stimulating saprotrophic microbes to mineralize soil organic matter (Cheng et al., 2012). Thus although AM fungi have no saprotrophic capacity of their own, they can strongly influence decomposition through their effects on bacteria and other saprotrophs that are associated with their hyphae.

Availability of N and P generally controls rates of primary production and decomposition in most ecosystems (Vitousek et al., 2010). The N cycle is tightly linked to the carbon cycle through many biotic processes, including assimilation and mineralization (McGill and Cole, 1981). In contrast, weathering of parent material ultimately controls P availability, and it cycles independently from C and N (Vitousek et al., 2010; Yang and Post, 2011). Mycorrhizas operate at the nexus of the C, N, and P cycles by actively linking the processes of assimilation, mineralization, and weathering.

4.3 MYCORRHIZAL INTERACTIONS WITH PARENT MATERIAL

The primary source of nutrients for plant growth is the decomposition and recycling of organic debris from the dead remains of plants, animals, and microorganisms. Because soils are open systems, essential plant nutrients are continually lost through erosion, leaching, and removal of vegetation. In addition, some nutrients become sequestered in insoluble compounds that are not accessible to plants. Consequently, with the exception of N (which is primarily fixed from the atmosphere), the nutrients required to maintain plant growth and production ultimately require prolonged and continual weathering from rock or minerals in parent material. In natural ecosystems with little soil organic matter, mycorrhizal fungi may be critical for the maintenance of plant communities because of their ability to sense nutrient-rich substrates and increase nutrient acquisition from raw rock-forming minerals (Chapters 3 and 8).

Parent material is the consolidated or unconsolidated geological material on which soil horizons form. Consolidated material may consist of igneous, sedimentary, or metamorphic rock. Unconsolidated material has been transported to a location by water (alluvium), gravity (colluvium), wind (eolian), or ice (glacial till). Soils inherit their minerals and much of their structure from their parent material, and soils derived from different parent materials have distinctive properties. A better understanding of mycorrhizal responses to the physical and chemical properties of different types of soil parent material could help predict the distribution patterns of different types of mycorrhizas and even the distributions of specific taxa of plant hosts and their associated fungi. This knowledge could be very helpful for incorporating mycorrhizas onto soil maps.

4.3.1 Soil Phosphorus Dynamics

Soil orders differ in their total P content because of interactions among soil parent material, weathering, and other pedogenic processes (Yang et al., 2013). In general, total P content is low in strongly weathered Ultisols and Oxisols and high in young soil orders such as Entisols and Inceptisols that have experienced little or no weathering, or Histosols with very low

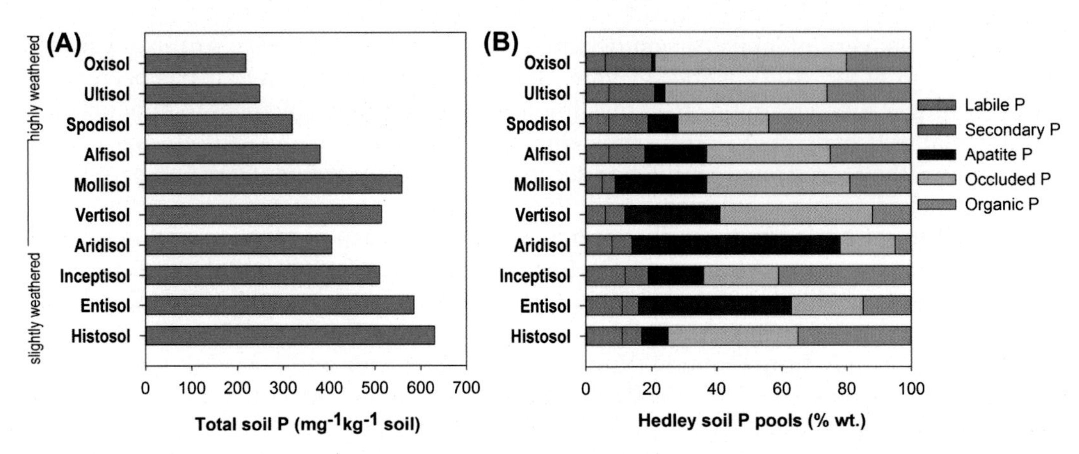

FIGURE 4.3 Soil orders arranged from slightly weathered to highly weathered, showing a (A) decrease in total soil P and (B) changes in the distribution of P fractions based on the Hedley procedure (Hedley and Stewart, 1982). *Data reproduced with permission from Yang, X., Post, W.M., 2011. Phosphorus transformations as a function of pedogenesis: a synthesis of soil phosphorus data using Hedley fractionation method. Biogeosciences 8, 2907–2916.*

decomposition (Fig. 4.3(A); Yang and Post, 2011). Mollisols also have relatively high total P content because they are often formed on unconsolidated P-rich glacial till and loess deposited during Quaternary glaciations (Ruhe, 1984).

The classic conceptual model of soil P dynamics during soil development tracks the transformation of mineral apatite through the weathering process (Walker and Syers, 1976). Over time, labile P is released from apatite and converted to organic forms of P by plants, mycorrhizas, and other biota. There is a gradual decrease in the fraction of P in apatite and an increase in P sorbed onto secondary minerals, which is eventually occluded by physical encapsulation by iron and aluminum oxides. The amount of P in these different fractions is roughly measured through a sequential extraction process called the Hedley fractionation procedure (Hedley and Stewart, 1982). A compilation of Hedley data shows that the most highly weathered soils, Oxisols and Ultisols, have very little apatite remaining and larger fractions of occluded and organic P (Fig. 4.3(B); Yang and Post, 2011). The Walker and Sayers model predicted that very little labile and secondary P (also referred to as "nonoccluded") should remain in the most highly weathered soil orders. However, Yang and Post's analysis showed significant amounts of labile and secondary P throughout all soil orders (Fig. 4.3B; Yang and Post, 2011). Walker and Sayers had assumed that occluded P is unavailable to plants and microbes, but Crews et al. (1995) challenged this assumption and hypothesized that mycorrhizas may dissolve certain occluded P minerals and reintroduce labile and secondary P into the actively cycling pool. Additional research is needed to test this hypothesis and gain a better understanding of the role that mycorrhizal fungi play in the P cycle.

4.3.2 Physicochemical Properties of Soil Parent Material

Parent material is an important driver of the chemical and physical properties of soil, and plants often adapt to these soil properties with their mycorrhizal symbioses. The availability of essential nutrients such as P, potassium, and calcium are influenced by parent material

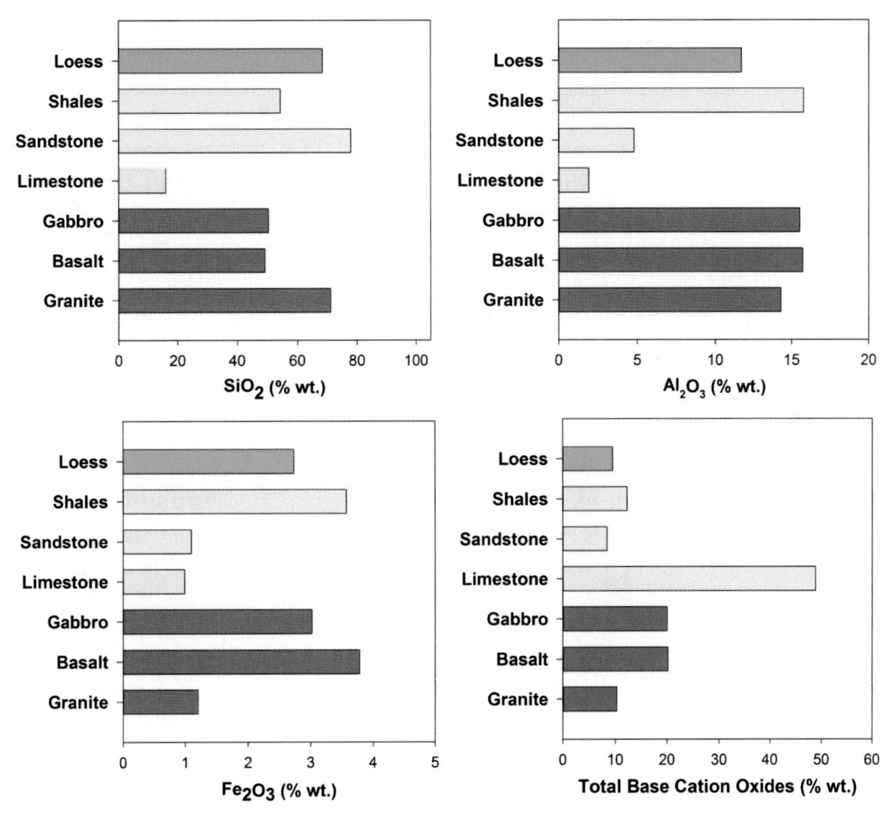

FIGURE 4.4 Chemical composition (% wt) of rock classes characteristic of parent material from Hartmann et al., 2012. The color of bars indicates different classes of parent material: *green* represents unconsolidated loess, *yellow* represents consolidated sedimentary rock, and *brown* represents consolidated igneous rock. Total base cation oxides comprise $K_2O + Na_2O + CaO + MgO$.

(Vitousek et al., 2010). The distinctive mineral composition of different parent materials may control resource availability and the potential role played by mycorrhizas. For example, limestone-derived soils are rich in base cations that bind tightly to P and remove it from the labile pool that is easily accessible to plant roots. Thus, it is not surprising that the genotype of big bluestem grass from a native prairie that developed on limestone parent material was shown to be much more dependent on AM symbioses than the genotype from a prairie that developed on glacial outwash with high levels of labile P (Johnson et al., 2010). Furthermore, the composition of mycorrhizal fungal communities varies with soil parent material. For example, in the Serengeti grasslands, AM fungal communities differ considerably in ancient sandstone-derived soils compared with soils with eolian inputs of volcanic ash (Antoninka et al., 2015; Stevens, 2016). In addition, EcM fungal communities have been shown to differ on soils derived from limestone and basalt (Gehring et al., 1998).

Parent materials have distinctive mineral compositions, and these differences can be used to generate testable hypotheses about the expected types of mycorrhizas that will form on them (Fig. 4.4). Among sedimentary parent materials, limestone is rich in base cations and relatively depleted in silica, aluminum, and iron oxides whereas shale shows the opposite pattern. Among igneous parent materials, granite has a much lower iron content compared with gabbro and basalt. One might hypothesize that in areas with "iron-poor" parent material, mycorrhizas may be more important for plant iron nutrition compared with areas with "iron-rich" parent material. The effect of parent material on soil pH may also be important. Calcium-rich soils derived from limestone should have a much higher pH compared with soils derived from other parent materials, and this may have a strong influence on the structure and function of mycorrhizal symbioses because AM, EcM, and ErM fungi are known to be sensitive to soil pH (Mitchell and Gibson, 2006; Jansa et al., 2014; Soudzilovskaia et al., 2015; Zhang et al., 2016).

The texture of soil parent material is also an important determinant of mycorrhizal mediation of soil properties. The large, weather-resistant quartz particles in soils derived from sandstone affect soil structure by influencing soil aggregation (Chapter 14) and pore size, which in turn will determine the aeration potential and water relations, both potentially important factors for structuring mycorrhizal fungal communities (Dickie et al., 2013; Jansa et al., 2014). The texture of soil parent material may also influence the weathering dynamics of soil because weathering rates are determined by the amount of exposed surface area and consequently soils with large particles sizes weather more slowly than those with small particle sizes (Vitousek et al., 2010).

4.3.3 Mycorrhizas as a Weathering Agent

Mycorrhizas influence soil parent material as well as respond to it (Fig. 4.1). Jenny defines parent material as the initial state of the soil system (Jenny, 1941). However, such a definition is almost impossible to quantify (Birkeland, 1999; Retallack, 2001) because the chemical and physical imprint from parent material diminishes with age so that the properties of surface horizons are generally very different from the properties of the soil parent material (Chesworth, 1973). Over time, the identity and initial state of parent material is increasingly difficult to determine. Nevertheless, in many cases parent material can be inferred from soil history or time functions of properties known to be modified by climate, topography, and biota (Dokuchaev, 1883; Jenny, 1941). In this regard, a more mechanistic understanding of the influence of mycorrhizas on the weathering of soil parent material will help advance the field of pedogenesis.

As described in Chapter 3, mycorrhizas may contribute to weathering processes by physically breaking apart rock and by accelerating the dissolution of parent material. Growth of roots and mycorrhizal hyphae contribute to the fragmentation and dissolution of parent material through acidification and excretion of low-molecular–weight organic acids (Robert and Berthelin, 1986; Neaman et al., 2005; Koele et al., 2014). Because of their reliance on carbon acquired from host photosynthate for growth, mycorrhizal fungi are energetically at an advantage over other microbes for accessing nutrients and essential metals attached to mineral surfaces (Hoffland et al., 2004). The importance of mycorrhizal-induced weathering for plant nutrition will vary in different soil orders. Mycorrhizal weathering is likely to be critical for P nutrition in Oxisols, but of no importance in Histosols, where instead mycorrhizas can acquire ample P from organic compounds.

4.4 MYCORRHIZAL INTERACTIONS WITH TOPOGRAPHY

Geomorphology generates local-scale variation in soil properties and biotic communities. At landscape and regional scales, soil formation is strongly influenced by climate and parent material; however, at local scales, topography and biological (including human) activities may be the most important factors controlling soil development (Jenny, 1980). The inclination (steepness), length, and aspect of slopes influence microclimates, local drainage capacity, water potential, erosion, and deposition (López et al., 2003; Dengiz and Baskan, 2010). The composition and productivity of plant communities are shaped by small-scale gradients in temperature, water, and nutrient availability driven by topography (Tromp-van Meerveld and McDonnell, 2006; Tietjen, 2016). AM (e.g., Gibson and Hetrick, 1988; Henkel et al., 1989; Murray et al., 2010; Zhang et al., 2013; Bonfim et al., 2016) and EcM (Yao et al., 2013) fungal diversity and community composition have been reported to vary in response to environmental gradients generated by geomorphic features. An awareness of these small-scale gradients may provide opportunities to design informative field experiments to test hypotheses that link edaphic factors to mycorrhizal structure and function.

Topographic relief generates environmental heterogeneity and influences soil properties and biological communities in many predictable ways. Differences in the temperature and moisture regimes of north- and south-facing slopes are caused by variation in the intensity and timing of solar gain. Gravity generates variation in the dynamics of erosional and depositional processes and creates distinctive physical and chemical signatures in soil profiles at the top and bottom of slopes. In addition, slope and aspect influence disturbances caused by biotic and abiotic forces. We can gain insights about the distribution of mycorrhizas by exploring the ways that topography influences microclimate, soil texture, soil chemistry, and disturbance agents.

4.4.1 Topography Influences on Physical Conditions and Processes

Topographic relief causes spatial variation of microclimates because the aspect and inclination of slopes control the solar gain of an area. In northern latitudes, air temperature may be as much as 5°C higher and soil temperature (6cm deep) as much as 10°C higher on north-facing versus south-facing slopes (Holch, 1931; Cottle, 1932; Fu and Rich, 2002). In addition, in mountainous regions, temperature systematically decreases with increasing elevation because of adiabatic cooling caused by decreasing air pressure (Körner, 1999). Compared with plants grown at lower elevations, higher elevation plants have a reduced effective temperature sum and shorter growing season, which ultimately results in less photosynthate to be shared with symbiotic fungi. Temperature has been shown to influence patterns of mycorrhizal association across latitudinal and elevation gradients (Bentivenga and Hetrick, 1992; Liu et al., 2015). Elevation gradients in mountainous regions often recapitulate latitudinal gradients with shifts from AM-dominated vegetation at low elevations and tropical latitudes to EcM- and ErM-dominated vegetation at high elevations and latitudes. As discussed earlier, EcM and ErM trees and shrubs with symbiotic Basidiomycota and Ascomycota can successfully acquire essential nutrients from the cold soils with very slow decomposition rates whereas warm-season grasses in association with Glomeromycota thrive in warm, wet environments.

Landscape position determines the movement of water, solutes, and particulates through the soil profile, which ultimately controls the formation and composition of soils. Hillslopes can be separated into distinct regions: ridge (or summit), crest, midslope, and toeslope, each with characteristic soil properties (Ludwig et al., 2005; Ovalles and Collins, 1986). Gravitational forces transport materials down hillslopes, and the steeper the slope the higher the rate of transport. Crest and midslope regions tend to have drier coarse-textured soils whereas toeslopes typically have wetter fine-textured soil (Guzman and Al-Kaisi, 2011). In addition, toeslopes typically have higher soil organic matter and nutrient levels because of inputs received from upper slopes (Liu et al., 2007; Baskan et al., 2016). Thus, lowland areas typically support a greater biomass of plants and mycorrhizal fungi compared with adjacent upland sites (Lugo et al., 2012; Li et al., 2014; Xu and Wan, 2008). Hillslope gradients in soil texture, moisture, and the amount and composition of organic compounds in soils can strongly influence mycorrhizas (Azcón-Aguilar and Barea, 2015; Gunina and Kuzyakov, 2015).

4.4.2 Topography Influences Disturbance Regimes

Geomorphology influences abiotic and biotic disturbances such as fire and herbivory. Fire is a natural disturbance in most ecosystems, with potentially profound effects on vegetation and mycorrhizas. The effects of fire are strongly controlled by topographic position (Di Folco and Kirkpatrick, 2011). Topography can control fire frequency and intensity by influencing fuel production, soil temperature, and moisture (Flatley et al., 2011; Collins and Calabrese, 2012). Fires are often more intense at toeslopes and lowlands compared with ridges and uplands because greater plant biomass generates a higher accumulation of litter. High-intensity fires have been shown to reduce the abundance and diversity of AM fungi whereas low-intensity fires may increase it (Korb et al., 2004; Zhang et al., 2013). Fire effects on EcM fungi also vary with topographic position because of variation in fire severity (Certini, 2005; Rincón and Pueyo, 2010).

Herbivory by large mammals can influence mycorrhizas, and topography may mediate the effects of this disturbance, particularly in grasslands and savannas. Herbivores can directly influence the relationship between host plants and mycorrhizal fungi because they alter plant nutrient requirements, nutrient availability, and the quantity of C supplied to the fungi (Gehring and Whitman, 2002). Because mycorrhizal fungi consume photosynthate and at the same time enhance mineral acquisition and growth capacity, the relationships among mycorrhizal fungi, herbivores, and plants is complex (Eom et al., 2001; Hartnett and Wilson, 2002). Topography influences animal behavior and distribution of grazing animals, thereby influencing grazing intensity and the redistribution of ingested nutrients from grazed areas to areas where the animals rest and ruminate. This results in spatially heterogeneous grazing intensity and deposition of dung and urine, leading to substantially greater loads of excreta N and P deposited in lowland areas (Iyyemperumal et al., 2007; Hoogendoorn et al., 2016). In upland areas that experience low to moderate herbivory, and lower mineral nutrient deposition, mycorrhizal symbiosis may increase grazing tolerance of their host plants by increasing nutrient acquisition. However, in lowland sites experiencing intense herbivory and nutrient inputs, plant regrowth of aboveground tissue and mycorrhizal fungi may be competing for limited photosynthate. This switch from nutrient limitation to carbon limitation as herbivory increases could result in a switch in the cost-benefit balance so that mycorrhizal symbioses may become detrimental rather than beneficial to plants (Hartnett and Wilson, 2002; Johnson et al., 2015).

4.4.3 Mycorrhizas Mediate Geomorphology

Mycorrhizas influence topography and respond to it. Dietrick and Perron (2006) note that the effects of living organisms on topography can be easily discerned over short timescales. Biotic processes can mediate chemical reactions, decrease soil density, disrupt the soil surface, and stabilize soil structure. The net-like structure of fine roots and mycorrhizal fungi may stabilize slopes and add internal structure to soils (Miller and Jastrow, 2000; Bast et al., 2016). This stabilization may influence geomorphic features. Process models and empirical evidence demonstrate that in the absence of biotic processes, slopes generally steepen and become concave with increasing drainage. Without the contributions of vegetation and mycorrhizas, the establishment of convex slopes may be very unlikely (Dietrick and Perron, 2006). Recognizing the roles of mycorrhizas in structuring and stabilizing soils may contribute to a more mechanistic understanding of geomorphology.

4.5 CONCLUSIONS

Mycorrhizal symbioses are integral components of soil; they are part of a complex adaptive system, and they respond to and influence the properties and functioning of soil. The process of evolution through natural selection has generated a diverse array of symbioses between plant roots and fungi to help plants adapt to their environment. Distinctive mycorrhizal syndromes can be identified that help plants overcome resource limitation and extreme environmental conditions from the poles to the equator. Important insights can be gained from considering how mycorrhizas interact with climate, soil parent material, and topography. This knowledge will help advance our understanding of the dynamic interplay between living and nonliving components of the Earth.

References

Aerts, R., 1997. Climate, leaf litter chemistry and leaf litter decompositiion in terrestrial ecosystems: a triangular relationship. Oikos 79, 439–449.

Antoninka, A., Ritchie, M.E., Johnson, N.C., 2015. The hidden Serengeti mycorrhizal fungi respond to environmental gradients. Pedobiologia—Journal of Soil Ecology 58, 165–176.

Atul-Nayyar, A., Hamel, C., Hanson, K., Germida, J., 2009. The arbuscular mycorrhizal symbiosis links N mineralization to plant demand. Mycorrhia 19, 239–246.

Azcón-Aguilar, C., Barea, J.M., 2015. Nutrient cycling in the mycorrhizosphere. Journal of Soil Science and Plant Nutrition 25 (2), 372–396.

Baskan, O., Dengiz, O., Gunturk, A., 2016. Effects of toposequence and land use-land cover on the spatial distribution of soil properties. Environmental Earth Sciences 75 (5), 1–10.

Bast, A., Wilcke, W., Graf, F., Lu¨scher, P., Gärtner, H., 2016. Does mycorrhizal inoculation improve plant survival, aggregate stability, and fine-root development on a coarse-grained soil in an alpine eco-engineering field experiment? Journal of Geophysical Research Biogeosciences 121, 2158–2171.

Bentivenga, S.P., Hetrick, B.A.D., 1992. Seasonal and temperature effects on mycorrhizal activity and dependence of cool-season and warm-season tallgrass prairie grasses. Canadian Journal of Botany 70, 1596–1602.
Bethlenfalvay, G.J., Dakessian, S., Pacovsky, R.S., 1984. Mycorrhizae in a southern California desert: ecological implications. Canadian Journal of Botany 62, 519–524.
Birkeland, P., 1999. Soils and Geomorphology. Oxford University Press, New York.
Bonfim, J.A., Vasconcellos, R.L.F., Gumiere, T., Mescolotti, D.D.L.C., Oehl, F., Cardoso, E.J.B.N., 2016. Diversity of arbuscular mycorrhizal fungi in a Brazilian Atlantic forest toposequence. Microbial Ecology 71 (1), 164–177.
Brady, N.C., 1984. The Nature and Properties of Soils. Macmillan Publishing Company, New York.
Buee, M., Courty, P.E., Mignot, D., Garbaye, J., 2007. Soil niche effect on species diversity and catabolic activities in an ectomycorrhizal fungal community. Soil Biology and Biochemistry 39, 1947–1955.
Bunn, R., Lekberg, Y., Zabinski, C., 2009. Arbuscular mycorrhizal fungi ameliorate temperature stress in thermophilic plants. Ecology 90 (5), 1378–1388.
Certini, G., 2005. Effects of fire on properties of forest soils: a review. Oecologia 143 (1), 1–10.
Chapela, I.H., Osher, L.J., Horton, T.R., Henn, M.R., 2001. Ectomycorrhizal fungi introduced with exotic pine plantations induce soil carbon depletion. Soil Biology &and Biochemistry 33, 1733–1740.
Chaudhary, V.B., Bowker, M.A., O'Dell, T.E., Grace, J.B., Redman, A.E., Rillig, M.C., et al., 2009. Untangling the biological contributions to soil stability in semiarid shrublands. Ecological Applications 19, 110–122.
Chen, W., Koide, R.T., Adams, T.S., DeForest, J.L., Cheng, L., Eissenstat, D.M., 2016. Root morphology and mycorrhizal symbioses together shape nutrient foraging strategies of temperate trees. Proceedings National Academy of Science USA 113, 8741–8746.
Cheng, L., Booker, F.L., Tu, C., Burkey, K.O., Zhou, L., Shew, H.D., et al., 2012. Arbuscular mycorrhial fungi increase organic carbon decomposition under elevated CO_2. Science 337, 1084–1087.
Chesworth, W., 1973. The parent rock effect in the genesis of soil. Geoderma 10, 215–225.
Collins, S.L., Calabrese, L.B., 2012. Effects of fire, grazing and topographic variation on vegetation structure in tallgrass prairie. Journal of Vegetation Science 23 (3), 563–575.
Connell, J.H., Lowman, M.D., 1989. Low-diversity tropical rain forests: some possible mechanisms for their existence. American Naturalist 134, 88–119.
Corrales, A., Mangan, S.A., Turner, B.L., Dalling, J.W., 2016. An ectomycorrhizal nitrogen economy facilitates monodominance in a neotropical forest. Ecology Letters 19, 383–392.
Cottle, H.J., 1932. Vegetation on north and south slopes of mountains in south-western Texas. Ecology 13, 121–134.
Crews, T., Kitayama, K., Fownes, J., Riley, R., Herbert, D., Mueller-Dombois, D., et al., 1995. Changes in soil phosphorus fractions and ecosystem dynamics across a long chronosequence in Hawaii. Ecology 76, 1407–1424.
Cripps, C.L., Eddington, L.H., 2005. Distribution of mycorrhizal types among alpine vascular plant families in the Beartooth Plateau, Rocky Mountains, USA, in Reference to large-scale patterns in arctic-alpine habitats. Arctic, Antarctic, and Alpine Research 37, 177–188.
Cui, M., Nobel, P.S., 1992. Nutrient status, water uptake and gas exchange for three desert succulents infected with mycorrhizal fungi. New Phytologist 122, 643–649.
Dengiz, O., Baskan, O., 2010. Characterization of soil profile development on different landscapes in semi-arid region of Turkey a case study; Ankara-Soğulca catchment. Anadolu Tarim Bilimleri Dergisi 25 (2), 106–112.
Di Folco, M.B., Kirkpatrick, J.B., 2011. Topographic variation in burning-induced loss of carbon from organic soils in Tasmanian moorlands. Catena 87 (2), 216–225.
Dickie, I.A., Martínez-García, B., Koele, N., Grelet, G.A., Tylianakis, J.M., Peltzer, D.A., et al., 2013. Mycorrhizas and mycorrhizal fungal communities throughout ecosystem development. Plant and Soil 367, 11–39.
Dietrick, W.E., Perron, J.T., 2006. The search for a topographic signature of life. Nature 439, 411–418.
Dokuchaev, V.V., 1883. Russian Chernozem. Selected Works of V.V. Dokuchaev. Israel Program for Scientific Translations, Jerusalem, p. 1. translated in 1967.
Eom, A.H., Wilson, G.W., Hartnett, D.C., 2001. Effects of ungulate grazers on arbuscular mycorrhizal symbiosis and fungal community structure in tallgrass prairie. Mycologia 93 (2), 233–242.
Flatley, W.T., Lafon, C.W., Grissino-Mayer, H.D., 2011. Climatic and topographic controls on patterns of fire in the southern and central Appalachian Mountains, USA. Landscape Ecology 26 (2), 195–209.
Fu, P., Rich, P.M., 2002. A geometric solar model with applications in agriculture and forestry. Computers and Electronics in Agriculture 37, 25–35.
Gavito, M.E., Schweiger, P., Jakobsen, I., 2003. P uptake by arbuscular mycorrhizal hyphae: effect of soil temperature and atmospheric CO_2 enrichment. Global Change Biology 9, 106–116.

Gehring, C.A., Theimer, T.C., Whitham, T.G., Keim, P., 1998. Ectomycorrhizal fungal community structure of pinyon pines growing in two environmental extremes. Ecology 79 (5), 1562–1572.

Gehring, C.A., Whitman, T.G., 2002. Mycorrhizae-herbivore interactions: population and community consequences. In: van der Heijden, M.G.A., Sanders, I.R. (Eds.), Mycorrhizal Ecology. Ecological Studies, vol. 157. Springer, New York, pp. 295–320.

Gibson, D.J., Hetrick, B.A.D., 1988. Topographic and fire effects on the composition and abundance of VA-mycorrhizal fungi in tallgrass prairie. Mycologia 80, 433–441.

Gunina, A., Kuzyakov, Y., 2015. Sugars in soil and sweets for microorganisms: review of origin, content, composition and fate. Soil Biology and Biochemistry 90, 87–100.

Guzman, J.G., Al-Kaisi, M.M., 2011. Landscape position effect on selected soil physical properties of reconstructed prairies in southcentral Iowa. Journal of Soil and Water Conservation 66 (3), 183–191.

Hartmann, J., Dürr, H.H., Moosdorf, N., Meybeck, M., Kempe, S., 2012. The geochemical composition of the terrestrial surface (without soils) and comparison with the upper continental crust. International Journal of Earth Sciences 101, 365–376.

Hartnett, D.C., Wilson, G.W.T., 2002. The role of mycorrhizas in plant community structure and dynamics: lessons from the grasslands. Plant and Soil 244, 319–331.

Hawkes, C.V., Hartley, I.P., Ineson, P., Fitter, A.H., 2008. Soil temperature affects carbon allocation within arbuscular mycorrhizal networks and carbon transport from plant to fungus. Global Change Biology 14, 1181–1190.

Hedley, M., Stewart, J., 1982. Method to measure microbial phosphate in soils. Soil Biology and Biochemistry 14, 377–385.

Henkel, T.W., Smith, W.K., Christensen, M., 1989. Infectivity and effectivity of indigenous vesicular-arbuscular mycorrhizal fungi from contiguous soils in southwestern Wyoming, USA. New Phytologist 112, 205–214.

Herman, D.J., Firestone, M.K., Nuccio, E., Hodge, A., 2012. Interactions between an arbuscular mycorrhizal fungus and a soil microbial communtiy mediating litter decomposition. FEMS Microbial Ecology 80, 236–247.

Hoffland, E., Kuyper, T.W., Wallander, H., Plassard, C., Gorbushina, A.A., Haselwandter, K., et al., 2004. The role of fungi in weathering. Frontiers in Ecology Environment 2, 258–564.

Holch, E.E., 1931. Development of roots and shoots of certain deciduous tree seedings in different forest sites. Ecology 12, 259–299.

Hoogendoorn, C.J., Newton, P.C.D., Devantier, B.P., Rolle, B.A., Theobald, P.W., Lloyd-West, C.M., 2016. Grazing intensity and micro-topographical effects on some nitrogen and carbon pools and fluxes in sheep-grazed hill country in New Zealand. Agriclture, Ecosystems and Environment 217, 22–32.

Iyyemperumal, K., Israel, D.W., Shi, W., 2007. Soil microbial biomass, activity and potential nitrogen mineralization in a pasture: Impact of stock camping activity. Soil Biology and Biochemistry 39 (1), 149–157.

Janos, D.P., 1980. Vesicular-arbuscular mycorrhizae affect lowland tropical rainforest plant growth. Ecology 61, 151–162.

Jansa, J., Erb, A., Oberholzer, H.R., Šmilauer, P., Egli, S., 2014. Soil and geography are more important determinants of indigenous arbuscular mycorrhizal communities than management practices in Swiss agricultural soils. Molecular Ecology 23, 2118–2135.

Jemo, M., Souleymanou, A., Frossard, E., Jansa, J., 2014. Cropping enhances mycorrhizal benefits to maize in a tropical soil. Soil Biology and Biochemistry 79, 117–124.

Jenny, H., 1941. Factors of Soil Formation: A System of Quantitative Pedology. McGraw-Hill Book Company, Inc, New York.

Jenny, H., 1980. The Soil Resource: Origin and Behavior. Springer-Verlag, New York.

Johnson, N.C., Rowland, D.L., Corkidi, L., Allen, E.B., 2008. Plant winners and losers duing grassland eutrophication differ in biomass allocation and mycorrhizas. Ecology 89, 2868–2878.

Johnson, N.C., Wilson, G.W.T., Bowker, M.A., Wilson, J., Miller, R.M., 2010. Resource limitation is a driver of local adaptation in mycorrhizal symbioses. Proceedings National Academy of Science USA 107, 2093–2098.

Johnson, N.C., Wilson, G.W.T., Wilson, J.A., Miller, R.M., Bowker, M.A., 2015. Mycorrhizal phenotypes and the Law of the minimum. New Phytologist 205, 1473–1484.

Klironomos, J.N., Hart, M.M., 2001. Animal nitrogen swap for plant carbon. Nature 410, 651–652.

Koele, N., Dickie, I.A., Blum, J.D., Gleason, J.D., de Graaf, L., 2014. Ecological significance of mineral weathering in ectomycorrhizal and arbuscular mycorrhizal ecosystems from a field-based comparison. Soil Biology and Biochemistry 69, 63–70.

Koide, R., Sharda, J.N., Herr, J.R., Malcolm, G.M., 2008. Ectomycorrhizal fungi and the biotrphy-saprotrophy continuum. New Phytologist 178 (2), 230–233.

Korb, J.E., Covington, W.W., Johnson, N.C., 2004. Vectors for non-native plants? Slash pile burning effects on soil properties, mycorrhizae and plant establishment: Recommendations for amelioration. Restoration Ecology 12, 52–62.

Körner, C., 1999. Alpine Plant Life: Functional Plant Ecology of High Mountain Ecosystems. Springer, Berlin.

Koske, R.E., 1987. Distribution of vesicular-arbuscular mycorrhizal fungi along a latitudinal temperature gradient. Mycologia 79, 55–68.

Lehto, T., Brosinsky, A., Heinonen-Tanski, H., Repo, T., 2008. Freezing tolerance of ectomycorrhizal fungi in pure culture. Mycorrhia 18, 385–392.

Lekberg, Y., Meadow, J., Rohr, J.R., Redecker, D., Zabinski, C., 2011. Importance of dispersal and thermal environment for mycorrhizal communities: lessons from Yellowstone National Park. Ecology 92 (6), 1292–1302.

Li, X., Gai, J., Cai, X., Li, X., Christie, P., Zhang, F., et al., 2014. Molecular diversity of arbuscular mycorrhizal fungi associated with two co-occurring perennial plant species on a Tibetan altitudinal gradient. Mycorrhiza 24 (2), 95–107.

Lindahl, B.D., Taylor, A.F.S., Finlay, R.D., 2002. Defining nutritional constraints on carbon cycling in boreal forests—towards a less 'phytocentric' perspective. Plant and Soil 242, 123–135.

Lindahl, B.D., Tunlid, A., 2015. Ectomycorrhizal fungi—potential organic matter decomposers, yet not saprotrophs. New Phytologist 205 (4), 1443–1447.

Liu, L., Hart, M.M., Zhang, J., Cai, X., Gai, J., Christie, P., et al., 2015. Altitudinal distribution patterns of AM fungal assemblages in a Tibetan alpine grassland. FEMS Microbial Ecology 91 (7), fiv078.

Liu, S.L., Guo, X.D., Fu, B.J., Lian, G., Wang, J., 2007. The effect of environmental variables on soil characteristics at different scales in the transition zone of the Loess Plateau in China. Soil Use and Management 23 (1), 92–99.

López, I.F., Lambert, M.G., Mackay, A.D., Valentine, I., 2003. The influence of topography and pasture management on soil characteristics and herbage accumulation in hill pasture in the North Island of New Zealand. Plant and Soil 255 (2), 421–434.

Ludwig, J.A., Wilcox, B.P., Breshears, D.D., Tongway, D.J., Imeson, A.C., 2005. Vegetation patches and runoff-erosion as interacting ecohydrological processes in semiarid landscapes. Ecology 86 (2), 288–297.

Lugo, M.A., Negritto, M.A., Jofré, M., Anton, A., Galetto, L., 2012. Colonization of native Andean grasses by arbuscular mycorrhizal fungi in Puna: a matter of altitude, host photosynthetic pathway and host life cycles. FEMS Microbial Ecology 81 (2), 455–466.

McGill, W.B., Cole, C.V., 1981. Comparative aspects of cycling of C, N, S and P through soil organic matter. Geoderma 26, 267–294.

McGuire, K.L., Allison, S.D., Fierer, N., Treseder, K.K., 2013. Ectomycorrhizal-dominated boreal and tropical forests have distinct fungal communities, but analogous spatial patterns across soil horizons. PLoS ONE 8 (7), e68278.

McGuire, K.L., Zak, D.R., Edwards, I.P., Blackwood, C.B., Upchurch, R., 2010. Slowed decomposition is biotically mediated in ectomycorrhizal, tropical rain forest. Oecologia 164, 785–795.

Miller, R.M., Jastrow, J.D., 2000. Mycorrhizal fungi influence soil structure. In: Kapulnik, Y., Douds, D.D.J. (Eds.), Arbuscular Mycorrhizas: Physiology and Function. Kluwer Academic Publishers, Dordrecht, The Netherlands, pp. 3–18.

Miller, R.M., Reinhardt, D.R., Jastrow, J.D., 1995. External hyphal production of vesicular-arbuscular mycorrhizal fungi in pasture and tallgrass prairie communities. Oecologia 103, 17–23.

Miller, R.M., Wilson, G.W.T., Johnson, N.C., 2012. Arbuscular mycorrhizae and grassland ecosystems. In: Southwood, D. (Ed.), Plant-Fungal Interactions. John Wiley & Sons, New York, pp. 59–84.

Mitchell, D.T., Gibson, B.R., 2006. Ericoid mycorrhizal associations: ability to adapt to a broad range of habitats. Mycologist 20, 2–9.

Mosca, E., Montecchio, L., Scattolin, L., Garbaye, J., 2007. Enzymatic activities of three ectomycorrhizal types of *Quercus robur* L. in relation to tree decline and thinning. Soil Biology & Biochemistry 39, 2897–2904.

Murray, T.R., Frank, D.A., Gehring, C.A., 2010. Ungulate and topographic control of arbuscular mycorrhizal fungal spore community composition in a temperate grassland. Ecology 91 (3), 815–827.

Neaman, A., Chorover, J., Brantley, S.L., 2005. Implications of the evolution of organic acid moieties for basalt weathering over geologic time. American Journal of Science 305, 147–185.

NRCS, 1999. Soil taxonomy: a basic system of soil classification for making and interpreting soil surveys. In: Agriculture NRCSUSDo.

Olsson, P.A., Thingstrup, I., Jakobsen, I., Baath, E., 1999. Estimation of the biomass of arbuscular mycorrhizal fungi in a linseed field. Soil Biology and Biochemistry 31, 1879–1887.

Ovalles, F.A., Collins, M.E., 1986. Soil-landscape relationships and soil variability in north central Florida. Soil Science Society of America Journal 50 (2), 401–408.

Perry, D.A., 1994. Forest Ecosystems. The Johns Hopkins University Press, Baltimore.

Phillips, R.P., Brzostek, E., Midgley, M.G., 2013. The mycorrhizal-associated nutrient economy: a new framework for predicting carbon-nutrient couplings in temperate forests. New Phytologist 199, 41–51.

Read, D.J., 1991. Mycorrhizas in ecosystems. Experimenta 47, 376–391.

Read, D.J., Perez-Moreno, J., 2003. Mycorrhizas and nutrient cycling in ecosystems—a journey towards relevance? New Phytologist 157, 475–492.

Retallack, G.J., 2001. Soils of the Past. Blackwell, Oxford.

Rincón, A., Pueyo, J.J., 2010. Effect of fire severity and site slope on diversity and structure of the ectomycorrhizal fungal community associated with post-fire regenerated *Pinus pinaster* Ait. seedlings. Forest Ecology and Management 260 (3), 361–369.

Robert, M., Berthelin, J., 1986. Role of biological and biochemical factors in soil mineral weathering. In: Huang, P.M., Schnitzer, M. (Eds.), Interactions of Soil Minerals with Natural Organics and Microbes. Soil Science Society of America, Madison, WI.

Ruhe, R.V., 1984. Soil-climate system across the prairies in Midwestern USA. Geoderma 34, 201–219.

Soudzilovskaia, N.A., Douma, J.C., Akhmetzhanova, A.A., van Bodegom, P.M., Cornwell, W.K., Moens, E.J., et al., 2015. Global patterns of plant root colonization intensity by mycorrhizal fungi explained by climate and soil chemistry. Global Ecology and Biogeography 24, 371–382.

Staddon, P.L., Ramsey, C.B., Ostle, N., Ineson, P., Fitter, A.H., 2003. Rapid turnover of hyphae of mycorrhizal fungi determined by AMS microanalysis of 14C. Science 300, 1138–1140.

Stevens, B.M., 2016. Soil Properties, Precipitation, Grazing, and Host Plant Species Influence Arbuscular Mycorrhizal Fungal Communities in the Serengeti. Master of Science. Northern Arizona University Flagstaff, Arizona.

Talbot, J.M., Allison, S.D., Treseder, K.K., 2008. Decomposers in disguise: mycorrhizal fungi as regulators of soil C dynamics in ecosystems under global change. Functional Ecology 22, 955–963.

Tedersoo, L., Smith, M.E., 2013. Lineages of ectomycorrhial fungi revisited: foraging strategies and novel lineages revealed by sequences from belowground. Fungal Biology Reviews 27, 83–99.

Tedersoo, L., Suvi, T., Beaver, K., Saar, I., 2007. Ectomycorrhizas of coltricia and coltriciella (hymenochaetales, basidiomycota) on caesalpiniaceae, dipterocarpaceae and myrtaceae in seychelles. Mycological Progress 6, 101–107.

Terrer, C., Vicca, S., Hungate, B.A., Phillips, R.P., Prentice, I.C., 2016. Mycorrhizal association as a primary control of the CO_2 fertilization effect. Science 353, 72–74.

Tibbett, M., Cairney, J.W.G., 2007. The cooler side of mycorrhizas: their occurrence and functioning at low temperatures. Canadian Journal of Botany 85, 51–62.

Tietjen, B., 2016. Same rainfall amount different vegetation-How environmental conditions and their interactions influence savanna dynamics. Ecological Modelling 326, 13–22.

Toljander, J.F., Santos-Gonzalez, J.C., Tehler, A., Finlay, R.D., 2008. Community analysis of arbuscular mycorrhizal fungi and bacteria in the maize mycorrhizosphere in a long-term fertilization trial. FEMS Microbial Ecology 65, 323–338.

Torti, S.D., Coley, P.D., Janos, D.P., 1997. Vesicular-arbuscular mycorrhizae in two tropical monodominant trees. Journal of Tropical Ecology 13, 623–629.

Tromp-van Meerveld, H.J., McDonnell, J.J., 2006. On the interrelations between topography, soil depth, soil moisture, transpiration rates and species distribution at the hillslope scale. Advances in Water Resources 29 (2), 293–310.

Tu, C., Booker, F.L., Watson, D.M., Chen, X., Rufty, T.W., Shi, W., et al., 2006. Mycorrhizal mediation of plant N acquisition and residue decomposition: Impact of mineral N inputs. Global Change Biology 12, 793–803.

van der Heijden, M.G.A., Sanders, I.R., 2002. Mycorrhizal ecology: synthesis and perspectives. In: Van der Heijden, M.G.A., Sanders, I.R. (Eds.), Mycorrhizal Ecology: Ecological Studies, vol. 157. Springer, New York.

Vitousek, P., Porder, S., Houlton, B.Z., Chadwick, O.A., 2010. Terrestrial phosphorus limitation: mechanisms, implications, and nitrogen-phosphorus interactions. Ecological Applications 20, 5–15.

Walker, T.W., Syers, J.K., 1976. The fate of phosphorus during pedogenesis. Geoderma 15, 1–19.

Wang, B., Qiu, Y.-L., 2006. Phylogenetic distribution and evolution of mycorrhizas in land plants. Mycorrhiza 16, 299–363.

Whiteside, M.D., Digman, M.A., Gratton, E., Treseder, K.K., 2012. Organic nitrogen uptake by arbuscular mycorrhizal fungi in a boreal forest. Soil Biology and Biochemistry 55, 7–13.

Wilson, G., Hartnett, D., 1998. Interspecific variation in plant responses to mycorrhizal colonization in tall grass prairie. American Journal of Botany 85, 1732–1738.

Worchel, E.R., Giauque, H.E., Kivlin, S.N., 2013. Fungal symbionts alter plant drought response. Microbial Ecology 65, 671–678.

Wright, J.S., 2002. Plant diversity in tropical forests: a review of mechanisms of species coexistence. Oecologia 130, 1–14.

Wu, T., 2011. Can ectomycorrhizal fungi circumvent the nitrogen mineralization for plant nutrition in termperate forest ecosystems? Soil Biology and Biochemistry 43, 1109–1117.

Xu, W., Wan, S., 2008. Water-and plant-mediated responses of soil respiration to topography, fire, and nitrogen fertilization in a semiarid grassland in northern China. Soil Biology and Biochemistry 40 (3), 679–687.

Yang, X., Post, W.M., 2011. Phosphorus transformations as a function of pedogenesis: a synthesis of soil phosphorus data using Hedley fractionation method. Biogeosciences 8, 2907–2916.

Yang, X., Post, W.M., Thornton, P.E., Jain, A., 2013. The distribution of soil phosphorus for global biogeochemical modeling. Biogeosciences 10, 2525–2537.

Yao, F., Vik, U., Brysting, A.K., Carlsen, T., Halvorsen, R., Kauserud, H., 2013. Substantial compositional turnover of fungal communities in an alpine ridge-to-snowbed gradient. Molecular Ecology 22 (19), 5040–5052.

Zhang, N., Xu, W., Yu, X., Lin, D., Wan, S., Ma, K., 2013. Impact of topography, annual burning and nitrogen addition on soil microbial communities in a semiarid grassland. Soil Science Society of America Journal 77 (4), 1214–1224.

Zhang, T., Wang, N.-F., Liu, H.-F., Zhang, Y.-Q., L-Y, Y., 2016. Soil pH is a key determinant of soil fungal community composition in the Ny-Alesund Region, Svalbard (high Arctic). Frontiers in Microbiology 7, 227.

Zhu, Y.-G., Miller, R.M., 2003. Carbon cycling by arbuscular mycorrhiza fungi in soil-plant systems. Trends in Plant Science 8, 407–409.

CHAPTER

5

Mycorrhizas Across Successional Gradients

F.P. Teste[1,2], *I.A. Dickie*[3]

[1]Grupo de Estudios Ambientales, IMASL-CONICET, Universidad Nacional de San Luis, San Luis, Argentina; [2]The University of Western Australia, Crawley (Perth), WA, Australia; [3]Lincoln University, Lincoln, New Zealand

5.1 SUCCESSION

Succession can be broadly defined as a change of biotic communities that occurs over time. Most plants rely on one or more root symbioses, but most research on succession in ecology has focused on plant communities without considering the important interacting roles of their root symbioses or soil microbial communities (Fierer et al., 2010; Meiners et al., 2015). Only in the last 20 years or so has the interaction between plant and soil microbial communities been incorporated into some models of succession that track differences in plant species performance due to plant–soil feedback (van der Putten et al., 2013, 2016; Meiners et al., 2015). We suggest that a better integration of mycorrhizal fungal communities into general models of succession would not only increase our understanding of the mycorrhizal symbiosis but also benefit terrestrial ecology in general.

Because plant ecologists have been studying ecological succession for decades, mycologists have adopted well-established concepts and approaches from plant-based studies to study fungal succession. Whether this adopted framework is well suited to understanding fungal communities remains untested. Nevertheless, some concepts remain true for plant and fungal succession. Firstly, what drives changes in plant or fungal communities during primary, secondary, and cyclic succession (i.e., succession that results from regular disturbances such as insect outbreaks which can "reset" the community to another stage (Watt, 1947)) differs because the starting conditions and resulting gradients are different. Secondly, the concepts of "climax" plant communities or "late-stage" fungal communities are no longer widely accepted as endpoint communities but rather seen as part of a continuum of change (Fierer et al., 2010; Meiners et al., 2015). Thirdly, in the light of the concept of long-term ecosystem development and retrogression (*sensu* Peltzer et al., 2010), determining

Mycorrhizal Mediation of Soil
http://dx.doi.org/10.1016/B978-0-12-804312-7.00005-X

what drives these communities to changes over long-term pedogenesis is likely to inform us about what are the fundamental drivers of plant and fungal communities (Dickie et al., 2013). Nonetheless, many concepts from plant succession do not appear entirely adoptable for studying mycorrhizal fungal succession, and we focus on outlining some of these within this chapter. In brief, we suggest that dispersal limitation, host dependency and specificity, plant–soil feedback, effects of mycorrhizal networks, and vertical distributions of fungal communities are concepts that require different approaches more " tailored" to fungi to accelerate our understanding of what drives mycorrhizal fungal succession.

Research on mycorrhizal fungal succession has steadily increased as a result of a growing recognition of the important role that mycorrhizal fungi play in structuring plant communities (Klironomos et al., 2011) and recent methodological improvements (Horton and Bruns, 2001; Peay et al., 2008; Lindahl et al., 2013; Hart et al., 2015). We highlight that changes in fungal communities over time are best conceptualized within the different stages of ecosystem

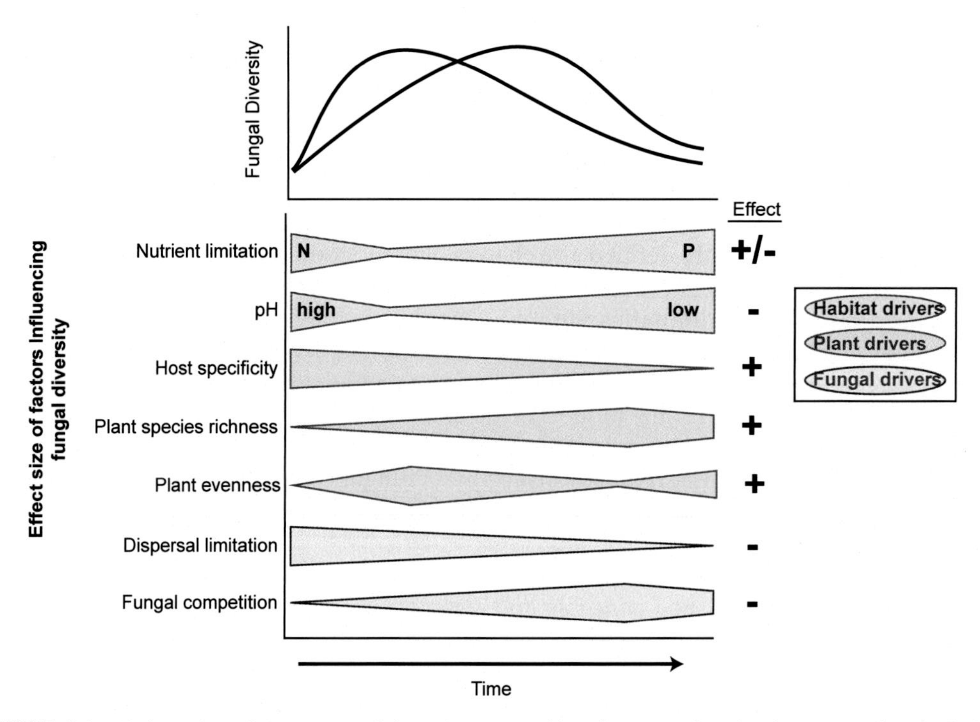

FIGURE 5.1 Shifts in fungal diversity and the importance of key factors within the three main drivers (habitat, plant, and fungal) with time. Observed patterns of fungal diversity over ecosystem development vary in the time of peak diversity (shown as two lines, reflecting patterns in Martinez-Garcia et al., 2015 and Krüger et al., 2015) but concur in diversity increasing and then decreasing (top). Major habitat (red), plant (green), and fungal (blue) drivers vary in their effect size in diversity with time (shown as width in bottom). Summarized from text: nutrient limitation shifts from N to P; pH can be somewhat high in early succession, becomes more neutral, and in old systems tends toward very acidic; host specificity tends to decline with succession at least in ectomycorrhizal fungi; plant communities become more diverse up until peak biomass mature ecosystems, but they can become dominated by large individuals (reducing plant evenness); fungal dispersal is most limiting early in succession whereas competitive exclusion becomes more likely in mature ecosystems. Many other possible drivers are not shown.

development (Peltzer et al., 2010; Dickie et al., 2013) using trends in fungal diversity over time (Fig. 5.1). There are potentially hundreds of secondary or cyclic succession events that can occur within each stage of ecosystem development, and this fundamental feature needs to be recognized to better understand what drives mycorrhizal fungal communities facing global environmental change.

In this chapter we first give a general overview of changes in ectomycorrhizal (EcM) and arbuscular mycorrhizal (AM) fungal communities. We focus on the ecology of mycorrhizal fungal communities in natural systems and emphasize the results of studies that have used soil chronosequences, which offer a good "space-for-time" replacement approach (Dickie et al., 2013) where sites only differ in age whereas parent material, climate, and the regional species pool are mostly invariant. We tend to focus more on succession of EcM fungal communities because more data are available compared with AM systems. We follow Zobel and Öpik (2014) in grouping the multiple factors into three "drivers" that are known to shape fungal communities and plant-fungal interactions over time: habitat, plant, and fungal drivers (Fig. 5.1). We conclude that the main drivers interact and that all drivers have considerable temporal and spatial dependencies. Finally, we highlight promising avenues for future research and outstanding questions.

5.2 SUCCESSION IN MYCORRHIZAL FUNGAL COMMUNITIES

Documenting changes in mycorrhizal fungal communities and diversity across gradients through time has value well beyond the quest to "catalog" fungal species for biodiversity surveying and conservation priorities. Gaining a better understanding of what drives these changes is highly relevant to better predicting host plant community dynamics and productivity, nutrient cycling, ecosystem functioning, and ultimately to better understand how the belowground subsystem may respond to global environmental change (Wardle and Lindahl, 2014). For example, changes in the dominance or functional traits of mycorrhizal fungi with time can affect the nutritional status and performance of the host plants (Erland and Taylor, 2002; Piotrowski and Rillig, 2008). Changes in fungal communities can also increase bedrock weathering (Gadd, 2007) and promote carbon (C) sequestration during secondary succession (Clemmensen et al., 2013).

The evidence for systematic changes in mycorrhizal fungal communities with succession was reviewed by Dickie et al. (2013), who found that EcM fungal diversity tended to accumulate over time during primary and secondary succession. The important exception was the loss of fungal genera with high host specificity to early successional plant species (e.g., *Alnicola* on *Alnus*, *Suillus* on *Pinus*). Also noteworthy is the steady decline of EcM fungal richness observed during ecosystem retrogression in a recent study using a 2-million-year soil chronosequence (Albornoz et al., 2016). AM fungal communities showed less clear patterns (Dickie et al., 2013), but Martinez-Garcia et al. (2015) later found that AM fungal communities tended to lose diversity after early succession, with fungi in the *Glomeraceae* persisting and becoming the long-term dominants in a study of 120,000 years of postglacial ecosystem development. Soon after, Krüger et al. (2015) documented an increase and decrease of AM fungal communities over much longer time scales (i.e., >2-million-year dune chronosequence) and on severely P-poor soils. They found evidence for strong environmental filtering as the main driver of AM fungal diversity where soil P availability was the most important edaphic variable (Krüger et al., 2015). In striking similarity to the results from Martinez-Garcia et al. (2015), Krüger and colleagues found Glomeraceae, specifically *Rhizophagus* and *Glomus* species, dominated across the sequence, although Krüger and colleagues also found Diversisporaceae in the oldest stages of retrogression. Before this work only a few studies measuring long-term changes in AM fungal abundance (via hyphal lengths and phospholipid fatty acids data) during soil development have been conducted (Piotrowski and Rillig, 2008). It is interesting to note that a general "hump-shaped" pattern, similar to what Krüger et al. (2015) showed, was found after major disturbances such as volcanoes, fire, and flooding (Piotrowski and Rillig, 2008). The AM fungal abundance tended to peak toward the middle of the gradient and then declined (Piotrowski and Rillig, 2008). An exception to this was the peak in diversity earlier in succession reported by Martinez-Garcia et al. (2015); however, this may reflect changes in plant evenness (with a few large tree species dominating root samples in mature sites) rather than a change in AM fungal diversity per se (Fig. 5.1).

Changes in mycorrhizal fungal communities during succession can be understood as being driven by the habitat, responding as passengers to plant community changes or driving the changes themselves (Fig. 5.1; Öpik et al., 2006). Martinez-Garcia et al. (2015) found close links of plant communities with fungal communities in an AM succession, supporting the passenger and driver hypotheses of Hart et al. (2001), but others have suggested a stronger role of soil habitat drivers in AM and EcM fungal communities (Krüger et al., 2015; Albornoz et al., 2016). The division of these three drivers provides a convenient, if somewhat contrived, way to categorize different factors controlling fungal communities throughout succession. Therefore we consider each driver in turn, but then we consider interactions among drivers.

5.3 HABITAT DRIVERS

5.3.1 The Changing Soil Abiotic Environment during Primary Succession

At the start of primary succession, soil development begins. Freshly formed soils are typically impoverished in biologically derived nutrients, particularly N, but rich in mineral-derived nutrients, such as P. During primary succession, vegetation patches appear, with nucleation, in part because of patchy inoculum of mycorrhizal fungi (Nara, 2006), contributing to an often highly spatially heterogeneous plant community. Over the first few years to decades, plant litter accumulates, soil pH decreases, and associated changes from inorganic to organic forms of soil nutrients occur. There are simultaneously drastic increases in mycorrhizal fungal diversity and rapid succession (Nara et al., 2003).

The type of disturbance (e.g., volcanic eruption, glacier retreat, sand dune formation) and the parent material (e.g., volcanic loess, sand, glacial till) produce different initial soil conditions and in turn produce different fungal successional trajectories. Therefore we discuss changes in fungal communities during primary succession within a context-dependent framework first and then attempt to find generalities. The considerable influence of soil conditions on fungal community assembly during primary succession has been well documented in recent work on glacial chronosequences (Davey et al., 2015). Still earlier studies found high patchiness within root systems, indicating that stochastic processes remain important determinants of the structure of fungal communities during primary succession

(Blaalid et al., 2012). Long-term research on fungal community change after glacier retreat has not shown any clear successional trajectory during primary succession (Jumpponen et al., 2012). That work also points to the important role of stochastic processes, but the authors could not completely dismiss the potentially influential role of soil types (i.e., bare soil or soil under established plants) and indirect effects of plant establishment in "driving" fungal community composition. Jumpoponen and colleagues suggest niche preferences related to organic legacies associated with soil development as important driving habitat characteristics.

5.3.2 The Changing Soil Abiotic Environment during Secondary Succession

Secondary successions occur within the context of ecosystem development, and depending on the ecosystem state, they can either rejuvenate an ecosystem or accelerate retrogression (Peltzer et al., 2010). Unlike primary succession, biologically derived nutrients and substrates such as N and organic C may already be present in high levels, and mineral nutrients including P may remain depleted. The other key difference between secondary and primary succession is that propagules of mycorrhizal fungi are likely to already be present in high densities depending on previous vegetation. For example, postagricultural succession is unlikely to be limited by a lack of AM fungal inoculum (Piotrowski and Rillig, 2008).

In general, many environmental conditions such as resource availability, light, and moisture levels change during secondary succession, and these have considerable effects on fungal community structure (Dickie et al., 2013). General secondary successional patterns in fungal communities are difficult to find despite the relatively large number of studies. Indeed, shifts in EcM fungal communities in tropical forests are not well explained by habitat drivers because soils are poorly differentiated across successional gradients (Tedersoo et al., 2011). In subtropical EcM forests, different fungal community assembly rules were identified and were grossly determined by the stage of secondary succession, in which habitat drivers were less important in older forests (Gao et al., 2015). In temperate forests, integrating factors such as stand development and canopy closure best described the important shifts observed in EcM fungal communities (Twieg et al., 2007). In pine forests from different continents, shifts in EcM fungal communities were clearly found, but drivers of change were not clear; nonetheless, soil (e.g., N availability) and stand characteristics (e.g., canopy closure) appeared important (Visser, 1995; Kipfer et al., 2011; LeDuc et al., 2013; Buscardo et al., 2015). These studies suggest that single factors or drivers do not fully explain fungal community dynamics; instead, interactions among soil factors, plant community characteristics, and fungal dispersal factors all show strong influences on EcM fungal community composition (Peay et al., 2010; Gao et al., 2015). Similar to primary succession, context dependencies such as the age of the dominant plant species and the remaining unexplained stochasticity challenges our capacity to better predict fungal shifts during secondary succession.

Shifts in mycorrhizal fungal communities during cyclic succession are topics in their infancy; hence, we do not discuss cyclic succession in any detail. However, we do acknowledge the work of Karst et al. (2015) that highlights that "habitat" (e.g., changes in litter input and light intensity), "plant" (i.e., changes in secondary chemistry of plants), and "fungal" (i.e., reduced EcM fungal diversity of species pool) drivers appear to interact and affect the composition and the role of EcM fungal communities found on emerging seedlings.

5.3.2.1 Key Soil Abiotic Factors that Regulate Mycorrhizal Community Composition

5.3.2.1.1 SOIL NUTRIENTS: NITROGEN AND PHOSPHORUS

During ecosystem development there are strong shifts in nutrient limitation (Peltzer et al., 2010). Early successional ecosystems have very high available P from fractured rock surfaces, which expose the P-containing mineral apatite to weathering (Walker and Syers, 1976). These initially high P levels are lost through leaching and through movement into unavailable forms, leading to increasing P limitation through time. Furthermore, chemical and biological transformations continue to reduce available soil P by converting primary mineral phosphate into organic forms and occluding P within secondary minerals (Turner and Condron, 2013). In contrast, N is initially very limited in early succession, but it increases rapidly after colonization by N-fixing plants (Walker and Syers, 1976). Thus an ecosystem moves through time from N to P limitation; in very old soils (i.e., during ecosystem retrogression) P limitation can be very strong, constrain plant productivity (Laliberté et al., 2012), and influence the distribution and diversity of fungal communities (Lambers et al., 2008; Krüger et al., 2015; Albornoz et al., 2016).

The industrial revolution has created "seminatural" soil N gradients via atmospheric N deposition. From these soil N gradients, studies typically show a decrease in EcM fungal species and considerable changes in community composition (Erland and Taylor, 2002; Cox et al., 2010). Fertilization experiments generally show similar trends, with a decrease in EcM fungal diversity and important changes in fungal community structure (Teste et al., 2012). So-called nutrient-source "specialist" fungal species are more negatively affected than "generalist" fungal species; furthermore, "nitrophilous" species (*Paxillus involutus* and *Lactarius rufus, Russula amoenolens*) can be found dominating the high end of the gradient (Wallenda and Kottke, 1998; Avis et al., 2008). Arbuscular mycorrhizal fungi have also been shown to be responsive to soil N, with one study finding a loss of the larger spore genera *Scutellospora* and *Gigaspora* in higher N sites whereas several *Glomus* were more tolerant of high N (Egerton-Warburton and Allen, 2000). Even within *Glomus,* some taxa appear to prefer higher N whereas others decline (Van Diepen et al., 2011).

Availability of P is a fundamental soil condition that affects the functioning of mycorrhizas (Smith and Read, 2008). Thus the influence of soil P might be expected to have stronger effects on the mycorrhizal communities (e.g., colonization and composition) than soil N (Treseder, 2004; Twieg et al., 2007; Dickie et al., 2009). Nonetheless, studies using buried substrate bags supplying P as a source of mineral apatite have found little response of fungal community composition (Hedh et al., 2008; Berner et al., 2012; Koele et al., 2014). Results from robust studies using long-term soil chronosequences have produced mixed conclusions. On the one hand, using an AM-dominated 120,000-year-old soil chronosequence with strong differences in soil chemistry between the soil ages, soil P availability was not an important factor (Martínez-García et al., 2015). On the other hand, with a much older (>2 million years) and nutrient-poor soil chronosequence, available P offered good predictive power of what structures AM and EcM fungal communities (Krüger et al., 2015; Albornoz et al., 2016).

Some studies have found that soil environmental gradients are more important than host plant effects (Schechter and Bruns, 2013; Sikes et al., 2014; Krüger et al., 2015) whereas others have found the opposite (Martinez-Garcia et al., 2015). This may, in part, reflect the strength of the gradient of soil nutrients and change in plant communities. Recent studies provide good evidence that diversity and functioning of mycorrhizal fungal communities appear to be limited, such as the host plants, by extremely low levels of soil available P (Krüger et al., 2015; Teste et al., 2016). More extreme soil gradients, such as those studied by Krüger et al. that encompass P limitation of fungi, may find a stronger habitat effect than studies of less extreme gradients.

5.3.2.1.2 MOISTURE

Soil moisture can undergo large changes during succession while also varying on a diurnal and seasonal basis. As vegetation biomass increases, the removal of moisture from the soil can inhibit decomposition (Koide and Wu, 2003). Major disturbances, particularly removal of trees, can result in raised water tables (i.e., paludification), limiting forest regeneration and increasing the reliance on deadwood as seedling establishment sites (Mallik, 2003). Conversely, the absence of disturbance can also lead to an excess of moisture where acidic leaf litter causes the formation of an impermeable iron pan in older forest sites and hence impeded drainage (Gaxiola et al., 2010).

Early studies on the effects of soil moisture on EcM fungal communities clearly showed changes in community composition and diversity (Worley and Hacskaylo, 1959; Fogel, 1980). Typically, fungal diversity decreases along a drying soil moisture gradient. Species with thick and melanized cell walls, such as *Cenococcum* spp., may increase in dominance in drier sites (Erland and Taylor, 2002; Fernandez and Koide, 2013). Excess soil moisture conditions (e.g., flooding, waterlogging) generally reduce mycorrhizal fungal biomass (Unger et al., 2009), and important shifts in EcM fungal composition and diversity are often likely observed as a result (Robertson et al., 2006; Sumorok et al., 2008; Wang et al., 2011). Shifts in AM fungal communities are not as drastic; fungal diversity tends to decrease in waterlogged sites, but the effects are highly dependent on the time the communities remain under excessive water conditions (Wang et al., 2011; Yang et al., 2016). Overall, fungal communities are responsive to soil moisture gradients; however, the direct effects of soil moisture in natural systems is not fully resolved because of interactive effects of plant host species and soil types (Erland and Taylor, 2002; Robertson et al., 2006). Effects of soil moisture gradients remain an area deserving more research given the more common occurrences of longer dry spells and excessive precipitation events due to climate change.

5.3.2.1.3 PH

Soil pH strongly declines during ecosystem development and can range from a pH of approximately 9 in the young developing soil to a pH of approximately 4 in the old nutrient-impoverished soils during ecosystem retrogression (Turner and Laliberté, 2015). Within each main successional type (primary and secondary) soil pH can be even more variable; however, in general a declining pattern remains as soil and plant communities age. The effect of soil pH is perhaps the most influential yet the most complex and interactive factor because it is strongly linked to other soil characteristics. Plant ecologists have been attempting to separate the complex effects of soil pH for some time, but they have only recently been able to elucidate the relative importance of soil pH in environmental filtering of plant communities using advanced statistical approaches such as structural equation modeling (SEM; Grace, 2006; Laliberté et al., 2014).

Mycorrhizal ecologists have not been as successful as plant ecologists in drawing firm conclusions on the direct effect of soil pH on fungal communities. We have mostly relied on in vitro studies or soil acidification/liming experiments to draw conclusions on the negative effect of low soil pH on growth of fungi and species diversity (Erland and Taylor, 2002, and references therein). Liming acidic forest soils has produced the opposite effects of acidification experiments, increasing EcM fungal species diversity and creating important shifts in community composition in favor of fungal species with abundant extraradical hyphae (Bakker et al., 2000). This effect may reach an optimum at moderately acidic levels because liming is also commercially used to reduce fungal diversity in the production of truffles (García-Montero et al., 2009).

There are a few studies using natural gradients in soil pH, and these have also shown marked negative effects on EcM fungal diversity with increasing soil acidity (Kumpfer and Heyser, 1986; Lu et al., 1999). In the case of AM fungi, soil pH has been shown to structure these fungal communities (Fitzsimons et al., 2008; Dumbrell et al., 2010; Lekberg et al., 2011) with strong interactive effects of the plant host community characteristics (Meadow and Zabinski, 2012). Drawing cause-and-effect conclusions about the direct effect of soil pH on mycorrhizal fungal communities still remains elusive and requires more studies using SEM or similar approaches.

5.4 PLANT DRIVERS

5.4.1 Plant Community Assembly

Plant succession is driven by three broad drivers: site conditions and history, species availability, and species performance (Meiners et al., 2015). After disturbance, all of these broad drivers directly and indirectly (via the host plants) affect the succession of the associated mycorrhizal fungal communities (Hart et al., 2001; Dickie et al., 2013; Zobel and Öpik, 2014). As plants establish during primary succession, competition among co-occurring plants for resources (e.g., light and nutrients) intensifies with time and is a major driver of plant community dynamics (Fargione and Tilman, 2002), which ultimately affect the mycorrhizal fungal communities (Kennedy, 2010; LeDuc et al., 2013). This basic premise has resulted in the "passenger" hypothesis—that changes in mycorrhizal fungal communities are simply a by-product of the considerable changes that occur to the hosting plant community (Hart et al., 2001). Although the passenger hypothesis has more empirical support during primary succession compared with secondary succession (Zobel and Öpik, 2014), more complex and rigorous experiments are needed to elucidate the validity of this theory during ecosystem development and over different spatial scales. There is also a need to determine the relative importance of the "passenger" hypothesis on harsh/nutrient-poor substrates found during other successional stages (e.g., ecosystem retrogression) that limit plant establishment (i.e., plant establishment is not simply dispersal limited).

5.4.2 Do Changing Plant Communities Drive Fungal Communities?

Plants, independent of the "habitat" drivers or "fungal" drivers (see Section 5.5), have the ability to structure mycorrhizal fungal communities via plant-driven processes such as plant biogeography, dispersal limitations of the host plant, and plant community dynamics. If "everything is everywhere for fungi," it makes sense that the assembly rules for plant communities drive the assembly rules for associated fungal communities. However, current thinking suggests that "everything is everywhere for fungi" is a misconception because fungi are not as ubiquitous as previously thought (Gladieux et al., 2015).

Plant traits such as size of root system and plant age have been identified as important factors that drive fungal succession (Twieg et al., 2007; Blaalid et al., 2012; Wang et al., 2012; Hart et al., 2014). Plants in older ecosystems have increased root length and branching, thinner root diameters, and lower nutrient concentrations than plants in younger ecosystems (Holdaway et al., 2011). Root density alone can explain a reasonable amount of variation in EcM fungal community structure, particularly during primary succession and presumably after severe disturbances causing secondary succession (Peay et al., 2011; LeDuc et al., 2013). Interactions between root density and the kind of EcM fungal exploration types have been proposed as overlooked drivers of EcM fungal successional shifts (Peay et al., 2011). Likewise, in AM fungal systems the interaction between root strategies and AM fungal traits are suggested to determine fungal and plant community structure and succession (Chagnon et al., 2013). The AM trait-based approach of Chagnon et al. (2013) that proposes to use interactions between "plant" and "fungal" traits more explicitly will help better predict the dynamics of AM fungal succession and species associations.

In at least one study the type of host plant community better predicted the structure of mycorrhizal communities than soil pH (Meadow and Zabinski, 2012). However, they did not find soil pH to be unimportant; it was a leading soil factor explaining the structure of the fungal communities, simply not to the same extent as the "plant" driver. Studies that have attempted to determine the relative importance of plant host and a key soil factor on mycorrhizal communities are overall rare and remain a research priority for mycorrhizal ecology. However, we can predict that "habitat" and "plant" drivers interact, and when they are included together we begin to better appreciate the integrative nature of the mycorrhizal symbioses.

5.4.3 Plant Host Specificity As a Driver of Changes in Fungal Communities

Community-level experiments have consistently shown that certain plant species mixtures (i.e., community types) influence the composition of mycorrhizal fungal communities (Hausmann and Hawkes, 2009, 2010; Pendergast et al., 2013). This outcome makes sense because strong evidence supports the idea that plants can reward fungal partners that are most "carbohydrate effective" (Kiers et al., 2011; Werner and Kiers, 2015). Furthermore, there is good evidence that plants can select or suppress certain mycorrhizal fungal species as exemplified with the local effects of invasive plant species on the abundance and composition of AM fungal communities (Mummey and Rillig, 2006; Callaway et al., 2008; Moora et al., 2011; Wilson et al., 2012; Lekberg et al., 2013). These studies suggest that plant host identity is another important factor; furthermore, plant host identity also has solid empirical support as potentially being the most influential factor on mycorrhizal fungal communities (Jumpponen et al., 2002, 2012; Öpik et al., 2009; Moora et al., 2011; Becklin et al., 2012; Martínez-García et al., 2015).

5.4.3.1 Host-Specific Ectomycorrhizal Fungal Species: A Paradox?

During primary succession there are important shifts in plant community composition such as the dominance and later replacement of EcM host-specific trees such as *Alnus*. Indeed, *Alnus* species almost exclusively inhabit disturbed, early successional, or riparian habitats (Chen and Li, 2004). The genus *Alnus* hosts a surprisingly low diversity (~40 species) of EcM fungal species (Tedersoo et al., 2009). However, the fungal species (e.g., *Alnicola*) are highly host specific and community structure is only in part shaped by abiotic factors, mainly organic matter concentration and soil pH (Tedersoo et al., 2009). Furthermore, variation in alpha and beta diversity of *Alnus*-EcM communities remains poorly understood because abiotic factors only had a minor influence at a regional scale and recent work can only speculate on the main drivers of these unique EcM communities (Roy et al., 2013). We require more empirical studies at a regional scale that integrate the potential drivers (e.g., host identity, habitat, geographical isolation) to comprehend what shapes *Alnus*-EcM communities but also to better understand other EcM fungal species showing specificity.

It is also notable that *Pinus*, although broader in fungal associates than *Alnus*, frequently associates with highly host-specific *Suillus* and/or *Rhizopogon* in early succession, both in the native range (Terwilliger and Pastor, 1999; Collier and Bidartondo, 2009) and when invasive (Hayward et al., 2015; Wood et al., 2015). Having highly host-specific plant-fungal associations dominate in early succession is somewhat unexpected because it would seem to potentially increase symbiont limitation during establishment. This remains an unresolved area of research.

5.5 FUNGAL DRIVERS

5.5.1 Fungal Community Assembly

Most studies assume that mycorrhizal fungal community assembly rules are similar to the assembly rules found in plant communities. In general, experimental designs and data analyses used for fungal community analysis have been derived from well-established plant successional models (Fierer et al., 2010). We have mentioned that there are aspects of fungal biology that suggest that the assumptions of plant models may apply to fungal systems but should be treated with caution. In particular, fungal communities utilize a much broader range of nutrient sources than plants (increasing potential for niche differentiation), show strong assembly history effects (Fukami et al., 2010), and interact in a complex three-dimensional space that lacks the strong asymmetric competition that typifies light competition in plants. The vast majority of mycorrhizal fungal species are dependent on host plants for survival and growth; thus what is best for the fungus is not necessarily best for the plant and vice versa (Smith and Read, 2008). This dependency adds complexity and increases the importance of interactive factors relative to other factors such as habitat drivers alone. Mechanisms underlying fungal succession include dispersal, resource availability, and plant succession (Fierer et al., 2010). However, there still remains substantial unexplained stochasticity in how fungal communities assemble, and predicting the composition will remain a formidable challenge if we simply "recycle" plant successional models (Schmidt et al., 2014).

A more integrative and mycocentric approach to predict changes in mycorrhizal fungal communities is to use a trait-based approach that incorporates the plant-fungal interactions in a tractable way (Peay et al., 2011; Chagnon et al., 2013). The modified Grime's C-S-R (competitor, stress tolerator, ruderal) framework (Chagnon et al., 2013) may best explain mycorrhizal fungal successional dynamics, spatial structure of the communities, and ultimately diversity. However, many questions remain, such as which trait(s) do fungal species share that allow fungal occurrence during primary succession or which traits are needed to remain during ecosystem retrogression? When the trait-based approach proposed by Chagnon et al. (2013) uses key functional traits, it also appears promising that it will increase our understanding of the interplay between plant and fungi across gradients and highlight the implications for ecosystem functioning. However, using traits such as hyphal exploration types as drivers of EcM fungal succession (Peay et al., 2011) ignores the likely possibility that such traits are phylogenetically constrained. Thus it is difficult to tease apart if an increase in one functional type is due to the increase demand in the trait's function or because the functional type is correlated phylogenetically with another trait.

5.5.1.1 Dispersal Limitations

Limitations in dispersal capabilities of mycorrhizal fungi imply that not all fungi from the local species pool arrive efficiently into a novel habitat. This is contrary to the early view of microbes as "everything is everywhere" (Baas-Becking, 1934), but it is supported by strong evidence of dispersal limitation of EcM fungi at local scales (Telford et al., 2006; Peay et al., 2010). This evidence is also supported by careful studies on the actual effective dispersal distances of EcM fungi, which found that most species only travel a few meters away from the sporocarps (Li, 2005). Some consistent patterns have emerged from work in primary succession that demonstrated that *Laccaria* and *Hebeloma* are good dispersers, capable of colonizing plants in bare nutrient-poor soils but then are typically replaced in later successional stages (Jumpponen et al., 2002; Dickie et al., 2013; Davey et al., 2015).

AM fungal species occurrence and composition in early primary successional landscapes appear to be more stochastic than EcM fungi. In particular, after volcanic eruption or glacier retreat, movement of soil containing AM fungal inoculum is likely an important factor related to the establishment of AM fungal species (Allen et al., 1992; Jumpponen et al., 2012). The stochastic dispersal of AM fungal spores is likely responsible for the patterns of establishment of the AM fungi in these systems. However, in dune systems harboring an AM fungal diversity "hotspot," most species were from the *Acaulospora* and *Glomus* and overall AM fungal richness was invariant across the vegetation gradient studied (Alves da Silva et al., 2015). However, in most other primary-succession dune systems, AM fungal communities tend to start off species rich and change significantly with time. It still remains unclear if these changes are a product of different dispersal capabilities or other plant-related processes (Sikes et al., 2012); however, soil characteristics certainly appear to select AM fungal communities (Sikes et al., 2014; Krüger et al., 2015) whereas in other systems host identity was clearly the main factor (Martínez-García et al., 2015).

We propose that the concept of "everything is everywhere" is scale and group dependent. For example, even within AM fungi, which generally appear globally distributed (Öpik et al., 2013), several empirical studies do conclude that there is some level of dispersal limitation (Dickie et al., 2013). At the foreland of the retreating Morteratsch glacier in the Central Alps, it was found that pioneering AM fungal species were mainly distributed by wind whereas other species (e.g., *Glomus rubiforme* and *Glomus aureum*) appear to need established mycorrhizal networks to colonize new areas (Oehl et al., 2011). There are too few studies to draw any firm conclusions, in particular during secondary succession, but dispersal strategies do appear to drive AM fungal community structure to some degree under primary succession.

5.5.1.2 *Rapid Root Colonization, Fungal Competition, and Priority Effects*

Mycorrhizal fungal species capable of colonizing roots rapidly can gain considerable competitive advantage by establishing and exploiting resources ahead of competing fungal species (Kennedy et al., 2009). This often results in "priority effects" because the species that establish first negatively affect the latter arriving species through preemption of shared resources or direct inhibition (Kennedy, 2010). The species that can establish rapidly before others may be aided by long-distance dispersal processes such as wind and mammal dispersion (Allen et al., 1992) or by rapid mycorrhization of roots via well-established nearby mycorrhizal networks (Nara, 2006; Simard et al., 2012). As result of these fungal-driven effects, distinct shifts in EcM community composition occur, and these shifts are spatially constrained (Dickie et al., 2002; Teste et al., 2009).

5.6 INTERACTING DRIVERS

5.6.1 Toward a General Model of Mycorrhizal Fungal Succession: Integrating Drivers

Currently, three main hypotheses on changes in mycorrhizal plants and fungi along environmental gradients have been proposed and tested: (1) the "habitat" hypothesis (i.e., both mycorrhizal fungi and plant communities are strongly influenced by changes in abiotic conditions; Zobel and Öpik, 2014), (2) the "passenger" hypothesis (i.e., plants shape fungal communities; Hart et al., 2001), and (3) the "fungal driver" hypothesis (i.e., fungi shape plant communities; Hart et al., 2001). From the mycocentric view of this chapter, these correlate with habitat drivers, plant drivers, and fungal drivers of fungal communities.

Nonetheless, we suggest that treating each of these drivers independently is not an approach that will render the best predictions. Rather, we propose the "interacting drivers" hypothesis, a more general hypothesis to "rule them all," which postulates that all of the main drivers (i.e., habitat, plant, and fungal) interact through time and space (Fig. 5.2). Integrating these three working hypotheses should provide us with better predictions of what shapes mycorrhizal fungal communities from primary to secondary succession and beyond into ecosystem retrogression. The interacting drivers hypothesis is derived from our conceptual framework (Fig. 5.2) and from the pioneering studies that suggested it (Johnson et al., 1991). Furthermore, our recent work along long-term soil chronosequences in Australia and New Zealand (Krüger et al., 2015; Martínez-García et al., 2015; Albornoz et al., 2016) also gives some level of support. However, the hypothesis requires more empirical studies to assess its predictive power, and we expect that it will be applicable to most ecological conditions encountered in natural plant communities.

5.6.1.1 *Nitrogen and the Interacting Drivers Hypothesis*

Limited N availability characterizes many EcM ecosystems (Read, 1991; Vitousek and Howarth, 1991). This may reflect either preexisting edaphic conditions selecting for EcM

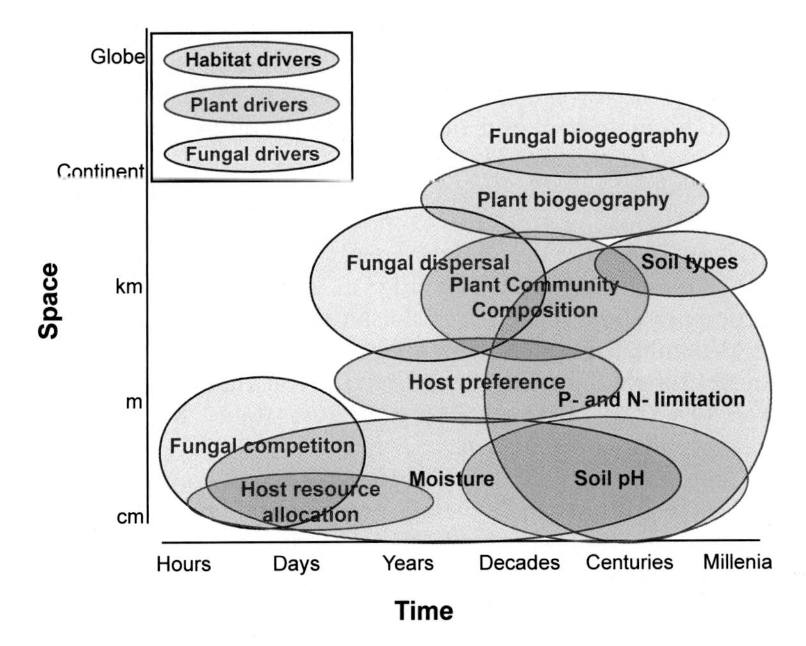

FIGURE 5.2 Conceptual synthesis of the ways all three main drivers (habitat, plant, and fungi) act at varying times and spatial scales, with examples of each. Colors follow Fig. 5.1.

dominance (Read, 1993; Read and Perez-Moreno, 2003) or the effect of EcM fungi on N cycling (Orwin et al., 2011; Näsholm et al., 2013; Averill et al., 2014; Dickie et al., 2014; Corrales et al., 2016). Plants also play a dominant role in soil N through hosting N-fixing symbionts, organic litter input, and N uptake. As such, soil N can be considered as representing all three drivers: habitat, plant, and fungi. In part, this reflects the perspective of the researcher. A short-term study might consider soil N availability to represent a preexisting habitat condition whereas a longer term perspective might view the same soil N as an outcome of a particular plant or fungal community.

5.6.1.2 Interacting Drivers Hypothesis Linking Plant and Fungal Communities

There are many basic plant community characteristics that change rapidly during primary succession. For example, canopy closure coincides with major shifts in EcM (Last et al., 1987; Visser, 1995; DeBellis et al., 2006; Twieg et al., 2007; LeDuc et al., 2013; Gao et al., 2015) and AM fungal communities (da Silva et al., 2014; Hart et al., 2014). Resulting shifts in plant communities that result from canopy closure are then seemingly reflected in shifts in mycorrhizal fungal communities. However, these changes in fungal communities during canopy closure are more likely the result of interactions between important factors from the "habitat" and "plant" drivers.

However, shifts in plant communities may represent the effect of fungal drivers because mycorrhizal fungi influence plant–plant interactions (Smith and Read, 2008). These effects are typically taxon specific because the mycorrhizal symbiosis exerts a continuum from mutualism to parasitism depending on the species involved and soil conditions (Klironomos, 2003; Hoeksema et al., 2010). Therefore plant communities are, at least in part, determined by fungal communities, as embodied in the "driver" hypothesis (Hart et al., 2001).

More specifically, during primary succession, EcM fungal species such as *Inocybe, Laccaria,* and *Scleroderma* spp. appear key to helping plants recolonize volcanic deserts (Nara et al., 2003). Suilloid fungi (*Rhizopogon* and *Suillus* spp.) have spore dispersal and longevity traits that allow them to be prominent and play important ecological roles in the early successional habitats (Ashkannejhad and Horton, 2006; Nguyen et al., 2012), which incidentally also help them become invasive and promote pine-tree invasion in their non-native ranges (Hayward et al., 2015; Wood et al., 2015). At the forefront of retreating glaciers, *Cortinarius, Inocybe, Laccaria,* and *Thelephora* spp. are commonly found as the most abundant groups at the forefront soil of retreating glaciers (Blaalid et al., 2012; Jumpponen et al., 2012). A lack of AM fungal inoculum can limit AM plant establishment into early successional habitat (Allen, 1987) or into EcM-dominated systems (Spence et al., 2011). Even if AM inoculum is present, changes within AM fungal communities can have strong effects on plant communities (van der Heijden et al., 1998, 2004), although Sikes et al. (2012) found that changes in AM fungi did not influence plant growth in succession.

Fungal communities interact with their host plants in a complex way and ultimately strongly influence the diversity, structure, and dynamics of plant communities (Bever, 2003; Bever et al., 2010; van der Putten et al., 2016). Variation in fungal species composition and specificity of interactions likely has two major effects: (1) regulates plant species performances (Moora et al., 2004; Johnson et al., 2008) and (2) further alters the fungal community composition. However, fungal community effects on plant performances are not general and require a more explicit incorporation of pedogenesis (i.e., "habitat" driver in a temporal context) as a key underlying mechanism independent of host plant identity and growth requirements (Sikes et al., 2012, 2014).

There are some clear examples that highlight the driving role of fungal communities in plant succession. The absence of EcM fungi can limit tree seedling establishment and invasion (Jones et al., 2003; Nuñez and Dickie, 2014). These will clearly alter the composition and dynamics of plant communities if trees cannot establish as rapidly as herbaceous or AM tree species. Once established, EcM fungal communities can have more indirect yet far-reaching effects on plant communities. For example, EcM fungi can have strong effects on the accumulation of organic substrate in soil and lead to considerable C sequestration (Orwin et al., 2011; Clemmensen et al., 2013; Averill et al., 2014). Such indirect effects can lead to exclusion of AM plants. Indeed, long-term replacement of AM plants by EcM vegetation (e.g., Cass pollen core data from New Zealand) has been documented (Dickie et al., 2014). Once AM plants are absent, a self-reinforcing pattern occurs in which a lack of AM fungi prevents reestablishment (Spence et al., 2011). Many of these processes are likely to be highly variable at large scales and steadily increasing in importance with time.

5.6.1.3 Shifts in Fungal Communities Depend on Scale

The interacting drivers hypothesis suggests that all main drivers interact to create the structure of fungal communities. The relative importance of different habitat, plant, and fungal drivers changes with spatial and temporal scale, but examples of all three occur at most scales. There is a growing recognition that factors that are important at explaining biological communities at local scales are not necessarily important at larger spatial scales (Willis and Whittaker, 2002; Tedersoo et al., 2014b; Bahram et al., 2015). For example, factors such as fungal and plant biogeography will determine the potential species pool for a site but have little predictive ability for finer scale patterns.

At large scales, EcM fungal communities may be driven by gradients in soil properties, particularly N availability and pH (Cox et al., 2010; Suz et al., 2014). However, EcM fungal communities can also be driven by plant and fungal biogeography. For example, *Alnus* tree host species are absent from the southern hemisphere; therefore *Alnicola* fungal species do not naturally occur, suggesting that plant drivers also operate at continental scales. Likewise, *Desceola* fungal species do not occur naturally in the northern hemisphere, indicating that fungal biogeography is also a factor at continental scales. The composition of AM fungal communities can be spatially structured at different scales (Wolfe et al., 2007). The influence of "scale" is far reaching (Willis and Whittaker, 2002), and if it is not taken into account it can restrict the scope of inference. For example, soil pH is often identified as a good predictor of fungal community dynamics, but attempts to determine the relative importance of such factors among others and to propose that some "drivers" are more important than others have often failed. However, the study of Fitzsimons et al. (2008) has identified temporal effects (i.e., time since disturbance) to have the largest effect on fungal community composition. These studies clearly support our current thinking that multiple drivers should be integrated along temporal and spatial scales.

5.6.1.3.1 TIME SCALE: PEDOGENESIS

Habitat drivers and environmental filtering are the most influential in shaping mycorrhizal fungal communities along long-term soil chronosequences within old, climatically buffered, infertile landscapes (Lambers et al., 2008, 2010; Krüger et al., 2015; Albornoz et al., 2016) found in Western Australia. Conversely, plant drivers were found to be more important in a shorter term retrogressive chronosequence in New Zealand (Martínez-García et al., 2015). When ecosystems are given more time to develop and enter retrogression, stronger soil gradients [e.g., a ~60-fold decline in total soil P concentration (Australian chronosequence) compared to a ~7-fold decrease (New Zealand chronosequence)] result. Thus the strength of an influential gradient (e.g., soil P) relative to other potentially influential gradients (e.g., host plant community composition) determines which main driver will be the most important in structuring the fungal communities.

5.6.1.3.2 SPATIAL SCALE: FROM ROOT TIPS TO CONTINENTAL SCALES

At large scales, increased N inputs have been consistently found to be a major factor responsible for altered EcM fungal community composition (Cox et al., 2010; Suz et al., 2014). These studies highlight that habitat drivers are more important at large scales (i.e., intercontinental scales) compared with local scales (i.e., 0.25-ha plots) where "plant drivers" (e.g., root density, stand age, and productivity) were most influential. However, there are also studies suggesting fine habitat effects (e.g., vertical niche partitioning) and plant drivers as being equally important at continental scales. Clearly the apparent inconsistencies among these studies support our ideas that drivers should be integrated and they all depend on the "scale." Other studies show different types of spatial-scale dependencies; for example, at local scales, "fungal drivers" (i.e., dispersal limitation of fungi) were found to be important factor in explaining the composition of EcM fungal communities (Peay et al., 2010). The study also suggests that such conclusions are tentative because accurate extrapolations over stronger soil gradients and over longer temporal scales are needed.

5.6.1.4 Advancing Mycorrhizal Community Ecology with Well-Suited Approaches and Tools

Chronosequences in general but particularly soil chronosequences should continue to be used to better our understanding of what drives mycorrhizal fungal communities. Although the use of soil chronosequences can certainly help determine the relative importance of soil factors, the resulting data require careful analysis as many soil factors can interact. Furthermore, two potential weaknesses of chronosequences can pose difficulties even if advanced data analyses are used: (1) they may not be a representative sample of the landscapes and (2) the regional species pool may not include all of the mycorrhizal types. Therefore these potential caveats can guide future studies in selecting the best sites as possible.

Because the main drivers interact and many important abiotic and biotic factors do not change independently from each other, advanced quantitative techniques that can incorporate this complexity should be used more frequently in mycorrhizal ecology. One promising approach is the use of SEM, which is well suited for explicitly quantifying simultaneous influences and strives toward proposing causation and ultimately allowing for strong inferences (Grace, 2006). Antoninka et al. (2009) have demonstrated the utility of SEM in AM-dominated systems. The main limitation to applying the technique has been a need for very high replication. Unfortunately the current trend toward using next-generation sequencing to achieve ever greater sequencing depth on ever fewer samples does little to alleviate this.

Meta-analysis provides compelling evidence-based conclusions about ecological hypotheses, effect size estimation of the main drivers, and comparison of effects across different temporal and spatial scales (Koricheva and Gurevitch, 2014). In mycorrhizal ecology, it has not been used as often as in plant ecology, but it has many of the same benefits if used correctly (Archmiller et al., 2015). The emerging controversy about whether host plant richness predicts EcM fungal diversity (Dickie, 2007; Tedersoo et al., 2012, 2014a; Gao et al., 2013, 2014) can serve as a good example of the potential power that meta-analyses can offer to mycorrhizal ecology but also how the statistical approach can be misused or not used to its full potential (Archmiller et al., 2015).

5.6.2 Outstanding Questions and Conclusions

In summary, we reviewed the changes that occur to mycorrhizal fungal communities during ecosystem development and what are the main drivers for this change. On the basis of this review, we think that the following questions should be answered:

- Can functional traits (response and effect traits) integrate successional drivers of plant and EcM fungal communities (Bruns and Kennedy, 2009; Koide et al., 2014)? Plant ecologists have used response and effect traits for better understanding what drives community structure and ecosystem function, however this approach has not been fully appreciated or applied to EcM fungal communities, although see Chagnon et al. (2013) for models about AM-dominated systems. We suspect that because of the large effects of mycorrhizal fungi on ecosystem function that such a functional-based approach will help better predict effects of fungal communities on ecosystem services during climate change and ecosystem degradation. This is particularly true because climatic and edaphic variables explain much of the fungal richness and community composition at the global scale (Tedersoo et al., 2014b). Nonetheless, accounting for phylogenetic autocorrelation is essential in any studies of mycorrhizal traits (e.g., Koele et al. (2012) for mycorrhizal status of plants).

- Can a better mechanistic understanding of the roles of the dominant fungal species in communities help us better understand what shapes host plant and fungal communities? We need to go beyond cataloging fungal species and describing diversity patterns to determining functional diversity and the roles of fungal species within the communities. Who is extracting P and N most efficiently and who is capable of scavenging deep in the soil profile, leaving behind recalcitrant fungal tissue? Answers to such mechanistic questions remains a priority to improve on the prediction of models of C sequestration and help land managers improve their efforts to mitigate climate change.

We conclude that there is no one driver that rules them all; instead, it is a matter of scale in space and time. We propose that the division of drivers into habitat, plant, and fungal, and particularly any suggestion that one is more important than the other, overlooks the fact that all three types of drivers operate and interact at multiple scales. Rather than focusing on "which driver is more important?", we conclude that the focus should be on the interaction of the drivers and better understand "at what spatial and temporal scales do each driver operate and how?".

References

Albornoz, F., Teste, F., Lambers, H., Bunce, M., Murray, D., White, N., Laliberté, E., 2016. Changes in ectomycorrhizal fungal community composition and declining diversity along a 2-million year soil chronosequence. Molecular Ecology. http://www.dx.doi.org/10.1111/mec.13778.

Allen, M.F., 1987. Re-establishment of mycorrhizas on Mount St Helens: migration vectors. Transactions of the British Mycological Society 88, 413–417.

Allen, M.F., Crisafulli, C., Friese, C.F., Jeakins, S.L., 1992. Re-formation of mycorrhizal symbioses on Mount St Helens, 1980–1990: interactions of rodents and mycorrhizal fungi. Mycological Research 96, 447–453.

Alves da Silva, D.K., de Souza, R.G., de Alencar Velez, B.A., da Silva, G.A., Oehl, F., Maia, L.C., 2015. Communities of arbuscular mycorrhizal fungi on a vegetation gradient in tropical coastal dunes. Applied Soil Ecology 96, 7–17.

Antoninka, A., Wolf, J.E., Bowker, M., Classen, A.T., Johnson, N.C., 2009. Linking above- and belowground responses to global change at community and ecosystem scales. Global Change Biology 15, 914–929.

Archmiller, A.A., Bauer, E.F., Koch, R.E., Wijayawardena, B.K., Anil, A., Kottwitz, J.J., Munsterman, A.S., Wilson, A.E., 2015. Formalizing the definition of meta-analysis in molecular ecology. Molecular Ecology 24, 4042–4051.

Ashkannejhad, S., Horton, T.R., 2006. Ectomycorrhizal ecology under primary succession on coastal sand dunes: interactions involving *Pinus contorta*, suilloid fungi and deer. New Phytologist 169, 345–354.

Averill, C., Turner, B.L., Finzi, A.C., 2014. Mycorrhiza-mediated competition between plants and decomposers drives soil carbon storage. Nature 505, 543–545.

Avis, P., Mueller, G., Lussenhop, J., 2008. Ectomycorrhizal fungal communities in two North American oak forests respond to nitrogen addition. New Phytologist 179, 472–483.

Baas-Becking, L., 1934. Geobiologie of inleiding tot de milieukunde. WP Van Stockum & Zoon NV.

Bahram, M., Peay, K.G., Tedersoo, L., 2015. Local-scale biogeography and spatiotemporal variability in communities of mycorrhizal fungi. New Phytologist 205, 1454–1463.

Bakker, M., Garbaye, J., Nys, C., 2000. Effect of liming on the ecto-mycorrhizal status of oak. Forest Ecology and Management 126, 121–131.

Becklin, K.M., Hertweck, K.L., Jumpponen, A., 2012. Host identity impacts rhizosphere fungal communities associated with three alpine plant species. Microbial Ecology 63, 682–693.

Berner, C., Johansson, T., Wallander, H., 2012. Long-term effect of apatite on ectomycorrhizal growth and community structure. Mycorrhiza 22, 615–621.

Bever, J.D., 2003. Soil community feedback and the coexistence of competitors: conceptual frameworks and empirical tests. New Phytologist 157, 465–473.

Bever, J.D., Dickie, I.A., Facelli, E., Facelli, J.M., Klironomos, J., Moora, M., Rillig, M.C., Stock, W.D., Tibbett, M., Zobel, M., 2010. Rooting theories of plant community ecology in microbial interactions. Trends in Ecology & Evolution 25, 468–478.

Blaalid, R., Carlsen, T., Kumar, S., Halvorsen, R., Ugland, K.I., Fontana, G., Kauserud, H., 2012. Changes in the root-associated fungal communities along a primary succession gradient analysed by 454 pyrosequencing. Molecular Ecology 21, 1897–1908.

Bruns, T.D., Kennedy, P.G., 2009. Individuals, populations, communities and function: the growing field of ectomycorrhizal ecology. New Phytologist 182, 12–14.

Buscardo, E., Rodriguez-Echeverria, S., Freitas, H., De Angelis, P., Pereira, J.S., Muller, L.A.H., 2015. Contrasting soil fungal communities in Mediterranean pine forests subjected to different wildfire frequencies. Fungal Diversity 70, 85–99.

Callaway, R.M., Cipollini, D., Barto, K., Thelen, G.C., Hallett, S.G., Prati, D., Stinson, K., Klironomos, J., 2008. Novel weapons: invasive plant suppresses fungal mutualists in America but not in its native Europe. Ecology 89, 1043–1055.

Chagnon, P.-L., Bradley, R.L., Maherali, H., Klironomos, J.N., 2013. A trait-based framework to understand life history of mycorrhizal fungi. Trends in Plant Science 18, 484–491.

Chen, Z., Li, J., 2004. Phylogenetics and biogeography of *Alnus* (Betulaceae) inferred from sequences of nuclear ribosomal DNA ITS Region. International Journal of Plant Science 165, 325–335.

Clemmensen, K.E., Bahr, A., Ovaskainen, O., Dahlberg, A., Ekblad, A., Wallander, H., Stenlid, J., Finlay, R.D., Wardle, D.A., Lindahl, B.D., 2013. Roots and associated fungi drive long-term carbon sequestration in boreal forest. Science 339, 1615–1618.

Collier, F.A., Bidartondo, M.I., 2009. Waiting for fungi: the ectomycorrhizal invasion of lowland heathlands. Journal of Ecology 97, 950–963.

Corrales, A., Mangan, S.A., Turner, B.L., Dalling, J.W., 2016. An ectomycorrhizal nitrogen economy facilitates monodominance in a neotropical forest. Ecology Letters 19, 383–392.

Cox, F., Barsoum, N., Lilleskov, E.A., Bidartondo, M.I., 2010. Nitrogen availability is a primary determinant of conifer mycorrhizas across complex environmental gradients. Ecology Letters 13, 1103–1113.

da Silva, I.R., Aragao de Mello, C.M., Ferreira Neto, R.A., Alves da Silva, D.K., de Melo, A.L., Oehl, F., Maia, L.C., 2014. Diversity of arbuscular mycorrhizal fungi along an environmental gradient in the Brazilian semiarid. Applied Soil Ecology 84, 166–175.

Davey, M., Blaalid, R., Vik, U., Carlsen, T., Kauserud, H., Eidesen, P.B., 2015. Primary succession of *Bistorta vivipara* (L.) Delabre (Polygonaceae) root-associated fungi mirrors plant succession in two glacial chronosequences. Environmental Microbiology 17, 2777–2790.

DeBellis, T., Kernaghan, G., Bradley, R., Widden, P., 2006. Relationships between stand composition and ectomycorrhizal community structure in boreal mixed-wood forests. Microbial Ecology 52, 114–126.

Dickie, I.A., 2007. Host preference, niches and fungal diversity. New Phytologist 174, 230–233.

Dickie, I.A., Koele, N., Blum, J.D., Gleason, J.D., McGlone, M.S., 2014. Mycorrhizas in changing ecosystems. Botany 92, 149–160.

Dickie, I.A., Koide, R.T., Steiner, K.C., 2002. Influences of established trees on mycorrhizas, nutrition, and growth of *Quercus rubra* seedlings. Ecological Monographs 72, 505–521.

Dickie, I.A., Martinez-Garcia, L.B., Koele, N., Grelet, G.A., Tylianakis, J.M., Peltzer, D.A., Richardson, S.J., 2013. Mycorrhizas and mycorrhizal fungal communities throughout ecosystem development. Plant and Soil 367, 11–39.

Dickie, I.A., Richardson, S.J., Wiser, S.K., 2009. Ectomycorrhizal fungal communities and soil chemistry in harvested and unharvested temperate *Nothofagus* rainforests. Canadian Journal of Forest Research-Revue Canadienne De Recherche Forestiere 39, 1069–1079.

Dumbrell, A.J., Nelson, M., Helgason, T., Dytham, C., Fitter, A.H., 2010. Relative roles of niche and neutral processes in structuring a soil microbial community. ISME Journal 4, 337–345.

Egerton-Warburton, L.M., Allen, E.B., 2000. Shifts in arbuscular mycorrhizal communities along an anthropogenic nitrogen deposition gradient. Ecological Applications 10, 484–496.

Erland, S., Taylor, A.F.S., 2002. Diversity of ecto-mycorrhizal fungal communities in relation to the abiotic environment. In: van der Heijden, M.G.A., Sanders, I.R. (Eds.), Mycorrhizal Ecology. Springer, Berlin, Germany, pp. 163–200.

Fargione, J., Tilman, D., 2002. Competition and coexistence in terrestrial plants. In: Sommer, U., Worm, B. (Eds.), Competition and Coexistence. Springer-Verlag, Berlin, pp. 165–206.

Fernandez, C.W., Koide, R.T., 2013. The function of melanin in the ectomycorrhizal fungus *Cenococcum geophilum* under water stress. Fungal Ecology 6, 479–486.

Fierer, N., Nemergut, D., Knight, R., Craine, J.M., 2010. Changes through time: integrating microorganisms into the study of succession. Research in Microbiology 161, 635–642.

Fitzsimons, M.S., Miller, R.M., Jastrow, J.D., 2008. Scale-dependent niche axes of arbuscular mycorrhizal fungi. Oecologia 158, 117–127.

Fogel, R., 1980. Mycorrhizae and nutrient cycling in natural forest ecosystems. New Phytologist 86, 199–212.

Fukami, T., Dickie, I.A., Paula Wilkie, J., Paulus, B.C., Park, D., Roberts, A., Buchanan, P.K., Allen, R.B., 2010. Assembly history dictates ecosystem functioning: evidence from wood decomposer communities. Ecology Letters 13, 675–684.

Gadd, G.M., 2007. Geomycology: biogeochemical transformations of rocks, minerals, metals and radionuclides by fungi, bioweathering and bioremediation. Mycological Research 111, 3–49.

Gao, C., Shi, N.-N., Liu, Y.-X., Peay, K.G., Zheng, Y., Ding, Q., Mi, X.-C., Ma, K.-P., Wubet, T., Buscot, F., et al., 2013. Host plant genus-level diversity is the best predictor of ectomycorrhizal fungal diversity in a Chinese subtropical forest. Molecular Ecology 22, 3403–3414.

Gao, C., Shi, N.-N., Liu, Y.-X., Zheng, Y., Ding, Q., Mi, X.-C., Ma, K.-P., Wubet, T., Buscot, F., Guo, L.-D., 2014. Host plant richness explains diversity of ectomycorrhizal fungi: response to the comment of Tedersoo et al. (2014). Molecular Ecology 23, 996–999.

Gao, C., Zhang, Y., Shi, N.-N., Zheng, Y., Chen, L., Wubet, T., Bruelheide, H., Both, S., Buscot, F., Ding, Q., et al., 2015. Community assembly of ectomycorrhizal fungi along a subtropical secondary forest succession. New Phytologist 205, 771–785.

García-Montero, L.G., Quintana, A., Valverde-Asenjo, I., Díaz, P., 2009. Calcareous amendments in truffle culture: a soil nutrition hypothesis. Soil Biology and Biochemistry 41, 1227–1232.

Gaxiola, A., McNeill, S., Coomes, D., 2010. What drives retrogressive succession? Plant strategies to tolerate infertile and poorly drained soils. Functional Ecology 24, 714–722.

Gladieux, P., Feurtey, A., Hood, M.E., Snirc, A., Clavel, J., Dutech, C., Roy, M., Giraud, T., 2015. The population biology of fungal invasions. Molecular Ecology 24, 1969–1986.

Grace, J.B., 2006. Structural Equation Modeling and Natural Systems. Cambridge University Press.

Hart, M.M., Aleklett, K., Chagnon, P.L., Egan, C., Ghignone, S., Helgason, T., Lekberg, Y., Öpik, M., Pickles, B.J., Waller, L., 2015. Navigating the labyrinth: a guide to sequence-based, community ecology of arbuscular mycorrhizal fungi. New Phytologist.

Hart, M.M., Reader, R.J., Klironomos, J.N., 2001. Life-history strategies of arbuscular mycorrhizal fungi in relation to their successional dynamics. Mycologia 93, 1186–1194.

Hart, M.M., Gorzelak, M., Ragone, D., Murch, S.J., 2014. Arbuscular mycorrhizal fungal succession in a long-lived perennial. Botany-Botanique 92.

Hausmann, N.T., Hawkes, C.V., 2009. Plant neighborhood control of arbuscular mycorrhizal community composition. New Phytologist 183, 1188–1200.

Hausmann, N.T., Hawkes, C.V., 2010. Order of plant host establishment alters the composition of arbuscular mycorrhizal communities. Ecology 91, 2333–2343.

Hayward, J., Horton, T.R., Pauchard, A., Nuñez, M.A., 2015. A single ectomycorrhizal fungal species can enable a *Pinus* invasion. Ecology 96, 1438–1444.

Hedh, J., Wallander, H., Erland, S., 2008. Ectomycorrhizal mycelial species composition in apatite amended and non-amended mesh bags buried in a phosphorus-poor spruce forest. Mycological Research 112, 681–688.

Hoeksema, J.D., Chaudhary, V.B., Gehring, C.A., Johnson, N.C., Karst, J., Koide, R.T., Pringle, A., Zabinski, C., Bever, J.D., Moore, J.C., et al., 2010. A meta-analysis of context-dependency in plant response to inoculation with mycorrhizal fungi. Ecology Letters 13, 394–407.

Holdaway, R.J., Richardson, S.J., Dickie, I.A., Peltzer, D.A., Coomes, D.A., 2011. Species- and community-level patterns in fine root traits along a 120 000-year soil chronosequence in temperate rain forest. Journal of Ecology 99, 954–963.

Horton, T.R., Bruns, T.D., 2001. The molecular revolution in ectomycorrhizal ecology: peeking into the black-box. Molecular Ecology 10, 1855–1871.

Johnson, N.C., Rowland, D.L., Corkidi, L., Allen, E.B., 2008. Plant winners and losers during grassland N-eutrophication differ in biomass allocation and mycorrhizas. Ecology 89, 2868–2878.

Johnson, N.C., Zak, D.R., Tilman, D., Pfleger, F.L., 1991. Dynamics of vesicular-arbuscular mycorrhizae during old field succession. Oecologia 86, 349–358.
Jones, M.D., Durall, D.M., Cairney, W.G., 2003. Ectomycorrhizal fungal communities in young forest stands regenerating after clearcut logging. New Phytologist 157, 399–422.
Jumpponen, A., Brown, S.P., Trappe, J.M., Cazares, E., Strommer, R., 2012. Twenty years of research on fungal-plant interactions on Lyman Glacier forefront – lessons learned and questions yet unanswered. Fungal Ecology 5, 430–442.
Jumpponen, A., Trappe, J.M., Cazares, E., 2002. Occurrence of ectomycorrhizal fungi on the forefront of retreating Lyman Glacier (Washington, USA) in relation to time since deglaciation. Mycorrhiza 12, 43–49.
Karst, J., Erbilgin, N., Pec, G.J., Cigan, P.W., Najar, A., Simard, S.W., Cahill, J.F., 2015. Ectomycorrhizal fungi mediate indirect effects of a bark beetle outbreak on secondary chemistry and establishment of pine seedlings. New Phytologist 208, 904–914.
Kennedy, P., 2010. Ectomycorrhizal fungi and interspecific competition: species interactions, community structure, coexistence mechanisms, and future research directions. New Phytologist 187, 895–910.
Kennedy, P.G., Peay, K.G., Bruns, T.D., 2009. Root tip competition among ectomycorrhizal fungi: are priority effects a rule or an exception? Ecology 90, 2098–2107.
Kiers, E.T., Duhamel, M., Beesetty, Y., Mensah, J.A., Franken, O., Verbruggen, E., Fellbaum, C.R., Kowalchuk, G.A., Hart, M.M., Bago, A., et al., 2011. Reciprocal rewards stabilize cooperation in the mycorrhizal symbiosis. Science 333, 880–882.
Kipfer, T., Moser, B., Egli, S., Wohlgemuth, T., Ghazoul, J., 2011. Ectomycorrhiza succession patterns in *Pinus sylvestris* forests after stand-replacing fire in the Central Alps. Oecologia 167, 219–228.
Klironomos, J., Zobel, M., Tibbett, M., Stock, W.D., Rillig, M.C., Parrent, J.L., Moora, M., Koch, A.M., Facelli, J.M., Facelli, E., et al., 2011. Forces that structure plant communities: quantifying the importance of the mycorrhizal symbiosis. New Phytologist 189, 366–370.
Klironomos, J.N., 2003. Variation in plant response to native and exotic arbuscular mycorrhizal fungi. Ecology 84, 2292–2301.
Koele, N., Dickie, I., Blum, J., Gleason, J., de Graaf, L., 2014. Ecological significance of mineral weathering in ectomycorrhizal and arbuscular mycorrhizal ecosystems from a field-based comparison. Soil Biology and Biochemistry 69, 63–70.
Koele, N., Dickie, I.A., Oleksyn, J., Richardson, S.J., Reich, P.B., 2012. No globally consistent effect of ectomycorrhizal status on foliar traits. New Phytologist 196, 845–852.
Koide, R.T., Fernandez, C., Malcolm, G., 2014. Determining place and process: functional traits of ectomycorrhizal fungi that affect both community structure and ecosystem function. New Phytologist 201, 433–439.
Koide, R.T., Wu, T., 2003. Ectomycorrhizas and retarded decomposition in a *Pinus resinosa* plantation. New Phytologist 158, 401–407.
Koricheva, J., Gurevitch, J., 2014. Uses and misuses of meta-analysis in plant ecology. Journal of Ecology 102, 828–844.
Krüger, M., Teste, F.P., Laliberté, E., Lambers, H., Coghlan, M., Zemunik, G., Bunce, M., 2015. The rise and fall of arbuscular mycorrhizal fungal diversity during ecosystem retrogression. Molecular Ecology 24, 4912–4930.
Kumpfer, W., Heyser, W., 1986. Effects of stemflow on the mycorrhiza of beech (*Fagus sylvatica* L.). In: Physiological and Genetical Aspects of Mycorrhizae. Aspects physiologiques et genetiques des mycorhizes, Dijon (France), July 1–5, 1985. INRA.
Laliberté, E., Turner, B.L., Costes, T., Pearse, S.J., Wyrwoll, K.H., Zemunik, G., Lambers, H., 2012. Experimental assessment of nutrient limitation along a 2-million-year dune chronosequence in the south-western Australia biodiversity hotspot. Journal of Ecology 100, 631–642.
Laliberté, E., Zemunik, G., Turner, B.L., 2014. Environmental filtering explains variation in plant diversity along resource gradients. Science 345, 1602–1605.
Lambers, H., Brundrett, M.C., Raven, J.A., Hopper, S.D., 2010. Plant mineral nutrition in ancient landscapes: high plant species diversity on infertile soils is linked to functional diversity for nutritional strategies. Plant and Soil 334, 11–31.
Lambers, H., Raven, J.A., Shaver, G.R., Smith, S.E., 2008. Plant nutrient-acquisition strategies change with soil age. Trends in Ecology & Evolution 23, 95–103.
Last, F., Dighton, J., Mason, P., 1987. Successions of sheathing mycorrhizal fungi. Trends in Ecology & Evolution 2, 157–161.
LeDuc, S.D., Lilleskov, E.A., Horton, T.R., Rothstein, D.E., 2013. Ectomycorrhizal fungal succession coincides with shifts in organic nitrogen availability and canopy closure in post-wildfire jack pine forests. Oecologia 172, 257–269.

Lekberg, Y., Gibbons, S.M., Rosendahl, S., Ramsey, P.W., 2013. Severe plant invasions can increase mycorrhizal fungal abundance and diversity. The ISME journal 7, 1424–1433.

Lekberg, Y., Meadow, J., Rohr, J.R., Redecker, D., Zabinski, C.A., 2011. Importance of dispersal and thermal environment for mycorrhizal communities: lessons from Yellowstone National Park. Ecology 92, 1292–1302.

Li, D.W., 2005. Release and dispersal of basidiospores from *Amanita muscaria* var. *alba* and their infiltration into a residence. Mycological Research 109, 1235–1242.

Lindahl, B.D., Nilsson, R.H., Tedersoo, L., Abarenkov, K., Carlsen, T., Kjøller, R., Kõljalg, U., Pennanen, T., Rosendahl, S., Stenlid, J., et al., 2013. Fungal community analysis by high-throughput sequencing of amplified markers – a user's guide. New Phytologist 199, 288–299.

Lu, X., Malajczuk, N., Brundrett, M., Dell, B., 1999. Fruiting of putative ectomycorrhizal fungi under blue gum (*Eucalyptus globulus*) plantations of different ages in Western Australia. Mycorrhiza 8, 255–261.

Mallik, A., 2003. Conifer regeneration problems in boreal and temperate forests with ericaceous understory: role of disturbance, seedbed limitation, and keytsone species change. Critical Reviews in Plant Sciences 22, 341–366.

Martínez-García, L.B., Richardson, S.J., Tylianakis, J.M., Peltzer, D.A., Dickie, I.A., 2015. Host identity is a dominant driver of mycorrhizal fungal community composition during ecosystem development. New Phytologist 205, 1565–1576.

Meadow, J.F., Zabinski, C.A., 2012. Linking symbiont community structures in a model arbuscular mycorrhizal system. New Phytologist 194, 800–809.

Meiners, S.J., Cadotte, M.W., Fridley, J.D., Pickett, S.T.A., Walker, L.R., 2015. Is successional research nearing its climax? New approaches for understanding dynamic communities. Functional Ecology 29, 154–164.

Moora, M., Berger, S., Davison, J., Öpik, M., Bommarco, R., Bruelheide, H., Kühn, I., Kunin, W.E., Metsis, M., Rortais, A., 2011. Alien plants associate with widespread generalist arbuscular mycorrhizal fungal taxa: evidence from a continental-scale study using massively parallel 454 sequencing. Journal of Biogeography 38, 1305–1317.

Moora, M., Öpik, M., Sen, R., Zobel, M., 2004. Native arbuscular mycorrhizal fungal communities differentially influence the seedling performance of rare and common *Pulsatilla* species. Functional Ecology 18, 554–562.

Mummey, D.L., Rillig, M.C., 2006. The invasive plant species *Centaurea maculosa* alters arbuscular mycorrhizal fungal communities in the field. Plant and Soil 288, 81–90.

Nara, K., 2006. Ectomycorrhizal networks and seedling establishment during early primary succession. New Phytologist 169, 169–178.

Nara, K., Nakaya, H., Wu, B.Y., Zhou, Z.H., Hogetsu, T., 2003. Underground primary succession of ectomycorrhizal fungi in a volcanic desert on Mount Fuji. New Phytologist 159, 743–756.

Näsholm, T., et al., 2013. Are ectomycorrhizal fungi alleviating or aggravating nitrogen limitation of tree growth in boreal forests?. New Phytologist 198, 214–221.

Nguyen, N.H., Hynson, N.A., Bruns, T.D., 2012. Stayin' alive: survival of mycorrhizal fungal propagules from 6-yr-old forest soil. Fungal Ecology 5, 741–746.

Nuñez, M.A., Dickie, I.A., 2014. Invasive belowground mutualists of woody plants. Biological Invasions 16, 645–661.

Oehl, F., Schneider, D., Sieverding, E., Burga, C.A., 2011. Succession of arbuscular mycorrhizal communities in the foreland of the retreating Morteratsch glacier in the Central Alps. Pedobiologia 54, 321–331.

Öpik, M., Metsis, M., Daniell, T.J., Zobel, M., Moora, M., 2009. Large-scale parallel 454 sequencing reveals host ecological group specificity of arbuscular mycorrhizal fungi in a boreonemoral forest. New Phytologist 184, 424–437.

Öpik, M., Moora, M., Liira, J., Zobel, M., 2006. Composition of root-colonizing arbuscular mycorrhizal fungal communities in different ecosystems around the globe. Journal of Ecology 94, 778–790.

Öpik, M., Zobel, M., Cantero, J., Davison, J., Facelli, J., Hiiesalu, I., Jairus, T., Kalwij, J., Koorem, K., Leal, M., et al., 2013. Global sampling of plant roots expands the described molecular diversity of arbuscular mycorrhizal fungi. Mycorrhiza 23, 411–430.

Orwin, K.H., Kirschbaum, M.U.F., St John, M.G., Dickie, I.A., 2011. Organic nutrient uptake by mycorrhizal fungi enhances ecosystem carbon storage: a model-based assessment. Ecology Letters 14, 493–502.

Peay, K.G., Garbelotto, M., Bruns, T.D., 2010. Evidence of dispersal limitation in soil microorganisms: isolation reduces species richness on mycorrhizal tree islands. Ecology 91, 3631–3640.

Peay, K.G., Kennedy, P.G., Bruns, T.D., 2008. Fungal community ecology: a hybrid beast with a molecular master. Bioscience 58, 799–810.

Peay, K.G., Kennedy, P.G., Bruns, T.D., 2011. Rethinking ectomycorrhizal succession: are root density and hyphal exploration types drivers of spatial and temporal zonation? Fungal Ecology 4, 233–240.

Peltzer, D.A., Wardle, D.A., Allison, V.J., Baisden, W.T., Bardgett, R.D., Chadwick, O.A., Condron, L.M., Parfitt, R.L., Porder, S., Richardson, S.J., et al., 2010. Understanding ecosystem retrogression. Ecological Monographs 80, 509–529.

Pendergast, T.H., Burke, D.J., Carson, W.P., 2013. Belowground biotic complexity drives aboveground dynamics: a test of the soil community feedback model. New Phytologist 197, 1300–1310.

Piotrowski, J.S., Rillig, M.C., 2008. Succession of arbuscular mycorrhizal fungi: patterns, causes, and considerations for organic agriculture. In: Sparks, D.L. (Ed.)Sparks, D.L. (Ed.), Advances in Agronomy, vol. 97, pp. 111–130.

Read, D., 1993. Mycorrhiza in plant communities. Advances in Plant Pathology 9, 1–31.

Read, D., Perez-Moreno, J., 2003. Mycorrhizas and nutrient cycling in ecosystems – a journey towards relevance? New Phytologist 157, 475–492.

Read, D.J., 1991. Mycorrhizas in ecosystems. Experientia 47, 376–391.

Robertson, S.J., Tackaberry, L.E., Egger, K.N., Massicotte, H.B., 2006. Ectomycorrhizal fungal communities of black spruce differ between wetland and upland forests. Canadian Journal of Forest Research 36, 972–985.

Roy, M., Rochet, J., Manzi, S., Jargeat, P., Gryta, H., Moreau, P.-A., Gardes, M., 2013. What determines *Alnus*-associated ectomycorrhizal community diversity and specificity? A comparison of host and habitat effects at a regional scale. New Phytologist 198, 1228–1238.

Schechter, S., Bruns, T., 2013. A common garden test of host-symbiont specificity supports a dominant role for soil type in determining AMF assemblage structure in *Collinsia sparsiflora*. PLoS One 8, e55507.

Schmidt, S.K., Nemergut, D.R., Darcy, J.L., Lynch, R., 2014. Do bacterial and fungal communities assemble differently during primary succession? Molecular Ecology 23, 254–258.

Sikes, B.A., Maherali, H., Klironomos, J.N., 2014. Mycorrhizal fungal growth responds to soil characteristics, but not host plant identity, during a primary lacustrine dune succession. Mycorrhiza 24, 219–226.

Sikes, B.A., Maherali, H.Z., Klironomos, J.N., 2012. Arbuscular mycorrhizal fungal communities change among three stages of primary sand dune succession but do not alter plant growth. Oikos 121, 1791–1800.

Simard, S.W., Beiler, K.J., Bingham, M.A., Deslippe, J.R., Philip, L.J., Teste, F.P., 2012. Mycorrhizal networks: mechanisms, ecology and modelling. Fungal Biology Reviews 26, 39–60.

Smith, S.E., Read, D.J., 2008. Mycorrhizal Symbiosis. Academic Press, New York, NY, USA.

Spence, L.A., Dickie, I.A., Coomes, D.A., 2011. Arbuscular mycorrhizal inoculum potential: a mechanism promoting positive diversity-invasibility relationships in mountain beech forests in New Zealand? Mycorrhiza 21, 309–314.

Sumorok, B., Kosiński, K., Michalska-Hejduk, D., Kiedrzyńska, E., 2008. Distribution of ectomycorrhizal fungi in periodically inundated plant communities on the Pilica River floodplain. Ecohydrology & Hydrobiology 8, 401–410.

Suz, L.M., Barsoum, N., Benham, S., Dietrich, H.-P., Fetzer, K.D., Fischer, R., Garcia, P., Gehrman, J., Kristoefel, F., Manninger, M., et al., 2014. Environmental drivers of ectomycorrhizal communities in Europe's temperate oak forests. Molecular Ecology 23, 5628–5644.

Tedersoo, L., Bahram, M., Dickie, I.A., 2014a. Does host plant richness explain diversity of ectomycorrhizal fungi? Re-evaluation of Gao et al. (2013) data sets reveals sampling effects. Molecular Ecology 23, 992–995.

Tedersoo, L., Bahram, M., Jairus, T., Bechem, E., Chinoya, S., Mpumba, R., Leal, M., Randrianjohany, E., Razafimandimbison, S., Sadam, A., et al., 2011. Spatial structure and the effects of host and soil environments on communities of ectomycorrhizal fungi in wooded savannas and rain forests of Continental Africa and Madagascar. Molecular Ecology 20, 3071–3080.

Tedersoo, L., Bahram, M., Põlme, S., Kõljalg, U., Yorou, N.S., Wijesundera, R., Ruiz, L.V., Vasco-Palacios, A.M., Thu, P.Q., Suija, A., et al., 2014b. Global diversity and geography of soil fungi. Science 346, 1052–1053.

Tedersoo, L., Bahram, M., Toots, M., Diédhiou, A.G., Henkel, T.W., Kjoller, R., Morris, M.H., Nara, K., Nouhra, E., Peay, K.G., et al., 2012. Towards global patterns in the diversity and community structure of ectomycorrhizal fungi. Molecular Ecology 21, 4160–4170.

Tedersoo, L., Suvi, T., Jairus, T., Ostonen, I., Põlme, S., 2009. Revisiting ectomycorrhizal fungi of the genus *Alnus*: differential host specificity, diversity and determinants of the fungal community. New Phytologist 182, 727–735.

Telford, R.J., Vandvik, V., Birks, H.J.B., 2006. Dispersal limitations matter for microbial morphospecies. Science 312, 1015.

Terwilliger, J., Pastor, J., 1999. Small mammals, ectomycorrhizae, and conifer succession in beaver meadows. Oikos 83–94.

Teste, F.P., Laliberté, E., Lambers, H., Auer, Y., Kramer, S., Kandeler, E., 2016. Mycorrhizal fungal biomass and scavenging declines in phosphorus-impoverished soils during ecosystem retrogression. Soil Biology and Biochemistry 92, 119–132.

Teste, F.P., Lieffers, V.J., Strelkov, S.E., 2012. Ectomycorrhizal community responses to intensive forest management: thinning alters impacts of fertilization. Plant and Soil 360, 333–347.

Teste, F.P., Simard, S.W., Durall, D.M., 2009. Role of mycorrhizal networks and tree proximity in ectomycorrhizal colonization of planted seedlings. Fungal Ecology 2, 21–30.

Treseder, K.K., 2004. A meta-analysis of mycorrhizal responses to nitrogen, phosphorus, and atmospheric CO_2 in field studies. New Phytologist 164, 347–355.

Turner, B.L., Condron, L.M., 2013. Pedogenesis, nutrient dynamics, and ecosystem development: the legacy of T.W. Walker and J.K. Syers. Plant and Soil 367, 1–10.

Turner, B.L., Laliberté, E., 2015. Soil development and nutrient availability along a 2 million-year coastal dune chronosequence under species-rich mediterranean shrubland in southwestern Australia. Ecosystems 18, 287–309.

Twieg, B.D., Durall, D.M., Simard, S.W., 2007. Ectomycorrhizal fungal succession in mixed temperate forests. New Phytologist 176, 437–447.

Unger, I.M., Kennedy, A.C., Muzika, R.-M., 2009. Flooding effects on soil microbial communities. Applied Soil Ecology 42, 1–8.

van der Heijden, M.G.A., 2004. Arbuscular mycorrhizal fungi as support systems for seedling establishment in grassland. Ecology Letters 7, 293–303.

van der Heijden, M.G.A., Klironomos, J.N., Ursic, M., Moutoglis, P., Streitwolf-Engel, R., Boller, T., Wiemken, A., Sanders, I.R., 1998. Mycorrhizal fungal diversity determines plant biodiversity, ecosystem variability and productivity. Nature 396, 69–72.

van der Putten, W.H., Bardgett, R.D., Bever, J.D., Bezemer, T.M., Casper, B.B., Fukami, T., Kardol, P., Klironomos, J.N., Kulmatiski, A., Schweitzer, J.A., et al., 2013. Plant–soil feedbacks: the past, the present and future challenges. Journal of Ecology 101, 265–276.

van der Putten, W.H., Bradford, M.A., Pernilla Brinkman, E., van de Voorde, T.F.J., Veen, G.F.C., 2016. Where, when and how plant-soil feedback matters in a changing world. Functional Ecology 30, 1109–1121.

Van Diepen, L.T., Lilleskov, E.A., Pregitzer, K.S., 2011. Simulated nitrogen deposition affects community structure of arbuscular mycorrhizal fungi in northern hardwood forests. Molecular Ecology 20, 799–811.

Visser, S., 1995. Ectomycorrhizal fungal succession in jack pine stands following wildfire. New Phytologist 129, 389–401.

Vitousek, P.M., Howarth, R.W., 1991. Nitrogen limitation on land and in the sea: how can it occur? Biogeochemistry 13, 87–115.

Walker, T., Syers, J., 1976. The fate of phosphorus during pedogenesis. Geoderma 15, 1–19.

Wallenda, T., Kottke, I., 1998. Nitrogen deposition and ectomycorrhizas. New Phytologist 139, 169–187.

Wang, Q., He, X.H., Guo, L.-D., 2012. Ectomycorrhizal fungus communities of *Quercus liaotungensis* Koidz of different ages in a northern China temperate forest. Mycorrhiza 22, 461–470.

Wang, Y., Huang, Y., Qiu, Q., Xin, G., Yang, Z., Shi, S., 2011. Flooding greatly affects the diversity of arbuscular mycorrhizal fungi communities in the roots of wetland plants. PLoS One 6, e24512.

Wardle, D.A., Lindahl, B.D., 2014. Disentangling global soil fungal diversity. Science 346, 1052–1053.

Watt, A., 1947. Pattern and process in the plant community. Journal of Ecology 35, 1–22.

Werner, G.D.A., Kiers, E.T., 2015. Partner selection in the mycorrhizal mutualism. New Phytologist 205, 1437–1442.

Willis, K.J., Whittaker, R.J., 2002. Species diversity – scale matters. Science 295, 1245–1248.

Wilson, G.W., Hickman, K.R., Williamson, M.M., 2012. Invasive warm-season grasses reduce mycorrhizal root colonization and biomass production of native prairie grasses. Mycorrhiza 22, 327–336.

Wolfe, B.E., Mummey, D.L., Rillig, M.C., Klironomos, J.N., 2007. Small-scale spatial heterogeneity of arbuscular mycorrhizal fungal abundance and community composition in a wetland plant community. Mycorrhiza 17, 175–183.

Wood, J.R., Dickie, I.A., Moeller, H.V., Peltzer, D.A., Bonner, K.I., Rattray, G., Wilmshurst, J.M., Gibson, D., 2015. Novel interactions between non-native mammals and fungi facilitate establishment of invasive pines. Journal of Ecology 103, 121–129.

Worley, J., Hacskaylo, E., 1959. The effects of available soil moisture on the mycorrhizal associations of Virginia pine. Forest Science 5, 267–268.

Yang, H., Koide, R.T., Zhang, Q., 2016. Short-term waterlogging increases arbuscular mycorrhizal fungal species richness and shifts community composition. Plant and Soil 1–12.

Zobel, M., Öpik, M., 2014. Plant and arbuscular mycorrhizal fungal (AMF) communities – which drives which? Journal of Vegetation Science 25, 1133–1140.

SECTION II

MYCORRHIZAL MEDIATION OF SOIL FERTILITY

Nancy Collins Johnson, Lead Editor

CHAPTER

6

Introduction: Perspectives on Mycorrhizas and Soil Fertility

L.K. Abbott[1], *N.C. Johnson*[2]

[1]University of Western Australia, Perth, WA, Australia; [2]Northern Arizona University, Flagstaff, AZ, United States

6.1 INTRODUCTION

The term *soil fertility* has ancient origins and has been consistently used over centuries to refer to the capability of soil to support plant production in agricultural contexts. Historically, the most common use of *soil fertility* has focused on provisioning mineral nutrients for plant growth (e.g. Foth and Ellis, 1997; Tisdale et al., 1985). An emphasis on fertilizer-based nutrient amendment has continued despite the well-understood contributions of soil organisms to nutrient availability (Lavelle et al., 1994) and soil physical conditions for plant growth (Evans, 1948; Six and Paustian, 2014). Abbott and Murphy (2003) explored the separation of biological, chemical, and physical soil fertility to provide clarity on the unique contributions and interactions among these components of soil fertility (Fig. 6.1). This was necessary to ensure that the contributions of soil organisms were considered within a soil fertility framework and not overshadowed by a focus solely on plant production responses to chemical fertility of soil. A further contribution of this conceptual model is that it provides insights about the interacting mechanisms that control soil fertility. Biological processes have the potential to contribute significantly to chemical and physical processes that influence soil fertility.

Mycorrhizas function at the nexus of biological, chemical, and physical components of soil fertility. Indeed, arbuscular mycorrhizal (AM) fungi are in unique positions to bridge these components of soil fertility (Abbott and Manning, 2015; Jeffries et al., 2003; Miller and Jastrow, 1992), and the same applies for ectomycorrhizal (EcM) fungi (Haselwandter and Bowen, 1996; Janos, 1980; Trappe, 1977) and ericoid mycorrhizal (ErM) fungi (Cairney and Meharg, 2003; Mitchell and Gibson, 2006). There is a risk that mycorrhizal contributions to soil fertility will be diminished without careful management of inputs that build chemical and physical fertility. This chapter introduces subsequent chapters that provide insights into how mycorrhizas can be managed to optimize their contributions to all facets of soil fertility.

Mycorrhizal Mediation of Soil
http://dx.doi.org/10.1016/B978-0-12-804312-7.00006-1

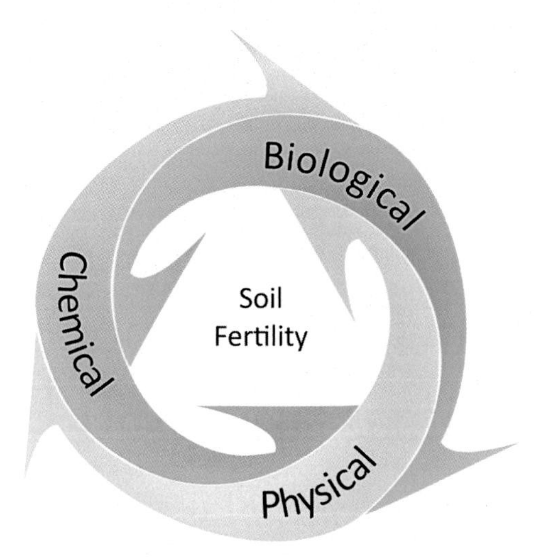

FIGURE 6.1 Soil fertility is generated from the interactions among physical, chemical, and biological processes (Abbott and Murphy, 2003). This chapter examines how mycorrhizas influence the individual and interactive effects of these processes. *Illustration created by Kara Skye Gibson.*

The three-component perspective helps highlight the fact that soil fertility is context specific. In an agricultural context, a "fertile soil" is primarily supported by applications of nutrients that increase its chemical fertility. High-input row crop production systems often override biological fertility with mechanical tillage and chemical fertilizers, which may diminish the contributions of AM fungal hyphae to soil structure (Degens, 1997; Six and Paustian, 2014; Tisdall and Oades, 1980) and biotic communities to N_2 fixation (Hardarson et al., 1984; Peoples et al., 2012; Rose et al., 2016). Likewise, practices that reduce contributions of EcM fungi in plantation forestry (Egnell et al., 2015) may lead to severe reductions in soil fertility due to reduced support of biological processes (Aucina et al., 2014; Cambi et al., 2015; Tikvic et al., 2012; Wu, 2011). Furthermore, compared with row crop systems, perennial forests have less opportunity for compensation by increasing chemical fertility through nutrient augmentation, except perhaps via pollution (Högberg et al., 2011; Tikvic et al., 2012). When soils are managed to support contributions of mycorrhizal fungi, processes involved in biological fertility will be favored and there will be greater balance in contributions from soil biological, chemical, and physical components of fertility (Abbott and Lumley, 2014; Abbott and Manning, 2015; Ekbald et al., 2013; Kalliokoski et al., 2010; Marschner and Dell, 1994; Näsholm et al., 2013).

Southwestern Australia is a global biodiversity hotspot where severely weathered soils have low chemical and physical fertility (Gibson et al., 2010; Lambers et al., 2010), but they support diverse plant communities through mycorrhizal symbioses and other microbial associations that facilitate plant nutrient absorption (Laliberte et al., 2012; Lambers et al., 2013). Plant production in these soils is extremely slow (Huston, 2013; Labiberte et al., 2013). Some plant species rely on mycorrhizal associations, but others do not because they have other strategies for accessing nutrients (Brundrett, 2006; Lambers et al., 2010). The diverse plant communities of southwestern Australia thrive on low levels of soil chemical fertility but high levels of soil biological fertility despite their slow growth rates. In contrast, the same soils are considered to be very infertile for fast-growing agricultural plants (Hingston et al., 1980; Bolan et al., 1983). Application of fertilizers to these soils for agriculture (Abbott and Robson, 1977) or forest management (Stoneman et al., 1995) may increase production but override mycorrhizal contributions to soil biological fertility. In certain situations, this may be undesirable. For example, reclamation of vegetation after mining operations requires the recovery of microbial processes, but biological fertility can take many years to develop because of limitations in plant growth associated with chemical (Daws et al., 2013, 2015) and physical constraints to soil fertility (Santini and Banning, 2016). Adding inappropriate levels of fertilizers may further delay the establishment of mycorrhizas in severely disturbed ecosystems such as mine tailings.

6.2 CONTRIBUTIONS OF MYCORRHIZAL FUNGI TO SOIL FERTILITY

The outcomes of interactions among mycorrhizal fungi in roots and in soil influence the extent to which they contribute to soil fertility. Mycorrhizal fungi occur in heterogeneous habitats in the soil, rhizosphere, and roots. Although the communities of mycorrhizal fungi as a whole have particular traits (van der Heijden and Scheublin, 2007), they are composed of individual taxa that express a range of responses to their immediate environment (Powell et al., 2009). In general, for AM fungi, these taxa concurrently co-colonize the same roots (Chagnon et al., 2012). For EcM fungi, root tips are commonly colonized by one dominant fungus, but co-colonization of root systems is common (Toju et al., 2016; Korkama et al., 2006). Niche differentiation among soil substrates within a soil profile has also been demonstrated to influence the spatial separation of different EcM hyphae within soil (Dickie et al., 2002). The result for AM fungi is the formation of a common mycorrhizal network (CMN; Barto et al., 2012; Kytovita et al., 2003; Robinson and Fitter, 1999) and communities of AM fungi can contribute via their CMNs to the mineral nutrition of neighboring plants (Bücking et al., 2016; Weremijewicz et al., 2016). EcM fungi act in a similar manner (e.g. He et al., 2010; Selosse et al., 2006) and vary in their effectiveness in contributing to soil fertility. Interactions among complex communities of mycorrhizal fungi and host plants influence the biological, chemical, and physical processes that define soil fertility (Fig. 6.2).

6.2.1 Contributions of Mycorrhizal Fungi to Soil Biological Fertility

Soil biological fertility encompasses the qualities of living components of soil that contribute to the nutritional requirements of plants (Abbott and Murphy, 2003; Fig. 6.2). From their central location at the soil–root interface, mycorrhizal fungi form a conduit for carbon from roots into soil that is independent of direct release into the rhizosphere from roots (Pringle, 2016). Hyphae of mycorrhizal fungi have the capacity to deliver root-derived carbon beyond the rhizosphere, with potential to influence soil biota involved in degradation of organic matter (Herman et al., 2012; Klironomos and Hart, 2001) and the resilience of fungal and bacterial communities (Veresoglou et al., 2016). This resource of carbon includes hyphae and spores that can be grazed by soil fauna (Cairney, 2012; Klironomos and Hart, 2001). Other

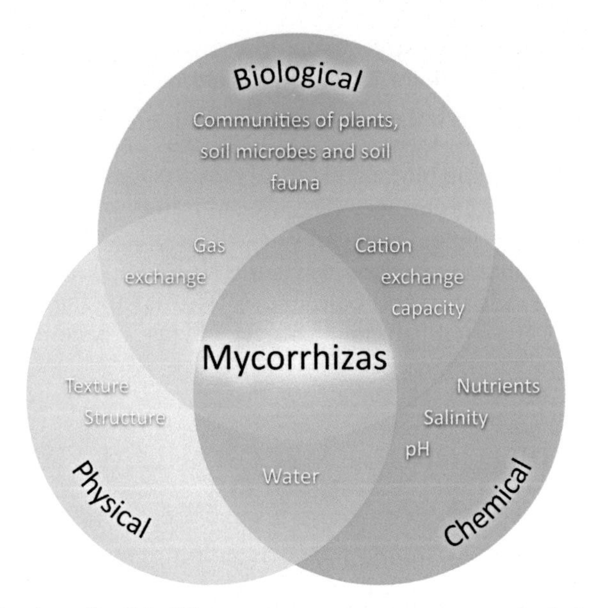

FIGURE 6.2 A holistic view of soil fertility encompasses synergy among physical, chemical, and biological factors that influence the capacity of soil to influence mycorrhizal function and for mycorrhizal fungi to influence soil conditions for plant growth. *Illustration created by Kara Skye Gibson.*

contributions involve complex interactions that influence the potential to prevent or minimize colonization of roots by pathogens, although such interactions may depend on the AM fungus (Sikes, 2010) and the susceptibility of the plant to pathogens (Ronsheim, 2016).

The capacity of mycorrhizal fungi to supply carbon and nutrients for use by other organisms (Chapter 9) can have a significant influence on soil biological fertility, assessed as the capability of soil biota to enable plants to resist disease and drought and possibly other soil constraints. Their capacity to access and trade plant carbon (Chapters 21 and 22) leads to multitrophic feedback loops, which can influence plant productivity, plant diversity, and the stabilization of organic matter in soil (Chapter 9). This highlights the potential of mycorrhizal fungi to play a significant role in regulation of microbial and faunal communities in soil, but the extent to which this occurs depends on the mycorrhizal taxa present, their abundance, their infectivity, and their capacity to access plant carbon. There clearly are differences between functional attributes of AM and EcM fungi, but there are also similarities in the overall beneficial contributions each group makes to soil processes that enhance soil biological fertility.

6.2.2 Contributions of Mycorrhizal Fungi to Soil Chemical Fertility

Soil chemical fertility refers to the capacity of the soil to provide the nutrients required by plants (Abbott and Murphy, 2003, Fig. 6.2). It is well known that mycorrhizal fungi have the potential to make significant contributions to plant nutrition by accessing nutrients that are not otherwise readily available to plants from inorganic and organic sources (Chapter 8; Courty et al., 2010). Mycorrhizas also ameliorate the effects of soil salinity (Chapter 15).

Complex feedback signals have evolved to control the physiological processes that enable fungi and plants to efficiently trade nutrients for carbon (Chapter 7; Bücking et al., 2012; Veresoglou et al., 2016). As plants approach nutrient adequacy, mycorrhizal colonization of roots may diminish (Thomson et al., 1986). This may or may not reduce the biomass of mycorrhizal fungi in the root system or in the soil (Carvalho et al., 2015). Rather, it is likely to shift the emphasis from mycorrhizal contributions to soil chemical fertility (Schweiger et al., 2007) or to those associated with soil physical fertility (Rillig and Mummey, 2006).

The net contributions of EcM and AM symbioses to soil chemical fertility are generally similar (Smith and Read, 2008), but there is greater potential for EcM fungi to contribute to nitrogen nutrition and AM fungi to phosphorus nutrition of plants (Johnson et al., 2015; Corrales et al., 2016). Although there is no question that AM and EcM fungi access organic and inorganic forms of nitrogen and phosphorus (Chapter 8), AM fungi have a high demand for nitrogen, and the economics of their "carbon-for-nitrogen" trade prevent a net biomass gain for their host plant (Püschel et al., 2016). EcM fungi are polyphyletic, and their diverse repertoire of enzymes helps them efficiently access nitrogen from a range of organic compounds (Courty et al., 2010). In contrast, all AM fungi belong to one phylum, the Glomeromycota (Chapter 1, Table 1.1), and this group of fungi has evolved to be very efficient in acquiring phosphorus and trading it for plant carbon (Parniske, 2008). In forest systems AM-associated trees have been shown to predominate in mineral soils and EcM trees in organic soils (Phillips et al., 2013).

6.2.3 Contributions of Mycorrhizal Fungi to Soil Physical Fertility

Soil physical fertility refers to the structural components of the soil that facilitate the movement of water and gases to plant roots (Abbott and Murphy, 2003; Fig. 6.2). These physical characteristics are dependent on soil type and land use (Bachelot et al., 2016; Williams et al., 2016) and are strongly influenced by biotic processes (Williams et al., 2016). The mycelium of AM and EcM fungi contributes to soil physical processes in the surface layers of soil (Chapters 14 and 17). Mycorrhizal hyphae interact with soil particles to form aggregates that can contribute to the protection of soil organic matter (Beare et al., 1997; Clemmensen et al., 2015; Six and Paustian, 2014) and help retention of soil water, even in sandy soils (Mickan, 2014; Mickan et al., 2016). All of these processes are associated with mycorrhizal networks and improved water relations (Chapters 16 and 18). Soil carbon sequestration associated with mycorrhizal fungi would contribute to soil physical fertility in association with processes associated with soil structure (Degens, 1997; Tisdall and Oades, 1982) by improving soil gas exchange (aeration) and water penetrability. All of these processes have the potential to improve soil conditions for plant growth and reduce the risk of soil structure decline, including resilience to soil erosion (Mardhiah et al., 2016), although under some circumstances mats of EcM hyphae may be responsible for soil water repellence (Unestam, 1991).

6.3 SOIL FERTILITY INFLUENCES MYCORRHIZAL FUNGI

Fertility-related soil properties can sometimes interfere with fungal growth in soil or in roots and alter mycorrhizal function, which in turn can limit its contribution to soil fertility. As earlier indicated, mycorrhizal fungi are diverse, and they can individually and collectively contribute to biological, chemical, and physical components of soil fertility in various ways. These contributions vary spatially and temporarily. However, mycorrhizal fungal taxa with known potential to enhance specific components of soil fertility may only do so when they occur at levels above a threshold level of abundance (Abbott and Robson, 1981a,b). "Effective" fungi (Pellegrino et al., 2011) cannot make a contribution if they occur at extremely low levels in soil and roots. This means they also need to be competitive with other mycorrhizal fungi (Cano and Bago, 2005; Knegt et al., 2016), and they need to overcome environmental constraints.

A range of soil factors can reduce the contribution of mycorrhizas to soil fertility if they limit colonization of roots. These include constraints such as salinity (Juniper and Abbott, 1993), soil compaction (Nadian et al., 1998), extreme pH (Robson and Abbott, 1989), water deficit (Mickan, 2014), high soil phosphorus (e.g. Thomson et al., 1986; Johnson et al., 2015), high soil nitrogen (e.g. Nouri et al., 2014), or adverse biological conditions such as selective grazing of fungal hyphae by soil fauna (Klironomos et al., 2004). Humans can inadvertently affect mycorrhizal fungal communities through land management practices, air pollution, and other unintended consequences of development. Agricultural systems affect AM fungi through imposition of soil disturbance in the form of tillage, fertilization, plant rotation, pest management, or grazing (Chapter 10). Disturbances in natural ecosystems can influence communities of AM and EcM fungi. This may be severe, as in the case of large-scale stockpiling of topsoil during mine site operations (Jasper et al., 1992), or local, such as tree removal, fire, or off-road vehicle traffic. Regional- and global-scale anthropogenic changes such as nitrogen eutrophication, CO_2 enrichment, and climate change can also affect mycorrhizas (Chapter 12). A better understanding of these effects will help develop management strategies to optimize mycorrhizal contributions to soil fertility in a changing world.

6.3.1 Mycorrhizal Function in Agricultural Ecosystems

Agricultural systems are primarily dominated by AM symbioses. Agricultural management practices can have various degrees of effect on individual AM fungi as well as on the community of AM fungi as a whole. Some agricultural practices (e.g. tillage, compaction plant rotation, irrigation, soil amendments such as lime, gypsum, compost) alter the soil's physical fertility and lead to changes in chemical fertility. For example, tillage can increase mineralization of nitrogen from soil organic matter. Other agricultural practices (e.g. fertilization, soil amendments such as compost, chemical sprays for weed and pest control) directly alter soil chemical fertility. Likewise, aspects of soil biological fertility can be influenced by agricultural management practices (e.g. disease control, stubble burning, application of compost and manure). In each case changes in soil conditions alter AM fungal biomass, their relative abundance, and their capacity to overcome plant nutrition deficiency. Other components of soil fertility may be indirectly influenced.

Tillage can break up the AM mycelium network, altering the extent and rate of root colonization of AM fungi. Some fungi may be more tolerant of this form of disturbance than others, but the abundance of "infective inocula" overall can also have an effect (Jasper et al., 1991). Therefore, to predict a likely alteration in the extent of mycorrhizal contribution to soil physical fertility in response to soil disturbance, knowledge of the extent to which dominant taxa are affected needs to be acquired. Ongoing tillage will select for AM fungi that are more resilient to disturbance. Conversely, reduced tillage will enable fungi that are less resilient to disturbance to be maintained to some degree within the AM fungal community.

The biomass of AM fungi in soil and roots is often significantly reduced by the addition of fertilizers to soil, thereby reducing a component of soil chemical fertility (Liu et al., 2012). Although the proportion of roots colonized by AM fungi may be reduced, the total biomass of AM fungi may simultaneously increase. This may be caused by a rapid expansion of roots (Carvaldo et al., 2015). In other cases higher levels of fertilizer application may be more detrimental to colonization of roots by some individual AM fungal taxa than to other taxa, reducing their capacity to contribute to soil chemical fertility. Well-managed application of fertilizer has the potential to complement AM fungal acquisition of poorly soluble nutrients located beyond the nutrient depletion zone. Establishing a balance between the functioning of the AM fungal community in nutrient access with the requirement of nutrients for commercial crop production is an ongoing challenge.

Fallowing land has the potential to reduce populations of AM fungi (Thompson et al., 2013), with potential for reduced contributions to all three components of soil fertility. On the other hand, rotations of crops, or crops and pastures can be used to maintain AM fungal communities at potentially functional levels. Applications of pesticides and herbicides have the potential to directly influence communities of AM fungi and decrease their capacity, at least temporally, to contribute to soil fertility. Although individual AM fungi may differ in their susceptibility to these chemicals, interactions between the plants and chemical applications also have potential to indirectly alter communities of AM fungi. As for other agricultural practices mentioned above, grazing of pastures or crops will influence root growth and carbon flow to AM fungi, leading to a change in soil biological fertility as well as components of soil chemical and physical fertility (Gehring and Whitham, 2002).

6.3.2 Mycorrhizal Function in Forest Ecosystems

EcM and AM symbioses occur in forest systems managed for production or fire reduction or they are unintentionally managed (Brundrett, 1991; Chen et al., 2000; Chapter 11). Various management practices used in agricultural systems [e.g. tillage, plant rotation (long term), soil amendment (e.g. fertilizer, lime, organic matter)] are also practiced in managed plantation ecosystems and during restoration of land disturbed by mining or other practices. Effects of disturbance on mycorrhizal fungi are similar in forestry to those discussed previously for agriculture, and here also there is considerable potential for them to contribute to all three components of soil fertility.

Inoculation with EcM fungi is practiced in plantation forestry, especially if forests are planted on land used previously for agricultural production or highly disturbed sites with no history of EcM plant species. Restoration of disturbed natural ecosystems is generally concerned with AM and EcM fungi. Effective reestablishment of AM fungi is commonly achieved by using recently collected top soil or by using appropriately managed stockpiled top soil (Miller et al., 1985; Jasper et al., 1987). There is relatively little opportunity to reestablish selected AM fungi into restored environments, although this has been attempted with various levels of success (e.g. Bell et al., 2003). Although there are many constraints to establishment of inoculant AM fungi during restoration, decisions about inoculation need to be considered carefully to avoid unintended consequences (Schwartz et al., 2006).

6.4 PRINCIPLES FOR MANAGEMENT OF MYCORRHIZAL FUNGI FOR SOIL FERTILITY

A key principle for effective management of AM fungal communities is that local knowledge about mycorrhizal fungi is important to maximize their contribution to soil fertility. Local knowledge can account for variability in complex systems. It can also provide a local "reference" framework within which to establish goals for mycorrhizal management for soil fertility benefits. At a minimum, there is a need for local knowledge of (1) the soil physical, chemical, and biological environment; (2) the state and amount of organic matter; (3) the types of mycorrhizal fungi already present; (4) the types of plant communities (or monocultures) in place or planned; (5) synchrony or asynchrony of plant growth cycles with life cycles of mycorrhizal fungi; and (6) location (e.g. north-facing versus south-facing slope, elevation, etc.).

There is a spectrum of possible plant responses to colonization by mycorrhizal fungi. It has been demonstrated than the influence of a single AM fungus on the biomass of different plant species can range from negative to positive (Klironomos, 2003). This finding accentuates the point that the effects of mycorrhizas are dependent on context. For agricultural crops or plantation forests, the beneficial services of mycorrhizas include not only increased crop yield or production but also increased drought tolerance, pathogen resistance, and competitive ability against weeds. In contrast, in the context of ecosystem restoration increased plant diversity is desirable, and in this case it is not desirable to have a mycorrhizal symbiosis that makes a single plant species a supercompetitor that dominates the community. Thus there is not a "one-size-fits-all" solution to mycorrhizal management; in fact, just the contrary—good management for production systems may be bad management for ecosystem restoration.

The relationship between "abundance" versus "function" for mycorrhizal fungi is not uniform for different plant species. There can be temporal factors associated with mycorrhizal "function" that may be more or less expressed for different combinations of plants and fungi. The relative importance of diversity of mycorrhizal fungi in the soil community may differ for different plant species and seasons. In addition, if fungal communities differ in their resistance or resilience to disturbances, this will influence their capacity to contribute to soil fertility, especially to soil chemical fertility under conditions of plant nutrient deficiency. The processes underlying these mycorrhizal effects are not yet understood well enough to generate predictable outcomes from mycorrhizal inoculation. There is a great need to develop mechanistic theories of mycorrhizal function that can be used to guide mycorrhizal management (Johnson and Graham, 2013).

Ecological stoichiometry and optimal foraging theory have been proposed as principles that may help determine the energetically most favorable strategy for nutrient acquisition by plants and their associated fungi (Chapter 12; Johnson, 2010). Ecological stoichiometry seeks to understand interactions between organisms and their environment based on the laws of conservation of matter and thermodynamics (Sterner and Elser, 2002), and optimal foraging theory assumes that evolution through natural selection will tend to select for behaviors and physiologies that optimize energy gained per unit energy invested in foraging (Pyke, 1984). Nutrients that are in short supply within the system will be immobilized, and those in abundant supply will be mineralized. Nutrient uptake strategies that provide access to the most limiting soil resource can be expected to have a strong selective advantage. Basic ecological studies of mycorrhizal fungi in natural and agricultural habitats are needed to link the range of functional traits expressed by different mycorrhizal fungi to different environmental conditions (van der Heijden and Scheublin, 2007). Surprisingly little is currently known about the ecology of mycorrhizal fungi; however, recent technological developments make this knowledge increasingly attainable.

6.5 LOOKING FORWARD

This is a golden age for microbiology because innovations in molecular genetics are providing the tools necessary to see the microbial world like never before. The genomic revolution has made it possible to identify the composition of mycorrhizal fungal communities inside plant roots and in the surrounding environment so that the true diversity of fungi and their associated microorganisms can be measured. This information is a first step toward elucidating the structure of mycorrhizal fungal communities and linking it to environmental conditions (biotic and abiotic). The next step is to link fungal community structure with symbiotic function using the tools of metatranscriptomics and proteomics, which facilitate the identification of the genes and proteins that generate desirable mycorrhizal services. However, the application of these new technologies to mycorrhizal research will be in vain unless they are guided by a holistic understanding of the role of mycorrhizas in ecosystems and agroecosystems. The next section of this book contributes this knowledge and helps advance mycorrhizal science from a descriptive to a predictive endeavor.

References

Abbott, L.K., Lumley, S.E., 2014. Mycorrhizal Fungi as a Potential Indicator of Soil Health in Mycorrhizal Fungi: Use in Sustainable Agriculture and Land Restoration. Springer, Heidelberg, Germany, pp. 17–31.

Abbott, L.K., Manning, D.A.C., 2015. Soil health and related ecosystem services in organic agriculture. Sustainable Agriculture Research 4, 116–125.

Abbott, L.K., Murphy, D.V., 2003. Soil Biological Fertility: A Key to Sustainable Land Use in Agriculture. Kluwer Academic, The Netherlands, p. 264 (Reprinted Springer, 2007).

Abbott, L.K., Robson, A.D., 1977. Growth stimulation of subterranean clover with vesicular arbuscular mycorrhizas. Australian Journal of Agricultural Research 28, 639–649.

Abbott, L.K., Robson, A.D., 1981a. Infectivity and effectiveness of five endomycorrhizal fungi: competition with indigenous fungi in field soils. Australian Journal of Agricultural Research 32, 621–630.

Abbott, L.K., Robson, A.D., 1981b. Infectivity and effectiveness of vesicular arbuscular mycorrhizal fungi: effect of inoculum type. Australian Journal of Agricultural Research 32, 631–639.

Aučina, A., Rudawska, M., Leski, T., Skridaila, A., Pašakinskiene, I., Riepšas, E., 2014. Forest litter as the mulch improving growth and ectomycorrhizal diversity of bare-root Scots pine (*Pinus sylvestris*) seedlings. iForest 8, 394–400. http://dx.doi.org/10.3832/ifor1083-008.

Bachelot, B., Uriarte, M., Zimmerman, J.K., Thompson, J., Leff, J.W., Asiaii, A., Koshner, J., McGuire, K., 2016. Long-lasting effects of land use history on soil fungal communities in second-growth tropical rain forests. Ecological Applications 26, 1881–1895.

Barto, E.K., Weidenhamer, J.D., Cipollini, D., Rillig, M.C., 2012. Fungal superhighways: do common mycorrhizal networks enhance below ground communication? Trends in Plant Science 17, 633–637.

Beare, M.H., Vikram Reddy, M., Tian, G., Srivastava, S.C., 1997. Agricultural intensification, soil biodiversity and agroecosystem function in the tropics: the role of decomposer biota. Applied Soil Ecology 6, 87–108.

Bell, J., Wells, S., Jasper, D.A., Abbott, L.K., 2003. Field inoculation with arbuscular mycorrhizal fungi in rehabilitation of mine sites with native vegetation, including *Acacia* spp. Australian Systematic Botany 16, 131–138.

Bolan, N.S., Robson, A.D., Barrow, N.J., 1983. Plant and soil factors including mycorrhizal infection causing sigmoidal response of plants to applied phosphorus. Plant and Soil 73, 187–201.

Brundrett, M.C., 1991. Mycorrhizas in natural ecosystems. Advances in Ecological Research 21, 171–313.

Brundrett, M.C., 2006. Understanding the roles of multifunctional mycorrhizal and endophytic fungi. In: Schults, B., Boyle, C., Sieber, T.N. (Eds.), Microbial Root Endophytes. Soil Biology, vol. 9. Springer-Verlag, Berlin, Heidelberg, pp. 281–298.

Bücking, H., Liepold, E., Ambilwade, P., 2012. The role of mycorrhizal symbiosis in nutrient uptake of plants and the regulatory mechanisms underlying these transport processes. Plant Science. 107–138. http://creativecommons.org/licenses/by/3.0.

Bücking, H., Mensah, J.A., Fellbaum, C.R., 2016. Common mycorrhizal networks and their effect on the bargaining power of the fungal partner in the arbuscular mycorrhizal symbiosis. Communicative & Integrative Biology. 9, e1107684. http://dx.doi.org/10.1080/19420889.2015.1107684.

Cairney, J.W.G., 2012. Extramatrical mycelia of ectomycorrhizal fungi as moderators of carbon dynamics in forest soil. Soil Biology and Biochemisty 47, 198–208.

Cairney, J.W.G., Meharg, A.A., 2003. Ericoid mycorrhiza: a partnership that exploits harsh edaphic conditions. European Journal of Soil Science 54, 735–740.

Cambi, M., Certini, G., Neri, F., Marchi, E., 2015. The Impact of heavy traffic on forest soils: a review. Forest Ecology and Management 338, 124–138.

Cano, C., Bago, A., 2005. Competition and substrate colonization strategies of three polyxenically grown arbuscular mycorrhizal fungi. Mycologia 97, 1201–1204.

Carvalho, M., Brito, I., Alho, L., Goss, M.J., 2015. Assessing the progress of colonization by arbuscular mycorrhiza of four plant species under different temperature regimes. Journal of Plant Nutrition and Soil Science 178, 515–522.

Chagnon, P.-L., Bradley, R.L., Klironomos, J.N., 2012. Using ecological network theory to evaluate the causes and consequences of arbuscular mycorrhizal community structure. New Phytologist 194, 307–312.

Chen, Y.L., Brundrett, M.C., Dell, B., 2000. Effects of ectomycorrhias and vesicular-arbuscular mycorhizas, alone and in competition, on root colonization and growth of *Eucalyptus globulus* and *E. urophylla*. New Phytologist 146, 545–556.

Clemmensen, K.E., Finlay, R.D., Dahlberg, A., Stenlid, J., Wardle, D.A., Lindahl, B.D., 2015. Carbon sequestration is related to mycorrhizal fungal community shifts during long-term succession in boreal forests. New Phytologist 205, 1525–1536.

Corrales, A., Mangan, S.A., Turner, B.K., Dalling, J.W., 2016. An ectomycorrhizal nitrogen economy facilitates monodominance in a neotropical forest. Ecology Letters 19, 383–392.

Courty, P.-E., Buée, M., Diedhiou, A.G., Frey-Klett, P., Le Tacon, F., Rineau, F., Turpault, M.-P., Uroz, S., Garbaye, J., 2010. The role of ectomycorrhizal communities in forest ecosystem processes: new perspectives and emerging concepts. Soil Biology and Biochemistry 42, 679–698.

Daws, M.I., Standish, R.J., Koch, J.M., Morald, T.K., 2013. Nitrogen and phosphorus fertilizer regime affect jarrah forest restoration after bauxite mining in Western Australia. Applied Vegetation Science 16, 610–618.

Daws, M.I., Standish, R.J., Koch, J.M., Morald, T.K., Tibbett, M., Hobbs, R.K., 2015. Forest Fertilisation and large legume species affect jarrah forest restoration after bauxite mining. Forest Ecology and Management 354, 10–17.

Degens, B.P., 1997. Macro-aggregation of soils by biological bonding and binding mechanisms and the factors affecting these: a review. Australian Journal of Soil Research 35, 431–459.

Dickie, I.A., IAlexander, I., Lennon, S., Opik, M., Selosse, M.A., van der Heijden, M.G.A., Martin, F.M., 2002. Evolving insights to understand mycorrhizas. New Phytologist 205, 1369–1374.

Egnell, G., Jurevics, A., Peichl, M., 2015. Negative effects of stem and stump harvest and deep soil cultivation on the soil carbon and nitrogen pools are mitigated by enhanced tree growth. Forest Ecology and Management 338, 57–67.

Ekbald, A., Wallander, H., Godbold, D.L., Cruz, C., Johnson, D., Baldrian, P., Bjork, R.G., Epron, E., Kielisewska-Rokicka, B., Kjoller, R., Kraigher, H., Matzner, E., Neumann, J., Plassard, C., 2013. The production and turnover of extramatical mycelium of ectomycorrhizal fungi in forest soils: role in carbon cycling. Plant and Soil 366, 1–27.

Evans, A.C., 1948. Earthworms and soil fertility. II Some effects of earthworms on soil structure. Applied Biology 35, 1–13.

Foth, H.D., Ellis, B.G., 1997. Soil Fertility. CRC Press, Inc., Boca Ratton, Florida.

Gehring, C.A., Whitham, T.G., 2002. Mycorrhizae-herbivore interactions: population and community consequences. In: van der Heijden, M.G.A., Sanders, I. (Eds.), Ecological Studies, vol. 157, pp. 295–320.

Gibson, N., Yates, C.J., Dillon, R., 2010. Plant communities of the ironstone ranges of South Western Australia: hotspots for plant diversity and mineral deposits. Biodiversity Conservation 19, 3961–3962.

Hardarson, G., Zapata, F., 1984. Effect of plant genotype and nitrogen fertilizer on symbiotic nitrogen fixation by soybean cultivars. Plant and Soil 82, 397–405.
Haselwandter, K., Bowen, G.D., 1996. Mycorrhizal relations in trees for agroforestry and land rehabilitation. Forest Ecology and Management 81, 1–17.
He, X.-H., Critchley, C., Bledsoe, C., 2010. Nitrogen transfer within and between plants through common mycorrhizal networks (CMNs). Critical Reviews in Plant Sciences 22, 531–567.
Herman, D.J., Firestone, M.K., Nuccio, E., Hodge, A., 2012. Interactions between an arbuscular mycorrhizal fungus and a soil microbial community mediating litter decomposition. FEMS Microbial Ecology 80, 236–247.
Hingston, F.J., Dimmock, G.M., Turton, A.G., 1980. Nutrient distribution in a jarrah (*Eucalyptus marginata* Donn Ex Sm.) ecosystem in south-western Australia. Forest Ecology and Management 3, 183–207.
Högberg, P., Johannisson, C., Yarwood, S., 2011. Recovery of ectomycorrhiza after 'nitrogen saturation' of a conifer forest. New Phytologist 198, 515–525.
Huston, M.A., 2013. Precipitation, soils NPP, and biodiversity: resurrection of Albrecht's curve. Ecological Monographs 82, 277–296.
Janos, D.P., 1980. Vesicular-arbuscular mycorrhizas affect lowland rain forest plant growth. Ecology 61, 151–162.
Jasper, A.D., Abbott, L.K., Robson, A.D., 1991. The effect of soil disturbance on VA mycorrhizal fungi, in soils from different vegetation types. New Phytologist 118, 471–476.
Jasper, D.A., Abbott, L.K., Robson, A.D., 1992. Soil disturbance in native ecosystems - the decline and recovery of infectivity of VA mycorrhizal fungi. In: Read, D.J., et al. (Ed.), Mycorrhizas in Ecosystems. CAB International, Wallingford, pp. 151–155.
Jasper, D.A., Robson, A.D., Abbott, L.K., 1987. The effect of surface mining on the infectivity of vesicular-arbuscular mycorrhizal fungi. Australian Journal of Botany 35, 641–652.
Jeffries, P., Gianinazzi, S., Perotto, S., Turnau, K., Barea, J.-M., 2003. The contribution of arbuscular mycorrhizal fungi in sustainable maintenance of plant health and soil fertility. Biology and Fertility of Soils 37, 1–16.
Johnson, N.C., 2010. Resource stoichiometry elucidates the structure and function of arbuscular mycorrhizas across scales. New Phytologist 185, 631–647.
Johnson, N.C., Graham, J.M., 2013. The continuum concept remains a useful framework for studying mycorhrrizal functioning. Plant and Soil 363, 411–419.
Johnson, N.C., Wilson, G.W.T., Wilson, J.A., Miller, R.M., Bowker, M.A., 2015. Mycorrhizal phenotypes and the law of the minimum. New Phytologist 205, 1473–1484.
Juniper, S., Abbott, L.K., 1993. Vesicular-arbuscular mycorrhizas and soil salinity. Mycorrhiza 4, 45–57.
Kalliokoski, T., Pennanen, T., Nygren, P., Sievänen, R., Helmisaari, H.-S., 2010. Belowground interspecific competition in mixed boreal forests: fine root and ectomycorrhiza characteristics along stand developmental stage and soil fertility gradients. Plant and Soil 330, 73–89.
Klironomos, J.N., Hart, M.M., 2001. Food-web dynamics: animal nitrogen swap for plant carbon. Nature 410, 651–652.
Klironomos, J.N., McCune, J., Moutoglis, P., 2004. Species of arbuscular mycorrhizal fungi affect mycorrhizal responses to simulated herbivory. Applied Soil Ecology 26, 133–141.
Klironomos, J.N., 2003. Variation in plant response to native and exotic arbuscular mycorrhizal fungi. Ecology 84, 2292–2301.
Knegt, B., Jansa, J., Franken, O., Engelmoer, D.J.P., Werner, G.D.A., Bucking, H., Kiers, E.T., 2016. Host plant quality mediates competition between mycorrhizal fungi. Fungal Ecology 20, 233–240.
Korkama, T., Pakkanen, A., Pennanen, T., 2006. Ectomycorrhizal community structure varies among Norway spruce (*Picea abies*) clones. New Phytologist 171, 815–824.
Kytovita, M.-M., Vestberg, M., Toumi, J., 2003. A test of mutual aid in common mycorrhizal networks: established vegetation negates benefit in seedlings. Ecology 84, 906–989.
Labiberte, E., Grace, J.B., Huston, M.A., Lambers, H., Tester, F.P., Turner, B.L., Wardle, D.A., 2013. How does pedogenesis drive plant diversity? Trends in Ecology and Evolution 28, 331–340.
Laliberte, E., Turner, B.L., Coates, T., Pearse, S.J., 2012. Experimental assessment of nutrient limitation along a 2-million-year dune chronosequence in the south-western Australia biodiversity hotspot. Journal of Ecology 100, 631–642.
Lambers, H., Ahmedi, I., Berkowitz, O., Dunne, C., Finnegan, P.M., Hardy, G.E.St.J., Jost, R., Laliberté, E., Pearse, S.J., Teste, F.P., 2013. Phosphorus nutrition of phosphorus-sensitive Australian native plants: threats to plant communities in a global biodiversity hotspot. Conservation Physiology 1, 1–21.
Lambers, H., Brundrett, M.C., Raven, J.A., Hopper, S.D., 2010. Plant mineral nutrition in ancient landscapes: high plant diversity on infertile soils is linked to functional diversity for nutritional strategies. Plant and Soil 334, 11–31.

Lavelle, P., Dangerfield, M., Fragoso, C., Eshenbrenner, V., Lopez-Hernandez, D., Pashanasi, B., Brussaard, L., 1994. The relationship between soil macrofauna and tropical soil fertility. In: Woomer, P.L., Swift, M.J. (Eds.), The Biological Management of Tropical Soil Fertility. Wiley, Chichester, pp. 137–169.

Liu, Y., Shi, G., Chen, G., Jiang, S., Ma, Z., An, L., Du, G., Johnson, N.C., Feng, H., 2012. Direct and indirect influences of 8 yr of nitrogen and phosphorus fertilization on Glomeromycota in an alpine meadow ecosystem. New Phytologist 194, 523–535.

Mardhiah, U., Caruso, T., Gurnell, A., Rillig, M.C., 2016. Arbuscular mycorrhizal fungal hyphae reduce soil erosion by surface flow in a greenhouse experiment. Applied Soil Ecology 99, 137–140.

Marschner, H., Dell, B., 1994. Nutrient uptake in mycorrhizal symbiosis. Plant and Soil 159, 89–201.

Mickan, B., 2014. Mechanisms for alleviating water stress involving arbuscular mycorrhizal fungi. In: Solaiman, Z.A., Abbott, L.K., Varma, A. (Eds.), Mycorrhizal Fungi: Use in Sustainable Agriculture and Land Restoration. Soil Biology, vol. 41, pp. 225–239.

Mickan, B.S., Abbott, L.K., Stefanova, K., Solaiman, Z.M., 2016. Demonstrated mechanisms for interactions between biochar and mycorrhizal fungi in water-deficient agricultural soil. Mycorrhiza 26, 565–574.

Miller, R.M., Jastrow, J.D., 1992. The role of mycorrhizal fungi in soil conservation. In: Bethlenfalvay, G.J., Linderman, R.G. (Eds.), Mycorrhizae in Sustainable Agriculture. Americal Society of Agronomy, vol. 54. Wisconsin, Madison, pp. 45–70. Special Publication.

Miller, R.M., Carnew, B.A., Moorman, T.B., 1985. Factors influencing survival of vesicular-arbuscular mycorrhizal propagules during topsoil storage. Journal of Applied Ecology 22, 259–266.

Mitchell, D.T., Gibson, B.R., 2006. Ericoid mycorrhizal association: ability to adapt to a broad range of habitats. Mycologist 20, 2–9.

Nadian, H., Smith, S.E., Am, A., Murray, R.S., Siebert, B.D., 1998. Effects of soil compaction on phosphorus uptake and growth of *Trifolium subterraneum* colonized by four species of vesicular-arbuscular mycorrhizal fungi. New Phytologist 139, 155–165.

Nasholm, T., Hogberg, P., Franklin, O., Metcalfe, D., Keel, S.G., Campbell, C., Hurry, V., Linder, S., Hogberg, M.N., 2013. Are ectomycorrjoza; fungi allegiating or aggravating nitrogen limitation of tree growth in boreal forests? New Phytologist 198, 214–221.

Nouri, E., Breuillin-Sessoms, F., Feller, U., Reinhardt, D., 2014. Phosphorus and nitrogen regulate abuscular mycorrhizal symbiosis in *Petuna hybrid*. PLoS One 9 (3), e90841. http://dx.doi.org/10.1371/journal.pone.0090841.

Parniske, M., 2008. Arbuscular mycorrhiza: the mother of plant root endosymbioses. Nature Reviews Microbiology 6, 763–775.

Pellegrino, E., Bedini, S., Avio, L., Bonari, E., Giovannetti, M., 2011. Field inoculation effectiveness of native and exotic arbuscular mycorrhizal fungi in a Mediterranean agricultural soil. Soil Biology and Biochemistry 43, 367–376.

Peoples, M.B., Brockwell, J., Hunt, J.R., Swan, A.D., Watson, L., Haynes, R.C., Li, G.D., Hackney, B., Nuttall, J.G., Davies, S.L., Fillery, I.R.P., 2012. Factors affecting the potential contributions of N_2 fixation by legumes in Australian pasture systems. Crop and Pasture Science 63, 759–786.

Phillips, R.P., Brzostek, E., Midgley, M.G., 2013. The mycorrhizal-associated nutrient economy: a new framework for predicting carbon-nutrient couplings in temperate forests. New Phytologist 199, 41–51.

Powell, J.R., Parrent, J.L., Hart, M.M., Klironomos, J.N., Rillig, M.C., Maherali, H., 2009. Phylogenetic trait conservatism and the evolution of functional trade-offs in arbuscular mycorrhizal fungi. Proceedings of the Royal Society B 276, 4237–4245.

Pringle, E.G., 2016. Integrating plant carbon dynamics with mutualism ecology. New Phytologist 210, 71–75.

Püschel, D., Janouskova, M., Hujslova, M., Slavıkova, R., Gryndlerova, H., Jansa, J., 2016. Plant-fungus competition for nitrogen erases mycorrhizal growth benefits of *Andropogon gerardii* under limited nitrogen supply. Ecology and Evolution 6, 4332–4346.

Pyke, G.H., 1984. Optimal foraging theory: a critical review. Annual Review of Ecology and Systematics 15, 523–575.

Rillig, M.C., Mummey, D.L., 2006. Mycorrhizas and soil structure. New Phytologist 171, 41–53.

Robinson, D., Fitter, A., 1999. The magnitude and control of carbon transfer between plants linked by a common mycorrhizal network. Journal of Experimental Botany 50, 9–13.

Robson, A.D., Abbott, L.K., 1989. The effect of soil acidity on microbial activity in soil. In: Robson, A.D. (Ed.), Soil Acidity and Plant Growth. Academic Press, Sydney, pp. 139–165.

Ronsheim, M.L., 2016. Plant genotype influences mycorrhiza benefits and susceptibility to a soil pathogen. The American Midland Naturalist 175, 103–112.

Rose, T., Julia, C.C., Shepherd, M., Rose, M.T., van Zwieten, L., 2016. Faba bean is less susceptible to fertiliser N impacts on biological N_2 fixation than chickpea in monoculture and intercropping systems. Biology and Fertility of Soils 52, 271–276.

Santini, T.C., Banning, N.C., 2016. Alkaline tailings as novel soil forming substrates: reframing perspectives on mining and refining wastes. Hyrdometallurgy 164, 38–47.

Schwartz, M.W., Hoeksema, J.D., Gehring, C.A., Johnson, N.C., Klironomos, J.N., Abbott, L.K., Pringle, A., 2006. The promise and the potential consequences of the global transport of mycorrhizal fungal inoculum. Ecology Letters 9, 501–515.

Schweiger, P.F., Robson, A.D., Barrow, N.J., Abbott, L.K., 2007. Arbuscular mycorrhizal fungi from three genera induce two-phase plant growth responses on a high P-fixing soil. Plant and Soil 292, 181–192.

Selosse, M.-A., Richard, F., He, X., Simard, S.W., 2006. Mycorrhizal networks: des liaisons dangereuses? Trends in Ecology and Evolution 21, 621–628.

Sikes, B., 2010. When do arbuscular mycorrhizal fungi protect plant roots from pathogens? Plant Signaling and Behavior 5, 763–765.

Six, J., Paistian, K., 2014. Aggregate-associated soil organic matter as an ecosystem property and a measurement tool. Soil Biology and Biochemistry 68, A4–A9.

Smith, S.E., Read, D., 2008. Mycorrhizal Symbiosis, third ed. Academic Press.

Sterner, R.W., Elser, J.J., 2002. Ecological Stoichiometry: The Biology of Elements from Molecules to the Biosphere. Princeton University Press, Princeton, New Jersey, USA.

Stoneman, G.L., Dell, B., Turner, N.C., 1995. Growth of *Eucalyptus marginata* (jarrah) seedlings in mediterranean-climate forest in south-west Australia in response to overstorey, site and fertilizer application. Forest Ecology and Management 79, 173–184.

Thompson, J.P., Clewett, T.G., Fiske, M.L., 2013. Field inoculation with arbuscular-mycorrhizal fungi overcomes phosphorus and zinc deficiencies of linseed (*Linum usitatissiumum*) in a vertisol subject to long-fallow disorder. Plant and Soil 371, 117–137.

Thomson, B.D., Robson, A.D., Abbott, L.K., 1986. Effects of phosphorus on the formation of mycorrhizas by *Gigaspora calospora* and *Glomus fasciculatum* in relation to root carbohydrates. New Phytologist 103, 751–765.

Tikvic, I., Ugarkovic, D., Kobasic, Z., October 8–12, 2012. Ecophysical disturbances of mycorrhiza caused by the application of forest operations in forest ecosystems – review. Forest Engineering Dubrovnik (Cavtat).

Tisdale, S.L., Nelson, W.L., Beaton, J.D., 1985. Soil Fertility and Fertilisers. . ISBN: 0-02-420830-2, p. 754.

Tisdall, J.M., Oades, J.M., 1980. The effect of crop rotation on aggregation in a red-brown Earth. Australian Journal of Soil Research 18, 423–433.

Tisdall, J.M., Oades, J.M., 1982. Organic matter and water-stable aggregates in soils. Journal of Soil Science 33, 141–163.

Toju, H., Yamamoto, S., Tanabe, A.S., Hayakawa, T., Ishii, H.S., 2016. Network modules and hubs in plant-root fungal biomes. Journal of Royal Society Interface. 13, 20151097. http://dx.doi.org/10.1098/rsif.2015.1097.

Trappe, J.M., 1977. Selection of fungi for ectomycorrhizal inoculation in nurseries. Annual Review of Phytopathology 151, 203–222.

Unestam, T., 1991. Water repellency, mat formation, and leaf-stimulated growth of some ectomycorrhizal fungi. Mycorrhiza 1, 13–20.

van der Heijden, M.G.A., Scheublin, T.R., 2007. Functional traits in mycorrhizal ecology: their use for predicting the impact of arbuscular mycorrhizal fungal communities on plant growth and ecosystem functioning. New Phytologist 174, 244–250.

Veresoglou, S.D., Anderson, I.C., de Sousa, N.M.F., Hempel, S., Rillig, M.C., 2016. Resilience of fungal communities to elevated CO_2. Microbial Ecology 72, 493–495.

Weremijewicz, J., Sternberg, L.S.L.O.R., Janos, D.P., 2016. Common mycorrhizal networks amplify competition by preferential mineral nutrient allocation to large host plants. New Phytologist. http://dx.doi.org/10.1111/nph.14041.

Williams, A., Kane, D.A., Ewing, P.M., Atwood, L.W., Jilling, A., Li, M., Lou, Y., Davis, A.S., Grandy, A.S., Huerd, S.C., Hunter, M.C., Koide, R.T., Mortensen, D.A., Smith, R.G., Snapp, S.S., Spokas, K.A., Yannarell, A.C., Jordan, N.R., 2016. Soil functional zone management: a vehicle for enhancing production and soil ecosystem services in row-crop agroecosystems. Frontiers in Plant Science 7. http://dx.doi.org/10.3389/fpls.2016.00065 Article 65.

Wu, T., 2011. Can ectomycorrhizal fungi circumvent the nitrogen mineralization for plant nutrition in temperate forest ecosystems? Soil Biology and Biochemistry 43, 1109–1117.

CHAPTER

7

Fungal and Plant Tools for the Uptake of Nutrients in Arbuscular Mycorrhizas: A Molecular View

M. Giovannetti[1,2,a], *V. Volpe*[1,a], *A. Salvioli*[1], *P. Bonfante*[1]

[1]Università degli Studi di Torino, Torino, Italy; [2]Gregor Mendel Institute (GMI), Austrian Academy of Sciences, Vienna Biocenter (VBC), Vienna, Austria

7.1 INTRODUCTION

One of the most ancient and successful strategies that plants have adopted for nutrient acquisition is the formation of symbiotic associations with arbuscular mycorrhizal (AM) fungi. Extensive networks of AM fungal hyphae reach unexplored soil niches and efficiently mine nutrients that are otherwise not accessible to plant roots. In return, plant hosts provide AM fungi with up to 20% of their assimilated carbon, allowing the fungi to complete their life cycle with the production of new spores. The AM symbiosis has been the default association for plants since their appearance on Earth, with the exception of plant taxa that have lost this capacity during their evolution, such as *Arabidopsis thaliana* and other plants belonging to the order Brassicales (Bravo et al., 2016).

The key structure of nutrient exchange between the two symbiotic partners is a highly branched hypha called the *arbuscule* because of its similarities to a tiny tree. Arbuscules are completely surrounded by the invagination of the plant plasma membrane, called the *periarbuscular membrane* (PAM), and plant nutrient transporters, which are induced by the presence of AM fungi, are localized in this membrane. Comparing fossils from 450-million-year-old arbuscules (Remy et al., 1994) and confocal micrographs of the same structures inside of contemporary plant roots shows how arbuscules have been dramatically conserved, suggesting that mechanisms controlling fungal morphogenesis inside plant cells have been maintained throughout plant evolution (Bonfante and Genre, 2008). Despite the high level of conservation of the arbuscule across time, it should be remembered that it constitutes an ephemeral

[a]These authors contributed equally to this work.

Mycorrhizal Mediation of Soil
http://dx.doi.org/10.1016/B978-0-12-804312-7.00007-3

structure lasting approximately 5 days within a single plant cortical cell (Kobae and Hata, 2010). The mechanisms underlying the collapse of arbuscules are still to be disentangled: a possible weekly checkpoint ensures that the symbiosis is still efficient to the partners. It is known that, among different partners, a fine-tuned molecular mechanism regulates the establishment and maintenance of the symbiosis, and the monitoring and understanding of the exchange of resources is a major tool to assess it, as described by the biological market theory (Kiers et al., 2011). In-depth research on nutrient uptake within AM symbioses will have two outcomes: (1) better understanding of the role of AM fungi in plant nutrition, improving their use in a more sustainable agriculture and (2) the elucidation of molecular drivers that generate and control the establishment of this highly conserved relationship.

In this chapter we will provide a detailed description of our current knowledge about fungal and plant molecular mechanisms involved in nutrient uptake within AM symbioses. In particular, three major macronutrients—nitrogen (N), phosphorus (P), and sulfur (S) (Fig. 7.1)—will be discussed with a special focus on old and new paradigms, future perspectives, and open questions. Apart from the localized direct effect of mycorrhizal interactions, we will also focus on systemic changes happening at the whole-plant level. Thanks to new available techniques, it is now known that the effect of the AM symbiosis is not limited to changes in plant nutrient homeostasis; rather, a whole set of systemic plant responses is happening.

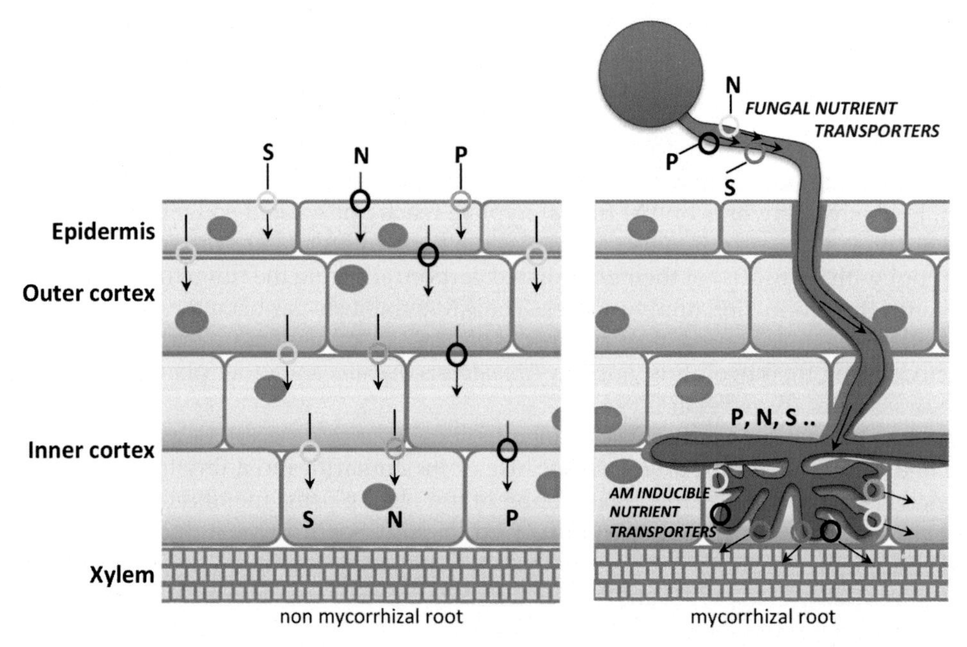

FIGURE 7.1 Direct and symbiotic plant nutrient uptake. Plants can absorb nutrients directly from the soil and through arbuscular mycorrhizal (AM) fungal mycelia. It is known that P is mainly absorbed throughout the fungus whereas the knowledge for the other macronutrient is still scattered. In the direct pathway, S, N, and P are absorbed by plant nutrient transporters, which accumulate in the epidermis and cortex (represented by *white*, *gray*, and *black circles*). In the mycorrhizal pathway, S, N, and P are taken up into AM fungal hyphae by fungal nutrient transporters (Tisserant et al., 2012) and are translocated to intracellular fungal structures in root cortical cells.

7.2 NITROGEN NUTRITION WITHIN ARBUSCULAR MYCORRHIZAS

Constituting up to 2% of plant dry weight, N is above all required for the biosynthesis of amino acids, nucleic acids, and chlorophyll. It usually represents the limiting factor in plant growth, and most agricultural practices aim to increase N soil contents through crop rotation and/or the massive use of N-based fertilizers. The high economic and environmental costs of N fertilizers (McSwiney and Robertson, 2005) are great incentives to explore more sustainable practices. Because N can be directly fixed from the atmosphere by some bacteria and Archaea, which possess the enzyme nitrogenase, and some plant species, as legumes, live in symbiosis with N-fixing bacteria, an enticing project is to engineer cereal crops with the capability to fix their own N (Rogers and Oldroyd, 2014). A more amenable approach is to understand and exploit the capacity of AM fungi to absorb N from different substrates. However, the actual contribution of AM fungi to plant N status is still under debate (Smith and Smith, 2011). As a first point, fungi—as plants and animals—need N not only for protein building but also for assembling chitin, which is a crucial component of their cell wall. Therefore AM fungi uptake N, as well as P and S, from the soil for their own metabolism, and only as a second step during the symbiosis functioning will they release the nutrients to their hosts.

In the past AM symbiosis has been associated with negative (George et al., 1995), neutral (Hawkins and George, 1999), and positive (Saia et al., 2014; Mensah et al., 2015) effects on plant N status (for up-to-date reviews see Corrêa et al., 2015; Bücking and Kafle, 2015), depending on set-up conditions, and several authors postulated that the effect of AM symbiosis on plant N status is merely a consequence of improved P nutrition (Reynolds et al., 2005). It has also been shown that the effect of AM symbiosis on plant N content and growth response is highly influenced by the AM fungal morphospecies used (Mensah et al., 2015). Indeed, it is clear that AM fungi are able to absorb different forms of N and pass it to the plant, similar to what happens with P and other nutrients (Toussaint et al., 2004; Tanaka and Yano, 2005; Jin et al., 2005). The kinetics of ammonium (NH_4^+) acquisition by AM fungi show 5 times higher affinity compared with plants, therefore improving plant N nutrition even in soils with low N levels (Pérez-Tienda et al., 2012).

The uptake of organic and inorganic N sources from the soil via specialized transporters has been described for ectomycorrhizal (EcM), ericoid (Grelet et al., 2009), and AM fungi (Gobert and Plassard, 2008). It has been proposed that the release from the fungus to the plant of these two different N forms depends on the nutritional and photosynthetic status of the plant (Chalot et al., 2006). The "traditional view" suggests an organic N transfer under high C availability, whereas under C depletion the synthesis of organic N would be downregulated in the fungus, leading to the transfer of inorganic N.

Inorganic N is present in the soil as either nitrate (NO_3^-) or ammonia (NH_4^+). Although both forms are absorbed by AM fungal hyphae, studies suggest that NH_4^+ is generally the preferential form because it is considered energetically more efficient than NO_3^- (Govindarajulu et al., 2005; Jin et al., 2005). However, preference for a specific N form depends on the soil where the symbiosis occurs: NO_3^- is abundant in most agricultural soils whereas NH_4^+ is dominant in many undisturbed or very acidic soils (Bücking and Kafle, 2015). NO_3^- uptake is an active process, which is coupled to an H^+-symport mechanism, whereas NH_4^+ uptake involves an antiport mechanism with an H^+ efflux (Bago et al., 1996; Bago and Azcón-Aguilar,

1997). Several NH_4^+ and NO_3^- transporters have been identified in AM fungi, and they are localized in extraradical or intraradical mycelium (IRM) (Tisserant et al., 2012). In *Rhizophagus irregularis* (previously *Glomus intraradices*) three sequences have been annotated as NH_4^+ transporters whereas only one NO_3^- transporter has been identified. The expression of NH_4^+ transporter *GintAMT1* was higher in the extraradical mycelium (ERM) and was induced by low levels of NH_4^+ (López-Pedrosa et al., 2006), and it has been suggested to play a preferential role in acquisition of N from the soil (López-Pedrosa et al., 2006). By contrast, the other NH_4^+ transporter of *R. irregularis*, *GintAMT2*, was induced in the intraradical fungal structures, indicating a preferential role in N acquisition during the symbiotic association (Pérez-Tienda et al., 2011). The localization of both GintAMT1 and GintAMT2 transporters via laser microdissection in the colonized cortical cells revealed that both transporters might have overlapping physiological functions in the symbiotic interface (Pérez-Tienda et al., 2011). A fungal NO_3^- transporter has been characterized in the ERM of *R. irregularis* (Tian et al., 2010); its expression was induced by the presence of NO_3^- but suppressed by an increase of NH_4^+ or glutamine, a downstream metabolite, in the internal hyphae (Fellbaum et al., 2012).

Arbuscular mycorrhizal fungi are also involved in the acquisition of organic N from the ERM (Gachomo et al., 2009). The amino acid permease of *Funnelliformis mosseae* (previously *Glomus mosseae*), *GmosAAP1*, was characterized, and its expression was detected in the extraradical mycelium and was induced by high levels of organic N (Cappellazzo et al., 2008). The authors hypothesize that GmosAAP1 plays a role in the first steps of amino acid acquisition, allowing direct amino acid uptake from the soil and extending the molecular tools by which AM fungi exploit soil resources (Cappellazzo et al., 2008). Recently, Belmondo and coauthors have characterized the dipeptide transporter, *RiPTR2*, from the AM fungus *R. irregularis*. They reported that it is expressed in the ERM but also upregulated in the intraradical phase, suggesting a dual role of this transporter in the uptake of organic N from soil and in the reabsorption of peptides from the interfacial apoplast (Belmondo et al., 2014).

The transfer of N from the soil to the plant via the fungus follows the same mechanisms as other nutrients: N uptake from the soil by fungal nutrient transporter into ERM, translocation along the hyphae and arbuscules, and finally uptake from the mycorrhizal interface by AM-inducible plant transporters in the root cortex (Smith and Smith, 2011). When NO_3^- is absorbed by AM fungi, it is reduced to nitrite by a nitrate reductase and then the nitrite is converted into NH_4^+ by a nitrite reductase. Excess NH_4^+ in the cells is considered to be toxic, thus NH_4^+ is rapidly assimilated into amino acids involving two pathways: the NAD(P)-glutamate dehydrogenase or the glutamine synthetase–glutamate synthase (GS-GOGAT) pathway. The latter is that predominately used by AM fungi. The GS-GOGAT cycle is used to convert the absorbed inorganic N (NO_3^- and NH_4^+) to arginine and then transported into the IRM (Ngwene et al., 2013) according to the following reactions: (1) inorganic N availability induces the expression of N transporters; (2) accumulation of NH_4^+ stimulates the GS-GOGAT pathway with an increase of glutamate (Glu) and glutamine (Gln) and consequently an upregulation of genes involved in arginine (Arg) synthesis; (3) Arg is then transferred from the ERM to IRM; (4) in the IRM Arg is converted to ornithine (Orn) by the increase of arginase activity or to urea by the activity of urease; (5) the NH_4^+ produced from the urea is released in the interfacial apoplast and then to the host, where the NH_4^+ availability induces the plant ammonium transporters; (6) in the host, near to the site where N is released, the plant glutamine synthetases (GS) are upregulated (Fig. 7.2). Tian and coauthors demonstrated that the arginine in the mycorrhizal plants increased

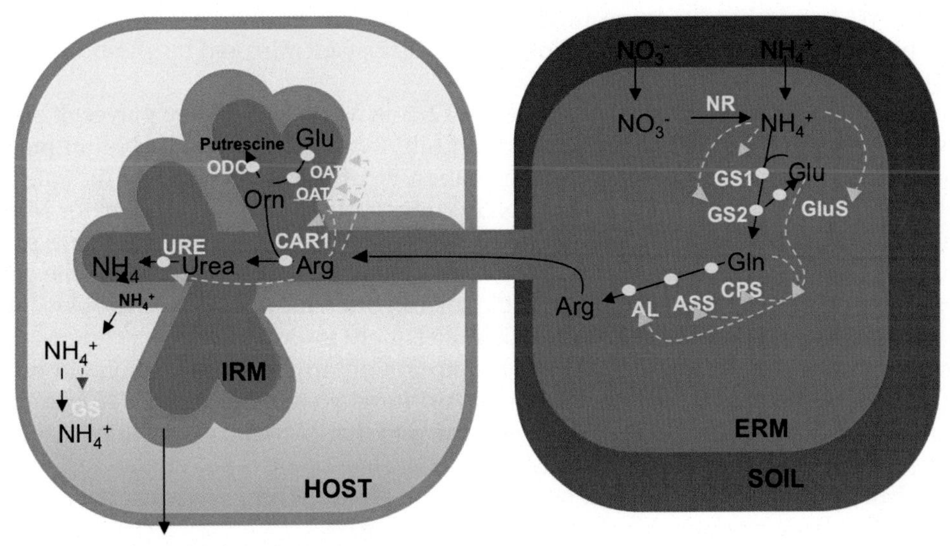

FIGURE 7.2 Model of N transport from soil to plant host through AM fungi. N as NH_4^+ and NO_3^- is absorbed from the soil by the extraradical mycelium (ERM) and is converted to arginine (Arg) involving the glutamine synthetase-glutamate synthase (GS-GOGAT) pathway. Arg is then transported to the arbuscule where it is broken down and the subsequent NH_4^+ is transferred first into the interfacial apoplast and then to the host cell. Activities of the enzymes are indicated in *light blue* and *circles* represent their action sites. *Gray dotted lines* represent the hypothetical regulation of N metabolites and transporter genes involved in the pathway. *AL*, Argininosuccinate lyase; *ASS*, argininosuccinate synthase; *CAR*, arginase; *CPS*, carbamoyl-phosphate synthetase; *Glu*, glutamate; *GluS*, glutamate synthase; *GS*, glutamine synthetase; *IRM*, extraradical mycelium; *NR*, nitrate/nitrite reductase; *OAT*, ornithine aminotransferase; *ODC*, ornithine decarboxylase; *Orn*, ornithine; *URE*, urease. *Modified from Tian, C., Kasiborski, B., Koul, R., Lammers, P.J., Bücking, H., Shachar-Hill, Y., 2010. Regulation of the nitrogen transfer pathway in the arbuscular mycorrhizal symbiosis: gene characterization and the Coordination of expression with nitrogen flux. Plant Physiology 153, 1175–1187.*

threefold and is the most abundant amino acid because of the fungal presence inside roots (Tian et al., 2010). However, other studies have shown that in addition to Arg, other amino acids, such as Gln and Glu, are mainly involved in N transport from the ERM to IRM compartment (George et al., 1992).

Many plant N transporters are induced during AM colonization. Among 47 nutrient-induced transporters observed in *Lotus japonicus* roots after 28 days of colonization (Guether et al., 2009a), a high-affinity amino acid transporter, the LjLHT1.2, was identified and characterized (Guether et al., 2011). *LjLHT1.2* transcripts were preferentially detected in arbusculated cells, suggesting that the LjLHT1.2 transporter is involved in the reuptake and in the recycle of amino acids from the mycorrhizal interface and cortical cells. Several AM-inducible plant NH_4^+ transporters have been identified in different plant species such as *L. japonicus*, *Sorghum bicolor*, and *Medicago truncatula*. In *Lotus*, the NH_4^+ transporter LjAMT2; 2 is reported as a high-affinity transporter that is preferentially expressed in arbusculated cells (Guether et al., 2009b). Similar induction was shown also in arbusculated cells of *M. truncatula* (Gomez et al., 2009), *Glycine max* (Kobae et al., 2010), and *Sorghum* (Koegel et al., 2013). The expression of NH_4^+ transporters in *Oryza sativa* is influenced by different factors: a group of these genes are overexpressed at lower N levels, such as OsAMT1; 1, OsAMT1; 3, and OsAMT3; 3, whereas some other transporters are exclusively influenced by the fungal presence, such as OsAMT3; 1 (Fiorilli et al., 2015), well mirrored by glutamate synthase expression (Pérez-Tienda et al., 2014).

Recently, it has been demonstrated that AMT2; 3 in *Medicago* is a key player of signaling cross-talk between P and N metabolism. Specifically, *mtpt4* mutants under N-limiting conditions are not showing any premature degeneration of arbuscules because of the role played by MtAMT2; 3. The functional role of the NH_4^+ transporter was only reported for MtAMT2; 4, suggesting that MtAMT2; 3 could be more involved in the N-sensing/signaling pathway (Breuillin-Sessoms et al., 2015). A further confirmation of this interconnection is the observation that a condition of low P and low N in the soil can dramatically increase the percentage of AM colonization (Bonneau et al., 2013) and, interestingly, in *Petunia*, N starvation is partially overruling the suppressive effect of high P nutrition on arbuscule formation (Nouri et al., 2014). Therefore there is strong evidence that AM fungi are able to acquire N from organic and inorganic sources and transfer it to the plant, and that N itself is equally contributing to the fine-tuned regulation of AM establishment and functioning. The cellular and molecular mechanism regulating the cross-talk between N and P within the metabolism of AM roots is still largely unknown and requires further investigation.

7.3 PHOSPHATE TRANSPORT IN ARBUSCULAR MYCORRHIZAL SYMBIOSIS

Phosphorus is an essential macronutrient for the proper growth and functioning of plants. The predominant form of P in the soil is the dihydrogen phosphate ion ($H_2PO_4^-$, Pi; Nussaume et al., 2011). This anion is very immobile in the soil because it strongly interacts with iron, aluminum, and calcium, limiting its availability for plants and leading to very low Pi concentrations ($<10\,\mu M$) at the root–soil interface (Smith et al., 2011). Moreover, another factor that restricts Pi availability in the soil is its assimilation by microbes.

To overcome scarce levels of Pi, plants have evolved a range of strategies to increase Pi uptake and Pi availability in the soil (Marschner, 1995), and AM symbiosis represents an evolutionarily persistent and efficient adaptation to address this function. AM fungi uptake Pi in a highly efficient way and convert it in polyphosphate (polyPi) molecules for movement through the mycelia. The polyPis are then hydrolyzed, probably by phosphatase activities during the transport from the fungal structures, as intercellular hyphae or arbuscules, into root cortical cells. It has been demonstrated that the fungal contribution in plant Pi acquisition ranges from a small percentage to almost all of the acquired Pi (Smith et al., 2003, 2004). Consequently, AM fungi provide a very effective and indirect pathway, called "the AM pathway," which interacts with "the direct pathway," where root epidermal cells and root hairs are involved in the Pi uptake directly from the soil (Smith et al., 2011). The interplay between the direct and AM pathways for Pi uptake was recently studied (Watts-Williams et al., 2015). The authors investigated the distal and local effects of AM colonization on the direct root Pi uptake, suggesting that AM fungi reduce the direct root Pi uptake activity locally but not in a distal and noncolonized patch of root (Watts-Williams et al., 2015). Some fungal Pi transporters appear to be responsible for the first step of this symbiotic Pi transport. Four AM Pi transporters have been characterized so far on the basis of a transcriptomic data set: GmosPT from *G. mosseae*, GvPT from *Glomus versiforme*, GiPT from *G. intraradices*, and one from *Gigaspora margarita* (Salvioli et al., 2016). They are all expressed in the extraradical hyphae, and GmosPT also showed an expression in the intraradical hyphae (Benedetto et al., 2005).

AM fungi are able to efficiently absorb Pi from the soil and accumulate it as polyPi. In addition, the rapid and massive accumulation of polyPi in fungal mycelia is accompanied by near-synchronous and near-equivalent uptake of Na^{+}, K^{+}, Ca^{2+}, and Mg^{2+}, with a parallel regulation of gene expression responsible for cation uptake, Pi and N metabolism, and maintenance of cellular homeostasis (Kikuchi et al., 2014). The release of fungal Pi within the root arbusculated cells is accompanied by a significant upregulation of plant Pi transporters, which have been identified in the following plants: *M. truncatula* (Harrison et al., 2002; Javot et al., 2007), *O. sativa* (Paszkowski et al., 2002), *L. japonicus* (Guether et al., 2009a; Volpe et al., 2016), *Astragalus sinicus* (Xie et al., 2013), *Solanum tuberosum* (Nagy et al., 2005), *Zea mays* L. (Nagy et al., 2006), various cereals (Glassop et al., 2005), *Lycopersicon esculentum* (Nagy et al., 2005; Xu et al., 2007; Balestrini et al., 2007), *Populus trichocarpa* (Loth-Pereda et al., 2011), *S. bicolor*, and *Linum usitatissimum* (Walder et al., 2015). These transporters belong to the PHT1 family of Pi transporters and show an amino acid sequence conserved from fungi to plants. They are 500–600 amino acids long, and they share a common structure with 12 predicted membrane-spanning domains of 17–25 amino acids residues that are arranged in a helix. The membrane-spanning regions are separated in two groups of six domains by a hydrophilic and charged loop (Nussaume et al., 2011; Johri et al., 2015). Furthermore, these transporters are specific to AM symbiosis because they are expressed strictly in response to AM fungi, being localized in the PAM (Fig. 7.3). This is the site where nutrient exchange between the two symbionts is located (Harrison et al., 2002). One of the best characterized AM-induced Pht1

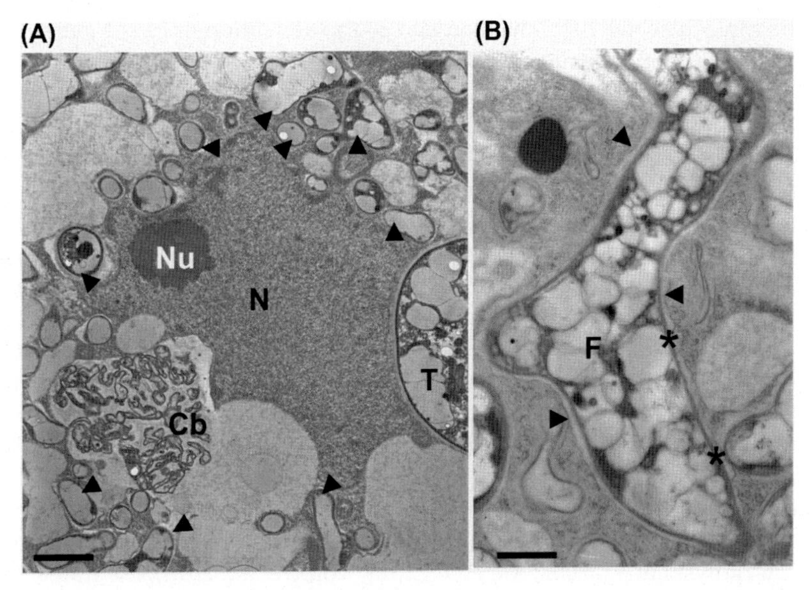

FIGURE 7.3 (A) Ultrastructural features of an arbusculated cell of *Lotus japonicus* colonized by *Gigaspora margarita* after 4 weeks. The cell shows an enlarged central nucleus surrounded by the fungal branches of the arbuscule. Bar corresponds to 2 mm. *Cb*, Collapsed arbuscular branches; *N*, nucleus; *Nu*, nucleolus; T, arbuscule main trunk. *Arrowheads* indicate arbuscular branches. (B) The magnification illustrates the detail of the interface area where the exchange processes between partners are localized. Bar corresponds to 0.7 mm. *F*, hyphal branch. *Asterisks* indicate fungal cell wall and *arrowheads* indicate periarbuscular membrane.

genes is *MtPT4* of *M. truncatula* (Harrison et al., 2002). MtPT4 protein was immunolocalized around the fine arbuscular branches, suggesting that the MtPT4 is exclusively present in the PAM. Moreover, MtPT4 was detected only in developing and mature arbuscules (Harrison et al., 2002; Javot et al., 2007). The heterologous expression of MtPT4 in yeast revealed a lower affinity for Pi, with an apparent K_m between 493 and 668 μM (Harrison et al., 2002). The translocation of Pi from the periarbuscular space throughout the PAM requires an H^+-energy gradient that is provided by H^+-ATPases (Krajinski et al., 2014; Wang et al., 2014b).

During AM colonization, the Pi transporter genes that are involved in the direct pathway can be downregulated independently of the Pi status. In *M. truncatula*, *MtPT1*, *MtPT2*, and *MtPT3* belong to PHT1 family (Bucher et al., 2001; Liu et al., 2008), but their mRNA level is reduced in mycorrhizal roots. Different from MtPT4, this gene expression pattern excludes their involvement in Pi transfer at the arbuscular interface (Chiou et al., 2001). In *M. truncatula* two other PHT1 genes of the direct pathway have been identified: *MtPT5* and *MtPT6*. Both genes are expressed in noncolonized root but they are also affected in AM plants (Grunwald et al., 2009). The *MtPT5* transcript level was normally downregulated in the presence of AM fungus, but not when the roots are colonized by *Gi. margarita* and *G. versiforme*. In this case the expression of the *MtPT5* gene was localized in the cortex of the colonized regions. In contrast, *MtPT6* was repressed during AM colonization, regardless of fungal species.

Expression of the direct Pi transporter genes in nonmycorrhizal plants is regulated by the Pi starvation signaling pathway. In *Arabidopsis*, which is never colonized by AM fungi, a complex network controls the regulation of genes involved in P uptake. The transcription factor (TF) PHR1 binds to the *cis*-regulatory P1BS motif that is present in the promoter region of most Pi starvation-induced genes (Rubio et al., 2001). Components of this network are conserved in plants. The molecular mechanisms that lead to the activation of PHT1 genes in response to AM symbiosis have been investigated (Chen et al., 2011). Two conserved motifs were identified as AM-induced PT promoters: one is P1BS and the other one is MYCS (mycorrhiza transcription factor binding sequence) and both are involved in the activation of AM-responsive PT genes (Chen et al., 2011). In *Arabidopsis* plants, four WRKY proteins have been reported to be involved in the Pi starvation signaling pathway (Devaiah et al., 2007; Chen et al., 2009; Wang et al., 2014a; Su et al., 2015) as positive regulators of *PHT1;1* genes (Devaiah et al., 2007; Su et al., 2015). Up to now, in the AM association, the WRKY proteins have been reported to regulate, during the presymbiotic phase, the plant defense responses genes but their involvement in the Pi starvation signaling pathway was not shown (Gallou et al., 2012).

Pi levels and Pi transporters are crucial regulators of AM symbiosis. When plants are grown at high levels of Pi, the AM colonization is drastically reduced or completely impaired, depending on the plant and on the Pi concentration of the nutrient solution (Breuillin et al., 2010; Balzergue et al., 2011). This effect is milder in the case of preformed arbuscules: when high Pi solution is applied to settled mycorrhizal roots, arbuscules can remain intact and their longevity is not altered (Kobae et al., 2016). Moreover, when AM-inducible transporters are nonfunctional or downregulated by RNA interference, the formation of arbuscules is impaired (Harrison et al., 2002; Paszkowski et al., 2002; Xie et al., 2013; Volpe et al., 2016); the transfer of the Pi ion itself across the PAM has been proposed to act as a cell signal triggering the accomplishment of the AM symbiosis (Yang et al., 2012). However, the suppressive effect of high Pi on root colonization by AM fungi as well as the degeneration of arbuscules in *mtpt4* mutants are partially overruled by N starvation (Nouri et al., 2014; Breuillin-Sessoms et al., 2015). It appears that a complex interplay between nutrient homeostasis, nutrient acquisition, and nutrient sensing is able to regulate the AM colonization and arbuscule morphogenesis, but its mechanism still needs to be elucidated. Recently, Volpe et al. (2016) demonstrated that the expression of *MtPT4* gene and another AM-induced PT gene, *LjPT4* from *L. japonicus*, was also found in the root tips of noncolonized plants, suggesting that both genes could act upstream of the Pi-sensing cellular machinery as potential membrane Pi transceptors. The authors suggested that both AM-responsive Pi transporters play a role in root tips, creating a link among Pi-perception, root branching, and Pi signaling. A similar role for Pi sensing was also proposed for OsPT13, a rice AM-responsive gene that shows no function in Pi transfer but is necessary for the correct formation of arbuscules (Yang et al., 2012).

To date the concept of transceptor has been described only in two plants, *A. thaliana* and *L. japonicus*, in which the NO_3^- transporter AtNRT1.1 and the NH_4^+ transporter LjAMT1; 3, respectively, have been proposed to act as sensor (Remans et al., 2006; Rogato et al., 2010). AtNRT1.1 is involved in root growth and development, stimulating the lateral root elongation (Remans et al., 2006). The authors suggest that this response is not a nutritional effect but results from the activation of the NO_3^- -sensing/signaling pathway. Taken together, these works are opening an interesting scenario in which the multiple functions of these nutrient transporters could represent an important link among plant root development, nutrient sensing, and AM formation.

7.4 SULFUR METABOLISM AND ARBUSCULAR MYCORRHIZAL SYMBIOSIS

Together with P, N, K, Ca, and Mg, S is an essential macronutrient for plants. Within a plant cell, S is needed as a backbone for cell wall and cell membrane formation, as a necessary part of sulfolipids, but also as a main component of essential amino acids such as cysteine and methionine (see Takahashi et al., 2011 for a review). Glutathione, an important S-containing redox controller, is a key metabolite in a plethora of plant cell developmental processes (Foyer and Noctor, 2011). In addition to basic metabolism, S plays important roles in plant secondary processes, and in *Arabidopsis* it is present in more than 400 metabolites (Gläser et al., 2014). Although most of the functions of these metabolites are still unknown, some have been the focus of many studies because of their pivotal action in plant immunity (i.e., camalexin; Nawrath and Métraux, 1999) or flavor quality (i.e., organosulfur compounds from *Allium* species; Kyung and Lee, 2001).

Throughout the 19th century S was abundant in the soil and not limiting plant growth because S-rich air pollutants from heavy industry were oxidized to sulfate in the atmosphere and deposited to the biosphere through the rainfall (Lehmann et al., 2008). However, during the last few decades S limitation is increasingly becoming an issue for many farmers, with a dramatic effect on final products. For example, bread leavening relies on a high concentration of proteins in wheat flour that are highly correlated with soil S levels (Granvogl et al., 2007). When plants are deprived of S, they show strong phenotypes such as yellowing of the leaves and reduced growth and seed number. Furthermore, during sulfate deficiency the biosynthesis of hormones, such as auxin and jasmonate, is induced and sulfate uptake and assimilation are regulated by ethylene, jasmonate, abscisic acid, and other phytohormones (Koprivova and Kopriva, 2016). Notwithstanding the major roles that S plays in plant health and metabolism, mycorrhizal researchers have only recently started to examine the effect of mycorrhization on plant S nutrition and the genetic determinants for S uptake by mycorrhizas.

In addition to preliminary and pioneering experiments comparing the capability of mycorrhizal and nonmycorrhizal plants to absorb ^{35}S from the soil 8 cm distant from the root (Rhodes and Gerdemann, 1978), the demonstration that AM fungi are able to absorb organic and inorganic S forms and transfer them to the plant root was elegantly provided by Allen and Shachar-Hill (2009) using transformed carrot roots and *R. irregularis* cultures. Later, it was confirmed in a two-compartment system with entire *Medicago* plants (Sieh et al., 2013). Mycorrhizal colonization can positively improve plant sulfate content, especially, but not exclusively, when plants are grown in low-sulfate conditions. This effect has been shown for *M. truncatula* (Casieri et al., 2012; Sieh et al., 2013), *L. japonicus* (Giovannetti et al., 2014), and maize (Gerlach et al., 2015; Sawers et al., 2016) in semicontrolled environments and field experiments.

Target and nontarget transcriptomic approaches have been used to investigate which genes are responsible for sulfate transfer in AM symbioses. Using quantitative reverse-transcriptase polymerase chain reaction to compare AM and non-AM *Medicago* grown at high and low levels of S, it has been shown that the presence of AM fungi induces the transcription of many different transporters, although this is not always consistent among studies (Casieri et al., 2012; Sieh et al., 2013). The challenge and intricacy of these kinds of studies is in the discrimination between direct effects caused by the fungal presence and indirect effects caused by modified nutrient status in mycorrhizal roots. Moreover, Sieh et al. (2013) reported that the efficiency of sulfate uptake in plant roots was dependent on Pi availability, observing an increase of S transfer when soil Pi content was low (Sieh et al., 2013).

In recent years the use of microarrays and RNA sequencing of plant and fungal transcriptomes has allowed a deeper characterization of key candidate genes potentially involved in mycorrhizal sulfate uptake. In *L. japonicus*, a strong regulation of the sulfate transporter, *LjSultr1;2*, in roots colonized by either *Gi. margarita* or *R. irregularis* was shown in two different studies (Guether et al., 2009a; Handa et al., 2015), and subsequent molecular characterization demonstrated its two distinct roles. On one hand, *LjSultr1;2* mediates direct sulfate uptake from soil when plants are under S starvation, and on the other hand it is strongly activated in arbuscule-containing cells once the symbiosis is formed (Giovannetti et al., 2014). *LjSultr1;2* is not the only sulfate transporter regulated during AM symbiosis. In fact, plants possess dozens of sulfate transporters (SULTRs), and a fine-tuned regulation allows the balance among storage, catabolism, and root-to-shoot movement of the sulfate pool.

SULTRs can be distinguished into four functional groups based on their sequences and their activation (for a review see Gigolashvili and Kopriva, 2014). Mainly from *Arabidopsis*, we know that group one encodes high-affinity SULTRs that usually mediate the direct uptake of sulfate from the soil, group two are low-affinity transporters and responsible for long-distance translocation of sulfate (Takahashi et al., 2000), group three is still largely diverse and its function is unknown, and group four encodes vacuolar sulfate exporters (Buchner et al., 2004; Takahashi et al., 2011). The structures and transcriptional regulation of SULTRs have also been characterized in other plant species, such as *Brassica oleracea*, poplar, wheat, and *M. truncatula* (Buchner et al., 2004, 2010; Dürr et al., 2010). A recent microarray study of *Medicago* root and leaf transcriptome responses to S deficiency showed that several hundred genes are regulated by the S status of plants and by their interactions with the AM fungus *R. irregularis*. In root tissues the low-S treatment induced downregulation of several genes, including genes coding for thiosulfate and glutathione sulfur transferases, gibberellin oxidases, and quinone oxidoreductases. It is interesting to note that this effect was less evident in mycorrhizal plants compared with nonmycorrhizal plants (Wipf et al., 2014). Understanding which genes regulate S metabolism in mycorrhizal plants is still a long way off, but further characterization of genes coming from –omics studies will better elucidate this complex system.

It is interesting to note that both in yeast (Kankipati et al., 2015) and in the model plant *A. thaliana*, SULTRs have been proposed to act not only in transferring sulfate through cell membranes but also in sensing sulfate, as shown by their capacity to regulate gene transcription of sulfate metabolism independent of sulfate transfer (Zhang et al., 2014). A parallel role in AM symbiosis is worth investigating because nutrient exchange is a key regulator of an established symbiosis and, so far, exclusively Pi transporters have been demonstrated to play such a role in the formation of a functional arbuscule (Harrison et al., 2002; Volpe et al., 2016).

7.5 FROM ROOT TO SHOOT AND BACK: EVIDENCE FOR A SYSTEMIC SIGNALING AND GENE REGULATION IN MYCORRHIZAL PLANTS

There is a fairly good understanding of the molecular basis for AM establishment and nutrient exchange dynamics, but little is known about the systemic spread of mycorrhizal signals. Mycorrhizal symbioses generate a plethora of effects in plant hosts, including improved growth and biomass, tolerance to biotic and abiotic stresses, and changes in metabolite production (van der Heijden et al., 2006; Pozo and Azcón-Aguilar, 2007; Smith and Smith, 2011). The idea that these diverse outcomes are not solely because of improved nutritional status is largely acknowledged. In recent years many authors have addressed the question of whether the establishment of mycorrhizas in isolated root compartments may elicit long-distance molecular effects, and this interest was initially directed toward mycorrhizal functioning in nutrient exchange. Although it is well acknowledged that expression of nutrient transporter genes (mainly P and N) is regulated in mycorrhizal roots, whether this regulation is systemic or not remains an open question. Using the split-root system that simultaneously allows comparison of a mycorrhizal and a nonmycorrhizal root from the same experimental plant provides a useful approach to address this question. In 2013, Koegel and colleagues analyzed the expression of several transporters in *S. bicolor* and found that two specific NH_4^+ transporters were locally but not systemically induced by AM colonization whereas an AM-inducible Pi transporter was overexpressed also in nonmycorrhizal roots of mycorrhizal plants (Koegel et al., 2013). This patchy situation has been confirmed by other studies. Recent research on wheat revealed that root colonization by AM fungi locally but not systemically upregulated the expression of AM-inducible Pi transporters, whereas distal effects on the expression of NO_3^- and NH_4^+ transporter genes varied depending on the specific gene and the AM species (Duan et al., 2015). Likewise, another study indicated that colonization by AM fungi decreased expression of direct Pi transporter genes (that mediate mycorrhizal-independent

Pi uptake) locally, but not systemically, and direct root Pi uptake was only weakly influenced on distal root portions (Watts-Williams et al., 2015). On the whole, these studies suggest that the colonization by AM fungi might produce a signal that is spread to the noncolonized roots, even if, probably depending on the specific plant–fungus couple and experimental conditions, different transporters (P or N) were found to be systemically regulated. It is well known that a systemic bidirectional signaling is involved in regulating the establishment of arbuscular mycorrhiza. For example, the root colonization by an AM fungus is able to negatively regulate further AM colonization, and this phenomenon is known as autoregulation (Staehelin et al., 2011). Taking advantage of the split-root system, many studies demonstrated that initial colonization of one root half is able to suppress the AM colonization on the other half (Vierheilig et al., 2000; Catford et al., 2003). In analogy with what has already been established during nodulation, the signaling molecules involved in the mycorrhizal autoregulation are proposed to migrate in the xylem and in the phloem to transfer the signal from the root to the shoot and then back to the belowground compartment (Staehelin et al., 2011). Furthermore, other researches demonstrate that the suppression of mycorrhization observed under high Pi availability is mainly dependent on the shoot P levels rather than the P status experienced by the roots, thus implying the involvement of a shoot-derived signal (Breuillin et al., 2010; Balzergue et al., 2011). These data prove that not only the suppressive signal works systemically, but such a feedback also requires the activity of signaling and sensing components localized in plant districts other than roots.

The examples provided so far demonstrate that molecular signaling is involved in events that are instrumental in the symbiosis itself. However, does this signaling also play a role in inducing the long-distance beneficial effects that characterize a mycorrhizal plant? Emerging evidence supports this view as in the case of systemic resistance to disease.

The ability of AM fungi to protect plants from pathogen attack through the induction of a systemic resistance has long been recognized (Cordier et al., 1998). It is interesting to note that nutrient supply experiments have revealed that this effect is not a consequence of the improved nutritional status (Fritz et al., 2006). In early 2002, Pozo and colleagues studied the influence of AM colonization on the biochemical defense of tomato against *Phytophtora parasitica* and found that AM fungi reduced disease symptoms. A systemic alteration of the activity of defense-related enzymes was also observed in nonmycorrhizal roots of mycorrhizal plants (Pozo et al., 2002). This "mycorrhiza-induced resistance" acts as a systemic priming of defense against a wide range of attacks, and its spread seems to involve complex hormonal signaling in which jasmonic acid (JA) and salicylic acid play a pivotal role (Cameron et al., 2013). In a recent work, mycorrhizal tomato plants challenged with the fungal pathogen *Alternaria solani* responded to the attack with a strong induction of pathogenesis-related genes, as well as other defense-related genes in leaves. However, pathogen infection did not induce these genes and enzymes in a jasmonate-biosynthesis defective plant, indicating that JA signaling is an essential component of the priming mechanism triggered by mycorrhizas (Song et al., 2015). At the same time these data also confirm that the long-distance protective effect involves the modulation of transcription of defense-related genes in plant districts far from the root. The same class of defense-related genes is also triggered when plants are challenged with AM spore exudates or the purified molecules contained in those exudates, such as short chitin oligomers (Giovannetti et al., 2015).

In this respect, pioneering research was performed by Maria Harrison's group, who used a microarray and quantitative polymerease chain reaction-based strategy to explore local and systemic transcriptional responses to AM fungi in *M. truncatula* roots and shoots (Liu et al., 2007). They observed a transcriptome remodeling in shoots as a consequence of root colonization by AM fungi, and in particular, genes dealing with stress and defense responses were upregulated. Accordingly, *M. truncatula* plants showed increased resistance against a virulent bacterial pathogen. Another point of novelty of this work is that the experiment was designed to avoid a significant enhancement of plant Pi nutrition by the AM fungus. The fact that under these conditions a transcriptome change could still be observed strengthens the idea that alterations in gene expression are not necessarily linked to the improvement of Pi nutrition.

After the work in Harrison's laboratory, numerous other studies analyzed changes in gene expression that are systemically induced by AM symbiosis by using microarrays and more recently through deep sequencing (Fiorilli et al., 2009; Salvioli et al., 2012; Zouari et al., 2014; Cervantes-Gámez et al., 2015). It is interesting to note that tomato was the plant species studied in all of the papers; indeed, in the last decade much attention has been paid to dissect the mechanisms that feature the mycorrhizal status of crop plants, with a particular focus on agronomically relevant traits. The advent of more refined molecular techniques enabled researchers to reveal the fine-scale physiological mechanisms shaping the generically defined "growth effect" triggered by AM fungi, revealing that it implies deep molecular and metabolic rearrangement in distal parts of the plant. These investigations clearly indicate that mycorrhizal establishment can improve the nutritional value of different crops. After a biochemical approach, Baslam and colleagues demonstrated that enhanced levels of carotenoids, chlorophylls, and tocopherols are induced in leaves from mycorrhizal lettuce plants (Baslam et al., 2011). Mycorrhizal colonization improved the nutrient quality of tomato fruits because concentrations of lycopene, carotenoid, and volatile compounds were significantly increased (Giovannetti et al., 2012; Hart et al., 2015). In recent works from our group, changes in gene expression in tomato fruit were accompanied by phenological modifications as an accelerated flowering and fruiting time, an overall increase in fruit yield, and qualitative and quantitative changes in amino acid profile (Salvioli et al., 2012). It is interesting to note that when the experiment was designed to avoid nutrient starvation in control plants, Pi and SULTRs turned up to be upregulated in fruits from AM plants, suggesting that this regulation is mycorrhizal specific and not only related to the improved nutritional status (Zouari et al., 2014).

Scaling up molecular results to the phenotypical level is necessary and a major challenge in this kind of research. An integrated approach that applies multiprofiling analysis that includes comparative ionomic, transcriptomic, metabolomic, and phenotypic analyses appears to be a promising approach for reconstructing the complexity of whole systems. Gerlach and colleagues recently dissected systemic responses of maize to AM symbiosis at the single-leaf level and were able to link plant growth enhancement with robust changes in elemental leaf composition, an increase in anthocyanin and lipid metabolism, a reprogramming of leaf C:N ratio, and the induction of defense-related genes (Gerlach et al., 2015).

The evidence so far points to the involvement of long distance "mycorrhizal signals" that move from root to shoot and back. To date, the major candidates for this role are hormones and micro-RNAs (miRNAs). The involvement of phytohormones in the establishment and functionality of the AM symbiosis is already well acknowledged, and more data consistently support their role in the spread of mycorrhiza-induced resistance (Cameron et al., 2013).

In addition, there is evidence suggesting that miRNAs also play a regulatory role during AM symbiosis (Devers et al., 2011). miRNAs are 20- to 25-nucleotide-long phloem-mobile RNAs that can regulate the plant physiology in response to diverse external stimuli, including nutrient availability (Kulcheski et al., 2015). It is interesting to note that an in silico analysis on solanaceous genomes identified Pi starvation and AM-related *cis*-elements in the putative promoter regions of some miRNA genes (Gu et al., 2014), strengthening the hypothesis that miRNA could play a role in mediating the plant response to mycorrhization.

In conclusion, the well acknowledged growth effect exerted by AM fungi is accompanied by molecular changes that imply a mycorrhiza-specific signaling and a fine tuning of metabolic functions of the distal part of the plant host. Such systemic outcome is challenging to study because it is influenced by many factors (i.e., the specific plant–fungus couple, the surrounding environment, the interaction with other microbes, etc.), but the fine understanding of how AM fungi shape plant physiology represents a frontier toward exploiting AM symbioses in sustainable agriculture (Box 7.1; Chapter 10).

SUMMARY POINTS

Better understanding nutrient transfer within AM symbiosis is a milestone for future sustainable and environmentally viable agriculture.

- Nitrogen (N)

AM fungi absorb inorganic and organic nitrogen forms and transfer them to the plant, where specific ammonium transporters are included.

- Phosphorus (P)

Phosphorus is a nonrenewable resource, and AM fungi significantly improve plant P nutrition. Phosphate itself and plant phosphate transporters are key regulators of AM symbiosis.

- Sulfur (S)

AM fungi transfer sulfur and improve plant sulfur status, especially when it represents the growth-limiting factor.

- Systemic effect

AM symbioses provide deep beneficial effects for host plants, including uncolonized regions of roots, shoots, and fruits.

FUTURE ISSUES

- Which **nutrient signals** regulate and execute inhibition or induction of AM development?
- Within the AM interface how much **nutrient reabsorption** is generated by symbiotic AM fungi?
- What is the role of **plant nutrient transceptors** in the establishment and the maintenance of arbusculated cells?
- How does plant and fungal **natural variation** affect AM fitness and effects?

7.6 PERSPECTIVES AND CONCLUSIONS

Notwithstanding their biotrophic nature, AM fungi possess amazing capacities to extract mineral nutrients from the soils, and genome sequencing (Tisserant et al., 2013) clearly demonstrates that their enzyme repertoire is rich in transporters that differ from pathogenic biotrophs that live on leaves and are fully dependent on their host for nutrients as well as C. In this context AM fungi release a part of these nutrients to the plants inducing the expression of many plant transporters and activating the so-called symbiotic pathway. However, the amount of nutrients that AM fungi retain for their own metabolism and how much is transferred to the plant hosts is still unclear, as well as the environmental conditions that control these exchanges. Laser microdissection studies showing the transcriptional profile of arbusculated cells (Balestrini et al., 2007; Fiorilli et al., 2013) strongly suggest that AM fungi control the release and absorption of nutrients at the arbuscule level. Understanding these issues could be crucial for a better exploitation of AM symbiosis and modern techniques, such as nanoscale secondary ion mass spectrometry (NanoSIMS) (Kaiser et al., 2015), will help to reveals such nutrient traffics in vivo.

AM-inducible plant transporters are well characterized but new functionalities have been recently unveiled, such as nutrient sensing, even in the absence of the fungal partner, opening the question of their role in an evolutionary context. During their evolution, land plants had to cope with mineral nutrient deprivation (Bonfante and Genre, 2008); therefore nutrient sensing could have represented a driving force in different nutrient uptake strategies (Volpe et al., 2016). This function may also exert a control on the root architecture, creating a strict link among plant development, environmental conditions, and AM establishment. Deciphering this network could help in understanding factors that limit the AM colonization in agricultural ecosystems.

Nevertheless, to increase plant productivity and agricultural sustainability, there is a need to fully understand and domesticate AM effects at a systemic level with open field studies using a multidisciplinary approach. One of the open questions is the reciprocal effect of plant photosynthesis within AM symbiosis: on the one hand, AM fungi act as C sinks, but on the other they can stimulate plant photosynthesis (Ruiz-Sánchez et al., 2010; Birhane et al., 2012). Finally, it becomes increasingly clear that plant and fungal ecotypes evolve specific adaptations to their local environments and, as a consequence, specific alleles regulate plant developmental processes in peculiar ways. Genome-wide association studies in legumes will be a key tool that can lead to the identification of natural alleles and their effects on plant phenotypes (Gentzbittel et al., 2015). In this context, as described for example in maize (Sawers et al., 2016), different plant ecotypes showed different mycorrhizal effects and nutrient levels and the same is valid from the fungal side, where different isolates induced various effects on the plant (Ceballos et al., 2013). Altogether new approaches will be the key to decipher systemic responses to AM symbioses and exploit them for a more sustainable agriculture through reducing the need for high-cost fertilizers without hampering yield.

Acknowledgments

The authors sincerely thank Nancy Johnson for her kind suggestions and help and Dr. Mara Novero for kindly providing the electron microscope picture. Research in Bonfante's laboratory was supported by the projects Mycoplant (Progetto di Ateneo and CSP 2012–2016), Mycoceres (CARIPLO- Agripole Project 2014–2017), and Green–Rice (FACCE; MIUR).

References

Allen, J.W., Shachar-Hill, Y., 2009. Sulfur transfer through an arbuscular mycorrhiza. Plant Physiology 149, 549–560.

Bago, B., Chamberland, H., Goulet, A., Vierheilig, H., Lafon- taine, J.G., Pichè, Y., 1996. Effect of nikkomycin Z, a chitin synthase inhibitor, on hyphal growth and cell wall structure of two arbuscular mycorrhizal fungi. Protoplasma 192, 80–92.

Bago, B., Azcón-Aguilar, C., 1997. Changes in the rhizospheric pH induced by arbuscular mycorrhiza formation in onion (*Allium cepa* L.). Zeitschrift für Pflanzenernährung und Bodenkunde 160, 333–339.

Balestrini, R., Gómez-Ariza, J., Lanfranco, L., Bonfante, P., 2007. Laser microdissection reveals that transcripts for five plant and one fungal phosphate transporter genes are contemporaneously present in arbusculated cells. Molecular plant-microbe interactions: MPMI 20, 1055–1062.

Balzergue, C., Puech-Pagès, V., Bécard, G., Rochange, S.F., 2011. The regulation of arbuscular mycorrhizal symbiosis by phosphate in pea involves early and systemic signalling events. Journal of Experimental Botany 62, 1049–1060.

Baslam, M., Garmendia, I., Goicoechea, N., 2011. Arbuscular mycorrhizal fungi (AMF) improved growth and nutritional quality of greenhouse-grown lettuce. Journal of Agricultural and Food Chemistry 59, 5504–5515.

Belmondo, S., Fiorilli, V., Pérez-Tienda, J., Ferrol, N., Marmeisse, R., Lanfranco, L., 2014. A dipeptide transporter from the arbuscular mycorrhizal fungus *Rhizophagus irregularis* is upregulated in the intraradical phase. Plant Traffic and Transport 5, 436.

Benedetto, A., Magurno, F., Bonfante, P., Lanfranco, L., 2005. Expression profiles of a phosphate transporter gene (GmosPT) from the endomycorrhizal fungus *Glomus mosseae*. Mycorrhiza 15, 620–627.

Birhane, E., Sterck, F.J., Fetene, M., Bongers, F., Kuyper, T.W., 2012. Arbuscular mycorrhizal fungi enhance photosynthesis, water use efficiency, and growth of frankincense seedlings under pulsed water availability conditions. Oecologia 169, 895–904.

Bonfante, P., Genre, A., 2008. Plants and arbuscular mycorrhizal fungi: an evolutionary-developmental perspective. Trends in Plant Science 13, 492–498.

Bonneau, L., Huguet, S., Wipf, D., Pauly, N., Truong, H.-N., 2013. Combined phosphate and nitrogen limitation generates a nutrient stress transcriptome favorable for arbuscular mycorrhizal symbiosis in *Medicago truncatula*. The New Phytologist 199, 188–202.

Bravo, A., York, T., Pumplin, N., Mueller, L.A., Harrison, M.J., 2016. Genes conserved for arbuscular mycorrhizal symbiosis identified through phylogenomics. Nature Plants 2, 15208.

Breuillin, F., Schramm, J., Hajirezaei, M., Ahkami, A., Favre, P., Druege, U., et al., 2010. Phosphate systemically inhibits development of arbuscular mycorrhiza in *Petunia hybrida* and represses genes involved in mycorrhizal functioning. The Plant Journal: For Cell and Molecular Biology 64, 1002–1017.

Breuillin-Sessoms, F., Floss, D.S., Gomez, S.K., Pumplin, N., Ding, Y., Levesque-Tremblay, V., et al., 2015. Suppression of arbuscule degeneration in *Medicago truncatula* phosphate transporter4 mutants is dependent on the ammonium transporter 2 family protein AMT2;3. The Plant Cell 27, 1352–1366.

Bucher, M., Rausch, C., Daram, P., 2001. Molecular and biochemical mechanisms of phosphorus uptake into plants. Journal of Plant Nutrition and Soil Science 164, 209–217.

Buchner, P., Parmar, S., Kriegel, A., Carpentier, M., Hawkesford, M.J., 2010. The sulfate transporter family in wheat: tissue-specific gene expression in relation to nutrition. Molecular Plant 3, 374–389.

Buchner, P., Takahashi, H., Hawkesford, M.J., 2004. Plant sulphate transporters: co-ordination of uptake, intracellular and long-distance transport. Journal of Experimental Botany 55, 1765–1773.

Bücking, H., Kafle, A., 2015. Role of arbuscular mycorrhizal fungi in the nitrogen uptake of plants: current knowledge and research Gaps. Agronomy 5, 587–612.

Cameron, D.D., Neal, A.L., van Wees, S.C.M., Ton, J., 2013. Mycorrhiza-induced resistance: more than the sum of its parts? Trends in Plant Science 18, 539–545.

Cappellazzo, G., Lanfranco, L., Fitz, M., Wipf, D., Bonfante, P., 2008. Characterization of an amino acid permease from the endomycorrhizal fungus *Glomus mosseae*. Plant Physiology 147, 429–437.

Casieri, L., Gallardo, K., Wipf, D., 2012. Transcriptional response of *Medicago truncatula* sulphate transporters to arbuscular mycorrhizal symbiosis with and without sulphur stress. Planta 235, 1431–1447.

Catford, J.-G., Staehelin, C., Lerat, S., Piché, Y., Vierheilig, H., 2003. Suppression of arbuscular mycorrhizal colonization and nodulation in split-root systems of alfalfa after pre-inoculation and treatment with Nod factors. Journal of Experimental Botany 54, 1481–1487.

Ceballos, I., Ruiz, M., Fernández, C., Peña, R., Rodríguez, A., Sanders, I.R., 2013. The in vitro mass-produced model mycorrhizal fungus, *Rhizophagus irregularis*, significantly increases yields of the globally important food security crop cassava. PloS One 8, e70633.

Cervantes-Gámez, R.G., Bueno-Ibarra, M.A., Cruz-Mendívil, A., Calderón-Vázquez, C.L., Ramírez-Douriet, C.M., Maldonado-Mendoza, I.E., et al., 2015. Arbuscular mycorrhizal symbiosis-induced expression changes in Solanum lycopersicum leaves revealed by RNA-seq analysis. Plant Molecular Biology Reporter 34, 89–102.

Chalot, M., Blaudez, D., Brun, A., 2006. Ammonia: a candidate for nitrogen transfer at the mycorrhizal interface. Trends in Plant Science 11, 263–266.

Chen, A., Gu, M., Sun, S., Zhu, L., Hong, S., Xu, G., 2011. Identification of two conserved *cis*-acting elements, MYCS and P1BS, involved in the regulation of mycorrhiza-activated phosphate transporters in eudicot species. The New Phytologist 189, 1157–1169.

Chen, Y.-F., Li, L.-Q., Xu, Q., Kong, Y.-H., Wang, H., Wu, W.-H., 2009. The WRKY6 transcription factor modulates PHOSPHATE1 expression in response to low Pi stress in Arabidopsis. The Plant Cell 21, 3554–3566.

Chiou, T.J., Liu, H., Harrison, M.J., 2001. The spatial expression patterns of a phosphate transporter (MtPT1) from *Medicago truncatula* indicate a role in phosphate transport at the root/soil interface. The Plant Journal: For Cell and Molecular Biology 25, 281–293.

Cordier, C., Pozo, M.J., Barea, J.M., Gianinazzi, S., Gianinazzi-Pearson, V., 1998. Cell defense responses associated with localized and systemic resistance to *Phytophthora parasitica* induced in tomato by an arbuscular mycorrhizal fungus. Molecular Plant-Microbe Interactions 11, 1017–1028.

Corrêa, A., Cruz, C., Ferrol, N., 2015. Nitrogen and carbon/nitrogen dynamics in arbuscular mycorrhiza: the great unknown. Mycorrhiza 25, 499–515.

Devaiah, B.N., Karthikeyan, A.S., Raghothama, K.G., 2007. WRKY75 transcription factor is a modulator of phosphate acquisition and root development in Arabidopsis. Plant Physiology 143, 1789–1801.

Devers, E.A., Branscheid, A., May, P., Krajinski, F., 2011. Stars and symbiosis: microRNA- and microRNA*-mediated transcript cleavage involved in arbuscular mycorrhizal symbiosis. Plant Physiology 156, 1990–2010.

Duan, J., Tian, H., Drijber, R.A., Gao, Y., 2015. Systemic and local regulation of phosphate and nitrogen transporter genes by arbuscular mycorrhizal fungi in roots of winter wheat (*Triticum aestivum* L.). Plant physiology and biochemistry: PPB/Société française de physiologie végétale 96, 199–208.

Dürr, J., Bücking, H., Mult, S., Wildhagen, H., Palme, K., Rennenberg, H., et al., 2010. Seasonal and cell type specific expression of sulfate transporters in the phloem of Populus reveals tree specific characteristics for SO(4)(2-) storage and mobilization. Plant Molecular Biology 72, 499–517.

Fellbaum, C.R., Gachomo, E.W., Beesetty, Y., Choudhari, S., Strahan, G.D., Pfeffer, P.E., et al., 2012. Carbon availability triggers fungal nitrogen uptake and transport in arbuscular mycorrhizal symbiosis. Proceedings of the National Academy of Sciences of the United States of America 109, 2666–2671.

Fiorilli, V., Catoni, M., Miozzi, L., Novero, M., Accotto, G.P., Lanfranco, L., 2009. Global and cell-type gene expression profiles in tomato plants colonized by an arbuscular mycorrhizal fungus. The New Phytologist 184, 975–987.

Fiorilli, V., Lanfranco, L., Bonfante, P., 2013. The expression of GintPT, the phosphate transporter of *Rhizophagus irregularis*, depends on the symbiotic status and phosphate availability. Planta 237, 1267–1277.

Fiorilli, V., Vallino, M., Biselli, C., Faccio, A., Bagnaresi, P., Bonfante, P., 2015. Host and Non-Host roots in rice: cellular and molecular approaches reveal differential responses to arbuscular mycorrhizal fungi. Frontiers in Plant Science 6, 636.

Foyer, C.H., Noctor, G., 2011. Ascorbate and glutathione: the heart of the redox hub. Plant Physiology 155, 2–18.

Fritz, M., Jakobsen, I., Lyngkjaer, M.F., Thordal-Christensen, H., Pons-Kühnemann, J., 2006. Arbuscular mycorrhiza reduces susceptibility of tomato to *Alternaria solani*. Mycorrhiza 16, 413–419.

Gachomo, E., Allen, J.W., Pfeffer, P.E., Govindarajulu, M., Douds, D.D., Jin, H., et al., 2009. Germinating spores of *Glomus intraradices* can use internal and exogenous nitrogen sources for de novo biosynthesis of amino acids. New Phytologist 184, 399–411.

Gallou, A., Declerck, S., Cranenbrouck, S., 2012. Transcriptional regulation of defence genes and involvement of the WRKY transcription factor in arbuscular mycorrhizal potato root colonization. Functional and Integrative Genomics 12 (1), 183–198.

Gentzbittel, L., Andersen, S.U., Ben, C., Rickauer, M., Stougaard, J., Young, N.D., 2015. Naturally occurring diversity helps to reveal genes of adaptive importance in legumes. Plant Genetics and Genomics 6, 269.

George, E., Heorge, K.-U., Vetterlein, D., Gorgus, E., Marschner, H., 1992. Water and nutrient translocation by hyphae of *Glomus mosseae*. Canadian Journal of Botany 70, 2130–2137.

George, E., Marschner, H., Jakobsen, I., 1995. Role of arbuscular mycorrhizal fungi in uptake of phosphorus and nitrogen from soil. Critical Reviews in Biotechnology 15, 257–270.

Gerlach, N., Schmitz, J., Polatajko, A., Schlüter, U., Fahnenstich, H., Witt, S., et al., 2015. An integrated functional approach to dissect systemic responses in maize to arbuscular mycorrhizal symbiosis. Plant, Cell and Environment 38, 1591–1612.

Gigolashvili, T., Kopriva, S., 2014. Transporters in plant sulfur metabolism. Frontiers in Plant Science 5.

Giovannetti, M., Avio, L., Barale, R., Ceccarelli, N., Cristofani, R., Iezzi, A., et al., 2012. Nutraceutical value and safety of tomato fruits produced by mycorrhizal plants. The British Journal of Nutrition 107, 242–251.

Giovannetti, M., Mari, A., Novero, M., Bonfante, P., 2015. Early *Lotus japonicus* root transcriptomic responses to symbiotic and pathogenic fungal exudates. Frontiers in Plant Science 6, 480.

Giovannetti, M., Tolosano, M., Volpe, V., Kopriva, S., Bonfante, P., 2014. Identification and functional characterization of a sulfate transporter induced by both sulfur starvation and mycorrhiza formation in *Lotus japonicus*. New Phytologist 204, 609–619.

Gläser, K., Kanawati, B., Kubo, T., Schmitt-Kopplin, P., Grill, E., 2014. Exploring the Arabidopsis sulfur metabolome. The Plant Journal: For Cell and Molecular Biology 77, 31–45.

Glassop, D., Smith, S.E., Smith, F.W., 2005. Cereal phosphate transporters associated with the mycorrhizal pathway of phosphate uptake into roots. Planta 222, 688–698.

Gobert, A., Plassard, C., 2008. The beneficial effect of mycorrhizae on N utilization by the host-plant: myth or reality? In: Varma, A., (Ed.), Mycorrhiza. Springer-Verlag, Berlin, pp. 209–240.

Gomez, S.K., Javot, H., Deewatthanawong, P., Torres-Jerez, I., Tang, Y., Blancaflor, E.B., et al., 2009. *Medicago truncatula* and *Glomus intraradices* gene expression in cortical cells harboring arbuscules in the arbuscular mycorrhizal symbiosis. BMC Plant Biology 9, 10.

Govindarajulu, M., Pfeffer, P.E., Jin, H., Abubaker, J., Douds, D.D., Allen, J.W., et al., 2005. Nitrogen transfer in the arbuscular mycorrhizal symbiosis. Nature 435, 819–823.

Granvogl, M., Wieser, H., Koehler, P., Von Tucher, S., Schieberle, P., 2007. Influence of sulfur fertilization on the amounts of free amino acids in wheat. Correlation with baking properties as well as with 3-Aminopropionamide and acrylamide generation during baking. Journal of Agricultural and Food Chemistry 55, 4271–4277.

Grelet, G.-A., Meharg, A.A., Duff, E.I., Anderson, I.C., Alexander, I.J., 2009. Small genetic differences between ericoid mycorrhizal fungi affect nitrogen uptake by Vaccinium. New Phytologist 181, 708–718.

Grunwald, U., Guo, W., Fischer, K., Isayenkov, S., Ludwig-Müller, J., Hause, B., et al., 2009. Overlapping expression patterns and differential transcript levels of phosphate transporter genes in arbuscular mycorrhizal, Pi-fertilised and phytohormone-treated *Medicago truncatula* roots. Planta 229, 1023–1034.

Guether, M., Balestrini, R., Hannah, M., He, J., Udvardi, M.K., Bonfante, P., 2009a. Genome-wide reprogramming of regulatory networks, transport, cell wall and membrane biogenesis during arbuscular mycorrhizal symbiosis in *Lotus japonicus*. The New Phytologist 182, 200–212.

Guether, M., Neuhäuser, B., Balestrini, R., Dynowski, M., Ludewig, U., Bonfante, P., 2009b. A mycorrhizal-specific ammonium transporter from *Lotus japonicus* acquires nitrogen released by arbuscular mycorrhizal fungi. Plant Physiology 150, 73–83.

Guether, M., Volpe, V., Balestrini, R., Requena, N., Wipf, D., Bonfante, P., 2011. LjLHT1.2—a mycorrhiza-inducible plant amino acid transporter from *Lotus japonicus*. Biology and Fertility of Soils 47, 925–936.

Gu, M., Liu, W., Meng, Q., Zhang, W., Chen, A., Sun, S., et al., 2014. Identification of microRNAs in six solanaceous plants and their potential link with phosphate and mycorrhizal signaling. Journal of Integrative Plant Biology 56, 1164–1178.

Handa, Y., Nishide, H., Takeda, N., Suzuki, Y., Kawaguchi, M., Saito, K., 2015. Rna-seq transcriptional profiling of an arbuscular mycorrhiza provides insights into regulated and coordinated gene expression in *Lotus japonicus* and *Rhizophagus irregularis*. Plant and Cell Physiology 56, 1490–1511.

Harrison, M.J., Dewbre, G.R., Liu, J., 2002. A phosphate transporter from *Medicago truncatula* involved in the acquisition of phosphate released by arbuscular mycorrhizal fungi. The Plant Cell 14, 2413–2429.

Hart, M., Ehret, D.L., Krumbein, A., Leung, C., Murch, S., Turi, C., et al., 2015. Inoculation with arbuscular mycorrhizal fungi improves the nutritional value of tomatoes. Mycorrhiza 25, 359–376.

Hawkins, H.-J., George, E., 1999. Effect of plant nitrogen status on the contribution of arbuscular mycorrhizal hyphae to plant nitrogen uptake. Physiologia Plantarum 105, 694–700.

van der Heijden, M.G.A., Streitwolf-Engel, R., Riedl, R., Siegrist, S., Neudecker, A., Ineichen, K., et al., 2006. The mycorrhizal contribution to plant productivity, plant nutrition and soil structure in experimental grassland. The New Phytologist 172, 739–752.

Javot, H., Penmetsa, R.V., Terzaghi, N., Cook, D.R., Harrison, M.J., 2007. A *Medicago truncatula* phosphate transporter indispensable for the arbuscular mycorrhizal symbiosis. Proceedings of the National Academy of Sciences 104, 1720–1725.

Jin, H., Pfeffer, P.E., Douds, D.D., Piotrowski, E., Lammers, P.J., Shachar-Hill, Y., 2005. The uptake, metabolism, transport and transfer of nitrogen in an arbuscular mycorrhizal symbiosis. The New Phytologist 168, 687–696.

Johri, A.K., Oelmüller, R., Dua, M., Yadav, V., Kumar, M., Tuteja, N., et al., 2015. Fungal association and utilization of phosphate by plants: success, limitations, and future prospects. Frontiers in Microbiology 6, 984.

Kaiser, C., Kilburn, M.R., Clode, P.L., Fuchslueger, L., Koranda, M., Cliff, J.B., et al., 2015. Exploring the transfer of recent plant photosynthates to soil microbes: mycorrhizal pathway vs direct root exudation. The New Phytologist 205, 1537–1551.

Kankipati, H.N., Rubio-Texeira, M., Castermans, D., Diallinas, G., Thevelein, J.M., 2015. Sul1 and Sul2 sulfate transceptors signal to protein kinase A upon exit of sulfur starvation. The Journal of Biological Chemistry 290, 10430–10446.

Kiers, E.T., Duhamel, M., Beesetty, Y., Mensah, J.A., Franken, O., Verbruggen, E., et al., 2011. Reciprocal rewards stabilize cooperation in the mycorrhizal symbiosis. Science (New York, N.Y.) 333, 880–882.

Kikuchi, Y., Hijikata, N., Yokoyama, K., Ohtomo, R., Handa, Y., Kawaguchi, M., et al., 2014. Polyphosphate accumulation is driven by transcriptome alterations that lead to near-synchronous and near-equivalent uptake of inorganic cations in an arbuscular mycorrhizal fungus. The New Phytologist 204, 638–649.

Kobae, Y., Hata, S., 2010. Dynamics of periarbuscular membranes visualized with a fluorescent phosphate transporter in arbuscular mycorrhizal roots of rice. Plant and Cell Physiology 51, 341–353.

Kobae, Y., Ohmori, Y., Saito, C., Yano, K., Ohtomo, R., Fujiwara, T., 2016. Phosphate treatment strongly inhibits new arbuscule development but not the maintenance of arbuscule in mycorrhizal rice roots. Plant Physiology 00127 2016.

Kobae, Y., Tamura, Y., Takai, S., Banba, M., Hata, S., 2010. Localized expression of arbuscular mycorrhiza-inducible ammonium transporters in soybean. Plant and Cell Physiology 51, 1411–1415.

Koegel, S., Ait Lahmidi, N., Arnould, C., Chatagnier, O., Walder, F., Ineichen, K., et al., 2013. The family of ammonium transporters (AMT) in *Sorghum bicolor*: two AMT members are induced locally, but not systemically in roots colonized by arbuscular mycorrhizal fungi. The New Phytologist 198, 853–865.

Koprivova, A., Kopriva, S., 2016. Hormonal control of sulfate uptake and assimilation. Plant Molecular Biology 1–11.

Krajinski, F., Courty, P.-E., Sieh, D., Franken, P., Zhang, H., Bucher, M., et al., 2014. The H^+-ATPase HA1 of *Medicago truncatula* is essential for phosphate transport and plant growth during arbuscular mycorrhizal symbiosis. The Plant Cell Online 26, 1808–1817.

Kulcheski, F.R., Côrrea, R., Gomes, I.A., de Lima, J.C., Margis, R., 2015. NPK macronutrients and microRNA homeostasis. Frontiers in Plant Science 6, 451.

Kyung, K.H., Lee, Y.C., 2001. Antimicrobial activities of sulfur compounds derived from S-Alk (En) Yl-L-cysteine sulfoxides in Allium and Brassica. Food Reviews International 17, 183–198.

Lehmann, J., Solomon, D., Zhao, F.-J., McGrath, S.P., 2008. Atmospheric SO_2 emissions since the late 1800s change organic sulfur forms in humic substance extracts of soils. Environmental Science and Technology 42, 3550–3555.

Liu, J., Maldonado-Mendoza, I., Lopez-Meyer, M., Cheung, F., Town, C.D., Harrison, M.J., 2007. Arbuscular mycorrhizal symbiosis is accompanied by local and systemic alterations in gene expression and an increase in disease resistance in the shoots. The Plant Journal: For Cell and Molecular Biology 50, 529–544.

Liu, J., Versaw, W.K., Pumplin, N., Gomez, S.K., Blaylock, L.A., Harrison, M.J., 2008. Closely related members of the *Medicago truncatula* PHT1 phosphate transporter gene family encode phosphate transporters with distinct biochemical activities. The Journal of Biological Chemistry 283, 24673–24681.

López-Pedrosa, A., González-Guerrero, M., Valderas, A., Azcón-Aguilar, C., Ferrol, N., 2006. GintAMT1 encodes a functional high-affinity ammonium transporter that is expressed in the extraradical mycelium of *Glomus intraradices*. Fungal genetics and biology: FG and B 43, 102–110.

Loth-Pereda, V., Orsini, E., Courty, P.-E., Lota, F., Kohler, A., Diss, L., et al., 2011. Structure and expression profile of the phosphate Pht1 transporter gene family in mycorrhizal *Populus trichocarpa*. Plant Physiology 156, 2141–2154.

Marschner, H., 1995. Mineral Nutrition of Higher Plants. Academic Press, London.

McSwiney, C.P., Robertson, G.P., 2005. Nonlinear response of N2O flux to incremental fertilizer addition in a continuous maize (*Zea mays* L.) cropping system. Global Change Biology 11, 1712–1719.

Mensah, J.A., Koch, A.M., Antunes, P.M., Kiers, E.T., Hart, M., Bücking, H., 2015. High functional diversity within species of arbuscular mycorrhizal fungi is associated with differences in phosphate and nitrogen uptake and fungal phosphate metabolism. Mycorrhiza 25, 533–546.

Nagy, R., Karandashov, V., Chague, V., Kalinkevich, K., Tamasloukht, M., Xu, G., et al., 2005. The characterization of novel mycorrhiza-specific phosphate transporters from *Lycopersicon esculentum* and *Solanum tuberosum* uncovers functional redundancy in symbiotic phosphate transport in solanaceous species. The Plant Journal: For Cell and Molecular Biology 42, 236–250.

Nagy, R., Vasconcelos, M.J.V., Zhao, S., McElver, J., Bruce, W., Amrhein, N., et al., 2006. Differential regulation of five Pht1 phosphate transporters from maize (*Zea mays* L.). Plant Biology (Stuttgart, Germany) 8, 186–197.

Nawrath, C., Métraux, J.P., 1999. Salicylic acid induction-deficient mutants of Arabidopsis express PR-2 and PR-5 and accumulate high levels of camalexin after pathogen inoculation. The Plant Cell 11, 1393–1404.

Ngwene, B., Gabriel, E., George, E., 2013. Influence of different mineral nitrogen sources (NO3(-)-N vs. NH4(+)-N) on arbuscular mycorrhiza development and N transfer in a *Glomus intraradices*-cowpea symbiosis. Mycorrhiza 23, 107–117.

Nouri, E., Breuillin-Sessoms, F., Feller, U., Reinhardt, D., 2014. Phosphorus and nitrogen regulate arbuscular mycorrhizal symbiosis in *Petunia hybrida*. PloS One 9, e90841.

Nussaume, L., Kanno, S., Javot, H., Marin, E., Pochon, N., Ayadi, A., et al., 2011. Phosphate import in plants: focus on the PHT1 transporters. Frontiers in Plant Science 2, 83.

Paszkowski, U., Kroken, S., Roux, C., Briggs, S.P., 2002. Rice phosphate transporters include an evolutionarily divergent gene specifically activated in arbuscular mycorrhizal symbiosis. Proceedings of the National Academy of Sciences 99, 13324–13329.

Pérez-Tienda, J., Corrêa, A., Azcón-Aguilar, C., Ferrol, N., 2014. Transcriptional regulation of host NH_4^+ transporters and GS/GOGAT pathway in arbuscular mycorrhizal rice roots. Plant physiology and biochemistry: PPB/Société française de physiologie végétale 75, 1–8.

Pérez-Tienda, J., Testillano, P.S., Balestrini, R., Fiorilli, V., Azcón-Aguilar, C., Ferrol, N., 2011. GintAMT2, a new member of the ammonium transporter family in the arbuscular mycorrhizal fungus *Glomus intraradices*. Fungal genetics and biology: FG and B 48, 1044–1055.

Pérez-Tienda, J., Valderas, A., Camañes, G., García-Agustín, P., Ferrol, N., 2012. Kinetics of NH (4) (+) uptake by the arbuscular mycorrhizal fungus *Rhizophagus irregularis*. Mycorrhiza 22, 485–491.

Pozo, M.J., Azcón-Aguilar, C., 2007. Unraveling mycorrhiza-induced resistance. Current Opinion in Plant Biology 10, 393–398.

Pozo, M.J., Cordier, C., Dumas-Gaudot, E., Gianinazzi, S., Barea, J.M., Azcón-Aguilar, C., 2002. Localized versus systemic effect of arbuscular mycorrhizal fungi on defence responses to Phytophthora infection in tomato plants. Journal of Experimental Botany 53, 525–534.

Remans, T., Nacry, P., Pervent, M., Filleur, S., Diatloff, E., Mounier, E., et al., 2006. The Arabidopsis NRT1.1 transporter participates in the signaling pathway triggering root colonization of nitrate-rich patches. Proceedings of the National Academy of Sciences of the United States of America 103, 19206–19211.

Remy, W., Taylor, T.N., Hass, H., Kerp, H., 1994. Four hundred-million-year-old vesicular arbuscular mycorrhizae. Proceedings of the National Academy of Sciences of the United States of America 91, 11841–11843.

Reynolds, H.L., Hartley, A.E., Vogelsang, K.M., Bever, J.D., Schultz, P.A., 2005. Arbuscular mycorrhizal fungi do not enhance nitrogen acquisition and growth of old-field perennials under low nitrogen supply in glasshouse culture. The New Phytologist 167, 869–880.

Rhodes, L.H., Gerdemann, J.W., 1978. Influence of phosphorus nutrition on sulfur uptake by vesicular-arbuscular mycorrhizae of onion. Soil Biology and Biochemistry 10, 361–364.

Rogato, A., D'Apuzzo, E., Chiurazzi, M., 2010. The multiple plant response to high ammonium conditions: the *Lotus japonicus* AMT1; 3 protein acts as a putative transceptor. Plant Signaling and Behavior 5, 1594–1596.

Rogers, C., Oldroyd, G.E.D., 2014. Synthetic biology approaches to engineering the nitrogen symbiosis in cereals. Journal of Experimental Botany eru098.

Rubio, V., Linhares, F., Solano, R., Martín, A.C., Iglesias, J., Leyva, A., et al., 2001. A conserved MYB transcription factor involved in phosphate starvation signaling both in vascular plants and in unicellular algae. Genes and Development 15, 2122–2133.

Ruiz-Sánchez, M., Aroca, R., Muñoz, Y., Polón, R., Ruiz-Lozano, J.M., 2010. The arbuscular mycorrhizal symbiosis enhances the photosynthetic efficiency and the antioxidative response of rice plants subjected to drought stress. Journal of Plant Physiology 167, 862–869.

Saia, S., Benítez, E., García-Garrido, J.M., Settanni, L., Amato, G., Giambalvo, D., 2014. The effect of arbuscular mycorrhizal fungi on total plant nitrogen uptake and nitrogen recovery from soil organic material. The Journal of Agricultural Science 152, 370–378.

Salvioli, A., Ghignone, S., Novero, M., Navazio, L., Venice, F., Bagnaresi, P., et al., 2016. Symbiosis with an endobacterium increases the fitness of a mycorrhizal fungus, raising its bioenergetic potential. The ISME journal 10, 130–144.

Salvioli, A., Zouari, I., Chalot, M., Bonfante, P., 2012. The arbuscular mycorrhizal status has an impact on the transcriptome profile and amino acid composition of tomato fruit. BMC plant biology 12, 44.

Sawers, R.J.H., Svane, S.F., Quan, C., Gronlund, M., Wozniak, B., Nigussie, M., et al., 2016. Outcome variation in maize interaction with arbuscular mycorrhizal fungi is correlated with extent of extra-radical mycelium. bioRxiv 042028.

Sieh, D., Watanabe, M., Devers, E.A., Brueckner, F., Hoefgen, R., Krajinski, F., 2013. The arbuscular mycorrhizal symbiosis influences sulfur starvation responses of *Medicago truncatula*. New Phytologist 197, 606–616.

Smith, S.E., Jakobsen, I., Grønlund, M., Smith, F.A., 2011. Roles of arbuscular mycorrhizas in plant phosphorus nutrition: interactions between pathways of phosphorus uptake in arbuscular mycorrhizal roots have important implications for understanding and manipulating plant phosphorus acquisition. Plant Physiology 156, 1050–1057.

Smith, S.E., Smith, F.A., 2011. Roles of arbuscular mycorrhizas in plant nutrition and growth: new paradigms from cellular to ecosystem scales. Annual Review of Plant Biology 62, 227–250.

Smith, S.E., Smith, F.A., Jakobsen, I., 2004. Functional diversity in arbuscular mycorrhizal (AM) symbioses: the contribution of the mycorrhizal P uptake pathway is not correlated with mycorrhizal responses in growth or total P uptake. New Phytologist 162, 511–524.

Smith, F.A., Smith, S.E., Timonen, S., 2003. Mycorrhizas. In: de Kroon, P.D.H., Visser, D.E.J.W. (Eds.), Ecological Studies. Root Ecology. Springer, Berlin Heidelberg, pp. 257–295.

Song, Y., Chen, D., Lu, K., Sun, Z., Zeng, R., 2015. Enhanced tomato disease resistance primed by arbuscular mycorrhizal fungus. Frontiers in Plant Science 6, 786.

Staehelin, C., Xie, Z.-P., Illana, A., Vierheilig, H., 2011. Long-distance transport of signals during symbiosis: are nodule formation and mycorrhization autoregulated in a similar way? Plant Signaling and Behavior 6, 372–377.

Su, T., Xu, Q., Zhang, F.-C., Chen, Y., Li, L.-Q., Wu, W.-H., et al., 2015. WRKY42 modulates phosphate homeostasis through regulating phosphate translocation and acquisition in Arabidopsis. Plant Physiology 167, 1579–1591.

Takahashi, H., Kopriva, S., Giordano, M., Saito, K., Hell, R., 2011. Sulfur assimilation in photosynthetic organisms: molecular functions and regulations of transporters and assimilatory enzymes. Annual Review of Plant Biology 62, 157–184.

Takahashi, H., Watanabe-Takahashi, A., Smith, F.W., Blake-Kalff, M., Hawkesford, M.J., Saito, K., 2000. The roles of three functional sulphate transporters involved in uptake and translocation of sulphate in *Arabidopsis thaliana*. The Plant Journal: For Cell and Molecular Biology 23, 171–182.

Tanaka, Y., Yano, K., 2005. Nitrogen delivery to maize via mycorrhizal hyphae depends on the form of N supplied. Plant, Cell and Environment 28, 1247–1254.

Tian, C., Kasiborski, B., Koul, R., Lammers, P.J., Bücking, H., Shachar-Hill, Y., 2010. Regulation of the nitrogen transfer pathway in the arbuscular mycorrhizal symbiosis: gene characterization and the coordination of expression with nitrogen flux. Plant Physiology 153, 1175–1187.

Tisserant, E., Kohler, A., Dozolme-Seddas, P., Balestrini, R., Benabdellah, K., Colard, A., et al., 2012. The transcriptome of the arbuscular mycorrhizal fungus *Glomus intraradices* (DAOM 197198) reveals functional tradeoffs in an obligate symbiont. The New Phytologist 193, 755–769.

Tisserant, E., Malbreil, M., Kuo, A., Kohler, A., Symeonidi, A., Balestrini, R., et al., 2013. Genome of an arbuscular mycorrhizal fungus provides insight into the oldest plant symbiosis. Proceedings of the National Academy of Sciences 110, 20117–20122.

Toussaint, J.-P., St-Arnaud, M., Charest, C., 2004. Nitrogen transfer and assimilation between the arbuscular mycorrhizal fungus *Glomus intraradices* Schenck & Smith and Ri T-DNA roots of *Daucus carota* L. in an in vitro compartmented system. Canadian Journal of Microbiology 50, 251–260.

Vierheilig, H., Maier, W., Wyss, U., Samson, J., Strack, D., Piché, Y., 2000. Cyclohexenone derivative- and phosphate-levels in split-root systems and their role in the systemic suppression of mycorrhization in precolonized barley plants. Journal of Plant Physiology 157, 593–599.

Volpe, V., Giovannetti, M., Sun, X.-G., Fiorilli, V., Bonfante, P., 2016. The phosphate transporters LjPT4 and MtPT4 mediate early root responses to phosphate status in non mycorrhizal roots. Plant, Cell and Environment 39, 660–671.

Walder, F., Brulé, D., Koegel, S., Wiemken, A., Boller, T., Courty, P.-E., 2015. Plant phosphorus acquisition in a common mycorrhizal network: regulation of phosphate transporter genes of the Pht1 family in sorghum and flax. The New Phytologist 205, 1632–1645.

Wang, H., Xu, Q., Kong, Y.-H., Chen, Y., Duan, J.-Y., Wu, W.-H., et al., 2014a. Arabidopsis WRKY45 transcription factor activates PHOSPHATE TRANSPORTER1;1 expression in response to phosphate starvation. Plant Physiology 164, 2020–2029.

Wang, E., Yu, N., Bano, S.A., Liu, C., Miller, A.J., Cousins, D., et al., 2014b. A H^+-ATPase that Energizes nutrient uptake during mycorrhizal symbioses in rice and *Medicago truncatula*. The Plant Cell 26, 1818–1830.

Watts-Williams, S.J., Jakobsen, I., Cavagnaro, T.R., Grønlund, M., 2015. Local and distal effects of arbuscular mycorrhizal colonization on direct pathway Pi uptake and root growth in *Medicago truncatula*. Journal of Experimental Botany erv202.

Wipf, D., Mongelard, G., van Tuinen, D., Gutierrez, L., Casieri, L., 2014. Transcriptional responses of *Medicago truncatula* upon sulfur deficiency stress and arbuscular mycorrhizal symbiosis. Frontiers in Plant Science 5.

Xie, X., Huang, W., Liu, F., Tang, N., Liu, Y., Lin, H., et al., 2013. Functional analysis of the novel mycorrhiza-specific phosphate transporter AsPT1 and PHT1 family from *Astragalus sinicus* during the arbuscular mycorrhizal symbiosis. The New Phytologist 198, 836–852.

Xu, G.-H., Chague, V., Melamed-Bessudo, C., Kapulnik, Y., Jain, A., Raghothama, K.G., et al., 2007. Functional characterization of LePT4: a phosphate transporter in tomato with mycorrhiza-enhanced expression. Journal of Experimental Botany 58, 2491–2501.

Yang, S.-Y., Grønlund, M., Jakobsen, I., Grotemeyer, M.S., Rentsch, D., Miyao, A., et al., 2012. Nonredundant regulation of rice arbuscular mycorrhizal symbiosis by two members of the phosphate transporter1 gene family. The Plant Cell 24, 4236–4251.

Zhang, B., Pasini, R., Dan, H., Joshi, N., Zhao, Y., Leustek, T., et al., 2014. Aberrant gene expression in the Arabidopsis SULTR1;2 mutants suggests a possible regulatory role for this sulfate transporter in response to sulfur nutrient status. The Plant Journal: For Cell and Molecular Biology 77, 185–197.

Zouari, I., Salvioli, A., Chialva, M., Novero, M., Miozzi, L., Tenore, G.C., et al., 2014. From root to fruit: RNA-Seq analysis shows that arbuscular mycorrhizal symbiosis may affect tomato fruit metabolism. BMC Genomics 15, 221.

CHAPTER

8

Accessibility of Inorganic and Organic Nutrients for Mycorrhizas

A. Hodge
University of York, York, United Kingdom

8.1 INTRODUCTION

Nitrogen (N) and phosphorus (P) are essential for plant growth, but in many ecosystems one or both of these key nutrients are limiting. The controls that determine the availability of N and P for the plant differ. Microorganisms play a central role in transformation and accessibility of N and can be thought of as the gatekeepers of the N cycle. In contrast, the supply of P tends to be dominated by inorganic equilibria. As mycorrhizal associations are known to benefit the nutrition of the host plant, a role for the fungal symbiont in enhancing N and P uptake may be expected. In an influential article, Read (1991) challenged the then conventional view that mycorrhizal symbiosis was primary involved in the acquisition of phosphate. Rather, he proposed that because certain mycorrhizal associations are dominant in ecosystems and soil environments with distinctive conditions, their associated fungal symbiont may make contrasting contributions to host plant nutrition (Read, 1991). This proposal resulted in a change of focus away from P and toward N for the ectomycorrhizal (EcM) association, whereas P remained the major nutritional focus of the arbuscular mycorrhizal (AM) association. Although N and P remain the main limiting nutrients in most (but not all) environments dominated by EcM or AM associations respectively, it has also become increasingly realized that both these mycorrhizal associations may aid acquisition of N and P and that the relationship between the two may also be important.

8.1.1 Nitrogen Availability

Plant roots capture N largely in inorganic form relying on microbes to release ammonium (NH_4^+) during the mineralization of organic matter. NH_4^+ can subsequently be oxidized to highly mobile nitrate (NO_3^-) via nitrification, the first step of which is driven by autotrophic bacteria and archaea. Plant roots and microorganisms must compete for these inorganic N

Mycorrhizal Mediation of Soil
http://dx.doi.org/10.1016/B978-0-12-804312-7.00008-5

sources (Hodge et al., 2000). In most natural systems, however, mineralization rates tend to be low and N occurs largely as organic forms. Some plants can take up simple soluble organic N compounds including amino acids and even small peptides, but because microorganisms can also access these N sources, competition also occurs (Hodge et al., 2000; Farrell et al., 2013). Moreover, the ecological importance of this N uptake pathway is uncertain (Jones et al., 2005; Farrell et al., 2013). Other plants rely on associations with ericoid (ErM) or EcM fungi to acquire N directly from organic sources, although these associations are limited to only about 1% (ErM) and 5% (EcM) of plant species. However, as the EcM association forms with woody plants, it is of economic as well as ecological importance. AM fungi, which form mycorrhizal associations with around two-thirds of all land species, were previously thought not to play any role in N acquisition for their host plant, but as discussed in this chapter, this view now has to be revaluated.

8.1.2 Phosphorus Availability

As for N, P in soil can be in inorganic or organic form. Whereas the total P content of a soil may be relatively high, its bioavailability is low with concentrations in the soil solution (from which the plant acquires P) only about 0.1–10 μM (Schachtman et al., 1998; Hinsinger, 2001). Phosphate ions react with charged surfaces such as organic matter and the principal cations in most soils [e.g., aluminum (Al^{3+}), iron (Fe^{3+}), and calcium (Ca^{2+})], and it is this adsorption that leads to sequestration of P in labile and non-labile pools (Fig. 8.1). Thus, most P is held in the solid phase in a wide range of inorganic and organic forms of varying (and not easily defined) chemical complexity. This solid phase can be thought of as comprising non-labile and labile phases. Labile P can be considered as more weakly held or "adsorbed" so is more likely to come into the solution phase and is in equilibrium with it. In contrast, non-labile P, which is usually the larger phase of the two, can be thought of as much less likely to come into solution. Over time, however, some non-labile P will react to become labile P. Thus there is a continuous equilibrium among the different phases; they are not separate per se, even though it may be convenient to consider them in this way.

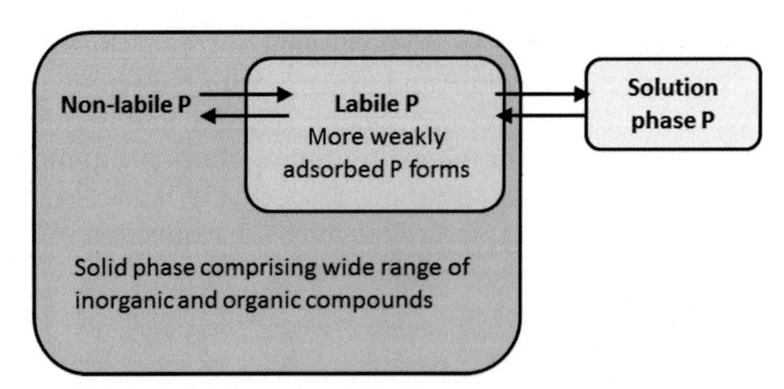

FIGURE 8.1 Diagram showing the solid and solution phases of phosphorus (P) in soil. The solid phases can be further thought of as comprising labile and non-labile phases. Phosphate forms in the labile soil phase tend to be less strongly held or "adsorbed"; thus they are more likely to come into the solution phase, and are in equilibrium with it. Phosphate forms in the non-labile phase are much less likely to come into solution. However, over time some of these P forms will come into the labile phase (i.e., when P is withdrawn from the solution phase, causing P to move from labile to solution) and these two phases are also in equilibrium.

8.2 MOVEMENT OF PHOSPHATE AND NITRATE IONS TO ROOTS

Roots predominantly obtain their nutrients from the soil solution and ions move to the root via mass flow or diffusion. Mass flow is driven by the transpiration pull of the plant, resulting in a water-potential gradient. Water that moves to the plant to satisfy this transpiration demand will also contain soluble ions and these will be available for root acquisition. Mass flow is generally a fast process and the amount reaching roots by this pathway (i.e., the product of water flux and the concentration of ions in the bulk soil solution) may be sufficient to meet plant demand. If this is not the case, then the flux across the root cell membranes will exceed that through soil to the root surface. A concentration gradient will then develop from the root surface out to the bulk soil, leading to a depletion zone (i.e., the volume of soil the root effectively can acquire the ion from) around the root (Fig. 8.2). Ions will move down this concentration gradient to the root via diffusion, a slower ion movement pathway.

NO_3^- barely interacts with charged surfaces in soil and is universally soluble in water. Thus, NO_3^- can move via mass flow or, if present at low concentrations such that the mass flux fails to meet plant demand, by diffusion. If NO_3^- is supplied by diffusion, then the resulting depletion zone tends to be both large and shallow, meaning NO_3^- can still be effectively acquired. Consequently, the issue for NO_3^- acquisition is not movement to the root per se, but amounts present in the soil. In contrast, the amounts of phosphate delivered to the plant root via mass flow are negligible because the soil solution concentration is so low, and instead the root has to predominantly rely on diffusion. The reaction of phosphate ions with charged surfaces and cations (Section 8.1.2) reduces availability in the soil solution and mobility through soil is severely restricted. Therefore depletion zones that develop as a result of diffusion of phosphate are narrow and sharply defined; they are extremely "depleted" in phosphate (Fig. 8.2). Trying to acquire sufficient phosphate is therefore a real problem for most plants.

8.3 INORGANIC PHOSPHORUS AND NITROGEN ACQUISITION BY ARBUSCULAR MYCORRHIZAL FUNGI

The AM association plays a key role in the acquisition of P for its associated host plant, a function that the AM fungal symbiont has most likely performed for 400 million years, given the contemporaneous origin of the symbiosis with the first land plants (Remy et al., 1994). The AM fungal extraradical mycelium (ERM) can explore a larger volume of soil, thus acquiring poorly mobile phosphate ions (mainly in the form of orthophosphate, $H_2PO_4^-$) beyond the P depletion zone (Sanders and Tinker, 1973; Fig. 8.2). Given the fine diameter of AM fungal hyphae (typically ≈2–20 μm), this permits P acquisition at a much reduced cost compared with new root construction (Harley, 1989). However, when inorganic P does arrive at the AM hyphal (or indeed the root) surface there are considerable issues to overcome to enable uptake. Concentrations of inorganic P in the soil are much lower (≈1000-fold) than cellular concentrations and the inside of the membrane has a more negative charge than the outside. Therefore negatively charged inorganic phosphate ions have to be taken up against this steep electrochemical gradient. Hydrogen (H_4^+)-adenosine triphosphatases (ATPases) on the plasma membrane generate the proton-motive force to enable H^+–phosphate co-transport by high-affinity phosphate transporters in the ERM (Harrison and van Buuren, 1995; Lei et al., 1991; Smith and

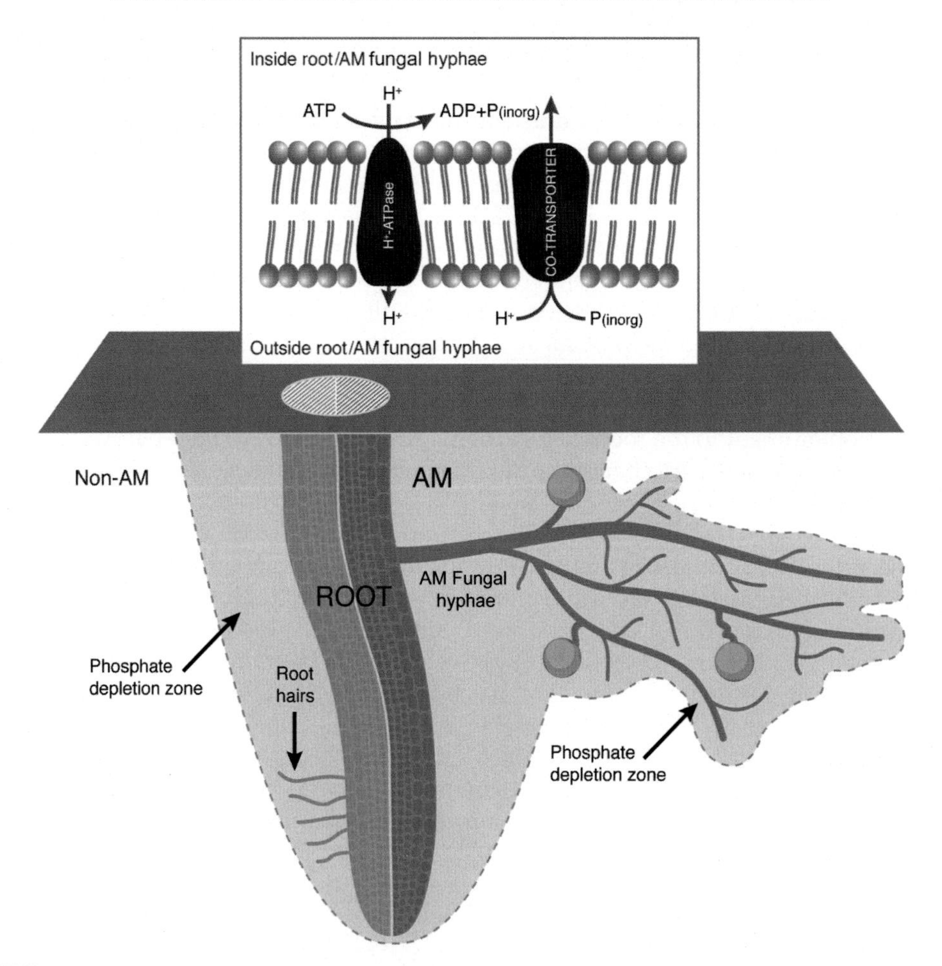

FIGURE 8.2 Schematic representation of the phosphate depletion zones that develop around a non-mycorrhizal root with root hairs (left) and a root colonized by arbuscular mycorrhizal (AM) fungi (right), where the depletion zone is larger as a result of the AM fungal hyphae exploring a larger soil volume. The box shows the uptake of phosphate across the plasma membrane and is common to both AM fungal hyphae and roots. Hydrogen protons (H_4^+) are pumped out the cell by H_4^+ adenosine triphosphatases (ATPases) on the plasma membrane to generate the proton-motive force to enable H^+–phosphate co-transport by high affinity phosphate transporters at the root surface or extraradical mycelium (ERM) of the AM fungal hyphae. *ADP*, Adenosine diphosphate; *ATP*, adenosine triphosphate; *CO*, carbon monoxide, $P_{(inorg)}$, inorganic phosphorus.

Smith, 2011; Fig. 8.2). Phosphate is then likely transported within the AM fungus as long chains of polyphosphate (polyP), which once in the intraradical hyphae (i.e., AM fungal hyphae inside the root) are hydrolyzed, thereby facilitating subsequent transfer to the host plant (Ohtomo and Saito, 2005). Transfer from fungus to plant occurs principally at the arbuscule interface, but likely also at the site of hyphal coils as demonstrated by the expression of P transporters at these sites (Karandashov et al., 2004). It is now widely accepted that acquisition of P via the AM symbiotic pathway down-regulates direct P uptake by the plant, although the controls on the regulation between these two pathways are yet to be discovered (Smith and Smith, 2011).

So what about inorganic N acquisition by AM fungi? Because the AM association tends to dominate in ecosystems where inorganic N sources (including highly mobile NO_3^- ions; see Section 8.2) represent both a significant and dynamic N pool, Read (1991) proposed that the host plant would be able to acquire sufficient N without AM fungal assistance. Yet there was some experimental evidence (Ames et al., 1983) suggesting AM fungi may play a role in N transfer, including from inorganic N (in this case, as NH_4^+) sources. Subsequent studies on an AM fungal N transfer role confirm that inorganic N can be acquired and transferred to their associated host, but values vary considerably, from a few percentage points to around 74% shoot N (Hodge and Storer, 2015). In general, providing plants are not water stressed; thus ion mobility is not restricted. AM fungi appear to contribute very low amounts of N to their host plant's nutrition when NO_3^- is the dominant N source applied, but higher amounts when the less mobile NH_4^+ is added (Hodge and Storer, 2015).

Although NH_4^+ is more mobile than phosphate, it can be held on negatively charged clay and organic matter particles, restricting availability, whereas NO_3^- ions are much more mobile (Section 8.2). However, similar to phosphate ions, NO_3^- uptake must occur against a steep electrochemical gradient. This is followed by subsequent reduction to NH_4^+. Thus for AM fungi (and indeed plants), uptake of NH_4^+ is energetically more attractive. Studies using root-organ culture, however, have shown that both NH_4^+ and NO_3^- can be taken up by AM fungal hyphae (Govindarajulu et al., 2005; Johansen et al., 1996), and a transporter gene (*GintAMT1*) in the ERM of *Rhizophagus irregularis*, which showed high sequence similarity with other fungal NH_4^+ transporters, has been characterized (López-Pedrosa et al., 2006). After uptake (and conversion to NH_4^+ in the case of NO_3^-), the N is likely translocated in the form of arginine from the ERM to the intraradical mycelium (IRM), where it is subsequently broken down to urea then to NH_4^+, the likely form in which the N is transferred to the host plant (Govindarajulu et al., 2005; Hodge and Storer, 2015). Arginine carries a net positive charge, leading to suggestions that it may bind to the negatively charged polyP, allowing N and P translocation with the AM fungal hyphae to be coupled (Govindarajulu et al., 2005). However, there are still considerable difficulties in the proposed AM-fungal N translocation pathway and subsequent transfer to the host plant that need to be resolved. These include how electrochemical charge balance is achieved after arginine formation, translocation and subsequent breakdown in the IRM and at the IRM-root interface after N transfer, whether arginine is the only form in which N is translocated and the total carbon (C) costs associated with the AM-fungal–N translocation pathway (Smith and Smith, 2011; Hodge and Storer, 2015). Moreover, whether N acquisition via the AM symbiotic pathway has an effect upon the direct (plant) uptake pathway, similar to that demonstrated for phosphate uptake, is also unknown (Hodge and Storer, 2015).

8.4 INORGANIC PHOSPHORUS AND NITROGEN ACQUISITION BY ECTOMYCORRHIZAL FUNGI

In EcM associations, a fungal sheath or mantle can often enclose the colonized root surface. A role for EcM fungi in P acquisition for their host may therefore be anticipated. Early work, largely using excised roots (Section 8.6.3), supported this view and also demonstrated accumulation of P in the fungal mantle, suggesting fungal control on the amounts transferred (Smith and Read, 2008). Similar to AM associations, under high P levels, EcM formation can be reduced (Bechem and Alexander, 2012) and EcM fungi also acquire inorganic P from outside the P depletion zone around the EcM root surface, which is then translocated in polyP form (Cairney, 2011). EcM fungi are, however, much more diverse than AM fungi and it is well established that they can exhibit different mycelial foraging strategies, including the production of rhizomorphs (linear aggregates of hyphae), which will likely impact P acquisition. A number of genes encoding P transporters have been detected in EcM fungi, most of which belong to the *Pht1* subfamily (proton– phosphate co-transport), but some cluster with sodium–phosphate transporters, suggesting P uptake may be influenced by external pH (Casieri et al., 2013). However, only a few transporters have thus far been characterized using yeast complementation (Tatry et al., 2009; Wang et al., 2014). Acidifying the hyphosphere soil by EcM fungal hyphae can also increase P availability through dissolution of insoluble inorganic P forms such as $AlPO_4$ (Cumming and Weinstein, 1990) in the same way that roots can (Hinsinger, 2001).

Also analogous to plant roots, many EcM fungi release low–molecular-weight organic acids, including the divalent and trivalent organic anions malate, citrate, and oxalate (Wallander, 2000; Zhang et al., 2014). However, with some notable exceptions (e.g., citrate release from cluster root species), organic acid release from most plant roots may be insufficient to have a sizable impact upon P availability and instead, may be caused by membranes becoming more leaky under conditions of P deficiency (Jones, 1998). In contrast, oxalate is an important metabolite of many fungi and a number of different functional roles have been implicated, including acting in metal and mineral transformations, acting as an electron donor in lignocellulose degradation, and increasing plant susceptibility to fungal infection (Gadd et al., 2014). Oxalate is also commonly reported as the dominant organic acid released by a number of EcM fungal species (Courty et al., 2010; Zhang et al., 2014), although not all EcM fungi share this capacity (Casarin et al., 2004). However, it appears likely that the secretion of organic acids by EcM fungi contribute to increasing P availability from insoluble P sources (Courty et al., 2010). Wallander (2000) reported a correlation between the concentration of oxalic acid and P in the soil solution when EcM fungal hyphae had access to a root-free compartment containing apatite as a P source. Moreover, Landeweert et al. (2001) suggested enhanced weathering of minerals caused by EcM organic acid release allowed the EcM fungi to access apatite inclusions, in addition to mobilizing other nutrients such as potassium, Ca, and magnesium. The production of oxalic acid by the EcM fungus *Paxillus involutus* was found to be influenced by the N source present; mycelia supplied with NO_3^- produced large quantities, whereas very little was produced when NH_4^+ was the N source (Lapeyrie et al., 1987).

Studies investigating the capability of EcM fungi to use NH_4^+ and/or NO_3^- have shown that the energetically attractive NH_4^+, which is also the dominate inorganic N form in most forest soils (Read, 1991), is generally preferred over NO_3^- under both in vitro (France and Reid, 1984; Finlay et al., 1992; Anderson et al., 1999) and field conditions (Clemmensen et al., 2008; Kranabetter et al., 2015). However, for some EcM fungi, a preference for NO_3^- over NH_4^+ has been reported (Scheromm et al., 1990) and considerable variation among EcM fungal species (France and Reid, 1984; Finlay et al., 1992; Nygren et al., 2008) and even among strains of the same species (Finlay et al., 1992; Anderson et al., 1999) in their ability to use NO_3^- has been found mainly from investigations carried out under in vitro conditions. Such interspecific and intraspecific variation in nutrient accessibility is a common theme in EcM fungal research, although the reported intraspecific variation does highlight the need for caution when conclusions are being drawn if multiple strains of the same species are not also considered. Moreover, data obtained from in vitro experiments may not necessary reflect actual N acquisition pathways operating under field conditions.

The acquisition and subsequent assimilation of NH_4^+ by EcM associations has been reasonably well characterized, although progress on NO_3^- assimilation has also been made (Smith and Read, 2008). A number of ammonium transporters in the external mycelium of EcM fungi and other EcM structures (e.g., the Hartig net and the fungal sheath/mantle) have been characterized; the expression of which can be differentially affected by both N form and availability, whereas nitrate reductase activity and NO_3^- assimilation has generally been found to be reduced by the presence of NH_4^+ (Smith and Read, 2008; Courty et al., 2015). For some EcM fungi the ability to acquire inorganic N forms may be more important than others if they are unable to access other, more complex N forms. The identification of certain EcM fungi that expressed high proteolytic activity (the so-called "protein" fungi), which had the ability to acquire N from complex organic form, coupled with the demonstration that their associated host plant achieved yields similar to those obtained with NH_4^+ (Abuzinadah and Read, 1986), suggested that these EcM fungi were less reliant on N mineralization processes. In contrast, other EcM fungi, such as *Pisolithus tinctorius* and *Laccaria laccata*, were found not to express such high proteolytic activities (the so-called "non-protein" fungi), implying inorganic N pools were an important source of N for these fungi (Read, 1991). The genome sequence of *Laccaria bicolor* revealed that this fungus has numerous genes encoding glycoside hydrolases and proteinases, which Martin et al. (2008) suggested a capacity to degrade proteins from decomposing leaf litter as well as from soil animal origin. The genome sequence also revealed multiple ammonium transporters (although only a single nitrate permease), which indicated a greater potential for NH_4^+ uptake compared with other basidiomycete fungi (Martin et al., 2008). EcM fungi as a group are evolutionary diverse and functionally complex; they occur in N-limited tundra and boreal ecosystems with slowed nutrient cycling to more N-rich temperate forests (Read, 1991). The diversity in N acquisition pathways, which reflects the form of resource in their environment, is perhaps not entirely unexpected.

8.5 ARBUSCULAR MYCORRHIZAL FUNGI AND ORGANIC NUTRIENT FORMS

Early observations that AM fungi had a tendency to proliferate hyphae in zones or "patches" of organic matter initially led to the suggestion that these fungi had saprotroph capabilities (Warner and Mosse, 1980). Although the latter has not been substantiated, the ability of AM fungi to proliferate, sometimes spectacularly so, in organic matter patches has (Hodge and Fitter, 2010). However, this may depend upon the type of organic substrate added. For example, the addition of cellulose often fails to evoke a proliferation response, whereas that of plant shoot material does (Hodge, 2014). The lack of response to cellulose might be expected given the apparent lack of saprotrophic lifestyle, but that AM fungi can proliferate hyphae in organic materials, which often promotes hyphal growth elsewhere (Hodge et al., 2001; Leigh et al., 2009), intuitively suggests a nutritional benefit to the fungus. The question therefore is, what is this benefit?

Enhanced P acquisition seemed an obvious candidate. Despite some evidence that AM fungi may be able to hydrolyze organic P forms under gnotobiotic conditions (Joner et al., 2000a), and that AM fungi may either directly release or influence root phosphatase activities (Joner et al., 2000b), there is little evidence that AM fungi access P sources that the nonmycorrhizal plant cannot (Hayman and Mosse, 1972; Joner et al., 2000b). However, AM fungi may benefit through their spatial placement in the soil and through acquiring inorganic P released via the mineralization of organic P forms by other microorganisms (Joner et al., 2000b; Zhang et al., 2016). Given their spatial proximity to sites of decomposition, it might similarly be expected that AM fungi would benefit in terms of N acquisition. The decomposition of organic materials can result in simple organic (such as amino acids) and inorganic N forms being released both spatially and temporally through the decomposing patch. Location at the sites of such N release therefore becomes important, as demonstrated for plant root proliferation responses (Hodge, 2004). Although a few reports of AM fungi acquiring amino acids directly exist, this seems surprisingly uncommon in comparison to both other fungal species and their EcM fungal counterparts, and N is most likely acquired in inorganic N form (Barrett et al., 2014; Leigh et al., 2011; Hodge and Fitter, 2010). However, even though in some cases the amounts of N transferred have been large, the contribution to plant N status has generally been low (typically $<7\%$; but see Leigh et al., 2009), leading to questions about the ecological significance of the AM-fungal N acquisition pathway (Smith and Smith, 2011).

AM fungal hyphae are N rich, and proliferation of hyphae in organic matter patches was shown to be an important N acquisition pathway for these fungi (Hodge and Fitter, 2010). It is therefore likely that AM fungi, similar to plants, have a large N requirement relative to P, and thus are perhaps less willing to transfer N than P to their host. This is supported by the findings of Johnson and colleagues using AM fungal communities from different prairie soils of contrasting N or P limitation. Whereas AM fungi could eliminate P limitation for the plant, they did not ameliorate N limitation (Johnson et al., 2015), even though the AM fungi increased N uptake from an N-limited soil (Johnson et al., 2010). As the AM fungal biomass would still acquire C for this N, it was not so adversely affected (Johnson et al., 2010). In contrast, the addition of inorganic N fertilizer (Johnson et al., 2015) or an N-rich organic patch (Thirkell et al., 2016) to a P-limited system results in both AM fungi and host plants showing a mutual benefit. Strikingly, Hodge and Fitter (2010) found shading plants, thus reducing C supply, did not affect the ability of the AM fungi to proliferate hyphae in an organic patch or acquire N from it, whereas Johnson et al. (2015) found shading did not influence ERM lengths when fertilized with N alone, but P and N fertilization resulted in reduced ERM. Thus, the impact of AM fungi on N transfer to their host likely depends on the stoichiometry of N and P in the system (including both that of the AM fungi and the plant), as well as potentially other nutrients that may be colimiting with N such as zinc (Corrêa et al., 2014).

A rather unexpected finding of the impact of AM fungi on organic nutrient sources has been their apparent ability to increase organic matter decomposition (Hodge et al., 2001; Cheng et al., 2012). There could be several mechanisms by which AM fungi influence organic matter decomposition (Fig. 8.3) (Hodge, 2014), including through hyphal exudation and turnover, export of inorganic N from the sites of decomposition, and physical and chemical modifications within the organic matter substrate.

Enhanced decomposition is not found in all studies. The addition of a simple (Hodge, 2001) or more complex (Verbruggen et al., 2016) organic substrate resulted in an increase

FIGURE 8.3 Proposed main pathways by which arbuscular mycorrhizal (AM) fungi may enhance decomposition of organic matter (OM) in soil. The image in the middle shows an AM fungus proliferating hyphae around a decomposing patch of dead root material. *C*, carbon; H^+, hydrogen; HCO_3^-, bicarbonate; *N*, nitrogen; NH_4^+, ammonium; NO_3^-, nitrate; OH^-, hydroxide. *Adapted from Hodge, A., 2014. Interactions between arbuscular mycorrhizal fungi and organic material substrates. Advances in Applied Microbiology 89, 47–99 and Nuccio, E.E., Hodge, A., Pett-Ridge, J., Herman, D.J., Weber, P.K., Firestone, M.K., 2013. An arbuscular mycorrhizal significantly modifies the soil bacterial community and nitrogen cycling during litter decomposition. Environmental Microbiology 15, 1870–1881. Image reproduced by permission of John Wiley and Sons.*

in C retention (but no impact on N) when AM fungi were present. Given the lack of evidence suggesting any substantial extracellular enzyme release by AM fungi, their impact on organic decomposition is most likely mediated through their influence on the rest of the microbial community via nutrient competition or altered C dynamics. Results from studies on AM fungal presence on microbial community composition are not consistent, with no discernible effect (Herman et al., 2012), a decrease in fungal decomposers (Verbruggen et al., 2016), or, depending on the bacterial taxa, both an increase and decrease in bacterial communities (Nuccio et al., 2013). Although differences in how microbial communities are assessed or experimental systems employed explain some of these differing results, the underlying nutrient dynamics in the system and their interaction no doubt also play a role as discussed.

8.6 ECTOMYCORRHIZAL FUNGI AND ORGANIC NUTRIENT FORMS

In the EcM, similar to the AM, association the fungus acquires C from its host plant in return for nutrients. Although some EcM fungi are capable of being grown in pure culture, there is no evidence that these fungi can complete their life cycle in the absence of a host, although the true extent of fungi capable of forming EcM associations is likely still to be realized. The fungi involved in the EcM symbiosis have evolved multiple times from saprotrophic ancestors and multiple lineages of fungi (although just how many times remains disputed) (Bruns and Shefferson, 2004; Tedersoo et al., 2010). Their ancestral origins have consequently given rise to much debate as to what extent, if any, a saprotrophic legacy remains in the EcM fungi, mirroring the earlier debate held on AM fungi (Section 8.5). Previous suggestions that some EcM fungi may have even reverted back from the symbiotic to the free-living saprotrophic condition owing to retention of genes necessary for plant cell-wall degradation (Hibbett et al., 2000), however, now seem unlikely (Wolfe et al., 2012) and the data from which this was based has been subsequently demonstrated to be subject to sampling bias and model dependency (Bruns and Shefferson, 2004).

EcM fungi have been shown to access nutrients from complex, yet model, organic forms including proteins (Abuzinadah and Read, 1986), chitin (Hodge et al., 1995), and inositol hexaphosphate or "phytate" (Dighton, 1983) under controlled conditions in the laboratory or greenhouse. However, compared with saprotrophic fungi and also ErM fungi, the capability of EcM fungi to decompose more complex compounds such as litter is generally much reduced (Colpaert and van Laere, 1996; Wu et al., 2005). This has largely been attributed to the relative inability of EcM fungi to depolymerize complex C sources, possibly as a result of catabolite repression of cellulase production caused by C flux (in the form of glucose) from their associated host plant or, more simply, because of reduced cellulolytic and ligninolytic activities in comparison with their saprotrophic or ErM counterparts (Read et al., 2004). Arguably retaining a substantial capacity to degrade plant structural compounds such as lignin and cellulose may not be a particularly desirable trait for a root-inhabiting symbiotic organism, because levels would have to be closely controlled within the living root environment. Equally, however, retaining some enzymatic activity for plant cell-wall degradation, or at least modification, from their saprotrophic ancestors may be beneficial in facilitating colonization of the root. The organic horizons that many EcM fungi occupy are rich in phenolic compounds derived from plant litter that is complex with nutrient sources including organic N forms, and may severely restrict the accessibility of these nutrients (Handley, 1961). Phenolics can be broadly characterized into lignin and soluble phenolics including polyphenols. Some simple phenolics have been suggested to enhance soil organic matter decomposition rates by acting as a C source for microbes (Inderjit and Mallik, 1997), but this may subsequently reduce net soil N availability through microbial immobilization of N (Kanerva et al., 2006), whereas more complex phenolic forms are often reported to retard litter decomposition (Palm and Sanchez, 1991). Hence having some capacity to "unlock" complex nutrient forms would be beneficial to enable nutrient acquisition by EcM fungal hyphae.

Decoding the genome sequences of EcM fungi has shed some light on this capability, but has also renewed debate upon the saprotrophic potential of EcM fungi. Definitions of 'saprotroph' vary considerably, but underpinning all is the notion that the organism is utilizing dead or decaying organic matter to support growth. Thus by implication (although not always explicitly stated) organic matter serves as the principal source of C for metabolic function. Earlier suggestions of saprotrophic behavior by EcM fungi were based on measured enzyme activities expressed during periods when host C supply was low (Courty et al., 2007; Cullings et al., 2008), resulting in the suggestion that C could even flow from the EcM fungi to the host tree during periods when host C demand was high but photosynthetic C supply was low (i.e., during bud break) (Courty et al., 2007). However, whether these enzyme activities were measured and interpreted correctly remains disputed (Baldrian, 2009). Other suggestions include viewing EcM fungi along a biotroph–saprotroph continuum rather than applying an "either/or" distinction (Koide et al., 2008), or that EcM fungi may be facultative saprotrophs. Although the latter suggestion is not entirely new, it has been developed to encompass the notion that EcM fungi may have a dual lifestyle, acting as symbionts within the plant roots and at the same time as "transitory" or facultative saprotrophs in soil (Martin and Nehls, 2009; Bonfante and Genre, 2010). In contrast, others have questioned the idea that EcM fungi have any capacity to function as saprotrophs, facultative or otherwise (Baldrian, 2009; Lindahl and Tunlid, 2015), highlighting instead a role for EcM fungi in decomposition processes. So what does the evidence suggest?

8.6.1 Evidence From Genome Analysis

Analysis of genome sequences from *L. bicolor* (Martin et al., 2008) and *Tuber melanosporum* (Martin et al., 2010), the first two EcM fungal genomes sequenced, revealed a lack of ligninolytic capability and a paucity in genes required for cellulose degradation. Similarly, Wolfe et al. (2012) found that *Amanita* species capable of forming EcM associations had lost two key genes in the cellulose degradation pathway and with it, the capacity to grow on media containing sterile litter as the sole C source. Meanwhile Treseder et al. (2006) found that EcM fungi failed to assimilate C from ^{14}C decomposing labeled litter added in a temperate deciduous forest. Collectively, this work may suggest that EcM fungi have lost their ability to decompose plant litter to any extent, perhaps surprisingly so, given their spatial distribution among litter horizons, and argues against saprotroph capability. Yet both *Cortinarius glaucopus* (Bödeker et al., 2014) and *Hebeloma cylindrosporum* (Kohler et al., 2015) retain a number of class II peroxidase genes encoding manganese (Mn) peroxidases, which oxidize Mn^{2+} to Mn^{3+}. Class II peroxidases are used by the so-called "white-rot" fungi (saprotrophs renowned for their ability to degrade lignin) to depolymerize the lignin matrix and reveal the more tractable cellulose C embedded within (Eastwood et al., 2011). As Mn^{3+} can be unstable in aqueous media, white-rot fungi release organic acids, notably oxalic acid, which chelates with the Mn^{3+}. The resulting complex then interacts with a wide range of compounds (including phenolics), resulting in structure disruption and subsequent degradation (Martínez, 2002). Many EcM fungi including *C. glaucopus* (Courty et al., 2010), but interestingly not *H. cylindrosporum* (Casarin et al., 2004), are known to release oxalic acid (Section 8.4) so in *C. glaucopus*, at least it may serve this additional function.

Not all saprotrophic fungi can degrade lignin; the saprotrophic brown-rot fungi cannot, yet are still able to breakdown cellulose and hemicellulose structures in wood to acquire C. Fenton chemistry, where reactive oxygen species (such as hydroxyl radicals) produced non-enzymatically during the oxidation of Fe^{2+} by hydrogen peroxide initially attack the cellulose structure, is usually believed to occur, although other possibilities, including a role for fungal-produced oxalic acid, may also operate (Green and Highley, 1997). Rineau et al. (2012) concluded that a similar hydroxyl radical attack mechanism was employed by the EcM fungus *P. involutus* to degrade organic matter. Whereas a similar set of enzymes involved in the oxidative degradation of wood by brown rot fungi were expressed by *P. involutus*, importantly, the transcripts encoding the extracellular enzymes for metabolizing the released C were absent. Furthermore, in *P. involutus*, unlike saprotrophic fungi, the addition of external glucose up-regulated gene expression encoding oxidative degradation, N transporters, and enzymes in the N and C metabolism pathways (Rineau et al., 2013).

The evidence in the previous paragraph demonstrates that as a group, EcM fungi are very diverse in their ability to decompose litter sources, which could aid in N (and P) acquisition. However, evidence that the EcM fungi utilize C sources from the decomposition process to any great extent is still lacking. Why then do some EcM fungi retain expensive enzymatic machinery to degrade complex compounds that could potentially result in C sources from organic material being released, and how does this compare with other mycorrhizal associations? In the orchid mycorrhizal association, C does move from the fungus to the host plant, at least in the juvenile achlorophyllous stage. However, in the ErM association, a greater capacity for organic matter decomposition has been demonstrated (Read et al., 2004). Genome analysis of two orchid mycorrhizal species (*Sebacina vermifera* and *Tulasnella calospora*) and one ErM fungal species (*Oidiodendron maius*) revealed an abundance of genes encoding for carbohydrate-active enzymes and proteins with a cellulose-binding domain (CBM1) compared with the EcM fungi sequenced (Kohler et al., 2015). Proteins with a cellulose-binding domain are often found to enhance the activities of exoenzymes and endoenzymes involved in the breakdown of cellulose. In contrast, these were rarely detected in the EcM fungal genomes, further suggesting that the EcM fungi are more reliant on host C than other potential C sources. In addition to a nutrient acquisition function, the differing expression of activities among the mycorrhizal fungi screened may be related to their colonization extents, allowing for host cell modifications by EcM fungi but penetration of cell walls by orchid and ErM fungi (Kohler et al., 2015).

8.6.2 Evidence From Field and Microcosm Studies

The distribution of EcM fungal communities throughout the soil profile may be expected to yield information on their capacity to decompose organic materials. However, results on niche partitioning are often contradictory (Dickie et al., 2002; Lindahl et al., 2007; Talbot et al., 2013). This likely reflects the wide range of other factors (i.e., seasonality, environmental conditions) that will impact EcM spatial distribution. In contrast, Lindahl et al. (2007) found the spatial location of saprotrophic, EcM, and ErM fungi in a *Pinus sylvestris* forest soil differed according to the age and quality of the litter. Saprotrophic fungi were confined to recent (<4 years old) litter near the surface where organic C sources were mineralized, but N was retained, resulting in a decrease in the litter C:N ratio. The mycorrhizal fungi dominated underneath in the more decomposed litter and humus layer (fermentation horizon; average age ≈10 years), where they mobilized N sources from the litter and transferred N to their associated host plant. Similarly, in a microcosm study to which fermentation horizon organic matter (FHOM) from a pine forest soil was added as discrete zones or patches, N concentrations of the FHOM decreased by 13% when colonized by *Thelephora terrestris* and by 23% when colonized by *Suillus bovinus*, respectively,

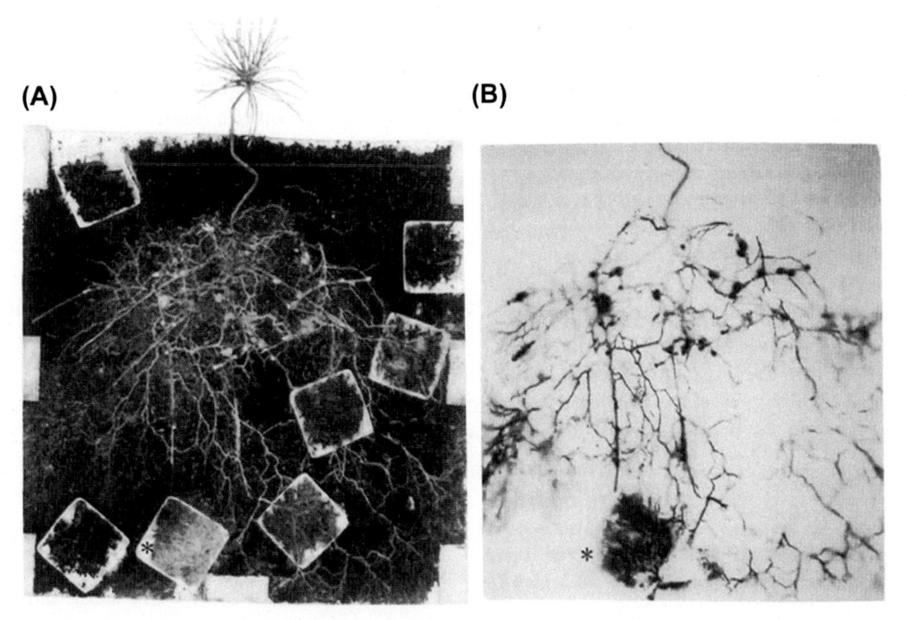

FIGURE 8.4 (A) Microcosm unit showing mycelium of the ectomycorrhizal fungus *Suillus bovinus* with trays of fermentation horizon organic matter (FHOM) at different stages of colonization. (B) Autoradiograph of the same microcosm unit after labeling of the plant shoot with ^{14}C-labelled carbon dioxide ($^{14}CO_2$). The allocation of recent photoassimilate to the tray containing FHOM soil colonized for less than 40 days is clearly evident and marked with an *asterisk. Reproduced from Bending, G.D., Read, D.J., 1995. The structure and function of the vegetative mycelium of ectomycorrhizal plants. V. Foraging behaviour and translocation of nutrients from exploited litter. New Phytologist 130, 401–409 with permission.*

resulting in an increased C:N ratio of the FHOM patch (Bending and Read, 1995). In contrast, P concentrations were only reduced when *S. bovinus* colonized the FMOM patches. Furthermore, recent photoassimilate from a *P. sylvestris* seedling was only allocated to FHOM patches colonized by the EcM fungi for less than 40 days, which coincided with the time of initial colonization of the FHOM patches by the EcM fungal hyphae to their early senescence (Fig. 8.4). Both of these studies (Bending and Read, 1995; Lindahl et al., 2007) suggest the mycorrhizal fungi were mainly utilizing the available N sources (rather than the C, which instead was supplied by the plant). Although no spatial distinction in EcM and saprotrophic species richness was observed in a *Pinus muricata* Californian forest soil, EcM community species richness could be explained by variation in the activity of enzymes targeting recalcitrant N sources (protease and peroxidase activities) in the bulk soil. In contrast, saprotrophic community structure was correlated with activity of carbohydrate and organic P–targeting enzymes, measured as acid phosphatase activity (Talbot et al., 2013), again suggesting N forms are the primary target of EcM fungi.

8.6.3 Ectomycorrhizal Fungi and Organic Phosphorus Sources

Despite the findings by Talbot et al. (2013), several other studies have reported EcM fungi have the capacity to acquire P from organic P sources, at least under controlled conditions (Dighton, 1983; Perez-Moreno and Read, 2001a). Forms of organic P in soils include phosphomonoesters (e.g., sugar-phosphates, inositol hexaphosphates, and mononucleotides) and phosphate diesters (phospholipids and nucleic acids), whereas organic polyphosphates (e.g., ATP) usually occur at lower amounts (Turner, 2008). Although inositol phosphates (which include phytate, a P reserve in plants) are thought to be the dominant form of organic P compounds in soils (Turner et al., 2002), some caution must be exercised in the interpretation of this data if based on ^{31}P nuclear magnetic resonance spectroscopy analysis; this method has subsequently been demonstrated to potentially overestimate the amount of phytate (Doolette et al., 2009; Smernik and Dougherty, 2007) and to underestimate the amount of phospholipids (Doolette et al., 2009) present, unless spiking by model organic P compounds is also included to aid in compound identification. Despite this, the ability of EcM fungi to potentially release phytase enzymes has been investigated extensively using both EcM roots and hyphae, but with highly variable results and no consensus of the global importance of such activities (Antibus et al., 1997; Colpaert et al., 1997; Louche et al., 2010).

Although a large number of studies have reported phosphomonoesterase activities from both EcM fungal hyphae and EcM root tips, phosphodiesterase activities tend to be less well documented. This may simply reflect the fact that phosphodiester compounds tend not to persist in soils for long periods (Cairney, 2011). Evidence suggests that at least some EcM fungi do have the capacity for acquiring P from organic sources, although again variation among EcM fungi (and EcM strains) to produce such enzyme capacities is usually large (Burke et al., 2014; Jones et al., 2010; Courty et al., 2010). Some of these studies have used excised root tips (Jones et al., 2010; Tedersoo et al., 2012), which readily allows (usually field) collection of sufficient uniform material for analyses. There are technical issues with using such material (e.g., the ERM network is severed, the excised tissue may become starved of organic C, and the transpiration stream is eliminated) and so results should be interpreted with a degree of caution. Nevertheless, using such an approach, Tedersoo et al. (2012) found that enzyme activities, including that of acid phosphatase, in excised EcM root tips were generally higher in those EcM fungi associated with production of a more extensive ERM network. Moreover, phosphomonoesterase activities by both EcM fungal mycelium and EcM roots have been shown in numerous studies to be repressed when inorganic P availability is high, suggesting a link between expression and P availability. However, it is also well established that EcM phosphomonoesterase activities are influenced by a wide range of environmental parameters including season, temperature, soil type, and N availability (Cairney, 2011).

Given that substantial amounts of P are held in complex organic forms in soil, a link between N and P acquisition is potentially significant. The mycelium of EcM fungi can acquire both N and P from various organic substrates including pollen (Perez-Moreno and Read, 2001a), litter from fermentation horizons (Perez-Moreno and Read, 2000), and nematodes (Perez-Moreno and Read, 2001b). In these studies, enzyme activities were not monitored, nor was the organic matter labeled with radioactive or stable isotopes to track the flow from substrate through the plant–soil system. Nevertheless, by observing P and N contents in the plant and remaining in the substrate, nutrient flow and transfer could be estimated. In most cases, N acquisition exceeded that of P acquisition. Interestingly, although EcM plants did acquire both N and P from the nematode necromass, the amount of N acquired was greater than the loss of N from the nematode substrate, suggesting the EcM fungi were also acquiring N from other sources in the clay–peat media. This is similar to the findings by Herman et al. (2012), but using an AM fungal system and decomposing root material added to soil. Proliferation of EcM and

AM fungal hyphae in response to organic substrates can therefore result in a general enhancement of nutrient acquisition by the mycobiont. Saprotrophic microorganisms were present in both systems and likely initiated the decomposition process (Perez-Moreno and Read, 2001b; Herman et al., 2012), yet both the EcM and AM fungi were able to effectively acquire the released nutrients for their host plant. As a result of the decomposition process, some P may also become available for capture as complex organic structures become modified. This may release the EcM or AM fungal partner from P limitation, allowing more effective N acquisition (Perez-Moreno and Read, 2001b; Thirkell et al., 2016). EcM roots tips inoculated by EcM fungi from an N-rich soil had higher capability to produce phosphomonoesterase activities than those inoculated by EcM fungi from an N-poor soil (Taniguchi et al., 2008), further suggesting a link between N and P availability and mycorrhizal function.

8.7 CONCLUSIONS

Both EcM and AM fungi can respond to and acquire nutrients from inorganic and organic nutrient sources. In the case of AM fungi, however, nutrients from organic forms are most likely acquired after decomposition by other microorganisms. The ability of EcM fungi and AM fungi to acquire N and P, respectively, for their host plant is confirmed, but a new appreciation of a role for EcM fungi in P acquisition, and for AM fungi in N acquisition, has been realized. Early work showed that EcM fungi accumulated P in their fungal mantle, whereas more recently, the high N demand and the differing ability of the AM fungal mycobiont to down-regulate root P uptake all point to a level of fungal control on the nutrient transfer process. Despite these findings, unfortunate phrases such as "AM fungi act as mere extensions of plant roots" are still prevalent in the literature. In contrast, EcM research has never been prone to such a poor appreciation of the fungal partner, most likely because of an early recognition of the wider enzymatic repertoire of these fungi. Recent debate on the EcM symbiosis has, instead, focused on a saprotroph/decomposer distinction, a situation not helped by these terms often being used interchangeably and being ill-defined in the literature. It is perhaps interesting to note that the early evidence of AM fungi proliferating hyphae in organic matter was initially suggested to indicate a potential ability to exist saprophytically (Nicolson, 1959) or a role in organic matter breakdown (Mosse, 1959). EcM fungi, however, are evolutionally diverse and functionally complex compared with AM fungi. Some EcM fungi have been demonstrated to operate oxidative decomposition pathways to release nutrients, whereas others do not. Nevertheless, EcM fungi with lesser decomposer capabilities may still be able to compete effectively for nutrients released through the action of other EcM fungi and microorganisms in much the same way as their AM fungi counterparts. The evidence largely argues for the host plant as a reliable source of C for the EcM fungi, which enables the EcM fungi to acquire N and P from sources in the soil including organic matter. Yet EcM fungi will encounter C-containing compounds (e.g., amino acids) as part of their nutrient acquisition role, potentially leading to ambiguity over the distinction between saprotroph and decomposer among the population of hyphae, some of which may be operating at considerable distance from their associated host plants. Therefore, perhaps we should focus less on semantics and instead view these systems as a series of nutrient pools with fluctuating spatiotemporal pool dynamics. The ability to effectively exploit such pools in a competitive environment becomes paramount, even if the fungus has not been intimately involved in the decomposition process. This may help explain some of the differing results found (e.g., whether AM fungi enhance or retard the decomposition of organic materials). The ability to explore the soil volume through an extensive mycelial network enables fungi to exploit such spatially and temporally distributed nutrient-rich zones and arguably to exploit plants that provide different "pools" of C.

Acknowledgments

I thank Alastair Fitter, Jon Pitchford, Roger Koide, and Nancy Johnson for providing comments on this chapter.

References

Abuzinadah, R.A., Read, D.J., 1986. The role of proteins in the nitrogen nutrition of ectomycorrhizal plants. II. Utlization of protein by mycorrhizal plants of *Pinus contorta*. New Phytologist 103, 495–506.

Ames, R.N., Reid, C.P.P., Porter, L.K., Cambardella, C., 1983. Hyphal uptake and transport of nitrogen from two ^{15}N-labelled sources by *Glomus mosseae*, a vesicular-arbuscular mycorrhizal fungus. New Phytologist 95, 381–396.

Anderson, I.C., Chambers, S.M., Cairney, J.W.G., 1999. Intra- and interspecific variation in patterns of organic and inorganic nitrogen utilization by three Australian *Pisolithus* species. Mycological Research 103, 1579–1587.

Antibus, R.K., Bower, D., Dighton, J., 1997. Root surface phosphatase activities and uptake of ^{32}P-labelled inositol phosphate in field-collected gray birch and red maple roots. Mycorrhiza 7, 39–46.

Baldrian, P., 2009. Ectomycorrhizal fungi and their enzymes in soils: is there enough evidence for their role as facultative soil saprotrophs? Oecologia 161, 657–660.

Barrett, G., Campbell, C.D., Hodge, A., 2014. The direct response of the external mycelium or arbuscular mycorrhizal fungi to temperature and the implications for nutrient transfer. Soil Biology & Biochemistry 78, 109–117.

Bechem, E.E.T., Alexander, I.J., 2012. Phosphorus nutrition of ectomycorrhizal *Gnetum africanum* plantlets from Cameroon. Plant and Soil 353, 379–393.

Bending, G.D., Read, D.J., 1995. The structure and function of the vegetative mycelium of ectomycorrhizal plants. V. Foraging behaviour and translocation of nutrients from exploited litter. New Phytologist 130, 401–409.

Bödeker, I.T.M., Clemmensen, K.E., de Boer, W., Martin, F., Olson, Å., Lindahl, B.D., 2014. Ectomycorrhizal *Cortinarius* species participate in enzymatic oxidation of humus in northern forest ecosystems. New Phytologist 203, 245–256.

Bonfante, P., Genre, A., 2010. Mechanisms underlying beneficial plant-fungus interactions in mycorrhizal symbiosis. Nature Communications 1, 48. http://dx.doi.org/10.1038/ncomms1046.

Bruns, T.D., Shefferson, R.P., 2004. Evolutionary studies of ectomycorrhizal fungi: recent advances and future directions. Canadian Journal of Botany 82, 1122–1132.

Burke, D.J., Smemo, K.A., Hewins, C.R., 2014. Ectomycorrhizal fungi isolated from old-growth northern hardwood forest display variability in extracellular enzyme activity in the presence of plant litter. Soil Biology & Biochemistry 68, 219–222.

Cairney, J.W.G., 2011. Ectomycorrhizal fungi: the symbiotic route to the root for phosphorus in forest soils. Plant and Soil 344, 51–71.

Casarin, V., Plassard, C., Hinsinger, P., Arvieu, J.-C., 2004. Quantification of ectomycorrhizal fungal effects on the bioavailability and mobilization of soil P in the rhizosphere of *Pinus pinaster*. New Phytologist 163, 177–185.

Casieri, L., Lahmidi, N.A., Doidy, J., Veneault-Fourrey, C., Migeon, A., Bonneau, L., Courty, P.E., Garcia, K., Charbonnier, M., et al., 2013. Biotrophic transportome in mutualistic plant-fungal interactions. Mycorrhiza 23, 597–625.

Cheng, L., Booker, F.L., Tu, C., Burkey, K.O., Zhou, L., Shew, H.D., Rufty, T.W., Hu, S.J., 2012. Arbuscular mycorrhizal fungi increase organic carbon decomposition under elevated CO_2. Science 337, 1084–1087.

Clemmensen, K.E., Sorensen, P.L., Michelsen, A., Jonasson, S., Ström, L., 2008. Site-dependent N uptake from N-form mixtures by artic plants, soil microbes and ectomycorrhizal fungi. Oecologia 155, 771–783.

Colpaert, J.V., van Laere, A., 1996. A comparison of the extracellular enzyme activities of two ectomycorrhizal and a leaf-saprotrophic basidiomycete colonising beech leaf litter. New Phytologist 133, 133–141.

Colpaert, J.V., van Laere, A., Tichelen, K.K., van Assche, J.A., 1997. The use of inositol hexaphosphate as a phosphorus source by mycorrhizal and non-mycorrhizal Scots pine (*Pinus sylvestris*). Functional Ecology 11, 407–415.

Corrêa, A., Cruz, C., Pérez-Tienda, J., Ferrol, N., 2014. Shedding light onto nutrient responses of arbuscular mycorrhizal plants: nutrient interactions may lead to unpredicted outcomes of the symbiosis. Plant Science 221–222, 29–41.

Courty, P.-E., Bréda, N., Garbaye, J., 2007. Relation between oak tree phenology and the secretion of organic matter degrading enzymes by *Lactarius quietus* ectomycorrhizas before and during bud break. Soil Biology & Biochemistry 39, 1655–1663.

Courty, P.-E., Buée, M., Diedhiou, A.G., Frey-Klett, P., Le Tacon, F., Rineau, F., Turpault, M.-P., Uroz, S., Garbaye, J., 2010. The role of ectomycorrhizal communities in forest ecosystem processes: new perspectives and emerging concepts. Soil Biology & Biochemistry 42, 679–698.

Courty, P.-E., Smith, P., Koegel, S., Redecker, D., Wipf, D., 2015. Inorganic nitrogen uptake and transport in beneficial plant root-microbe interactions. Critical Reviews in Plant Sciences 34, 4–16.

Cullings, K., Ishkhanova, G., Henson, J., 2008. Defoliation effects on enzyme activities of the ectomycorrhizal fungus *Suillus granulatus* in a *Pinus contorta* (lodgepole pine) stand in Yellowstone National Park. Oecologia 158, 77–83.

Cumming, J.R., Weinstein, L.H., 1990. Utilization of $AlPO_4$ as a phosphorus source by ectomycorrhizal *Pinus rigida* Mill. seedlings. New Phytologist 116, 99–106.

Dickie, I.A., Xu, B., Koide, R.T., 2002. Vertical niche differentiation of ectomycorrhizal hyphae in soil as shown by T-RFLP analysis. New Phytologist 156, 527–535.

Dighton, J., 1983. Phosphatase production by mycorrhizal fungi. Plant and Soil 71, 455–462.

Doolette, A.L., Smernik, R.J., Dougherty, W.J., 2009. Spiking improved solution phosphorus-31 nuclear magnetic resonance identification of soil phosphorus compounds. Soil Science Society of America Journal 73, 919–927.

Eastwood, D.C., Floudas, D., Binder, M., Majcherczyk, A., Schneider, P., Aerts, A., Asiegbu, F.O., Baker, S.E., Barry, K., Bendiksby, M., et al., 2011. The plant cell wall – decomposing machinery underlies the functional diversity of forest fungi. Science 333, 762–765.

Farrell, M., Hill, P.W., Farrar, J., DeLuca, T.H., Roberts, P., Kielland, K., Dahlgren, R., Murphy, D.V., Hobbs, P.J., Bardgett, R.D., et al., 2013. Oligopeptides represent a preferred source of organic N uptake: a global phenomenon? Ecosystems 16, 133–145.

Finlay, R.D., Frostegård, Å., Sonnerfeldt, A.-M., 1992. Utilisation of organic and inorganic nitrogen sources by ectomycorrhizal fungi in pure culture and in symbiosis with *Pinus contorta* Dougl. ex Loud. New Phytologist 120, 105–115.

France, R.C., Reid, C.P.P., 1984. Pure culture growth of ectomycorrhizal fungi on inorganic nitrogen sources. Microbial Ecology 10, 187–195.

Gadd, G.M., Bahri-Esfahani, J., Li, Q., Rhee, Y.J., Wei, Z., Fomina, M., Liang, X., 2014. Oxalate production by fungi: significance in geomycology, biodeterioration and bioremediation. Fungal Biology Reviews 28, 36–55.

Govindarajulu, M., Pfeffer, P.E., Jin, H., Abubaker, J., Douds, D.D., Allen, J.W., Bücking, H., Lammers, P.J., Shachar-Hill, Y., 2005. Nitrogen transfer in the arbuscular mycorrhizal symbiosis. Nature 435, 819–823.

Green, F., Highley, T.L., 1997. Mechanism of brown-rot decay: paradigm or paradox. International Biodeterioration & Biodegradation 39, 113–124.

Handley, W.R.C., 1961. Further evidence for the importance of residual leaf protein complexes in litter decomposition and the supply of nitrogen for plant growth. Plant and Soil 15, 37–73.

Harley, J.L., 1989. The significance of mycorrhiza. Mycological Research 92, 129–139.

Harrison, M.J., van Buuren, M.L., 1995. A phosphate transporter from the mycorrhizal fungus *Glomus versiforme*. Nature 378, 626–629.

Hayman, D.S., Mosse, B., 1972. Plant growth responses to vesicular-arbuscular mycorrhiza. III. Increased uptake of labile P from soil. New Phytologist 71, 41–47.

Herman, D.J., Firestone, M.K., Nuccio, E., Hodge, A., 2012. Interactions between an arbuscular mycorrhizal fungus and a soil microbial community mediating litter decomposition. FEMS Microbiology Ecology 80, 236–247.

Hibbett, D.S., Gllbert, L.-B., Donoghue, M.J., 2000. Evolutionary instability of ectomycorrhizal symbiosis in basidiomycetes. Nature 407, 506–508.

Hinsinger, P., 2001. Bioavailability of soil inorganic P in the rhizosphere as affected by root-induced chemical changes: a review. Plant and Soil 237, 173–195.

Hodge, A., Fitter, A.H., 2010. Substantial nitrogen acquisition by arbuscular mycorrhizal fungi from organic material has implications for N cycling. Proceedings of the National Academy of Sciences, United States of America 107, 13754–13759.

Hodge, A., Storer, K., 2015. Arbuscular mycorrhiza and nitrogen: implications for individual plants through to ecosystems. Plant and Soil 386, 1–19.

Hodge, A., Alexander, I.J., Gooday, G.W., 1995. Chitinolytic enzymes of pathogenic and ectomycorrhizal fungi. Mycological Research 99, 935–941.

Hodge, A., Robinson, D., Fitter, A., 2000. Are microorganisms more effective than plants at competing for nitrogen? Trends in Plant Science 5, 304–308.

Hodge, A., Campbell, C.D., Fitter, A.H., 2001. An arbuscular mycorrhizal fungus accelerates decomposition and acquires nitrogen directly from organic material. Nature 413, 297–299.

Hodge, A., 2001. Arbuscular mycorrhizal fungi influence decomposition of, but not plant nutrient capture from, glycine patches in soil. New Phytologist 151, 725–734.

Hodge, A., 2004. The plastic plant: root responses to heterogeneous supplies of nutrients. New Phytologist 162, 9–24.

Hodge, A., 2014. Interactions between arbuscular mycorrhizal fungi and organic material substrates. Advances in Applied Microbiology 89, 47–99.

Inderjit, Mallik, A.U., 1997. Effect of phenolic compounds on selected soil properties. Forest Ecology and Management 92, 11–18.

Johansen, A., Finlay, R.D., Olsson, P.A., 1996. Nitrogen metabolism of external hyphae of the arbuscular mycorrhizal fungus *Glomus intraradices*. New Phytologist 133, 705–712.

Johnson, N.C., Wilson, G.W.T., Bowker, M.A., Wilson, J., Miller, R.M., 2010. Resource limitation is a driver of local adaptation in mycorrhizal symbiosis. Proceedings of the National Academy of Sciences, United States of America 107, 2093–2098.

Johnson, N.C., Wilson, G.W.T., Wilson, J.A., Miller, R.M., Bowker, M.A., 2015. Mycorrhizal phenotypes and the Law of the Minimum. New Phytologist 205, 1473–1484.

Joner, E.J., Ravnskov, S., Jakobsen, I., 2000a. Arbuscular mycorrhizal phosphate transport under monoxenic conditions using radio-labelled inorganic and organic phosphate. Biotechnology Letters 22, 1705–1708.

Joner, E.J., van Aarle, I.M., Vosatka, M., 2000b. Phosphatase activity of extra-radical arbuscular mycorrhizal hyphae: a review. Plant and Soil 226, 199–210.

Jones, D.L., Healy, J.R., Willett, V.B., Farrar, J.F., Hodge, A., 2005. Dissolved organic nitrogen uptake by plants – an important N uptake pathway? Soil Biology & Biochemistry 37, 413–423.

Jones, M.D., Twieg, B.D., Ward, V., Barker, J., Durall, D.M., Simard, S., 2010. Functional complementarity in Douglas-fir ectomycorrhizas for extracellular enzyme activity after wildfire or clearcut logging. Functional Ecology 24, 1139–1151.

Jones, D.L., 1998. Organic acids in the rhizosphere – a critical review. Plant and Soil 205, 25–44.

Kanerva, S., Kitunen, V., Kiikkilä, O., Loponen, J., Smolander, A., 2006. Response of soil C and N transformations to tannin fractions originating from Scots pine and Norway spruce needles. Soil Biology & Biochemistry 38, 1364–1374.

Karandashov, V., Nagy, R., Wegmuller, S., Amrhein, N., Bucher, M., 2004. Evolutionary conservation of a phosphate transporter in the arbuscular mycorrhizal symbiosis. Proceedings of the National Academy of Sciences of the United States of America 101, 6285–6290.

Kohler, A., Kuo, A., Nagy, L.G., Morin, E., Barry, K.W., Buscot, F., Canbäck, B., Choi, C., Cichocki, N., Clum, A., et al., 2015. Convergent losses of decay mechanisms and rapid turnover of symbiosis genes in mycorrhizal mutualists. Nature Genetics 47, 410. http://dx.doi.org/10.1038/ng.3223 U176.

Koide, R.T., Sharda, J.N., Herr, J.R., Malcolm, G.M., 2008. Ectomycorrhizal fungi and the biotrophy-saprotrophy continuum. New Phytologist 178, 230–233.

Kranabetter, J.M., Hawkins, B.J., Jones, M.D., Robbins, S., Dyer, T., Li, T., 2015. Species turnover (β-diversity) in ectomycorrhizal fungi linked to NH_4^+ uptake capacity. Molecular Ecology 24, 5992–6005.

Landeweert, R., Hoffland, E., Finlay, R.D., Kuyper, T.W., van Breemen, N., 2001. Linking plants to rocks: ectomycorrhizal fungi mobilize nutrients from minerals. Trends in Ecology and Evolution 16, 248–254.

Lapeyrie, F., Chilvers, G.A., Bhem, C.A., 1987. Oxalic acid synthesis by the mycorrhizal fungus *Paxillus involutus* (Batsch. ex Fr.) Fr. New Phytologist 106, 139–146.

Lei, J., Bécard, G., Catford, J.G., Piché, Y., 1991. Root factors stimulate ^{32}P uptake and plasmalemma ATPase activity in vesicular-arbuscular mycorrhizal fungus *Gigaspora margarita*. New Phytologist 118, 289–294.

Leigh, J., Hodge, A., Fitter, A.H., 2009. Arbuscular mycorrhizal fungi can transfer substantial amounts of nitrogen to their host plant from organic material. New Phytologist 181, 199–207.

Leigh, J., Fitter, A.H., Hodge, A., 2011. Growth and symbiotic effectiveness of an arbuscular mycorrhizal fungus in organic matter in competition with soil bacteria. FEMS Microbiology Ecology 76, 428–438.

Lindahl, B.D., Tunlid, A., 2015. Ectomycorrhizal fungi – potential organic decomposers, yet not saprotrophs. New Phytologist 205, 1443–1447.

Lindahl, B., Ihrmark, K., Boberg, J., Trumbore, S.E., Högberg, P., Stenlid, J., Finlay, R.D., 2007. Spatial separation of litter decomposition and mycorrhizal nitrogen uptake in a boreal forest. New Phytologist 173, 611–620.

López-Pedrosa, A., González-Guerrero, M., Valderas, A., Acón-Aguilar, C., Ferrol, N., 2006. *GintAMT1* encodes a functional high-affinity ammonium transporter that is expressed in the extraradical mycelium of *Glomus intraradices*. Fungal Genetics and Biology 43, 102–110.

Louche, J., Ali, M.A., Cloutier-Huteau, B., Sauvage, F.-X., Quiquampoix, H., Plassard, C., 2010. Efficiency of acid phosphatases secreted from the ectomycorrhizal fungus *Hebeloma cylindrosporum* to hydrolyse organic phosphorus in podzols. FEMS Microbiology Ecology 73, 323–335.

Martin, F., Nehls, U., 2009. Harnessing ectomycorrhizal genomics for ecological insights. Current Opinion in Plant Biology 12, 508–515.

Martin, F., Aerts, A., Ahrén, D., Brun, A., Danchin, E.G.J., Duchaussoy, F., Gibon, J., Kohler, A., Lindquist, E., Pereda, V., et al., 2008. The genome of *Laccaria biocolor* provides insights into mycorrhizal symbiosis. Nature 452, 88–92.

Martin, F., Kohler, A., Murat, C., Balestrini, R., Coutinho, P.M., Jaillon, O., Montanini, B., Morin, E., Noel, B., Percudani, R., et al., 2010. Périgord black truffle genome uncovers evolutionary origins and mechanisms of symbiosis. Nature 464, 1033–1038.

Martínez, A.T., 2002. Molecular biology and structure-function of lignin-degrading heme peroxidases. Enzyme and Microbial Technology 30, 425–444.

Mosse, B., 1959. Observations on the extra-matrical mycelium of a vesicular-arbuscular endophyte. Transactions of the British Mycological Society 42, 439–448.

Nicolson, T.H., 1959. Mycorrhiza in the Gramineae. I. Vesicular-arbuscular endophytes, with special reference to the external phase. Transactions of the British Mycological Society 42, 421–438.

Nuccio, E.E., Hodge, A., Pett-Ridge, J., Herman, D.J., Weber, P.K., Firestone, M.K., 2013. An arbuscular mycorrhizal significantly modifies the soil bacterial community and nitrogen cycling during litter decomposition. Environmental Microbiology 15, 1870–1881.

Nygren, C.M.R., Eberhardt, U., Karlsson, M., Parrent, J.L., Lindahl, B.D., Taylor, A.F.S., 2008. Growth on nitrate and occurrence of nitrate reductase-encoding genes in a phylogenetically diverse range of ectomycorrhizal fungi. New Phytologist 180, 875–889.

Ohtomo, R., Saito, M., 2005. Polyphosphate dynamics in mycorrhizal roots during colonization of an arbuscular mycorrhizal fungus. New Phytologist 167, 571–578.

Palm, C.A., Sanchez, P.A., 1991. Nitrogen release from the leaves of some tropical legumes as affected by their lignin and polyphenolic contents. Soil Biology & Biochemistry 23, 83–88.

Perez-Moreno, J., Read, D.J., 2000. Mobilization and transfer of nutrients from litter to tree seedlings via the vegetative mycelium of ectomycorrhizal plants. New Phytologist 145, 301–309.

Perez-Moreno, J., Read, D.J., 2001a. Exploitation of pollen by mycorrhizal mycelial systems with special reference to nutrient recycling in boreal forest. Proceedings of the Royal Society of London B 268, 1329–1335.

Perez-Moreno, J., Read, D.J., 2001b. Nutrient transfer from soil nematodes to plants: a direct pathway provided by the mycorrhizal mycelial network. Plant, Cell and Environment 24, 1219–1226.

Read, D.J., Leake, J.R., Perez-Moreno, J., 2004. Mycorrhizal fungi as drivers of ecosystem processes in heathland and boreal forest biomes. Canadian Journal of Botany 82, 1243–1263.

Read, D.J., 1991. Mycorrhizas in ecosystems. Experientia 47, 376–391.

Remy, W., Taylor, T.N., Haas, H., Kerp, H., 1994. Four hundred-million-year-old vesicular-arbuscular mycorrhizae. Proceedings of the National Academy of Sciences of the United States of America 91, 11841–11843.

Rineau, F., Roth, D., Shah, F., Smits, M., Johansson, T., Canbäck, B., Olsen, P.B., Persson, P., Grell, M.N., Lindquist, E., et al., 2012. The ectomycorrhizal fungus *Paxillus involutus* converts organic matter in plant litter using a trimmed brown-rot mechanism involving Fenton chemistry. Environmental Microbiology 14, 1477–1487.

Rineau, F., Shah, F., Smits, M.M., Persson, P., Johansson, T., Carleer, R., Troein, C., Tunlid, A., 2013. Carbon availability triggers the decomposition of plant litter and assimilation of nitrogen by an ectomycorrhizal fungus. ISME Journal 7, 2010–2022.

Sanders, F.E., Tinker, P.B., 1973. Phosphate inflow into mycorrhizal roots. Pesticide Science 4, 385–395.

Schachtman, D.P., Reid, R.J., Ayling, S.M., 1998. Phosphorus uptake by plants: from soil to cell. Plant Physiology 116, 447–453.

Scheromm, P., Plassard, C., Salsac, L., 1990. Effect of nitrate and ammonium nutrition on the metabolism of the ectomycorrhizal basidiomycete, *Heleloma cylindrosporum* Romagn. New Phytologist 114, 227–234.

Smernik, R.J., Dougherty, W.J., 2007. Identification of phytate in phosphorus-31 nuclear magnetic resonance spectra: the need for spiking. Soil Science Society of America Journal 71, 1045–1050.

Smith, S.E., Read, D.J., 2008. Mycorrhizal Symbiosis, third ed. Academic Press, London, UK.

Smith, S.E., Smith, F.A., 2011. Roles of arbuscular mycorrhizas in plant nutrition and growth: new paradigms from cellular to ecosystem scales. Annual Review of Plant Biology 62, 227–250.

Talbot, J.M., Bruns, T.D., Smith, D.P., Branco, S., Glassman, S.I., Erlandson, S., Vilgalys, R., Peay, K.G., 2013. Independent roles of ectomycorrhizal and saprotrophic communities in soil organic matter decomposition. Soil Biology & Biochemistry 57, 282–291.

Taniguchi, T., Kataoka, R., Futai, K., 2008. Plant growth and nutrition in pine (*Pinus thunbergii*) seedlings and dehydrogenase and phosphatase activity of ectomycorrhizal root tips inoculated with seven individual ectomycorrhizal fungal species at high and low nitrogen conditions. Soil Biology & Biochemistry 40, 1235–1243.

Tatry, M.-V., El Kassis, E., Lambilliotte, R., Corratgé, C., van Aarle, I., Amenc, L.K., Alary, R., Zimmermann, S., Sentenac, H., et al., 2009. Two differentially regulated phosphate transporters from the symbiotic fungus *Hebeloma cylindrosporum* and phosphorus acquisition by ectomycorrhizal *Pinus pinaster*. The Plant Journal 57, 1092–1102.

Tedersoo, L., May, T.W., Smith, M.E., 2010. Ectomycorrhizal lifestyle in fungi: global diversity, distribution and evolution of phylogenetic linkages. Mycorrhiza 20, 217–263.

Tedersoo, L., Naadel, T., Bahram, M., Pritsch, K., Buegger, F., Leal, M., Kõljalg, U., Põldmaa, K., 2012. Enzymatic activities and stable isotope patterns of ectomycorrhizal fungi in relation to phylogeny and exploration types in afrotropical rain forest. New Phytologist 195, 832–843.

Thirkell, T.J., Cameron, D.D., Hodge, A., 2016. Resolving the 'nitrogen paradox' of arbuscular mycorrhizas: fertilization with organic matter brings considerable benefits for plant nutrition and growth. Plant, Cell & Environment. http://dx.doi.org/10.1111/pce.12667.

Treseder, K.K., Torn, M.S., Masiello, C.A., 2006. An ecosystem-scale radiocarbon tracer to test use of litter carbon by ectomycorrhizal fungi. Soil Biology & Biochemistry 38, 1077–1082.

Turner, B.L., Papházy, M.J., Haygarth, P.M., McKelvie, I.D., 2002. Inositol phosphates in the environment. Philosophical Transactions of the Royal Society Series B Biological Sciences 357, 449–469.

Turner, B.L., 2008. Resource partitioning for soil phosphorus: a hypothesis. Journal of Ecology 96, 698–702.

Verbruggen, E., Jansa, J., Hammer, E.C., Rillig, M.C., 2016. Do arbuscular mycorrhizal fungi stabilize litter-derived carbon in soil? Journal of Ecology 104, 261–269.

Wallander, H., 2000. Uptake of P from apatite by *Pinus sylvestris* seedlings colonised by different ectomycorrhizal fungi. Plant and Soil 218, 249–256.

Wang, J., Li, T., Wu, X., Zhao, Z., 2014. Molecular cloning and functional analysis of a H^+-dependent phosphate transporter gene from the ectomycorrhizal fungus *Boletus edulis* in southwest China. Fungal Biology 118, 453–461.

Warner, A., Mosse, B., 1980. Independent spread of vesicular-arbuscular mycorrhizal fungi in soil. Transactions of the British Mycological Society 74, 407–445.

Wolfe, B.E., Tulloss, R.E., Pringle, A., 2012. The irreversible loss of a decomposition pathway marks the single origin of an ectomycorrhizal symbiosis. PLoS One 7, e39597.

Wu, T., Kabir, Z., Koide, R.T., 2005. A possible role for saprotrophic microfungi in the N nutrition of ectomycorrhizal *Pinus resinosa*. Soil Biology & Biochemistry 37, 965–975.

Zhang, L., Wang, M.-X., Li, H., Yuan, L., Huang, J.-G., Penfold, C., 2014. Mobilization of inorganic phosphorus from soils by ectomycorrhizal fungi. Pedosphere 24, 683–689.

Zhang, L., Xu, M., Liu, Y., Zhang, F., Hodge, A., Feng, G., 2016. Carbon and phosphorus exchange may enable cooperation between an arbuscular mycorrhizal fungus and a phosphate-solubilizing bacterium. New Phytologist 210, 1022–1032.

CHAPTER

9

Mycorrhizas as Nutrient and Energy Pumps of Soil Food Webs: Multitrophic Interactions and Feedbacks

P.M. Antunes, A. Koyama
Algoma University, Sault Ste. Marie, ON, Canada

9.1 INTRODUCTION

Mycorrhizal fungi form bridges among the rhizospheres of different plants through which organic carbon (C), nutrients [e.g., nitrogen (N) and phosphorus (P)], allelochemicals, bacteria, and viruses can travel either inside the hyphae or attached to the hyphosphere. Plant roots can be a quantitatively more important source of C for soil food webs than aboveground litter in temperate forests (Pollierer et al., 2009, 2012) and agroecosystems (Kramer et al., 2012), and mycorrhizal networks represent an important channel for this C. In this chapter we will review how the most abundant types of mycorrhizal fungi [i.e., ectomycorrhizal (EcM) and arbuscular mycorrhizal (AM) fungi] directly and indirectly influence the soil food web through multitrophic feedback loops (Fig. 9.1). We highlight the "state of the art," identify knowledge gaps, and propose future research directions.

We postulate three major channels through which photosynthetically derived C is transferred into the soil food web: via bacteria, fungi, and herbivores (Hunt et al., 1987). The fungal channel consists of saprotrophs in addition to mycorrhizal fungi. Although there is some debate that some types of mycorrhizal fungi (e.g., EcM fungi) may also function as saprotrophs (Chapter 20), recent evidence suggests that mycorrhizal fungi do not assimilate C from organic matter (Lindahl and Tunlid, 2015) and should be considered a distinct C channel of the food web. Because more than 80% of terrestrial vascular plant species are associated with mycorrhizal fungi and up to 30% of photosynthate is allocated to these symbionts (Wang and Qiu, 2006; Smith and Read, 2008), it is plausible to assume that soil food webs in any given terrestrial ecosystem derive energy via one or multiple types of mycorrhizal fungi to some degree.

Mycorrhizal Mediation of Soil
http://dx.doi.org/10.1016/B978-0-12-804312-7.00009-7

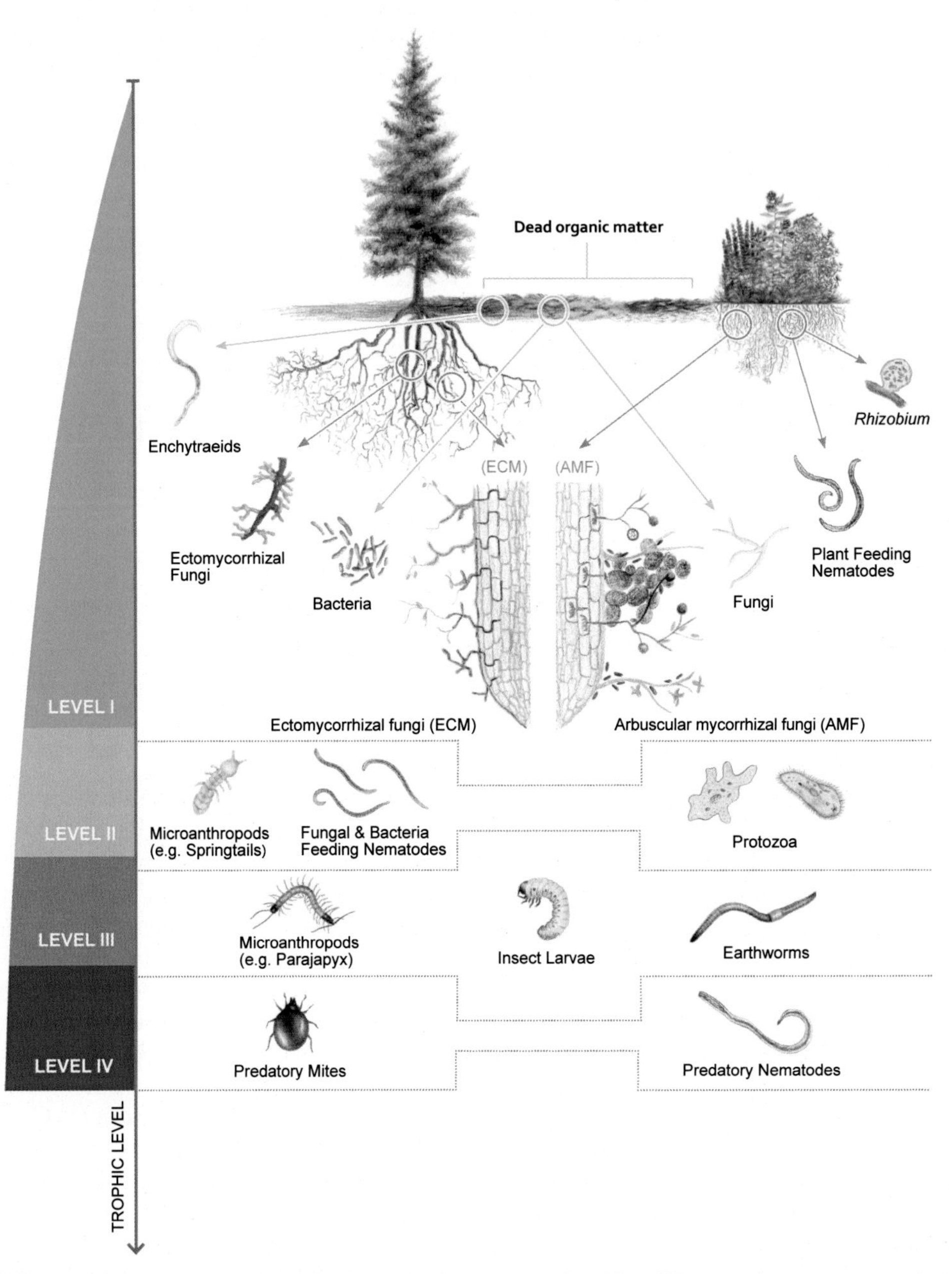

FIGURE 9.1 Simplified food web including the dead (*green arrows*) and live (*blue arrows*) organic matter channels. For simplicity, examples of organisms that directly interact with plants or plant litter, although not necessarily primary producers (i.e., Enchytraeids, plant feeding nematodes, bacteria, and fungi), were placed in level 1. *Illustrations by Angeline Castilloux (Algoma University Fine Arts Student).*

Mycorrhizal fungi penetrate the roots of plants to extract carbohydrates through a mutualistic symbiosis. Mutualism was favored in the process of mycorrhizal evolution in a way that the fungal partner must provide mineral nutrients and potentially other benefits in exchange for the C received (Bever et al., 2009; Kiers et al., 2011; Bever, 2015; Argüello et al., 2016). Consequently, mycorrhizal fungi are intrinsically tied to the host plant at the base of the trophic chain, having direct and indirect effects on soil food webs. Second trophic-level organisms in the rhizosphere and hyphosphere can directly consume mycorrhizal biomass. Indirect effects on the food web include biotic interactions associated with the stimulation of net primary productivity through enhanced plant nutrition, physiology, and facilitation of other plant-associated symbioses. Indirect effects also extend to trophic levels III and IV rhizosphere and hyphosphere biota that do not directly feed on mycorrhizal roots or fungal hyphae (Fig. 9.1). These organisms may access C compounds in exudates and promote plant (Lugtenberg and Kamilova, 2009) and fungal growth responses (Frey-Klett et al., 2007). Another indirect effect is that mycorrhizal fungi promote soil aggregation (Miller and Jastrow, 2000; Wilson et al., 2009; Rillig et al., 2010) that provides diverse habitats for soil biota. This chapter explores these indirect and direct effects in the context of the soil food web, including multitrophic soil feedback loops. We start by focusing on mycorrhizal effects on net primary productivity, including specific effects on plant growth and physiology with consequences on the soil food web.

9.1.1 Mycorrhizas and Net Primary Productivity

"Net primary production (NPP) is the difference between total photosynthesis and total respiration in an ecosystem" and is estimated by quantifying the new organic matter formed and retained by living plants in a given time (Clark et al., 2001). Even though primary production transferred below-ground is often larger than that kept above-ground (Jackson et al., 1997), few studies account for the below-ground components of NPP, including mycorrhizal fungi (Long and Hutchin, 1991; Hobbie, 2006; Litton and Giardina, 2008; Clemmensen et al., 2015; Fernandez et al., 2016). Mycorrhizal fungi may affect processes such as the shedding of leaves, flowers, and fruits above-ground and root turnover below-ground. These processes are key to understanding the structure and function of the soil food web but there is little experimental data available because most studies restrict the quantification of mycorrhizal effects to total host biomass.

Mycorrhizas affect the soil food web through changes in root architecture, lifespan, and turnover. They are associated with fine root promotion (e.g., Klein et al., 2016). However, the turnover of fine roots (<2 mm in diameter) can represent a particularly significant C loss, exceeding that of leaves (Jackson et al., 1997). Plant biomass is also lost through herbivory, and likely via volatile organic compounds and leached organic compounds (e.g., allelopathic compounds). Therefore to account for the net effect of mycorrhizal fungi on NPP there is a need for studies capable of simultaneously accounting for these multiple processes associated with biomass losses. In addition, studies must account for total mycorrhizal fungal biomass, including the different types of mycorrhizal fungi co-occurring in ecosystems [consult Orwin et al. (2011) for a modeling approach and Soudzilovskaia et al. (2015) for a proposed methodology to account for the contribution of AM and EcM fungi to C cycling]. Typically studies take into consideration inorganic or organic nutrient uptake via mycorrhizal fungi but fail to concurrently consider the C transfers between hosts and mycorrhizal fungi (see Chapter 21). An exception is the study by Orwin et al. (2011) taking into account C transfers between hosts and EcM fungi and ericoid mycorrhizas while considering mycorrhizal access to organic nutrients. They demonstrated that under nutrient-limiting conditions, mycorrhizal mobilization of organic nutrients can result in increased below-ground NPP through greater host transfer of C to mycorrhizal fungi. Concurrently, increased competition with saprotrophs may further lead to overall reduced soil respiration (see Section 9.2 and Chapters 23 and 24).

Although it is well established that mycorrhizal colonization affects root biomass (Jonsson et al., 2001; Lerat et al., 2002), fewer studies have focused on combined effects on root architecture, turnover, and lifespan. Root lifespan varies according to plant species and is tightly connected to root architecture. It has been shown that root lifespan varies considerably even within small root size classes (Wells and Eissenstat, 2001; Tierney and Fahey, 2002). Taking this into consideration, novel approaches that use branch order have emerged to determine root turnover rates more accurately (Guo et al., 2008a). Host responses to mycorrhizal fungi include reductions in root/shoot ratio (Azcon and Ocampo, 1981 as cited by Johnson, 2010) and, consequently, to potential net reductions in the turnover and decomposition of roots (Langley and Hungate, 2003). In these cases, mycorrhizal biomass replaces root biomass, which may change decomposition dynamics. Furthermore, AM fungi have been shown to significantly reduce root lifespan. This may be the result of the mycorrhizal stimulation of more branched root systems, which are thinner and therefore more susceptible to degradation; however, they may also be more suitable to further mycorrhizal colonization (Guo et al., 2008b). Although the stimulation of root branching by AM fungi has been shown in many plant species; this is not a consistent response (Péret et al., 2009 and review by McCormack and Guo (2014)). Moreover, mycorrhizal protection against pathogens and herbivores among other factors (e.g., increased stress tolerance) may contribute to counteract reductions in root/shoot ratio (Lewandowski et al., 2013), thereby further complicating calculations on the effect of mycorrhizas on NPP and, consequently, the soil food web. This is an area requiring further research.

9.1.2 Mycorrhizas and Plant–Soil Feedback

Plant–soil feedback results from the abiotic and biotic changes imposed by a plant on its rhizosphere that in turn provide feedback to the plant community (Ehrenfeld et al., 2005; van der Putten et al., 2013). Plant community diversity depends on the balance between mutualistic, competitive, and antagonistic biotic interactions. Concurrently, the diversity of soil biotic communities depends on plant community diversity (Eisenhauer et al., 2012). Mycorrhizal fungi have the potential to increase plant diversity and productivity in communities (van der Heijden et al., 1998; Klironomos et al., 2000; van der Heijden et al., 2008). This may result from alleviation of plant–plant competition (Wagg et al., 2011) and effects on the temporal dynamics of plant–soil feedback (Bever et al., 2012). In contrast, mycorrhizal fungi may reduce plant community diversity when the community is dominated by highly mycorrhizal-responsive species (Hartnett and Wilson, 2002).

AM fungi are not host specific, but soil and plant traits determine AM fungal community structure, which may lead either to positive or negative feedback. Positive feedback contributes to dominance in plant communities and, consequently, to lower plant species diversity. In contrast, negative feedback is necessary for increased plant diversity in communities. Negative feedback caused by hosts preferentially associating with relatively poor mycorrhizal growth promoters relative to other co-existing plants is possible (Bever, 2002; but see Argüello et al., 2016). However, interactions between mycorrhizal fungi, pathogens, and potentially other enemies in the context of soil and plant traits are key to understanding feedback responses (Wehner et al., 2010; Reininger et al., 2015). Newsham et al. (1995) showed that the sole function of an AM fungus can be pathogen protection of its host. Since then, many other studies have shown that host pathogen protection may be an important trait of mycorrhizal fungi, likely driven by isolate specific traits (Wehner et al., 2010, 2011; Jung et al., 2012; Lewandowski et al., 2013). Furthermore, pathogen accumulation in rarer plants may act as a selective pressure for the promotion of positive feedback loops between mycorrhizal fungi and other groups of beneficial soil biota (Latz et al., 2012). Among these are endophytes (Brundrett, 2006; Herre et al., 2007; Hardoim et al., 2015; but Mack and Rudgers, 2008), plant growth promoting bacteria (Artursson et al., 2006; Miransari, 2011), and N_2 fixing bacteria (Bauer et al., 2012; Nguyen and Bruns, 2015). The field of virology is also starting to explore not only the role of mycorrhizal symbioses on disease-causing plant viruses (Rúa et al., 2013) but also on the detection of mycoviruses in mycorrhizas (Kitahara et al., 2014; Ezawa et al., 2015). Recently Ke et al. (2015) used a trait-based approach linking plant–soil feedback to the interplay between pathogens and mycorrhizal fungi while taking into consideration plant and microbial traits, including litter quality. They found that the importance of litter decomposability as a driver of plant–soil feedback was greater when the relative abundance of mycorrhizal fungi increased because of their positive feedback on litter production. Future empirical studies should integrate trait-based approaches in the context of food webs to enable greater predictability regarding the direction of plant–soil feedback (Kardol et al., 2015).

9.2 MYCORRHIZAS AND SAPROTROPHS

In this section we focus on the interactions between mycorrhizal fungi, bacteria, and fungal saprotrophs (i.e., organisms that obtain their source of metabolic C from dead organic matter) (Fig. 9.1). This is critical, given its implications for C sequestration in light of climate warming. Saprotrophs not only represent a large component of the soil biomass but also play a key role in soil organic matter (SOM) decomposition and, given that soil represents the largest terrestrial pool of C, the effect of mycorrhizal fungi on SOM decomposition/stabilization is extremely important (Averill et al., 2014). There are two contrasting hypotheses with regard to the role that mycorrhizal fungi play on decomposition through interactions with saprotrophs. They can either interact positively through increased C allocation via mycorrhizas into the rhizosphere and hyphosphere or negatively because of competition for resources such as nutrients, water, and space. The latter hypothesis is gaining more support (see below), which is consistent with the idea that mycorrhizal fungi contribute to C stabilization in soils. Furthermore, the more nutrients mycorrhizal fungi are able of accessing from soil, the greater the SOM stabilization effect via competition with saprotrophs.

AM fungi do not directly decompose or take up any significant amounts of organic material (Joner et al., 2000; Nottingham et al., 2013). However, the circumstances whereby AM fungi may indirectly affect decomposition by increasing or decreasing saprotroph activity is still an open question in the literature(Chapters 20 and 24). The interplay between photosynthetic rate, C allocation below-ground, and macronutrient availability in soil is key to understanding the relationship between AM fungi and saprotrophic activity. Although AM fungi contribute to the formation of water-stable soil aggregates that may protect SOM from decomposing (Rillig and Mummey, 2006; Chapter 14), they also grow profusely on patches of SOM, exerting a physical action as they grow through it that results in a larger surface area exposed to saprotrophs, which can be carried along the hyphosphere. Voříšková and Baldrian (2013), have shown that AM fungi increase over time in the leaf litter of a temperate forest, especially 8 months after the litter was deposited. The greater the C allocation from hosts into AM fungi, the greater the priming effect (i.e., transient decomposition of relatively more recalcitrant SOM) via greater saprotrophic activity, which results from the exudation of energy-rich photosynthates primarily by the roots and turnover of fine roots and hyphae, leading to the concomitant release of ammonium (and C losses) required to compensate for the increased plant–AM fungal demand (Cheng et al., 2012; Nuccio et al., 2013). In turn, the export of ammonium may lead to positive feedback mechanisms for saprotrophs through AM fungal–mediated changes in enzymatic activity and pH reduction (Bago et al., 1996; Geisseler et al., 2010). It appears, however, that the priming effect may be transient and driven primarily by root exudates rather than the direct action of AM fungi (e.g., Hodge, 2014; Koller et al., 2013a,b; Shahzad et al., 2015; Verbruggen et al., 2016). The importance of this feedback loop involving ammonium acquisition and transport to hosts in the relationship between AM fungi and saprotrophs is highlighted by evidence supporting that it has been a strong factor of evolutionary selection (Cappellazzo et al., 2007; Hodge, 2014).

Whereas increases in C allocation by plants below-ground may contribute to promote decomposition, AM fungi have been found to reduce the decomposition of certain organic materials (Leifheit et al., 2015; Verbruggen et al., 2016). This may be linked not only to SOM recalcitrance and/or high C:N ratio but also to a variety of other potential mechanisms, including competition for limiting nutrients, localized reductions in water availability caused by mycorrhizal uptake, the structure of microbial communities present, and temporal aspects either linked to priority effects of microbial introduction or successional processes associated with the decomposition process(Chapters 20 and 24). For instance, using high-throughput molecular methods, Nuccio et al. (2013) demonstrated that the presence of *Glomus hoi* hyphae hosted by *Plantago lanceolata* altered the relative abundances of about 10% of bacterial taxa known as active decomposers (Tanahashi et al., 2005); i.e., Firmicutes increased and Actinobacteria decreased.

Most of the studies focusing on the interaction between AM fungi and saprotrophic fungi primarily investigated how their interaction affected the host plants. Once again, effects of mycorrhizal–saprotrophic interactions can vary in this context. McAllister et al. (1994, 1997) investigated interactions between the AM fungus *Glomus mosseae* and four species of saprotrophic fungi, and their resulting effects on host plants, maize (*Zea mays*) and lettuce (*Lactuca sativa*). In both studies, they found that presence of saprotrophic fungi alone did not affect the biomass of nonmycorrhizal host plants. However, with AM fungal inoculation, the presence of saprotrophic fungi resulted in smaller plants compared with the mycorrhizal plants without the saprotrophic fungi. This effect appears to have resulted from AM fungi–saprotroph competition for nutrients. However, when the AM fungi were inoculated before introducing saprotrophic fungi, no such significant difference in host plant biomass was detected and the growth of saprotrophic fungi was smaller. These older studies in combination with more recent findings (Shahzad et al., 2015; Verbruggen et al., 2016) emphasize the importance of temporary dynamic interactions between mycorrhizal and saprotrophic fungi and their host plants.

Understanding the effect of AM fungi on soil microbial community composition is important to determine which microbial groups either decompose or stabilize C through AM fungi–saprobe interactions. Recent studies have established bidirectional compositional effects between AM fungi and microbial communities (Welc et al., 2010; Leigh et al., 2011). We propose that the way forward is to adhere to an approach similar to that by Nuccio et al. (2013); using experimental designs with either individual taxa or plant and indigenous microbial communities and $^{13}C/^{15}N$–labeled SOM (e.g., from different natural ecosystems) in a dual-compartment system (i.e., with a root-free compartment). The use of stable isotopes is necessary to determine the fate of SOM in the plant host and soil biota. The growth rates of microbial communities (i.e., those promoted by indigenous AM fungi and utilizing the SOM) in this dual compartment system need to be determined during the decomposition process and succession using different molecular incorporation methods (Rousk and Bååth, 2011). At the same time, the use of metagenomic tools can be geared toward providing insight into ecological function; for example, using both ribosomal and functional genes and taking advantage of emerging functional databases such as FUNGuild (Voříšková and Baldrian, 2013; Nguyen et al., 2015). Furthermore, in light of increasing evidence on the importance of local adaptation, using sympatric combinations of soil, mycorrhizal fungi, and plant hosts is important for ecological relevance or to take advantage of more significant responses that can emerge from coevolved adaptations between symbiotic partners (Revillini et al., 2016; Rúa et al., 2016). Given the obligate nature of the mycorrhizal symbiosis and the associated challenges to establish negative controls in the field, perfecting the use of "rotative cores" to sever (or not) the hyphae either in the greenhouse or possibly in the field may be one among other useful approaches to be used in this context (Verbruggen et al., 2016; Brito et al., 2009). Furthermore, bottom-up and top-down effects need to be taken into consideration. For instance, the presence of AM fungi was found to significantly change the structure of the saprotrophic fungal community, and the change was amplified when fungivorous Collembola were present (Tiunov and Scheu, 2005).

Recently, opposing views have emerged in the literature regarding EcM fungi either being facultative saprotrophs, falling along a biotrophy–saprotrophy continuum (Koide et al., 2008), or that they are instead decomposers owing to their oxidizing primarily nitrogenous organic compounds (Lindahl et al., 2007; Chapters 20 and 24). The foundational theory by Frank (1894) that mycorrhizal fungi play a role as decomposers but are not saprotrophs is gaining increasing support. Evidence indicates that the evolution of symbiosis has led to the loss of genes encoding for enzymes involved in the release of metabolically readily available C compounds in EcM fungi. However, enzymes capable of oxidizing N from organic pools have been retained. Indeed, there have been a number of in vitro tests demonstrating the production of lignin-decomposing enzymes by ericoid and EcM fungi (Burke and Cairney, 2002; Courty et al., 2005). Ectomycorrhizal fungi can thus acquire nutrients, primarily N, from organic material with energy provided by their hosts (Lindahl and Tunlid, 2015). The fate of C obtained from the hosts and in compounds resulting from EcM fungi decomposition is an important literature gap, especially considering climate change and C sequestration. This includes the interactions between mycorrhizal fungi and saprotrophs. By stimulating photosynthesis, particularly in organic soils (where inorganic N is low), and acting as active N scavengers, EcM fungi may suppress saprotrophic decomposition and contribute to C sequestration to a much greater extent than AM fungi (Orwin et al., 2011; Averill et al., 2014). However, future research should focus on revealing neutral, synergistic, competitive, and antagonistic interactions between mycorrhizal fungi and different groups of saprotrophs (e.g., brown rot fungi, white rot fungi, ligninolytic and nonligninolytic bacteria) either directly or through modifications of litter quality.

9.3 MYCORRHIZAS AND HERBIVORES

Many soil animals feed directly on live plant roots. They include nematodes (Denton et al., 1998), Collembola (Endlweber et al., 2009), and insects such as weevils (Gange, 2001), craneflies (Treonis et al., 2005), and click beetles (Sonnemann et al., 2012). Root-feeding nematodes and Collembola live below-ground throughout their lives, whereas most of the root-feeding insects are soil dwellers only in their juvenile stages (Brown and Gange, 1990). Therefore direct effects of root-feeding insects on soil food webs are limited to their juvenile stages. Root-feeding insects can, however, indirectly affect the soil food web by changing plant physiology with resulting effects on mycorrhizal fungi (Johnson and Rasmann, 2015).

Interactions between AM fungi and root-feeding nematodes have been well studied, mostly for economically important agricultural plants (Hussey and Roncadorl, 1982; Ingham, 1988; Smith, 1988; Pinochet et al., 1996; Roncadori, 1997; Borowicz, 2001; Hol and Cook, 2005; Schouteden et al., 2015). Most of the studies have focused on beneficial effects of AM fungi on their host plants against damage by nematodes. Results were, however, mixed and dependent on nematode identity (Ingham, 1988). Root-feeding nematodes can be categorized into three groups based on their foraging strategies (Schouteden et al., 2015): ectoparasitic nematodes stay in the rhizosphere, migratory endoparasitic nematodes migrate and feed inside of roots, and sedentary endoparasitic nematodes create feeding sites in roots to stay sedentary. In a meta-analysis, Borowicz (2001) found that nematodes in general reduce AM fungal colonization regardless of their feeding types. Conversely, AM fungi tended to have negative effects on sedentary endoparasitic nematodes but improved the growth of migratory endoparasitic nematodes. Among the sedentary endoparasitic nematodes, AM fungi reduced the abundance of root-knot nematodes more than that of cyst nematodes (Hol and Cook, 2005). The authors also reported that AM fungi tended to increase the damage caused to roots by ectoparasitic nematodes compared with nonmycorrhizal control plants. In one of the few available studies conducted on nonagricultural plants, De La Peña et al. (2006) demonstrated that AM fungi inoculation of the sand dune grass *Ammophila arenaria* decreased nematode colonization and reproduction. This suppression of nematode activity was not caused by induced systemic resistance (ISR) but by local mechanisms. On the other hand, ISR mediated by AM fungi has been suggested in some studies as a mechanism to suppress root-feeding nematodes (Elsen et al., 2008; Hao et al., 2012; Vos et al., 2012a,b).

Interactions between root herbivores and EcM fungi have not been well studied because most likely, few commercially important agricultural plants host EcM fungi. However, the increasing number of invasive insects that cause significant mortality of commercially important tree species may trigger research in this area (Treu et al., 2014), even if the insects are primarily leaf-eaters.

9.4 MYCORRHIZAS AND FUNGIVORES

Fungivores include nematodes (Hussey and Roncadori, 1981; Ingham et al., 1985; Hua et al., 2014), protozoa (Petz et al., 1985; Hekman et al., 1992), earthworms (Heděnec et al., 2013), and microarthropods such as Collembola (Gange, 2000) and mites (Mitchell and Parkinson, 1976; Schneider et al., 2005) (Fig. 9.1). Soil fungivores are considered the main drivers of fungal community structure and fungal decomposition processes (Tordoff et al., 2008; Bardgett et al., 1993; Klironomos and Kendrick, 1995; Filser, 2002). Most research has concentrated on the negative effect of grazing on mycorrhizal hyphae (Boerner and Harris, 1991; Kaiser and Lussenhop, 1991) and spores (Bakonyi et al., 2002). However, there is some evidence that mycorrhizal fungi may also actively engage in "chemical warfare" with fungivores to gain privileged access to their nutrients (Klironomos and Hart, 2001; Perez-Moreno and Read, 2001).

The direct effects of mycorrhizal-feeding fungivores on mycorrhizal fungi and their host plants depends on their abundance (Finlay, 1985; Harris and Boerner, 1990; Klironomos and Ursic, 1998; Gange, 2000; Crowther et al., 2012) and can be positive (Lussenhop, 1996), neutral (Larsen and Jakobsen, 1996; Wurst et al., 2004) or negative (Warnock et al., 1982). In some cases, effects may be negative for one symbiotic partner and positive for the other (Lussenhop, 1996; Bakhtiar et al., 2001; Bakonyi et al., 2002; Hua et al., 2014), and the mechanisms for this may result from complex feedback loops. For example, Lussenhop (1996) interpreted the stimulated growth of mycorrhizal roots by fungivorous Collembola as compensatory growth for the reduction of mycorrhizas. Alternatively, preferential feeding by Collembola and other fungivores on saprotrophic fungi over mycorrhizal fungi can provide indirect positive feedback to mycorrhizal fungi (Green et al., 1999; Gange, 2000; Lindahl et al., 2001). For example, Crowther et al. (2013) showed that the presence of fungivorous isopods that preferentially fed on basidiomycete fungi increased the relative abundances of AM and other fungi. The increased fungal diversity associated with the presence of isopods promoted enzymatic activity linked to mineralization. Similarly to AM fungi, under certain conditions (e.g., at relatively low densities) fungivores can have positive effects on the host (Ek et al., 1994). In addition, Kaneda and Kaneko (2004) found that a Collembola species, *Folsomia candida*, preferred EcM fungal hyphae cut from the root 56h before feeding than freshly cut hyphae. This result indicates that Collembola may preferentially feed on dead mycorrhizal hyphae, thereby enhancing host-nutrient availability.

Understanding the community structure of fungivores combined with their feeding preferences for either mycorrhizal fungi or saprotrophs is important to predict SOM turnover caused by the competitive interactions between these two groups. Studies on these interactions are still sparse and have emphasized AM fungi (but Hiol et al., 1994; Kanters et al., 2015). Many studies have demonstrated that Collembola (Bardgett et al., 1993; Hiol et al., 1994; Klironomos and Ursic, 1998; Klironomos et al., 1999; Gange, 2000; Scheu and Simmerling, 2004; Heděnec et al., 2013), nematodes (Hasna et al., 2007), mites (Maraun et al., 2011), and earthworms (Bonkowski et al., 2000) have preferential feeding for saprotrophic fungal taxa rather than mycorrhizal fungi, which may in some cases only serve as choice of last resort (e.g., Schreiner and Bethlenfalvay, 2003). Nevertheless, mycorrhizal-feeding fungivores may contribute to regulate the flow of photosynthates below-ground from mycorrhizal plants (Johnson et al., 2005). $^{13}C/^{15}N$-labeling methods have suggested that photosynthesized C enters fungivores via grazing on EcM fungal hyphae (Pollierer et al., 2012). But again, this C pathway may not be the most important part of their diet (Potapov and Tiunov, 2016). Conversely, Schneider et al. (2005) demonstrated that the most preferred fungal types by three species of oribatid mites (*Carabodes femoralis*, *Nothrus silvestris* and *Oribatula tibialis*) included ericoid mycorrhizal fungi (*Hymenoscyphus ericae*), EcM fungi (*Boletus badius*), and saprotrophic fungi (*Alternaria alternata*) when given choices of 10 fungal species, including six EcM fungi, one ericoid fungus, and three saprotrophic fungal species. However, feeding preferences and selectiveness significantly differed among the three mite species, indicating that it is challenging to generalize feeding preferences for fungivorous mites.

A general limitation of fungivore feeding-preference experiments is that many species are polyphagous and can even feed on live roots and litter (Endlweber et al., 2009; Schreiner and Bethlenfalvay, 2003). Therefore the array of resources presented can be limiting by not representing the full spectrum of choices available in natural conditions. For example, some studies employed only mycorrhizal fungi, leaving no other feeding choices, which can potentially overestimate the negative effect of fungivores on mycorrhizal fungi (e.g., Warnock et al., 1982; Harris and Boerner, 1990). Schreiner and Bethlenfalvay (2003) observed that *Isotoma* species of Collembola preferred AM fungal hyphae of mixed *Glomus* species to mixed species of saprotrophic fungi. However, when crop residue was added as an alternative choice, the Collembola preferentially fed on the crop residue over the two types of mycorrhizal hyphae. Therefore, when assessing the feeding effects of fungivores on mycorrhizal fungi it is important to implement mixed diets and to consider that resources change in space and time (Scheu and Simmerling, 2004; Staaden et al., 2010).

Mechanisms that modulate fungivore foraging behavior include the use of olfactory cues to select preferred fungal taxa, alert for toxicity, and as indicators of prior hyphal grazing (e.g., Staaden et al., 2012). Klironomos et al. (1999) suggested thickness of hyphae and nutritional values as two potential explanations. Collembola and mites prefer thin rather than thick hyphal segments of AM fungi (Klironomos and Kendrick, 1996), and AM fungal hyphae tend to be thick and multilayered (Klironomos et al., 1999). Mycorrhizal hyphae and spores may be less nutritious than those of saprotrophic fungi, which has not been tested. Another possible explanation is the transfer of secondary metabolites from host plants to mycorrhizal fungi. Duhamel et al. (2013) demonstrated that when host plants (*P. lanceolata*) were exposed to fungivorous Collembola, catalpol, a secondary metabolite, was transferred from the plants to AM fungal hyphae and that hyphal biomass was not significantly affected. Additional data indeed show that mycorrhizal networks are conduits for the transfer of allelopathic compounds in soil (Barto et al., 2011; Achatz et al., 2014). Some mycorrhizal fungi have evolved diffusible compounds that are toxic to plants and soil biota (Kempken and Rohlfs, 2010). Streiblová et al. (2012) described the volatile organic compounds emitted by truffles such as *Tuber melanosporum*, which allow them to exclude competing biota from areas defined as *brûlé* (i.e., burnt), meaning that it is lacking plants and many members of soil biota. Martin et al. (2008) found a family of proteins, including one with a snake toxin–like domain, in a genome analysis of the EcM fungi *Laccaria bicolor*. Using microcosms, Klironomos and Hart (2001) demonstrated that *F. candida* Collembola died after exposure to *L. bicolor* and that the carcasses were internally infected by the fungus, which extracted N from the carcasses and mobilized it to their host plant, *Pinus strobus*. Conversely, another EcM fungus, *Cenococcum geophilum*, isolated from the same study site did not show any evidence of toxicity for the Collembola. How widespread toxicity is among mycorrhizal fungi is an area requiring more studies.

The feeding choices of fungivores can affect their reproductive ability (Klironomos and Ursic, 1998; Staaden et al., 2010; Hedĕnec et al., 2013). Larsen et al. (2008) investigated how fungal diets affected reproductive performances of *F. candida* and *Folsomia fimetaria*. The fungal diets included two AM fungi (*Glomus intraradices* and *Glomus invermaium*), root pathogenic fungi (*Rhizoctonia solani* and *Fusarium culmorum*), and saprotrophic fungi (*Penicillium hordei* and *Trichoderma harzianum*). Overall, AM fungi provided less reproductive success to the Collembola species than other fungi. For *F. fimetaria*, the two AM fungi provided the least reproductive success, but *F. candida* feeding on *G. intraradices* had intermediate reproductive success. These dietary preferences are consistent with findings by Klironomos et al. (1999); AM fungal species tended to provide less reproductive success to a species of Collembola, *A. alternata*, compared with saprotrophic fungal species. Mixed diets, even with low- and high-quality components, generally benefit Collembola reproduction (Scheu and Folger, 2004). However, such tendency in Collembola reproductive success with saprotrophic over mycorrhizal fungi may not apply to other soil fungivores. Ruess and Dighton (1996) and Ruess et al. (2000) found that a fungivorous nematode, *Aphelenchoides* spp., generally reproduced better on EcM fungal diet than with saprotrophic fungi extracted from a spruce forest soil. However, Giannakis and Sanders (1990) reported that EcM fungus *Laccaria laccata* enhanced the reproduction rates as much as the common mushroom (*Agaricus bisporus*) for three species of fungivorous nematodes, but this was not the case for the other five species of EcM fungi.

We propose that future research on mycorrhizal–fungivore interactions should continue to take advantage of high-throughput metagenomic identification tools (e.g., Anslan et al., 2016) combined with stable isotope tracking and/or the differential N and C isotope fractionation signatures associated with the different foraging habits of fungi (Griffith, 2004; Johnson et al., 2005; Mayor et al., 2009). In addition, little is known about certain groups. For instance, little is known about fungivorous protozoa feeding on mycorrhizal fungi and only few studies have focused on mycorrhizal interactions with protozoa in general. Nutrient flows through bacterivorous protozoa have been deemed more important than through fungivorous protozoa, but this view is changing as studies on the diversity and feeding preferences of fungivorous protozoa emerge (Geisen et al., 2016). In terms of mycorrhizal feedbacks, it is known that some protozoa can significantly contribute to mobilize N from SOM into AM fungi, whereas other protozoa in the rhizosphere stimulate photosynthesis through hormonal effects (Koller et al., 2013b). Future research needs to consider protozoa diversity, including unculturable species and their differential feeding preference among bacteria and fungal guilds.

One of the most notable implications of mycorrhizas–fungivore interactions is mycorrhizal dispersal. It has been long recognized that mycorrhizal propagules (i.e., hyphae and spores) have been found in gut contents, casts, and feces of Collembola (Klironomos and Moutoglis, 1999), mites (Lilleskov and Bruns, 2005), earthworms (Reddell and Spain, 1991; Gange, 1993; Zaller et al., 2011), and even fungal-feeding mammals (Mangan and Adler, 2002). Mycorrhizal propagules in casts and feces can be viable to infect potential host trees (Reddell and Spain, 1991; Mangan and Adler, 2002). Klironomos and Moutoglis (1999) observed that Collembola dispersed AM fungi several centimeters away from the rhizosphere and beyond what AM fungi could disperse on their own. Mangan and Adler (2002) demonstrated that spiny rats (*Proechimys semispinosus*) from islands in Panama feed on AM fungal spores as an alternative food source when fruits are not available. It is likely that other fungal-consuming soil macrofauna can disperse AM fungal propagules over long distances at different temporal scales. Further study of the mechanisms of mycorrhizal dispersal is important to elucidate both small and large scale patterns of mycorrhizal distribution (Davison et al., 2015).

9.5 MYCORRHIZAS AND BACTERIVORES

Soil bacterivores are represented mostly by protozoa and nematodes (Rønn et al., 2012). They interact with mycorrhizal fungi indirectly, and only few studies have investigated these interactions and effects on host plants (Trap et al., 2016). Bacterivores in the rhizosphere often increase nutrient availability (comprehensive list in Trap et al., 2016). This increased nutrient availability can stimulate plant growth (e.g., Ingham et al., 1985; Herdler et al., 2008; Ekelund et al., 2009) and reproduction (Bonkowski et al., 2001a,b; Krome et al., 2009). Bonkowski (2004) suggested that rhizosphere bacterivores increase N availability via the "microbial loop" (Clarholm, 1985) involving complex interactions among plant, bacteria, and bacterivores (but Ekelund et al., 2009). Through meta-analysis, Trap et al. (2016) found that, overall, the presence of bacterivores in the rhizosphere increased shoot and root biomass by over 20% compared with controls without bacterivores, with no significant change in shoot/root ratio.

Studies investigating interactions among host plants, bacterivores, AM fungi (Herdler et al., 2008; Koller et al., 2013a,b; but Wamberg et al., 2003a,b), and EcM fungi (Jentschke et al., 1995; Setälä et al., 1999; Bonkowski et al., 2001a,b; Irshad et al., 2012) indicate that the presence of both mycorrhizal fungi and bacterivores results in increased above-ground production of the host plants (Trap et al., 2016). The presence of AM fungi can reduce bacterivore populations (e.g., Wamberg et al., 2003a) by decreasing root exudation (Meier et al., 2013) and/or lateral root production (Jentschke et al., 1995; Bonkowski et al., 2001a,b; Herdler et al., 2008), which in turn can reduce bacterial populations. However, the potential counteracting effects between mycorrhizal fungi and bacterivores on root physiology, biomass, and architecture appear to still lead to a net positive effect in regard to stimulation of plant above-ground performance when both groups are present (Trap et al., 2016). Bonkowski et al. (2001a,b) postulated that plants might allocate their resources to optimize simultaneous exploitation of plant–mycorrhizal mutualism as well as beneficial effects of bacterivores via the microbial loop. Indeed, Koller et al. (2013a,b) demonstrated that the presence of both bacteria-feeding protozoa and AM fungi created an additive positive effect, resulting in significantly higher above-ground plant growth than presence of either protozoa or AM fungi alone.

Only few studies have specifically focused on reciprocal effects between EcM fungi and bacterivores, and the results are inconclusive; Jentschke et al. (1995) reported that the presence of EcM fungi increased the number of bacterivores. In contrast, Irshad et al. (2012) found that EcM fungi caused reductions in both bacteria and bacterivore nematodes.

In conclusion, positive feedback loops among plant host, mycorrhizal fungi, and bacterivores appear to be prevalent. However, more ecologically relevant studies are needed to reveal how mycorrhizal fungi and bacterivores affect each other.

9.6 MYCORRHIZAS AND HIGHER TROPHIC LEVELS

Soil food webs are considered to be donor-controlled systems; the resource density controls consumer density, but the reverse does not occur, especially for fungi (Persson et al., 1996); thus basal resources, including mycorrhizal fungi, are not considered to be affected by soil animals at higher trophic levels. For instance, Mikola and Setälä (1998) could not find evidence of top-down trophic cascades on microbial biomass in a soil microcosm experiment using soil food webs up to the third level with a predatory nematode species. However, Bengtsson et al. (1996) suggest that there are empirical and theoretical grounds to challenge the concept. More and more studies support that top-down trophic cascades can be significant enough to influence biomass and structure of soil microbes, including mycorrhizal fungi as well as plants. One example was demonstrated by Bradford et al. (2002) using different animal size classes to manipulate functional groups to partly reflect their trophic levels (Turnbull et al., 2014). Their objective was to investigate food web effects on various ecological functions, including mycorrhizal and plant community composition. Bradford et al. (2002) found that microcosms including soil animals in all the size classes significantly reduced AM fungal colonization and root biomass compared with those without macrofauna, indicating a top-down trophic cascade to basal resources including roots and associated mycorrhizal fungi. However, the manipulation of soil animals by size did not affect NPP or net ecosystem productivity, most likely because positive and negative effects of the manipulations canceled each other out. This indicates that interactions among plants, mycorrhizal fungi, and soil animals in multiple trophic levels in a food web can be complex, and specific interactions between trophic levels involving mycorrhizal fungi described in the previous sections can be obscured by other interactions caused by different groups in the soil food web. This was suggested by Bradford et al. (2002) and Ladygina et al. (2010); effects of one functional group can be canceled out by another group in complex trophic interactions of soil food webs, resulting in similar overall plant productivity among treatments with different combinations of functional groups.

Effects of soil food web community on plant performance (e.g., NPP) is often a primary interest in many soil food web studies, especially those related to agricultural plants (e.g., McAllister et al., 1997). However, many other ecosystem processes, including nutrient mineralization, SOM formation, greenhouse gas fluxes, and soil moisture retention, are also essential ecosystem services to humanity (Wall et al., 2012). Thus biodiversity in a soil food web and associated complex interactions and feedbacks among various players, including mycorrhizal fungi and animals at higher trophic levels, in the web should contribute to "multifunctionality" (Maestre et al., 2012); an ecosystem can maintain various functions, and its below-ground food web is a critical component for all the functions (Barrios, 2007). The importance of diversity in soil food webs, including mycorrhizal fungi, for multifunctionality was demonstrated by Wagg et al. (2014). By manipulating soil food web structure using different size classes of soil organisms in a manner similar to that by Bradford et al. (2002), Wagg et al. (2014) showed that biodiversity of soil food webs controlled many ecosystem functions, and the biodiversity and multifunctionality indices were significantly correlated with each other. This is consistent with the result of a study by de Vries et al. (2013), which demonstrated that soil food web properties were significant predictors of ecosystem processes, including C and N cycling across European land use systems. Given the evidence to support that mycorrhizal fungi play major roles in soil food webs, mycorrhizal fungi must significantly contribute to ecosystem multifunctionality.

9.7 THE WAY FORWARD

More than a decade ago, Scheu (2002) suggested two lines of research as important to advance soil food web ecology in the future: the adoption of new methodologies and better experimentation, including more food web components. The first suggestion has been well implemented; new technologies in molecular and stable isotope methods have contributed to our understanding of soil food web ecology (Traugott et al., 2013). For instance, advances in molecular methods such as high-throughput sequencing and bioinformatics have helped us better understand the community structure of soil microbes, including mycorrhizal fungi, which are the basal resources in soil food webs (e.g., Nuccio et al., 2013; Crowther et al., 2013). One future application of high-throughput sequencing can be the investigation of microbivore diets. For instance, it can be used to investigate gut contents of fungivores (e.g., Collembola and earthworms) to explore the relative importance between saprotrophic and mycorrhizal fungi as food sources (Remén et al., 2010; Greenstone et al., 2012).

Advances and popularity of stable isotope analyses for C and N have helped us understand energy and nutrient flow through soil food webs (e.g., Pollierer et al., 2007) and assess food web compartmentalization and trophic levels (Pollierer et al., 2009; Crotty et al., 2012). Notable techniques include compound-specific stable isotope analyses of biomarkers such as phospholipids and neutral and lipid fatty acids (e.g., Olsson and Johnson, 2005; Drigo et al., 2010; Pollierer et al., 2012). Developing stable isotope instruments such as nanoscale secondary ion mass spectrometry (nanoSIMS) (Wagner, 2009; Hatton et al., 2012) can be applied for soil food web ecology in the future. For instance, nanoSIMS can quantify incorporation of substrates labeled with stable isotopes (e.g., ^{13}C and ^{15}N) of single microbes and animals (Musat et al., 2012) such as bacteria (Lechene et al., 2006), cyanobacteria (Ploug et al., 2010, 2011), and zooplankton (Eybe et al., 2009). Thus nanoSIMS can be a promising technology to reveal energy and nutrient transfer in soil food webs.

In regard to the second suggestion by Scheu (2002), i.e., better experimentation including more food web components, it has only been implemented to a limited extent. Two studies using manipulation of functional groups in soil food webs via body sizes by Bradford et al. (2002) and Wagg et al. (2014) described above (Section 9.6), and a full-factorial experiment with three different functional groups by Ladygina et al. (2010) are three of the few studies which strived to assess effects of multiple functional groups on ecosystem processes. Most of the studies investigating the relationship between mycorrhizal fungi and soil organisms at higher trophic levels have used simple microcosms or mesocosms with reduced factors. This reductionist approach is often necessary for theoretical and practical reasons to investigate specific cause-and-effect relationships among well-defined groups in trophic levels, keeping experiments manageable with limited resources. However, in the real world, multifaceted complexity exists in any given soil food web, such as many plant species (both mycorrhizal and nonmycorrhizal) interacting with multiple mycorrhizal fungal species, various plant species of different functional groups at different developmental stages and many more species of soil organisms at different trophic levels (Beare et al., 1995). A soil food web in a real world is truly a "tangled bank," as Charles Darwin (1859) described to emphasize the complexity of ecosystems in *On the Origin of Species*. Therefore the interactions and feedbacks among mycorrhizal fungi, their host plants, and specific groups in the soil food webs should be interpreted and viewed in the context of interactivity (Moore et al., 2003). To advance our understanding of complex interactions among multiple trophic levels in soil food webs and their effects on ecosystem processes, we need more food web experiments using innovative manipulation means for functional groups, such as body sizes as employed by Bradford et al. (2002).

An alternative approach to account for complexity in soil food webs is to use casual correlations based on a priori knowledge for observational data, such as de Vries et al. (2013), who found that food web characteristics were significant predictors for C and N biogeochemistry in soils. Similar approaches in combination with analytical methods requiring high computational power such as structural equation models (Grace et al., 2012) will be a future direction for soil food web ecology.

Another way to contend with the complex nature of dynamics of soil food webs and their resulting ecosystem functions is modeling. This approach has been used since the 1980s (e.g., Hunt et al., 1987; Moore and William Hunt, 1988; Moore et al., 1993; Ke et al., 2015) and continues to contribute to our understanding of energy and nutrient flows through soil food webs. Modern computational capacity helps us simulate food web dynamics with reasonable costs and time and enhance our understanding of complex interactions and feedbacks among players in a soil food web, including mycorrhizal fungi. For instance, Moore et al. (2014) added extracellular enzymes and root structure in two-dimensional space in detrital food webs with two trophic levels to simulate energy dynamics and system stability. Similar simulation models with more trophic levels are feasible with increasing computational power with reasonable costs in the future.

Acknowledgments

This work was funded by an NSERC Discovery Grant and Canada Research Chair to PMA. We are grateful to Angeline Castilloux and Tiina Keranen for their support designing Fig. 9.1. We are also thankful to Dr. Robert Koller for his thorough review of our manuscript and valuable feedback.

References

Achatz, M., Morris, E.K., Müller, F., Hilker, M., Rillig, M.C., 2014. Soil hypha-mediated movement of allelochemicals: arbuscular mycorrhizae extend the bioactive zone of juglone. Functional Ecology 28 (4), 1020–1029.

Anslan, S., Bahram, M., Tedersoo, L., 2016. Temporal changes in fungal communities associated with guts and appendages of Collembola as based on culturing and high-throughput sequencing. Soil Biology and Biochemistry 96, 152–159.

Argüello, A., O'Brien, M.J., van der Heijden, M.G.A., Wiemken, A., Schmid, B., Niklaus, P.A., 2016. Options of partners improve carbon for phosphorus trade in the arbuscular mycorrhizal mutualism. Ecology Letters 19 (6), 648–656.

Artursson, V., Finlay, R.D., Jansson, J.K., 2006. Interactions between arbuscular mycorrhizal fungi and bacteria and their potential for stimulating plant growth. Environmental Microbiology 8 (1), 1–10.

Averill, C., Turner, B.L., Finzi, A.C., 2014. Mycorrhiza-mediated competition between plants and decomposers drives soil carbon storage. Nature 505 (7484), 543–545.

Azcon, R., Ocampo, J., 1981. Factors affecting the vesicular-arbuscular infection and mycorrhizal dependency of thirteen wheat cultivars. New Phytologist 87 (4), 677–685.

Bago, B., Vierheilig, H., Piche, Y., Azcon-Aguilar, C., 1996. Nitrate depletion and pH changes induced by the extraradical mycelium of the arbuscular mycorrhizal fungus *Glomus intraradices* grown in monoxenic culture. New Phytologist 133 (2), 273–280.

Bakhtiar, Y., Miller, D., Cavagnaro, T., Smith, S., 2001. Interactions between two arbuscular mycorrhizal fungi and fungivorous nematodes and control of the nematode with fenamifos. Applied Soil Ecology 17 (2), 107–117.

Bakonyi, G., Posta, K., Kiss, I., Fabian, M., Nagy, P., Nosek, J., 2002. Density-dependent regulation of arbuscular mycorrhiza by collembola. Soil Biology and Biochemistry 34 (5), 661–664.

Bardgett, R., Whittaker, J., Frankland, J., 1993. The diet and food preferences of *Onychiurus procampatus* (Collembola) from upland grassland soils. Biology and Fertility of Soils 16 (4), 296–298.

Barrios, E., 2007. Soil biota, ecosystem services and land productivity. Ecological Economics 64 (2), 269–285.

Barto, E.K., Hilker, M., Müller, F., Mohney, B.K., Weidenhamer, J.D., Rillig, M.C., 2011. The fungal fast lane: common mycorrhizal networks extend bioactive zones of allelochemicals in soils. PLoS One 6 (11), e27195.

Bauer, J.T., Kleczewski, N.M., Bever, J.D., Clay, K., Reynolds, H.L., 2012. Nitrogen-fixing bacteria, arbuscular mycorrhizal fungi, and the productivity and structure of prairie grassland communities. Oecologia 170 (4), 1089–1098.

Beare, M., Coleman, D., Crossley Jr., D., Hendrix, P., Odum, E., 1995. A Hierarchical Approach to Evaluating the Significance of Soil Biodiversity to Biogeochemical Cycling. The Significance and Regulation of Soil Biodiversity. Springer, pp. 5–22.

Bengtsson, J., Setälä, H., Zheng, D.W., 1996. Food webs and nutrient cycling in soils: interactions and positive feedbacks. In: Food Webs. Springer, pp. 30–38.

Bever, J.D., 2002. Negative feedback within a mutualism: host-specific growth of mycorrhizal fungi reduces plant benefit. Proceedings of the Royal Society of London Series B-Biological Sciences 269 (1509), 2595–2601.

Bever, J.D., 2015. Preferential allocation, physio-evolutionary feedbacks, and the stability and environmental patterns of mutualism between plants and their root symbionts. New Phytologist 205 (4), 1503–1514.

Bever, J.D., Richardson, S.C., Lawrence, B.M., Holmes, J., Watson, M., 2009. Preferential allocation to beneficial symbiont with spatial structure maintains mycorrhizal mutualism. Ecology Letters 12 (1), 13–21.

Bever, J.D., Platt, T.G., Morton, E.R., 2012. Microbial population and community dynamics on plant roots and their feedbacks on plant communities. Annual Review of Microbiology 66, 265–283.

Boerner, R.E.J., Harris, K.K., 1991. Effects of collembola (arthropoda) and relative germination date on competition between mycorrhizal *Panicum virgatum* (Poaceae) and nonmycorrhizal *Brassica nigra* (Brassicaceae). Plant and Soil 136 (1), 121–129.

Bonkowski, M., 2004. Protozoa and plant growth: the microbial loop in soil revisited. New Phytologist 162 (3), 617–631.

Bonkowski, M., Geoghegan, I.E., Birch, A.N.E., Griffiths, B.S., 2001a. Effects of soil decomposer invertebrates (protozoa and earthworms) on an above-ground phytophagous insect (cereal aphid) mediated through changes in the host plant. Oikos 95 (3), 441–450.

Bonkowski, M., Griffiths, B.S., Ritz, K., 2000. Food preferences of earthworms for soil fungi. Pedobiologia 44 (6), 666–676.

Bonkowski, M., Jentschke, G., Scheu, S., 2001b. Contrasting effects of microbial partners in the rhizosphere: interactions between Norway Spruce seedlings (*Picea abies* Karst.), mycorrhiza (*Paxillus involutus* (Batsch) Fr.) and naked amoebae (protozoa). Applied Soil Ecology 18 (3), 193–204.

Borowicz, V.A., 2001. Do arbuscular mycorrhizal fungi alter plant-pathogen relations? Ecology 82 (11), 3057–3068.

Bradford, M.A., Tordoff, G.M., Eggers, T., Jones, T.H., Newington, J.E., 2002. Microbiota, fauna, and mesh size interactions in litter decomposition. Oikos 99 (2), 317–323.

Brito, I., de Carvalho, M., Goss, M.J., 2009. Techniques for arbuscular mycorrhiza inoculum reduction. In: Symbiotic Fungi. Springer, Berlin, Heidelberg, pp. 307–318.

Brown, V.K., Gange, A.C., 1990. Insect herbivory below ground. Advances in Ecological Research 20, 1–58.

Brundrett, M.C., 2006. Understanding the roles of multifunctional mycorrhizal and endophytic fungi. In: Microbial Root Endophytes. Springer, pp. 281–298.

Burke, R., Cairney, J., 2002. Laccases and other polyphenol oxidases in ecto-and ericoid mycorrhizal fungi. Mycorrhiza 12 (3), 105–116.

Cappellazzo, G., Lanfranco, L., Bonfante, P., 2007. A limiting source of organic nitrogen induces specific transcriptional responses in the extraradical structures of the endomycorrhizal fungus Glomus intraradices. Current Genetics 51 (1), 59–70.

Cheng, L., Booker, F.L., Tu, C., Burkey, K.O., Zhou, L., Shew, H.D., Rufty, T.W., Hu, S., 2012. Arbuscular mycorrhizal fungi increase organic carbon decomposition under elevated CO_2. Science 337 (6098), 1084–1087.

Clarholm, M., 1985. Possible roles for roots, bacteria, protozoa and fungi in supplying nitrogen to plants. *Special Publications Series of the British Ecological Society*. In: Fitter, A.H., Atkinson, D., Read, D.J., Usher, M.B. (Eds.), Ecological interactions in soil, Blackwell Scientific Publications, pp. 355–365.

Clark, D.A., Brown, S., Kicklighter, D.W., Chambers, J.Q., Thomlinson, J.R., Ni, J., 2001. Measuring net primary production in forests: concepts and field methods. Ecological Applications 11 (2), 356–370.

Clemmensen, K.E., Finlay, R.D., Dahlberg, A., Stenlid, J., Wardle, D.A., Lindahl, B.D., 2015. Carbon sequestration is related to mycorrhizal fungal community shifts during long-term succession in boreal forests. New Phytologist 205 (4), 1525–1536.

Courty, P.E., Pritsch, K., Schloter, M., Hartmann, A., Garbaye, J., 2005. Activity profiling of ectomycorrhiza communities in two forest soils using multiple enzymatic tests. New Phytologist 167 (1), 309–319.

Crotty, F.V., Adl, S.M., Blackshaw, R.P., Murray, P.J., 2012. Using stable isotopes to differentiate trophic feeding channels within soil food webs. Journal of Eukaryotic Microbiology 59 (6), 520–526.

Crowther, T.W., Boddy, L., Jones, T.H., 2012. Functional and ecological consequences of saprotrophic fungus–grazer interactions. The ISME Journal 6 (11), 1992–2001.

Crowther, T.W., Stanton, D.W., Thomas, S.M., A'Bear, A.D., Hiscox, J., Jones, T.H., Voříšková, J., Baldrian, P., Boddy, L., 2013. Top-down control of soil fungal community composition by a globally distributed keystone consumer. Ecology 94 (11), 2518–2528.

Darwin, C., 1859. On the Origin of Species (John Murray, London), Mentor ed. New American Library, New York City.

Davison, J., Moora, M., Öpik, M., Adholeya, A., Ainsaar, L., Ba, A., Burla, S., Diedhiou, A., Hiiesalu, I., Jairus, T., 2015. Global assessment of arbuscular mycorrhizal fungus diversity reveals very low endemism. Science 349 (6251), 970–973.

De La Peña, E., Echeverría, S.R., Van Der Putten, W.H., Freitas, H., Moens, M., 2006. Mechanism of control of root-feeding nematodes by mycorrhizal fungi in the dune grass *Ammophila arenaria*. New Phytologist 169 (4), 829–840.

de Vries, F.T., Thebault, E., Liiri, M., Birkhofer, K., Tsiafouli, M.A., Bjornlund, L., Jorgensen, H.B., Brady, M.V., Christensen, S., de Ruiter, P.C., et al., 2013. Soil food web properties explain ecosystem services across European land use systems. Proceedings of the National Academy of Sciences 110 (35), 14296–14301.

Denton, C.S., Bardgett, R.D., Cook, R., Hobbs, P.J., 1998. Low amounts of root herbivory positively influence the rhizosphere microbial community in a temperate grassland soil. Soil Biology and Biochemistry 31 (1), 155–165.

Drigo, B., Pijl, A.S., Duyts, H., Kielak, A.M., Gamper, H.A., Houtekamer, M.J., Boschker, H.T., Bodelier, P.L., Whiteley, A.S., van Veen, J.A., 2010. Shifting carbon flow from roots into associated microbial communities in response to elevated atmospheric CO_2. Proceedings of the National Academy of Sciences 107 (24), 10938–10942.

Duhamel, M., Pel, R., Ooms, A., Bucking, H., Jansa, J., Ellers, J., van Straalen, N.M., Wouda, T., Vandenkoornhuyse, P., Kiers, E.T., 2013. Do fungivores trigger the transfer of protective metabolites from host plants to arbuscular mycorrhizal hyphae? Ecology 94 (9), 2019–2029.

Ehrenfeld, J.G., Ravit, B., Elgersma, K., 2005. Feedback in the plant-soil system. Annual Review of Environment and Resources 30, 75–115.

Eisenhauer, N., Reich, P.B., Scheu, S., 2012. Increasing plant diversity effects on productivity with time due to delayed soil biota effects on plants. Basic and Applied Ecology 13 (7), 571–578.

Ek, H., Sjögren, M., Arnebrant, K., Söderström, B., 1994. Extramatrical mycelial growth, biomass allocation and nitrogen uptake in ectomycorrhizal systems in response to collembolan grazing. Applied Soil Ecology 1 (2), 155–169.

Ekelund, F., Saj, S., Vestergård, M., Bertaux, J., Mikola, J., 2009. The "soil microbial loop" is not always needed to explain protozoan stimulation of plants. Soil Biology and Biochemistry 41 (11), 2336–2342.

Elsen, A., Gervacio, D., Swennen, R., De Waele, D., 2008. AMF-induced biocontrol against plant parasitic nematodes in *Musa* sp.: a systemic effect. Mycorrhiza 18 (5), 251–256.

Endlweber, K., Ruess, L., Scheu, S., 2009. Collembola switch diet in presence of plant roots thereby functioning as herbivores. Soil Biology and Biochemistry 41 (6), 1151–1154.

Eybe, T., Bohn, T., Audinot, J., Udelhoven, T., Cauchie, H., Migeon, H., Hoffmann, L., 2009. Uptake visualization of deltamethrin by NanoSIMS and acute toxicity to the water flea *Daphnia magna*. Chemosphere 76 (1), 134–140.

Ezawa, T., Ikeda, Y., Shimura, H., Masuta, C., 2015. Detection and characterization of mycoviruses in arbuscular mycorrhizal fungi by deep-sequencing. Plant Virology Protocols: New Approaches to Detect Viruses and Host Responses 171–180.

Fernandez, C.W., Langley, J.A., Chapman, S., McCormack, M.L., Koide, R.T., 2016. The decomposition of ectomycorrhizal fungal necromass. Soil Biology and Biochemistry 93, 38–49.

Filser, J., 2002. The role of Collembola in carbon and nitrogen cycling in soil. Pedobiologia 46 (3), 234–245.

Finlay, R., 1985. Interactions between soil micro-arthropods and endomycorrhizal associations of higher plants. Special Publications Series of the British Ecological Society. In: Fitter, A.H. (Ed.), Ecological Interactions in Soil, Blackwell Scientific Publications, Oxford, UK, pp 319–332.

Frank, A., 1894. Die bedeutung der mykorrhizapilze für die gemeine Kiefer. Forstwissenschaftliches Centralblatt 16, 185–190.

Frey-Klett, P., Garbaye, J., Tarkka, M., 2007. The mycorrhiza helper bacteria revisited. New Phytologist 176 (1), 22–36.

Gange, A., 2000. Arbuscular mycorrhizal fungi, Collembola and plant growth. Trends in Ecology & Evolution 15 (9), 369–372.

Gange, A.C., 1993. Translocation of mycorrhizal fungi by earthworms during early succession. Soil Biology and Biochemistry 25 (8), 1021–1026.

Gange, A.C., 2001. Species-specific responses of a root- and shoot-feeding insect to arbuscular mycorrhizal colonization of its host plant. New Phytologist 150 (3), 611–618.

Geisen, S., Koller, R., Hünninghaus, M., Dumack, K., Urich, T., Bonkowski, M., 2016. The soil food web revisited: diverse and widespread mycophagous soil protists. Soil Biology and Biochemistry 94, 10–18.

Geisseler, D., Horwath, W.R., Joergensen, R.G., Ludwig, B., 2010. Pathways of nitrogen utilization by soil microorganisms–a review. Soil Biology and Biochemistry 42 (12), 2058–2067.

Giannakis, N., Sanders, F.E., 1990. Interactions between mycophagous nematodes, mycorrhizal and other soil fungi. Agriculture, Ecosystems & Environment 29 (1–4), 163–167.

Grace, J.B., Schoolmaster, D.R., Guntenspergen, G.R., Little, A.M., Mitchell, B.R., Miller, K.M., Schweiger, E.W., 2012. Guidelines for a graph-theoretic implementation of structural equation modeling. Ecosphere 3 (8), 1–44.

Green, H., Larsen, J., Olsson, P.A., Jensen, D.F., Jakobsen, I., 1999. Suppression of the biocontrol agent *Trichoderma harzianum* by mycelium of the arbuscular mycorrhizal fungus *Glomus intraradices* in root-free soil. Applied and Environmental Microbiology 65 (4), 1428–1434.

Greenstone, M.H., Weber, D.C., Coudron, T.A., Payton, M.E., Hu, J.S., 2012. Removing external DNA contamination from arthropod predators destined for molecular gut-content analysis. Molecular Ecology Resources 12 (3), 464–469.

Griffith, G.W., 2004. The use of stable isotopes in fungal ecology. Mycologist 18 (04), 177–183.

Guo, D., Mitchell, R.J., Withington, J.M., Fan, P.-P., Hendricks, J.J., 2008a. Endogenous and exogenous controls of root life span, mortality and nitrogen flux in a longleaf pine forest: root branch order predominates. Journal of Ecology 96 (4), 737–745.

Guo, D., Xia, M., Wei, X., Chang, W., Liu, Y., Wang, Z., 2008b. Anatomical traits associated with absorption and mycorrhizal colonization are linked to root branch order in twenty-three Chinese temperate tree species. New Phytologist 180 (3), 673–683.

Hao, Z., Fayolle, L., van Tuinen, D., Chatagnier, O., Li, X., Gianinazzi, S., Gianinazzi-Pearson, V., 2012. Local and systemic mycorrhiza-induced protection against the ectoparasitic nematode *Xiphinema* index involves priming of defence gene responses in grapevine. Journal of Experimental Botany 63 (10), 3657–3672.

Hardoim, P.R., van Overbeek, L.S., Berg, G., Pirttilä, A.M., Compant, S., Campisano, A., Döring, M., Sessitsch, A., 2015. The hidden world within plants: ecological and evolutionary considerations for defining functioning of microbial endophytes. Microbiology and Molecular Biology Reviews 79 (3), 293–320.

Harris, K.K., Boerner, R., 1990. Effects of belowground grazing by collembola on growth, mycorrhizal infection, and P uptake of *Geranium robertianum*. Plant and Soil 129 (2), 203–210.

Hartnett, D.C., Wilson, G.W., 2002. The Role of Mycorrhizas in Plant Community Structure and Dynamics: Lessons from Grasslands. In Diversity and Integration in Mycorrhizas. Springer, Netherlands, pp. 319–331.

Hasna, M., Insunza, V., Lagerlöf, J., Rämert, B., 2007. Food attraction and population growth of fungivorous nematodes with different fungi. Annals of Applied Biology 151 (2), 175–182.

Hatton, P.J., Remusat, L., Zeller, B., Derrien, D., 2012. A multi-scale approach to determine accurate elemental and isotopic ratios by nano-scale secondary ion mass spectrometry imaging. Rapid Communications in Mass Spectrometry 26 (11), 1363–1371.

Heděnec, P., Radochová, P., Nováková, A., Kaneda, S., Frouz, J., 2013. Grazing preference and utilization of soil fungi by *Folsomia candida* (Isotomidae: Collembola). European Journal of Soil Biology 55, 66–70.

Hekman, W.E., Boogert, P.J., Zwart, K.B., 1992. The physiology and ecology of a novel, obligate mycophagous flagellate. FEMS Microbiology Letters 86 (3), 255–265.
Herdler, S., Kreuzer, K., Scheu, S., Bonkowski, M., 2008. Interactions between arbuscular mycorrhizal fungi (*Glomus intraradices*, Glomeromycota) and amoebae (*Acanthamoeba castellanii*, Protozoa) in the rhizosphere of rice (*Oryza sativa*). Soil Biology and Biochemistry 40 (3), 660–668.
Herre, E.A., Mejia, L.C., Kyllo, D.A., Rojas, E., Maynard, Z., Butler, A., Van Bael, S.A., 2007. Ecological implications of anti-pathogen effects of tropical fungal endophytes and mycorrhizae. Ecology 88 (3), 550–558.
Hiol, F.H., Dixon, R.K., Curl, E.A., 1994. The feeding preference of myophagous collembola varies with the ectomycorrhizal symbiont. Mycorrhiza 5 (2), 99–103.
Hobbie, E.A., 2006. Carbon allocation to ectomycorrhizal fungi correlates with belowground allocation in culture studies. Ecology 87 (3), 563–569.
Hodge, A., 2014. Interactions between arbuscular mycorrhizal fungi and organic material substrates. Advances in Applied Microbiology 89, 47.
Hol, W.G., Cook, R., 2005. An overview of arbuscular mycorrhizal fungi–nematode interactions. Basic and Applied Ecology 6 (6), 489–503.
Hua, J., Jiang, Q., Bai, J., Ding, F., Lin, X., Yin, Y., 2014. Interactions between arbuscular mycorrhizal fungi and fungivorous nematodes on the growth and arsenic uptake of tobacco in arsenic-contaminated soils. Applied Soil Ecology 84, 176–184.
Hunt, H., Coleman, D., Ingham, E., Ingham, R.E., Elliott, E., Moore, J., Rose, S., Reid, C., Morley, C., 1987. The detrital food web in a shortgrass prairie. Biology and Fertility of Soils 3 (1–2), 57–68.
Hussey, R., Roncadori, R., 1981. Influence of *Aphelenchus avenae* on vesicular-arbuscular endomycorrhizal growth response in cotton. Journal of Nematology 13 (1), 48.
Hussey, R., Roncadorl, R., 1982. Vesicular-arbuscular mycorrhizae may limit nematode activity and improve plant growth. Plant Disease 66 (1), 9–14.
Ingham, R., 1988. Interactions between nematodes and vesicular-arbuscular mycorrhizae. Agriculture, Ecosystems & Environment 24 (1), 169–182.
Ingham, R.E., Trofymow, J., Ingham, E.R., Coleman, D.C., 1985. Interactions of bacteria, fungi, and their nematode grazers: effects on nutrient cycling and plant growth. Ecological Monographs 55 (1), 119–140.
Irshad, U., Brauman, A., Villenave, C., Plassard, C., 2012. Phosphorus acquisition from phytate depends on efficient bacterial grazing, irrespective of the mycorrhizal status of *Pinus pinaster*. Plant and Soil 358 (1–2), 155–168.
Jackson, R.B., Mooney, H.A., Schulze, E.-D., 1997. A global budget for fine root biomass, surface area, and nutrient contents. Proceedings of the National Academy of Sciences 94 (14), 7362–7366.
Jentschke, G., Bonkowski, M., Godbold, D.L., Scheu, S., 1995. Soil protozoa and forest tree growth: non-nutritional effects and interaction with mycorrhizae. Biology and Fertility of Soils 20 (4), 263–269.
Johnson, D., Krsek, M., Wellington, E.M., Stott, A.W., Cole, L., Bardgett, R.D., Read, D.J., Leake, J.R., 2005. Soil invertebrates disrupt carbon flow through fungal networks. Science 309 (5737), 1047.
Johnson, N.C., 2010. Resource stoichiometry elucidates the structure and function of arbuscular mycorrhizas across scales. New Phytologist 185 (3), 631–647.
Johnson, S.N., Rasmann, S., 2015. Root-feeding insects and their interactions with organisms in the rhizosphere. Annual Review of Entomology 60, 517–535.
Joner, E., Briones, R., Leyval, C., 2000. Metal-binding capacity of arbuscular mycorrhizal mycelium. Plant and Soil 226 (2), 227–234.
Jonsson, L.M., Nilsson, M.-C., Wardle, D.A., Zackrisson, O., 2001. Context dependent effects of ectomycorrhizal species richness on tree seedling productivity. Oikos 93 (3), 353–364.
Jung, S.C., Martinez-Medina, A., Lopez-Raez, J.A., Pozo, M.J., 2012. Mycorrhiza-induced resistance and priming of plant defenses. Journal of Chemical Ecology 38 (6), 651–664.
Kaiser, P.A., Lussenhop, J., 1991. Collembolan effects on establishment of vesicular – arbuscular mycorrhizae in soybean (*Glycine max*). Soil Biology & Biochemistry 23 (3), 307–308.
Kaneda, S., Kaneko, N., 2004. The feeding preference of a collembolan (*Folsomia candida* Willem) on ectomycorrhiza (*Pisolithus tinctorius* (Pers.)) varies with mycelial growth condition and vitality. Applied Soil Ecology 27 (1), 1–5.
Kanters, C., Anderson, I.C., Johnson, D., 2015. Chewing up the wood-wide web: selective grazing on ectomycorrhizal fungi by Collembola. Forests 6 (8), 2560–2570.
Kardol, P., Veen, G.F., Teste, F.P., Perring, M.P., 2015. Peeking into the black box: a trait-based approach to predicting plant–soil feedback. New Phytologist 206 (1), 1–4.

Ke, P.-J., Miki, T., Ding, T.-S., 2015. The soil microbial community predicts the importance of plant traits in plant–soil feedback. New Phytologist 206 (1), 329–341.

Kempken, F., Rohlfs, M., 2010. Fungal secondary metabolite biosynthesis–a chemical defence strategy against antagonistic animals? Fungal Ecology 3 (3), 107–114.

Kiers, E.T., Duhamel, M., Beesetty, Y., Mensah, J.A., Franken, O., Verbruggen, E., Fellbaum, C.R., Kowalchuk, G.A., Hart, M.M., Bago, A., et al., 2011. Reciprocal rewards stabilize cooperation in the mycorrhizal symbiosis. Science 333 (6044), 880–882.

Kitahara, R., Ikeda, Y., Shimura, H., Masuta, C., Ezawa, T., 2014. A unique mitovirus from Glomeromycota, the phylum of arbuscular mycorrhizal fungi. Archives of Virology 159 (8), 2157–2160.

Klein, T., Siegwolf, R.T.W., Körner, C., 2016. Belowground carbon trade among tall trees in a temperate forest. Science 352 (6283), 342–344.

Klironomos, J., Bednarczuk, E., Neville, J., 1999. Reproductive significance of feeding on saprobic and arbuscular mycorrhizal fungi by the collembolan, *Folsomia candida*. Functional Ecology 13 (6), 756–761.

Klironomos, J., Ursic, M., 1998. Density-dependent grazing on the extraradical hyphal network of the arbuscular mycorrhizal fungus, *Glomus intraradices*, by the collembolan, *Folsomia candida*. Biology and Fertility of Soils 26 (3), 250–253.

Klironomos, J.N., Hart, M.M., 2001. Food-web dynamics: animal nitrogen swap for plant carbon. Nature 410 (6829), 651–652.

Klironomos, J.N., Hart, M.M., Neville, J., 2000. The influence of arbuscular mycorrhizae on the relationship between plant diversity and productivity. Ecology Letters 3 (2), 137–141.

Klironomos, J., Kendrick, B., 1995. Relationships among microarthropods, fungi, and their environment. Plant and Soil 170 (1), 183–197.

Klironomos, J.N., Kendrick, W.B., 1996. Palatability of microfungi to soil arthropods in relation to the functioning of arbuscular mycorrhizae. Biology and Fertility of Soils 21 (1–2), 43–52.

Klironomos, J.N., Moutoglis, P., 1999. Colonization of nonmycorrhizal plants by mycorrhizal neighbours as influenced by the collembolan, Folsomia candida. Biology and Fertility of Soils 29 (3), 277–281.

Koide, R.T., Sharda, J.N., Herr, J.R., Malcolm, G.M., 2008. Ectomycorrhizal fungi and the biotrophy–saprotrophy continuum. New Phytologist 178 (2), 230–233.

Koller, R., Rodriguez, A., Robin, C., Scheu, S., Bonkowski, M., 2013a. Protozoa enhance foraging efficiency of arbuscular mycorrhizal fungi for mineral nitrogen from organic matter in soil to the benefit of host plants. New Phytologist 199 (1), 203–211.

Koller, R., Scheu, S., Bonkowski, M., Robin, C., 2013b. Protozoa stimulate N uptake and growth of arbuscular mycorrhizal plants. Soil Biology and Biochemistry 65, 204–210.

Kramer, S., Marhan, S., Ruess, L., Armbruster, W., Butenschoen, O., Haslwimmer, H., Kuzyakov, Y., Pausch, J., Scheunemann, N., Schoene, J., et al., 2012. Carbon flow into microbial and fungal biomass as a basis for the belowground food web of agroecosystems. Pedobiologia 55 (2), 111–119.

Krome, K., Rosenberg, K., Bonkowski, M., Scheu, S., 2009. Grazing of protozoa on rhizosphere bacteria alters growth and reproduction of *Arabidopsis thaliana*. Soil Biology and Biochemistry 41 (9), 1866–1873.

Ladygina, N., Henry, F., Kant, M.R., Koller, R., Reidinger, S., Rodriguez, A., Saj, S., Sonnemann, I., Witt, C., Wurst, S., 2010. Additive and interactive effects of functionally dissimilar soil organisms on a grassland plant community. Soil Biology and Biochemistry 42 (12), 2266–2275.

Langley, J.A., Hungate, B.A., 2003. Mycorrhizal controls on belowground litter quality. Ecology 84 (9), 2302–2312.

Larsen, J., Jakobsen, I., 1996. Interactions between a mycophagous Collembola, dry yeast and the external mycelium of an arbuscular mycorrhizal fungus. Mycorrhiza 6 (4), 259–264.

Larsen, J., Johansen, A., Larsen, S.E., Heckmann, L.H., Jakobsen, I., Krogh, P.H., 2008. Population performance of collembolans feeding on soil fungi from different ecological niches. Soil Biology and Biochemistry 40 (2), 360–369.

Latz, E., Eisenhauer, N., Rall, B.C., Allan, E., Roscher, C., Scheu, S., Jousset, A., 2012. Plant diversity improves protection against soil-borne pathogens by fostering antagonistic bacterial communities. Journal of Ecology 100 (3), 597–604.

Lechene, C., Hillion, F., McMahon, G., Benson, D., Kleinfeld, A.M., Kampf, J.P., Distel, D., Luyten, Y., Bonventre, J., Hentschel, D., 2006. High-resolution quantitative imaging of mammalian and bacterial cells using stable isotope mass spectrometry. Journal of Biology 5 (6), 1.

Leifheit, E., Verbruggen, E., Rillig, M., 2015. Arbuscular mycorrhizal fungi reduce decomposition of woody plant litter while increasing soil aggregation. Soil Biology and Biochemistry 81, 323–328.

Leigh, J., Fitter, A.H., Hodge, A., 2011. Growth and symbiotic effectiveness of an arbuscular mycorrhizal fungus in organic matter in competition with soil bacteria. FEMS Microbiology Ecology 76 (3), 428–438.

Lerat, S., Lapointe, L., Gutjahr, S., Piché, Y., Vierheilig, H., 2002. Carbon partitioning in a split-root system of arbuscular mycorrhizal plants is fungal and plant species dependent. New Phytologist 157, 589–595.

Lewandowski, T.J., Dunfield, K.E., Antunes, P.M., 2013. Isolate identity determines plant tolerance to pathogen attack in assembled mycorrhizal communities. PLoS One 8 (4), e61329.

Lilleskov, E.A., Bruns, T.D., 2005. Spore dispersal of a resupinate ectomycorrhizal fungus, *Tomentella sublilacina*, via soil food webs. Mycologia 97 (4), 762–769.

Lindahl, B., Stenlid, J., Finlay, R., 2001. Effects of resource availability on mycelial interactions and 32P transfer between a saprotrophic and an ectomycorrhizal fungus in soil microcosms. FEMS Microbiology Ecology 38 (1), 43–52.

Lindahl, B.D., Ihrmark, K., Boberg, J., Trumbore, S.E., Hogberg, P., Stenlid, J., Finlay, R.D., 2007. Spatial separation of litter decomposition and mycorrhizal nitrogen uptake in a boreal forest. New Phytologist 173 (3), 611–620.

Lindahl, B.D., Tunlid, A., 2015. Ectomycorrhizal fungi – potential organic matter decomposers, yet not saprotrophs. New Phytologist 205 (4), 1443–1447.

Litton, C., Giardina, C., 2008. Below-ground carbon flux and partitioning: global patterns and response to temperature. Functional Ecology 22 (6), 941–954.

Long, S.P., Hutchin, P.R., 1991. Primary production in grasslands and coniferous forests with climate change: an overview. Ecological Applications 139–156.

Lugtenberg, B., Kamilova, F., 2009. Plant-growth-promoting rhizobacteria. Annual Review of Microbiology 63, 541–556.

Lussenhop, J., 1996. Collembola as mediators of microbial symbiont effects upon soybean. Soil Biology & Biochemistry 28 (3), 363–369.

Mack, K.M.L., Rudgers, J.A., 2008. Balancing multiple mutualists: asymmetric interactions among plants, arbuscular mycorrhizal fungi, and fungal endophytes. Oikos 117 (2), 310–320.

Maestre, F.T., Quero, J.L., Gotelli, N.J., Escudero, A., Ochoa, V., Delgado-Baquerizo, M., García-Gómez, M., Bowker, M.A., Soliveres, S., Escolar, C., 2012. Plant species richness and ecosystem multifunctionality in global drylands. Science 335 (6065), 214–218.

Mangan, S.A., Adler, G.H., 2002. Seasonal dispersal of arbuscular mycorrhizal fungi by spiny rats in a neotropical forest. Oecologia 131 (4), 587–597.

Maraun, M., Erdmann, G., Fischer, B.M., Pollierer, M.M., Norton, R.A., Schneider, K., Scheu, S., 2011. Stable isotopes revisited: their use and limits for oribatid mite trophic ecology. Soil Biology and Biochemistry 43 (5), 877–882.

Martin, F., Aerts, A., Ahrén, D., Brun, A., Danchin, E., Duchaussoy, F., Gibon, J., Kohler, A., Lindquist, E., Pereda, V., 2008. The genome of *Laccaria bicolor* provides insights into mycorrhizal symbiosis. Nature 452 (7183), 88–92.

Mayor, J.R., Schuur, E.A., Henkel, T.W., 2009. Elucidating the nutritional dynamics of fungi using stable isotopes. Ecology Letters 12 (2), 171–183.

McAllister, C., Garcia-Garrido, J., Garcia-Romera, I., Godeas, A., Ocampo, J., 1997. Interaction between *Alternaria alternata* or *Fusarium equiseti* and *Glomus mosseae* and its effects on plant growth. Biology and Fertility of Soils 24 (3), 301–305.

McAllister, C.B., Garcia-Romera, I., Godeas, A., Ocampo, J., 1994. Interactions between *Trichoderma koningii, Fusarium solani* and *Glomus mosseae*: effects on plant growth, arbuscular mycorrhizas and the saprophyte inoculants. Soil Biology and Biochemistry 26 (10), 1363–1367.

McCormack, M.L., Guo, D., 2014. Impacts of environmental factors on fine root lifespan. Frontiers in Plant Science 5, 205.

Meier, I.C., Avis, P.G., Phillips, R.P., 2013. Fungal communities influence root exudation rates in pine seedlings. FEMS Microbiology Ecology 83 (3), 585–595.

Mikola, J., Setälä, H., 1998. No evidence of trophic cascades in an experimental microbial-based soil food web. Ecology 79 (1), 153–164.

Miller, R., Jastrow, J., 2000. Mycorrhizal fungi influence soil structure. In: Arbuscular Mycorrhizas: Physiology and Function. Springer, pp. 3–18.

Miransari, M., 2011. Interactions between arbuscular mycorrhizal fungi and soil bacteria. Applied Microbiology and Biotechnology 89 (4), 917–930.

Mitchell, M.J., Parkinson, D., 1976. Fungal feeding or oribatid mites (Acari: Cryptostigmata) in an aspen woodland soil. Ecology 302–312.

Moore, J.C., Boone, R.B., Koyama, A., Holfelder, K., 2014. Enzymatic and detrital influences on the structure, function, and dynamics of spatially-explicit model ecosystems. Biogeochemistry 117 (1), 205–227.

Moore, J.C., De Ruiter, P.C., Hunt, H.W., 1993. Influence of productivity on the stability of real and model ecosystems. Science 261, 906–908.

Moore, J.C., McCann, K., Setälä, H., De Ruiter, P.C., 2003. Top-down is bottom-up: does predation in the rhizosphere regulate aboveground dynamics? Ecology 84 (4), 846–857.

Moore, J.C., William Hunt, H., 1988. Resource compartmentation and the stability of real ecosystems. Nature 333 (6170), 261–263.

Musat, N., Foster, R., Vagner, T., Adam, B., Kuypers, M.M.M., 2012. Detecting metabolic activities in single cells, with emphasis on nanoSIMS. FEMS Microbiology Reviews 36 (2), 486–511.

Newsham, K.K., Fitter, A.H., Watkinson, A.R., 1995. Arbuscular mycorrhiza protect an annual grass from root pathogenic fungi in the field. Journal of Ecology 83 (6), 991–1000.

Nguyen, N.H., Bruns, T.D., 2015. The microbiome of *Pinus muricata* ectomycorrhizae: community assemblages, fungal species effects, and *Burkholderia* as important bacteria in multipartnered symbioses. Microbial Ecology 69 (4), 914–921.

Nguyen, N.H., Song, Z., Bates, S.T., Branco, S., Tedersoo, L., Menke, J., Schilling, J.S., Kennedy, P.G., 2015. FUNGuild: an open annotation tool for parsing fungal community datasets by ecological guild. Fungal Ecology 20, 241–248.

Nottingham, A.T., Turner, B.L., Winter, K., Chamberlain, P.M., Stott, A., Tanner, E.V., 2013. Root and arbuscular mycorrhizal mycelial interactions with soil microorganisms in lowland tropical forest. FEMS Microbiology Ecology 85 (1), 37–50.

Nuccio, E.E., Hodge, A., Pett-Ridge, J., Herman, D.J., Weber, P.K., Firestone, M.K., 2013. An arbuscular mycorrhizal fungus significantly modifies the soil bacterial community and nitrogen cycling during litter decomposition. Environmental Microbiology 15 (6), 1870–1881.

Olsson, P.A., Johnson, N.C., 2005. Tracking carbon from the atmosphere to the rhizosphere. Ecology Letters 8 (12), 1264–1270.

Orwin, K.H., Kirschbaum, M.U., St John, M.G., Dickie, I.A., 2011. Organic nutrient uptake by mycorrhizal fungi enhances ecosystem carbon storage: a model-based assessment. Ecology Letters 14 (5), 493–502.

Péret, B., Svistoonoff, S., Laplaze, L., 2009. When plants socialize: symbioses and root development. Annual Plant Reviews 37, 209–238.

Perez-Moreno, J., Read, D., 2001. Nutrient transfer from soil nematodes to plants: a direct pathway provided by the mycorrhizal mycelial network. Plant, Cell & Environment 24 (11), 1219–1226.

Persson, L., Bengtsson, J., Menge, B.A., Power, M.E., 1996. Productivity and consumer regulation—concepts, patterns, and mechanisms. In: Food Webs. Springer, pp. 396–434.

Petz, W., Foissner, W., Adam, H., 1985. Culture, food selection and growth rate in the mycophagous ciliate *Grossglockneria acuta* Foissner, 1980: first evidence of autochthonous soil ciliates. Soil Biology and Biochemistry 17 (6), 871–875.

Pinochet, J., Calvet, C., Camprubi, A., Fernandez, C., 1996. Interactions between migratory endoparasitic nematodes and arbuscular mycorrhizal fungi in perennial crops: a review. Plant and Soil 185 (2), 183–190.

Ploug, H., Musat, N., Adam, B., Moraru, C.L., Lavik, G., Vagner, T., Bergman, B., Kuypers, M.M., 2010. Carbon and nitrogen fluxes associated with the cyanobacterium *Aphanizomenon* sp. in the Baltic Sea. The ISME Journal 4 (9), 1215–1223.

Ploug, H., Adam, B., Musat, N., Kalvelage, T., Lavik, G., Wolf-Gladrow, D., Kuypers, M.M., 2011. Carbon, nitrogen and O_2 fluxes associated with the cyanobacterium *Nodularia spumigena* in the Baltic Sea. The ISME Journal 5 (9), 1549–1558.

Pollierer, M.M., Dyckmans, J., Scheu, S., Haubert, D., 2012. Carbon flux through fungi and bacteria into the forest soil animal food web as indicated by compound-specific ^{13}C fatty acid analysis. Functional Ecology 26 (4), 978–990.

Pollierer, M.M., Langel, R., Körner, C., Maraun, M., Scheu, S., 2007. The underestimated importance of belowground carbon input for forest soil animal food webs. Ecology Letters 10 (8), 729–736.

Pollierer, M.M., Langel, R., Scheu, S., Maraun, M., 2009. Compartmentalization of the soil animal food web as indicated by dual analysis of stable isotope ratios ($^{15}N/^{14}N$ and $^{13}C/^{12}C$). Soil Biology and Biochemistry 41 (6), 1221–1226.

Potapov, A.M., Tiunov, A.V., 2016. Stable isotope composition of mycophagous collembolans versus mycotrophic plants: do soil invertebrates feed on mycorrhizal fungi? Soil Biology and Biochemistry 93, 115–118.

van der Putten, W.H., Bardgett, R.D., Bever, J.D., Bezemer, T.M., Casper, B.B., Fukami, T., Kardol, P., Klironomos, J.N., Kulmatiski, A., Schweitzer, J.A., 2013. Plant–soil feedbacks: the past, the present and future challenges. Journal of Ecology 101 (2), 265–276.

Reddell, P., Spain, A.V., 1991. Earthworms as vectors of viable propagules of mycorrhizal fungi. Soil Biology and Biochemistry 23 (8), 767–774.

Reininger, V., Martinez-Garcia, L.B., Sanderson, L., Antunes, P.M., 2015. Composition of fungal soil communities varies with plant abundance and geographic origin. AoB Plants 7, plv110.

Remén, C., Krüger, M., Cassel-Lundhagen, A., 2010. Successful analysis of gut contents in fungal-feeding oribatid mites by combining body-surface washing and PCR. Soil Biology and Biochemistry 42 (11), 1952–1957.

Revillini, D., Gehring, C.A., Johnson, N.C., 2016. The role of locally adapted mycorrhizas and rhizobacteria in plant–soil feedback systems. Functional Ecology 30, 1086–1098. http://dx.doi.org/10.1111/1365-2435.12668.

Rillig, M.C., Mardatin, N.F., Leifheit, E.F., Antunes, P.M., 2010. Mycelium of arbuscular mycorrhizal fungi increases soil water repellency and is sufficient to maintain water-stable soil aggregates. Soil Biology & Biochemistry 42 (7), 1189–1191.

Rillig, M.C., Mummey, D.L., 2006. Mycorrhizas and soil structure. New Phytologist 171 (1), 41–53.

Roncadori, R., 1997. Interactions between arbuscular mycorrhizas and plant parasitic nematodes in agro-ecosystems. Multitrophic interactions in terrestrial systems. In: The 36th Symposium of the British Ecological Society. Blackwell Science, Oxford, pp. 101–113.

Rønn, R., Vestergård, M., Ekelund, F., 2012. Interactions between bacteria, protozoa and nematodes in soil. Acta Protozoologica 51 (3), 223–235.

Rousk, J., Bååth, E., 2011. Growth of saprotrophic fungi and bacteria in soil. FEMS Microbiology Ecology 78 (1), 17–30.

Rúa, M.A., Antoninka, A., Antunes, P.M., Chaudhary, V.B., Gehring, C., Lamit, L.J., Piculell, B.J., Bever, J.D., Zabinski, C., Meadow, J.F., et al., 2016. Home-field advantage? Evidence of local adaptation among plants, soil, and arbuscular mycorrhizal fungi through meta-analysis. BMC Evolutionary Biology 16 (1), 1–15.

Rúa, M.A., Umbanhowar, J., Hu, S., Burkey, K.O., Mitchell, C.E., 2013. Elevated CO_2 spurs reciprocal positive effects between a plant virus and an arbuscular mycorrhizal fungus. New Phytologist 199 (2), 541–549.

Ruess, L., Dighton, J., 1996. Cultural studies on soil nematodes and their fungal hosts. Nematologica 42 (3), 330–346.

Ruess, L., Zapata, E.J.G., Dighton, J., 2000. Food preferences of a fungal-feeding Aphelenchoides species. Nematology 2 (2), 223–230.

Scheu, S., 2002. The soil food web: structure and perspectives. European Journal of Soil Biology 38 (1), 11–20.

Scheu, S., Folger, M., 2004. Single and mixed diets in Collembola: effects on reproduction and stable isotope fractionation. Functional Ecology 18 (1), 94–102.

Scheu, S., Simmerling, F., 2004. Growth and reproduction of fungal feeding Collembola as affected by fungal species, melanin and mixed diets. Oecologia 139 (3), 347–353.

Schneider, K., Renker, C., Maraun, M., 2005. Oribatid mite (Acari, Oribatida) feeding on ectomycorrhizal fungi. Mycorrhiza 16 (1), 67–72.

Schouteden, N., De Waele, D., Panis, B., Vos, C.M., 2015. Arbuscular mycorrhizal fungi for the biocontrol of plant-parasitic nematodes: a review of the mechanisms involved. Frontiers in Microbiology 6.

Schreiner, R.P., Bethlenfalvay, G., 2003. Crop residue and Collembola interact to determine the growth of mycorrhizal pea plants. Biology and Fertility of Soils 39 (1), 1–8.

Setälä, H., Kulmala, P., Mikola, J., Markkola, A.M., 1999. Influence of ectomycorrhiza on the structure of detrital food webs in pine rhizosphere. Oikos 113–122.

Shahzad, T., Chenu, C., Genet, P., Barot, S., Perveen, N., Mougin, C., Fontaine, S., 2015. Contribution of exudates, arbuscular mycorrhizal fungi and litter depositions to the rhizosphere priming effect induced by grassland species. Soil Biology and Biochemistry 80, 146–155.

Smith, G., 1988. The role of phosphorus nutrition in interactions of vesicular-arbuscular mycorrhizal fungi with soilborne nematodes and fungi. Phytopathology (USA) 78 (3), 371–374.

Smith, S.E., Read, D.J., 2008. Mycorrhizal Symbioses. Academic Press, London, U.K.

Sonnemann, I., Baumhaker, H., Wurst, S., 2012. Species specific responses of common grassland plants to a generalist root herbivore (Agriotes spp. larvae). Basic and Applied Ecology 13 (7), 579–586.

Staaden, S., Milcu, A., Rohlfs, M., Scheu, S., 2010. Fungal toxins affect the fitness and stable isotope fractionation of Collembola. Soil Biology and Biochemistry 42 (10), 1766–1773.

Staaden, S., Milcu, A., Rohlfs, M., Scheu, S., 2012. Olfactory cues associated with fungal grazing intensity and secondary metabolite pathway modulate Collembola foraging behaviour. Soil Biology and Biochemistry 43 (7), 1411–1416.

Streiblová, E., Gryndlerová, H., Gryndler, M., 2012. Truffle brûlé: an efficient fungal life strategy. FEMS Microbiology Ecology 80 (1), 1–8.

Soudzilovskaia, N.A., Heijden, M.G., Cornelissen, J.H., Makarov, M.I., Onipchenko, V.G., Maslov, M.N., Akhmetzhanova, A.A., Bodegom, P.M., 2015. Quantitative assessment of the differential impacts of arbuscular and ectomycorrhiza on soil carbon cycling. New Phytologist 208 (1), 280–293.

Tierney, G.L., Fahey, T.J., 2002. Fine root turnover in a northern hardwood forest: a direct comparison of the radiocarbon and minirhizotron methods. Canadian Journal of Forest Research 32 (9), 1692–1697.

Tanahashi, T., Murase, J., Matsuya, K., Hayashi, M., Kimura, M., Asakawa, S., 2005. Bacterial communities responsible for the decomposition of rice straw compost in a Japanese rice paddy field estimated by DGGE analysis of amplified 16S rDNA and 16S rRNA fragments. Soil Science & Plant Nutrition 51 (3), 351–360.

Tiunov, A.V., Scheu, S., 2005. Arbuscular mycorrhiza and Collembola interact in affecting community composition of saprotrophic microfungi. Oecologia 142 (4), 636–642.

Tordoff, G.M., Boddy, L., Jones, T.H., 2008. Species-specific impacts of collembola grazing on fungal foraging ecology. Soil Biology and Biochemistry 40 (2), 434–442.

Trap, J., Bonkowski, M., Plassard, C., Villenave, C., Blanchart, E., 2016. Ecological importance of soil bacterivores for ecosystem functions. Plant and Soil 398 (1–2), 1–24.

Traugott, M., Kamenova, S., Ruess, L., Seeber, J., Plantegenest, M., 2013. Empirically characterising trophic networks: what emerging DNA-based methods, stable isotope and fatty acid analyses can offer. Advances in Ecological Research 49, 177–224.

Treonis, A.M., Grayston, S.J., Murray, P.J., Dawson, L.A., 2005. Effects of root feeding, cranefly larvae on soil microorganisms and the composition of rhizosphere solutions collected from grassland plants. Applied Soil Ecology 28 (3), 203–215.

Treu, R., Karst, J., Randall, M., Pec, G.J., Cigan, P.W., Simard, S.W., Cooke, J.E.K., Erbilgin, N., Cahill, J.F., 2014. Decline of ectomycorrhizal fungi following a mountain pine beetle epidemic. Ecology 95 (4), 1096–1103.

Turnbull, M.S., George, P.B., Lindo, Z., 2014. Weighing in: size spectra as a standard tool in soil community analyses. Soil Biology and Biochemistry 68, 366–372.

Verbruggen, E., Jansa, J., Hammer, E.C., Rillig, M.C., 2016. Do arbuscular mycorrhizal fungi stabilize litter-derived carbon in soil? Journal of Ecology 104 (1), 261–269.

Voříšková, J., Baldrian, P., 2013. Fungal community on decomposing leaf litter undergoes rapid successional changes. The ISME Journal 7 (3), 477–486.

van der Heijden, M.G., Klironomos, J.N., Ursic, M., Moutoglis, P., Streitwolf-Engel, R., Boller, T., Wiemken, A., Sanders, I.R., 1998. Mycorrhizal fungal diversity determines plant biodiversity, ecosystem variability and productivity. Nature 396 (6706), 69–72.

van der Heijden, M.G., Bardgett, R.D., Van Straalen, N.M., 2008. The unseen majority: soil microbes as drivers of plant diversity and productivity in terrestrial ecosystems. Ecology letters 11 (3), 296–310.

Vos, C., Tesfahun, A., Panis, B., De Waele, D., Elsen, A., 2012a. Arbuscular mycorrhizal fungi induce systemic resistance in tomato against the sedentary nematode *Meloidogyne incognita* and the migratory nematode *Pratylenchus penetrans*. Applied Soil Ecology 61, 1–6.

Vos, C., Van den Broucke, D., Lombi, F.M., De Waele, D., Elsen, A., 2012b. Mycorrhiza-induced resistance in banana acts on nematode host location and penetration. Soil Biology & Biochemistry 47, 60–66.

Wagg, C., Bender, S.F., Widmer, F., van der Heijden, M.G., 2014. Soil biodiversity and soil community composition determine ecosystem multifunctionality. Proceedings of the National Academy of Sciences 111 (14), 5266–5270.

Wagg, C., Jansa, J., Stadler, M., Schmid, B., Van Der Heijden, M.G., 2011. Mycorrhizal fungal identity and diversity relaxes plant-plant competition. Ecology 92 (6), 1303–1313.

Wagner, M., 2009. Single-cell ecophysiology of microbes as revealed by Raman microspectroscopy or secondary ion mass spectrometry imaging. Annual Review of Microbiology 63, 411–429.

Wall, D.H., Bardgett, R.D., Behan-Pelletier, V., Herrick, J.E., Jones, T.H., Ritz, K., Six, J., 2012. Soil Ecology and Ecosystem Services. OUP, Oxford.

Wamberg, C., Christensen, S., Jakobsen, I., 2003a. Interaction between foliar-feeding insects, mycorrhizal fungi, and rhizosphere protozoa on pea plants. Pedobiologia 47 (3), 281–287.

Wamberg, C., Christensen, S., Jakobsen, I., Müller, A.K., Sørensen, S.J., 2003b. The mycorrhizal fungus (*Glomus intraradices*) affects microbial activity in the rhizosphere of pea plants (*Pisum sativum*). Soil Biology and Biochemistry 35 (10), 1349–1357.

Wang, B., Qiu, Y.L., 2006. Phylogenetic distribution and evolution of mycorrhizas in land plants. Mycorrhiza 16 (5), 299–363.

Warnock, A., Fitter, A., Usher, M., 1982. The influence of a springtail *Folsomia candida* (Insecta, Collembola) on the mycorrhizal association of leek *Allium porrum* and the vesicular-arbuscular mycorrhizal endophyte *Glomus fasciculatus*. New Phytologist 90 (2), 285–292.

Wehner, J., Antunes, P.M., Powell, J.R., Caruso, T., Rillig, M.C., 2011. Indigenous arbuscular mycorrhizal fungal assemblages protect grassland host plants from pathogens. PLoS One 6 (11).

Wehner, J., Antunes, P.M., Powell, J.R., Mazukatow, J., Rillig, M.C., 2010. Plant pathogen protection by arbuscular mycorrhizas: a role for fungal diversity? Pedobiologia 53 (3), 197–201.

Welc, M., Ravnskov, S., Kieliszewska-Rokicka, B., Larsen, J., 2010. Suppression of other soil microorganisms by mycelium of arbuscular mycorrhizal fungi in root-free soil. Soil Biology and Biochemistry 42 (9), 1534–1540.

Wells, C.E., Eissenstat, D.M., 2001. Marked differences in survivorship among apple roots of different diameters. Ecology 82 (3), 882–892.

Wilson, G.W., Rice, C.W., Rillig, M.C., Springer, A., Hartnett, D.C., 2009. Soil aggregation and carbon sequestration are tightly correlated with the abundance of arbuscular mycorrhizal fungi: results from long-term field experiments. Ecology Letters 12 (5), 452–461.

Wurst, S., Dugassa-Gobena, D., Langel, R., Bonkowski, M., Scheu, S., 2004. Combined effects of earthworms and vesicular–arbuscular mycorrhizas on plant and aphid performance. New Phytologist 163 (1), 169–176.

Zaller, J.G., Heigl, F., Grabmaier, A., Lichtenegger, C., Piller, K., Allabashi, R., Frank, T., Drapela, T., 2011. Earthworm-mycorrhiza interactions can affect the diversity, structure and functioning of establishing model grassland communities. PLoS One 6 (12), e29293.

CHAPTER

10

Implications of Past, Current, and Future Agricultural Practices for Mycorrhiza-Mediated Nutrient Flux

C. Hamel[1], *C. Plenchette*[2]

[1]Agriculture and Agri-Food Canada, Quebec, QC, Canada; [2]Consultant in Agronomy and Soil Biology and Microbiology, Quetigny, France

10.1 INTRODUCTION

Consumed directly or after being transformed into animal proteins, plants have always been the main food source for humans. Plants provide the basic materials for shelter, clothing, and medicine, and supply numerous industries. In exploiting plants, humans have significantly modified the natural environment.

Terrestrial plants appeared at the same time as arbuscular mycorrhizal (AM) fungi, some 450 million years ago (Pirozynski and Dalpé, 1989). It is well accepted that AM fungi were instrumental in the emergence of terrestrial plants, given that soil as we know it did not exist at the time and that rootless plants needed such fungal partners to take up nutrients from the rocky substrate that covered the continents. Mycorrhizal symbioses contributed to the sustainable colonization of terrestrial environments by plants, and when humans appeared, some million years ago, plants and their mycorrhizal symbionts had already shared a long history.

10.2 AGRICULTURE IN THE PAST

For a long time, the human population was sparse and large regions of Earth remained uninhabited. The vegetation was dependent on mycorrhizas. The productivity of native plants was low but sufficient for a modest and dispersed human population. Humans were nomadic hunter-gatherers for a long time, traveling according to the seasons to search for food. According to anthropologists, the first signs of agricultural activities (seeding plants

Mycorrhizal Mediation of Soil
http://dx.doi.org/10.1016/B978-0-12-804312-7.00010-3

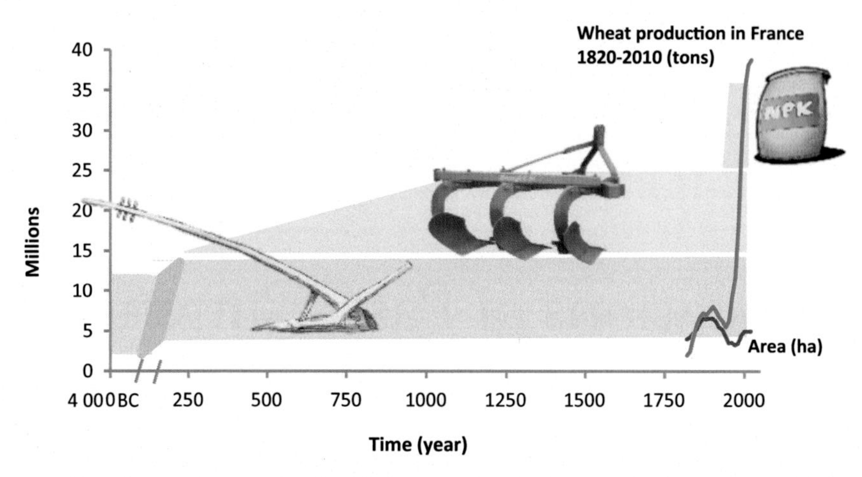

FIGURE 10.1 Time frame of two major historical events that fundamentally transformed global crop production in improving crop yield with impact on the population dynamics of arbuscular mycorrhizal fungi. The switch from the ard to the moldboard plow for seedbed preparation began early in our era, and the rise of the fertilizer industry, which led to the selection of high-yielding crop genotypes as well as water and air pollution, took place after the Second World War (1939–1945). The evolution of wheat productivity relative to the size of production area in France from 1820 to 2010 illustrates the dramatic effect of fertilization on food production. *Data on wheat productivity and area in production in France since 1820 are from Ausubel, J.H., Wernick, I.K., Waggoner, P.E., February 2013. Peak farmland and the prospect for land sparing. Population and Development Review, 38 (s1), 221–242. http://dx.doi.org/10.1111/j.1728–4457.2013.00561.x. http://onlinelibrary.wiley.com/doi/10.1111/j.1728-4457.2013.00561.x/abstract.*

and raising animals) appeared in the Neolithic period, some 12,000 years ago. The wild plants that were being domesticated had arbuscular mycorrhizas. By that time, soil had developed as a result of the alteration of the parent material by soil-forming factors and was providing a slow flux of mineral nutrients in plant-available form. Mycorrhizas were fundamental organs for the capture of this nutrient flux by plants.

Wheat and barley were the first plants to benefit from the application of wood ash and animal manure. The soil was untilled and thus in a favorable state for mycorrhizal formation and function, although as we now know, small-grain cereals have a lower mycorrhizal dependency than plants with a high nutrient demand such as the potato (Plenchette et al., 1983). At this early stage in agricultural development, the mycorrhizal symbiosis was unchanged. However, plant productivity was very low (Fig. 10.1), because cropping practices remained rudimentary and essentially unchanged for century. This is why wheat production in France and most developed countries was below 1 t ha^{-1} at the beginning of the 20th century.

This subsistence agriculture was depleting the soil despite the low productivity of cropping practices, and a piece of land could not be continuously cultivated. Exhausted soils were fallowed for various lengths of time while soil fertility was being restored. There was little export of nutrients at that time. The food produced was consumed locally and mineral nutrients were returned to local soils. Thus the transit of mineral nutrients through mycorrhizas was low but consistent.

10.3 MODERN AGRICULTURE

According to experts, the world's population was about 5 million around 8000 to 10,000 BC, when agriculture began. In the thousands of years separating this epoch from the beginning of our era, humanity grew to at least 200 million people (Durand, 1974). This is a very modest number on the scale of the planet. During antiquity, humanity was naturally sustained by rudimentary agriculture, despite reports of numerous famines. Since then, the human population has grown exponentially to reach 4 billion in 1974 (McEvedy and Jones, 1978). Agriculture evolved with the ever-increasing demand for food and industrial crops. This evolution had a great impact on AM fungi.

10.3.1 Soil Tillage

Humans learned the importance of the conditions in the seeds' environment after seeding. The first instrument for loosening the soil before seeding was the ard, also called the *scratch plow*, which was used in Mesopotamia around 4000 BC. This pointed instrument was drawn by humans or animals through the soil to dig a furrow. The ard did not disturb the soil between the rows, leaving the root system of the previous crop and the extraradical networks of AM hyphae intact. The ard in all its forms was gradually replaced by the moldboard plow, which was reported in Gaul as early as the 2nd century (Marbach, 2004). The moldboard plow could work the soil to greater depths and modified the distribution of AM fungi in the soil (Smith, 1978). Both instruments coexisted for a very long time, and even today the ard is still used in some developing countries. The moldboard plow brings soil mineral nutrients to the surface by soil inversion and favors root development by loosening the soil matrix. This fundamental change in the practice of crop production was quite disruptive to the AM fungal community, because the architecture of the hyphal networks of the previous crop is broken up. The moldboard plow also buries the superficial soil horizons, where AM fungal spores are most abundant, more deeply. Extensive root development in soil loosened by the moldboard plow reduced the dependency of crops on mycorrhizas for nutrient and water uptake (Baylis, 1975).

The mycorrhizal fungi associated with an establishing crop fuse with the existing extraradical hyphal networks in undisturbed soil, favoring the nutrition of the crop. Consequently in the absence of tillage, direct-seeded plants can take up more phosphorus (P) than the same plants seeded in tilled soil (e.g., maize; O'Halloran et al., 1986), and the more intensive the soil tillage, the less developed the mycorrhizas (Kabir, 2005).

The replacement of the horse by the tractor in developed countries and extraordinary advances in tillage equipment (e.g., chisel plow, disk harrow, spring-tooth harrow, rotary hoe) contributed to the decline of AM fungal communities in soils under intensive production. However, agriculture has not been industrialized to the same degree in all countries of the world, and traditional cultural practices are still used in some areas. In those soils, AM fungal communities have evolved differently (Jefwa et al., 2012).

10.3.2 Cultural Systems

The evolution of plowing equipment in medieval Europe brought major change to the landscape. In the past, half of the land on a traditional farm was in pasture, and the other half was divided in two, with one quarter in the production of cereals (essentially wheat) and the other quarter in fallow. Cereals were thus grown on only 25% of the landscape. The emergence of advanced plows led to the adoption of the horse as a draft animal. Grass was supplemented with cereals in the diet of working horses, and the area under cereals expanded from 25% to 33% of the landscape. Cereals, a group of plants with reduced mycorrhizal dependency (Baylis, 1975; Plenchette et al., 1983), became dominant in the landscape.

Another negative factor for AM fungi was the replacement of the practice of fallow in developed countries. Horticultural crops came to replace part of fallowed land and pastures, reducing the size of undisturbed land, disfavoring AM fungal diversity (Vestberg et al., 1999; Trejo et al., 2016). During the 20th century, cropping systems were initially complex, combining animal production and crop rotation systems with up to six crops, but diversity was reduced as farms became increasingly specialized. Plant diversity decreased as rotations dominated by two or three crop species were implemented. Monocultures became prevalent in certain areas such as the North American Great Plains, where wheat monoculture was often used in rotation with summer fallow, increasing retention of soil moisture for the following year. These systems and their associated agronomic practices significantly modified both the landscape and the soil. The AM fungi of the Great Plains demonstrated a surprising level of resilience to the application of these practices over several decades (Dai et al., 2013; Bainard et al., 2015), but in areas such as Europe where the climate is not limiting and intensive production is practiced, AM fungi have almost disappeared from certain soils (Gianinazzi-Pearson et al., 1985; Plenchette, 1989; Baon et al., 1992).

10.3.3 Changes in Soils and Plants

Plants were initially thought to feed directly on soil organic matter, according to the humus theory of plant nutrition supported by Albrecht Thaër (1809) and Sir Humphry Davy (1813). Despite the assertions of Bernard Palissy, a ceramist who in 1563 reported that it is the minerals rather than the carbonaceous compounds in manure that make plants grow (Feller et al., 2001), and of Nicolas-Théodore de Saussure, a chemist and botanist who studied plant constituents and stated in 1800 that plants feed on mineral nutrients (Robin and Blondel-Mégrelis, 2001), the humus theory prevailed for a long time before being gradually abandoned. Another revolutionary concept that was difficult for the scientific community to accept came after Albert Bernhard Frank, introduced the term *mycorrhiza* in 1885 (Frank, 2005).

It is now well accepted that the principle of fertilization is the application of minerals to the soil. Fertilization consists of augmenting the existing soil nutrients to enable a yield objective to be attained that is determined by plant genetics, climate, and economic considerations. The uptake of mineral nutrients by plants is mediated by roots and thus by mycorrhizas. Mycorrhizal fungi are particularly effective for the uptake of P (Javot et al., 2007), a macronutrient that readily binds to soil constituents and is difficult to extract from soil. The important role of AM fungi in P extraction has been well demonstrated but P fertilization progressively decreased the importance of AM symbioses in croplands.

Over time, it was necessary to design production systems that could sustain an ever-increasing population. After the Second World War (1939–1945), probably the most important and remarkable advance in terms of agricultural productivity, the Green Revolution, began in temperate regions and later expanded to tropical areas (Craswell and Karjalainen, 1990). The Green Revolution resulted from a combination of factors, including improvements in crop genetics and farm equipment and, most importantly, the rise of the agrochemical industry. The major driver of the Green Revolution was the transformation of the war industry, which produced powder for weapons, into the chemical industry, which produced fertilizers and pesticides.

As mineral fertilizer use became common practice, it appeared that soils receiving regular fertilizer inputs were more productive and responsive to fertilization. The concept of fertility-building was born (Bosc, 1988). It was thought necessary to add enough phosphate ions to the soil to build soil reserves for the coming years and saturate phosphate-binding sites in the soil. Once such a level was reached, all subsequent P inputs would be readily available to plants. P fertilization was important up to the end of the 20th century, and P levels of cropped soils often became excessive. Field trials conducted over 40 years in France showed that P fertilization could be withdrawn for at least 10 years before a decline in yield was encountered (Gachon, 1988). It is well known that arbuscular mycorrhizas are particularly effective in low-P soils and that the AM potential of soils declines as the soil P level increases (Duvert et al., 1990). Thus the loss of contribution of AM fungi to croplands in developed countries was largely the result of high soil P levels.

Other factors negatively impact the communities of AM fungi in soil. The full potential of new cultivars cannot be reached without the use of fertilizers and pesticides. In developing countries, cropping practices such as plant and sod burning are also detrimental to AM fungal communities (Rasoamampionona et al., 2008) through direct destruction and indirect impact on soil quality. Thus modern agriculture has adversely affected arbuscular mycorrhizas and mycorrhizal-mediated nutrient fluxes.

10.4 AGRICULTURE IN THE FUTURE

Modern legumes such as alfalfa and soybean, and the Haber–Bosch process now produce about 50% of the atmospheric nitrogen (N) that is fixed each year, and this has an impact on global warming (Kissel, 2014). Since 1974, the world's population has almost doubled. As much as 37.6% of the land area around the world is classified as agricultural land by the Food and Agriculture Organization of the United Nations (FAO, 2013; Roser, 2016), and 10.51% of the Earth's surface is in crop production. The world's population is expected to exceed 9 billion by 2050; by that time, with the projected improvement in the economic situation in the developing world, experts predict a 50% increase in the demand for bioproducts from agriculture (Farsund et al., 2015). Producing more food and bioproducts on the same area of land is often considered the best way forward to meet global population needs, given that farming already puts pressure on the environment and that the conversion of forests to cropland would increase greenhouse gas emissions and reduce ecosystem services (Phalan et al., 2011; Farsund et al., 2015).

The intensification of crop production with agrochemicals and improved plant genetics has sustained population growth. However, the benefits of the Green Revolution have come with environmental impacts. High N and P fertilization rates combined with inefficient uptake of

N and P fertilizers by crop plants negatively impact water and air quality. Furthermore, the concentration of limited P deposits in only a few countries places global food security at risk. Clearly, intensified crop production must be N and P efficient to be sustainable.

In the face of environmental degradation and the need to feed an increasing population, the intensification of sustainable agricultural production is rising on the priority list of countries around the world. As yield gains achieved from the adoption of improved agronomic practices and genetic improvement of major crops plateau (Long and Ort, 2010), there is renewed interest in ecologically intensifying crop production through the management of the root microbiome. Because it is the cornerstone of efficient nutrient cycling (Cavagnaro et al., 2015), the AM symbiosis has a prime role to play in the agriculture of the future. Facilitation of nutrient flux through mycorrhizas must be prioritized, if agricultural production is to be sustainably intensified.

In the Anthropocene era, environmental conditions are changing so rapidly that the adaptive capacity of terrestrial ecosystems has been overwhelmed. The temperature of the Earth increased by 0.85°C between 1850 and 2012; atmospheric carbon dioxide (CO_2) levels increased by 40% from a preindustrial value of 280ppm (IPCC, 2014) to more than 400ppm in 2015 (Dlugokencky and Tans, 2015); and atmospheric N inputs into the soil have doubled in the last century (Kissel, 2014). Rising atmospheric CO_2 levels influence plant physiology (Lenka and Lal, 2012) and the altered soil nutrient balance in cropped fields affects soil biology (Herrera-Peraza et al., 2011). Increased soil P levels under intensive fertilization programs and shifts in soil P pools (Tran and N'Dayegamiye, 1995; McDowell and Condron, 2012) have negatively impacted on the AM symbiosis of crop plants (Hoeksema et al., 2010; Johnson, 2010).

The environment is rapidly changing, influencing the biosphere in ways that are difficult to predict (Barnosky et al., 2012; Brook et al., 2013; Reichstein et al., 2013; Anderegg et al., 2015) and agricultural practices have also changed the soil environment considerably. The dominance of *Funneliformis mosseae* in agricultural soils around the world (Rosendahl et al., 2009; Dai et al., 2013) indicates that agricultural soils have shifted significantly from their natural state. The agriculture industry and policies are increasingly recognizing the need for science to take the lead in modifying the agroecosystem and adapting agronomic practices to promote soil biology and the root microbiome in ways that improve the efficiency of crop production. Plant-available nutrient levels in agricultural soils must be managed to minimize environmental losses, and practices that augment the functionality of AM symbioses are integral to achieving sustainable intensification of crop production.

10.4.1 Inoculants

The emerging technology of AM fungal inoculants is attracting interest in agriculture even though at this stage of technological development the effect of these products on crop production is difficult to predict. However, the possibility of enhancing plant nutrition and productivity through the introduction of AM fungi by means of crop inoculation has been demonstrated (Pellegrino et al., 2012; Pellegrino and Bedini, 2014). Inoculants often contain AM fungal strains selected for their ability to improve plant growth and nutrition in a wide range of conditions, but inoculants can also be developed for use in specific environments where they perform particularly well (Rivera et al., 2007; Herrera-Peraza et al., 2011; Pellegrino and Bedini, 2014; Labidi et al., 2015). Synergistic consortia of AM fungal strains (Jin et al., 2013) and diverse plant-growth–promoting microorganisms are being considered for the development of second-generation inoculants (Adesemoye and Kloepper, 2009; Reddy and Saravanan, 2013; Abd-Alla et al., 2014). The agronomic practices that are required for the best use of AM inoculants are slowly being refined (Rivera et al., 2007; Maiti et al., 2011).

The effectiveness of inoculants containing AM fungi will improve as we gain more knowledge of the ecology of these fungi (Verbruggen et al., 2013; Kiers and Denison, 2014; Rodriguez and Sanders, 2015). To be effective, the AM fungi introduced into the agroecosystems through inoculation have to become established rapidly in crop roots or in hyphal networks and have to function effectively in the soil environment, in addition to bearing superior symbiotic traits of interest. Methods to rapidly assess the value of resident AM fungal communities are being sought (Formey et al., 2012; Thonar et al., 2012). There is no doubt that emerging (Pellegrino and Bedini, 2014; Stockinger et al., 2014; Nadimi et al., 2016) and future technologies for AM community assessment will allow farmers to consider the AM fungal community as one component of agroecosystems to be managed in precision agriculture, in terms of whether or not there is a need to improve this community through inoculation or other means. Precision agriculture involves using information on the crop being grown, the properties of the soil, and the conditions of the environment for the precise management of soils, crops, and pests through variable application rates and types of inputs. The goal of precision agriculture is to maximize the ratio of yield to input costs. Precision agriculture relies on high technology and is evolving at a rapid pace as developments are being made in remote sensing (Anisi et al., 2015; Behmann et al., 2015; Gago et al., 2015) and envirotyping (Xu, 2016).

The role of AM fungi in reducing nutrient loss from soil is well established (Cavagnaro et al., 2015). Extraradical AM hyphal networks, which can finely enmesh the soil matrix, are powerful organs for the extraction of sparse soil nutrients. One way to improve the capacity of AM fungi to extract soil nutrients is to increase the extent of the extraradical AM hyphal network. The currency used by plants to pay for the nutrient provision service offered by AM fungi is photosynthates. Fortunately, gains in photosynthesis efficiency could be made (Long and Ort, 2010; Long and Zhu, 2014). The net conversion efficiency of intercepted solar energy into biomass energy in a growing season averages around 0.5%, whereas the biological limit is 4.5–6% for C3 and C4 plants (Long and Zhu, 2014). Dense extraradical AM hyphal networks must be built and maintained for the powerful extraction of soil nutrients by highly productive and nutrient-efficient crops.

10.4.2 Bioactive Molecules

Plants have evolved with the ability to modify their environment, particularly their biological environment. Plants use a wide range of shapes, colors, and chemicals to attract pollinators, seed dispersal agents, biocontrol agents, and symbionts, and to deter pests (Hadacek, 2002). Bioactive phytochemicals have a role to play in the ecological intensification of crop production. The microbiome of crop roots can be engineered through the skillful use of phytochemicals. Certain flavonoids, peptides, strigolactones, and volatile organic compounds of plant origin can attract, stimulate, or inhibit AM fungi (Gianinazzi-Pearson et al., 1989; Besserer et al., 2006; Horii et al., 2009; Bazghaleh et al., 2016) and other inhabitants of the root microbiome (Birkett et al., 2004; Steinkellner et al., 2007; Dor et al., 2011; Cruz et al., 2012; Kretzschmar et al., 2012). Some of the bioactive phytochemicals produced by roots are selective and can promote the growth of certain beneficial fungal endophytes while suppressing pathogens (Bazghaleh et al., 2016).

It has been shown with chickpea plants that intraspecific variation in the phytochemical profile of plants (Cruz et al., 2012) coincides with intraspecific variation in the composition of their root microbiome, particularly in the composition of the root microbiome's AM fungal component (Bazghaleh et al., 2015). Such intraspecific variation indicates the possibility of using traditional plant breeding methods to select crop genotypes that preferentially associate with highly effective AM fungi. Genes coding for the production of phytochemicals that specifically attract or stimulate highly effective AM fungal species could also be inserted into the genome of crop plants to make their roots more desirable to AM fungi.

There is potential for using bioactive phytochemicals in the management of AM symbioses in crop plants. Phytochemicals that stimulate AM fungal growth could be added to bioreactors to increase the yield of AM fungi to facilitate AM inoculant production. Certain phytochemicals from mycorrhizal roots may selectively inhibit the germination of spores of some AM fungi but not others (Ellouze et al., 2012). There may be potential to develop AM inoculants formulated with bioactive molecules that selectively inhibit certain ineffective resident AM fungal species in order to mitigate undesirable effects, such as prior colonization of roots by ineffective AM fungi, which block subsequent access to roots (Verbruggen et al., 2013). The goal would be to maximize colonization of crop roots by more effective AM fungi introduced into soil through inoculation. Precision seeding equipment could be used to selectively place plant genotypes, inoculants, and bioactive molecules in different field zones according to the spatial heterogeneity of soil properties and the topography of the field, with the goal of increasing the sustainability of agricultural production.

10.5 CONCLUSION

The golden age of nutrient fluxes through mycorrhizas occurred during prehistoric times, although arbuscular mycorrhizas would have made major and consistent contributions to crop production for much of the history of humanity. Two main technological developments have had significant negative impacts on the function of AM symbioses in agriculture. First, the evolution of tillage equipment negatively impacted mycorrhizas while improving productivity. Second, the Green Revolution brought about largely by the fertilizer industry essentially reduced the relevance of mycorrhizas in agriculture, especially in countries with intensive crop production. As the contribution of AM symbioses to nutrient fluxes in agricultural soil declined, negative impacts of agriculture on the environment increased. It is likely that if it is to be sustainable, future agricultural systems will depend on nutrient fluxes through AM networks.

References

Abd-Alla, M.H., El-Enany, A.-W.E., Nafady, N.A., Khalaf, D.M., Morsy, F.M., 2014. Synergistic interaction of *Rhizobium leguminosarum* bv. viciae and arbuscular mycorrhizal fungi as a plant growth promoting biofertilizers for faba bean (*Vicia faba* L.) in alkaline soil. Microbiological Research 169, 49–58.

Adesemoye, A.O., Kloepper, J.W., 2009. Plant–microbes interactions in enhanced fertilizer-use efficiency. Applied Microbiology and Biotechnology 85, 1–12.

Anderegg, W.R.L., Hicke, J.A., Fisher, R.A., Allen, C.D., Aukema, J., Bentz, B., et al., 2015. Tree mortality from drought, insects, and their interactions in a changing climate. New Phytologist 208, 674–683.

Anisi, M.H., Abdul-Salaam, G., Abdullah, A.H., 2015. A survey of wireless sensor network approaches and their energy consumption for monitoring farm fields in precision agriculture. Precision Agriculture 16, 216–238.

Bainard, L.D., Dai, M., Gomez, E.F., Torres-Arias, Y., Bainard, J.D., Sheng, M., et al., 2015. Arbuscular mycorrhizal fungal communities are influenced by agricultural land use and not soil type among the Chernozem great groups of the Canadian Prairies. Plant and Soil 387, 351–362.

Baon, J.B., Smith, S.E., Alston, A.M., Wheeler, R.D., 1992. Phosphorus efficiency of three cereals as related to indigenous mycorrhizal infection. Australian Journal of Agricultural Research 43, 479–491.

Barnosky, A.D., Hadly, E.A., Bascompte, J., Berlow, E.L., Brown, J.H., Fortelius, M., et al., 2012. Approaching a state shift in Earth's biosphere. Nature 486, 52–58.

Baylis, G.T.S., 1975. The magniolioid mycorrhiza and mycotrophy in root systems derived from it. In: Mosse, B., Sanders, F.E., Tinker, P.B. (Eds.), Endomycorrhizas. Academic Press, pp. 373–389.

Bazghaleh, N., Hamel, C., Gan, Y., Knight, J.D., Vujanovic, V., Cruz, A.F., et al., 2016. Phytochemicals induced in chickpea roots selectively and non-selectively stimulate and suppress fungal endophytes and pathogens. Plant and Soil. http://dx.doi.org/10.1007/s11104-016-2977-z.

Bazghaleh, N., Hamel, C., Gan, Y., Tar'an, B., Knight, J.D., 2015. Genotype-specific variation in the structure of root fungal communities is related to chickpea plant productivity. Applied and Environmental Microbiology 81, 2368–2377.

Behmann, J., Mahlein, A.-K., Rumpf, T., Römer, C., Plümer, L., 2015. A review of advanced machine learning methods for the detection of biotic stress in precision crop protection. Precision Agriculture 16, 239–260.

Besserer, A., Puech-Pagès, V., Kiefer, P., Gomez-Roldan, V., Jauneau, A., Roy, S., et al., 2006. Strigolactones stimulate arbuscular mycorrhizal fungi by activating mitochondria. PLoS Biology 4, e226.

Birkett, M.A., Bruce, T.J.A., Martin, J.L., Smart, L.E., Oakley, J., Wadhams, L.J., 2004. Responses of female orange wheat blossom midge, *Sitodiplosis mosellana*, to wheat panicle volatiles. Journal of Chemical Ecology 30, 1319–1328.

Bosc, M., 1988. Enseignements fournis par des essais de longue durée sur la fumure phosphatée et potassique. Essais sur la fertilisation potassique. In: Gachon, L. (Ed.), Phosphore et Potassium Dans Les Relations Sol-Plante : Conséquences sur la Fertilisation. INRA, Paris, France, pp. 403–466.

Brook, B.W., Ellis, E.C., Perring, M.P., Mackay, A.W., Blomqvist, L., 2013. Does the terrestrial biosphere have planetary tipping points? Trends in Ecology and Evolution 28, 396–401.

Cavagnaro, T.R., Bender, S.F., Asghari, H.R., van der Heijden, M.G.A., 2015. The role of arbuscular mycorrhizas in reducing soil nutrient loss. Trends in Plant Science 20, 283–290.

Craswell, E.T., Karjalainen, U., 1990. Recent research on fertilizer problems in Asian agriculture. Fertilizer Research 26, 243–248.

Cruz, A.F., Hamel, C., Yang, C., Matsubara, T., Gan, Y., Singh, A.K., et al., 2012. Phytochemicals to suppress Fusarium head blight in wheat–chickpea rotation. Phytochemistry 78, 72–80.

Dai, M., Bainard, L.D., Hamel, C., Gan, Y., Lynch, D., 2013. Impact of land use on arbuscular mycorrhizal fungal communities in rural Canada. Applied and Environmental Microbiology 79, 6719–6729.

Davy, H., 1813. Elements of Agricultural Chemistry, in a Course of Lectures for the Board of Agriculture. Longman, Hurst, Rees, Orne and Brown, London, UK.

Dlugokencky, E., Tans, P., 2015. Trends in Atmospheric Carbon Dioxide. ESRL Global Monitoring Division. Earth System Research Laboratory. National Oceanic & Atmospheric Administration. [WWW document] URL http://www.esrl.noaa.gov/gmd/ccgg/trends/global.html. [accessed 20 July 2016].

Dor, E., Joel, D.M., Kapulnik, Y., Koltai, H., Hershenhorn, J., 2011. The synthetic strigolactone GR24 influences the growth pattern of phytopathogenic fungi. Planta 234, 419–427.

Durand, J.D., 1974. Historical Estimates of World Population An evaluation. PSC Analytical and Technical Reports No. 10. University of Pennsylvania Population Studies Center, Philadelphia, PA, USA.

Duvert, P., Perrin, R., Plenchette, C., 1990. Soil receptiveness to VA mycorrhizal association: concept and method. Plant and Soil 124, 1–6.

Ellouze, W., Hamel, C., Cruz, A.F., Ishii, T., Gan, Y., Bouzid, S., et al., 2012. Phytochemicals and spore germination: at the root of AMF host preference? Applied Soil Ecology 60, 98–104.

FAO, 2013. Table 4: land. In: FAO Statistical Yearbook 2013. Food and Agriculture Organization of the United Nations, Rome, Italy, pp. 34–37. [WWW document] URL http://www.fao.org/docrep/018/i3107e/i3107e00.htm. [accessed 20 July 2016].

Farsund, A.A., Daugbjerg, C., Langhelle, O., 2015. Food security and trade: reconciling discourses in the food and agriculture organization and the world trade organization. Food Security 7, 383–391.

Feller, C., Boulaine, J., Pedro, G., 2001. Indicateurs de fertilité et durabilité des systèmes de culture au début du XIXe siècle : L'approche de Albrecht THAËR (1752-1828). Étude et Gestion des Sols 8, 33–46.

Formey, D., Molès, M., Haouy, A., Savelli, B., Bouchez, O., Bécard, G., et al., 2012. Comparative analysis of mitochondrial genomes of *Rhizophagus irregularis*—syn. *Glomus irregulare*—reveals a polymorphism induced by variability generating elements. New Phytologist 196, 1217–1227.

Frank, B., 2005. On the nutritional dependence of certain trees on root symbiosis with belowground fungi (an English translation of A.B. Frank's classic paper of 1885). Mycorrhiza 15, 267–275.

Gachon, L., 1988. Phosphore et Potassium Dans les Relations Sol-Plante : Conséquences Pour la Fertilisation. INRA, Paris, France.

Gago, J., Douthe, C., Coopman, R.E., Gallego, P.P., Ribas-Carbo, M., Flexas, J., et al., 2015. UAVs challenge to assess water stress for sustainable agriculture. Agricultural Water Management 153, 9–19.

Gianinazzi-Pearson, V., Gianinazzi, S., Trouvelot, A., 1985. Evaluation of the infectivity and effectiveness of indigenous vesicular-arbuscular fungal populations in some agricultural soils in Burgundy. Canadian Journal of Botany 63, 1521–1524.

Gianinazzi-Pearson, V., Branzanti, B., Gianinazzi, S., 1989. In vitro enhancement of spore germination and early hyphal growth of a vesicular-arbuscular mycorrhizal fungus by host root exudates and plant flavonoids. Symbiosis 7, 243–255.

Hadacek, F., 2002. Secondary metabolites as plant traits: current assessment and future perspectives. Critical Reviews in Plant Sciences 21, 273–322.

Herrera-Peraza, R.A., Hamel, C., Fernández, F., Ferrer, R.L., Furrazola, E., 2011. Soil–strain compatibility: the key to effective use of arbuscular mycorrhizal inoculants? Mycorrhiza 21, 183–193.

Hoeksema, J.D., Chaudhary, V.B., Gehring, C.A., Johnson, N.C., Karst, J., Koide, R.T., et al., 2010. A meta-analysis of context-dependency in plant response to inoculation with mycorrhizal fungi. Ecology Letters 13, 394–407.

Horii, S., Matsumura, A., Kuramoto, M., Ishii, T., 2009. Tryptophan dimer produced by water-stressed bahia grass is an attractant for *Gigaspora margarita* and *Glomus caledonium*. World Journal of Microbiology and Biotechnology 25, 1207–1215.

IPCC, 2014. Core writing team. In: Pachauri, R.K., Meyer, L.A. (Eds.), 2014. Climate Change 2014: Synthesis Report. Contribution of Working Groups I, II and III to the Fifth Assessment Report of the Intergovernmental Panel on Climate Change. IPCC, Geneva, Switzerland.

Jefwa, J.M., Okoth, S., Wachira, P., Karanja, N., Kahindi, J., Njuguini, S., et al., 2012. Impact of land use types and farming practices on occurrence of arbuscular mycorrhizal fungi (AMF) Taita-Taveta district in Kenya. Agriculture. Ecosystems and Environment 157, 32–39.

Jin, H., Germida, J.J., Walley, F.L., 2013. Impact of arbuscular mycorrhizal fungal inoculants on subsequent arbuscular mycorrhizal fungi colonization in pot-cultured field pea (*Pisum sativum* L.). Mycorrhiza 23, 45–59.

Javot, H., Pumplin, N., Harrison, M.J., 2007. Phosphate in the arbuscular mycorrhizal symbiosis: transport properties and regulatory roles. Plant, Cell and Environment 30, 310–322.

Johnson, N.C., 2010. Resource stoichiometry elucidates the structure and function of arbuscular mycorrhizas across scales. New Phytologist 185, 631–647.

Kabir, Z., 2005. Tillage or no-tillage: impact on mycorrhizas. Canadian Journal of Plant Science 85, 23–29.

Kiers, E.T., Denison, R.F., 2014. Inclusive fitness in agriculture. Philosophical Transactions of the Royal Society B: Biological Sciences 369, 20130367.

Kissel, D.E., 2014. The historical development and significance of the Haber Bosch Process. Better Crops With Plant Food 98 (2), 9–11.

Kretzschmar, T., Kohlen, W., Sasse, J., Borghi, L., Schlegel, M., Bachelier, J.B., et al., 2012. A petunia ABC protein controls strigolactone-dependent symbiotic signalling and branching. Nature 483, 341–344.

Labidi, S., Ben Jeddi, F., Tisserant, B., Yousfi, M., Sanaa, M., Dalpé, Y., et al., 2015. Field application of mycorrhizal bio-inoculants affects the mineral uptake of a forage legume (*Hedysarum coronarium* L.) on a highly calcareous soil. Mycorrhiza 25, 297–309.

Lenka, N.K., Lal, R., 2012. Soil-related constraints to the carbon dioxide fertilization effect. Critical Reviews in Plant Sciences 31, 342–357.

Long, S., Zhu, X.-G., 2014. Photosynthesis: the final frontier. Resource: Engineering and Technology for a Sustainable World 21 (6), 16.

Long, S.P., Ort, D.R., 2010. More than taking the heat: crops and global change. Current Opinion in Plant Biology 13, 241–248.

Maiti, D., Toppo, N.N., Variar, M., 2011. Integration of crop rotation and arbuscular mycorrhiza (AM) inoculum application for enhancing AM activity to improve phosphorus nutrition and yield of upland rice (*Oryza sativa* L.). Mycorrhiza 21, 659–667.

Marbach, A., 2004. Recherches sur les instruments aratoires et le travail du sol en Gaule Belgique. British Archaeological Report International Series, 1235. Archaeopress, Oxford, UK.

McDowell, R.W., Condron, L.M., 2012. Phosphorus and the Winchmore trials: review and lessons learnt. New Zealand Journal of Agricultural Research 55, 119–132.

McEvedy, C., Jones, R., 1978. Facts on file. In: Atlas of World Population History. Penguin Reference Books, New York, NY, USA, pp. 342–351.

Nadimi, M., Stefani, F.O.P., Hijri, M., 2016. The large (134.9 kb) mitochondrial genome of the glomeromycete *Funneliformis mosseae*. Mycorrhiza. http://dx.doi.org/10.1007/s00572-016-0710-7.

O'Halloran, I.P., Miller, M.H., Arnold, G., 1986. Absorption of P by corn (*Zea mays* L.) as influenced by soil disturbance. Canadian Journal of Soil Science 66, 287–302.

Pellegrino, E., Bedini, S., 2014. Enhancing ecosystem services in sustainable agriculture: biofertilization and biofortification of chickpea (*Cicer arietinum* L.) by arbuscular mycorrhizal fungi. Soil Biology and Biochemistry 68, 429–439.

Pellegrino, E., Turrini, A., Gamper, H.A., Cafà, G., Bonari, E., Young, J.P.W., et al., 2012. Establishment, persistence and effectiveness of arbuscular mycorrhizal fungal inoculants in the field revealed using molecular genetic tracing and measurement of yield components. New Phytologist 194, 810–822.

Phalan, B., Onial, M., Balmford, A., Green, R.E., 2011. Reconciling food production and biodiversity conservation: land sharing and land sparing compared. Science 333, 1289–1291.

Pirozynski, K.A., Dalpé, Y., 1989. Geological history of the Glomaceae with particular reference to mycorrhizal symbiosis. Symbiosis 7, 1–36.

Plenchette, C., 1989. Mycorhizal soil infectivity of Deherain's plots. Comptes Rendus de l'Academie d'Agriculture de France, 75, pp. 23–29.

Plenchette, C., Fortin, J.A., Furlan, V., 1983. Growth responses of several plant species to mycorrhizas in a soil of moderate P-fertility—I. Mycorrhizal dependency under field conditions. Plant and Soil 70, 199–209.

Rasoamampionona, B., Rabeharisoa, L., Andrianjaka, A., Duponnois, R., Plenchette, C., 2008. Arbuscular mycorrhizas in Malagasy cropping systems. Biological Agriculture and Horticulture 25, 327–337.

Reddy, C.A., Saravanan, R.S., 2013. Polymicrobial multi-functional approach for enhancement of crop productivity. Advances in Applied Microbiology 82, 53–113.

Reichstein, M., Bahn, M., Ciais, P., Frank, D., Mahecha, M.D., Seneviratne, S.I., et al., 2013. Climate extremes and the carbon cycle. Nature 500, 287–295.

Rivera, R., Fernández, F., Fernández, K., Ruis, L., Sánchez, C., Riera, M., 2007. Advances in the management of effective arbuscular mycorrhizal symbiosis in tropical ecosystems. In: Hamel, C., Plenchette, C. (Eds.), Mycorrhiza in Crop Production. The Haworth Press Inc, Binghamton, NY, USA, pp. 151–190.

Robin, P., Blondel-Mégrelis, M., 2001. 1800 et 1840, Physiologie végétale et chimie agricole. 1, Saussure, une publication à ressusciter. Comptes Rendus de l'Académie d'Agriculture de France 87 (4), 31–59.

Rodriguez, A., Sanders, I.R., 2015. The role of community and population ecology in applying mycorrhizal fungi for improved food security. ISME Journal 9, 1053–1061.

Rosendahl, S., McGee, P., Morton, J.B., 2009. Lack of global population genetic differentiation in the arbuscular mycorrhizal fungus *Glomus mosseae* suggests a recent range expansion which may have coincided with the spread of agriculture. Molecular Ecology 18, 4316–4329.

Roser, M., 2016. Land Use in Agriculture. Our World in Data. [WWW document] URL https://ourworldindata.org/land-use-in-agriculture/ [accessed 20 July 2016].

Smith, T.F., 1978. A note on the effect of soil tillage on the frequency and vertical distribution of spores of vesicular-arbuscular endophytes. Australian Soil Research 16, 359–361.

Steinkellner, S., Lendzemo, V., Langer, I., Schweiger, P., Khaosaad, T., Toussaint, J.-P., et al., 2007. Flavonoids and strigolactones in root exudates as signals in symbiotic and pathogenic plant-fungus interactions. Molecules 12, 1290–1306.

Stockinger, H., Peyret-Guzzon, M., Koegel, S., Bouffaud, M.-L., Redecker, D., 2014. The largest subunit of RNA polymerase II as a new marker gene to study assemblages of arbuscular mycorrhizal fungi in the field. PLoS ONE 9, e107783. http://dx.doi.org/10.1371/journal.pone.0107783.

Thaër, A.D., 1809. In: Prechoud, J.J. (Ed.), Principes Raisonnés D'agriculture. (Translated From German) Crud EVB (Paris, France).

Thonar, C., Erb, A., Jansa, J., 2012. Real-time PCR to quantify composition of arbuscular mycorrhizal fungal communities—Marker design, verification, calibration and field validation. Molecular Ecology Resources 12, 219–232.

Tran, T.S., N'Dayegamiye, A., 1995. Long-term effects of fertilizers and manure application on the forms and availability of soil phosphorus. Canadian Journal of Soil Science 75, 281–285.

Trejo, D., Barois, I., Sangabriel-Conde, W., 2016. Disturbance and land use effect on functional diversity of the arbuscular mycorrhizal fungi. Agroforestry Systems 90, 265–279.

Verbruggen, E., van der Heijden, M.G.A., Rillig, M.C., Kiers, E.T., 2013. Mycorrhizal fungal establishment in agricultural soils: factors determining inoculation success. New Phytologist 197, 1104–1109.

Vestberg, M., Cardoso, M., Mårtensson, A., 1999. Occurrence of arbuscular mycorrhizal fungi in different cropping systems at Cochabamba, Bolivia. Agricultural and Food Science in Finland 8, 309–318.

Xu, Y., 2016. Envirotyping for deciphering environmental impacts on crop plants. Theoretical and Applied Genetics 129, 653–673.

CHAPTER

11

Integrating Ectomycorrhizas Into Sustainable Management of Temperate Forests

M.D. Jones

University of British Columbia Okanagan Campus, Kelowna, BC, Canada

11.1 INTRODUCTION

Sustainable forest management is defined as "management that maintains and enhances the long-term health of forest ecosystems for the benefit of all living things while providing environmental, economic, social, and cultural opportunities for present and future generations" (Canadian Council of Forest Ministers, 2008; http://www.sfmcanada.org/images/Publications/EN/MeasureOurProgress_EN.pdf). Starting in 1992, many forest-dependent countries have come together regularly to create and refine criteria and indicators of sustainable forest management. This process has been dubbed the "Montréal Process" and the current Montréal Working Group comprises countries representing 60% of the world's forests, including 90% of the temperate and boreal forests (http://www.montrealprocess.org/). In 2015, the most recent iteration of the Process generated seven criteria, with five of them relevant to mycorrhizal associations: (1) conservation of biological diversity, (2) maintenance of productive capacity of forest ecosystems, (3) maintenance of forest ecosystem health and vitality, (4) conservation and maintenance of soil and water resources, and (5) maintenance of forest contributions to global carbon cycles. Chapters 13–26 consider criteria 4 and 5. The topics covered in this chapter relate primarily to criteria 1–3 and will focus on ectomycorrhizal (EcM) associations in temperate and boreal forests.

Ectomycorrhizal fungi contribute both directly and indirectly to biodiversity in forests. By influencing above-ground plant diversity and productivity (Kernaghan, 2005), as well as serving as food for mammals, above-ground invertebrates, and below ground meso- and macrofauna (Claridge and May, 1994; Krivtsov et al., 2003), they contribute to overall species diversity in forests (criterion 1). Ectomycorrhizal fungi are also reservoirs of high species

Mycorrhizal Mediation of Soil
http://dx.doi.org/10.1016/B978-0-12-804312-7.00011-5

diversity themselves (Molina et al., 1992). Because of their role in enhancing plant access to both organic and inorganic nutrients (Chapter 8), ectomycorrhizas are crucial to maintaining productivity of forests (criterion 2). Many EcM forest trees grow very poorly without EcM fungi and cannot survive without them unless extensively fertilized (e.g., Hacskaylo, 1972). Sporocarps of some species of EcM fungi also contribute to forest value as nontimber forest products. Criterion 3, maintenance of forest ecosystem health and vitality, depends on forest resistance and resilience in response to disturbance. In the mycorrhizal context, colonization of seedlings after stand-destroying disturbances, including commercial harvesting, contributes to forest resilience because of the role of EcM fungi in nutrient cycling (Courty et al., 2010) and suppression of greenhouse gas emissions (Holz et al., 2016). Mycorrhizal fungi may also influence the ability of stands to regenerate after insect attack (Karst et al., 2015).

Soil biota have a higher density of networked relationships in forest ecosystems than in other types of ecosystems, with fungi forming some of the most numerous relationships with other organisms (Creamer et al., 2016). Ectomycorrhizal fungal hyphae, in particular, form an important component of forest soil food webs, and their loss can decrease the numbers of the soil mesofauna that are dependent on them (Siira-Pietikainen et al., 2003). Furthermore, fungal species richness appears to be an important driver of carbon cycling. This chapter begins by reviewing how harvesting systems and immediate site treatment influence EcM fungal communities and subsequent colonization of seedlings. In the second section, site preparation and issues surrounding the inoculation of seedlings in nurseries is presented. Finally, the effects of ongoing stand management on EcM fungal communities are discussed.

11.2 HARVESTING SYSTEMS

The pattern of harvesting in managed forests affects EcM fungal communities in any small fragments of remaining forest, as well as in regenerating stands. Types of harvesting systems range from selection systems, where individual trees or small groups of trees are selected for harvest based on various criteria and result in reduced stem density or small openings in the forest (e.g., single-tree selection, patchcut, group selection), to clearcut systems, where large openings are created (O'Hara, 2014). Within clearcut systems, widely-spaced very young (advance regeneration) or mature individuals can be left behind (eg, Luoma et al., 2006). In aggregated retention systems, groups of green trees, forming patches with intact soils, remain within cutblocks (Gustaffson et al., 2012). In general, tree harvesting is thought to influence biological and physicochemical aspects of soil in the following order: clearcut > seed tree > shelter wood > aggregated retention > group selection > single-tree selection (Marshall, 2000). Because harvesting pattern affects ectomycorrhiza formation, both in the regenerating stand by altering the relative availability of different types of inoculum, as well in remnant stands through effects on EcM fungal diversity, it is a key consideration when managing for EcM symbioses.

11.2.1 Rotation Age: Frequency of Harvesting

The frequency at which commercial harvesting occurs affects the diversity of EcM fungi in mature stands or patches across the landscape. Over the first years to decades of stand development, pioneer EcM fungi become less abundant and the community of symbiotic fungi on roots becomes substantially more diverse (Richter and Bruhn, 1993; Visser, 1995; Jones et al., 1997; Twieg et al., 2007; LeDuc et al., 2013). Twieg et al. (2007) concluded that, for mixed interior Douglas-fir *(Pseudotsuga menziesii)* and paper birch *(Betula papyrifera)* stands in the interior of British Columbia, EcM fungal richness and species composition stabilizes by 65 years after disturbance. Similarly, in a coastal western hemlock *(Tsuga heterophylla)* forest, Lim and Berbee (2013) found that the species composition of EcM fungi differed substantially between 24-year-old stands and 100- or 300-year-old stands. In particular, some EcM fungi that dominated communities in the mature and old-growth stands were only rarely encountered in the younger stands. Crucially, a recent meta-analysis concluded that, in North America and Europe, it would take at least 90 years for forests to recover their EcM fungi to preharvest values, and some EcM fungi do not appear to return to preharvest levels by rotation age (Spake et al., 2015). For example, whereas *Piloderma fallax* strands were observed in a majority of old-growth (>400 years old) Douglas-fir stands in Oregon, they occurred in only 6% of plots in rotation-age (45–50 years old) stands (Smith et al., 2000). Together, these results indicate that, to retain a full complement of EcM fungi, especially rare species, a portion of the landscape must remain as old-growth stands. To do this, harvesting should be focused on stands that have regenerated after disturbances during normal rotation cycles (Simard et al., 2013).

11.2.2 Large Openings: Clearcuts

Although colonization by EcM fungi can be severely reduced on degraded sites or sites on which trees have not recently grown or do not regenerate (Amaranthus et al., 1989; Dickie and Reich, 2005), overall, colonization rates and EcM diversity of seedlings in clearcuts are often as high as for seedlings in forests (Jones et al., 2003). It is well established, however, that EcM fungi colonizing seedlings in clearcuts are different from those colonizing seedlings of similar ages in intact forests (Jones et al., 2003). This is not surprising, given that making large openings in the forest causes a major change in both soil physiochemical and biological properties and in environmental filters influencing EcM fungal community assembly (Parke et al., 1983; Cavender-Bares et al., 2009; Dickie et al., 2009). Soils in clearcuts show greater temperature and moisture extremes (Chen et al., 1993; Pennock and vanKessel, 1997), have redistribution of organic matter during harvesting (Marshall, 2000) and compaction in some parts of the cutblock, and can experience short-term flushes of nutrients (Prescott, 2002), including nitrate and soluble organic nitrogen (N) (Titus et al., 2006). By 5–10 years after harvest, organic matter content, total N, and soluble organic N tend to be lower in clearcuts than adjacent forests, especially if the forest floor is removed (Hartmann et al., 2009; Closa and Goicoechea, 2010). Soil food webs are often disrupted, with some groups of soil fauna being more affected than others (Marshall, 2000). Roots of harvested trees die, leaving only small, unharvested saplings and some shrubs as potential plant partners for EcM fungi remaining in the soil. This may explain the greater spatial structure for fungal hyphae in clearcuts than in adjacent conifer forests (Mummey et al., 2010).

Most importantly, the forms of EcM fungal inoculum available to seedlings change with clearcut logging (Jones et al., 2003). Seedlings germinating in established forests can be colonized by fungal hyphae extending vegetatively from EcM roots, spore banks, and newly dispersed spores (Tendersoo et al., 2008; Peay et al., 2012; Glassman et al., 2015). Immediately after clearcutting, EcM fungi associated with the dying roots of harvested trees can effectively colonize young seedlings (Ba et al., 1991; Hagerman and Durall, 2004), but the availability of this source decreases with time since harvest. In a cold, subalpine, spruce-fir forest in Canada, root systems including a diverse range of ectomycorrhizas remained intact for two summers after harvest (Hagerman et al., 1999a), whereas in warmer climates where decomposition rates are faster, few active ectomycorrhizas remained 9 months after a fall harvest (Harvey et al., 1980). As time from harvest increases, rates of colonization of seedlings regenerating in clearcuts go down (Perry et al., 1987). When regeneration develops through natural seeding, rather than by planting seedlings, good seed supply within 1 year of harvest or wildfire may be required in order for germination and establishment of well-colonized seedlings to occur (Nilsson et al., 1996; Purdy et al., 2002).

With the loss of active EcM fungal mycelia associated with living roots (Lazaruk et al., 2005), inoculum sources in large areas of the clearcut are restricted to spore banks, sclerotia, and newly dispersed spores (Fleming, 1984; Fox, 1986), so-called "resistant propagules" (Baar et al., 1999). Hence one explanation for the large shift in the community composition of EcM fungi in clearcuts is that only a subset of EcM fungal species are present in spore banks (Taylor and Bruns, 1999). In North America, spore banks are dominated by *Rhizopogon*, *Wilcoxina*, *Cenococcum*, *Thelephora*, *Amphinema*, *Tuber*, *Laccaria*, and *Suillus* species (Nguyen et al., 2012; Glassman et al., 2015). These are the same genera that form the majority of ectomycorrhizas on young conifer and hardwood seedlings in clearcuts, after wildfire, and in pot studies (Jones et al., 1997; Izzo et al., 2006; Barker et al., 2013). By definition, banked spores were deposited before harvest, so their density, although patchy, should not vary with distance from the surrounding forest (Fig. 11.1(A)).

By contrast, actively dispersing spores decrease in density with distance from the forest edge (Fig. 11.1(B)), because even though some hyphae may be present in clearcuts (Gordon and van Norman, 2014), EcM fungi do not fruit in the absence of their phytobiont (Romell, 1939). Modeling efforts indicate that 95% of spores will fall within 1 m of the sporocarps in the forest (Galante et al., 2011), yet even over 1000 m, dispersal can be sufficient to colonize the majority of seedlings (Peay et al., 2012). Only a small portion of EcM fungi are effective at long-distance dispersal; hence, in a bishop pine *(Pinus muricata)* ecosystem, 93% of pine seedlings exposed only to natural spore rain were colonized by only three species: *Thelephora terrestris*, *Suillus pungens*, and *Tomentella sublilacina* (Peay et al., 2012). In addition to air dispersal, newly-produced spores can be moved short distances by soil mesofauna and macrofauna (Lilleskov and Bruns, 2005) or larger distances by small and large mammals (Ashkannejhad and Horton, 2006; Wood et al., 2015). Overall, given that newly dispersed spores range in the distance they disperse from the sporocarp, with some of these distances being quite small, a greater diversity of EcM fungal spores would be expected within a few meters of the edges of the clearcut.

Many EcM fungi colonize more effectively via vegetative hyphae associated with living trees than from spores (Deacon and Fleming, 1992 and references therein). For these species, inoculum levels will be higher in the zone where roots of forest trees extend into the clearcut (Fig. 11.1(C)). Indeed, both colonization rates and diversity are higher within clearcuts for seedlings planted within the rooting zone of mature trees (Kranabetter and Wylie, 1998; Durall et al., 1999; Hagerman et al., 1999b). Roots are often assumed to be concentrated within the drip line of the tree crown (eg, Luoma et al., 2006). Indeed, tree root densities dropped

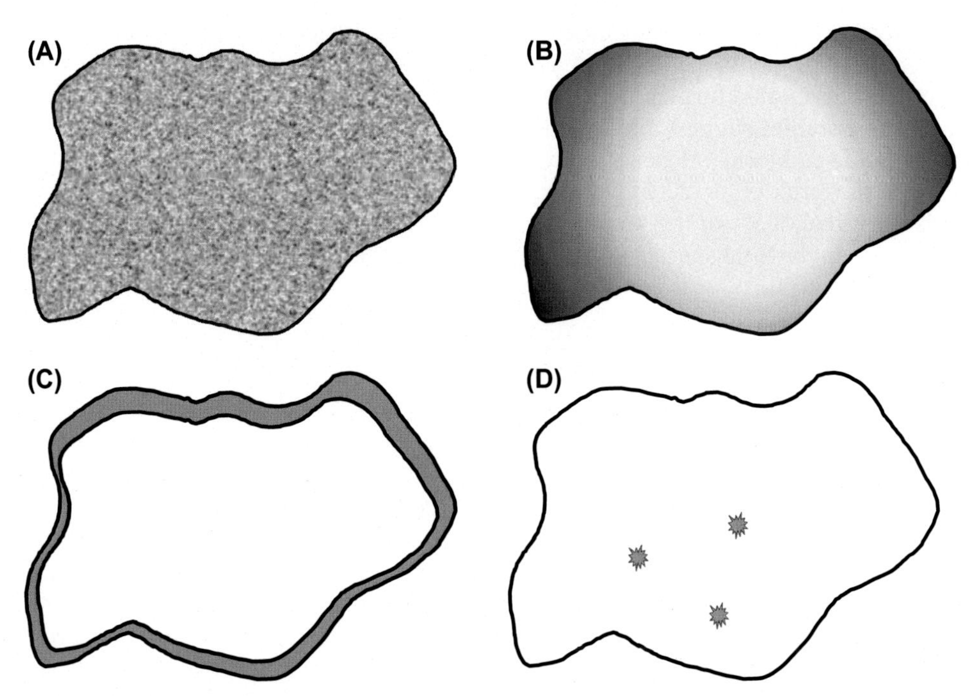

FIGURE 11.1 Distribution of ectomycorrhizal fungal inoculum types across a 20-ha clearcut. (A) Resistant propagules such as spore banks and sclerotia. (B) Actively dispersing spores. (C) Vegetative hyphae from ectomycorrhizal plants in the surrounding forest. (D) Vegetative hyphae from refuge plants.

substantially by 1.5 m from edges of Douglas-fir (*P. menziesii*)/western larch (*Larix occidentalis*) forests (Harvey et al., 1980), by 1.5 times the distance to the drip line of isolated trees in another Douglas-fir–dominated system (Luoma et al., 2006), and within 5–6 m of the forest edge in a lodgepole pine (*Pinus contorta*) clearcut (Parsons et al., 1994). The observation that EcM fungal colonization rates decrease substantially beyond 15 m in a range of ecosystems, including Engelmann spruce (*Picea engelmannii*) (Hagerman et al., 1999b), oak (*Quercus macrocarpa*) (Dickie and Reich, 2005), and southern beech (*Nothofagus* spp.) (Dickie et al., 2012) ecosystems, is evidence that symbiotic EcM fungal hyphae extend approximately that far. In summary, access to several kinds of EcM fungal inoculum, including active mycelia and newly dispersed spores, can be improved by increasing the ratio of forest edge to clearcut area. This can be accomplished by using smaller cutblocks, cutblocks with complex edges, or by including patches of intact forest (aggregated green-tree retention) within the cutblock.

11.2.3 Aggregated Retention

The best way to retain older trees and their EcM fungal symbionts after harvesting is through retention forestry (Gustafsson et al., 2012). Forest patches left after aggregated retention can be considered to behave as biological islands, and to act as "lifeboats" for many forest-dependent organisms through the regeneration phase (Gustafsson et al., 2012). These forest islands are important for retaining rare EcM fungi over the long term, but size of the patches is important because diversity decreases faster in small islands than large islands with time (MacArthur and Wilson, 1967; Peay et al., 2007). In addition, diversity decreases faster over time in forest retention patches located further away from contiguous forest (Peay et al., 2010). Both empirical and predictive studies suggest that for the greatest diversity of EcM fungi available to colonize seedlings in the cutblock, patches should be located no more than 20m from the next patch or forest edge (Lazaruk et al., 2005). Aggregated retention patches of approximately 0.2ha and comprising a total of 3% of the area of the cutblock were sufficient to retain most of the EcM fungal species in the local species pool of a coastal Douglas-fir (*P. menziesii* var. *menziesii*) forest (Kranabetter et al., 2013).

11.2.4 Dispersed Green-Tree Retention

From the perspective of managing for EcM fungi, harvesting trees in a dispersed pattern has the advantage of leaving sources of active hyphal inoculum evenly spread through the block. This may be more effective than leaving a few large islands of trees within the cutblock because both Lazaruk et al. (2005) and Jones et al. (2008) found that the distance to aggregated retention patches was more important that the size of the patch in terms of EcM diversity remaining as inoculum in the harvested area around the patch 1–3 years after harvest. Patches of one or two trees were as effective as larger groups. This was corroborated by Luoma et al. (2006), who found that the diversity of ectomycorrhizas decreased outside the drip line of evenly dispersed trees in three Douglas-fir–dominated forests where 15% of the basal area had been retained; richness decreased by 50% when more than 8m from trees. In white spruce (*Picea glauca*) stands exposed to dispersed retention, even 80% stem removal resulted in no change in the richness or diversity of ectomycorrhizas in the residual strips of the 5-m wide machine corridors 2years after harvest (Lazaruk et al., 2005). These studies all provide evidence that dispersed retention is an effective method of leaving widespread living hyphal inoculum in a harvested block. It was surprising, then, that Teste et al. (2009) found no evidence that mature interior Douglas fir (*P. menziesii* var. *glauca*) left behind after variable retention harvesting acted as a dominant inoculum source for Douglas fir seedlings planted within 5m of the tree. This may have been a result of slight differences among EcM fungi in preference for age of plant partners (Bradbury et al., 1998). Regardless, retention of dispersed trees is effective in retaining EcM fungal diversity on a site as a legacy from one rotation to the next.

11.2.5 Refuge Plants

In addition to trees of the same species left behind after harvest, other EcM host plants may remain in the cutblock and act as reservoirs of active EcM fungi (Fig. 11.1(D)). For example, depending on the management system, EcM alder, birch, aspen, willow, poplar, and *Lithocarpus* spp., or ericoid or arbutoid mycorrhizal shrub species, may remain and/or resprout after harvest. These species associate with a subset of the same EcM fungi that associates with harvested species (Massicotte et al., 1999; Kennedy et al., 2003; Twieg et al., 2007) and may therefore act as inoculum sources for seedlings. For example, Hagerman et al. (2001) found that roots of arbutoid species such as bearberry (*Arctostaphylos uva-ursi*) were as effective as EcM roots at maintaining a high diversity of EcM fungi in a clearcut area and at supplying inoculum to Douglas-fir seedlings in a greenhouse bioassay (Hagerman and Durall, 2004). Similarly, Douglas-fir seedlings planted within patches of manzanita

(*Arctostaphylos* spp.) were more likely to be colonized, and with a greater diversity of EcM fungi, than seedlings planted in plots containing only arbuscular mycorrhizal hosts (Horton et al., 1999). *Betula papyrifera* served as an excellent source of inoculum for birch seedlings planted within its rooting zone, helping to maintain high EcM fungal richness (Kranabetter, 1999). Harvesting practices that reduce disturbance of the forest floor will be most successful in retaining shrubby refuge plants as a source of EcM fungal diversity.

Little work has been done on the role of ericoid shrubs and their role as inoculum sources in stands regenerating after harvest, even though endophytic fungi that form ericoid mycorrhizas are often found in young conifer seedlings (Kernaghan et al., 2003) and some can form ectomycorrhizas (Grelet et al., 2010). Whether these fungi can ever enhance seedling establishment is unknown, but at least on nutrient-poor clearcut areas, ericaceous shrubs are associated with poor growth of conifer seedlings. The reduced growth seems to be a result of competition for nutrients rather than a negative effect of ericoid mycorrhizal fungi (Prescott and Sajedi, 2008).

11.2.6 Importance of Mycorrhizal Networks

The formation of mycorrhizas by EcM fungi growing from existing plants, either forest trees or refuge plants, creates a mycorrhizal network between the seedling and these other plants. Such EcM networks appear to increase survival (Teste and Simard, 2008), or growth (Booth, 2004), of seedlings in the majority of cases examined (van der Heijden and Horton, 2009). The networks allow for better distribution of water among connected plants (Bingham and Simard, 2011) and can allow some of the carbon needs of the fungus to be met by the larger host (Wu et al., 2002), thereby allowing young seedlings to have access to a large nutrient acquisition network without a concomitant allocation of photosynthate. Although seedlings in clearcut regions are unlikely to be carbon limited and can increase their photosynthetic rates to support the network (Simard et al., 1997), mycorrhizal networks are expected to increase the resistance of an EcM fungal community to disturbance by providing multiple plant partners for each fungus (Amaranthus and Perry, 1994; Simard and Durall, 2004).

As stands age, individual mycelia of some EcM fungi will increase in size, allowing mycorrhizal networks to become more extensive. For example, in one 30 × 30 m plot, the root systems of 55 of 56 sampled Douglas-fir trees were connected to at least one other tree by associating with a total of 27 genets of *Rhizopogon vinicolor* or *Rhizopogon vesiculosus* (Beiler et al., 2010). These two species of *Rhizopogon* are known to form extensive networks in unevenly aged Douglas-fir forests in the Pacific Northwest of North America (Kretzer et al., 2004). Little is known about the effect of selective harvesting on EcM fungi networks; however, one study indicates that trees can remain networked after diffuse retention harvesting. Van Dorp et al. (2016) found that *R. vinicolor* formed robust networks, almost to the exclusion of *R. vesiculosus*, in mature (few trees < 50 years old) interior Douglas-fir stands, but by 25 years after selective logging, *R. vesiculosus* had become more common and both species formed extensive networks (Fig. 11.2). If only *R. vinicolor* networks were considered in the unevenly aged regenerating stands, the networks appear susceptible to fragmentation with an additional round of selective logging; however, when *R. vesiculosus* links were added to the networks, they were extensive and robust (Van Dorp, 2016). Therefore it appears that *Rhizopogon* networks in interior Douglas-fir forests of British Columbia are resilient to selective logging by switching between fungal species. Whether this is the case for other forest ecosystems remains to be determined.

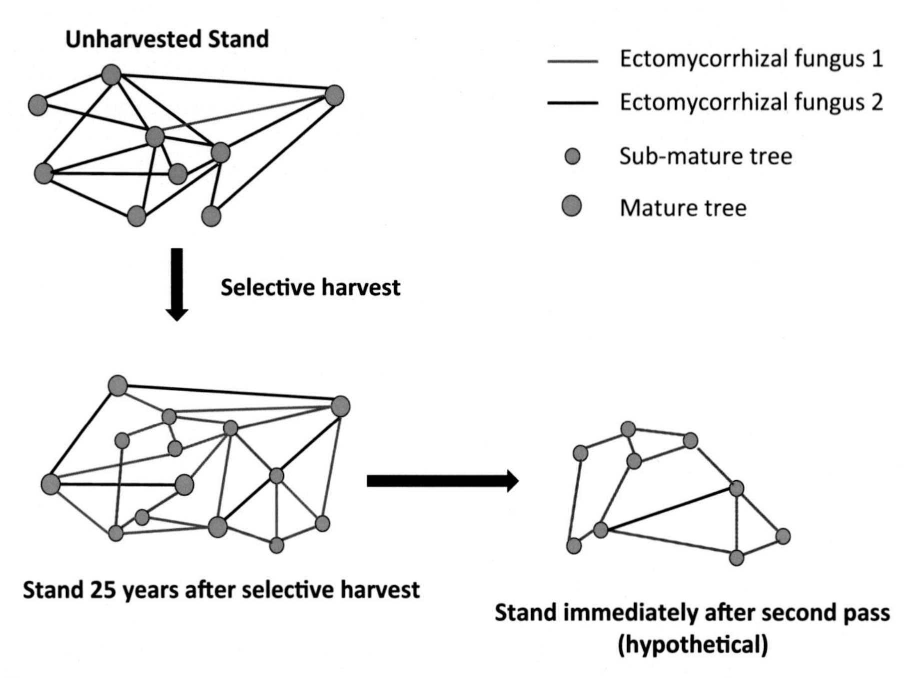

FIGURE 11.2 Mycorrhizal networks in stands of different ages, based on Van Dorp (2016). Before harvest, trees in interior Douglas-fir stands are highly linked by a strand-forming fungus. After selective harvest and a 25-year period of regeneration, a robust mycorrhizal network has reestablished, but is based on a different strand-forming fungus. Because of the robust nature of the network, it is appears to be quite resistant to fragmentation by a second round of harvesting, as long as both species of ectomycorrhizal fungi are considered.

11.2.7 Coarse Woody Debris

Coarse woody debris is defined as intact downed wood greater than 8–10cm in diameter (Stevens, 1997), and its presence in a stand is considered to be an indicator of low-intensity forest management (Bassler et al., 2014). Although its importance in providing habitat for a range of organisms has triggered regulations in some jurisdictions regarding the retention of downed wood at harvest commercial logging still typically results in reduced amounts of coarse woody debris (Bunnell and Houde, 2010). The increased use of whole-tree harvesting will exacerbate this situation. In mature natural forests, partially decomposed wood can be an important habitat for EcM roots, supporting high densities compared with mineral or organic soils (Harvey et al., 1979, 1987). Some species of EcM fungi are particularly vulnerable, either because they produce resupinate fruiting bodies on the underside of logs or because they preferentially colonize roots growing in rotting wood (Harvey et al., 1979; Goodman and Trofymow, 1998; Tedersoo et al., 2003; Walker et al., 2014; Poznanovic et al., 2015). For example, two common EcM fungi, *Amphinema byssoides* and *Tylospora fibrillosa*, whose presence is not apparent above ground in many heavily managed Swedish spruce (*Picea abies*) forests, formed sporocarps on a high proportion of spruce logs 5years after the logs were experimentally placed into stands (Olsson et al., 2011). Although Olsson et al. (2011) concluded that these EcM fungi are probably common in the soils of heavily managed stands but just do not fruit because of a lack of suitable substrate, downed wood is clearly important for reproduction, genetic recombination, and long-distance dispersal of these species.

Whereas the presence of resupinate fruit bodies does not necessarily reflect the EcM fungi that colonize roots in decaying wood (Tedersoo et al., 2009), there is some evidence that leaving large pieces of downed wood at harvest affects colonization in the regenerating stand. In 12–13-year-old subalpine clearcut areas, planting nonmycorrhizal Engelmann spruce seedlings into decayed wood or microsites next to retained logs had no effect on the EcM fungal species colonizing their roots over the first 2 years (Walker and Jones, 2013), even though the hyphae of *T. fibrillosa* and *Russula curtipes* occurred more commonly in decayed wood than in mineral soil (Walker et al., 2014). Over longer periods, however, retained wood affected the composition of the EcM fungal community of Engelmann spruce saplings at the plot scale (Walker et al., 2012), and resulted in increased colonization of individual Scots pine (*Pinus sylvestris*) roots after 15 years (Väre, 1989a). Therefore although downed wood may not be required to support initial colonization of very young seedlings, over the long term it is expected to increase EcM fungal genetic and species diversity in a stand.

11.2.8 Compaction

Soil compaction during mechanical harvesting increases bulk density and reduces water infiltration and nutrient availability; these changes can last at least 5 years (Page-Dumroese et al., 2006). Consequently, growth of trees on compacted soils is commonly inhibited (Kozlowski, 1999). Soil compaction can also reduce the number of ectomycorrhizas on root systems, depending on the tree species (Page-Dumroese et al., 1998). Because compaction can suppress root tip number in species such as Douglas-fir (Page-Dumroese et al., 1998), the effect may not imply a reduced ability to colonize roots, but rather just a change in root system size or architecture in response to increased bulk density. Nevertheless, compaction resulted in long-term changes in the species composition of soil fungal communities, including EcM fungi (Hartmann et al., 2012).

Landings and haul roads are extreme situations where the topsoil has been totally removed and compaction is severe. Although these areas can be rehabilitated sufficiently to allow good seedling growth through the incorporation of topsoil, formation of ectomycorrhizas may still be lower than surrounding clearcut areas (Campbell et al., 2008). Given that landings and roads can comprise 5% of mechanically harvested forests in some areas (Campbell et al., 2008), mechanisms for restoring EcM fungal inoculum potential and diversity in these areas should be a high priority.

11.3 STAND REESTABLISHMENT

11.3.1 Is a Change in Ectomycorrhizal Fungal Community Immediately After Commercial Harvesting Likely to Affect Forest Resilience?

Is sustainable forest management compromised because the species composition of the EcM fungi colonizing seedlings in clearcuts is different from seedlings of similar ages in mature forests? There are several pieces of evidence that suggest it is not. First, communities of clearcut EcM fungi are very similar to those found on seedlings after natural disturbance such as wildfire, especially when organic soil horizons have been removed from the clearcuts (Barker et al., 2013). These EcM fungal communities tend to be less diverse than in mature stands (Visser, 1995; Twieg et al., 2007; Barker et al., 2013), likely because only a subset of EcM fungi survive severe wildfire (Glassman et al., 2015). Whereas colonization of young seedlings by a high diversity of EcM fungi theoretically provides them with access to a greater range of organic nutrients (Jones et al., 2010), rapid colonization by a high diversity of EcM fungi is not necessarily correlated with seedling growth rates in the field (Hagerman et al., 1999b; Kranabetter, 2004; Barker et al., 2013, 2014; Velmala et al., 2013). Second, when transplanted into clearcuts, interior hybrid spruce (*P. engelmannii* × *P. glauca* (Kranabetter, 2004)); or subalpine fir (*Abies lasiocarpa*; Nicholson and Jones, unpublished data) seedlings colonized by these pioneer fungi grow as well or better than seedlings colonized by later successional EcM fungi. Pioneer fungal species appear to be perfectly capable of supplying N and phosphorus (P) to interior spruce seedlings growing in clearcuts (Jones et al., 2009) and are associated with the proliferation of fine roots (Kranabetter, 2004), which has been correlated with high growth rate in Norway spruce (*P. abies*) provenances (Velmala et al., 2013). Finally, it is important to consider the contribution of EcM fungi to organic matter turnover through their production of extracellular enzymes (Courty et al., 2010). Although activities of hydrolytic organic-matter–degrading enzymes are typically higher on the surfaces of ectomycorrhizas sampled from forests than from clearcuts (Walker et al., 2016), reciprocal transplant experiments have demonstrated that ectomycorrhizas formed by pioneer fungi increase their enzyme activities when transplanted into forests (Nicholson and Jones, unpublished). Therefore pioneer EcM fungal communities retain similar capacities for organic matter turnover as forest EcM fungal communities, which contribute to the resilience of the stand. Clearcut soils tend to have reasonably high nutrient availabilities in the first year or two after harvest (Prescott, 2002; Barker et al., 2014). This likely contributes to the ability of seedlings to survive and flourish when colonized by early successional fungi; therefore forest managers should not be concerned if seedlings are colonized primarily by a low diversity of pioneer fungi in the first year or two after establishment in clearcuts.

Although pioneer EcM fungi appear well adapted to support seedling growth after disturbance, it is still worth considering how the many aspects of postharvest site management influence colonization by EcM fungi in the developing stand. Notably, it is common for sites to be mechanically prepared for planting by scarifying the forest floor, inverting and mounding soil, ploughing or broadcast burning. These processes may be invoked to improve the seed bed, raise soil temperature and/or improve drainage, reduce competition from other plants, and/or stimulate nutrient mineralization. All of these can affect the interaction between roots and EcM fungal inoculum.

11.3.2 Timing of Planting

The timing of replanting after harvesting is crucial with respect to optimizing colonization by EcM inoculum on site. Dying ectomycorrhizas, which form a widely and evenly dispersed source of locally-adapted EcM fungi, persist for only a few years after harvest (Harvey et al., 1980; Dahlberg and Stenstrom, 1990; Ba et al., 1991; Hagerman et al., 1999a). If regeneration is through natural seed sources, then forest managers have little control over timing, but planted seedlings should be planted within 1–2 years of harvest to optimize the amount and diversity of inoculum. Clearcuts older than 10 years postharvest can have very low inoculum potentials (Amaranthus and Perry, 1994).

11.3.3 Site Preparation and Broadcast Burning

Even when initial seedling growth is improved, forest floor removal through scarification can result in a reduced initial percentage of colonization of some species, including interior Douglas-fir (Simard et al., 2003; Barker et al., 2013), lodgepole pine (Campbell et al., 2006), or eucalypts (Barry et al., 2015). Similarly broadcast burning can reduce EcM colonization and diversity (Jurgensen et al., 1997 and references therein), although this does not always occur (Mah et al., 2001) and probably depends on the intensity and completeness of the burn. These results are not surprising, because although EcM fungi occur throughout the mineral soil profile (Genney et al., 2006), the diversity and density of ectomycorrhizas in many coniferous forests tends to be highest in the forest floor (Harvey et al., 1979; Goodman and Trofymow, 1998; Tedersoo et al., 2008), or at the interface between the fermentation/humus layer and the mineral soil (Harvey et al., 1986). An exception can be found in very dry forest types such as ponderosa pine (*Pinus ponderosa*; Jurgensen et al., 1997). Even with roots removed, the inoculum potential of organic horizons is often higher than mineral horizons (Heinonsalo and Sen, 2007; Barry et al., 2015). This is likely why scarification or other mechanical site preparation can delay colonization by local fungi, leaving nursery fungi to dominate root systems of seedlings for the first several years after planting (Lazaruk et al., 2008). Overall, retention of forest floor should be a goal, not only because it increases availability of EcM fungal inoculum but also because it provides important nutrient capital and hosts soil food webs crucial for long-term site productivity and carbon storage.

Planting mounds can be made in several ways: (1) by scraping upper soil layers into piles or rows, leaving scarified areas in between (eg, Harvey et al., 1996); (2) by inverting a volume of soil using an excavator (Vyse, 1999; Menkis et al., 2010), e.g., leaving a hole beside each mound; or (3) by overturning the soil using a plow, so that a furrow is left adjacent to mounded row of inverted soil (eg, Väre, 1989b). These last two do not change the nutrient content of the soil, but result in double humus layers at the bottom of the mounds. The effects of mounding on root growth and colonization appears to vary on a site and species basis. Using the first approach, Harvey et al. (1996) found that the number of EcM roots per seedling did not differ between mounds and control areas with intact forest floor 3 years after planting of Douglas-fir and western white pine (*Pinus monticola*) seedlings. Using the second approach, with 10–20 cm inverted mounds, growth of new roots of planted Norway spruce was much greater than on untreated planting sites, and the percentage of those roots colonized was also greater (Pennanen et al., 2005). Similarly a 50% increase in colonization rate was observed for Engelmann spruce and subalpine fir planted on inverted mounds in a subalpine forest (MacKay, Vyse, Jones, unpublished). By contrast, 15-year-old Scots pine saplings grew poorly and had fewer short roots and lower EcM fungal colonization on ploughed mounds than in untreated areas (Väre, 1989b). Given the wide range of results, it is impossible to predict the effect of mounding on colonization for any particular species/site combination, although it appears more likely to have a positive effect on root growth and colonization in wetter, colder sites than in drier sites.

Stumps are removed during site preparation if there is concern about transmission of root pathogens, such as *Armillaria* spp. on a site. It is also an alternative to mounding as a mechanism for exposing mineral soil at the surface (Kataja-aho et al., 2012). Removal of slash and/or stumps reduced colonization rate and diversity of Norway spruce ectomycorrhizas in one study in spite of more vigorous growth (Menkis et al., 2010), but had no effect on colonization or EcM types of the same species in another (Kataja-aho et al., 2012). The response depends on species and stand type (Page-Dumroese et al., 1998); however, this may have been a result of the low number of root tips produced in destumped plots rather than a lower rate of colonization.

11.3.4 Herbicide or Heat Treatment

Herbicide application has an advantage over scarification for reducing vegetation competing with young seedlings in that it retains the forest floor and associated nutrients and inoculum. Seedling survival (Jurgensen et al., 1997) and growth (Harvey et al., 1997) after planting can improve with herbicide use compared with scarification, but colonization by EcM fungi may be reduced, depending on the tree species and herbicide. Simard et al. (2003) found no reduction in colonization of lodgepole pine with glyphosate in the field. By contrast, Franco et al. (2015) detected a reduction in colonization of stone pine (*P. pinea*) seedlings in a pot study with benfluralin. Even where colonization rates were not affected, implying that fungal activity was not reduced, recommended rates of triclopyr, imazapyr, and sulfometuron resulted in a lower number of fine roots in several conifer species (Busse et al., 2004). Consequently careful pretesting of herbicide types and doses should be carried out to avoid negative effects on ectomycorrhizas in the field. Competition can also be controlled through steam treatment, and one study found that it did not reduce the rate or diversity of EcM fungal colonization of Scots pine 3 years after germination on site (Jaderlund et al., 1998).

11.4 SEEDLING PRODUCTION

11.4.1 Inoculation: Is Worth it?

When deciding whether the expense of inoculation is warranted, it is worth considering colonization patterns of seedlings in the field. When seeds germinate after stand-replacing disturbances such as wildfire or clearcutting, many do not become colonized until late in the fall or even the next growing season (Miller et al., 1998; Hagerman et al., 1999b). Miller et al. (1998) speculated that only those seedlings that became mycorrhizal by the fall were able to overwinter, but the presence of as many as 20% nonmycorrhizal seedlings the following summer indicates that this is not always the case (Jones et al., 2009; Barker et al., 2013). The relationship between colonization and growth of young germinants can be complex. Jones et al. (2009) found that overall, naturally colonized Engelmann spruce seedlings did not differ in size from seedlings that had remained nonmycorrhizal for 13 months in the field. When examined more closely, however, seedlings colonized by some EcM fungal species were significantly larger than nonmycorrhizal seedlings, whereas seedlings dominated by other EcM fungi were significantly smaller. These differences were matched with the ability of the fungi to acquire nitrogen. Likewise, Purdy et al. (2002) found that seedling size was negatively correlated with the proportion of root tips colonized by some common pioneer fungi but not others. By contrast, the growth of bishop pine seedlings was positively correlated with the richness of spore-derived EcM fungi on their root systems (Peay et al., 2012). Overall, the evidence indicates that conifer seedlings are not dependent on EcM fungi for their survival over their first few months in the field, especially in low-stress sites. Nevertheless, at some point over the first several years, colonization is essential for proper tree growth and development (Vozzo and Hacskaylo, 1971; Hacskaylo, 1972).

As reviewed by Jones et al. (2003), inoculation of seedlings with individual EcM fungi in nurseries can result in a range of responses. Mycorrhizal seedlings may grow faster, at the same rate, or slower than nonmycorrhizal seedlings in the nursery, and these differences can be carried over into the field (Jones et al., 2003 and references therein). Response to inoculation in nurseries therefore appears to depend on site, fungal species, and tree species. Careful evaluation of these factors should be carried out before an inoculation program is established. Inoculation is most likely to be needed to afforest or reforest sites with degraded soils, harsh climatic conditions, or low EcM fungal inoculum potential (Querejeta et al., 1998; Polanco et al., 2008; Sebastiana et al., 2013) or when exotic tree species are planted in nonnative soils (Dell et al., 2002; Chen et al., 2006). These situations are beyond the mandate of this chapter, but in such situations, detailed tests should be performed on methods of inoculation and suitable species and isolates (eg, Brundrett et al., 2005). Where possible, it is ideal to use EcM fungal isolates native to the local area, because isolates from the site matched to tree genotypes from the area are most likely to maximize tree growth potential (Kranabetter et al., 2015).

Another factor to consider is that both hardwood and conifer EcM tree species often become spontaneously colonized in both container and bare-root nurseries (Rudawska et al., 2006; Fernandez et al., 2013; Pietras et al., 2013), although inoculation can increase the colonization rate substantially in some nurseries (e.g., Khasa et al., 2001; Franco et al., 2014). A reduction in fertilizer concentrations and decrease in the bulk density of potting substrate often increases colonization rates (Hunt, 1991; Dell and Malajczuk, 1994; Chen et al., 2006). The EcM fungi that spontaneously colonize seedlings in both bare-root and container nurseries are typically a subset of the fungi that colonize in disturbed sites, including logged sites. Although they may not be locally adapted to the sites at which they are planted, they often colonize via air-dispersed spores and hence are early successional EcM fungi native to the locality of the nursery. Pioneer EcM fungi, such as *A. byssoides*, *T. terrestris*, and *Laccaria laccata* are common both in nurseries and disturbed sites, and were associated with stronger growth by interior hybrid spruce in clearcuts than later successional EcM fungi (Kranabetter, 2004). As a result, and because these pioneer EcM fungi are gradually replaced by local fungi after outplanting (Jones et al., 1997; Klavina et al., 2013), spontaneous colonization by EcM fungi in nurseries should be viewed favorably.

11.4.2 Assisted Migration: Seed Sources

One of the biggest challenges to sustainable forest management is the changing climate. Many forested areas are becoming hotter and more drought-prone; hence some jurisdictions are changing seed guidelines so that harvested sites are replanted using seeds from lower latitude or elevation provenances (e.g., http://www.nrcan.gc.ca/forests/climate-change/adaptation/13121). We know very little about local adaptation of EcM associations within tree or fungal species, but work by Kranabetter et al. (2015) on coastal Douglas-fir populations planted up to 400km north of their site of origin found that dissimilarity in EcM fungal communities between local trees and distant populations was negatively correlated to the difference in mean annual temperature between the two locations. Furthermore, Kranabetter et al. (2015) estimated that 15% of the loss of height growth observed on transfer could be attributed to the local, unmatched EcM fungal communities. Similarly Pickles et al. (2015) found evidence for local adaptation of EcM fungi and tree genotypes in a glasshouse bioassay of six populations of interior Douglas-fir grown in eight soils, from over 6° latitude, 7° longitude, and 800 m elevation. Part of the local adaptation may be the result of the N status of a site, because EcM fungal community composition is correlated to natural gradients in the supply rate and form of N in the soil (Kranabetter et al., 2009). Consequently, assisted migration could trigger maladapted mutualisms between the imported tree genotypes, which are predicted to be better adapted to the new climate conditions, and local EcM fungi, which are adapted to soil nutrient status (Kranabetter et al., 2015). To add further complexity, the change in climate will likely alter mineralization processes so that the relative supply of amino acids/peptides versus ammonium versus nitrate changes from the current situation, resulting in the loss of local adaptation by the resident EcM fungal community. To counter the unpredictability of climate and soil factors over the course of a rotation, a diverse late-seral EcM fungal community using harvesting and management approaches discussed elsewhere in this chapter should be encouraged (Kranabetter, 2014). That will maximize the functional redundancy and phenotypic plasticity present in diverse EcM fungal communities (Mosca et al., 2007; Jones et al., 2010).

11.4.3 Planting of Single-Species Versus Mixed Stands

Sometimes foresters are interested in changing the composition of stands from mixed species, or species of less commercial interest, to monoculture stands of species with greater commercial value or that are faster growing (e.g., Roach et al., 2015). This may result in the loss of some EcM fungi from the site. Although many EcM fungi are host generalists, some are host specific (Molina et al., 1992). In addition, positive plant–soil feedbacks, such as local site adaptations discussed previously, will not be possible if tree species are changed. Planting mixtures of tree species adapted to the site, including broadleaf trees, is important for sustainable forest management. It will not only reduce the spread of pathogens (Baleshta et al., 2015) but will encourage a greater diversity of EcM fungi (Jones et al., 1997; DeBellis et al., 2006), and therefore is more likely to retain taxonomic and functional diversity of EcM fungi in the stand.

11.5 STAND MANAGEMENT

Once stands have been established and are being managed through a rotation, the diversity of EcM fungi is generally expected to increase over many decades (Visser, 1995; Twieg et al., 2007; Spake et al., 2015). Regardless, the intensity of stand management can affect community assembly processes for forest fungi. Bassler et al. (2014) used the amount and distribution of dead wood, together with species of saproxylic beetles, to generate indices of stand management for 69 plots of European beech forest in southern Germany. Like Goldmann et al. (2015), they found no effect of stand management intensity on richness of EcM fungal species, but environmental influences (filtering) became more influential in determining EcM fungal community composition in highly managed than in less managed stands.

11.5.1 Thinning

Thinning is used to prevent overcrowding and to encourage faster diameter growth in the remaining trees. In addition, it is predicted to maintain productivity in stands attacked by some forest insects (Russell et al., 2015). Although thinning removes host trees, it does not necessarily result in a reduction in EcM diversity. Buée et al. (2005) studied two adjacent European beech stands: one that had been thinned to 170 stems per ha every 6 years for 38 years, and one from which only dead trees had been removed (705 stems per ha). Richness of EcM fungi colonizing beech roots was slightly, but significantly, higher in samples from the thinned plot. Similarly, colonization of planted red oak (*Quercus rubra*) seedlings was actually higher when planted in thinned oak or red pine (*Pinus resinosa*) stands than in unthinned stands over the first year (Zhou et al., 1997). Thinning by approximately 50% per 2500 stems ha^{-1} had no effect on colonization, richness, or diversity of lodgepole pine stands (Teste et al., 2012). Based on sporocarp surveys, 50% thinning had only minor effects on the succession of EcM fungi over the first 4 years in Scots pine forests (Shaw et al., 2003). If thinning is not too extreme (e.g., up to 60% loss in basal area of beech in Gessler et al., 2005), the density of root tips may not be reduced, and this may explain the lack of negative effects of thinning on EcM fungi. Furthermore, retention of small diameter thinning residues can even increase the density of EcM roots from remaining trees (Dighton et al., 2012). Overall, thinning appears more likely to increase than decrease colonization by, and diversity of, EcM fungi in managed stands; therefore, it appears to be a safe practice as part of sustainable forest management.

11.5.2 Brushing

In addition to thinning of the commercial species, brushing or removal of unwanted tree species may also affect EcM fungal diversity. For example, removal of shrubs from montado (*Quercus suber*) woodlands, when combined with grazing or tilling, reduced the number of ectomycorrhizal sporocarps and changed the species composition of the sporocarp community (Azul et al., 2009). A similar reduction was not observed with sporocarps of saprotrophic fungi, suggesting that disturbance itself was not responsible for the change in EcM fungi, but rather that understory or codominant trees help support fruiting in a greater range of EcM fungal species. Very few studies have examined the effects of removal of shrubs on the ectomycorrhizas themselves, but Zhou et al. (1997) found removal of understory (mostly bracken and maple) caused a slight reduction in EcM density on roots, with no effect on colonization percentage. Because host diversity increases the diversity of EcM fungi in a stand (DeBellis et al., 2006), brushing should be reduced as much as possible.

11.5.3 Prescribed Burning

A range of mechanisms are used to prevent the build-up of fuel in fire-prone forest types, especially those found in Mediterranean or semiarid climates. These include mechanical mastication, thinning, and prescribed burning, often in combination. Prescribed burning is also used to encourage natural regeneration of pines in some ecosystems (Hancock et al., 2009). Although it reduces the risk of catastrophic fire, prescribed burns also kill some fine roots, thereby reducing the abundance of ectomycorrhizas (e.g., Hart et al., 2005). The effects of prescribed burns on EcM fungal community structure appear to be minimal for single (Curlevski et al., 2011; Southworth et al., 2011) or low-frequency fires (Bastias et al., 2006). By contrast, when burned every 2 years, EcM fungal communities in *Eucalyptus pilularis* forests were altered considerably compared with those burned every 4 years (Bastias et al., 2006). Nevertheless, given that fire return intervals in these types of forests are naturally high, and that functional traits of EcM fungi appear to be quite resilient (Jones et al., 2010), it is likely that prescribed burning will not impede ecosystem services provided by EcM fungi over the long term.

11.5.4 Fertilization

Anthropogenic enrichment of soils with nitrogen causes a reduction in EcM fungal colonization and species richness, and a change in the EcM fungal community (Treseder, 2004; Lilleskov et al., 2011). The same is true for high rates of fertilizer applications (Teste et al., 2012). These results should not be extrapolated to application of fertilizer to forests on an operational basis, however. Operational applications tend to be of a mixture of nutrients, and at a lower level than many experimental or pollution situations. They often result in relatively minor changes to EcM fungal communities (Jones et al., 2012; Hay et al., 2015), especially over the long term (Wright et al., 2009). When regenerating stands are nutrient deficient, fertilization may even result in EcM fungal communities that are more similar to mature stands than to unfertilized regenerating stands (Lim and Berbee, 2013). If fertilization is accompanied by thinning, so that plants allocate more photosynthate below ground, even fewer effects on ectomycorrhizas are observed (Teste et al., 2012). Although changes may not be substantial, there is general consensus that certain EcM fungi are sensitive to anthropogenic nitrogen enrichment. These include *Suillus* spp., some *Cortinarius* spp., *T. fibrillosa*, and *Tomentella* spp (Lilleskov et al., 2011; Jones et al., 2012; Teste et al., 2012). Studies of extracellular enzyme activities of ectomycorrhizas reveal that nutrient cycling capacity is not substantially affected by infrequent, balanced fertilization (Jones et al., 2012; Hay et al., 2015); hence operational levels of fertilization are unlikely to have a negative impact on EcM fungal function.

11.6 CONCLUSIONS

Although not all these observations will apply across all types of temperate forests, some generalizations regarding ectomycorrhizas and sustainable forest management are possible:

1. Retention of some old-growth stands will be essential in order to maintain rare EcM fungal species across landscapes.
2. Where possible, high edge-to-area ratios should be the goal in cutblocks, either through smaller cutblocks or the use of aggregated retention. This will facilitate the spread of EcM fungi through mycelial networks from the forest into the regenerating stand.
3. Most types of site preparation are unlikely to have long-term negative effects on EcM fungal communities. Treatments that increase root density tend to increase the diversity of EcM fungi on root systems.
4. Inoculation is not necessary to regenerate preexisting forests except in the most degraded sites. If planted within a year or two of harvest, seedlings will be colonized by pioneer species of EcM fungi, which in turn will be gradually replaced with later successional EcM fungi.

5. If inoculation is carried out, fungi adapted to local conditions should be used, so testing should be done in advance of large inoculation schemes.
6. Planting a diverse mixture of trees, including both broadleaf and coniferous trees if forests native to the area contained such mixtures, will help retain EcM fungal diversity.
7. Thinning or selective harvesting will not have a negative effect on EcM fungal diversity. Limited evidence suggests that mycorrhizal networks can recover within decades.
8. Infrequent fertilization with a balanced fertilizer may cause minor changes in the species composition of EcM fungal communities, but there does not appear to be any major loss of function.

Overall, there is fidelity of EcM fungal communities to previous communities (Walker et al., 2014), likely because of dispersal limitations (Peay et al., 2010), soil abiotic characteristics (Kranabetter et al., 2009; Kennedy et al., 2015), and priority effects (Kennedy et al., 2009). Consequently EcM fungal communities, although changed by harvesting, still show strong site-to-site differences that may reflect the local species pool (Pither and Aarssen, 2005; Goldmann et al., 2015) or differences in soil physicochemical properties (Walker et al., 2014; Kennedy et al., 2015) combined with environmental filtering (Courty et al., 2016). Especially for forests adapted to frequent stand-destroying disturbance, taxonomic and functional resilience of EcM fungal communities is generally high.

Acknowledgment

Melanie Jones gratefully acknowledges continuous support for her research program from the Natural Sciences and Engineering Research Council of Canada and from various programs of the British Columbia Ministry of Forests, Lands and Natural Resource Operations. Marty Kranabetter and Dan Durall provided valuable feedback on earlier versions of this chapter.

References

Amaranthus, M.P., Perry, D.A., 1994. The functioning of ectomycorrhizal fungi in the field: linkages in space and time. Plant and Soil 159 (1), 133–140.

Amaranthus, M.P., Trappe, J.M., Molina, R.J., 1989. Long-term forest productivity and the living soil. Maintaining the Long-Term Productivity of Pacific Northwest Forest Ecosystems 36–52.

Ashkannejhad, S., Horton, T.R., 2006. Ectomycorrhizal ecology under primary succession on coastal sand dunes: interactions involving *Pinus contorta*, suilloid fungi and deer. New Phytologist 169, 345–354.

Azul, A.M., Castro, P., Sousa, J.P., Freitas, H., 2009. Diversity and fruiting patterns of ectomycorrhizal and saprobic fungi as indicators of land-use severity in managed woodlands dominated by *Quercus suber*: a case study from southern Portugal. Canadian Journal of Forest Research-Revue Canadienne de Recherche Forestiere 39 (12), 2404–2417.

Ba, A.M., Garbaye, J., Dexheimer, J., 1991. Influence of fungal propagules during the early stage of the time sequence of ectomycorrhizal colonization on afzelia: African seedlings. Canadian Journal of Botany-Revue Canadienne de Botanique 69 (11), 2442–2447.

Baar, J., Horton, T.R., Kretzer, A.M., Bruns, T.D., 1999. Mycorrhizal colonization of *Pinus muricata* from resistant propagules after a stand-replacing wildfire. New Phytologist 143 (2), 409–418.

Baleshta, K.E., Simard, S.W., Roach, W.J., 2015. Effects of thinning paper birch on conifer productivity and understory plant diversity. Scandinavian Journal of Forest Research 30, 699–709.

Barker, J.S., Simard, S.W., Jones, M.D., Durall, D.M., 2013. Ectomycorrhizal fungal community assembly on regenerating Douglas-fir after wildfire and clearcut harvesting. Oecologia 172 (4), 1179–1189.

Barry, K.M., Janos, D.P., Nichols, S., Bowman, D., 2015. *Eucalyptus obliqua* seedling growth in organic vs. mineral soil horizons. Frontiers in Plant Science 6.

Bassler, C., Ernst, R., Cadotte, M., Heibl, C., Muller, J., 2014. Near-to-nature logging influences fungal community assembly processes in a temperate forest. Journal of Applied Ecology 51 (4), 939–948.
Bastias, B.A., Xu, Z.H., Cairney, J.W.G., 2006. Influence of long-term repeated prescribed burning on mycelial communities of ectomycorrhizal fungi. New Phytologist 172 (1), 149–158.
Beiler, K.J., Durall, D.M., Simard, S.W., Maxwell, S.A., Kretzer, A.M., 2010. Architecture of the wood-wide web: *Rhizopogon* spp. genets link multiple Douglas-fir cohorts. New Phytologist 185, 543–553.
Bingham, M.A., Simard, S.W., 2011. Do mycorrhizal network benefits to survival and growth of interior Douglas-fir seedlings increase with soil moisture stress? Ecology and Evolution 1 (3), 306–316.
Booth, M.G., 2004. Mycorrhizal networks mediate overstorey-understorey competition in a temperate forest. Ecology Letters 7 (7), 538–546.
Bradbury, S.M., Danielson, R.M., Visser, S., 1998. Ectomycorrhizas of regenerating stands of lodgepole pine (*Pinus contorta*). Canadian Journal of Botany-Revue Canadienne de Botanique 76, 218–227.
Brundrett, M., Malajczuk, N., Gong, M., Xu, D., Snelling, S., Dell, B., 2005. Nursery inoculation of eucalyptus seedlings in Western Australia and Southern China using spores and mycelial inoculum of diverse ectomycorrhizal fungi from different climatic regions. Forest Ecology and Management 209, 193–205.
Buée, M., Vairelles, D., Garbaye, J., 2005. Year-round monitoring of diversity and potential metabolic activity of the ectomycorrhizal community in a beech (*Fagus silvatica*) forest subjected to two thinning regimes. Mycorrhiza 15, 235–245.
Bunnell, F.L., Houde, I., 2010. Down wood and biodiversity: implications to forest practices. Environmental Reviews 18, 397–421.
Barker, J.S., Simard, S.W., Jones, M.D., 2014. Clearcutting and high severity wildfire have comparable effects on growth of direct-seeded interior Douglas-fir. Forest Ecology and Management 331, 188–195.
Busse, M.D., Fiddler, G.O., Ratcliff, A.W., 2004. Ectomycorrhizal formation in herbicide-treated soils of differing clay and organic matter content. Water Air and Soil Pollution 152 (1–4), 23–34.
Campbell, D.B., Bulmer, C.E., Jones, M.D., Philip, L.J., Zwiazek, J.J., 2008. Incorporation of topsoil and burn-pile debris substantially increases early growth of lodgepole pine on landings. Canadian Journal of Forest Research-Revue Canadienne De Recherche Forestiere 38 (2), 257–267.
Campbell, D.B., Kiiskila, S., Philip, L.J., Zwiazek, J.J., Jones, M.D., 2006. Effects of forest floor planting and stock type on growth and root emergence of *Pinus contorta* seedlings in a cold northern cutblock. New Forests 32 (2), 145–162.
Canadian Council of Forest Ministers, 2008. Measuring our progress: putting sustainable forest management into practice across Canada and beyond, Catalogue number: Fo4-26/2008E-PDF. ISBN: 978-1-100-11167-4.
Cavender-Bares, J., Izzo, A., Robinson, R., Lovelock, C.E., 2009. Changes in ectomycorrhizal community structure on two containerized oak hosts across an experimental hydrologic gradient. Mycorrhiza 19 (3), 133–142.
Chen, J.Q., Franklin, J.F., Spies, T.A., 1993. Contrasting microclimates among clear-cut, edge, and interior of old growth Douglas-fir forest. Agricultural and Forest Meteorology 63 (3–4), 219–237.
Chen, Y.L., Kang, L.H., Dell, B., 2006. Inoculation of *Eucalyptus urophylla* with spores of *Scleroderma* in a nursery in south China: comparison of field soil and potting. Forest Ecology and Management 222 (1–3), 439–449.
Claridge, A.W., May, T.W., 1994. Mycophagy among Australian mammals. Australian Journal of Ecology 19, 251–275.
Closa, I., Goicoechea, N., 2010. Seasonal dynamics of the physicochemical and biological properties of soils in naturally regenerating, unmanaged and clear-cut beech stands in northern Spain. European Journal of Soil Biology 46 (3–4), 190–199.
Courty, P.E., Buee, M., Diedhiou, A.G., Frey-Klett, P., Le Tacon, F., Rineau, F., et al., 2010. The role of ectomycorrhizal communities in forest ecosystem processes: new perspectives and emerging concepts. Soil Biology and Biochemistry 42 (5), 679–698.
Courty, P.E., Munoz, F., Selosse, M.A., Duchemin, M., Criquet, S., Ziarelli, F., Buée, M., Plassard, C., Taudière, A., Garbaye, J., Richard, F., 2016. Into the functional ecology of ectomycorrhizal communities: environmental filtering of enzymatic activities, Journal of Ecology. http://dx.doi.org/10.1111/1365-2745.12633.
Creamer, R.E., Hannula, S.E., Van Leeuwen, J.P., Stone, D., Rutgers, M., Schmelz, R.M., et al., 2016. Ecological network analysis reveals the inter-connection between soil biodiversity and ecosystem function as affected by land use across Europe. Applied Soil Ecology 97, 112–124.
Curlevski, N.J.A., Artz, R.R.E., Anderson, I.C., Cairney, J.W.G., 2011. Response of soil microbial communities to management strategies for enhancing Scots pine (*Pinus sylvestris*) establishment on heather (*Calluna vulgaris*) moorland. Plant and Soil 339 (1–2), 413–424.

Dahlberg, A., Stenstrom, E., 1990. The effect of clear-cutting on the ectomycorrhizal flora on Scots pine. Agriculture Ecosystems and Environment 28, 91–94.

Deacon, J.W., Fleming, L.V., 1992. Interactions of ectomycorrhizal fungi. In: Allen, M.F. (Ed.), Mycorrhizas in Ecosystems. Chapman & Hall, New York, USA, pp. 249–300.

DeBellis, T., Kernaghan, G., Bradley, R., Widden, P., 2006. Relationships between stand composition and ectomycorrhizal community structure in boreal mixed-wood forests. Microbial Ecology 52 (1), 114–126.

Dell, B., Malajczuk, N., Dunstan, W.A., 2002. Persistence of some Australian *Pisolithus* species introduced into eucalypt plantations in China. Forest Ecology and Management 169 (3), 271–281.

Dell, B., Malajczuk, N., 1994. Fertiliser requirements for ectomycorrhizal eucalypts in forest nurseries and field plantings in southern China. In: International Symposium and Workshop on Mycorrhizas for Plantation Forestry in Asia. Kaiping, Peoples R China, pp. 96–100.

Dickie, I.A., Reich, P.B., 2005. Ectomycorrhizal fungal communities at forest edges. Journal of Ecology 93 (2), 244–255.

Dickie, I.A., Davis, M., Carswell, F.E., 2012. Quantification of mycorrhizal limitation in beech spread. New Zealand Journal of Ecology 36 (2), 210–215.

Dickie, I.A., Richardson, S.J., Wiser, S.K., 2009. Ectomycorrhizal fungal communities and soil chemistry in harvested and unharvested temperate Nothofagus rainforests. Canadian Journal of Forest Research-Revue Canadienne de Recherche Forestiere 39 (6), 1069–1079.

Dighton, J., Helmisaari, H.S., Maghirang, M., Smith, S., Malcolm, K., Johnson, W., et al., 2012. Impacts of forest post thinning residues on soil chemistry, fauna and roots: implications of residue removal in Finland. Applied Soil Ecology 60, 16–22.

Durall, D.M., Jones, M.D., Wright, E.F., Kroeger, P., Coates, K.D., 1999. Species richness of ectomycorrhizal fungi in cutblocks of different sizes in the interior cedar-hemlock forests of northwestern British Columbia: sporocarps and ectomycorrhizae. Canadian Journal of Forest Research-Revue Canadienne de Recherche Forestiere 29 (9), 1322–1332.

Fernandez, N.V., Marchelli, P., Fontenla, S.B., 2013. Ectomycorrhizas naturally established in *Nothofagus* nervosa seedlings under different cultivation practices in a forest nursery. Microbial Ecology 66 (3), 581–592.

Fleming, L.V., 1984. Effects of soil trenching and coring on the formation of ectomycorrhizas on birch seedlings grown around mature trees. New Phytologist 98 (1), 143–153.

Fox, F.M., 1986. Ultrastructure and infectivity of sclerotia of the ectomycorrhizal fungus *Paxillus involutus* on Birch (*Betula* spp.). Transactions of the British Mycological Society 87, 627–631.

Franco, A.R., Pereira, S.I.A., Castro, P.M.L., 2015. Effect of benfluralin on *Pinus pinea* seedlings mycorrhized with *Pisolithus tinctorius* and *Suillus bellinii*: study of plant antioxidant response. Chemosphere 120, 422–430.

Franco, A.R., Sousa, N.R., Ramos, M.A., Oliveira, R.S., Castro, P.M.L., 2014. Diversity and persistence of ectomycorrhizal fungi and their effect on nursery: inoculated *Pinus pinaster* in a post-fire plantation in northern Portugal. Microbial Ecology 68 (4), 761–772.

Galante, T.E., Horton, T.R., Swaney, D.P., 2011. 95 % of basidiospores fall within 1 m of the cap: a field- and modeling-based study. Mycologia 103 (6), 1175–1183.

Genney, D.R., Anderson, I.C., Alexander, I.J., 2006. Fine-scale distribution of pine ectomycorrhizas and their extramatrical mycelium. New Phytologist 170 (2), 381–390.

Gessler, A., Jung, K., Gasche, R., Papen, H., Heidenfelder, A., Borner, E., et al., 2005. Climate and forest management influence nitrogen balance of European beech forests: microbial N transformations and inorganic N net uptake capacity of mycorrhizal roots. European Journal of Forest Research 124 (2), 95–111.

Glassman, S.I., Peay, K.G., Talbot, J.M., Smith, D.P., Chung, J.A., Taylor, J.W., et al., 2015. A continental view of pine-associated ectomycorrhizal fungal spore banks: a quiescent functional guild with a strong biogeographic pattern. New Phytologist 205 (4), 1619–1631.

Goldmann, K., Schoning, I., Buscot, F., Wubet, T., 2015. Forest management type influences diversity and community composition of soil fungi across temperate forest ecosystems. Frontiers in Microbiology 6.

Goodman, D.M., Trofymow, J.A., 1998. Distribution of ectomycorrhizas in micro-habitats in mature and old-growth stands of Douglas-fir on southeastern Vancouver Island. Soil Biology and Biochemistry 30 (14), 2127–2138.

Gordon, M., van Norman, K., 2014. Molecular monitoring of protected fungi: mycelium persistence in soil after timber harvest. Fungal Ecology 9, 34–42.

Grelet, G.-A., Johnson, D., Vralstad, T., Anderson, I.J., Anderson, I.C., 2010. New insights into the mycorrhizal *Rhizoscyphus ericae* aggregate: spatial structure and co-colonization of ectomycorrhizal and ericoid roots. New Phytologist 188, 210–222.

Gustafsson, L., Baker, S.C., Bauhus, J., Beese, W.J., Brodie, A., Kouki, J., et al., 2012. Retention forestry to maintain multifunctional forests: a world perspective. BioScience 62, 633–645.

Hacskaylo, E., 1972. Mycorrhiza: the ultimate in reciprocal parasitism? BioScience 22, 577–583.

Hagerman, S.M., Durall, D.M., 2004. Ectomycorrhizal colonization of greenhouse-grown Douglas-fir (*Pseudotsuga menziesii*) seedlings by inoculum associated with the roots of refuge plants sampled from a Douglas-fir forest in the southern interior of British Columbia. Canadian Journal of Botany-Revue Canadienne de Botanique 82 (6), 742–751.

Hagerman, S.M., Sakakibara, S.M., Durall, D.M., 2001. The potential for woody understory plants to provide refuge for ectomycorrhizal inoculum at an interior Douglas-fir forest after clear-cut logging. Canadian Journal of Forest Research-Revue Canadienne de Recherche Forestiere 31 (4), 711–721.

Hagerman, S.M., Jones, M.D., Bradfield, G.E., Gillespie, M., Durall, D.M., 1999a. Effects of clear-cut logging on the diversity and persistence of ectomycorrhizae at a subalpine forest. Canadian Journal of Forest Research-Revue Canadienne de Recherche Forestiere 29 (1), 124–134.

Hagerman, S.M., Jones, M.D., Bradfield, G.E., Sakakibara, S.M., 1999b. Ectomycorrhizal colonization of *Picea engelmannii* × *Picea glauca* seedlings planted across cut blocks of different sizes. Canadian Journal of Forest Research-Revue Canadienne De Recherche Forestiere 29 (12), 1856–1870.

Hancock, M.H., Summers, R.W., Amphlett, A., Willi, J., 2009. Testing prescribed fire as a tool to promote Scots pine *Pinus sylvestris* regeneration. European Journal of Forest Research 128, 319–333.

Hart, S.C., Classen, A.T., Wright, R.J., 2005. Long-term interval burning alters fine root and mycorrhizal dynamics in a ponderosa pine forest. Journal of Applied Ecology 42 (4), 752–761.

Hartmann, M., Howes, C.G., VanInsberghe, D., Yu, H., Bachar, D., Christen, R., et al., 2012. Significant and persistent impact of timber harvesting on soil microbial communities in northern coniferous forests (vol 6, pg 2199, 2012). ISME Journal 6 (12), 2320 –2320.

Hartmann, M., Lee, S., Hallam, S.J., Mohn, W.W., 2009. Bacterial, archaeal and eukaryal community structures throughout soil horizons of harvested and naturally disturbed forest stands. Environmental Microbiology 11 (12), 3045–3062.

Harvey, A.E., PageDumroese, D.S., Jurgensen, M.F., Graham, R.T., Tonn, J.R., 1997. Site preparation alters soil distribution of roots and ectomycorrhizae on outplanted western white pine and Douglas-fir. Plant and Soil 188 (1), 107–117.

Harvey, A.E., PageDumroese, D.S., Jurgensen, M.F., Graham, R.T., Tonn, J.R., 1996. Site preparation alters biomass, root and ectomycorrhizal development of outplanted western white pine and Douglas-fir. New Forests 11 (3), 255–270.

Harvey, A.E., Jurgensen, M.F., Larsen, M.J., Graham, R.T., 1987. Relationships among soil microsite, ectomycorrhizae, and natural conifer regeneration of old-growth forests in western Montana. Canadian Journal of Forest Research-Revue Canadienne de Recherche Forestiere 17 (1), 58–62.

Harvey, A.E., Jurgensen, M.F., Larsen, M.J., Schlieter, J.A., 1986. Distribution of active ectomycorrhizal short roots in forest soils of the inland northwest: effects of site and disturbance. USDA Forest Service Intermountain Research Station Research Paper 374, 1–8.

Harvey, A.E., Jurgensen, M.F., Larsen, M.J., 1980. Clear-cut harvesting and ectomycorrhizae: survival of activity on residual roots and influence on a bordering forest stand in western Montana. Canadian Journal of Forest Research-Revue Canadienne De Recherche Forestiere 10 (3), 300–303.

Harvey, A.E., Larsen, M.J., Jurgensen, M.F., 1979. Comparative distribution of ectomycorrhizae in soils of three western Montana forest habitat types. Forest Science 25 (2), 350–358.

Hay, T.N., Phillips, L.A., Nicholson, B.A., Jones, M.D., 2015. Ectomycorrhizal community structure and function in interior spruce forests of British Columbia under long term fertilization. Forest Ecology and Management 350, 87–95.

Heinonsalo, J., Sen, R., 2007. Scots pine ectomycorrhizal fungal inoculum potential and dynamics in podzol-specific humus, eluvial and illuvial horizons one and four growth seasons after forest clear-cut logging. Canadian Journal of Forest Research-Revue Canadienne de Recherche Forestiere 37 (2), 404–414.

Holz, M., Aurangojeb, M., Kasimir, A., Boeckx, P., Kuzyakov, Y., Klemedtsson, L., et al., 2016. Gross nitrogen dynamics in the mycorrhizosphere of an organic forest soil. Ecosystems 19 (2), 284–295.

Horton, T.R., Bruns, T.D., Parker, V.T., 1999. Ectomycorrhizal fungi associated with *Arctostaphylos* contribute to *Pseudotsuga menziesii* establishment. Canadian Journal of Botany-Revue Canadienne De Botanique 77 (1), 93–102.

Hunt, G.A., 1991. Ectomycorrhizal fungi in British Columbia container nurseries. FRDA Handbook, 9. Forestry Canada, Victoria, Canada. ISSN 0835–1929.

Izzo, A., Nguyen, D.T., Bruns, T.D., 2006. Spatial structure and richness of ectomycorrhizal fungi colonizing bioassay seedlings from resistant propagules in a Sierra Nevada forest: comparisons using two hosts that exhibit different seedling establishment patterns. Mycologia 98 (3), 374–383.

Jaderlund, A., Norberg, G., Zackrisson, O., Dahlberg, A., Teketay, D., Dolling, A., et al., 1998. Control of bilberry vegetation by steam treatment: effects on seeded Scots pine and associated mycorrhizal fungi. Forest Ecology and Management 108 (3), 275–285.

Jones, M.D., Phillips, L.A., Treu, R., Ward, V., Berch, S.M., 2012. Functional responses of ectomycorrhizal fungal communities to long-term fertilization of lodgepole pine (*Pinus contorta* Dougl. ex Loud. var. *latifolia* Engelm.) stands in central British Columbia. Applied Soil Ecology 60, 29–40.

Jones, M.D., Twieg, B.D., Ward, V., Barker, J., Durall, D.M., Simard, S.W., 2010. Functional complementarity of Douglas-fir ectomycorrhizas for extracellular enzyme activity after wildfire or clearcut logging. Functional Ecology 24 (5), 1139–1151.

Jones, M.D., Grenon, F., Peat, H., Fitzgerald, M., Holt, L., Philip, L.J., et al., 2009. Differences in (15)N uptake amongst spruce seedlings colonized by three pioneer ectomycorrhizal fungi in the field. Fungal Ecology 2, 110–120.

Jones, M.D., Twieg, B.D., Durall, D.M., Berch, S.M., 2008. Location relative to a retention patch affects the ECM fungal community more than patch size in the first season after timber harvesting on Vancouver Island, British Columbia. Forest Ecology and Management 255 (3–4), 1342–1352.

Jones, M.D., Durall, D.M., Cairney, J.W.G., 2003. Ectomycorrhizal fungal communities in young forest stands regenerating after clearcut logging. New Phytologist 157 (3), 399–422.

Jones, M.D., Durall, D.M., Harniman, S.M.K., Classen, D.C., Simard, S.W., 1997. Ectomycorrhizal diversity on *Betula papyrifera* and *Pseudotsuga menziesii* seedlings grown in the greenhouse or outplanted in single-species and mixed plots in southern British Columbia. Canadian Journal of Forest Research-Revue Canadienne de Recherche Forestiere 27 (11), 1872–1889.

Jurgensen, M.F., Harvey, A.E., Graham, R.T., PageDumroese, D.S., Tonn, J.R., Larsen, M.J., et al., 1997. Impacts of timber harvesting on soil organic matter, nitrogen, productivity, and health of Inland Northwest forests. Forest Science 43 (2), 234–251.

Karst, J., Erbilgin, N., Pec, G.J., Cigan, P.W., Najar, A., Simard, S.W., et al., 2015. Ectomycorrhizal fungi mediate indirect effects of a bark beetle outbreak on secondary chemistry and establishment of pine seedlings. New Phytologist 208 (3), 904–914.

Kataja-aho, S., Pennanen, T., Lensu, A., Haimi, J., 2012. Does stump removal affect early growth and mycorrhizal infection of spruce (*Picea abies*) seedlings in clear-cuts? Scandinavian Journal of Forest Research 27 (8), 746–753.

Kennedy, P.G., Peay, K.G., Bruns, T.D., 2009. Root tip competition among ectomycorrhizal fungi: are priority effects a rule or an exception? Ecology 90 (8), 2098–2107.

Kennedy, P.G., Izzo, A.D., Bruns, T.D., 2003. There is high potential for the formation of common mycorrhizal networks between understorey and canopy trees in a mixed evergreen forest. Journal of Ecology 91 (6), 1071–1080.

Kennedy, N.M., Robertson, S.J., Green, D.S., Scholefield, S.R., Arocena, J.M., Tackaberry, L.E., et al., 2015. Site properties have a stronger influence than fire severity on ectomycorrhizal fungi and associated N-cycling bacteria in regenerating post-beetle-killed lodgepole pine forests. Folia Microbiologica 60 (5), 399–410.

Kernaghan, G., 2005. Mycorrhizal diversity: cause and effect? Pedobiologia 49 (6), 511–520.

Kernaghan, G., Sigler, L., Khasa, D., 2003. Mycorrhizal and root endophytic fungi of containerized *Picea glauca* seedlings assessed by rDNA sequence analysis. Microbial Ecology 45, 128–136.

Khasa, P.D., Sigler, L., Chakravarty, P., Dancik, B.P., Erickson, L., Mc Curdy, D., 2001. Effect of fertilization on growth and ectomycorrhizal development of container-grown and bare-root nursery conifer seedlings. New Forests 22 (3), 179–197.

Klavina, D., Gaitnieks, T., Menkis, A., 2013. Survival, growth and ectomycorrhizal community development of container- and bare-root grown *Pinus sylvestris* and *Picea abies* seedlings outplanted on a forest clear-cut. Baltic Forestry 19 (1), 39–49.

Kozlowski, T.T., 1999. Soil compaction and growth of woody plants. Scandinavian Journal of Forest Research 14 (6), 596–619.

Kranabetter, J.M., 2014. Ectomycorrhizal fungi and the nitrogen economy of conifers: implications for genecology and climate change mitigation. Canadian Journal of Botany-Revue Canadienne de Botanique 92, 417–423.

Kranabetter, J.M., 2004. Ectomycorrhizal community effects on hybrid spruce seedling growth and nutrition in clearcuts. Canadian Journal of Botany-Revue Canadienne de Botanique 82, 983–991.

Kranabetter, J.M., 1999. The effect of refuge trees on a paper birch ectomycorrhiza community. Canadian Journal of Botany-Revue Canadienne de Botanique 77 (10), 1523–1528.

Kranabetter, J.M., Wylie, T., 1998. Ectomycorrhizal community structure across forest openings on naturally regenerated western hemlock seedlings. Canadian Journal of Botany-Revue Canadienne de Botanique 76 (2), 189–196.

Kranabetter, J.M., Stoehr, M., O'Neill, G.A., 2015. Ectomycorrhizal fungal maladaptation and growth reductions associated with assisted migration of Douglas-fir. New Phytologist 206 (3), 1135–1144.

Kranabetter, J.M., De Montigny, L., Ross, G., 2013. Effectiveness of green-tree retention in the conservation of ectomycorrhizal fungi. Fungal Ecology 6 (5), 430–438.

Kranabetter, J.M., Durall, D.M., MacKenzie, W.H., 2009. Diversity and species distribution of ectomycorrhizal fungi along productivity gradients of a southern boreal forest. Mycorrhiza 19 (2), 99–111.

Kretzer, A.M., Dunham, S., Molina, R., Spatafora, J.W., 2004. Microsatellite markers reveal the below ground distribution of genets in two species of *Rhizopogon* forming tuberculate ectomycorrhizas on Douglas fir. New Phytologist 161 (1), 313–320.

Krivtsov, V., Illian, J.B., Liddell, K., Garside, A., Bezginova, T., Salmond, R., et al., 2003. Some aspects of complex interactions involving soil mesofauna: analysis of the results from a Scottish woodland. Ecological Modelling 170, 441–452.

Lazaruk, L.W., Macdonald, S.E., Kernaghan, G., 2008. The effect of mechanical site preparation on ectomycorrhizae of planted white spruce seedlings in conifer-dominated boreal mixed wood forest. Canadian Journal of Forest Research-Revue Canadienne de Recherche Forestiere 38 (7), 2072–2079.

Lazaruk, L.W., Kernaghan, G., Macdonald, S.E., Khasa, D., 2005. Effects of partial cutting on the ectomycorrhizae of *Picea glauca* forests in northwestern Alberta. Canadian Journal of Forest Research-Revue Canadienne De Recherche Forestiere 35 (6), 1442–1454.

LeDuc, S.D., Lilleskov, E.A., Horton, T.R., Rothstein, D.E., 2013. Ectomycorrhizal fungal succession coincides with shifts in organic nitrogen availability and canopy closure in post-wildfire jack pine forests. Oecologia 172 (1), 257–269.

Lilleskov, E.A., Hobbie, E.A., Horton, T.R., 2011. Conservation of ectomycorrhizal fungi: exploring the linkages between functional and taxonomic responses to anthropogenic N deposition. Fungal Ecology 4 (2), 174–183.

Lilleskov, E.A., Bruns, T.D., 2005. Spore dispersal of a resupinate ectomycorrhizal fungus, *Tomentella sublilacina*, via soil food webs. Mycologia 97 (4), 762–769.

Lim, S., Berbee, M.L., 2013. Phylogenetic structure of ectomycorrhizal fungal communities of western hemlock changes with forest age and stand type. Mycorrhiza 23 (6), 473–486.

Luoma, D.L., Stockdale, C.A., Molina, R., Eberhart, J.L., 2006. The spatial influence of *Pseudotsuga menziesii* retention trees on ectomycorrhiza diversity. Canadian Journal of Forest Research-Revue Canadienne De Recherche Forestiere 36 (10), 2561–2573.

MacArthur, R.H., Wilson, E.O., 1967. The Theory of Island Biogeography. Princeton University Press, Princeton, USA.

Mah, K., Tackaberry, L.E., Egger, K.B., Massicotte, H.B., 2001. The impacts of broadcast burning after clear-cutting on the diversity of ectomycorrhizal fungi associated with hybrid spruce seedlings in central British Columbia. Canadian Journal of Forest Research-Revue Canadienne de Recherche Forestiere 31 (2), 224–235.

Marshall, V.G., 2000. Impacts of forest harvesting on biological processes in northern forest soils. Forest Ecology and Management 133 (1–2), 43–60.

Massicotte, H.B., Molina, R., Tackaberry, L.E., Smith, J.E., Amaranthus, M.P., 1999. Diversity and host specificity of ectomycorrhizal fungi retrieved from three adjacent forest sites by five host species. Canadian Journal of Botany-Revue Canadienne de Botanique 77, 1053–1076.

Menkis, A., Uotila, A., Arhipova, N., Vasaitis, R., 2010. Effects of stump and slash removal on growth and mycorrhization of *Picea abies* seedlings outplanted on a forest clear-cut. Mycorrhiza 20 (7), 505–509.

Miller, S.L., McClean, T.M., Stanton, N.L., Williams, S.E., 1998. Mycorrhization, physiognomy, and first-year survivability of conifer seedlings following natural fire in Grand Teton National Park. Canadian Journal of Forest Research-Revue Canadienne de Recherche Forestiere 28 (1), 115–122.

Molina, R., Massicotte, H.B., Trappe, J.M., 1992. Ecological role of specificity phenomena in ectomycorrhizal plant-communities: potentials for interplant linkages and guild development. In: Allen, M.F. (Ed.), Mycorrhizas in Ecosystems. Chapman & Hall, New York, USA, pp. 106–112.

Mosca, E., Montecchio, L., Scattolin, L., Garbaye, J., 2007. Enzymatic activities of three ectomycorrhizal types of *Quercus robur* L. in relation to tree decline and thinning. Soil Biology and Biochemistry 39 (11), 2897–2904.

Mummey, D.L., Clarke, J.T., Cole, C.A., O'Connor, B.G., Gannon, J.E., Ramsey, P.W., 2010. Spatial analysis reveals differences in soil microbial community interactions between adjacent coniferous forest and clearcut ecosystems. Soil Biology and Biochemistry 42 (7), 1138–1147.

Nguyen, N.H., Hynson, N.A., Bruns, T.D., 2012. Stayin' alive: survival of mycorrhizal fungal propagules from 6-yr-old forest soil. Fungal Ecology 5 (6), 741–746.

Nilsson, M.C., Steijlen, I., Zackrisson, O., 1996. Time-restricted seed regeneration of Scots pine in sites dominated by feather moss after clear-cutting. Canadian Journal of Forest Research-Revue Canadienne de Recherche Forestiere 26 (6), 945–953.

O'Hara, K.L., 2014. Multiaged Silviculture: Managing for Complex Forest Stand Structure. Oxford University Press, Oxford, UK.

Olsson, J., Jonsson, B.G., Hjalten, J., Ericson, L., 2011. Addition of coarse woody debris: the early fungal succession on *Picea abies* logs in managed forests and reserves. Biological Conservation 144 (3), 1100–1110.

Page-Dumroese, D.S., Jurgensen, M.F., Tiarks, A.E., Ponder, F., Sanchez, F.G., et al., 2006. Soil physical property changes at the North American Long-Term Soil Productivity study sites: 1 and 5 years after compaction. Canadian Journal of Forest Research-Revue Canadienne de Recherche Forestiere 36 (3), 551–564.

Page-Dumroese, D.S., Harvey, A.E., Jurgensen, M.F., Amaranthus, M.P., 1998. Impacts of soil compaction and tree stump removal on soil properties and outplanted seedlings in northern Idaho, USA. Canadian Journal of Soil Science 78 (1), 29–34.

Parke, J.L., Linderman, R.G., Trappe, J.M., 1983. Effect of root zone temperature on ectomycorrhiza and vesicular arbuscular mycorrhiza formation in disturbed and undisturbed forest soils of southwest Oregon. Canadian Journal of Forest Research-Revue Canadienne de Recherche Forestiere 13 (4), 657–665.

Parsons, W.F.J., Miller, S.L., Knight, D.H., 1994. Root-gap dynamics in a lodgepole pine forest: ectomycorrhizal and nonmycorrhizal fine root activity after experimental gap formation. Canadian Journal of Forest Research-Revue Canadienne de Recherche Forestiere 24, 1531–1538.

Peay, K.G., Schubert, M.G., Nguyen, N.H., Bruns, T.D., 2012. Measuring ectomycorrhizal fungal dispersal: macroecological patterns driven by microscopic propagules. Molecular Ecology 21 (16), 4122–4136.

Peay, K.G., Garbelotto, M., Bruns, T.D., 2010. Evidence of dispersal limitation in soil microorganisms: isolation reduces species richness on mycorrhizal tree islands. Ecology 91 (12), 3631–3640.

Peay, K.G., Bruns, T.D., Kennedy, P.G., Bergemann, S.E., Garbelotto, M., 2007. A strong species-area relationship for eukaryotic soil microbes: island size matters for ectomycorrhizal fungi. Ecology Letters 10 (6), 470–480.

Pennanen, T., Heiskanen, J., Korkama, U., 2005. Dynamics of ectomycorrhizal fungi and growth of Norway spruce seedlings after planting on a mounded forest clearcut. Forest Ecology and Management 213 (1–3), 243–252.

Pennock, D.J., vanKessel, C., 1997. Effect of agriculture and of clear-cut forest harvest on landscape-scale soil organic carbon storage in Saskatchewan. Canadian Journal of Soil Science 77 (2), 211–218.

Perry, D.A., Molina, R., Amaranthus, M.P., 1987. Mycorrhizae, mycorrhizosphere, and reforestation: current knowledge and research needs. Canadian Journal of Forest Research-Revue Canadienne de Recherche Forestiere 17, 929–940.

Pickles, B.J., Twieg, B.D., O'Neill, G.A., Mohn, W.W., Simard, S.W., 2015. Local adaptation in migrated interior Douglas-fir seedlings is mediated by ectomycorrhizas and other soil factors. New Phytologist 207 (3), 858–871.

Pietras, M., Rudawska, M., Leski, T., Karlinski, L., 2013. Diversity of ectomycorrhizal fungus assemblages on nursery grown European beech seedlings. Annals of Forest Science 70 (2), 115–121.

Pither, J., Aarssen, L.W., 2005. The evolutionary species pool hypothesis and patterns of freshwater diatom diversity along a pH gradient. Journal of Biogeography 32 (3), 503–513.

Polanco, M.C., Zwiazek, J.J., Jones, M.D., MacKinnon, M.D., 2008. Responses of mycorrhizal jack pine (*Pinus banksiana*) seedlings to NaCl and boron. Trees-Structure and Function 22 (6), 825–834.

Poznanovic, S.K., Lilleskov, E.A., Webster, C.R., 2015. Sharing rotting wood in the shade: ectomycorrhizal communities of co-occurring birch and hemlock seedlings. Mycorrhiza 25 (2), 153–164.

Prescott, C.E., 2002. The influence of the forest canopy on nutrient cycling. Tree Physiology 22, 1193–1200.

Prescott, C.E., Sajedi, T., 2008. The role of salal in forest regeneration problems in coastal British Columbia: problem or symptom? Forestry Chronicle 84 (1), 29–36.

Purdy, B.G., Macdonald, S.E., Dale, M.R.T., 2002. The regeneration niche of white spruce following fire in the mixed wood boreal forest. Silva Fennica 36 (1), 289–306.

Querejeta, J.I., Roldan, A., Albaladejo, J., Castillo, V., 1998. The role of mycorrhizae, site preparation, and organic amendment in the afforestation of a semi-arid Mediterranean site with *Pinus halepensis*. Forest Science 44 (2), 203–211.

Richter, D.L., Bruhn, J.N., 1993. Mycorrhizal fungus colonization of *Pinus resinosa* AIT transplanted on northern hardwood clearcuts. Soil Biology and Biochemistry 25 (3), 355–369.

Roach, W.J., Simard, S.W., Sachs, D.L., 2015. Evidence against planting lodgepole pine monocultures in the cedar-hemlock forests of southeastern British Columbia. Forestry 88 (3), 345–358.

Romell, L.G., 1939. The ecological problem of mycotrophy. Ecology 20 (2), 163–167.

Rudawska, M., Leski, T., Trocha, L.K., Gornowicz, R., 2006. Ectomycorrhizal status of Norway spruce seedlings from bare-root forest nurseries. Forest Ecology and Management 236 (2–3), 375–384.

Russell, M.B., D'Amato, A.W., Albers, M.A., Woodall, C.W., Puettmann, K.J., Saunders, M.R., et al., 2015. Performance of the forest vegetation simulator in managed white spruce plantations influenced by Eastern spruce budworm in northern Minnesota. Forest Science 61 (4), 723–730.

Sebastiana, M., Pereira, V.T., Alcantara, A., Pais, M.S., Silva, A.B., 2013. Ectomycorrhizal inoculation with *Pisolithus tinctorius* increases the performance of *Quercus suber* L. (cork oak) nursery and field seedlings. New Forests 44 (6), 937–949.

Shaw, P.J.A., Kibby, G., Mayes, J., 2003. Effects of thinning treatment on an ectomycorrhizal succession under Scots pine. Mycological Research 107, 317–328.

Siira-Pietikainen, A., Haimi, J., Fritze, H., 2003. Organisms, decomposition, and growth of pine seedlings in boreal forest soil affected by sod cutting and trenching. Biology and Fertility of Soils 37 (3), 163–174.

Simard, S., Martin, K., Vyse, A., Larson, B., 2013. Meta-networks of fungi, fauna and flora as agents of complex adaptive systems. In: Messier, C., Puettman, K.J., Coates, K.D. (Eds.), Managing Forests as Complex Adaptive Systems: Building Resilience to the Challenge of global Change. Routledge, Abingdon, UK, pp. 133–164.

Simard, S.W., Durall, D.M., 2004. Mycorrhizal networks: a review of their extent, function, and importance. Canadian Journal of Botany-Revue Canadienne de Botanique 82 (8), 1140–1165.

Simard, S.W., Jones, M.D., Durall, D.M., Hope, G.D., Stathers, R.J., Sorensen, N.S., et al., 2003. Chemical and mechanical site preparation: effects on *Pinus contorta* growth, physiology, and microsite quality on grassy, steep forest sites in British Columbia. Canadian Journal of Forest Research-Revue Canadienne de Recherche Forestiere 33 (8), 1495–1515.

Simard, S.W., Perry, D.A., Smith, J.E., Molina, R., 1997. Effects of soil trenching on occurrence of ectomycorrhizas on *Pseudotsuga menziesii* seedlings grown in mature forests of *Betula papyrifera* and *Pseudotsuga menziesii*. New Phytologist 136, 327–340.

Smith, J.E., Molina, R., Huso, M.M.P., Larsen, M.J., 2000. Occurrence of *Piloderma fallax* in young, rotation-age, and old-growth stands of Douglas-fir (*Pseudotsuga menziesii*) in the Cascade Range of Oregon, USA. Canadian Journal of Botany-Revue Canadienne de Botanique 78 (8), 995–1001.

Southworth, D., Donohue, J., Frank, J.L., Gibson, J., 2011. Mechanical mastication and prescribed fire in conifer-hardwood chaparral: differing responses of ectomycorrhizae and truffles. International Journal of Wildland Fire 20 (7), 888–896.

Spake, R., Ezard, T.H.G., Martin, P.A., Newton, A.C., Doncaster, C.P., 2015. A meta-analysis of functional group responses to forest recovery outside of the tropics. Conservation Biology 29, 1695–1703.

Stevens, V., 1997. The Ecological Role of Coarse Woody Debris: an Overview of the Ecological Role of CWD in BC Forests. Research Branch, British Columbia Ministry of Forests, Victoria, Canada. Working Paper 30/1997.

Taylor, D.L., Bruns, T.D., 1999. Community structure of ectomycorrhizal fungi in a *Pinus muricata* forest: minimal overlap between the mature forest and resistant propagule communities. Molecular Ecology 8 (11), 1837–1850.

Tedersoo, L., Gates, G., Dunk, C.W., Lebel, T., May, T.W., Koljalg, U., et al., 2009. Establishment of ectomycorrhizal fungal community on isolated *Nothofagus cunninghamii* seedlings regenerating on dead wood in Australian wet temperate forests: does fruit-body type matter? Mycorrhiza 19 (6), 403–416.

Tedersoo, L., Suvi, T., Jairus, T., Koljalg, U., 2008. Forest microsite effects on community composition of ectomycorrhizal fungi on seedlings of *Picea abies* and *Betula pendula*. Environmental Microbiology 10 (5), 1189–1201.

Tedersoo, L., Koljalg, U., Hallenberg, N., Larsson, K.H., 2003. Fine scale distribution of ectomycorrhizal fungi and roots across substrate layers including coarse woody debris in a mixed forest. New Phytologist 159 (1), 153–165.

Teste, F.P., Simard, S.W., 2008. Mycorrhizal networks and distance from mature trees alter patterns of competition and facilitation in dry Douglas-fir forests. Oecologia 158 (2), 193–203.

Teste, F.P., Lieffers, V.J., Strelkov, S.E., 2012. Ectomycorrhizal community responses to intensive forest management: thinning alters impacts of fertilization. Plant and Soil 360 (1–2), 333–347.

Teste, F.P., Simard, S.W., Durall, D.M., 2009. Role of mycorrhizal networks and tree proximity in ectomycorrhizal colonization of planted seedlings. Fungal Ecology 2, 21–30.

Titus, B.D., Prescott, C.E., Maynard, D.G., Mitchell, A.K., Bradley, R.L., Feller, M.C., et al., 2006. Post-harvest nitrogen cycling in clearcut and alternative silvicultural systems in a montane forest in coastal British Columbia. Forestry Chronicle 82 (6), 844–859.

Treseder, K.K., 2004. A meta-analysis of mycorrhizal responses to nitrogen, phosphorus, and atmospheric CO_2 in field studies. New Phytologist 164 (2), 347–355.

Twieg, B.D., Durall, D.M., Simard, S.W., 2007. Ectomycorrhizal fungal succession in mixed temperate forests. New Phytologist 176 (2), 437–447.

van der Heijden, M.G.A., Horton, T.R., 2009. Socialism in soil? The importance of mycorrhizal fungal networks for facilitation in natural ecosystems. Journal of Ecology 97, 1139–1150.

Van Dorp, C.H., 2016. Rhizopogon Mycorrhizae of Interior Douglas-Fir and their Mycorrhizal Networks in Selectively Harvested and Non-harvested Forests. MSc thesis, University of British Columbia Okanagan campus, Kelowna, BC, Canada. https://open.library.ubc.ca/cIRcle/collections/ubctheses/24/items/1.0304573.

Van Dorp, C., Beiler, K., Durall, D., 2016. Dominance of a *Rhizopogon* sister species corresponds to forest age structure. Mycorrhiza 26 (2), 169–175.

Väre, H., 1989a. Influence of decaying birch logs to Scots pine mycorrhizae at clear-cutted ploughed sites in northern Finland. Agriculture, Ecosystems and Environment 28, 539–545.

Väre, H., 1989b. The mycorrhizal condition of weakened Scots pine saplings grown on ploughed sites in northern Finland. Canadian Journal of Forest Research-Revue Canadienne de Recherche Forestiere 19, 341–346.

Velmala, S.M., Rajala, T., Haapanen, M., Taylor, A.F.S., Pennanen, T., 2013. Genetic host-tree effects on the ectomycorrhizal community and root characteristics of Norway spruce. Mycorrhiza 23 (1), 21–33.

Visser, S., 1995. Ectomycorrhizal fungal succession in jack pine stands following wildfire. New Phytologist 129 (3), 389–401.

Vozzo, J.A., Hacskaylo, E., 1971. Inoculation of *Pinus caribaea* with ectomycorrhizal fungi in Puerto Rico. Forest Science 17, 239–245.

Vyse, A., 1999. Is everything all right up there? A long-term interdisciplinary silvicultural systems project in a high elevation fir-spruce forest at Sicamous Creek, BC. Forestry Chronicle 75 (3), 467–472.

Walker, J.K.M., Jones, M.D., 2013. Little evidence for niche partitioning among ectomycorrhizal fungi on spruce seedlings planted in decayed wood versus mineral soil microsites. Oecologia 173 (4), 1499–1511.

Walker, J.K.M., Ward, V., Jones, M.D., 2016. Ectomycorrhizal fungal exoenzyme activity differs on spruce seedlings planted in forests versus clearcuts. Trees—Structure and Function 30, 497–508.

Walker, J.K.M., Phillips, L.A., Jones, M.D., 2014. Ectomycorrhizal fungal hyphae communities vary more along a pH and nitrogen gradient than between decayed wood and mineral soil microsites. Botany 92 (6), 453–463.

Walker, J.K.M., Ward, V., Paterson, C., Jones, M.D., 2012. Coarse woody debris retention in subalpine clearcuts affects ectomycorrhizal root tip community structure within fifteen years of harvest. Applied Soil Ecology 60, 5–15.

Wood, J.R., Dickie, I.A., Moeller, H.V., Peltzer, D.A., Bonner, K.I., Rattray, G., et al., 2015. Novel interactions between non-native mammals and fungi facilitate establishment of invasive pines. Journal of Ecology 103 (1), 121–129.

Wright, S.H.A., Berch, S.M., Berbee, M.L., 2009. The effect of fertilization on the below-ground diversity and community composition of ectomycorrhizal fungi associated with western hemlock (*Tsuga heterophylla*). Mycorrhiza 19 (4), 267–276.

Wu, B.Y., Nara, K., Hogetsu, T., 2002. Spatiotemporal transfer of carbon-14-labelled photosynthate from ectomycorrhizal *Pinus densiflora* seedlings to extraradical mycelia. Mycorrhiza 12 (2), 83–88.

Zhou, M.Y., Sharik, T.L., Jurgensen, M.F., Richter, D.L., 1997. Ectomycorrhizal colonization of *Quercus rubra* seedlings in response to vegetation removals in oak and pine stands. Forest Ecology and Management 93 (1–2), 91–99.

CHAPTER

12

Mycorrhizal Mediation of Soil Fertility Amidst Nitrogen Eutrophication and Climate Change

M.F. Allen, E.B. Allen

University of California, Riverside, CA, United States

12.1 INTRODUCTION

Global environmental change includes multiple complex changes beyond temperature (T) increases. The primary global directional shift is the increasing atmospheric carbon dioxide (CO_2). Through deforestation and fossil fuel burning, humans have dramatically increased transfer of carbon (C) from the lithosphere and biosphere to the atmosphere. Atmospheric CO_2 has increased from 320 ppm in the 1950s to nearly 410 ppm by the spring of 2016, with increasing rates of approximately 1 ppm annually in the decade of the 1960s to nearly 3 ppm annually in the 2010s (NOAA, 2016). Because C is the energy storage compound for all life forms, any change to source or sink alters the interactions of organisms and environment. In most terrestrial ecosystems, C is driven by the rate of photosynthesis (J_{CO_2}) and respiration (R). J_{CO_2} is a direct function of the atmospheric partial pressure of CO_2. As the partial pressure of CO_2 increases, the diffusion of CO_2 into the plant tissue increases per unit of surface area. If there is adequate photosynthetic machinery, then the fixation of C increases linearly with the atmospheric CO_2 levels. This drives an excess of C in comparison with resources derived from the soil [water (H_2O), nitrogen (N), phosphorus (P), other nutrients]. Thus increasing CO_2 drives a greater need for the resources that mycorrhizas provide.

Short-wave radiation passes through the atmosphere and directly affects solid materials such as organisms and soil. The result of this radiation is increasing atmospheric T and reradiation of long-wave radiation (infrared). Most of this reradiation occurs back into space, lowering the T of the globe. But infrared radiation also excites molecules of CO_2 and H_2O vapor in the atmosphere, some of which reradiate back, resulting in increasing Ts. Water vapor is dynamic spatially and temporally. However, CO_2 rapidly diffuses and homogenizes in the atmosphere, and its impact is especially pronounced in regions with low water vapor such as deserts, high elevations, and high latitudes (Beer et al., 2016). These areas have experienced the highest increases in T. Overall, a

Mycorrhizal Mediation of Soil
http://dx.doi.org/10.1016/B978-0-12-804312-7.00012-7

consequence of the increasing atmospheric CO_2 is increasing global Ts, observed since the advent of agriculture but especially since the industrial revolution over 2 centuries ago. Radiation also drives evaporation. Because the earth spins on its axis, evaporation and precipitation are affected by complex forcing functions that vary in time and space daily, seasonally, and in long-term climate drivers. Therefore both T and moisture are driven by elevated atmospheric CO_2.

A second consequence of the advancing agricultural and industrial revolution is a significant increase in the availability of biologically active forms of N. N is an abundant molecule, but most of it is in the form of atmospheric N_2 that cannot be used by organisms. It must be converted to ammonium (NH_4^+), organic N (N_o), or nitrogen oxide (NO_x). In nature that conversion occurs through biological N_2 fixation by prokaryotic microorganisms or through lightning ($N_2 \rightarrow NO_x$). High-input agriculture produces high amounts of available N that is applied to fields, lawns, and forests. Industrial activity also produces high levels of NO_x and ammonia (NH_3), in some areas comparable to agricultural application (Galloway et al., 2004). Mycorrhizal associations take up N and exchange that N with the plant for C. But, if the plant has N in excess of its need, the plant will reduce allocations of C to the fungal partner, in direct contrast to its response to elevated CO_2.

Molecular clock evidence suggests that fungi first appeared in the Ediacaran period and were likely aquatic organisms beginning to occupy marshy environments from the Ediacaran period to the Cambrian period (Droser et al., 2014). The groups that later formed mycorrhizal fungi (MF), including the Glomeromycota and Mucorales, were present by the late Silurian period. Arbuscular mycorrhizal (AM) symbioses first appeared in the Ordovician period when atmospheric CO_2 levels were far higher than today (Allen, 1996; see also Chapter 2), but P was limiting (e.g., Malloch et al., 1980). Atmospheric CO_2 then dropped as land plants and mycorrhizas co-colonized land (Allen, 1996). Effective decomposer fungi and ectomycorrhizal (EM) fungi proliferated some 200 million years ago, again during a period of high CO_2, drawing the C through leaves, roots, and microbes within soil and rock materials. Levels have been varying between 190 ppm (during the Pleistocene glacial maximum) and 1000 ppm since then (e.g., during the Eocene hothouse, Pearson, 2010). The dynamic interplay of atmospheric CO_2, climate, soil fertility, and mycorrhizas has regulated terrestrial life on earth since the beginning (see Chapter 2).

Importantly, we postulate that C and N in plant tissue have inversely covaried through recent Earth's history. As CO_2 increases, N becomes more limiting, and vice versa. One way that plants could deal with the increasing C is by relying more on MF to acquire the nutrients necessary to continue growth. Alternatively, as CO_2 declines, nutrients become more readily available and plants become less dependent on MF. This trade-off may well be reflected in the evolutionary history of vegetation communities (Becklin et al., 2012). Just as important, if this postulate is true, a question emerges: does the mycorrhizal relationship have the genetic mechanisms to cope with a world of increasing CO_2 and simultaneous increasing available N? We will explore this idea through this chapter.

12.2 MECHANISMS OF MYCORRHIZAL NUTRITION AND STOICHIOMETRY

The optimum relative composition of elements in a cell for optimal performance is called the *stoichiometry*. Stoichiometry is used to understand relationships from plant response to elevated CO_2 to shifting elements in food webs (Sterner and Elser, 2002). The Redfield (1934) ratio of 106 C:16 N:1 P was developed for oceanic phytoplankton. Plant tissue is between 43% and 50% C, with the remainder largely oxygen (O) and hydrogen (H). A stoichiometric ratio for terrestrial plants is more on the order of 250 C:7 N:1 P, or a C:N ratio of 35. This is because plants form structural elements (e.g., cellulose and lignins) with high C demands. In contrast to plants, fungi have chitin walls and a higher protein content and thus have a ratio more like 20 C:2 N:1 P, or a C:N ratio of 10 (Allen et al., 2003).

In the environment, both C (as CO_2 in the atmosphere) and nutrients (in the lithosphere) are largely taken up in inorganic forms against a concentration gradient. Although atmospheric CO_2 has been enriched over the past century from a concentration of 320 ppm to approximately 400 ppm, there is still a steep concentration gradient, ranging from 0.0400% CO_2 in the atmosphere to 50% C in the leaf. Leaves draw down inorganic C as CO_2 from the atmosphere (400 ppm) into the leaf cell (approaching 0 ppm) by fixing inorganic C into organic C through photosynthesis, an energetic process utilizing sunlight. Plants take up available N [NO_3^- or NH_4^+] from soil containing 0.1–20NH_4^+mg/kg. Plant leaf cells must take up N to 2% (or 20,000 mg/kg) against a gradient from soil of 3–3.5 orders of magnitude. The P gradient is in a similar range of magnitude.

A mycorrhizal symbiosis creates a paradox. MF have higher N and P concentrations than plants. The free energy gradient of transfer of C from plant to fungus, and N and P from fungus to plant are favorable for exchange. This was shown in the down-gradient in P from AM fungi to plant cells with the arbuscule, to adjacent plant cells, to the vascular tissue (Schoknecht and Hattingh, 1976). Also, the ratio of organic P to inorganic P (P_o:P_i) in leaves from AM plants was greater than 50%, whereas it was below 50% in leaves from nonmycorrhizal plants (Allen et al., 1981), indicating both greater photosynthesis and a reduced free energy gradient for P transport. However, to obtain that P and N the concentration gradient from soil to MF was even greater than for the plant tissue, requiring even more energy.

The mechanism is that the small diameter hyphal tips of MF can access both water and nutrients in 2-μm (diameter) soil pores that are too small for even the finest root hairs (Allen, 2007; see also Chapter 8). Hyphae rapidly integrate hyphal P into forms other than phosphate (e.g., P_o, polyphosphate granules) and transport it to the arbuscular cells or the Hartig net cells, creating a free energy gradient for uptake. Many of these fungi also have enzymatic capabilities for converting organic molecules into available inorganic forms that go beyond the ability of plant roots. Together these processes mean that mycorrhizas, despite the stoichiometric gradients, improve the uptake of nutrients. These concepts are quantifiable, and can be used to understand and model the role of mycorrhizas in plant nutrition (Allen et al., 2003) in the face of a changing environment.

12.3 NUTRIENT UPTAKE AND MYCORRHIZAL FUNGI: THE BASICS

To characterize the impacts of global change on mycorrhizas, we will utilize the two processes whereby MF can alter the physiology of the host plant. First we focus on the uptake of nutrients into plant leaves for photosynthesis, such as N for ribulose-1,5-bisphosphate carboxylase/oxygenase (RuBisCO), the dominant enzyme for photosynthesis, and P for energy storage compounds [adenosine triphosphate (ATP), nicotinamide adenine dinucleotide phosphate (NADP)]. For this process, we utilize the Agren and Bosatta (1996) model. We then must access those nutrients, which are distributed at very low concentrations in the lithosphere. To understand how the roots and fungi acquire these nutrients, we turn to the model of Treseder and Allen (2002) that is built on how nutrient uptake feeds back into acquisition.

12.3.1 Plant Uptake Model

We can build on the model of Agren and Bosatta (1996) (see Allen et al., 2003). In this model (Fig. 12.1), as the nutrient concentration (in this case, N or P) increases, growth rate also increases. Mycorrhizas increase access to the limiting nutrient, resulting in greater uptake toward maximal plant growth rate. The maximal mycorrhizal activity will occur where nutrient levels are limiting and CO_2 is adequate. When the plant has adequate nutrition for the CO_2, then C support for mycorrhizas declines. Elevated CO_2 drives plant nutrient concentrations down. Mycorrhizal activity then again increases as plants exhibit nutrient stress. As plant C is exchanged for fungal P or N, then plant growth rates again increase. This creates a dynamic interplay between plant and fungus, built around C:N or C:P ratios. As atmospheric CO_2 increases, the plant nutrient level declines, resulting in more allocation of C toward the optimal level for mycorrhizal formation. As P or N exchange from fungus to plant increases, more C is fixed, either maintaining the plant nutrient level within the optimal mycorrhizal formation range or even increasing the nutrient level to where the plant rejects additional infections or simply outgrows the fungal inoculum density with a consequent lowering of mycorrhizal formation. This model can result in a relatively simple but dynamic relationship. Mycorrhizas simply provide the nutrients but take C that otherwise would be used in plant growth directly.

Importantly, N deposition adds another complication. With N deposition, growth rate increases to a point where the plant does not need to rely on the fungus to obtain that N (e.g., Allen et al., 2010). N levels can even increase to the point of toxicity (a "burning" of plant leaf tissue) and cause reduced plant size or mortality (e.g., Padgett and Allen, 1999; Allen et al., 2010). In addition, P then

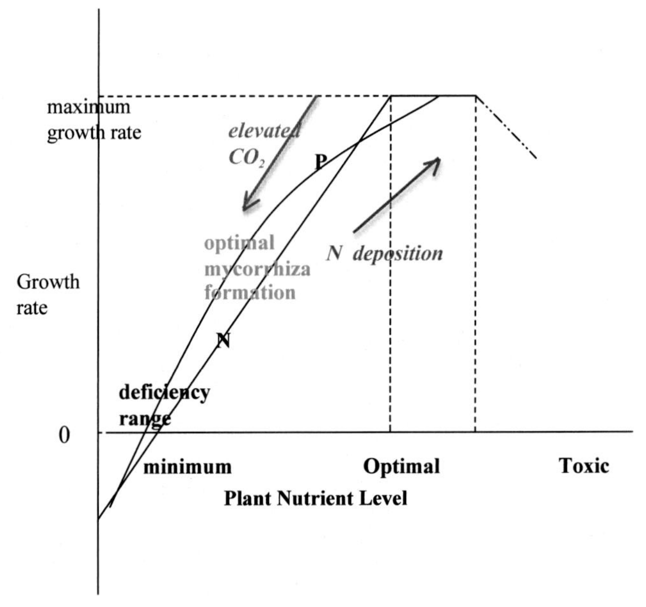

FIGURE 12.1 Growth rate versus plant nutrient levels, optimizing for mycorrhizal formation (infection, colonization, activity, or other measure), elevated CO_2, and N deposition. At extremely low levels of soil fertility where the high C:N or C:P ratio of the fungus requires N or P, immobilization by the fungus (if present) could well occur, a situation found in the extreme deficiencies of ancient soils in Australia created by high aluminum (Al), P-fixing soils. This process could also occur in new high silica soils such as the Mount St. Helens pumice, or even in some soils with moderate nutrition but high calcium (Ca) [or iron (Fe) or Al] P-fixing soils. *Redrawn from Agren, G.I., Bosatta, E., 1996. Theoretical Ecosystem Ecology: Understanding Element Cycles. Cambridge University Press, Cambridge, England.*

becomes limiting. This relatively simple two-dimensional dynamic model now becomes a highly complex, three-dimensional interactive model.

12.3.2 Soil Availability Model

Treseder and Allen (2002) defined a simple quadratic model between mycorrhizal activity and soil nutrient levels (Fig. 12.2). Although it is outwardly straight-forward, the implications are enormous and complex. There is an optimal supply level of nutrients that supports a mycorrhiza and the plant. This equates to the optimal formation range in Fig. 12.1. As nutrient supply goes above the optimal supply rate for supporting mycorrhizas, the plant simply stops allocating C to the fungus. Alternatively at the low end, there are not enough nutrients to support the fungal tissue. Here at the low end, the N or P required by the plant exceeds the rate that can be supplied by the fungus. Most research sites in central and eastern North America and Europe have soil nutrient supply rates that are in the optimal range, or exceeding that range with high use of superfertilizers. Western North America and other young soils tend to be in the upper nutrient supply range. Exceptions appear to include places like the Mount St. Helens pumice (Titus and del Moral, 1998), where soil nutrient levels are extremely low (MacMahon and Warner, 1984). In areas with highly leached, often ancient soils such as South Africa and Australia, many soils have supply rates below those to support the fungus (Propster and Johnson, 2015; Teste et al., 2016).

12.3.3 Nutrient Forms and Availability

A major challenge for understanding the contribution of mycorrhizas to plant nutrition is integrating the mechanisms of mycorrhizal nutrient uptake into the Agren and Bosatta model.

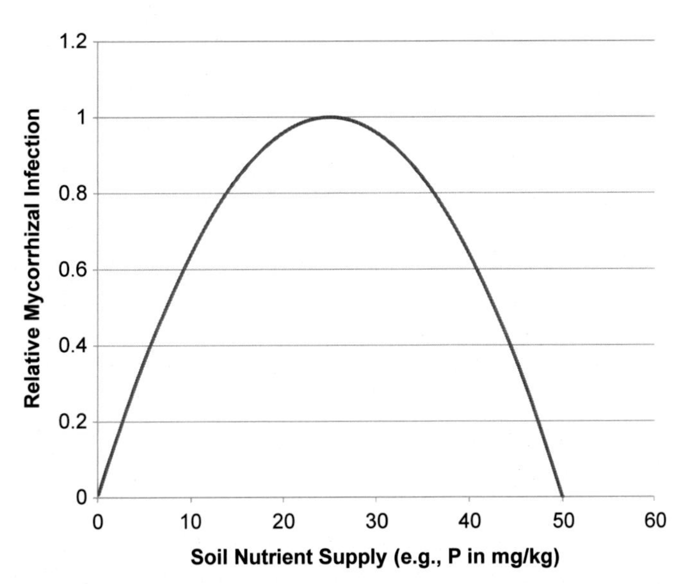

FIGURE 12.2 Relative mycorrhizal activity (e.g., infection, colonization) versus soil nutrient availability. The specific shape of this curve can vary, with the optimum moving up or down, but the general pattern conforms to a quadratic distribution model. *Redrawn from Treseder, K.K., Allen, M.F., 2002. Direct N and P limitation of arbuscular mycorrhizal fungi: a model and field test. New Phytologist 155, 507–515.*

The best documented mechanism for uptake of nutrients is simply that MF extend outward from either a constrained root tip in the case of ectomycorrhizas, or from the region of elongation in the case of AM associations. There are three mechanisms for increasing nutrient uptake: the increased direct uptake, including throughput of water that brings more nutrients to the root; the decomposition of organic matter with the subsequent transport of nutrients released by mineralization; or the shifting physical chemistry of soil inorganic elements that weathers nutrients.

12.3.3.1 Throughput

Hyphae of MF extend the absorption capacity for water and nutrients beyond the roots and root hairs (Hatch, 1937). The importance of the hyphal extension into the soil cannot be underestimated for uptake of nutrients in both AM (e.g., Hattingh et al., 1973) and EM (Hatch, 1937) symbioses. This is especially true for nutrients that bind to soil particles such as NH_4^+ and HPO_3^-. For these nutrients, the extension of the hyphae of MF to tap individual molecules from soil particles, especially in soil pores, is crucial to mycorrhizal functioning. Also, both EM and AM fungi scavenge N and P from high organic matter patches as those nutrients are released through decomposition (see Chapter 8; Hodge and Fitter, 2010; Herrera et al., 1978; St. John et al., 1983).

But there is more to the story than simple surface area in mesic soils. Frank (1885) first described a mycorrhiza as a wet-nurse to the plant, "die Amme des Baumes." Stahl (1900) further observed that water containing nutrients flows through the MF and assumed that this was the predominant mechanism for increasing nutrient uptake associated with a mycorrhiza. In wet soils, many nutrients such as HNO_3^- and many cations will flow with the transpiration water to the root surface in a process called *bulk flow.* This process provides plants with nutrients that can go into solution, sometimes in excess of that provided by the hyphae. But under high transpiration demands the hyphae can provide a significant portion of a plant's water (Allen, 1982; Hardie, 1985), and that water can be transported large distances (Duddridge et al., 1980).

As soil continues to dry, hyphae bind to soil particles, often holding those particles to the root, reducing the impacts of root shrinkage forming air gaps. Hyphae also bridge natural gaps in soil and individually tap water droplets, as can be seen in the depletion of the water droplets along hyphal surfaces (Fig. 12.3). Also important, as soil becomes wet, precipitation water first fills macropores and then fills micropores. However, as plants utilize that water and soil dries, macropores are depleted first via roots and root hairs. Then water shrinks back into micropores ($<80\,\mu m$) and eventually ultramicropores ($<5\,\mu m$). As the water evaporates from these pores, nutrients become ever more concentrated. Hyphae, with their diameters of as little as $2\,\mu m$, can penetrate even to the larger ultramicropores (Allen, 2007). These hyphae can tap this water, unavailable to the roots and root hairs, providing both water and nutrients during dry-down periods.

Another variant of throughput occurs when plant roots tap deep-water sources, such as ground water or perched water tables. If the surface soil is dry, a process called *hydraulic lift* occurs. During the day, when stomata are open, normal transpiration occurs using the deeper water. However, as stomata close in the evening, the water within the stem reverses direction (water only flows in response to an energy potential gradient) and flows out the roots, into fine roots, and into the hyphae of MF (Querejeta et al., 2003). This process sustains the hyphae even during very dry soil periods (Allen, 2006). Just as important, water flows through the hydrophobic coarse hyphae but can emerge through the hydrophilic hyphal tip (Fig. 12.4) into the soil. That water includes dissolved nutrients such as NH_4^+ and P. With daylight and

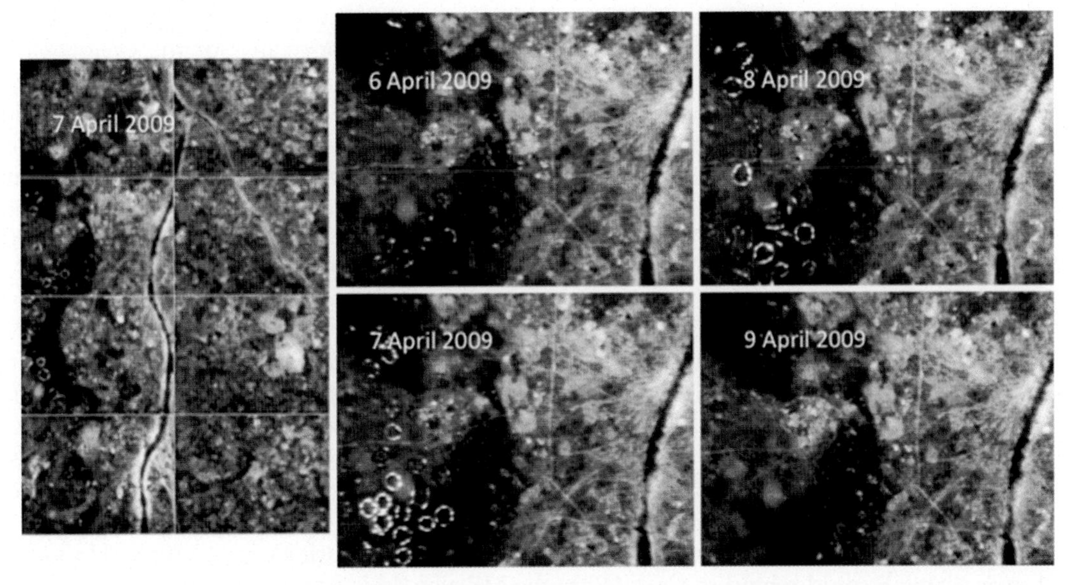

FIGURE 12.3 Acquisition of water across gaps by arbuscular mycorrhizal (AM) fungal hyphae. These images are taken by an automated minirhizotron from a forb/grass meadow at the James Reserve in the mountains of southern California (see Hernandez and Allen, 2013).

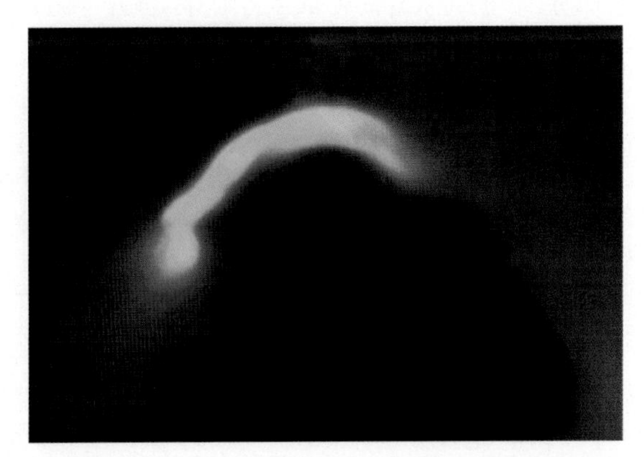

FIGURE 12.4 Hydraulically-lifted water emerging from a hydrophilic mycorrhizal root tip. *From Egerton-Warburton, L.M., Querejeta, J.I., Allen, M.F., 2008. Efflux of hydraulically lifted water from mycorrhizal fungal hyphae during imposed drought. Plant Signaling and Behavior 3, 68–71.*

stomatal opening, the water on the hyphal tips with the dissolved nutrients is reabsorbed and both the nutrients and water are transported to the leaves (Egerton-Warburton et al., 2008).

12.3.3.2 Mineralization

Supply rates can be limited by mineralization rates of organic nutrients or by the weathering rates of inorganic ones. Glomeromycota may not have the genes for breaking down plant cell walls either within living or dead cells (Tisserant et al., 2013), indicating that they are dependent upon the monosaccharides provided by the host plant for their C. But currently the functional genomic data are only available for the AM fungus *Rhizophagus irregularis* (formerly *Glomus intraradices*). More sequences are needed, especially for the *Gigasporales*. The physiological evidence for acquisition of organic P and N, especially in soils with high organic matter content is inconclusive (Hart et al., 2001). But AM fungi do have both acid and alkaline phosphatases in both intraradical and extraradical hyphae (Allen et al., 1981; Gianinazzi et al., 1979; van Aarle et al., 2002) and appear to have the ability to utilize P from phytates (Allen et al., 1981). EM fungi are more evolutionary advanced than AM fungi, and largely have the ability to break down organic nutrients into amino acid forms that can be readily transported to the plant (e.g., Vandeboncoeur et al., 2015).

12.3.3.3 Weathering

Mycorrhizas are also directly involved in elemental weathering, making soil nutrients available (see Chapter 3). This is especially crucial for P nutrition. MF increase soil R locally (e.g., Knight et al., 1989; Allen and Kitajima, 2014). In the presence of H_2O, the CO_2 preferentially binds with one H and the O, forming bicarbonate $(HCO_3^-)+H^+$, thereby increasing the soil H^+ content. Increasing H^+ weathers Ca-apatite, converting (hydroxyapatite) $Ca_5(PO_4)_3OH+7H^+$ to $5Ca^{2+}+3H_2PO_4^-+H_2O$ [where $H_2PO_4^-$ is a dihydrogen phosphate (PO_4) ion], converting unavailable P into an available form (Jurinak et al., 1986). However, that available P could rebind to Ca^{2+}. What the MF also do is release oxalates (e.g., $H_2C_2O_4$) and other organic acids along the hyphal surface (Graustein et al., 1977; Jurinak et al., 1986). Those organic acids preferentially bind the Ca^{2+} locally from:

$$Ca_5(PO_4)_3OH + 5H_2C_2O_4 = 5CaC_2O_4 + 3H_2PO_4^- + H^+ + H_2O,$$

or P in a form that can be taken up.

Artificial addition of $H_2C_2O_4$ increases P availability and uptake (Cannon et al., 1995). In theory, this same process would apply to P forms that are unavailable in ancient, highly weathered soils ($AlPO_4$) and wet, highly leached soils ($AlPO_4$ or $FePO_4$). Indeed, current efforts to acidify $AlPO_4$ (Zhao et al., 2013) as a source of P should only require high rates of R and mycorrhizas in a soil environment!

$FePO_4$ becomes of special interest in understanding the role of mycorrhizas in nutrition and environmental change. With the increase in H^+ ions associated with R of CO_2 into H_2O (and HCO_3^-), there would be dissociation of $FePO_4$, wherein the HPO_4^- could be absorbed Fe^{3+} would predominate. Fe is also an important limiting element for plant production. MF are known to produce siderophores, high-affinity organic compounds that bind Fe^{3+} (Haselwandter, 2008). On the cell surface, the Fe^{3+} is reduced to Fe^{2+}, wherein the Fe-siderophore complex is transported into the fungus. There are a number of these complex organic compounds that function across ecosystems. EM and ericoid MF are well known producers of a range of siderophores (e.g., Haselwandter et al., 2013). AM fungi are known to improve Fe uptake, suggesting siderophore activity, but these have not been identified or isolated. Together, MF have a dramatic impact on uptake of a number of unavailable forms of soil nutrients.

In extremely nutrient-deficient ecosystems, MF may likely require nutrients beyond the ability of the soils to provide. This appears to be the case especially with limiting P in extremely old, weathered soils. In extreme cases, some plants form cluster roots (Lambers, 2014). Cluster roots do not appear to support mycorrhizal formation. However, in some sites, such as newly burned or younger soils where the nutrient levels are slightly higher, mycorrhizas may form in purportedly nonmycotrophic families such as the Proteaceae (Bellgard, 1991; Bellgard et al., 1994). More research is needed into the nature of the ecology and evolution of cluster roots and the details of root–fungal interactions (Teste et al., 2016). It could well be that individual cluster roots are simply too small to support the internal structures of MF or that flow rates are constrained by the small cells, again precluding the formation of a viable mycorrhiza (Rillig et al., 1998).

Finally, whereas mycorrhizas generally act to increase the uptake of nutrients, in some cases AM fungi can actually reduce nutrient supply rate by immobilizing critical nutrients. Janos et al. (2013) found that hyphal networks of AM fungi inhibited growth of EM seedlings by sequestering Fe in glomalin but sustaining P uptake by AM. Clearly, there are multiple, interactive complex chemical processes occurring simultaneously within and across mycorrhizal systems.

12.3.4 Temporal Dynamics

Soils, plants, and microbes are not static entities. In our minirhizotron research, we measure soil CO_2 concentrations at 5-minute intervals because the variation is so high that rapid shifts can occur. Factors such as sunflecks can rapidly raise soil T (T_s), and a precipitation event can result in soil water channeling down soil cracks and old roots in a spatially and temporally complex pattern that rapidly shifts soil moisture (θ) (Allen, 2006; Allen et al., 2007). MF preferentially grow and die at diurnal time scales (Hernandez and Allen, 2013).

Seastedt and Knapp (1993) outlined their "transient maxima" hypothesis, in which the shifting nonequilibrium conditions allow for the uptake of more resources than a static limitation. As an example, they showed that rapid uptake of N when it is temporally available serves to increase productivity beyond that predicted by simply assessing soil nutrient availability. That same model can be applied to both N mineralization and uptake, and by responses such as rapid changes in soil respiration (R_s), shifting temporally H^+ and thus P availability and uptake as well as the other factors described in this chapter.

It is also important to think about these processes on a repeated temporal scale. Plants and their MF are not static, but have distinct phenologies at seasonal to annual cycles. R_s has repeated patterns at daily and annual scales (Vargas et al., 2010a), as well as predictable lags in R_s, CO_2 concentrations, and growth of fine roots and hyphae (Vargas and Allen, 2008; Vargas et al., 2010b). Variables can change in unpredicted ways. For example, we found that under high θ, diffusion of CO_2 out of the soil was restricted and CO_2 rapidly built up, both under snow and under high rainfall regimes. The added CO_2 and θ would consequently increase H^+ concentrations and increase availability of HPO_4^-.

12.4 MYCORRHIZAS AND GLOBAL CHANGE

Global change is commonly inferred to be climate change. However, it is important to remember that the change in T is driven by changes in atmospheric CO_2 (see Section 12.1). Atmospheric humidity is a function of T and wind flow patterns, which are influenced by topography. The ability of the atmosphere to hold moisture increases with increasing T. In coastal mountains, increasing ocean Ts will generate high-humidity air masses that will be blown up mountain sides, and with increasing elevation, clouds will form and precipitation will increase. As that drier air descends down the inland side, it becomes even drier and hotter. On the oceanic side of the mountains, the increasing T increases humidity, which is a greenhouse gas, trapping more heat. On the drier side, with low humidity, CO_2 acts as a stronger greenhouse gas. But, increasing the partial pressure of CO_2 in the atmosphere increases water-use efficiency, thereby reducing the impacts of drought. Together, these mean that global change is a complex interplay between negative and positive forces with significant impacts on mycorrhizal functioning.

12.4.1 Impacts of Increasing Carbon Dioxide

As atmospheric CO_2 increases, there is a greater pressure to fix C, and because the increasingly fixed C increases C-to-nutrient ratios, there will be a shift to allocate C to nutrient acquisition. The greatest beneficiaries are MF. This process should markedly increase the weathering process, because elevated CO_2 not only results in greater mycorrhizal activity but also greater soil aggregation (see Chapter 14; Rillig et al., 1999), exudation of a broad range in organic compounds, N_2 fixation (for those plants with N_2-fixing associations; Allen et al., 2005), and ultimately higher R_s (e.g., Allen et al., 2014) with higher T. The greater R, in turn, will lead to greater weathering and more organic nutrients, which creates multiple positive feedback mechanisms enhancing mycorrhizal activity and mycorrhizal-regulated nutrient availability.

12.4.2 Temperature and Soil Moisture

Broadly, T is predicted to increase globally, and the warming is expected to be more severe with increasing latitude, altitude, and aridity (IPCC, 2013). This is simply because water vapor is a stronger greenhouse gas than CO_2, so where humidity and cloudiness are high, the CO_2 impact is lessened. Where CO_2 does drive T, nighttime Ts are especially likely to increase, meaning higher T_s. A change of 2–5°C in the atmospheric T, the expected increase, is unlikely to drive major changes in mycorrhizal functioning. What will dramatically change are the extremes, because variation is also a likely outcome (Fig. 12.5). This means that there are likely to be more extreme high-T days and more extended-heat stress days, especially in arid to semiarid regions. This means greater evapotranspiration, thereby potentially driving a greater reliance on MF for drought tolerance, but also longer periods of dependency on processes such as hydraulic lift, as long as there is accessible ground water. At the lower end of the T curve, there would probably be less impact than on the upper end of the curve.

Changing θ will have a measurable impact on J_{CO_2} if θ drops below critical values for plants. Decreasing θ would mean less J_{CO2}, and under soil water potential (ψ) less than −1.5 MPa, permanent wilt could even occur for species that are less drought tolerant, resulting in mortality of plants and associated MF. Even for drought-tolerant species, there would still be a reduced J_{CO_2}, and therefore less C available for MF and for the rhizosphere processes associated with weathering described previously. We modeled potential changes in J_{CO_2}, C allocation to fine roots (hence MF), and R_s. Increasing T will likely decrease forest productivity by increasing leaf water stress and reducing the allocation of C below ground (Allen et al., 2014). We did not

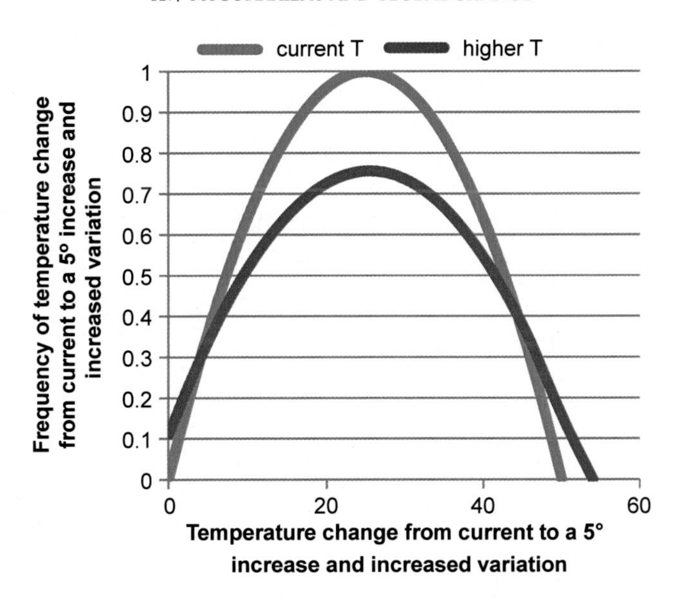

FIGURE 12.5 Impacts of a 5°C increase in global temperature (T), coupled with an increase in variation.

include all the potential nutritional implications that could also result in a dramatic impact in productivity. However, the dynamics noted are still relatively gradual. As we discovered with AM fungi and elevated CO_2, if the increase is multigenerational, adaptation may well occur such that the impacts could be less dramatic (Klironomos et al., 2005).

Altering enzymatic structure for adaptation to increasing T could be problematic. R_s, which integrates a large array of enzymes, increases with a T coefficient (Q_{10}) of 1.4 (Mahecha et al., 2010). Although there are variations with θ and feedbacks for other C forms, the likelihood is that R will simply continue to increase within the range of T changes that are expected. However, J_{CO2} could well be altered at the higher levels. RuBisCO, the enzyme responsible for photosynthesis, appears to peak about 30°C and declines above that value. Importantly, in some trees (e.g., black spruce) CO_2 assimilation declines above 20°C, whereas R continues to increase (Sage et al., 2008) This could result in a threshold of C loss both internal to the plant and in C allocation to MF at the same time that R_s is increasing. Thus C allocation and nutrient need would be mismatched, affecting all processes that are regulated by MF.

On a larger scale, it is likely that EM symbioses will be affected more than AM symbioses because the former tend to exist in areas with higher precipitation and a narrower range of environmental conditions. Thus a smaller T change has a greater impact on EM functioning (Vargas et al., 2010b). AM symbioses can be found from deserts to rainforests, but are often constrained in environments with extremely cold Ts. Environmental change, especially increasing aridity, is likely to result in a loss of EM and gain of AM symbioses, enduring a wide T range. This shift could have major effects on the patterns of nutrient supply, from organic nutrient mineralization to glomalin-binding of cations, as discussed previously. Together we envision a conversion of all of these multiple mechanisms, some of which may

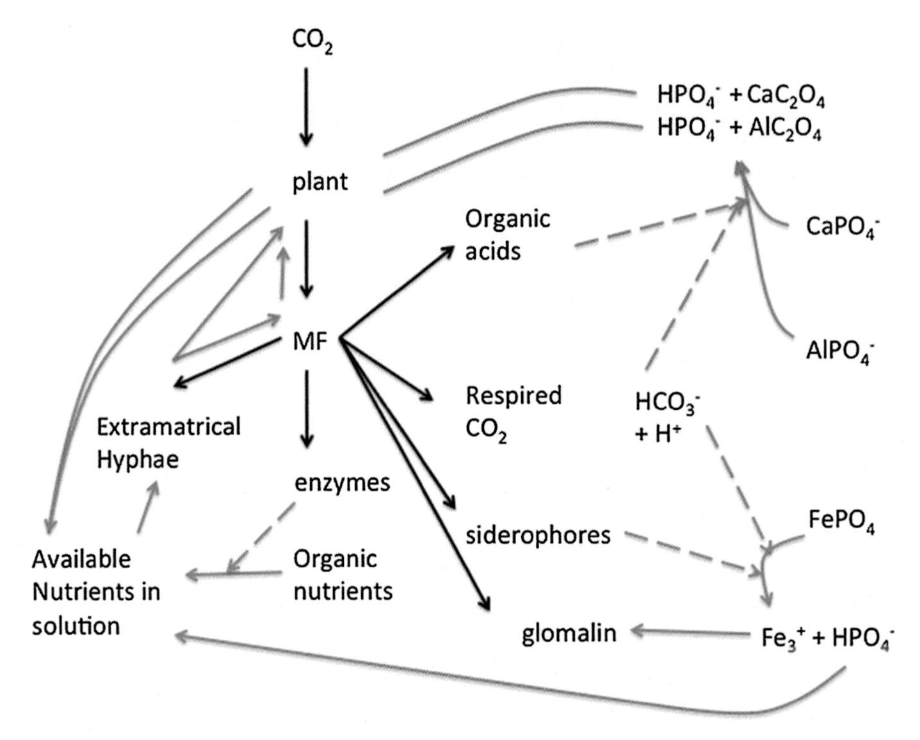

FIGURE 12.6 Conceptual model of interacting nutrient acquisition pathways.

well cancel each other out, and some of which may well result in entirely new trajectories in the local vegetation and nutrient cycling mechanisms (Fig. 12.6).

12.5 MYCORRHIZAS AND NITROGEN DEPOSITION

Deposition of reactive N has increased threefold to fourfold globally since 1860 (Galloway et al., 2004). Reactive N includes forms oxidized NO_x from combustion and reduced $NH_{3/4}$ from agricultural activities. Unlike CO_2, reactive N forms are deposited downwind of sources of emission and form gradients of N deposition ranging from greater than 30 kg N/ha year to background levels of less than 2 kg N/ha year across short distances of less than 100 km (Fenn et al., 2010; Simkin et al., 2016). The high deposition values are equivalent to amounts farmers apply to fertilize agricultural fields and thus have major impacts on agricultural as well as wildland vegetation. Activity of MF has been assessed across N deposition gradients and in N fertilization experiments in areas of low deposition to determine impacts of elevated N on their functioning.

The interaction of mycorrhizas with N uptake has been the focus of research since Frank's original studies (1888). Today, because of innovative studies of isotopic transfers and fractionation (e.g., Zimmer et al., 2007; Hobbie and Hobbie, 2006), we are in a position to begin to quantify that relationship and thus to project likely impacts of global change. In a pinôn pine stand in New Mexico, we found that the trees used 1.6 g N/m^2 for net primary production. EM fungi provided 0.4 g N/m^2 to the host plant while retaining 1.2 g N/m^2. Based on this estimate, then 25% of the host N was provided by the EM fungi. After a high fertilization treatment or adding 5 g of NO_3^- –N, we indeed found that relative EM fungal activity declined (Allen et al., 2010). The applied fertilization replaced that provided by the fungus at a lowered C cost. In areas of high N deposition (2 g N/m^2), that deposition alone can provide sufficient N for plant growth, reducing the value of the EM symbiosis. As per our models (see Section 12.3), the plant will then reduce C allocation to the fungus, thereby starving the fungus.

We know less about N use by AM fungi, because they appear to use largely the same N forms as the plants and do not appear to fractionate the N transferred (see Chapter 8, Hodge and Fitter, 2010; Nakano et al., 2001). However, we can make some general estimates. High levels of N deposition in *Artemisia californica*–dominated areas of southern California is causing the composition of vegetation to change (Cox et al., 2014; Allen et al., 2016). Using litter and soil from background conditions, decomposition provided approximately 1.2 g N/m^2 (Sirulnik et al., 2007), much of which was likely transported by AM fungi to the host (Yoshida and Allen, 2001, 2004). With N deposition, AM infection was reduced and the composition of the fungal community was altered (Egerton-Warburton and Allen, 2000).

Fertilization experiments and observations along N deposition gradients have enabled assessment of critical loads (CLs) of N for ecosystem changes, and in particular for MF (Allen et al., 2016). The CL for loss of AM fungal species' richness and reduced percentage of root infection was 10–11 kg N/ha year (Allen et al., 2016). The CL for vegetation type conversion from native shrubland to exotic annual grassland was also 11 kg N/ha year (Cox et al., 2014), suggesting that AM fungi and plants were responding similarly to elevated mineral N. Some species of AM fungi, such as *Glomus aggregatum* and *Glomus leptotichum*, increased with elevated N in the rhizospheres of remnant native shrubs (Egerton-Warburton and Allen, 2000). Other species of AM fungi that form small *Glomus*-type spores, *Claroideoglomus* (=*Glomus claroideum*) became common in invasive grasses sequenced in areas of high N deposition (O'Neill and Allen, unpublished data). Conversely, large-spored Gigasporaceae all declined with elevated N. These different responses may be related to physiology of the fungi, because Gigasporaceae tend to be slow-growing, later successional fungi that are especially active in soils with organic nutrient forms (Hart et al., 2001). In Europe, native grasses have been replacing native shrublands under N deposition (Bobbink et al., 1998). Interestingly, many species with small *Glomus*-type spores, now classified as *Claroideoglomus* spp. and some *Diversispora* spp. also increased under long-term N fertilization (Liu et al., 2015; Sochorova et al., 2016). Emphasis in Europe has been on losses of diversity and functioning of EM fungi under elevated N (Kjoller et al., 2012). N deposition values are often high in Europe, up to 40 kg N/ha year, and areas of low N deposition with background clean-air N levels are uncommon, so reported CL values tend to be higher in Europe than the United States (Pardo et al., 2011). Fertilization of 30 kg N/ha year resulted in increased piñon pine mortality during summer drought, because N promoted increased above-ground productivity without concomitant increases in below-ground EM fungal activity (Allen et al., 2010). The combined increases in CO_2 with global N deposition may be promoting shifts in the distribution and functioning of many species of plants and MF. An understanding of the CL of N under ambient CO_2 may be used in regulatory decisions for controlling nitrogenous air pollutants.

Research and regulation has dramatically reduced the detrimental impacts of acid rain for sulfuric acid ($H_2SO_4^-$) pollution. However, in many areas the acidification of soils and waters did not decline (Likens et al., 1996). Calculations indicated that simultaneous with the decline in $H_2SO_4^-$, there was a rise in the deposition of NO_x, resulting in an HNO_3^- acidification. This acidification would also act to weather Ca, Fe, or $AlPO_4$, increasing P availability, but also resulting in the release of Al (or other cations) to aquatic or soil sinks. The impacts of increasing both N and P availability would likely act to again reduce plant dependency on mycorrhizas.

12.6 WHAT IS NEEDED? A STOICHIOMETRIC CHALLENGE

In 2003, we developed a matrix model built around C:N:P stoichiometry to examine the impacts of multiple fungal species, each of which had slightly different nutrient acquisition strategies (Allen et al., 2003). Using that approach, we found a complex array of both positive and negative production outputs to differing combinations of AM and EM fungi. For modeling purposes, we gave the EM fungi differing N acquisition strategies; for example, fungus A had a strong enzymatic ability to mineralize organic N, versus fungus B with a larger exploration strategy. If these differing fungi were added in different sequences, such as a successional transition, or plant competition outcomes, then a vastly more complex array of productivity outputs could result. When compared with actual plant growth outcomes with an array of MF, this modeled outcome is more accurate in comparison to field studies than a more simplistic approach wherein all MF simply acquire more nutrients. This model was expanded for *Quercus agrifolia* (see Box 2, Johnson et al., 2006) resulting in a complex plane of nonlinear production over the course of a growing season, because different species of EM and AM fungi colonize newly growing fine roots. This represents a relatively accurate picture of the developing pattern of live oak, which forms EM symbioses with tens to hundreds of fungal species, and AM symbioses with at least tens of Glomeromycota.

Our challenge now is to begin parameterizing this type of model in four ways. First we need to vary atmospheric CO_2. Mature *Q. agrifolia* trees established when the atmosphere was between 280 and 300 ppm. Today, that value exceeds 400 ppm. This makes more C driving up the C-to-nutrient gradients, which must drive greater mycorrhizal dependency to reduce those gradients. This also means greater soil CO_2, with all of the chemistry changes described previously. Second, T and moisture drivers are shifting. As T increases, θ declines because of greater evaporational and transpirational demands. However, water-use efficiency also goes up because the partial pressure of CO_2 is higher. This trade-off is likely to be dynamic and variable spatially and temporally. Finally, the environment of the late 21st century is likely to be a novel one for mycorrhizas and ecosystems. Previously, when atmospheric CO_2 increased, N had to become more limiting to production. Therefore adjusting the C:N ratio meant more mycorrhizal activity to increase N uptake. When CO_2 decreased (i.e., during the Pleistocene period), nutrients and especially N would have become less limiting. But the environment of the next few decades is going to be one where both C and N simultaneously increase. How are mycorrhizas and vegetation types going to adapt?

We predict that:

1. There will be a greater response to AM symbioses, or even a relative increase in nonmycotrophic species where P is not limiting. In many cases, this means an environment more conducive for invasive plant and fungal species. We know little about invasive fungal species and not enough about the processes of invasion of plants. Understanding the role of mycorrhizal symbiosis in plant invasions becomes even more crucial.
2. Fine roots and MF will become even more dynamic as T_s increases, and daily variation in θ increases. Understanding mycorrhizal dynamics on a diurnal or even faster scale will become a more crucial issue (e.g., Hernandez and Allen, 2013).
3. There will likely be an increase in stable soil organic materials such as glomalin, potentially also impacting essential elements such as Fe and altering the vegetative communities (e.g., Janos et al., 2013).

In summary, we expect novel environmental stressors under the parameters that are shaping global environmental change during the 21st century. Mycorrhizal symbioses are not going to be come less important, but the specific species, types, and dynamics regulating plant production and interaction are going to shift in unpredictable and complex ways.

Acknowledgments

We acknowledge the support of the Agricultural Experiment Station Hatch Projects CA-R-PPA-6689-H and CA-R-BPS-5821-H, and the US National Science Foundation DEB-1442537.

References

Agren, G.I., Bosatta, E., 1996. Theoretical Ecosystem Ecology: Understanding Element Cycles. Cambridge University Press, Cambridge, England.

Allen, M.F., Allen, E.B., Lansing, J.L., Pregitzer, K.S., Hendrick, R.L., Ruess, R.W., Collins, S.L., 2010. Responses to chronic N fertilization of ectomycorrhizal piñon but not arbuscular mycorrhizal juniper in a piñon-juniper woodland. Journal of Arid Environments 74, 1170–1176.

Allen, E.B., Egerton-Warburton, L.M., Hilbig, B.E., Valliere, J.M., 2016. Interactions of arbuscular mycorrhizal fungi, critical load of nitrogen deposition, and shifts from native to invasive species in a southern California shrubland. Botany 94, 425–433. http://dx.doi.org/10.1139/cjb-2015-0266.

Allen, M.F., 1982. Influence of vesicular-arbuscular mycorrhizae on water movement through *Bouteloua gracilis*. New Phytologist 91, 191–196.

Allen, M.F., 1996. The ecology of arbuscular mycorrhizae: a look back into the 20th century and a peek into the 21st. Centenary Review Article, British Mycological Society. Mycological Research 100, 769–782.

Allen, M.F., 2006. Water dynamics of mycorrhizas in arid soils. In: Gadd, G.M. (Ed.), Fungi in Biogeochemical Cycles. Cambridge University Press, Cambridge, England, pp. 74–97.

Allen, M.F., 2007. Mycorrhizal fungi: highways for water and nutrients in arid soils. Vadose Zone Journal 6, 291–297.

Allen, M.F., Kitajima, K., 2014. Net primary production of ectomycorrhizas in a California forest. Fungal Ecology. 10, 81–90. http://dx.doi.org/10.1016/j.funeco.2014.01.007.

Allen, M.F., Sexton, J.C., Moore, T.S., Christensen, M., 1981. The influence of phosphate source on vesicular-arbuscular mycorrhizae of *Bouteloua gracilis*. New Phytologist 87, 687–694.

Allen, M.F., Swenson, W., Querejeta, J.I., Egerton-Warburton, L.M., Treseder, K.K., 2003. Ecology of mycorrhizae: A conceptual framework for complex interactions among plants and fungi. Annual Review of Phytopathology 41, 271–303.

Allen, M.F., Klironomos, J.N., Treseder, K.K., Oechel, W.C., 2005. Responses of soil biota to elevated CO_2 in a chaparral ecosystem. Ecological Applications 15, 1701–1711.

Allen, M.F., Kitajima, K., Hernandez, R.R., 2014. Mycorrhizae and global change. In: Tausz, M., Grulke, N.E. (Eds.), Trees in a Changing Environment. Springer – Plant Sciences, Dordrecht, The Netherlands, pp. 37–59.

Allen, M.F., Vargas, R., Graham, E., Swenson, W., Hamilton, M., Taggart, M., Harmon, T.C., Rat'ko, A., Rundel, P., Fulkerson, B., Estrin, D., 2007. Soil sensor technology: life within a pixel. BioScience 57, 859–867.

Becklin, K.M., Medeiros, J.S., Ward, J.K., 2012. Mycorrhizal functioning across the glacial-interglacial transition: evidence from stable isotopes. In: COS 90-3. 97th Annual Meeting of the Ecological Society of America.

Beer, C., et al., 2016. Terrestrial gross carbon dioxide uptake: global distribution and covariation with climate. Science 329, 834–838.

Bellgard, S.E., Whelan, R.J., Muston, R.M., 1994. The impact of wildfire on vesicular-arbuscular mycorrhizal fungi and their potential to influence the re-establishment of post-fire plant communities. Mycorrhiza 4, 139–146.

Bellgard, S.E., 1991. Mycorrhizal associations of plant species in Hawkesbury Sandstone vegetation. Australian Journal of Botany 39, 357–364.

Bobbink, R., Hornung, M., Roelofs, J.G.M., 1998. The effects of air-borne nitrogen pollutants on species diversity in natural and semi-natural European vegetation. Journal of Ecology 86, 717–738.

Cannon, J.P., Allen, E.B., Dudley, L.M., Jurinak, J.J., 1995. The effects of oxalates produced by *Salsola tragus* on the phosphorus nutrition of *Stipa pulchra*. Oecologia 102, 265–272.

Cox, R.D., Preston, K.L., Johnson, R.F., Minnich, R.A., Allen, E.B., 2014. Influence of landscape-scale variables on vegetation conversion in southern California, USA. Global Ecology and Conservation 2, 190–203.

Droser, M.L., Gehling, G.G., Dzaugis, M.E., Kennedy, M.J., Rice, D., Allen, M.F., 2014. A new Ediacaran fossil with a novel sediment displacive life habit. Journal of Paleontology 88, 145–151.

Duddridge, J.A., Malibari, A., Read, D.J., 1980. Structure and function of mycorrhizal rhizomorphs with special reference to their role in water transport. Nature (London) 287, 834–836.

Egerton-Warburton, L.M., Allen, E.B., 2000. Shifts in arbuscular mycorrhizal communities along an anthropogenic nitrogen deposition gradient. Ecological Applications 10, 484–496.

Egerton-Warburton, L.M., Querejeta, J.I., Allen, M.F., 2008. Efflux of hydraulically lifted water from mycorrhizal fungal hyphae during imposed drought. Plant Signaling and Behavior 3, 68–71.

Fenn, M.E., Allen, E.B., Weiss, S.B., Jovan, S., Geiser, L.H., Tonnesen, G.S., Johnson, R.F., Rao, L.E., Gimeno, B.S., Yuan, F., Meixner, T., Bytnerowicz, A., 2010. Nitrogen critical loads and management alternatives for N-impacted ecosystems in California. Journal of Environmental Management 91, 2404–2423.

Frank, A.B., 1885. Über die auf Wurzelsymbiose beruhende Ernährung gewisser Baume durch unterirdische Pilze. Berichte der Deutsche Botanische Gesellschaft 3, 128–145.

Galloway, J.N., Dentener, F.J., Capone, D.G., Boyer, E.W., Howarth, R.W., Seitzinger, S.P., Asner, G.P., Cleveland, C.C., Green, P.A., Holland, E.A., Karl, D.M., Michaels, A.F., Porter, J.H., Townsend, A.R., Vörösmarty, C.J., 2004. Nitrogen cycles: past, present, and future. Biogeochemistry 70, 153–226.

Galloway, J.N., Dentener, F.J., Capone, D.G., Boyer, E.W., Howarth, R.W., Seitzinger, S.P., Asner, G.P., Cleveland, C.C., Green, P.A., Holland, E.A., Karl, D.M., Michaels, A.F., Porter, J.H., Townsend, A.R., Vörösmarty, C.J., 2004. Nitrogen cycles: past, present, and future. Biogeochemistry 70, 153–226.

Gianinazzi, S., Gianinazzi-Pearson, V., Dexheimer, J., 1979. Enzymatic studies on the metabolism of vesicular-arbuscular mycorrhiza. III. Ultrastructural localization of acid and alkaline phosphatase in onion roots infected with *Glomus mosseae* (Nicol. & Gerd.). New Phytologist 82, 127–132.

Graustein, V.C., Cromack Jr.V.K., Sollins, P., 1977. Calcium oxalate: occurrence in soils and effect on nutrient geochemical cyclesGlomus mosseae. Science 198, 1252–1254.

Hardie, K., Cromack Jr.K., Sollins, P., 1985. The effect of removal of extraradical hyphae on water uptake by vesicular-arbuscular mycorrhizal plants. New Phytologist 101, 677–684.

Hart, M.M., Reader, R.J., Klironomos, J.N., 2001. Life-history strategies of arbuscular mycorrhizal fungi in relation to the successional dynamics. Mycologia 83, 1184–1196.

Haselwandter, K., Häninger, G., Ganzera, M., Haas, H., Nicholson, G., Winkelmann, G., 2013. Linear fusigen as the major hydroxamate siderophore of the ectomycorrhizal Basidiomycota *Laccaria laccata* and *Laccaria bicolor*. Biometals 26, 969–979.

Haselwandter, K., Häninger, G., Ganzera, M., Haas, H., Nicholson, G., Winkelmann, G., 2008. Structure and function of siderophores produced by mycorrhizal fungiLaccaria laccataLaccaria bicolor. Mineralogical Magazine 72, 61–64.

Hatch, A.B., 1937. The physical basis of mycotrophy in *Pinus*. The Black Rock Forest Bulletin 6, 1–168.

Hattingh, M.J., Gray, L.E., Gerdemann, J.W., 1973. Uptake and translocation of 32P-labelled phosphate to onion roots by endomycorrhizal fungiPinus. Soil Science 116, 383–387.

Hernandez, R.R., Allen, M.F., Gerdemann, J.W., 2013. Diurnal patterns of productivity of arbuscular mycorrhizal fungi revealed with the soil ecosystem observatory. New Phytologist 200, 547–557. http://dx.doi.org/10.1111/nph.12393.

Herrera, R., Merida, T., Stark, N., Jordan, C.F., 1978. Direct phosphorus transfer from leaf litter to roots. Naturwissenschaften 65, 208–209. http://dx.doi.org/10.1111/nph.12393.

Hobbie, J.E., Hobbie, E.A., Stark, N., Jordan, C.F., 2006. N-15 in symbiotic fungi and plants estimates nitrogen and carbon flux rates in arctic tundra. Ecology 87, 816–822.

Hodge, A., Fitter, A.H., 2010. Substantial nitrogen acquisition by arbuscular mycorrhizal fungi from organic material has implications for N cycling. Proceedings of the National Academy of Sciences of the United States 107, 13754–13759.

IPCC, A., Fitter, A.H., 2013. Climate change 2013: the physical science basis. Intergovernmental Panel on Climate Change. 107, 13754–13759http://www.ipcc.ch/report/ar5/wg1/.

Janos, D.P., Scott, J., Aristizábal, C., Bowman, D.M.J.S., 2013. Arbuscular-mycorrhizal networks inhibit *Eucalyptus tetrodonta* seedlings in rain forest microcosms. PLoS One. 8, e57716. http://dx.doi.org/10.1371/journal.pone.0057716.

Johnson, N.C., Hoeksema, J.D., Bever, J.D., Chaudhary, V.B., Gehring, C., Klironomos, J., Koide, R., Miller, R.M., Moore, J., Moutoglis, P., Schwartz, M., Simard, S., Swenson, W., Umbanhowar, J., Wilson, G., Zabinski, C., 2006. From Lilliput to Brobdingnag: extending models of mycorrhizal function across scalesEucalyptus tetrodonta. Bioscience 56, 889–900. http://dx.doi.org/10.1371/journal.pone.0057716.

Jurinak, J.J., Dudley, L.M., Allen, M.F., Knight, W.G., Gehring, C., Klironomos, J., Koide, R., Miller, R.M., Moore, J., Moutoglis, P., Schwartz, M., Simard, S., Swenson, W., Umbanhowar, J., Wilson, G., Zabinski, C., 1986. The role of calcium oxalate in the availability of phosphorus in soils of semiarid regions: a thermodynamic study. Soil Science 142, 255–261.

Kjoller, R., Nilsson, L.O., Hansen, K., Schmidt, I.K., Vesterdal, L., Gundersen, P., 2012. Dramatic changes in ectomycorrhizal community composition, root tip abundance and mycelial production along a stand-scale nitrogen deposition gradient. New Phytologist 194, 278–286.

Klironomos, J.N., Allen, M.F., Rillig, M.C., Piotrowski, J., Makvandi-Nejad, S., Wolfe, B.E., Powell, J.R., 2005. Abrupt rise in atmospheric CO_2 overestimates community response in a model plant-soil system. Nature 433, 621–624.

Knight, W.G., Allen, M.F., Jurinak, J.J., Dudley, L.M., 1989. Elevated carbon dioxide and solution phosphorus in soil with vesicular-arbuscular mycorrhizal western wheatgrass. Soil Science Society of America Journal 53, 1075–1082.

Lambers, H. (Ed.), 2014. Plant Life on the Sandplains in Southwest Australia: a Global Biodiversity Hotspot. Kwongan Matters. University of Western Australia Press. 350 p.

Likens, G.E., Driscoll, C.T., Buso, D.C., 1996. Long-term effects of acid rain: response and recovery of a forest ecosystem. Science 272, 244–246.

Liu, Y., Johnson, N.C., Mao, L., Shi, G., Jiang, S., Ma, X., Du, G., An, L., Feng, H., 2015. Phylogenetic structure of arbuscular mycorrhizal community shifts in response to increasing soil fertility. Soil Biology and Biochemistry 89, 196–205.

MacMahon, J.A., Warner, N., 1984. Dispersal of mycorrhizal fungi: processes and agents. In: Williams, S.E., Allen, M.F. (Eds.), VA Mycorrhizae and Reclamation of Arid and Semiarid Lands. University of Wyoming Agricultural Experiment Station, Laramie, pp. 28–41.

Mahecha, M.D., Reichstein, M., Carvalhais, N., Lasslop, G., Lange, H., Seneviratne, S.I., Vargas, R., Ammann, C., Altaf Arain, M., Alessandro Cescatti, A., et al., 2010. Global convergence in the temperature sensitivity of respiration at the ecosystem level. Science 329, 838–840.

Malloch, D.W., Pirozynski, K.A., Raven, P.H., 1980. Ecological and evolutionary significance of mycorrhizal symbiosis in vascular plants (a review). Proceedings of the National Academy of Sciences of the United States 77, 2113–2118.

Nakano, A., Takahashi, K., Koide, R.T., 2001. Determination of the nitrogen source for arbuscular mycorrhizal fungi by N-15 application to soil and plants. Mycorrhiza 10, 267–273.

NOAA, 2016. NOAA Global Monitoring Division. http://www.esrl.noaa.gov/gmd/.

Padgett, P.E., Allen, E.B., 1999. Differential responses to nitrogen fertilization in native shrubs and exotic annuals common to Mediterranean coastal sage scrub of California. Plant Ecology 144, 93–101.

Pardo, L.H., Fenn, M.E., Goodale, C.L., Geiser, L.H., Driscoll, C.T., Allen, E.B., Baron, J.S., Bobbink, R., Bowman, W.D., Clark, C.M., Emmett, B., Gilliam, F.S., Greaver, T.L., Hall, S.J., Lilleskov, E.A., Liu, L.L., Lynch, J.A., Nadelhoffer, K.J., Perakis, S.S., Robin-Abbott, M.J., Stoddard, J.L., Weathers, K.C., Dennis, R.L., 2011. Effects of nitrogen deposition and empirical nitrogen critical loads for ecoregions of the United States. Ecological Applications 21, 3049–3082.

Pearson, P.N., 2010. Increased atmospheric CO_2 during the middle Eocene. Science 330, 763–764. http://dx.doi.org/10.1126/science.1197894.

Propster, J.R., Johnson, N.C., 2015. Uncoupling the effects of phosphorus and precipitation on arbuscular mycorrhizas in the Serengeti. Plant and Soil 388, 21–34.

Querejeta, J.I., Egerton-Warburton, L.E., Allen, M.F., 2003. Direct nocturnal water transfer from oaks to their mycorrhizal symbionts during severe soil drying. Oecologia 134, 55–64.

Redfield, A.C., 1934. On the proportions of organic derivations in sea water and their relation to the composition of plankton. In: Daniel, R.J. (Ed.), James Johnstone Memorial Volume. University Press of Liverpool, pp. 176–192.

Rillig, M.C., Allen, M.F., Klironomos, J.N., Field, C.B., 1998. Arbuscular mycorrhizal percent root infection and infection intensity of *Bromus hordeaceus* grown in elevated atmospheric CO_2. Mycologia 90, 199–205.

Rillig, M.C., Wright, S.F., Allen, M.F., Field, C.B., 1999. Rise in carbon dioxide changes soil structure. Nature 400, 628.

Sage, R.F., Way, D.A., Kublen, D.S., 2008. Rubisco, Rubisco activase, and global climate change. Journal of Experimental Botany 15. http://dx.doi.org/10.1093/jxb/em053.

Schoknecht, J.O., Hattingh, M.J., 1976. X-ray microanalysis of elements of VA mycorrhizal and non-mycorrhizal onions. Mycologia 68, 296–303.

Seastedt, T.R., Knapp, A.K., 1993. Consequences of nonequilibrium resource availability across multiple time scales: the transient maxima hypothesis. American Naturalist 141, 621–633.

Simkin, S.M., Allen, E.B., Bowman, W.D., Clark, C.M., Belnap, J., Brooks, M.L., Cade, B.S., Collins, S.L., Geiser, L.H., Gilliam, F.S., Jovan, S.E., Pardo, L.H., Schulz, B.K., Stevens, C.J., Suding, K.N., Throop, H.L., Waller, D.M., 2016. Conditional vulnerability of plant diversity to atmospheric nitrogen deposition across the United States. Proceedings of the National Academy of Sciences of the United States of America 113, 4086–4091.

Sirulnik, A.G., Allen, E.B., Meixner, T., Allen, M.F., 2007. Impacts of anthropogenic N additions on nitrogen mineralization from plant litter in exotic annual grasslands. Soil Biology & Biochemistry 39, 24–32.

Sochorová, L., Jansa, J., Verbruggen, E., Hejcman, M., Schellberg, J., Kiers, E.T., Johnson, N.C., 2016. Long-term agricultural management maximizing hay production can significantly reduce belowground C storage. Agriculture, Ecosystems and Environment 220, 104–114.

Stahl, E., 1900. Der Sinn der mycorrhizenbildung. Jahrbucher fuer wissenschaftliche Botanik 34, 539–668.

Sterner, R.W., Elser, J.J., 2002. Ecological Stoichiometry. Princeton University Press, Princeton, New Jersey.

St. John, T.V., Coleman, D.C., Reid, C.P.P., 1983. Association of vesicular-arbuscular mycorrhizal hyphae with soil organic particles. Ecology 64, 957–959.

Teste, F.P., Laliberté, E., Lambers, H., Auer, Y., Kramer, S., Kandeler, E., 2016. Mycorrhizal fungal biomass and scavenging declines in phosphorus-impoverished soils during ecosystem retrogression. Soil Biology & Biochemistry 92, 119–132.

Tisserant, E., Malbreil, M., Kuo, A., Kohler, A., Symeonidi, A., Balestrini, R., Charron, P., Duensing, N., dit Frey, N.F., Gianinazzi-Pearson, V., et al., 2013. Genome of an arbuscular mycorrhizal fungus provides insight into the oldest plant symbiosis. Proceedings of the National Academy of Sciences of the United States 110, 20117–20122.

Titus, J.H., del Moral, R., 1998. Vesicular-arbuscular mycorrhizae influence Mount St. Helens pioneer species in greenhouse experiments. Oikos 81, 495–510.

Treseder, K.K., Allen, M.F., 2002. Direct N and P limitation of arbuscular mycorrhizal fungi: a model and field test. New Phytologist 155, 507–515.

van Aarle, I., Rouhier, H., Saito, M., 2002. Phosphatase activities of arbuscular mycorrhizal intraradical and extraradical mycelium, and their relation to phosphorus availability. Mycological Research 106, 1224–1229.

Vandeboncoeur, M.A., Ouimette, A.P., Hobbie, E.A., 2015. Mycorrhizal roots in a temperate forest take up organic nitrogen from C-13 and N-15-leveled organic matter. Plant and Soil 397, 303–315.

Vargas, R., Allen, M.F., 2008. Dynamics of fine root, fungal rhizomorphs and soil respiration in a mixed temperate forest: integrating sensors and observations. Vadose Zone Journal 7, 1055–1064.

Vargas, R., Baldocchi, D.D., Allen, M.F., Bahn, M., Black, T.A., Collins, S.L., Yuste, J.C., Hirano, T., Jassal, R.S., Pumpanen, J., Tang, J., 2010b. Looking deeper into the soil: biophysical controls and seasonal lags of soil CO_2 production and efflux. Ecological Applications 20, 1569–1582.

Vargas, R., Detto, M., Baldocchi, D.D., Allen, M.F., 2010a. Multiscale analysis of temporal variability of soil CO_2 production as influenced by weather and vegetation. Global Change Biology 16, 1589–1605.

Yoshida, L.C., Allen, E.B., 2001. Response to ammonium and nitrate by a mycorrhizal annual invasive grass and a native shrub in southern California. American Journal of Botany 88, 1430–1436.
Yoshida, L.C., Allen, E.B., 2004. ^{15}N uptake by mycorrhizal *Artemisia californica* and the invasive *Bromus madritensis* of a N-eutrophied shrubland. Biology and Fertility of Soil 39, 243–248.
Zhao, X.H., Zhao, Y.Q., Kearney, P., 2013. Phosphorus recovery as $AlPO_4$ from beneficially reused aluminium sludge arising from water treatment. Environmental Technology 34, 263–268.
Zimmer, K., Hynson, N.A., Gebauer, G., Allen, E.B., Allen, M.F., Read, D.J., 2007. Wide geographical and ecological distribution of nitrogen and carbon gains from fungi in pyroloids and monotropoids (Ericaceae) and in orchids. New Phytologist 175, 166–175.

S E C T I O N III

MYCORRHIZAL MEDIATION OF SOIL STRUCTURE AND SOIL-PLANT WATER RELATIONS

Catherine Gehring, Lead Editor

CHAPTER

13

Introduction: Mycorrhizas and Soil Structure, Moisture, and Salinity

C.A. *Gehring*
Northern Arizona University, Flagstaff, AZ, United States

13.1 INTRODUCTION

Mycorrhizal fungi are best known for their promotion of plant nutrition, but they also have well-documented influences on the physical structure of soil that can affect the availability of soil resources, including water, to plants and other organisms. In addition, associations with mycorrhizal fungi alter plant physiology directly and indirectly, facilitating the growth of plants in dry or saline soils with potential feedback effects on soil properties. Whereas many studies of the relationships among mycorrhizal fungi and soil structure, salinity, and water availability focus on measurements of plant performance, there is a growing consideration of the plant–soil–fungal system as an interacting unit. This chapter briefly introduces the topics of soil structure, soil–water relationships, and soil salinity to set the stage for subsequent chapters that examine how the mycorrhizal symbiosis influences those properties of soil.

13.2 SOIL STRUCTURE

Soil structure refers to the arrangement of individual soil particles, sand, silt and clay, into larger aggregates of varying sizes and shapes. Shapes include granular, columnar, or blocky forms; soils with no apparent structure are termed *massive* (Brady and Weil, 2008). Soil aggregates are often also characterized by size; macroaggregates are typically greater than 250 μm in size and consist of multiple smaller microaggregates joined by microbes, plants, and animals and the chemicals that they produce. Physical processes such as alternate wetting and drying or freezing and thawing also contribute to soil aggregate formation. Soil structure is important because it is one of the major factors determining how fast water and air enter and move through the soil, which in turn influences soil resource availability for plants and habitat for other organisms, including fungi (Brady and Weil, 2008). Measurements of soil aggregate stability assess the ability of soil aggregates to resist fragmentation in the face of disruptive forces such as wind,

Mycorrhizal Mediation of Soil
http://dx.doi.org/10.1016/B978-0-12-804312-7.00013-9

water, or physical disturbance (e.g., tillage). Soil texture, soil parent material, soil organic matter, and disturbance all influence soil aggregate stability (Bird et al., 2007). A higher proportion of macroaggregates to microaggregates can increase soil quality as a result of increased biological activity and nutrient cycling (Arshad et al., 1996). Land-use changes such as deforestation and conversion to agriculture have well-documented effects on soil structural properties. For example, conversion from forest to pasture in Amazonia resulted in changes in pore-size distribution that reduced water availability to plants (Young et al., 1998). In southwestern France, deforestation followed by intensive cultivation affected soil structure at the macroaggregate and microaggregate scales, possibly because of a reduction in the earthworm population, which can play an important role in soil aggregation (Besnard et al., 1996). Plant cover in arid lands of the southwestern United States contributed substantially to soil aggregate stability, but in this case the effect of disturbance was more complex, depending on soil depth and testing method (Bird et al., 2007). Biological soil crust cover had the strongest influence on surface soil stability in semiarid shrubland, with plants and arbuscular mycorrhizal (AM) fungi contributing more strongly to subsurface stability (Chaudhary et al., 2009).

Global changes such as warming temperatures, more frequent droughts and nitrogen (N) deposition are predicted to change soil structural properties. Hotter, drier conditions are likely to increase wind erosion, whereas water erosion could increase during the intense precipitation events predicted for some areas (Young et al., 1998). Changes in patterns of drying and rewetting are also likely to have consequences because of aggregate disintegration resulting from changes in microbial activity and the availability of organic carbon (Young et al., 1998). The extent of these soil changes is likely to be mediated by organisms. Along a climate and erosion gradient in a Mediterranean shrubland ecosystem, plant species' richness was strongly, positively associated with soil aggregate stability and water-holding capacity (Garcia-Fayos and Bochet, 2009).

Long-term enrichment of tallgrass prairie with nitrogen (17 years) increased the formation of water-stable macroaggregate formation, as did annual burning (Wilson et al., 2009), with corresponding decreases in microaggregate formation. Macroaggregate formation was strongly influenced by the abundance of AM fungi. These studies underscore the importance of gaining a deeper understanding of the forces that promote and disrupt aggregate formation in the soil, including the role of mycorrhizal fungi whose abundant hyphae can play pivotal roles.

In Chapter 14, Lehmann, Leifheit and Rillig review the literature examining how mycorrhizal fungi influence soil structure, describing patterns in relation to type of mycorrhizal association, soil properties, and experimental variables such as pot size and study length. They also explore the mechanisms of these effects in detail, including a discussion of interactions between mycorrhizal fungi and other soil biota. Lehmann et al. (Chapter 14) also suggest directions for future study and make a call for greater integration of this topic into global scale models of mycorrhizal function.

13.3 SOIL SALINITY

Salts, compounds formed by the joining of sodium (Na^+), potassium (K^+), calcium (Ca^{2+}), magnesium (Mg^{2+}), chlorine (Cl^-), and other ions, are a natural component of soil that are released as soil minerals weather. Salts can also be deposited in soil via dust and precipitation or through irrigation water, sea water, and some fertilizers. They can be leached from soils by precipitation (Brady and Weil, 2008), but leaching may be limited in arid and semiarid lands where salts are more likely to accumulate in soils. The amount and types of salts in soils vary and if smaller ions like Ca^{2+} and Mg^{2+} are common, increased salinity can promote soil aggregate formation. However, if larger Na^+ ions dominate, the forces that hold clay particles together can be disrupted, reducing soil aggregation and lowering water infiltration and hydraulic conductivity and increasing surface crusting (Miller and Donahue, 1995). Water availability to plants in these "sodic" soils can decline as sodium chloride levels increase because excess sodium reduces the amount of water entering the soil (infiltration) and the rate at which water moves through the soil (hydraulic conductivity) (Warrence et al., 2003).

Salt accumulation in soils is a significant concern, particularly in more arid areas of the world. It is challenging to quantify the land area affected by saline soils, but the Food and Agriculture Organization of the United Nations (FAO) estimated that globally the total area of saline soils was 397 million ha, and that of sodic soils, 434 million ha (FAO, 2015). Data from FAO's database, Aquastat, indicate that in some areas of the world as much as 50% of the drylands used for irrigated agriculture are area affected by salinity, with an estimated 12 million ha of irrigated land lost from agricultural production by 2001 (FAO, 2015). Salinization could result in up to 50% loss of arable land over the next 40 years (Porcel et al., 2012).

The accumulation of salts in soil negatively affects plants, and they in turn have evolved mechanisms for coping with salt stress. Both nutrient and water uptake by roots can be reduced in saline soils, resulting in osmotic stress, and ions such as Na^+ and Cl^- can interfere with cellular processes such as protein synthesis, photosynthesis, and respiration (Ruiz-Lozano et al., 2012). Salinity also leads to greater production of reactive oxygen species that can cause oxidative damage to cells (Ding et al., 2010). Plants respond to salinity through tolerance of osmotic stress, exclusion of Na^+ or Cl^-, or tolerance of their tissues to the accumulation of Na^+ or Cl^- (Munns and Tester, 2008). Species of plants vary in salt tolerance and recent phylogenetic analyses indicate that salt tolerance has evolved repeatedly, but is more likely to evolve in some plant lineages than others (Bromham, 2014). Comparisons of sugar beet cultivars (*Beta vulgaris* ssp. *vulgaris*) and their wild, salt-tolerant ancestor, the sea beet (*B. vulgaris* ssp. *maritima*) showed that domestication slightly reduced salt tolerance and that over 200 years of breeding for improved salt tolerance among cultivars had not been effective (Rozema et al., 2015).

Although some of the most salt-tolerant plants belong to nonmycorrhizal plant families such as the Chenopodiaceae, mycorrhizal fungi can significantly influence the salt tolerance of their host plants. In Chapter 15, Miransari briefly reviews the mechanisms by which plants tolerate saline soils and provides a detailed consideration of how AM fungi contribute to plant salt tolerance. AM fungi have been studied extensively because of their importance to crops grown in areas prone to salt accumulation, but ectomycorrhizal fungi (EcM) are also affected by soil salinity and can promote salt tolerance, although responses may be host plant- and fungal species-specific. The root colonization and species composition of the EcM of *Salix* and *Betula* trees in a saline meadow was related to the level of salinity, season of the year, and tree species (Hrynkiewicz et al., 2015). Inoculation of white spruce (*Picea glauca*) with the EcM fungus *Hebeloma crustiliniforme* increased nutrient uptake and root hydraulic conductance, and decreased sodium levels relative to nonmycorrhizal controls (Muhsin and Zwiazek, 2001). Different patterns were observed in aspen (*Populus tremuloides*) and birch

(*Betula papyrifera*) inoculated with the same species of EcM fungus (Yi et al., 2008). Miransari (Chapter 15) discusses similar patterns in AM fungi and makes recommendations for future study and enhancement of agricultural production through consideration of the combined salt tolerance of plants and their AM fungal associates.

13.4 SOIL MOISTURE

The water content of soils influences rates of soil formation and erosion, soil structure, and the availability of water for plant growth. Water in soil is held as films on the surface of particles and in small pores (Miller and Donahue, 1995). As a result, soil water-holding capacity is controlled primarily by soil texture and organic matter content (Miller and Donahue, 1995). Soils with smaller particles (silt and clay) have a higher water-holding capacity than coarse-textured sandy soils. Low amounts of clay or silt can result in poor soil aggregation, limiting the pore spaces that facilitate water storage. Very clay-rich soils have high water-holding capacity, but the water can be difficult for plants to access because of the energy required to extract water bonded to clay particles (Miller and Donahue, 1995). Soil organic matter influences rainwater infiltration and retention and reduces erosion (FAO, 2015). Hudson (1994) estimated that for every 1% increase in soil organic matter, the available water-holding capacity (the maximum amount of plant-available water a soil can provide) in the soil increased by 3.7%. Strong positive correlations between soil organic matter and available water-holding capacity were observed regardless of soil texture (Hudson, 1994).

A deeper understanding of the factors that influence soil–water dynamics is important given the changes in the water cycle predicted with climate change (IPCC, 2007). Droughts are predicted to intensify in many areas of the world (Trenberth et al., 2013), including temperate and tropical regions (Duffy et al., 2015). Drought is expected to shrink and fracture soil aggregates and increase soil–water repellency, changing water infiltration patterns (Hendrickx and Flury, 2001). Water exclusion experiments accompanied by dye tracers provided experimental support for these hypotheses (Gimbel et al., 2016). Water infiltration was altered by drought in soils of varying texture (clay, loam, sand), largely owing to changes in water repellency (Gimbel et al., 2016). Changes in soil properties with drought are likely to influence plant–soil–water relationships. Soil texture and associated variation in plant–soil–water relationships explained the distribution and decline with more arid conditions of overstory trees in western Australia (Poot and Veneklaas, 2013). Likewise, because of their abundance in the soil and intimate connection with plant roots, mycorrhizal fungi can influence and be influenced by variation in soil water availability in complex ways.

In Chapters 16–18, interactions among mycorrhizal fungi, their host plants, and soil moisture are explored from different perspectives. In Chapter 16, Gehring, Swaty, and Deckert review the effects of drought on mycorrhizal fungi and the influence of mycorrhizal fungi on host-plant drought tolerance. They also describe the potential consequences of differential drought-related host plant mortality to mycorrhizal fungi, soils, and plant communities. Querejeta (Chapter 17) takes a detailed, mechanistic look at how both AM fungi and EcM influence the hydraulic properties of soils. He considers these relationships from the hyphal to ecosystem scales and suggests avenues for future research, including experiments that provide a longer-term perspective than previous work. Pickles and Simard (Chapter 18) examine the roles of mycorrhizal networks in water uptake and hydraulic redistribution with a focus on EcM in drought-affected forests. They consider the management implications of changes in fungal networks and suggest diverse areas of new research, including studies of mycorrhizal networks in deeper soil and interactions among AM fungi and EcM networks.

Although seemingly covering a diverse array of topics, the chapters in this section of the book clearly illustrate the dynamic interplay between mycorrhizas and soils and their importance in a world of increasing stresses such as salinization and drought.

References

Arshad, M.A., Lowery, B., Grossman, B., 1996. Physical tests for monitoring soil quality. In: Doran, J.W., Jones, A.J. (Eds.), Methods for Assessing Soil Quality, pp. 123–141 Madison, WI.

Besnard, E., Chenu, C., Balesdent, J., Puget, P., Arrouays, D., 1996. Fate of particulate organic matter in soil aggregates during cultivation. European Journal Soil Science 47, 495–503.

Bird, S.B., Herrick, J.E., Wander, M.M., Murray, L., 2007. Multiscale variability in soil aggregate stability: implications for understanding semiarid grassland degradation. Geoderma 140, 106–118.

Brady, N.C., Weil, R., 2008. The Nature and Properties of Soils, Fourteenth ed. Pearson. 980 pp.

Bromham, L., 2014. Macroevolutionary patterns of salt tolerance in angiosperms. Annals of Botany 115, 333–341. http://dx.doi.org/10.1093/aob/mcu229.

Chaudhary, V.B., O'Dell, T.E., Bowker, M.A., Grace, J.B., Redman, A.E., Rillig, M.C., et al., 2009. Untangling the biological controls of soil stability in semi-arid shrublands. Ecological Applications 19, 110–122.

Ding, M., Hou, P., Shen, X., et al., 2010. Salt-induced expression of genes related to Na^+/K^+ and ROS homeostasis in leaves of salt resistant and salt-sensitive poplar species. Plant Molecular Biology 73, 251–269.

Duffy, P.B., Brando, P., Asner, G.P., Field, C.B., 2015. Projections of future meteorological drought and wet periods in the Amazon. Proceedings of the National Academy of Sciences 112, 13172–13177.

Food and Agriculture Organization of the United Nations, 2015. Status of the World's Soil Resources. 650 p. FAO.

Garcia-Fayos, P., Bochet, E., 2009. Indication of antagonistic interaction between climate change and erosion on plant species richness and soil properties in semiarid Mediterranean ecosystems. Global Change Biology 15, 306–318.

Gimbel, K., Puhlmann, H., Weiler, M., 2016. Does drought alter hydrological functions in forest soils? Hydrology and Earth Systems Sciences 20, 1301–1317.

Hendrickx, J.M.H., Flury, M., 2001. Uniform and preferential flow mechanisms in the vadose zone. In: Conceptual Models of Flow and Transport in the Fractured Vadose Zone, Panel on Conceptual Models of Flow and Transport in the Fractured Vadose Zone. US National Committee for Rock Mechanics, Board on Earth Sciences and Resources, National Research Council, National Academy Press, Washington, D.C, pp. 149–187. http://dx.doi.org/10.17226/10102.

Hrynkiewicz, K., Szymańska, S., Piernik, A., Thiem, D., 2015. Ectomycorrhizal community structure of *Salix* and *Betula* spp. at a saline site in central Poland in relation to the seasons and soil parameters. Water, Air, and Soil Pollution 226, 99–106. http://dx.doi.org/10.1007/s11270-015-2308-7.

Hudson, B.D., 1994. Soil organic matter and available water capacity. Journal of Soil and Water Conservation 49, 189–194.

IPCC, 2007. Climate Change 2007: the Physical Science Basis Contribution of Working Group I to the Fourth Assessment Report of the Intergovernmental Panel on Climate Change. Cambridge University Press, Cambridge, United Kingdom, and New York.

Miller, R.W., Donahue, R.L., 1995. Soils in Our Environment, Seventh ed. Prudence Hall, Englewood Cliffs, NJ. 323pp.

Muhsin, T., Zwiazek, J.J., 2001. Colonization with *Hebeloma crustuliniforme* increases water conductance and limits shoot sodium uptake in white spruce (*Picea glauca*) seedlings. Plant Soil 238, 217–225.

Munns, R., Tester, M., 2008. Mechanisms of salinity tolerance. Annual Review of Plant Biology 59, 651–681.

Poot, P., Veneklaas, E.J., 2013. Species distribution and crown decline are associated with contrasting water relations in four common sympatric eucalypt species in southwestern Australia. Plant Soil 364, 409. http://dx.doi.org/10.1007/s11104-012-1452-8.

Porcel, R., Aroca, R., Ruíz-Lozano, J.M., 2012. Salinity stress alleviation using arbuscular mycorrhizal fungi: a review. Agronomy for Sustainable Development 32, 181–200.

Rozema, J., Cornelisse, D., Zhang, Y., Li, H., Bruning, B., Katschnig, D., et al., 2015. Comparing salt tolerance of beet cultivars and their halophytic ancestor: consequences of domestication and breeding programs. AOB Plants 7. http://dx.doi.org/10.1093/aobpla/plu083.

Ruiz-Lozano, J.M., Porcel, R., Azcón, C., Aroca, R., 2012. Regulation by arbuscular mycorrhizae of the integrated physiological response to salinity in plants: new challenges in physiological and molecular studies. Journal of Experimental Botany 63, 4033–4044.

Trenberth, K.E., Dai, A., van der Schrier, G., Jones, P.D., Barichivich, J., Briffa, K.R., et al., 2013. Global warming and changes in drought. Nature Climate Change 4, 17–22.

Warrence, N.J., Bauder, J.W., Pearson, K.E., 2003. Basics of Salinity and Sodicity Effects on Soil Physical Properties. Land Resources and Environmental Sciences Department Montana State University.

Wilson, G.W.T., Rice, C.W., Rillig, M.C., Springer, A., Hartnett, D.C., 2009. Arbuscular mycorrhizal fungi control soil aggregation and carbon sequestration. Ecology Letters 12, 452–461.

Yi, H., Calvo Polanco, M., MacKinnon, M., Zwiazek, J.J., 2008. Responses of ectomycorrhizal *Populus tremuloides* and *Betula papyrifera* seedlings to salinity. Environmental and Experimental Botany 62, 357–363.

Young, I.M., Blanchart, E., Chenu, C., Dangerfield, M., Fragoso, C., Grimaldi, M., 1998. The interaction of soil biota and soil structure under global change. Global Change Biology 4, 703–712.

CHAPTER

14

Mycorrhizas and Soil Aggregation

A. Lehmann, E.F. Leifheit, M.C. Rillig
Freie Universität Berlin, Berlin, Germany

14.1 INTRODUCTION: SOIL AGGREGATION, ITS COMPONENT PROCESSES, AND SIGNIFICANCE OF SOIL STRUCTURE

Soil aggregation is a dynamic and complex ecosystem process consisting of the formation, stabilization, and disintegration of soil aggregates (and associated soil pore spaces). This process is influenced by physical forces, chemical bonds, environmental factors (e.g., soil texture, freeze–thaw cycles, wetting-drying events), and biological agents, e.g., soil microorganisms and their interactions (e.g., Tisdall and Oades, 1982; Golchin et al., 1994; Rillig and Mummey, 2006). The basic modules in this process are the soil aggregates, which are constructs made of organic material (e.g., plant, microbial or animal derived debris), primary soil particles and minerals, adhering to each other more strongly than to surrounding particles (Kemper and Rosenau, 1986).

Soil aggregation generally is suggested to follow a conceptual, hierarchical model where primary particles (<53 μm diameter) form microaggregates (<250 μm diameter) via physicochemical forces and persistent binding agents (e.g., humified organic matter and complexes of polyvalent metal cations), whereas microaggregates are assembled into macroaggregates (>250 μm diameter) together with organic debris. Microaggregates can also form within macroaggregates. Macroaggregates are bound by relatively more transient agents including fine roots, fungal hyphae, and soil microbes (Tisdall and Oades, 1982; Six et al., 2004). Macroaggregates, and to a smaller extent microaggregates, are exposed to disruptive forces (e.g., shear forces or erosion events) causing their break-down into fragments or even primary particles (Fig. 14.1). The process of disruption provides access to the otherwise encapsulated organic matter, which will either be further degraded or function as a nucleus or building block for new aggregates (Oades, 1984; Six et al., 2004; Jastrow et al., 2007).

Soil aggregates and the associated soil pores of different sizes and shapes, formed within and between aggregates, are essential components of soil structure that fundamentally affect soil quality, fertility, and sustainability. For example, soil pores maintain and facilitate soil gas exchange, water infiltration, habitats of soil microorganisms, and subsequently biogeochemical cycling. Furthermore, root penetration and growth in soil is facilitated by an established pore matrix. Soil erosion by water and wind is diminished in well aggregated soils. Each year 75,000 Mt of soil are

Mycorrhizal Mediation of Soil
http://dx.doi.org/10.1016/B978-0-12-804312-7.00014-0

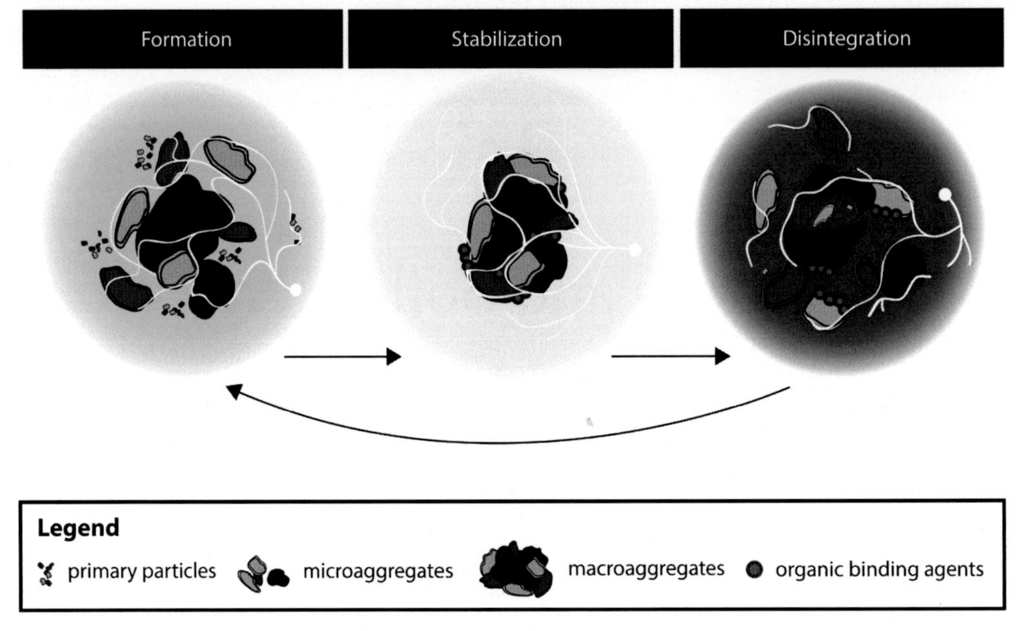

FIGURE 14.1 Soil aggregation is a key ecosystem process, and the formation, stabilization, and disintegration of soil aggregates is influenced by physical forces, chemical bonds, and biological agents as fungal hyphae (*white lines*) maintaining a dynamic aggregate turnover. In this hierarchical process, primary soil particles are assembled to microaggregates (20–250 μm), which in turn build up to macroaggregates (<250 μm). Microaggregates can also form within macroaggregates.

removed by erosion events, causing an estimated 17,000 million dollars of damage (Pimentel et al., 1995). Soil erosion ultimately leads to reduced soil fertility and sustainability, and hence losses in plant productivity (Frye et al., 1982; Mokma and Sietz, 1992). Soil organic matter losses in eroded soil inhibit the formation and stabilization of soil aggregates, thus fueling the cycle of soil loss.

Above all, soil structure is the stage upon which life in soil plays out, and as such it is a critical aspect of the soil habitat, the loss of which could give rise to local extinctions of biota (Veresoglou et al., 2015). Given the importance of soil structure for soil biota and ecosystem functioning, it comes as no surprise that considerable research efforts have been directed at describing and understanding the effects of mycorrhizal fungi, dominant players in many soil food webs, on soil aggregation processes.

14.2 EVIDENCE FOR INVOLVEMENT OF DIFFERENT TYPES OF MYCORRHIZAS IN SOIL AGGREGATION

14.2.1 Arbuscular Mycorrhizal Fungi

Among the different types of mycorrhizal associations the arbuscular mycorrhizal (AM) fungi are the best known for their improvement of soil structure and as one of the most ancient symbioses, being estimated to originate 460–600 Ma ago (Redecker et al., 2000;

Redecker and Raab, 2006). AM fungi are members of the phylum Glomeromycota (Schussler et al., 2001) and are ubiquitous root colonizers in the majority of land plants, including a broad variety of crops (Wang and Qiu, 2006; Smith and Smith, 2011). They form characteristic intercellular and intracellular structures, mediating the intimate symbiotic interactions; the most prominent are the arbuscules. Besides its intraradical portion consisting of hyphae and storage organs, the vesicles, AM fungi support an extraradical mycelium, which is mostly relevant for nutrient acquisition beyond the depletion zone of the associated host's root (Parniske, 2008). The key property of AM fungi for stabilizing soil aggregates most likely is their extensive extraradical hyphal growth, a factor that has often been found to correlate with soil stability parameters (e.g., Wilson et al., 2009; Barto et al., 2010). Rillig et al. (2010) showed in an in vitro study in which AM fungi grew in sterilized soil that there is a direct causal link between the amount of water-stable aggregates (WSAs) and the presence of AM fungal mycelium.

The extent to which soil aggregation is promoted depends on the AM fungal species involved, because members of different families can differentially produce soil hyphae (Hart and Reader, 2002). Several studies observed variations in the soil aggregation capacity between AM fungal species (Piotrowski et al., 2004; Enkhtuya and Vosatka, 2005; Klironomos et al., 2005). Furthermore, the combination of fungal species and host plant is important for the direction and strength of the relationship along the mutualism–parasitism continuum (Johnson et al., 1997), and likewise hyphal growth and thus soil aggregation depend on this interaction (e.g., Piotrowski et al., 2004). The role of AM fungal diversity and richness has been addressed in only a small number of studies. Schreiner and Bethlenfalvay (1997) found that a mix of three species was more beneficial to plant growth promotion and soil aggregation than the single species alone. More research is required to clarify the role of species diversity for soil structure. Apart from fungal and host identity, hyphal proliferation depends on environmental conditions such as nutrient levels, soil pH, and water content (Johnson et al., 2003; Parniske, 2008; Helgason and Fitter, 2009; Pietikainen et al., 2009).

The importance of AM fungi for soil structure was uncovered during the 1950s (Martin et al., 1955; review of Six et al., 2004) and has since been well documented in narrative reviews (e.g., Oades, 1993; Tisdall, 1994; Six et al., 2004; Rillig and Mummey, 2006), as well as in a quantitative review (Leifheit et al., 2014). A meta-analysis quantitatively synthesizing studies from 1986 to 2012 examining the AM fungal impact on soil aggregation revealed a positive overall effect (Fig. 14.2) (Leifheit et al., 2014). Furthermore, it could be shown that the selection of experimental parameters can have a crucial impact on the outcome of a study testing for AM fungal effects on water-stable macroaggregates. The highest effect sizes (highest influence of AM fungi on soil aggregation) were found for pot studies, sterilized soil, sandy texture, near neutral soil pH, a pot size smaller than 2.5 kg, and a duration between 2.2 and 5 months. Lowest effect sizes were found for field studies, nonsterilized soil, fine-textured soil, and soil with an acidic pH. The effect of AM fungi on soil aggregation was independent of the fungal species richness, the soil organic carbon content, or the selected laboratory procedures for determining soil aggregation. Potential effects on soil aggregation caused by the host plant identity do not seem to be strong enough to overrule the effect of AM fungi. To further validate these findings, research should focus on studies with various AM fungal genera, richness, and diversity levels in differing contexts (e.g., soil type).

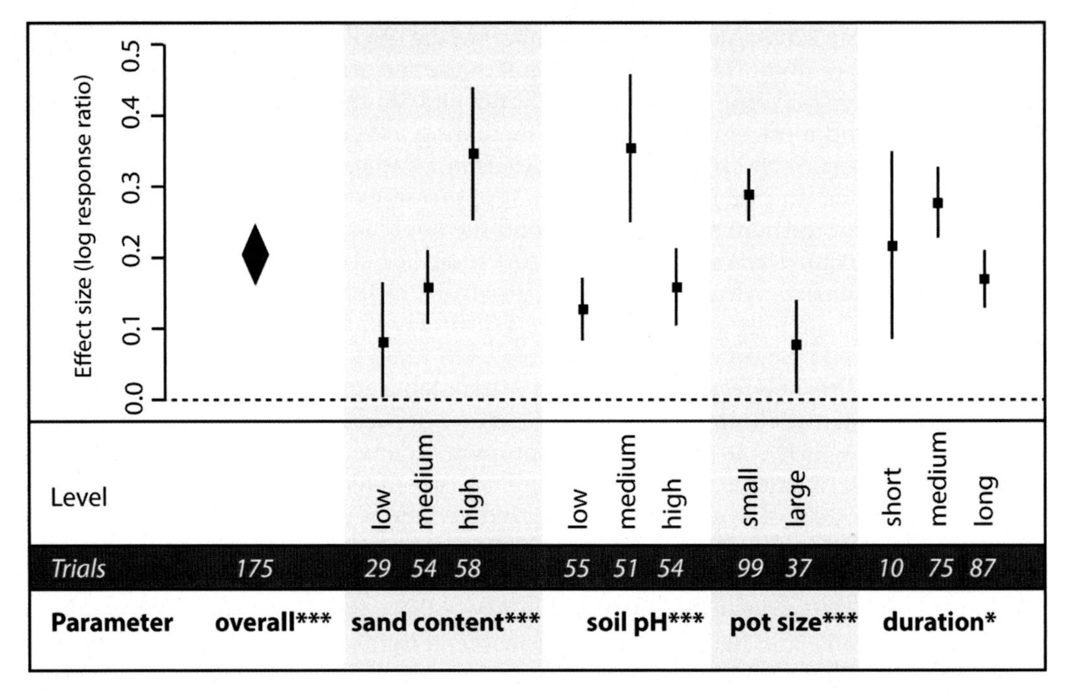

FIGURE 14.2 Overall effect of arbuscular mycorrhizal fungi on soil aggregation and the impact of soil derived (sand content, soil pH) and experiment-related variables (pot size, experimental duration). For the categorical variables (parameters), the level means and 95% confidence intervals were presented, which were estimated via a random-effects model with nonparametric bootstrapping with 3999 iterations. The level categorization was: sand content (low, 7–40%; medium, 41%; high, 42–82%); soil pH (low, 5–6.7; medium, 6.8–8; high, 8.1–8.9); pot size (small, <2.5 kg; large, ≥2.5 kg); and experimental duration (short, <2.2 month; medium, 2.2–5 month; long, >5 month). Trial numbers represent the number of individual experimental systems (not synonymous to independent studies) included in the analyses. *Asterisks* represent significance level (***, $p<.0001$; *, $p<.05$). *Redrawn from Leifheit, E.F., Veresoglou, S.D., Lehmann, A., Morris, E.K., Rillig, M.C., 2014. Multiple factors influence the role of arbuscular mycorrhizal fungi in soil aggregation: a meta-analysis. Plant and Soil 374 (1–2), 523–537.*

14.2.2 Ectomycorrhizal Fungi

Ectomycorrhizal fungi (EcM) are formed between a fungal root symbiont mainly belonging to the phylum Basidiomycota, but also to Ascomycota and Mucoromycotina (Tedersoo et al., 2010), and a plant host, which are almost all perennial woody plants. This type of symbiosis was estimated to originate 50–200 Ma ago (LePage et al., 1997; Berbee and Taylor, 2001; LePage, 2003). The main characteristics of EcM are the mantle encasing the host's roots, the Hartig net, and an extensive extraradical mycelium (Smith and Read, 2008).

The impact of EcM on soil aggregation is far less intensely studied than for AM fungi (Rillig and Mummey, 2006), possibly because the topic of soil aggregation is of inherently less interest in forests compared with grasslands or agricultural systems. However, there are observational and experimental studies demonstrating that they can improve soil structure. In an observational study, Thornton et al. (1956) found higher aggregation of a sandy soil under *Pinus radiata*, an EcM host, than under a *Leptospermum* species. Caesar-TonThat et al. (2013) showed that soil adjacent (40–50 cm distance) to *Agaricus lilaceps* fairy rings had higher mean weight diameter (MWD) of aggregates than soil sampled at a 1-m distance from outside the ring.

In experimental studies under controlled environmental conditions, the beneficial effect of EcM on soil aggregation was confirmed. Graf and Frei (2013) tested the EcM fungus *Melanogaster variegatus* forming mycorrhizas with *Alnus indica* and found higher water stability of macroaggregates. Similar effects could be detected by Caravaca et al. (2002) for *Pisolithus arhizus* and *Pinus halepensis*. Furthermore, Zheng et al. (2014) evaluated nine different EcM species forming mycorrhiza with *Pinus sylvestris* L. and showed that six fungal species, when in symbiosis, caused an increase in MWD and WSA, demonstrating that the effect is fungal-species specific.

14.2.3 Other Mycorrhizal Types

Soils may harbor a variety of mycorrhizal types besides the well-studied AM fungi and EcM, e.g., ericoid mycorrhizas (ERMs) or the newly discovered types formed by the Sebacinales and Mucoromycotina. There is a dearth of information regarding soil aggregation effects for all of these other types.

ErMs are formed exclusively between members of the Ericaceae and fungi belonging mainly to the Ascomycota but also Basidiomycota, namely to the order Sebacinales, which enter their hosts only at specialized distal roots. The ErMs can be seen as an adaptation to the often acidic and nutrient poor habitat of Ericaceae, growing in heathland, boreal forests, and bogs (Cairney and Meharg, 2003), and it was estimated that it originated 140 Ma ago (Cullings, 1996). Studies evaluating the impact of ERMs on soil aggregation are extremely rare. One example is the study by Tisdall et al. (1997), testing, among other saprotrophic fungi, the ErM fungus *Rhizoscyphus ericae* (synonymous with *Hymenoscyphus ericae*) but without a plant host; thus no mycorrhizal symbiosis could be established. However, in this in vitro experiment the fungus promoted the aggregation of a clay substrate.

For the Sebacinales forming EcM and ErMs (Selosse et al., 2002, 2007), no data on their effect on soil structure are available, although these fungi are ubiquitously present and may occur as endophytes in a large number of plants (Weiss et al., 2011). Because there is an increasing interest in this group of fungi owing to their strong beneficial impact on plant growth performance (e.g., Barazani et al., 2005), we expect that there will soon be studies evaluating the importance of Sebacinales species for soil aggregation.

A similar lack of data can be reported for the newly discovered liverwort–Mucoromycotina symbiosis, which was potentially formed between Haplomitriopsida and Endogonales (Field et al., 2015). Along with the Glomeromycota, these fungi likely associated with the earliest land plants and facilitated their conquest of the new terrestrial habitat. Their impact on soil aggregation needs to be tested in future studies, perhaps also in association with angiosperm hosts.

14.3 MECHANISMS OF SOIL AGGREGATION

The positive impact of mycorrhizal fungi on soil aggregation has been known since the early 1900s and although intense research followed and new insights were revealed by observational, experimental, correlational and data mining approaches, the precise mechanisms still remain unclear (e.g., Tisdall and Oades, 1982; Miller and Jastrow, 1990; Six et al., 2004; Rillig and Mummey, 2006; Leifheit et al., 2014).

The complex nature of soil aggregation with its simultaneously occurring process components (formation, stabilization, disintegration) is influenced by a multitude of interrelated factors and mechanisms mediated, among others, by soil biota. For mycorrhizal fungi, these can be loosely grouped into biophysical, biochemical, and biological mechanisms mediated directly or indirectly by their hyphae or mycelium traits (Fig. 14.3) (Rillig and Mummey, 2006). Even though these are discussed separately here, these mechanisms interact in reality.

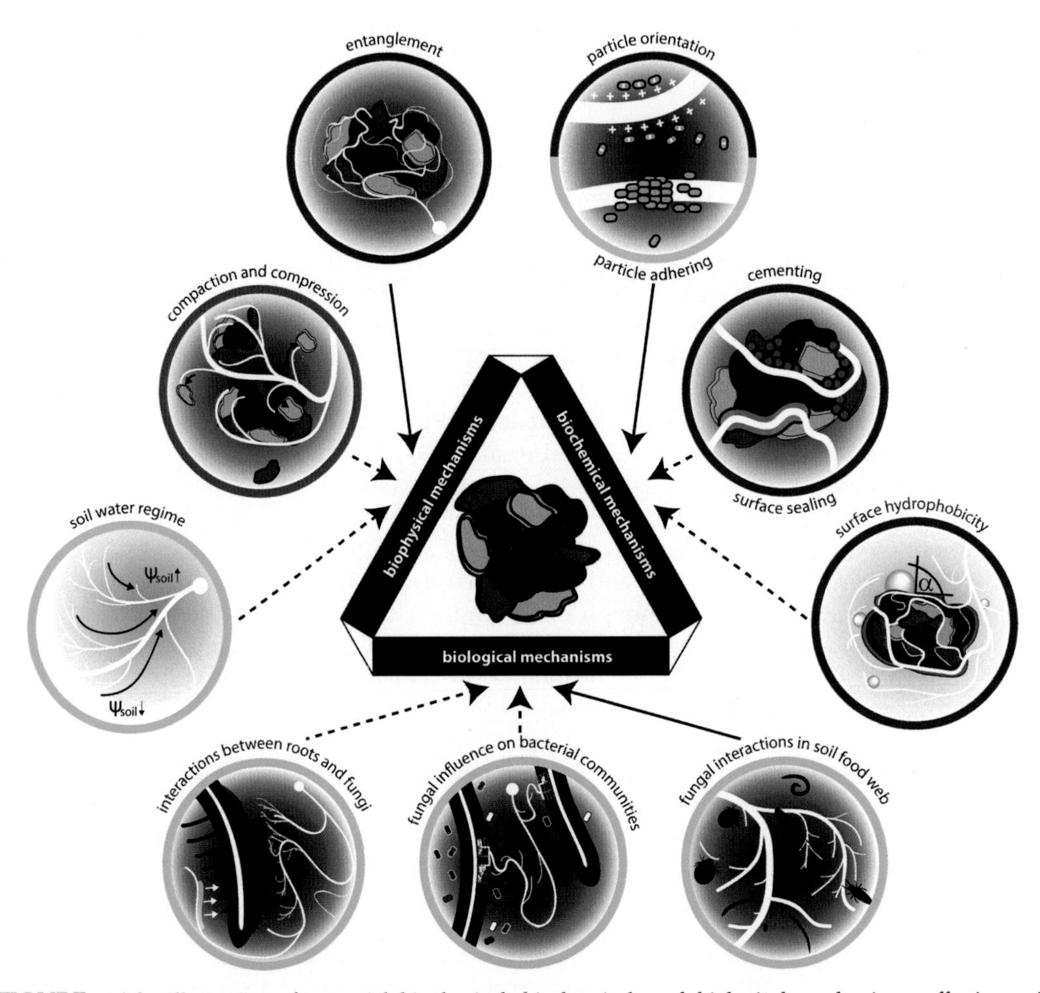

FIGURE 14.3 Illustration of potential biophysical, biochemical, and biological mechanisms affecting soil aggregation for which empirical evidence is either reported for mycorrhizal fungi (*green circles*) or for nonmycorrhizal fungi (*red circles*). For some potentially important mechanisms, no data was available (*gray circle*). The relevance of these mechanisms for soil aggregation was scarcely evaluated (*solid arrow*) but most often only hypothesized (*dashed arrow*). For further details on this topic see text of Section 14.3.

14.3.1 Biophysical Mechanisms

The biophysical mechanisms of soil aggregation are mediated by the direct physical interaction of individual hyphae, or the mycelium as a whole, with soil particles or aggregates. This set of mechanisms consists of enmeshment (or entanglement), alignment of particles by exertion of physical pressure, and changes in local water potential caused by the mycelium (Fig. 14.3). Each of these mechanisms is influenced by a number of mycelium traits, mostly architectural in nature.

The entanglement of soil particles and aggregates is one evident but less studied mechanism. There is correlational evidence demonstrating that with increasing length of the extraradical mycelium of AM fungi, the MWD is enhanced; thus it was suggested that with higher hyphal density per unit soil more particle surfaces can be affected and enmeshed (Tisdall and Oades, 1980; Miller and Jastrow, 1990). However, observational and experimental evidence is so far only available for nonmycorrhizal fungi (Gupta and Germida, 1988; Meadows et al., 1994; Daynes et al., 2012).

While growing, hyphae exert pressure on the surrounding soil (Money, 1994), potentially pushing and moving particles and aggregates into closer proximity and thus locally compressing the soil and aligning particles. There are no reports on experiments testing this mechanism in fungi, neither mycorrhizal nor nonmycorrhizal fungi.

Fungi can directly alter the water regime in the soil alongside their hyphae, in the so-called "hyphosphere." Mycorrhizal fungi can transport water from soil patches to their associated hosts, whereas plants as well can transfer water by nocturnal hydraulic lift to their symbionts (Augé, 2001; Querejeta et al., 2003; Egerton-Warburton et al., 2007; Marjanovic and Nehls, 2008). This water movement could result in dampening or drying the soil in the mycorrhizosphere, potentially affecting the process of soil aggregation.

A number of fungal traits (Rillig et al., 2014; Lehmann and Rillig, 2015) influence the ability of the mycelium (or hyphae) to perform one or more of these biophysical mechanisms; including mycelium extension rate, branching patterns and angles, anastomosis ability, hyphal tensile strength, lifespan, and healing ability. These are described and discussed in the following paragraphs, mostly with examples from AM fungi and EcM in mind. However, all of these points can be applied to other mycorrhizal fungal types, and in fact to fungi in general.

The mycelial extension rate (and volume covered) determines the soil volume being exposed to fungal influence. This is an important trait, including for soil aggregation, because it determines if a fungus can even influence a given volume of soil. Little is known about how far mycorrhizal fungi colonies can reach outside their plant host in the surrounding soil. This ability will drastically vary depending on the type of considered hyphae; for explorative hyphae (e.g., runner hyphae in AM fungi), fast mycelial extension is possible compared with slow-growing absorptive hyphal structures (Bago et al., 1998).

The impact of hyphal branching patterns and angles can define the intensity of soil–hyphal interactions; branching increases the hyphal surface area-to-volume ratio, whereas narrow angles could increase the potential to move soil particles toward each other (Lehmann and Rillig, 2015). AM fungi can have quite narrow branching angles of 30–40 degrees (Friese and Allen, 1991). However, branching patterns and angles vary strongly during fungal development and with environmental cues; e.g., AM hyphal branching intensifies at close proximity to host because of the establishment of the symbiosis or at nutrient rich soil patches for nutrient uptake (Paszkowski, 2006). The branching angle can also change during a lifetime, as was shown for the nonmycorrhizal fungus *Neurospora crassa*, for which the branching angle decreases after 22 hours of growth from 90 to 63 degrees (McLean and Prosser, 1987).

Resistance of the mycelial network against disturbances contributes to the stability of enmeshed soil particles and aggregates and is primarily determined by hyphal interlinkages and hyphal tensile strength. Hyphal linkages, or anastomoses, can be formed within the same or different hyphae of a genet or even between different genets of a fungal species, but no connections between different species are known for AM fungi and EcM species (de la Providencia et al., 2005; Sbrana et al., 2007; Wu et al., 2012). The intensity or number of anastomoses can vary among species; for AM fungal species, a higher number of anastomoses was found in Glomeraceae than Gigapsoraceae (Giovannetti et al., 1999; de la Providencia et al., 2005).

Hyphae can confer shear resistance and tensile strength to soil units, as demonstrated for nonmycorrhizal fungi (Tisdall et al., 2012). This is partially a consequence of the tensile strength of the hyphae themselves, which is a variable property influenced among others by fungal development (Li et al., 2002). However, to the best of our knowledge, hyphal tensile strength has never been measured in soil. Hyphal tensile strength can be understood and approximated as a function of hyphal diameter, wall thickness, and septation patterns/intensity. According to this view, tear resistance of hyphae could vary depending on fungal species, developmental stage, and location of hyphae in the colony. Hyphal diameter in AM fungal species was found to range from 1 to 18 μm (Dodd, 1994), whereas the thickness of hyphal walls varied between 1 and 4 μm; even further thickening of hyphal walls could be detected in AM and nonmycorrhizal fungi (McLean and Prosser, 1987; Dodd et al., 2000). Additionally, for other nonmycorrhizal fungi, wall thickness increased with increasing distance from colony margin (Trinci and Collinge, 1975) and decreased with higher branching order (McLean and Prosser, 1987).

The regenerative capability and longevity of the mycelium determines the period of functionality and integrity of the mycelial network potentially enmeshing and stabilizing soil aggregates. The lifespan of AM fungal extraradical hyphae can last 5–6 days, whereas the explorative runner hyphae were found to persist more than 30 days (Staddon et al., 2003; Olsson and Johnson, 2005), and most likely these hyphae last even much longer in soil. For EcM rhizomorphs, lifespan was estimated to average 11 months (Treseder et al., 2005).

The ability to repair disrupted hyphae after disturbance-induced injuries and breakages has been demonstrated for AM fungi (de la Providencia et al., 2005; de la Providencia et al., 2007), but such a mechanism has not been found for EcM (Taylor and Alexander, 2005). Briefly, the injury is sealed off by septa and subsequently, hyphal regrowth starts and branching from or behind the septa follows. The hyphal elongation and reorientation occurs until contact of regrown hyphae is established and hyphae can fuse. In AM fungi, the hyphal healing capability varies among species, especially for multiple injuries per hypha or long distance injuries; some species can reconnect cut hyphae with a 5-mm wide gap between the severed ends (de la Providencia et al., 2007).

14.3.2 Biochemical Mechanisms

The biochemical mechanisms are mainly attributed to fungal products (e.g., polysaccharides, glycoproteins, and hydrophobins) released from living or decomposing hyphae. They are thought to attract and align soil particles, function as cements stabilizing aggregates by filling cracks, or by covering aggregate/particle surfaces (Fig. 14.3). This knowledge derives mainly from research with nonmycorrhizal fungi showing that fungal products can decrease particle detachment from aggregates and increase water stability of artificial aggregates amended with fungal exudates (Griffiths and Jones, 1965; Caesar-TonThat and Cochran, 2000; Daynes et al., 2012; Tisdall et al., 2012). A variety of polysaccharides can be produced by AM fungi and EcM during their lifecycle (Bonfante Fasolo and Gianinazzi Pearson, 1982; Hooker et al., 2007); thus it seems plausible that for mycorrhizal fungi this mechanisms also applies to soil aggregation.

Observational studies also revealed that AM fungi have clay and sand particles adhering to their hyphae and thus increase particle attachment in colonized roots compared with uncolonized roots (Sutton and Sheppard, 1976; Tisdall and Oades, 1979). In nonmycorrhizal fungi, it could further be demonstrated that the zone of orientation of clay particles by fungal hyphae can be 1–5 μm thick and can be established within 3 days. Besides the hyphal influence on clay particles via surface charges, exudates as well can cause particle orientation, as demonstrated for the polysaccharide scleroglucan (Chenu, 1989); this exudate additionally formed fibers linking the surfaces of particles, further increasing the potential stabilization effect.

Among fungal exudates, hydrophobins are of special interest. They are ubiquitous proteins found in filamentous fungi (Wessels, 1997; Wosten, 2001) with highly conserved cysteine residues (Linder et al., 2005) that are essential for mediating interactions of hyphae and their hydrophobic/hydrophilic environment with aqueous/aerial phases. They are hypothesized to increase aggregate stability toward water as a disintegrating force by sealing of cracks or simply enhance their water repellency; hence hydrophobins can slow down or even prevent the wetting of aggregates or the filling of their small cracks and cavities, thus encapsulating the enclosed air and keeping the aggregate from breaking and falling apart (Sullivan, 1990; Piccolo and Mbagwu, 1999). The surface sealing by fungal mucilages and the resulting smoothing of the otherwise rough aggregates can be well studied by scanning electron micrography (Gupta and Germida, 1988; Daynes et al., 2012). Soil water repellency was further shown to increase in the presence of EcM and AM fungal hyphae (e.g., Schantz and Piemeisel, 1917; York and Canaway, 2000; Rillig et al., 2010); however, so far no hydrophobins could be identified in AM fungi, although analogues likely exist.

Other proteins of fungal origin may also be involved in soil aggregation (Rillig et al., 2007). For AM fungi, for example, glomalin-related soil protein (GRSP) (Rillig, 2004b) has been discussed as an agent involved in soil aggregation based on the assumption that it is mostly of AM fungal origin. However, it is unclear whether glomalin(s) are produced by AM fungi and to what extent extracted GRSP fractions really are of glomeromycotan origin (e.g., Rosier et al., 2006).

Mycorrhizal fungi that are not obligate biotrophs need enzymes for the degradation of extracellular organic compounds. ErMs and EcM produce extracellular enzymes capable of degrading organic substances such as cellulases, hemicellulases, and polyphenoloxidases (Read and Perez-Moreno, 2003; Read et al., 2004; Talbot et al., 2008); these enzymes could be needed for decomposition of plant litter (Read and Perez-Moreno, 2003) or might even potentially degrade mucilaginous substances on aggregates acting as cements. However, the amounts produced and the activity of these enzymes is much lower than, for example, in saprobic fungi (Read and Perez-Moreno, 2003); hence there is more research needed to evaluate what impact on aggregate destabilization mycorrhizal fungi really have compared with saprobic fungi.

Another potential, but so far not studied, (bio)chemical mechanism relates to mycorrhizal fungi changing the hyphosphere pH; AM fungi can reduce soil pH from 5.9 to 3.5 within 18 weeks (Bago et al., 1998). By locally acidifying the soil, AM fungi could prevent clay lixiviation and with it the loss of clay particles to lower soil layers; clay particles are important building blocks of soil aggregates. The process of lixiviation is characterized by dispersion of clays at alkaline conditions and/or reduced salt concentrations and the relocation of dispersed clay particles downwards in the soil profile. Finally, the process gets stopped by for example, flocculation, encapsulated air, or a stagnating water flow (Scheffer and Schachtschabel, 2002). Whether AM fungi or other mycorrhizal types can contribute to the reduction of this process by acidification of their hyphosphere remains unclear.

14.3.3 Biological Interaction Mechanisms

Mycorrhizal fungi do not exist in isolation, but are in intense relationships with roots of their associated host, are integrated in the soil food web, and also shape and interact with the soil microbial community (Fig. 14.3). For example, they can modify soil microbial communities in their hyphosphere and thus influence soil aggregation by indirect, strongly interrelated, biological mechanisms (Andrade et al., 1997).

Hyphosphere communities are dynamic systems determined by the growth and decay, fragmentation, and reconnection of the mycelium. The community composition depends on fungal species, hyphal vitality, and their developmental stage, affecting the attraction and attachment of soil microbes toward the fungal hyphae (Toljander et al., 2006; Scheublin et al., 2010). For bacteria, there is strong evidence that their abundance and richness is increased in the hyphosphere compared with bulk soil (Ames et al., 1984; Meyer and Linderman, 1986; Secilia and Bagyaraj, 1987; Warmink and van Elsas, 2008) and that their chemotactic responses are altered depending on the colonization state of plant roots (Sood, 2003). These effects were suggested to be driven by fungal exudates, but no hard evidence is available (Jansa et al., 2013). Hyphal-associated bacteria are known to affect soil aggregation by particle alignment (Caesar-Tonthat, 2002).

For fungi, the potential consequences of species interactions on soil aggregation are clearly underexplored, although they can represent a dominant fraction of soil microbial biomass in organic soil layers (Joergensen and Wichern, 2008). Depending on the biome, mycorrhizal fungi can comprise 30% of microbial and 80% of fungal biomass, respectively (Hogberg and Hogberg, 2002). Few studies investigated the interaction of mycorrhizal fungi with other fungi in terms of ripple-on effects on soil aggregation. In two studies testing the interaction of a mycorrhizal and a nonmycorrhizal Ascomycete for their effect on soil aggregation, a nonadditive increase in macroaggregate stability in the combined compared with the single treatments could be found for EcM but not AM fungi (Caravaca et al., 2004, 2005). Further research is needed to identify the mechanisms behind these soil structure improvements.

Another biological component affected by mycorrhizal fungi are roots; with those they maintain intense chemical "cross-talks" resulting in morphological and physiological changes in both symbiosis partners, indirectly influencing soil aggregation. There is evidence that this cross-talking affects specifically hyphal and root branching patterns. From the EcM fungus *Pisolithus tinctorius* an indole alkaloid (hypaphorine) can be isolated that can reduce lateral root development of the host (Beguiristain and Lapeyrie, 1997; Ditengou et al., 2000). Conversely, the host (*Eucalyptus globulus*) can increase the colony diameter or intensify the branching pattern of this EcM fungus by exuding flavanols and cytokinins, namely rutin and zeatin, respectively (Lagrange et al., 2001; Martin et al., 2001). In arbuscular mycorrhizas, both symbionts can induce intensified branching in their partners; for AM fungi, the mechanism is not known, but for plants this phenomenon is induced via the plant hormone strigolactone (Yano et al., 1996; Paszkowski and Boller, 2002; Paszkowski, 2006; Harris, 2008). Not surprisingly, the cross-talking also affects quantity and quality of exudates such as amino acids and carbohydrates (Leyval and Berthelin, 1993; Azaizeh et al., 1995). Thus the amount and quality of biochemical compounds acting as potential glues could also be affected by this biological mechanism.

Furthermore, mycorrhizal fungi are suggested to indirectly affect soil aggregation via their interplay with fungivores. Although AM fungi are of lower nutritional value compared with other soil fungi, their sheer abundance contributes to an extensive fungal-based energy channel in the soil food web (Holtkamp et al., 2011). Even though fungi are typically assumed to be not predominantly top-down controlled, i.e., controlled in their abundance by consumers, it is quite clear that fungal hyphae are differentially consumed, and that therefore such food web interactions could also have important consequences for mycorrhizal effects on soil aggregation (Rillig and Mummey, 2006). Fungivores can be found among all major soil biota taxa, including microarthropods, bacteria, mites, nematodes, and protists (Hunt et al., 1987; Geisen et al., 2016), where some taxa/species are found to be facultative or even obligate fungal feeders (e.g., Old and Darbyshire, 1978; Okada et al., 2005). There is very little research on the possible consequences of the interplay of fungi and fungal grazers for soil aggregation. Rillig and Mummey (2006) suggested that fungivores could have several possible effects: (1) they could alter fungal community composition by preferential feeding, thus potentially shifting the abundance of fungi with differential impact on soil aggregation; (2) they could induce shifts in fungal traits related to soil aggregation (e.g., mycelial architecture and exudate quantity/quality); and (3) fungal communities could shift fungivore abundance and communities. For microarthropods (e.g., collembola), it was hypothesized that they can positively affect fungal growth and respiration by grazing-induced stimulation at moderate densities of grazers (Fitter and Sanders, 1992; Lussenhop, 1992; Fitter and Garbaye, 1994); such an induced mycelial growth should have positive effect on soil aggregation. In the only two studies available, testing the combined impact of AM fungi and collembola, a nonadditive increase in the combination treatment compared with the two single treatments was found (Helgason and Fitter, 2009; Siddiky et al., 2012a,b). This was likely caused by complementarity in-soil aggregation mechanisms between fungi and collembola, and by the fact that collembola appeared to preferentially graze on saprobic fungal hyphae, preferring them over the AM fungal hyphae in these greenhouse experiments.

14.4 RELATIVE IMPORTANCE OF MYCORRHIZAS

Soil aggregation is a process that is influenced by a range of abiotic factors (e.g., wet–dry cycles, freezing–thawing), site characteristics (e.g., texture), but also by a range of biotic factors, of which mycorrhizas are but one. Studies that aimed at examining and quantifying mycorrhizal effects have often been carried out under circumstances in which their effects are expected to be largest, or in which effects by other players are mostly excluded. This has been and still is necessary in order to establish mechanism and causality. However, from an ecological perspective it is also important to ask about the relative importance of mycorrhizas, either in comparison to effects of other soil biota or site factors or, ideally, the effect of mycorrhizas given all other biotic and abiotic components.

14.4.1 In Relation to Other Biota

Effect sizes mediated by mycorrhizas can be modified by other soil biota, which necessitates interaction studies in a common experiment. Depending on the organism, species interactions are scarcely explicitly covered in experiments and hence our understanding of the relative importance of mycorrhizas on soil aggregation is quite limited.

An omnipresent organism group in experiments focusing on AM fungi is, of course, plant roots; this is because of the obligate status of this mycorrhizal symbiosis. Hence it is not surprising that mycorrhizal effects are typically conceptualized as including roots (e.g., Rillig and Mummey, 2006). As a matter of fact, only a limited number of studies have used root exclusion compartments to disentangle root and fungal contributions to soil aggregation. From such studies, it is known that the combination of root and mycorrhizal hyphae can lead to a nonadditive increase in WSA compared with root- and hyphae-only treatments (Thomas et al., 1993; Andrade et al., 1998). The application of hyphae-only (or hyphosphere) compartments or sterile in vitro growth systems (Rillig et al., 2010) revealed that the soil aggregation ability of mycorrhizal fungi can be comparable to that of roots (Thomas et al., 1993; Andrade et al., 1998).

Microbiota, including nonmycorrhizal fungi (e.g., saprobic fungi) and eubacteria, share the same habitat with mycorrhizal hyphae and influence each other in a number of ways, including competition and alteration of the surrounding soil and its structure (Andrade et al., 1998; Rillig, 2004a; Rillig and Mummey, 2006; Nottingham et al., 2013). Thus it is not straightforward to disentangle experimentally the relative importance of mycorrhizal fungi and associated microbiota, because this necessitates the existence of sterile culture systems. There is so far evidence that the co-occurrence of mycorrhizal fungi and bacteria promotes soil aggregation and that this effect is species specific (Rillig et al., 2005; Caesar-TonThat et al., 2013). Leifheit et al. (2015) went one step further and examined the soil aggregation potential of AM fungi and a natural soil microbial community. They found that both organism groups applied as individual treatments improved water stability of macroaggregates, whereas in the combination treatment a nonadditive increase was detected, i.e., the level of soil aggregation when the two treatments were combined was comparable to that of the single-additions. However, those experimental results are context dependent and many more studies are needed before general conclusions can be reached.

A well-studied organism group affecting soil aggregation are earthworms, whose effects on soil aggregation can mainly be related to disturbance and the production of casts and biopores (Shipitalo and Le Bayon, 2004). Depending on the species, soil conditions, and experimental duration, earthworms can process up to two-thirds of litter-amended artificial soil and convert it into macroaggregates (Ziegler and Zech, 1992), demonstrating their enormous potential impact on soil structure. There are only limited reports available on the combined examination of earthworms and mycorrhizal fungi. Milleret et al. (2009) found no significant change in water stability of macroaggregates for either the individual AM fungi treatment or the combined AM fungi and earthworm treatment. Thus further research is needed to disentangle both biotic components and to evaluate their individual contributions to soil aggregation.

Under natural conditions, mycorrhizal fungi are exposed to fungivores, which themselves affect soil aggregation—to what extent is rather unclear. Maaß et al. (2015) addressed this topic and summarized potential mechanisms by which microarthropods, with a focus on collembola, could affect soil aggregation. Beside production of organic materials (e.g., egg clutches, integuments, necromass, and fecal pellets) their main contribution could be related to the interaction with other soil biota (e.g., grazing on fungal hyphae). Grazing by microarthropods is one of the most researched fungal feeding interactions, but only two studies tested the impact of collembola on soil aggregation in combination with mycorrhizal fungi. Siddiky et al. (2012a) were the first to empirically test the effects of collembola on soil aggregation. They showed that collembola contribute comparably to soil aggregation as AM fungi when used as separated biota treatments; although in the combination treatment a nonadditive increase in WSA could be detected. Because of the nature of the experiment, non-AM fungi were also present in the soil, which was the preferred fungal food source for collembola. Hence further research is needed to evaluate the relative impact of mycorrhizal fungi on soil aggregation when grazed. This is also true for other groups of soil fungivores, including nematodes and protists, for which there are no studies available for consequences on soil aggregation.

14.4.2 Across Different Settings

Gauging the relative effect of mycorrhizal fungi on soil aggregation given all other factors (e.g., biotic and abiotic) is very difficult to achieve experimentally, and therefore research has so far relied on observational approaches. Here, data are typically collected in the field and subsequently used to disentangle mycorrhizal effects through tools such as structural equation modeling (Jastrow et al., 1998). This method is uniquely suited to this question, because it takes into account the hierarchical nature of predictor variables and different hypothesized causal pathways. With such approaches it was possible to reveal the overall importance of mycorrhizal fungal hyphae and roots for soil aggregation over a range of different systems. In both natural ecosystems (arid and mountain grasslands) and managed sites (managed grassland and woodland), roots and extraradical mycelia of AM fungi had a strong impact on aggregate stability (Miller and Jastrow, 1990; Chaudhary et al., 2009; Owen et al., 2009; Li et al., 2015).

Biotic effects, including those of mycorrhizal fungi, have also been examined at larger spatial scales (e.g., Germany), integrating across differences in soil characteristics, e.g., soil texture. Barto et al. (2010) showed in their study, which included 27 different sites across Germany with different land use intensity and site characteristics, that such abiotic factors can override the role of mycorrhizal fungi in soil aggregation. Many more studies are needed to arrive at a thorough understanding of the relative role of mycorrhizas across complex landscapes; of course, results are also prone to be very strongly influenced by the biotic and abiotic factors included in measurements, because only those can be considered in the statistical models.

14.5 AVENUES AND NEEDS FOR FUTURE RESEARCH

The mediation of soil aggregation by mycorrhizas is a critical process, especially given the ever-increasing demands on soils and threats to sustainable management of this resource in agroecosystems. Soil aggregation is clearly an under researched process compared with the more "classical" functions such as nutrient uptake. Therefore it is important to keep in mind clear paths for research, which we outline here.

14.5.1 Mechanisms

There is quite an appalling lack of mechanistic understanding of how mycorrhizal fungi or roots really build up and stabilize (or disintegrate) soil aggregates. We should not be satisfied with "just-so" stories, as appealing as they may be, but strive to replace this with a thorough process understanding. This requires a new approach that employs a combination of a variety of methods. These could include advanced imaging, direct observation in time-lapse videos, or novel soil-on-a-chip approaches (Stanley et al., 2016) coupled with more informed correlational approaches based on measurement of a number of traits in a larger set of fungi (and roots).

A promising approach to make inroads toward mechanistic understanding is to systematically adopt a trait-based approach, as has been proposed by Rillig et al. (2014) and Lehmann and Rillig (2015). A first step is to disentangle soil aggregation process components (Fig. 14.1) and to identify related fungal characteristics influencing them. Rillig et al. (2014) addressed this task for a broad range of potentially important fungal traits, as discussed previously. These potentially relevant traits have to be evaluated for their explanatory value. In order to do so, dedicated experiments enabling the separation of the soil aggregation process components have to be designed. As suggested by Lehmann and Rillig (2015) specific soil aggregate formation and stabilization assays could be utilized to test fungal ability to form new aggregates out of primary particles using soil powder or slurry (Tisdall et al., 1997) and to stabilize (artificially produced) macroaggregates (e.g., Caesar-TonThat and Cochran, 2000). Then correlations of trait and assay-derived data will reveal the important fungal characteristics for each soil aggregation process component.

Rigorously adopting such a trait-based approach for a large number of species of mycorrhizal fungi would be an ambitious task. Like trait-based approaches for plants (Cornelissen et al., 2003; Perez-Harguindeguy et al., 2013), this requires a collaboration of researchers. An important step toward an effective collaboration would be to agree on standardized measurement conditions and methods. This is necessary to ensure comparability of experimental results, because plant and mycorrhizal fungal growth strongly respond to environmental conditions; for example, in nutrient-rich soil patches, fungi show intense hyphal branching, whereas for plant hosts the dependence on their fungal partner can be diminished and carbon (C) translocation can be reduced (Olsson et al., 2002; Grant et al., 2005; Paszkowski, 2006). The vision would be to collate all these pieces of information in public, curated databases, which should also capture variability in the expression of mycorrhizal fungal traits and with it their effect on soil aggregation.

14.5.2 Relative Importance and Greater Coverage of Ecosystem Types

The systematic study of the relative importance of mycorrhizas, given other factors, requires a large research effort, especially if such data are to be integrated in global models.

It should include a wide parameter space, including various ecosystem types, not just the much-researched agricultural systems and grasslands. Variance partitioning and structural equation modeling can then be used to estimate the relative importance of mycorrhizal mycelium in these various settings, and lessons learned from this will be very important for management of mycorrhiza for this specific process.

Another related research focus should also be trying to better understand the interaction of mycorrhizas and other factors, especially abiotic factors such as wet–dry or freeze–thaw cycles. For obvious reasons, most studies have focused on understanding mycorrhizal contributions to soil aggregation in a given abiotic setting, keeping such factors constant. But efforts should move beyond demonstration of the fact that mycorrhizal fungi can aggregate soil, with particular choices of study parameters optimized for detection of such effects; this has been convincingly shown already. We now need to move to a new mode of experimentation that critically challenges our knowledge: under what conditions (biotic and abiotic) does the importance of mycorrhizas fade?

14.5.3 Conceptual Advances: Functions and Ecosystem Engineering

Even though soil aggregation is often thought of as a "function" of mycorrhiza, it is worth pondering whether this is really the case in the narrow sense; i.e., considering that the "proper function of a trait is the effect for which it was selected by natural selection" (Neander, 1991). Even though it is fully appropriate to regard soil aggregation (or its process components) as an ecosystem process probably notably under the influence of mycorrhizas, it seems unlikely that this has been selected for; rather it appears more probable that this is a process that arose as a consequence of functions such as nutrient uptake and foraging. However, this does not mean that there could not be interesting feedbacks between fungi and soil aggregation (e.g., Rillig and Steinberg, 2002), perhaps best viewed within the framework of ecosystem engineering. Work aimed at conceptually clarifying the soil aggregation function of mycorrhizas and other soil biota could therefore be enlightening.

14.5.4 Model Integration

Integration of soil aggregation and mycorrhizas together in process models is a very pressing need. Very few models consider soil aggregation from an organismic perspective (Caruso and Rillig, 2011), and this includes mycorrhizas. On the other hand, models of mycorrhizal function are focused on nutrient delivery and do not consider the soil aggregation effects. This is a large gap, and future effort fueled by the targeted collection of data is required to close it. Success in this endeavor is crucial, because otherwise soil aggregation risks being an overlooked process at larger scales (e.g., global models), at least as mediated by mycorrhizas.

References

Ames, R.N., Reid, C.P.P., Ingham, E.R., 1984. Rhizosphere bacterial population responses to root colonization by a vesicular arbuscular mycorrhizal fungus. New Phytologist 96 (4), 555–563.

Andrade, G., Mihara, K.L., Linderman, R.G., Bethlenfalvay, G.J., 1997. Bacteria from rhizosphere and hyphosphere soils of different arbuscular-mycorrhizal fungi. Plant and Soil 192 (1), 71–79.

Andrade, G., Mihara, K.L., Linderman, R.G., Bethlenfalvay, G.J., 1998. Soil aggregation status and rhizobacteria in the mycorrhizosphere. Plant and Soil 202 (1), 89–96.

Augé, R.M., 2001. Water relations, drought and vesicular-arbuscular mycorrhizal symbiosis. Mycorrhiza 11 (1), 3–42.

Azaizeh, H.A., Marschner, H., Romheld, V., Wittenmayer, L., 1995. Effects of a vesicular-arbuscular mycorrhizal fungus and other soil-microorganisms on growth, mineral nutrient acquisition and root exudation of soil-grown maize plants. Mycorrhiza 5 (5), 321–327.

Bago, B., Azcon-Aguilar, C., Piche, Y., 1998. Architecture and developmental dynamics of the external mycelium of the arbuscular mycorrhizal fungus *Glomus intraradices* grown under monoxenic conditions. Mycologia 90 (1), 52–62.

Barazani, O., Benderoth, M., Groten, K., Kuhlemeier, C., Baldwin, I.T., 2005. *Piriformospora indica* and *Sebacina vermifera* increase growth performance at the expense of herbivore resistance in *Nicotiana attenuata*. Oecologia 146 (2), 234–243.

Barto, E.K., Alt, F., Oelmann, Y., Wilcke, W., Rillig, M.C., 2010. Contributions of biotic and abiotic factors to soil aggregation across a land use gradient. Soil Biology and Biochemistry 42 (12), 2316–2324.

Beguiristain, T., Lapeyrie, F., 1997. Host plant stimulates hypaphorine accumulation in *Pisolithus tinctorius* hyphae during ectomycorrhizal infection while excreted fungal hypaphorine controls root hair development. New Phytologist 136 (3), 525–532.

Berbee, M.L., Taylor, J.W., 2001. Fungal molecular evolution: gene trees and geologic time. In: McLaughlin, D.J., McLaughlin, E.G., Lemke, P.A. (Eds.), The Mycota: Systematics and Evolution, vol. 7B, pp. 229–245.

Bonfante Fasolo, P., Gianinazzi Pearson, V., 1982. Ultrastructural aspects of endomycorrhiza in the Ericaceae. 3. Morphology of the dissociated symbionts and modifications occurring during their reassociation in axenic culture. New Phytologist 91 (4), 691–704.

Caesar-TonThat, T.C., Cochran, V.L., 2000. Soil aggregate stabilization by a saprophytic lignin-decomposing basidiomycete fungus - I. Microbiological aspects. Biology and Fertility of Soils 32 (5), 374–380.

Caesar-TonThat, T.C., Espeland, E., Caesar, A.J., Sainju, U.M., Lartey, R.T., Gaskin, J.F., 2013. Effects of *Agaricus lilaceps* fairy rings on soil aggregation and microbial community structure in relation to growth stimulation of western wheatgrass (*Pascopyrum smithii*) in eastern Montana rangeland. Microbial Ecology 66 (1), 120–131.

Caesar-Tonthat, T.C., 2002. Soil binding properties of mucilage produced by a basidiomycete fungus in a model system. Mycological Research 106, 930–937.

Cairney, J.W.G., Meharg, A.A., 2003. Ericoid mycorrhiza: a partnership that exploits harsh edaphic conditions. European Journal of Soil Science 54 (4), 735–740.

Caravaca, F., Garcia, C., Hernandez, M.T., Roldan, A., 2002. Aggregate stability changes after organic amendment and mycorrhizal inoculation in the afforestation of a semiarid site with *Pinus halepensis*. Applied Soil Ecology 19 (3), 199–208.

Caravaca, F., Alguacil, M.M., Azcon, R., Diaz, G., Roldan, A., 2004. Comparing the effectiveness of mycorrhizal inoculation and amendment with sugar beet, rock phosphate and *Aspergillus niger* to enhance field performance of the leguminous shrub *Dorycnium pentaphyllum* L. Applied Soil Ecology 25 (2), 169–180.

Caravaca, F., Alguacil, M.M., Azcon, R., Parlade, J., Torres, P., Roldan, A., 2005. Establishment of two ectomycorrhizal shrub species in a semiarid site after in situ amendment with sugar beet, rock phosphate, and *Aspergillus niger*. Microbial Ecology 49 (1), 73–82.

Caruso, T., Rillig, M.C., 2011. Direct, positive feedbacks produce instability in models of interrelationships among soil structure, plants and arbuscular mycorrhizal fungi. Soil Biology and Biochemistry 43 (6), 1198–1206.

Chaudhary, V.B., Bowker, M.A., O'Dell, T.E., Grace, J.B., Redman, A.E., Rillig, M.C., Johnson, N.C., 2009. Untangling the biological contributions to soil stability in semiarid shrublands. Ecological Applications 19 (1), 110–122.

Chenu, C., 1989. Influence of a fungal polysaccharide, scleroglucan, on clay microstructures. Soil Biology and Biochemistry 21 (2), 299–305.

Cornelissen, J.H.C., Lavorel, S., Garnier, E., Diaz, S., Buchmann, N., Gurvich, D.E., Reich, P.B., ter Steege, H., Morgan, H.D., van der Heijden, M.G.A., et al., 2003. A handbook of protocols for standardised and easy measurement of plant functional traits worldwide. Australian Journal of Botany 51 (4), 335–380.

Cullings, K.W., 1996. Single phylogenetic origin of ericoid mycorrhizae within the Ericaceae. Canadian Journal of Botany-Revue Canadienne De Botanique 74 (12), 1896–1909.

Daynes, C.N., Zhang, N., Saleeba, J.A., McGee, P.A., 2012. Soil aggregates formed in vitro by saprotrophic Trichocomaceae have transient water-stability. Soil Biology and Biochemistry 48, 151–161.

de la Providencia, I.E., de Souza, F.A., Fernandez, F., Delmas, N.S., Declerck, S., 2005. Arbuscular mycorrhizal fungi reveal distinct patterns of anastomosis formation and hyphal healing mechanisms between different phylogenic groups. New Phytologist 165 (1), 261–271.

de la Providencia, I.E., Fernandez, F., Declerck, S., 2007. Hyphal healing mechanism in the arbuscular mycorrhizal fungi *Scutellospora reticulata* and *Glomus clarum* differs in response to severe physical stress. FEMS Microbiology Letters 268 (1), 120–125.

Ditengou, F.A., Beguiristain, T., Lapeyrie, F., 2000. Root hair elongation is inhibited by hypaphorine, the indole alkaloid from the ectomycorrhizal fungus *Pisolithus tinctorius*, and restored by indole-3-acetic acid. Planta 211 (5), 722–728.

Dodd, J.C., Boddington, C.L., Rodriguez, A., Gonzalez-Chavez, C., Mansur, I., 2000. Mycelium of arbuscular mycorrhizal fungi (AMF) from different genera: form, function and detection. Plant and Soil 226 (2), 131–151.

Dodd, J.C., 1994. Approaches to the study of the extraradical mycelium of arbuscular mycorrhizal fungi. In: Gianinazzi, S., Schüepp, H. (Eds.), Impact of Arbuscular Mycorrhizas on Sustainable Agriculture and Natural Ecosystems. Birkhäuser Verlag, Basel, Switzerland, pp. 147–166.

Egerton-Warburton, L.M., Querejeta, J.I., Allen, M.F., 2007. Common mycorrhizal networks provide a potential pathway for the transfer of hydraulically lifted water between plants. Journal of Experimental Botany 58 (6), 1473–1483.

Enkhtuya, B., Vosatka, M., 2005. Interaction between grass and trees mediated by extraradical mycelium of symbiotic arbuscular mycorrhizal fungi. Symbiosis 38 (3), 261–276.

Field, K.J., Rimington, W.R., Bidartondo, M.I., Allinson, K.E., Beerling, D.J., Cameron, D.D., Duckett, J.G., Leake, J.R., Pressel, S., 2015. First evidence of mutualism between ancient plant lineages (Haplomitriopsida liverworts) and Mucoromycotina fungi and its response to simulated Palaeozoic changes in atmospheric CO_2. New Phytologist 205 (2), 743–756.

Fitter, A.H., Garbaye, J., 1994. Interactions between mycorrhizal fungi and other soil organisms. Plant and Soil 159 (1), 123–132.

Fitter, A.H., Sanders, I.R., 1992. Interactions with the soil fauna. In: Allen, M.F. (Ed.), Mycorrhizal Functioning: An Integrative Plant-Fungal Process. Chapman & Hall, London, UK, pp. 333–354.

Friese, C.F., Allen, M.F., 1991. The spread of VA mycorrhizal fungal hyphae in the soil - Inoculum types and external hyphal architecture. Mycologia 83 (4), 409–418.

Frye, W.W., Ebelhar, S.A., Murdock, L.W., Blevins, R.L., 1982. Soil-erosion on properties and productivity of 2 Kentucky soils. Soil Science Society of America Journal 46 (5), 1051–1055.

Geisen, S., Koller, R., Hünninghaus, M., Dumack, K., Urich, T., Bonkowski, M., 2016. The soil food web revisited: diverse and widespread mycophagous soil protists. Soil Biology and Biochemistry 94, 10–18.

Giovannetti, M., Azzolini, D., Citernesi, A.S., 1999. Anastomosis formation and nuclear and protoplasmic exchange in arbuscular mycorrhizal fungi. Applied and Environmental Microbiology 65 (12), 5571–5575.

Golchin, A., Oades, J.M., Skjemstad, J.O., Clarke, P., 1994. Study of free and occluded particulate organic matter in soils by solid state 13C P/MAS NMR spectroscopy and scanning electron microscopy. Australian Journal of Soil Research 32, 285–309.

Graf, F., Frei, M., 2013. Soil aggregate stability related to soil density, root length, and mycorrhiza using site-specific *Alnus incana* and *Melanogaster variegatus s.l.* Ecological Engineering 57, 314–323.

Grant, C., Bittman, S., Montreal, M., Plenchette, C., Morel, C., 2005. Soil and fertilizer phosphorus: effects on plant P supply and mycorrhizal development. Canadian Journal of Plant Science 85 (1), 3–14.

Griffiths, E., Jones, D., 1965. Microbiological aspects of soil structure - I. Relationships between organic amendments, microbial colonization and changes in aggregate stability. Plant and Soil 23 (1), 17–33.

Gupta, V.V.S.R., Germida, J.J., 1988. Distribution of microbial biomass and its activity in different soil aggregate size classes as affected by cultivation. Soil Biology and Biochemistry 20 (6), 777–786.

Harris, S.D., 2008. Branching of fungal hyphae: regulation, mechanisms and comparison with other branching systems. Mycologia 100 (6), 823–832.

Hart, M.M., Reader, R.J., 2002. Taxonomic basis for variation in the colonization strategy of arbuscular mycorrhizal fungi. New Phytologist 153 (2), 335–344.

Helgason, T., Fitter, A.H., 2009. Natural selection and the evolutionary ecology of the arbuscular mycorrhizal fungi (*Phylum Glomeromycota*). Journal of Experimental Botany 60 (9), 2465–2480.

Hogberg, M.N., Hogberg, P., 2002. Extramatrical ectomycorrhizal mycelium contributes one-third of microbial biomass and produces, together with associated roots, half the dissolved organic carbon in a forest soil. New Phytologist 154 (3), 791–795.

Holtkamp, R., van der Wal, A., Kardol, P., van der Putten, W.H., de Ruiter, P.C., Dekker, S.C., 2011. Modelling C and N mineralisation in soil food webs during secondary succession on ex-arable land. Soil Biology and Biochemistry 43 (2), 251–260.

Hooker, J.E., Piatti, P., Cheshire, M.V., Watson, C.A., 2007. Polysaccharides and monosaccharides in the hyphosphere of the arbuscular mycorrhizal fungi Glomus E3 and *Glomus tenue*. Soil Biology and Biochemistry 39 (2), 680–683.

Hunt, H.W., Coleman, D.C., Ingham, E.R., Ingham, R.E., Elliott, E.T., Moore, J.C., Rose, S.L., Reid, C.P.P., Morley, C.R., 1987. The detrital food web in a shortgrass prairie. Biology and Fertility of Soils 3 (1–2), 57–68.

Jansa, J., Bukovska, P., Gryndler, M., 2013. Mycorrhizal hyphae as ecological niche for highly specialized hypersymbionts - or just soil free-riders? Frontiers in Plant Science 4.

Jastrow, J.D., Miller, R.M., Lussenhop, J., 1998. Contributions of interacting biological mechanisms to soil aggregate stabilization in restored prairie. Soil Biology and Biochemistry 30 (7), 905–916.

Jastrow, J.D., Amonette, J.E., Bailey, V.L., 2007. Mechanisms controlling soil carbon turnover and their potential application for enhancing carbon sequestration. Climatic Change 80 (1–2), 5–23.

Joergensen, R.G., Wichern, F., 2008. Quantitative assessment of the fungal contribution to microbial tissue in soil. Soil Biology and Biochemistry 40 (12), 2977–2991.

Johnson, N.C., Graham, J.H., Smith, F.A., 1997. Functioning of mycorrhizal associations along the mutualism-parasitism continuum. New Phytologist 135 (4), 575–586.

Johnson, N.C., Rowland, D.L., Corkidi, L., Egerton-Warburton, L.M., Allen, E.B., 2003. Nitrogen enrichment alters mycorrhizal allocation at five mesic to semiarid grasslands. Ecology 84 (7), 1895–1908.

Kemper, W.D., Rosenau, R.C., 1986. Aggregate stability and size distribution. In: Lute, A. (Ed.), Methods of Soil Analysis. Part I - Physical and Mineralogical Methods. SSSA, Madison, USA, pp. 425–443.

Klironomos, J.N., Allen, M.F., Rillig, M.C., Piotrowski, J., Makvandi-Nejad, S., Wolfe, B.E., Powell, J.R., 2005. Abrupt rise in atmospheric CO_2 overestimates community response in a model plant-soil system. Nature 433 (7026), 621–624.

Lagrange, H., Jay-Allgmand, C., Lapeyrie, F., 2001. Rutin, the phenolglycoside from eucalyptus root exudates, stimulates *Pisolithus* hyphal growth at picomolar concentration. New Phytologist 149 (2), 349–355.

Lehmann, A., Rillig, M.C., 2015. Understanding mechanisms of soil biota involvement in soil aggregation: a way forward with saprobic fungi? Soil Biology and Biochemistry 88, 298–302.

Leifheit, E.F., Veresoglou, S.D., Lehmann, A., Morris, E.K., Rillig, M.C., 2014. Multiple factors influence the role of arbuscular mycorrhizal fungi in soil aggregation–a meta-analysis. Plant and Soil 374 (1–2), 523–537.

Leifheit, E.F., Verbruggen, E., Rillig, M.C., 2015. Arbuscular mycorrhizal fungi reduce decomposition of woody plant litter while increasing soil aggregation. Soil Biology and Biochemistry 81, 323–328.

LePage, B.A., Currah, R.S., Stockey, R.A., Rothwell, G.W., 1997. Fossil ectomycorrhizae from the middle Eocene. American Journal of Botany 84 (3), 410–412.

LePage, B.A., 2003. The evolution, biogeography and palaeoecology of the Pinaceae based on fossil and extant representatives. Acta Horticulturae 29–52.

Leyval, C., Berthelin, J., 1993. Rhizodeposition and net release of soluble organic-compounds by pine and beech seedlings inoculated with rhizobacteria and ectomycorrhizal fungi. Biology and Fertility of Soils 15 (4), 259–267.

Li, Z.J., Shukla, V., Wenger, K., Fordyce, A., Pedersen, A.G., Marten, M., 2002. Estimation of hyphal tensile strength in production-scale *Aspergillus oryzae* fungal fermentations. Biotechnology and Bioengineering 77 (6), 601–613.

Li, X., Zhang, J., Gai, J., Cai, X., Christie, P., Li, X., 2015. Contribution of arbuscular mycorrhizal fungi of sedges to soil aggregation along an altitudinal alpine grassland gradient on the Tibetan Plateau. Environmental Microbiology 17, 2841–2857.

Linder, M.B., Szilvay, G.R., Nakari-Setala, T., Penttila, M.E., 2005. Hydrophobins: the protein-amphiphiles of filamentous fungi. FEMS Microbiology Reviews 29 (5), 877–896.

Lussenhop, J., 1992. Mechanisms of microarthropod microbial interactions in soil. Advances in Ecological Research 23, 1–33.

Maaß, S., Caruso, T., Rillig, M.C., 2015. Functional role of microarthropods in soil aggregation. Pedobiologia 58 (2–3), 59–63.

Marjanovic, Z., Nehls, U., 2008. Ectomycorrhiza and water transport. In: Varma, A. (Ed.), Mycorrhiza - State of the Art, Genetics and Molecular Biology, Eco-Function, Biotechnology, Eco-Physiology, Structure and Systematics. Springer Verlag, Berlin, Germany.

Martin, J.P., Martin, W.P., Page, J.B., Raney, W.A., de Ment, J.D., 1955. Soil Aggregation - Advances in Agronomy, vol. 7. Academic Press, New York, pp. 1–37.

Martin, F., Duplessis, S., Ditengou, F., Lagrange, H., Voiblet, C., Lapeyrie, F., 2001. Developmental cross talking in the ectomycorrhizal symbiosis: signals and communication genes. New Phytologist 151 (1), 145–154.

McLean, K.M., Prosser, J.I., 1987. Development of vegetative mycelium during colony growth of *Neurospora crassa*. Transactions of the British Mycological Society 88, 489–495.

Meadows, A., Meadows, P.S., Wood, D.M., Murray, J.M.H., 1994. Microbiological effects on slope stability - an experimental analysis. Sedimentology 41 (3), 423–435.

Meyer, J.R., Linderman, R.G., 1986. Selective influence on populations of rhizosphere or rhizoplane bacteria and actinomycetes by mycorrhizas formed by *Glomus fasciculatum*. Soil Biology and Biochemistry 18 (2), 191–196.

Miller, R.M., Jastrow, J.D., 1990. Hierarchy of root and mycorrhizal fungal interactions with soil aggregation. Soil Biology and Biochemistry 22 (5), 579–584.

Milleret, R., Le Bayon, R.-C., Gobat, J.-M., 2009. Root, mycorrhiza and earthworm interactions: their effects on soil structuring processes, plant and soil nutrient concentration and plant biomass. Plant and Soil 316 (1–2), 1–12.

Mokma, D.L., Sietz, M.A., 1992. Effects of soil-erosion on corn yields on Marlette soils in south-central Michigan. Journal of Soil and Water Conservation 47 (4), 325–327.

Money, N.P., 1994. Osmotic adjustment and the role of turgor in mycelial fungi. In: Wessels, J.G.H., Meinhardt, F. (Eds.), The Mycota I - Growth, Differentiation and Sexuality. Springer Verlag, Berlin, Germany, pp. 67–87.

Neander, K., 1991. Functions as selected effects: the conceptual analyst's defense. Philosophy of Science 58 (2), 168–184.

Nottingham, A.T., Turner, B.L., Winter, K., Chamberlain, P.M., Stott, A., Tanner, E.V.J., 2013. Root and arbuscular mycorrhizal mycelial interactions with soil microorganisms in lowland tropical forest. FEMS Microbiology Ecology 85 (1), 37–50.

Oades, J.M., 1984. Soil organic-matter and structural stability - mechanisms and implications for management. Plant and Soil 76 (1–3), 319–337.

Oades, J.M., 1993. The role of biology in the formation, stabilization and degradation of soil structure. Geoderma 56 (1–4), 377–400.

Okada, H., Harada, H., Kadota, I., 2005. Fungal-feeding habits of six nematode isolates in the genus *Filenchus*. Soil Biology and Biochemistry 37 (6), 1113–1120.

Old, K.M., Darbyshire, J.F., 1978. Soil fungi as food for giant amoebae. Soil Biology and Biochemistry 10 (2), 93–100.

Olsson, P.A., Johnson, N.C., 2005. Tracking carbon from the atmosphere to the rhizosphere. Ecology Letters 8 (12), 1264–1270.

Olsson, P.A., van Aarle, I.M., Allaway, W.G., Ashford, A.E., Rouhier, H., 2002. Phosphorus effects on metabolic processes in monoxenic arbuscular mycorrhiza cultures. Plant Physiology 130 (3), 1162–1171.

Owen, S.M., Sieg, C.H., Gehring, C.A., Bowker, M.A., 2009. Above- and belowground responses to tree thinning depend on the treatment of tree debris. Forest Ecology and Management 259 (1), 71–80.

Parniske, M., 2008. Arbuscular mycorrhiza: the mother of plant root endosymbioses. Nature Reviews Microbiology 6 (10), 763–775.

Paszkowski, U., Boller, T., 2002. The growth defect of Irt1, a maize mutant lacking lateral roots, can be complemented by symbiotic fungi or high phosphate nutrition. Planta 214 (4), 584–590.

Paszkowski, U., 2006. A journey through signaling in arbuscular mycorrhizal symbioses 2006. New Phytologist 172 (1), 35–46.

Perez-Harguindeguy, N., Diaz, S., Garnier, E., Lavorel, S., Poorter, H., Jaureguiberry, P., Bret-Harte, M.S., Cornwell, W.K., Craine, J.M., Gurvich, D.E., et al., 2013. New handbook for standardised measurement of plant functional traits worldwide. Australian Journal of Botany 61 (3), 167–234.

Piccolo, A., Mbagwu, J.S.C., 1999. Role of hydrophobic components of soil organic matter in soil aggregate stability. Soil Science Society of America Journal 63 (6), 1801–1810.

Pietikainen, A., Mikola, J., Vestberg, M., Setala, H., 2009. Defoliation effects on *Plantago lanceolata* resource allocation and soil decomposers in relation to AM symbiosis and fertilization. Soil Biology and Biochemistry 41 (11), 2328–2335.

Pimentel, D., Harvey, C., Resosudarmo, P., Sinclair, K., Kurz, D., McNair, M., Crist, S., Shpritz, L., Fitton, L., Saffouri, R., et al., 1995. Environment and economic costs of soil erosion and conservation benefits. Science 267 (5201), 1117–1123.

Piotrowski, J.S., Denich, T., Klironomos, J.N., Graham, J.M., Rillig, M.C., 2004. The effects of arbuscular mycorrhizas on soil aggregation depend on the interaction between plant and fungal species. New Phytologist 164 (2), 365–373.

Querejeta, J.I., Egerton-Warburton, L.M., Allen, M.F., 2003. Direct nocturnal water transfer from oaks to their mycorrhizal symbionts during severe soil drying. Oecologia 134 (1), 55–64.

Read, D.J., Perez-Moreno, J., 2003. Mycorrhizas and nutrient cycling in ecosystems - a journey towards relevance? New Phytologist 157 (3), 475–492.
Read, D.J., Leake, J.R., Perez-Moreno, J., 2004. Mycorrhizal fungi as drivers of ecosystem processes in heathland and boreal forest biomes. Canadian Journal of Botany-Revue Canadienne De Botanique 82 (8), 1243–1263.
Redecker, D., Raab, P., 2006. Phylogeny of the Glomeromycota (arbuscular mycorrhizal fungi): recent developments and new gene markers. Mycologia 98 (6), 885–895.
Redecker, D., Kodner, R., Graham, L.E., 2000. Glomalean fungi from the Ordovician. Science 289 (5486), 1920–1921.
Rillig, M.C., Mummey, D.L., 2006. Mycorrhizas and soil structure. New Phytologist 171 (1), 41–53.
Rillig, M.C., Steinberg, P.D., 2002. Glomalin production by an arbuscular mycorrhizal fungus: a mechanism of habitat modification? Soil Biology and Biochemistry 34 (9), 1371–1374.
Rillig, M.C., Lutgen, E.R., Ramsey, P.W., Klironomos, J.N., Gannon, J.E., 2005. Microbiota accompanying different arbuscular mycorrhizal fungal isolates influence soil aggregation. Pedobiologia 49 (3), 251–259.
Rillig, M.C., Caldwell, B.A., Wosten, H.A.B., Sollins, P., 2007. Role of proteins in soil carbon and nitrogen storage: controls on persistence. Biogeochemistry 85 (1), 25–44.
Rillig, M.C., Mardatin, N.F., Leifheit, E.F., Antunes, P.M., 2010. Mycelium of arbuscular mycorrhizal fungi increases soil water repellency and is sufficient to maintain water-stable soil aggregates. Soil Biology and Biochemistry 42 (7), 1189–1191.
Rillig, M.C., Aguilar-Trigueros, C.A., Bergmann, J., Verbruggen, E., Veresoglou, S.D., Lehmann, A., 2014. Plant root and mycorrhizal fungal traits for understanding soil aggregation. New Phytologist 205 (4), 1385–1388.
Rillig, M.C., 2004a. Arbuscular mycorrhizae and terrestrial ecosystem processes. Ecology Letters 7 (8), 740–754.
Rillig, M.C., 2004b. Arbuscular mycorrhizae, glomalin, and soil aggregation. Canadian Journal of Soil Science 84 (4), 355–363.
Rosier, C.L., Hoye, A.T., Rillig, M.C., 2006. Glomalin-related soil protein: assessment of current detection and quantification tools. Soil Biology and Biochemistry 38 (8), 2205–2211.
Sbrana, C., Nuti, M.P., Giovannetti, M., 2007. Self-anastomosing ability and vegetative incompatibility of *Tuber borchii* isolates. Mycorrhiza 17 (8), 667–675.
Schantz, E.C., Piemeisel, F.J., 1917. Fungus fairy rings in Eastern Colorado and their effect on vegetation. Journal of Agricultural Research 11, 191–245.
Scheffer, F., Schachtschabel, P., 2002. Bodenentwicklung, Bodensystematik und Bodenverarbeitung. Lehrbuch der Bodenkunde. Spektrum Akademischer Verlag, Berlin, Germany.
Scheublin, T.R., Sanders, I.R., Keel, C., van der Meer, J.R., 2010. Characterisation of microbial communities colonising the hyphal surfaces of arbuscular mycorrhizal fungi. ISME Journal 4 (6), 752–763.
Schreiner, R.P., Bethlenfalvay, G.J., 1997. Plant and soil response to single and mixed species of arbuscular mycorrhizal fungi under fungicide stress. Applied Soil Ecology 7 (1), 93–102.
Schussler, A., Schwarzott, D., Walker, C., 2001. A new fungal phylum, the Glomeromycota: phylogeny and evolution. Mycological Research 105, 1413–1421.
Secilia, J., Bagyaraj, D.J., 1987. Bacteria and actinomycetes associated with pot cultures of vesicular arbuscular mycorrhizas. Canadian Journal of Microbiology 33 (12), 1069–1073.
Selosse, M.A., Bauer, R., Moyersoen, B., 2002. Basal hymenomycetes belonging to the Sebacinaceae are ectomycorrhizal on temperate deciduous trees. New Phytologist 155 (1), 183–195.
Selosse, M.-A., Setaro, S., Glatard, F., Richard, F., Urcelay, C., Weiss, M., 2007. Sebacinales are common mycorrhizal associates of Ericaceae. New Phytologist 174 (4), 864–878.
Shipitalo, M.J., Le Bayon, R.C., 2004. Quantifying the effects of earthworms on soil aggregation and porosity. In: Edwards, C.A. (Ed.), Earthworm Ecology. CRC Press, USA, pp. 183–194.
Siddiky, M.R.K., Kohler, J., Cosme, M., Rillig, M.C., 2012a. Soil biota effects on soil structure: interactions between arbuscular mycorrhizal fungal mycelium and collembola. Soil Biology and Biochemistry 50, 33–39.
Siddiky, M.R.K., Schaller, J., Caruso, T., Rillig, M.C., 2012b. Arbuscular mycorrhizal fungi and collembola non-additively increase soil aggregation. Soil Biology and Biochemistry 47, 93–99.
Six, J., Bossuyt, H., Degryze, S., Denef, K., 2004. A history of research on the link between (micro)aggregates, soil biota, and soil organic matter dynamics. Soil and Tillage Research 79 (1), 7–31.
Smith, S.E., Read, D., 2008. Mycorhizal Symbiosis. Academic Press, USA, pp. 189–387.
Smith, F.A., Smith, S.E., 2011. What is the significance of the arbuscular mycorrhizal colonisation of many economically important crop plants? Plant and Soil 348 (1–2), 63–79.

Sood, S.G., 2003. Chemotactic response of plant-growth-promoting bacteria towards roots of vesicular-arbuscular mycorrhizal tomato plants. FEMS Microbiology Ecology 45 (3), 219–227.
Staddon, P.L., Ramsey, C.B., Ostle, N., Ineson, P., Fitter, A.H., 2003. Rapid turnover of hyphae of mycorrhizal fungi determined by AMS microanalysis of C-14. Science 300 (5622), 1138–1140.
Stanley, C.E., Grossmann, G., Casadevalli Solvas, X., de Mello, A.J., 2016. Soil-on-a-chip: microfluidic platforms for environmental organismal studies. Lab on a Chip 16 (2), 228–241.
Sullivan, L.A., 1990. Soil organic-matter, air encapsulation and water-stable aggregation. Journal of Soil Science 41 (3), 529–534.
Sutton, J.C., Sheppard, B.R., 1976. Aggregation of sand-dune soil by endomycorrhizal fungi. Canadian Journal of Botany-Revue Canadienne De Botanique 54 (3–4), 326–333.
Talbot, J.M., Allison, S.D., Treseder, K.K., 2008. Decomposers in disguise: mycorrhizal fungi as regulators of soil C dynamics in ecosystems under global change. Functional Ecology 22 (6), 955–963.
Taylor, A.F.S., Alexander, I., 2005. The ectomycorrhizal symbiosis: life in the real world. Mycologist 19 (3), 102–112.
Tedersoo, L., May, T.W., Smith, M.E., 2010. Ectomycorrhizal lifestyle in fungi: global diversity, distribution, and evolution of phylogenetic lineages. Mycorrhiza 20 (4), 217–263.
Thomas, R.S., Franson, R.L., Bethlenfalvay, G.J., 1993. Separation of vesicular-arbuscular mycorrhizal fungus and root effects on soil aggregation. Soil Science Society of America Journal 57 (1), 77–81.
Thornton, R.H., Cowie, J.D., McDonald, D.C., 1956. Mycelial aggregation of sand under *Pinus radiata*. Nature 177 (4501), 231–232.
Tisdall, J.M., Oades, J.M., 1979. Stabilization of soil aggregates by the root systems of ryegrass. Australian Journal of Soil Research 17 (3), 429–441.
Tisdall, J.M., Oades, J.M., 1980. The effect of crop-rotation on aggregation in a red-brown earth. Australian Journal of Soil Research 18 (4), 423–433.
Tisdall, J.M., Oades, J.M., 1982. Organic-matter and water-stable aggregates in soils. Journal of Soil Science 33 (2), 141–163.
Tisdall, J.M., Smith, S.E., Rengasamy, P., 1997. Aggregation of soil by fungal hyphae. Australian Journal of Soil Research 35 (1), 55–60.
Tisdall, J.M., Nelson, S.E., Wilkinson, K.G., Smith, S.E., McKenzie, B.M., 2012. Stabilisation of soil against wind erosion by six saprotrophic fungi. Soil Biology and Biochemistry 50, 134–141.
Tisdall, J.M., 1994. Possible role of soil - microorganisms in aggregation in soils. Plant and Soil 159 (1), 115–121.
Toljander, J.F., Artursson, V., Paul, L.R., Jansson, J.K., Finlay, R.D., 2006. Attachment of different soil bacteria to arbuscular mycorrhizal fungal extraradical hyphae is determined by hyphal vitality and fungal species. FEMS Microbiology Letters 254 (1), 34–40.
Treseder, K.K., Allen, M.F., Ruess, R.W., Pregitzer, K.S., Hendrick, R.L., 2005. Lifespans of fungal rhizomorphs under nitrogen fertilization in a pinyon-juniper woodland. Plant and Soil 270 (1–2), 249–255.
Trinci, A.P.J., Collinge, A.J., December 1975. Hyphal wall growth in *Neurospora crassa* and *Geotrichum candidum*. Journal of General Microbiology 91, 355–361.
Veresoglou, S.D., Halley, J.M., Rillig, M.C., 2015. Extinction risk of soil biota. Nature Communications 6.
Wang, B., Qiu, Y.L., 2006. Phylogenetic distribution and evolution of mycorrhizas in land plants. Mycorrhiza 16 (5), 299–363.
Warmink, J.A., van Elsas, J.D., 2008. Selection of bacterial populations in the mycosphere of *Laccaria proxima*: is type III secretion involved? ISME Journal 2 (8), 887–900.
Weiss, M., Sykorova, Z., Garnica, S., Riess, K., Martos, F., Krause, C., Oberwinkler, F., Bauer, R., Redecker, D., 2011. Sebacinales everywhere: previously overlooked ubiquitous fungal endophytes. PLos One 6 (2), e16793.
Wessels, J.G., 1997. Hydrophobins: proteins that change the nature of the fungal surface. Advances in Microbial Physiology 38, 1–45.
Wilson, G.W.T., Rice, C.W., Rillig, M.C., Springer, A., Hartnett, D.C., 2009. Soil aggregation and carbon sequestration are tightly correlated with the abundance of arbuscular mycorrhizal fungi: results from long-term field experiments. Ecology Letters 12 (5), 452–461.
Wosten, H.A.B., 2001. Hydrophobins: multipurpose proteins. Annual Review of Microbiology 55, 625–646.
Wu, B., Maruyama, H., Teramoto, M., Hogetsu, T., 2012. Structural and functional interactions between extraradical mycelia of ectomycorrhizal *Pisolithus* isolates. New Phytologist 194 (4), 1070–1078.
Yano, K., Yamauchi, A., Kono, Y., 1996. Localized alteration in lateral root development in roots colonized by an arbuscular mycorrhizal fungus. Mycorrhiza 6 (5), 409–415.

York, C.A., Canaway, P.M., 2000. Water repellent soils as they occur on UK golf greens. Journal of Hydrology 231, 126–133.

Zheng, W., Morris, E.K., Rillig, M.C., 2014. Ectomycorrhizal fungi in association with *Pinus sylvestris* seedlings promote soil aggregation and soil water repellency. Soil Biology and Biochemistry 78, 326–331.

Ziegler, F., Zech, W., 1992. Formation of water-stable aggregates through the action of earthworms - implications from laboratory experiments. Pedobiologia 36 (2), 91–96.

CHAPTER

15

Arbuscular Mycorrhizal Fungi and Soil Salinity

M. Miransari
AbtinBerkeh Scientific Ltd. Company, Isfahan, Iran

15.1 INTRODUCTION

Soil salinity is among the most important stresses worldwide decreasing plant growth and agricultural production. The amount of arable land in the world is equal to 1.5 billion hectares, of which 77 million hectares are unsuitable for crop growth because of high salinity. The amount of land occupied by saline soils is increasing in many parts of the world, especially those under arid and semiarid conditions where precipitation is low and temperatures and rates of evaporation are high. Arid and semiarid lands cover more than 7% of the total land on the earth (Selvakumar et al., 2014; Miransari, 2016).

The presence of salt reduces plant growth as a result of its adverse effects on plant morphology and physiology. Under salinity stress the enhanced uptake of sodium (Na^+) and chloride (Cl^-) ions negatively affects plant growth, which is mainly the result of: (1) decreased uptake of essential nutrients such as potassium by the plant; (2) decreased water use efficiency; (3) toxic effects of Na^+ and Cl^- on plant morphological traits such as root system size, (4) toxic effects of Na^+ and Cl^- on plant physiology including the activity of enzymes, the function of cell membranes, and the production of plant hormones (Alam, 1994; Munns and James, 2003); and (5) increased oxidative stress caused by high Na^+ and Cl^- levels. Oxidative stress results from the production of reactive oxygen species (ROS) including OH (hydroxyl radical), hydroxide ion (OH^-), oxygen ions ($O_2^{\,-}$), and hydrogen peroxide (H_2O_2) (Munns, 2002; Dolatabadian et al., 2011; Golpayegani and Tilebeni, 2011; Bothe, 2012).

Plants use different mechanisms to alleviate the stress of salinity, with the most important one considered to be the adjustment of osmotic potential through the production of solutes in plant cells. Plants can accomplish this osmotic adjustment by the exclusion of Na^+ and $Cl^-$from plant leaves, and/or by increasing the concentration of Na^+ and Cl^- in plant cells to balance the soil salt concentration. However, these mechanisms must be strictly regulated in plant cells (Liu and Baird, 2003; Munns and Gilliham, 2015). Na^+ and Cl^- ions must

Mycorrhizal Mediation of Soil
http://dx.doi.org/10.1016/B978-0-12-804312-7.00015-2

be allocated to vacuoles to keep the concentration of Na^+ and Cl^- less than toxic levels; the increased concentration of potassium (K^+) and organic solutes follows to regulate the cellular ionic concentration (Munns and Tester, 2008; Conn and Gilliham, 2010; Flowers et al., 2015). Plants, especially the salt-tolerant species, also can alleviate salt stress by altering their leaf growth and angle, increasing root growth to access deeper sources of water, producing osmolytes and activating antioxidants and various stress genes (Munns, 2002; Liu and Baird, 2003; Miransari et al., 2015). The mechanisms used by plants at the tissue and cellular levels to tolerate salt stress are presented in Fig. 15.1.

More specifically, the following molecular processes can allow plants to tolerate salt stress: (1) transporters of Na^+/hydrogen (H^+) result in the localization of Na^+ from the cytoplasm to the vacuole or across the plasmalemma, and hence outside the cells, for the exchange of H^+ and overexpression of such transporters, allowing the plant to tolerate the stress (Xia et al., 2002; Jung et al., 2013); (2) the transport of water and small molecules such as glycerol, amino acids, urea, ions, and peptides by the aquaporin channels (Uehlein et al., 2007); (3) the modification of enzymatic activities, regulating the production of osmolytes; (4) the activation of

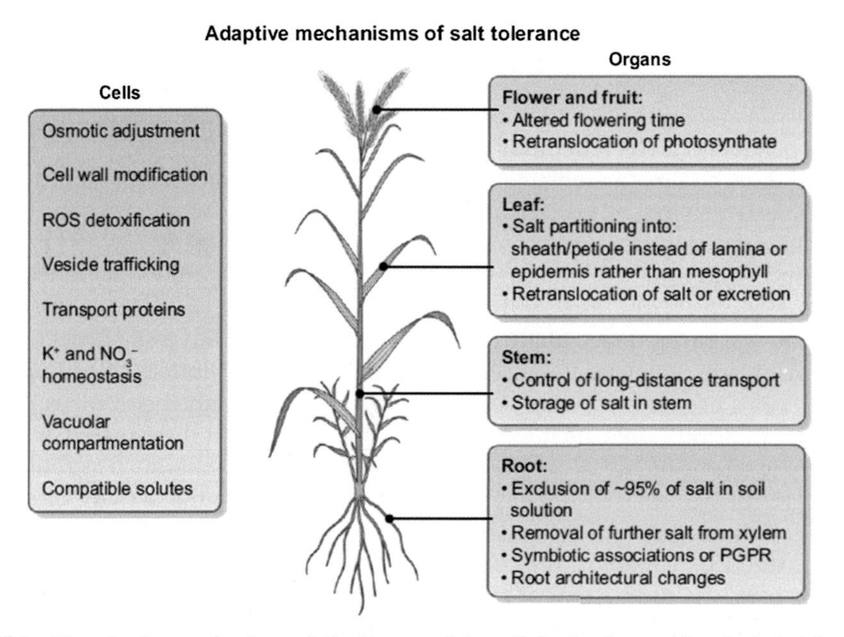

FIGURE 15.1 The adaptive mechanisms, at the tissue and the cellular levels, used by plants to tolerate salinity stress. At the tissue level the following mechanisms are used by plants: (1) altered time of different growth stages, (2) retranslocation of photosynthate, (3) salt partitioning and excretion to different plant parts such as leaf and stem or to the soil, (4) morphological alteration, and (5) symbiotic association with mycorrhizal fungi and rhizobium or nonsymbiotic association with PGPR. At the cellular level, the plant uses the following mechanisms to tolerate salinity stress: (1) modification of the cell wall, (2) scavenging of ROS, (3) adjustment of cellular osmotic potential, (4) protein functioning, (5) nutrient homeostasis such as K^+ and NO_3^-, and (6) salt compartmentation into vacuoles. *K^+*, potassium; *NO_3^-*, nitrate; *PGPR*, plant growth–promoting rhizobacteria; *ROS*, reactive oxygen species. *Munns, R., Gilliham, M., 2015. Salinity tolerance of crops – what is the cost? New Phytologist 208, 668–673, with kind permission from Wiley, License number: 3814870588841.*

genes, which control the production of antioxidants for scavenging ROS, and (5) altered root hydraulic conductivity and mycorrhizal colonization (Augé, 2001).

Several methods have been used to alleviate the effects of salinity on plant growth, including leaching the soil to reduce salt concentration, using salt-tolerant plant genotypes, and using soil microbes including arbuscular mycorrhizal (AM) fungi and plant growth–promoting rhizobacteria (PGPR). Soil microbes can allow host plants to tolerate the stress of high salt soils, producing greater crop yields (Egamberdieva and Lugtenberg, 2014). In this chapter, I will review studies exploring whether and how mycorrhizal fungi and PGPR can improve salt tolerance in their host plants. I focus on AM fungi because of their importance to many crop plants experiencing increased salinity in locations around the world. However, more has yet to be investigated about the mechanisms by which mycorrhizal fungi affect the growth of plants under salt stress. Accordingly, the main focus of this chapter is on the use of the important soil microbes, AM fungi, for the alleviation of salinity stress.

15.2 ARBUSCULAR MYCORRHIZAL FUNGI AND SALT STRESS

The importance of AM fungi for the alleviation of salt stress and promotion of plant growth and agricultural yield has been demonstrated by a large body of research. AM fungi have the potential to alleviate salt stress because of their ability to alter root morphology and expand the area scavenged by plants for nutrients through an extensive hyphal network. AM fungi also can increase the uptake of water and nutrients by the host plant, alleviating both abiotic and biotic stresses. AM fungi also produce the glycoprotein glomalin, which improves the structure and properties of soil. These fungi also alter plant physiology, increasing plant growth and tissue nutrients, allowing decreases in the use of chemical fertilizer (Gianinazzi et al., 2010; Estrada et al., 2013a,b,c).

Mycorrhizal fungi are able to enhance host plant tolerance of salinity by: (1) increasing the uptake of water and nutrients from the soil through an extensive hyphal network (Gopal et al., 2012; Smith et al., 2011; Bharti et al., 2014), (2) altering plant morphology and physiology in ways that allow the host plant to handle stress more efficiently (Miransari et al., 2008), (3) producing plant hormones, and (4) interacting with other soil microbes including PGPR and rhizobia (Miransari, 2010; Kang et al., 2014; Liu et al., 2015; Mardukhi et al., 2015). More details about these mechanisms are presented in the following paragraphs through a selection of case studies.

Bothe (2012) reviewed studies of AM fungal alleviation of salt stress from an ecological perspective, finding that whereas germination of AM fungal spores and the growth of their hyphae were adversely affected by salinity stress, the roots of salt-tolerant plants tended to be highly colonized by AM fungi. Spore communities of several saline soils were dominated by *Funneliformis geosporum* (formerly *Glomus geosporum*), whereas the roots of salt-tolerant plant species were commonly colonized by *Rhizophagus irregularis* (formerly *Glomus intraradices*).

Several studies of agricultural crops have shown that plant genotype and AM fungal species influence the ability of plants to tolerate high salinity. Daei et al. (2009) investigated the effects of different species of AM fungi on the growth and yield of several wheat genotypes under field saline conditions using salty water. The genotypes of wheat and the species of AM fungi responded differently to stress; *Claroideogloum etunicatum* (formerly *Glomus etunicatum*) was the most efficient fungal species, followed by *Funneliformis mosseae* and *R. irregularis*. The authors concluded that it is possible to alleviate the adverse effects of salinity on wheat growth under field saline conditions, especially if the right combination of wheat genotype and fungal species is used.

In a similar study, Mardukhi et al. (2015) investigated the effects of different species of AM fungi on the nutrient uptake of different wheat genotypes under salt stress, again finding that host genotype and AM fungal species both influence response to salinity. The AM fungal species studied were *F. mosseae*, *C. etunicatum*, and *R. irregularis* at salinity levels of 4, 8, and 12 dS/m. Although different species of AM fungi were able to alleviate the stress of salinity on the growth of wheat genotypes, the mixed inoculum treatment was the most effective, significantly increasing wheat growth as a result of increased rates of nutrient uptake. Among the wheat genotypes, genotype line 9 absorbed higher rates of nutrients in association with AM fungi, whereas genotype Chamran was able to alleviate salt stress by decreasing the uptake of Na^+ and Cl^- ions.

Similarly, Hajiboland et al. (2010) evaluated the effects of mycorrhization on the growth of tomato (*Solanum lycopersicum* L.) plants under salinity stress using salt-sensitive and salt-tolerant genotypes. The salinity treatments of control, low [electrical conductivity (EC) = 0.63 dS/m], medium (EC = 5 dS/m), and high (EC = 10 dS/m) levels of EC, and the mycorrhizal species *R. irregularis* used for the experiment. The stimulating effect of mycorrhizal fungi was higher in the salt-tolerant tomato genotype than the salt-sensitive genotype. The fungi increased the uptake of phosphorus (P), calcium (Ca), and K, enhanced the ratios of Ca/Na and K/Na, and also promoted carbon assimilation by increasing stomatal conductance and influencing photosystem II. There was also a higher activity of enzymes scavenging ROS, thereby reducing the rate of peroxide production. The differences in the tolerance of the genotypes under salinity stress were related to their ion balance, leaf gas exchange, and phytochemistry.

The alleviating effects of mycorrhizal fungi on the growth of tomato (*S. lycopersicum* L.) under salinity stress were also investigated by Huang et al. (2010), using a pot culture. Although salt stress significantly decreased plant growth, the fungi were able to alleviate the stress by increasing plant dry matter and leaf area. However, the effects of fungi were more pronounced on the dry matter of the shoots than the roots. Mycorrhization increased the activity of several antioxidants including superoxide dismutase (SOD), peroxidase (POD), ascorbate POD (ASA-POD), and catalase (Cat), especially in the early stages of the salt treatments. Both Hajiboland et al. (2010) and Huang et al. (2010) concluded that AM fungi were able to alleviate the oxidative stress resulting from high salinity by increasing the production of antioxidants and scavenging ROS.

In a similar study, Al-Karaki (2000) examined the effects of *F. mosseae* on the uptake of nutrients by tomato plants under greenhouse conditions using salty water treatments of 1.4 dS/m (control), 4.7 dS/m (medium), and 7.4 dS/m (high). The growth medium was a sterilized soil containing low phosphorus. Salinity stress decreased the rate of mycorrhization, but inoculation with *F. mosseae* increased the weight of plant roots and shoots as well as the leaf area. The mycorrhizal plants also contained higher levels of P, zinc (Zn), iron, and copper under the control and medium levels of stress. The fungi were able to decrease the rate of Na^+ in plant shoots compared with nonmycorrhizal plants. The authors concluded that the fungi were able to enhance the salinity tolerance of tomato plants by improving growth and nutrient uptake and that they can be used for planting tomato plants under saline conditions in arid and semiarid areas.

In a study aimed at identifying a genetic mechanism for AM fungal promotion of salt tolerance in the tomato, He and Huang (2013) investigated the effects of mycorrhization on plant growth, Na^+ and Cl^- content, and Na^+/H^+ vacuolar expression of the antiporter gene (*LeNHX1*) in the leaf and roots of tomato plants subjected to NaCl salinity (0.5% and 1%). The fungi enhanced salt tolerance and plant growth, decreased the concentration of Na^+ in plant but had no effect on plant Cl^- content. Although the expression of *LeNHX1* was induced in mycorrhizal and nonmycorrhizal tomato plants by NaCl stress, its expression did not increase in mycorrhizal plants. The authors indicated that the increased tolerance of mycorrhizal tomato plants under salinity is not a function of *LeNHX1*, which is able to translocate Na^+ from the cytosol to the vacuole across the tonoplast.

Using a pot experiment, Wu et al. (2010) investigated the effects of mycorrhizal fungal species *F. mosseae* and *Paraglomus occultum* on the growth, root morphology, photosynthesis, and ionic balance of citrus (*Citrus tangerine* Hort. Ex Tanaka) seedlings under salinity stress. The 85-day-old seedlings were subjected to the stress of salinity for 60 days using 100 mM NaCl. Salinity stress did not affect the colonization of seedlings by *P. occultum*; however, it significantly decreased the rate of colonization by *F. mosseae*. Plant growth and physiology including plant height, root and shoot growth, stem diameter, total dry matter, the rate of photosynthesis, the rate of transpiration, and stomatal conductance were improved by mycorrhization under the control and 100 mM concentration of NaCl compared with noninoculated seedlings. Root properties including root length and root surface area were also increased by mycorrhization; however, the seedlings inoculated with *F. mosseae* had a higher root volume. Mycorrhization by both species significantly decreased leaf Na^+ concentration and significantly increased leaf K^+ and magnesium (Mg^{2+}) concentration, as well as the K^+/Na^+ ratio. The concentration of Ca^{2+} increased in seedlings inoculated with *P. occultum* and in control plants; however, it decreased in seedlings inoculated with *F. mosseae*. Inoculation with *F. mosseae* also resulted in higher Ca^{2+}/Na^+ and Mg^{2+}/Na^+ ratios in control plants, although *P. occultum* just increased the ratio of Mg^{2+}/Na^+ under salinity stress. The authors indicated that mycorrhizal fungi are able to alleviate the stress of salinity on the citrus seedlings by increasing above-ground and below-ground growth, adjusting the ionic balance and improving the process of photosynthesis (Wu et al., 2010).

Although most studies of the effects of AM fungi on salt tolerance emphasize the consequences for host plant growth and physiology, Hammer and Rillig (2011) examined the effect of salt stress on glomalin production. Glomalin is a glycoprotein produced by AM fungi that plays an important role in soil aggregation (Chapter 14). Glomalin production has been hypothesized to represent a stress response of the fungus as its abundance is negatively correlated with hyphal length (Hammer and Rillig, 2011). These authors investigated the effects of salinity (using NaCl) and osmotic stress (using glycerol) on the growth of mycorrhizal fungi in vitro culture. Interestingly, salinity stress induced by NaCl significantly increased the production of glomalin by *R. irregularis*, whereas osmotic stress induced by glycerol did not (Hammer and Rillig 2011). This study suggests that changes in salinity might affect soil properties mediated by AM fungi (aggregation) as well as host plant physiology.

Feng et al. (2002) examined the interactive effects of salt stress, inoculation with an AM fungus, and fertilization with P on plant performance. They inoculated maize (*Zea mays* L.) with

F. mosseae and subjected plants to salinity stress using NaCl under greenhouse conditions. The effects of the experimental treatments, including two salt levels (control and 100 mM) and two P levels (0.05 and 0.1 mM) for 34 days, were examined after a 34-day nonsaline pretreatment. Inoculation with AM fungi increased plant root and shoot growth. Under certain combinations of salinity and P fertilization, the content of plant chlorophyll, soluble sugars and P were higher in mycorrhizal maize than nonmycorrhizal maize. However, the concentration of Na^+ was similar in mycorrhizal and nonmycorrhizal plants. Also, inoculation with AM fungi resulted in higher concentrations of electrolytes and a lower rate of electrolyte leakage in some combinations of salinity and P fertilization. There was a similar P content in mycorrhizal plants treated with lower levels of P as in nonmycorrhizal plants treated with higher levels of P; however, the fungi resulted in higher plant dry biomass, electrolyte concentrations, and soluble sugars in the roots. The authors indicated that a higher rate of soluble sugar and electrolyte concentration in mycorrhizal plants is indicative of higher osmoregulation potential by these plants. They also indicated that the higher demand for carbohydrates by the fungi increases the rate of soluble sugar in the roots of the host plant, which is not a function of plant P content, and enhances plant tolerance to the osmotic stress resulting from salinity in mycorrhizal plants.

Using a meta-analysis, Chandrasekaran et al. (2014) investigated the effects of mycorrhizal fungi on plant growth under saline conditions. Their results supported the results of many of the case studies detailed in the previous paragraphs, because mycorrhizal fungi significantly increased plant growth in saline conditions by enhancing nutrient uptake of N, P, and K, elevating the activity of antioxidant enzymes including Cat, SOD, POD, and ASA-POD, and decreasing the uptake of Na^+. However, the production of proline was not significantly affected. Proline accumulation is a commonly observed metabolic response to salinity stress in plants. Higher uptake of N, P, and K that resulted from the increased growth of roots and plant tolerance under stress increased because of the enhanced production of antioxidant enzymes.

15.3 SALINITY IN COMBINATION WITH DROUGHT AND WARMING

In many cases, plants and their associated mycorrhizal fungi experience salt stress and drought or high temperatures simultaneously, leading to further decreases in plant growth and revenue from agriculture. Cho et al. (2006) investigated the combined effects of drought and salinity stress on the growth of sorghum (*Sorghum bicolor*). They hypothesized that the response of mycorrhizal plants to drought stress would be more pronounced when the plant is subjected to drought stress under saline conditions. Using two greenhouse experiments, different plant properties were measured after inoculation by mycorrhizal fungi, including *R. irregulare* and *Gigaspora margarita,* and a mix of AM fungi from a semiarid location. The plants were subjected to drought and salinity (by NaCl stress and osmotic stress using concentrated nutrients) as well as soil leaching. The authors found that inoculation with AM fungi affected plant response to drought stress, but this effect was not consistently influenced by changes in salinity.

Using inoculation with *C. etunicatum*, Zhu et al. (2010) evaluated the effects of temperature stress on the growth and physiology of maize (*Z. mays* L.), finding that similar mechanisms involved in tolerance to salt were involved in tolerance of high temperatures. Zhu et al. (2010)

measured growth, osmotic potential, the peroxidation of membrane lipids, and the production and activity of antioxidants in maize using a pot experiment. The plants were grown under nonstressed conditions for 6 weeks and then exposed to temperatures of 5, 15, 25, 35, and 40°C for 1 week. The relative permeability of cell membranes and the content of malondialdehyde, a chemical marker for oxidative stress, in plant roots and shoots were decreased in maize inoculated by AM fungi. Although higher proline and soluble sugar content were found in the roots of inoculated plants, proline content was lower in the leaves of mycorrhizal plants than nonmycorrhizal plants. The activity of antioxidant enzymes including Cat, POD, and SOD were increased by the fungi in both roots and shoots. The authors concluded that AM fungi were able to alleviate the stress of temperature on maize growth by: (1) increasing the production of compounds essential for osmotic adjustment, (2) reducing the rate of lipid peroxidation, (3) decreasing the permeability of membranes, and (4) increasing the activity of antioxidant enzymes.

Also illustrating that some similar mechanisms may operate across temperature, drought, and salinity stress, Wu and Zou (2009) showed that AM fungi have direct effects on the production of ROS in drought-stressed citrus plants. The plants were subjected to water stress 1 year after planting by withholding irrigation for 12 days. During the stress period the rate of mycorrhizal colonization and arbuscule production decreased. The rate of hydrogen peroxide, superoxide anions, and malondialdehyde significantly decreased in mycorrhizal plants compared with nonmycorrhizal plants during the experiment. The activities of Cat, SOD, and guaiacol POD significantly increased in mycorrhizal plants during the drought-stress period. There was also a higher rate of glutathione and ascorbate content in the mycorrhizal plants during the stress period compared with nonmycorrhizal plants. Correlation analysis revealed that just total mycorrhizal colonization and the amount of arbuscules, and not the numbers of vesicles or entry points, significantly affected the metabolism of reactive oxygen. The authors concluded that the fungi were able to alleviate the effects of drought stress on citrus growth chiefly by enhancing the metabolism of ROS and hence decreasing oxidative stress.

15.4 STUDIES OF SALINITY RESPONSES OF INDIGENOUS ARBUSCULAR MYCORRHIZAL FUNGI

Because of methodological limitations, the functions of indigenous mycorrhizal fungi in soil have not been assessed as fully as the pot experiments described previously, but interesting results have emerged. Liu et al. (2016) evaluated the effects of indigenous mycorrhizal fungi on the growth of cotton and maize under field saline conditions. The alleviating effects of indigenous mycorrhizal fungi on the growth of these two crop species was the result of: (1) a greater concentration of proline, (2) higher K^+/Na^+ resulting from enhanced uptake of K^+, and (3) increased P uptake by the mycorrhizal plants. These mechanisms are similar to those detailed in the studies described in the previous section.

Guo and Gong (2014) investigated how the diversity and species composition of AM fungi were affected by different abiotic stresses (pH, salinity, and study location) and plant traits (tolerance under salinity stress, origin, and style of life). According to their analyses, among the abiotic stresses tested the highest rate of variation was related to soil salinity of 4.2% and pH of 6.9 significantly affecting the composition of the fungal community. The authors also found that plant traits including salinity tolerance, plant origin, and style of life did not significantly affect the fungal diversity in the host plant roots, but they did influence AM fungal species composition in saline ecosystems. This study points out the importance of considering the combined consequences of increased salinity for indigenous AM fungi and their host plants.

15.5 PLANT ROOT PROPERTIES, MYCORRHIZAL FUNGI AND SALINITY STRESS

The properties of plant roots may influence plant responses to salinity directly or through their interactions with root symbionts, so some of the dynamics of root growth will be reviewed here. Among the most important functions of plant roots are maintaining the position of plant in the soil, uptake of water and nutrients, interaction with soil microbes including mycorrhizal fungi, and affecting the properties of the rhizosphere (Bonfante and Anca, 2009; Miransari, 2011; el Zahar Haichar et al., 2014). Root growth is controlled by different plant signals. Although plant genetics is the most important parameter determining root and hence plant growth, variation in the environment can also affect root and plant responses. The response of plant roots to variation in environmental conditions is determined by the following roots traits: ion content, the architecture of roots, and the properties of root cells (Ristova and Busch, 2014).

Using a meta-analysis of about 1000 data sets, Yang et al. (2015) investigated how mycorrhizal responsiveness was affected by the properties of plant roots. According to their analysis, plants with tap roots were more responsive to mycorrhizal colonization than host plants with fibrous roots, as indicated by higher plant growth and P and nitrogen (N) uptake. However, this same pattern was not observed under biotic stresses, where no difference was observed among rooting strategies. These results suggest complex interactions among stress, root system properties, and inoculation with mycorrhizal fungi that could be pursued in the context of salinity stress in future studies.

15.6 SIGNALING, MYCORRHIZAL FUNGI, AND SALINITY STRESS

Mycorrhizal fungal spores are able to germinate under suitable conditions in the absence of their host plant, indicating that signal molecules produced by the host plant roots are not essential for the germination of the spores (Giovannetti, 2000). However, for the growth and production of hyphae and for symbiosis development, the presence of the host plant is necessary, indicating biochemical signaling between host plant and fungus (Miransari et al., 2014). However, this communication can be interrupted by unfavorable environmental conditions, including abiotic environmental stress as has been documented previously for symbiotic N fixation. For example, Miransari and Smith (2007, 2008, 2009) found that the initial stages of symbiotic association, including the exchange of the signal molecule, genistein, between the two symbionts were adversely affected by stress. Preincubation of the N-fixing bacterium, *Bradyrhizobium japonicum,* with genistein can alleviate this stress. Similar results might be expected for the symbiotic development of AM fungi with their host plant under stress. If the related molecules, the *myc* factors, are produced at suitable levels, they can result in the establishment of symbiotic development with the host plant. However, this communication between the fungus and the host plant could be interrupted if stress adversely affects signaling molecules (Miransari, 2014a; Kumar et al., 2015). Although the plant is able to alleviate salinity stress to some extent on its own using different hormonal signaling pathways, its symbiotic association with soil microbes such as mycorrhizal fungi and PGPR can result in the activation of different signaling pathways in the host plant and the associative microbes. Different signaling pathways are activated in plants under stress, including hormonal signaling, mitogen-activated protein kinase, and production of ROS (Miransari, 2012, 2014a,b,c,d).

Among the plant hormones that can affect the symbiosis between mycorrhizal fungi and the host plant are the strigolactones. These hormones are able to regulate plant growth and development including plant architecture, leaf senescence, secondary growth, and reproduction under nutrient deficient conditions (Kapulnik et al., 2011; Yamada et al., 2014). Strigolactones can also act as detection hormones for the development of the symbiosis between mycorrhizal fungi and the host plant. López-Ráez (2015) considered the effects of water-related stresses such as salinity and drought on strigolactones in the presence or absence of the AM symbiosis. Evidence from the literature supported the suggestion that water scarcity increased abscisic acid (ABA) levels and down-regulated strigolactone production in the absence of AM fungi. However, in the presence of AM fungi, plants induced strigolactone and ABA biosynthesis, promoting the symbiosis and increasing plant performance in stressful conditions (López-Ráez, 2015). The role of strigolactones in salt stress itself needs further study, because salt stress can adversely affect the production of strigolactones in plants in some cases, whereas other studies show that strigolactone production increases with salt stress, with production at its highest under severe salt stress (Abd-Allah et al., 2015; Kong et al., 2016).

15.7 TRIPARTITE INTERACTIONS AND SALINITY STRESS

Mycorrhizal fungi and bacteria commonly interact in soil and roots and these interactions may influence the tolerance of host plants to salt stress. In the soil, PGPR can directly influence plant responses to stress or do so indirectly through effects on AM fungi. For example, AM fungi improve plant growth in large part by increasing nutrient uptake, particularly P, which is poorly soluble in soil. Interactions between mycorrhizal fungi and P-solubilizing microorganisms can significantly affect plant growth by increasing the availability of P, and hence its higher uptake by the host plant. Such beneficial interactions can significantly decrease the use of P chemical fertilizers in the field, which is of economic and environmental significance. For example, P-solubilizing bacteria could be used with rock phosphate and AM fungi to increase the uptake and availability of P by the host plant (Bagyaraj et al., 2015). N-fixing bacteria, both free-living and symbiotic with plant roots, can make N more available to plants by converting N_2 into other forms. They often co-occur with AM fungi in soil or roots. A few examples illustrate the potential for these tripartite associations to influence salt tolerance.

Garg and Chandel (2011) investigated the interactive effects of the mycorrhizal fungus *F. mosseae* and salinity stress on the process of biological N fixation by chickpea (*Cicer arietinum* L.). Salt-sensitive and salt-tolerant genotypes of chickpea were used for the experiment and tested under different levels of salinity with and without mycorrhizal inoculation. Plant growth of both roots and shoots were adversely affected by high salinity. Mycorrhizal colonization resulted in significantly higher numbers of nodules, higher nodule dry weight and increased activity of an N-fixing enzyme under both stressful and nonstressful conditions. As salinity increased, mycorrhizal colonization decreased, but the mycorrhizal dependency of the host plant increased, particularly in the salt-sensitive genotype. The concentration of Na^+ and K^+ was less and higher, respectively, in the salt tolerant genotype compared with the salt sensitive genotype. Salinity decreased the rate of N and P uptake by both genotypes; however this effect was reduced in the salt-tolerant genotype. Based on these results, *F. mosseae* improved chickpea growth and N fixation under saline conditions, but this effect differed between plant genotypes.

Using a similar pot experiment, Kadian et al. (2014) examined the effects of inoculation with two species of mycorrhizal fungi, *F. mosseae* and *Acaulospora laevis*, and *B. japonicum* on the growth and nodulation of mung bean (*Vigna radiata* L.) under salinity stresses of 4, 8, and 12 dS/m generated using NaCl. The fungi were able to alleviate the stress on plant growth by increasing dry biomass and yield, nutrient uptake, chlorophyll content, protein, and proline content, and by decreasing electrolyte leakage compared with the nonmycorrhizal treatment. The fungi decreased the uptake of Na^+ by the host plant and increased the uptake of P, N, and K. The authors concluded that for the mung bean plants co-inoculated with the soil microbes tested in this research, tolerance to salinity stress increased mainly by increasing plant nutrient uptake (especially P), production of plant hormones, and improving the conditions of the plant rhizosphere and the soil, as well as by altering the physiological and biochemical properties of the host plant.

Franzini et al. (2010) evaluated the effects of single and double inoculation with two different mycorrhizal fungi and two different strains of rhizobium on the growth of four genotypes of bean (*Phaseolus vulgaris* L.) under a moderate level of salinity. Surprisingly, one AM fungal species and one rhizobium strain adversely affected the growth of the host plant. The authors found that at this moderate level of salinity, the mycorrhizal fungi decreased nodulation and the N fixation activity of the rhizobium, and hence bean growth. Also as shown above, plant genotype and AM fungal species interact in important ways, as does the strain of N-fixing bacteria.

Manaf and Zayed (2015) investigated the combined effects of mycorrhizal fungi and *Pseudomonas fluorescens* on the nodulation, biochemistry, growth, and yield of cowpea (*Vigna unguiculata* L.) under salt stress. The rate of mycorrhizal colonization decreased as the salinity of the irrigation water increased. The highest rate of nodulation, growth, and yield was found in the treatment including tap water, mycorrhizal fungi, and *P. fluorescens*. However, the treatment including 6000 ppm NaCl and *P. fluorescens* resulted in the highest weight of 100 seeds at 19.71 g. Similar results were obtained for the concentration of carotenoids, which was highest in the treatment containing 6000 ppm NaCl, mycorrhizal fungi, and *P. fluorescens* at 0.449 mg/g fresh weight. The activity of SOD was the highest in the treatment including 3000 ppm NaCl, at 7.25 unit/mg protein. The highest soluble sugars and proline concentrations were in the treatment containing 6000 ppm NaCl and mycorrhizal fungi at 1001.62 μg/g fresh weight and 2.56 mg/g fresh weight, respectively.

In an example of how studies of salinity stress and tripartite interactions can be applied, Rabie (2005) investigated the effects of seawater treatments on the growth of mung bean inoculated with *G. clarum* and/or treated with foliar application of the plant hormone, kinetin. Plant growth and physiological parameters including dry biomass, chlorophyll content, height, protein and sugar content, stomatal conductivity, rate of transpiration, N and P efficiencies, activity of the N fixation enzyme, and the activities of alkaline phosphatases was increased by mycorrhizal fungi in the presence or absence of kinetin compared with nonmycorrhizal plants. The fungi increased the uptake of nutrients including P, N, K, Mg, and Ca by the host plant and decreased the ratios of Na/P, Na/N, Na/K, Na/Mg, and Na/Ca. The growth of mung bean plants was highly dependent on AM fungi in the presence or absence of kinetin under high salinity. The fungi also increased the tolerance index of the host plant in the presence of kinetin significantly more than in the absence of kinetin. The author indicated that the fungi and kinetin were able to enhance the growth of mung bean under salinity stress, although the fungi were much more effective. It is thus possible to use mycorrhizal fungi and kinetin for planting mung bean under certain dilutions of seawater.

At the molecular level, some genes that are associated with the regulation of ROS production in plants affect symbioses as well. For example, if *PvRbohB*, a plant nicotinamide adenine dinucleotide phosphate oxidase (NADPH) gene in bean (*P. vulgaris*) roots, is overexpressed, the number of nodules increases but the rate of mycorrhizal colonization decreases. The reverse pattern is seen when the gene is down-regulated (Arthikala et al., 2015). Arthikala et al. (2015) investigated an alleviating mechanism, electrolyte leakage, by rhizobium and mycorrhizal fungi in a transgenic bean overexpressing *PvRbohB* under stress. They found that rhizobium and mycorrhizal fungi were able to alleviate the stress, although the effect of the latter was more significant.

15.8 AGRONOMICAL CONSEQUENCES OF USING MYCORRHIZAL FUNGI IN SALINE FIELDS

The studies discussed in this chapter illustrate the important effects of mycorrhizal fungi to plants growing under saline conditions. It is thus possible to enhance crop yield under salinity stress using mycorrhizal fungi. Studies of wheat illustrate the potential for improved agricultural production using AM fungi. A meta-analysis of studies conducted between 1975 and 2013 conducted by Pellegrino et al. (2015) documented the following consequences of mycorrhizal inoculation for wheat: (1) enhanced dry matter and the uptake of nutrients including N, P, and Zn (2) increased grain yield by 20% and harvest index by 25%; and (3) a positive correlation between the uptake of P and Zn by grain and grain yield with mycorrhizal root colonization. The most important parameters affecting wheat response to mycorrhizal colonization were soil pH, soil organic matter, total soil N, soil available P, soil texture, the species of mycorrhizal fungus, and climate. The authors concluded that using mycorrhizal inocula under field conditions is an efficient agronomic practice significantly increasing wheat yield, although the economical benefits of such a practice have yet to be tested under different cropping systems. As described previously, inoculation with AM fungi also increases the salt tolerance of wheat, potentially expanding the environments in which it can be grown.

15.9 CONCLUSIONS AND FUTURE PERSPECTIVES

Plants are subjected to different kind of stresses worldwide. Salinity is among the most important stresses significantly decreasing plant growth and yield. Plants, especially salt-tolerant species, are able to tolerate salt stress using a variety of morphological and physiological mechanisms. This review shows that AM fungi can facilitate tolerance of high salinity through a variety of important mechanisms, including the activation of different genes and signaling pathways, the production of osmolytes and antioxidants, and adjustments of the ratio of K^+/Na^+ in the host plant. Accordingly, the use of mycorrhizal fungi for the alleviation of soil stresses under field conditions should be practiced more efficiently and more widely. Future research should include investigation of strategies that allow more convenient use of mycorrhizal fungi in the field, including areas already experiencing salinity stress. Another priority should be finding and isolating salt tolerant fungal species, which can be produced and used at large scales under stressful plant growth conditions. Future research should also emphasize the species and genotypes of crop plants that are most responsive to mycorrhizal fungi under stressful conditions, including the combined stresses of salinity, drought and high temperatures. Methods that enhance the positive interactions of mycorrhizal fungi with rhizobium in a tripartite association can also increase the applicability of fungal species and rhizobium strains under field conditions.

References

Abd-Allah, E.F., Hashem, A., Alqarawi, A.A., Bahkali, A.H., Alwhibi, M.S., 2015. Enhancing growth performance and systemic acquired resistance of medicinal plant *Sesbania sesban* (L.) Merr using arbuscular mycorrhizal fungi under salt stress. Saudi Journal of Biological Sciences 22, 274–283.

Augé, R.M., 2001. Water relations, drought and vesicular-arbuscular mycorrhizal symbiosis. Mycorrhiza 11, 3–42.

Alam, S., 1994. Nutrient uptake by plants under stress condition. In: Pessarakli, M. (Ed.), Handbook of Plant and Crop Stress. Marcel Dekker, New York, pp. 227–243.

Al-Karaki, G., 2000. Growth of mycorrhizal tomato and mineral acquisition under salt stress. Mycorrhiza 10, 51–54.

Arthikala, M., Nava, N., Quinto, C., 2015. Effect of Rhizobium and arbuscular mycorrhizal fungi inoculation on electrolyte leakage in Phaseolus vulgaris roots overexpressing RbohB. Plant Signaling & Behavior 10, e1011932.

Bagyaraj, D., Sharma, M., Maiti, D., 2015. Phosphorus nutrition of crops through arbuscular mycorrhizal fungi. Current Science 108, 1288–1293.

Bharti, N., Barnawal, D., Awasthi, A., Yadav, A., Kalra, A., 2014. Plant growth promoting rhizobacteria alleviate salinity induced negative effects on growth, oil content and physiological status in *Mentha arvensis*. Acta Physiologiae Plantarum 36, 45–60.

Bonfante, P., Anca, L., 2009. Plants, mycorrhizal fungi, and bacteria: a network of interactions. Annul Reviews of Microbiology 63, 363–383.

Bothe, H., 2012. Arbuscular mycorrhiza and salt tolerance of plants. Symbiosis 58, 7–16.

Chandrasekaran, M., Boughattas, S., Hu, S.H., Oh, H., Sa, T., 2014. A meta-analysis of arbuscular mycorrhizal effects on plants grown under salt stress. Mycorrhiza 24, 611–625.

Cho, K., Toler, H., Lee, J., Ownley, B., Stutz, J., Moore, J., Augé, R., 2006. Mycorrhizal symbiosis and response of sorghum plants to combined drought and salinity stresses. Journal of Plant Physiology 163, 517–528.

Conn, S., Gilliham, M., 2010. Comparative physiology of elemental distribution in plants. Annals of Botany 105, 1081–1102.

Daei, G., Ardekani, M., Rejali, F., Teimuri, S., Miransari, M., 2009. Alleviation of salinity stress on wheat yield, yield components, and nutrient uptake using arbuscular mycorrhizal fungi under field conditions. Journal of Plant Physiology 166, 617–625.

Dolatabadian, A., ModarresSanavy, S., Ghanati, F., 2011. Effect of salinity on growth, xylem structure and anatomical characteristics of soybean. Notulae Scientia Biologicae 3, 41–45.

Egamberdieva, D., Lugtenberg, B., 2014. Use of plant growth-promoting rhizobacteria to alleviate salinity stress in plants. In: Miransari, M. (Ed.), Use of Microbes for the Alleviation of Soil Stresses, vol. 1. Springer, New York. ISBN: 978-1-4614-9465-2, pp. 73–96.

el Zahar Haichar, F., Santaella, C., Heulin, T., Achouak, W., 2014. Root exudates mediated interactions belowground. Soil Biology and Biochemistry 77, 69–80.

Estrada, B., Aroca, R., Maathuis, F., et al., 2013a. Arbuscular mycorrhizal fungi native from a Mediterranean saline area enhance maize tolerance to salinity through improved ion homeostasis. Plant, Cell & Environment 36, 1771–1782.

Estrada, B., Aroca, R., Barea, J.M., et al., 2013b. Native arbuscular mycorrhizal fungi isolated from a saline habitat improved maize antioxidant systems and plant tolerance to salinity. Plant Science 201, 42–51.

Estrada, B., Aroca, R., Azcón-Aguilar, C., Barea, J.M., et al., 2013c. Importance of native arbuscular mycorrhizal inoculation in the halophyte *Asteriscus maritimus* for successful establishment and growth under saline conditions. Plant and Soil 370 (1–2), 175–185.

Feng, G., Zhang, F., Li, X., Tian, C., Tang, C., Rengel, Z., 2002. Improved tolerance of maize plants to salt stress by arbuscular mycorrhiza is related to higher accumulation of soluble sugars in roots. Mycorrhiza 12, 185–190.

Flowers, T., Munns, R., Colmer, T., 2015. Sodium chloride toxicity and the cellular basis of salt tolerance in halophytes. Annals of Botany 115, 419–431.

Franzini, V., Azcón, R., Mendes, F., Aroca, R., 2010. Interactions between *Glomus* species and *Rhizobium* strains affect the nutritional physiology of drought-stressed legume hosts. Journal of Plant Physiology 167, 614–619.

Garg, N., Chandel, S., 2011. Effect of mycorrhizal inoculation on growth, nitrogen fixation, and nutrient uptake in *Cicer arietinum* (L.) under salt stress. Turkish Journal of Agriculture and Forestry 35, 205–214.

Gianinazzi, S., Gollotte, A., Binet, M., van Tuinen, D., Redecker, D., Wipf, D., 2010. Agroecology: the key role of arbuscular mycorrhizas in ecosystem services. Mycorrhiza 20, 519–530.

Giovannetti, M., 2000. Spore germination and pre-symbiotic mycelial growth. In: Kapulnik, Y., Douds, D. (Eds.), Arbuscular Mycorrhizas: Physiology and Function. Springer, Netherlands. ISBN: 978-90-481-5515-6, pp. 47–68.

Golpayegani, A., Tilebeni, H., 2011. Effect of biological fertilizers on biochemical and physiological parameters of basil (*Ociumum basilicm* L.) medicine plant. American-Eurasian Journal of Agricultural & Environmental Sciences 11, 411–416.

Gopal, S., Chandrasekaran, M., Shagol, C., Kim, K., Sa, T., 2012. Spore associated bacteria (SAB) of arbuscular mycorrhizal fungi (AMF) and plant growth promoting rhizobacteria (PGPR) increase nutrient uptake and plant growth under stress conditions. Korean Journal of Soil Science and Fertilizer 45, 582–592.

Guo, X., Gong, J., 2014. Differential effects of abiotic factors and host plant traits on diversity and community composition of root-colonizing arbuscular mycorrhizal fungi in a salt-stressed ecosystem. Mycorrhiza 24, 79–94.

Hajiboland, R., Aliasgharzadeh, N., Laiegh, S., Poschenrieder, C., 2010. Colonization with arbuscular mycorrhizal fungi improves salinity tolerance of tomato (*Solanum lycopersicum* L.) plants. Plant and Soil 331, 313–327.

Hammer, E., Rillig, M., 2011. The influence of different stresses on glomalin levels in an arbuscular mycorrhizal fungus—salinity increases glomalin content. PloS One 6, e28426.

He, Z., Huang, Z., 2013. Expression analysis of *LeNHX1* gene in mycorrhizal tomato under salt stress. Journal of Microbiology 51, 100–104.

Huang, Z., He, C., He, Z., Zou, Z., Zhang, Z., 2010. The effects of arbuscular mycorrhizal fungi on reactive oxyradical scavenging system of tomato under salt tolerance. Agricultural Sciences in China 9, 1150–1159.

Jung, H., Park, S., Kang, H., 2013. Regulation of RNA metabolism in plant development and stress responses. Journal of Plant Biology 56, 123–129.

Kadian, N., Yadav, K., Aggarwal, A., 2014. Application of AM fungi with *Bradyrhizobium japonicum* in improving growth, nutrient uptake and yield of *Vigna radiata* L. under saline soil. Journal of Stress Physiology & Biochemistry 10, 134–152.

Kang, S., Khan, A., Waqas, M., You, Y., Kim, J., Kim, J., Hamayun, M., Lee, I., 2014. Plant growth-promoting rhizobacteria reduce adverse effects of salinity and osmotic stress by regulating phytohormones and antioxidants in *Cucumis sativus*. Journal of Plant Interactions 9, 673–682.

Kapulnik, Y., Delaux, P.M., Resnick, N., Mayzlish-Gati, E., Wininger, S., Bhattacharya, C., Sejalon-Delmas, N., Combier, J.P., Becard, G., Belausov, E., Beeckman, T., Dor, E., Hershenhorn, J., Koltai, H., 2011. Strigolactones affect lateral root formation and root-hair elongation in Arabidopsis. Planta 233, 209–216.

Kong, X., Luo, Z., Dong, H., Eneji, A.E., Li, W., 2016. H_2O_2 and ABA signaling are responsible for the increased Na+ efflux and water uptake in *Gossypium hirsutum* L. roots in the non-saline side under non-uniform root zone salinity. Journal of Experimental Botany 67, 2247–2261.

Kumar, A., Dames, J., Gupta, A., Sharma, S., Gilbert, J., Ahmad, P., 2015. Current developments in arbuscular mycorrhizal fungi research and its role in salinity stress alleviation: a biotechnological perspective. Critical Reviews in Biotechnology 35, 461–474.

Liu, S., Guo, X., Feng, G., Maimaitiaili, B., Fan, J., He, X., 2016. Indigenous arbuscular mycorrhizal fungi can alleviate salt stress and promote growth of cotton and maize in saline fields. Plant and Soil 398, 195–206.

Liu, X., Baird, W., 2003. Differential expression of genes regulated in response to drought or salinity stress in sunflower. Crop Science 43, 678–687.

Liu, Y., Johnson, N.C., Mao, L., Shi, G., Jiang, S., Ma, X., Du, G., An, L., Feng, H., 2015. Phylogenetic structure of arbuscular mycorrhizal community shifts in response to increasing soil fertility. Soil Biology and Biochemistry 89, 196–205.

López-Ráez, J.A., 2015. How drought and salinity affect arbuscular mycorrhizal symbiosis and strigolactone biosynthesis? Planta 1–11.

Manaf, H., Zayed, M., 2015. Productivity of cowpea as affected by salt stress in presence of endomycorrhizae and Pseudomonas fluorescens. Annals of Agricultural Sciences 60, 219–226.

Mardukhi, B., Rejali, F., Daei, G., Ardakani, M.R., Malakouti, M.J., Miransari, M., 2015. Mineral uptake of mycorrhizal wheat (*Triticum aestivum* L.) under salinity stress. Communications in Soil Science and Plant Analysis 46, 343–357.

Miransari, M., Smith, D., 2007. Overcoming the stressful effects of salinity and acidity on soybean nodulation and yields using signal molecule genistein under field conditions. Journal of Plant Nutrition 30, 1967–1992.

Miransari, M., Smith, D., 2008. Using signal molecule genistein to alleviate the stress of suboptimal root zone temperature on soybean-Bradyrhizobium symbiosis under different soil textures. Journal of Plant Interactions 3, 287–295.

Miransari, M., Bahrami, H., Rejali, F., Malakouti, M., 2008. Using arbuscular mycorrhiza to alleviate the stress of soil compaction on wheat (*Triticum aestivum* L.) growth. Soil Biology and Biochemistry 40, 1197–1206.

Miransari, M., Smith, D., 2009. Alleviating salt stress on soybean (*Glycine max* (L.) Merr.)–*Bradyrhizobium japonicum* symbiosis, using signal molecule genistein. European Journal of Soil Biology 45, 146–152.

Miransari, M., 2010. Contribution of arbuscular mycorrhizal symbiosis to plant growth under different types of soil stress. Plant Biology 12, 563–569.

Miransari, M., 2011. Interactions between arbuscular mycorrhizal fungi and soil bacteria. Applied Microbiology and Biotechnology 89, 917–930.

Miransari, M., 2012. Signaling molecules in the arbuscular mycorrhizal fungi. In: Gupta, V., Ayyachamy, M. (Eds.), Biotechnology of Fungal Genes. Published by CRC Press, USA. ISBN: 9781578087877.

Miransari, M., 2014a. Plant growth promoting rhizobacteria. Journal of Plant Nutrition 37, 2227–2235.

Miransari, M., 2014b. Plant, mycorrhizal fungi, and bacterial network. In: Hakim, K., Rehman, R., Tahir, I. (Eds.), Plant Signaling: Understanding the Molecular Crosstalk. Springer, India. ISBN: 978-81-322-1541-7, pp. 315–325.

Miransari, M., 2014c. MAPK Signaling Cascade Affecting Plant Response Under Stress. AbtinBerkeh Ltd. Company, Isfahan, Iran, and Creative Space, an Amazon Company, USA. ISBN: 9781500114749.

Miransari, M., 2014d. Plant Hormonal Signaling Under Stress. Published By AbtinBerkeh Ltd. Company, Isfahan, Iran, and Creative Space, an Amazon Company, USA. ISBN: 9781500156039.

Miransari, M., Abrishamchi, A., Khoshbakht, K., Niknam, V., 2014. Plant hormones as signals in arbuscular mycorrhizal symbiosis. Critical Reviews in Biotechnology 34, 123–133.

Miransari, R., Miransari, A., Miransari, F., Miransari, A., Adham, S., Miransari, M., 2015. Plant and Stress: I. Wheat (*Triticum aestivum* L.). AbtinBerkeh Ltd. Company, Isfahan, Iran, and Creative Space, an Amazon Company, USA. ISBN: 978-1512370829.

Miransari, M. (Ed.), 2016. Abiotic and Biotic Stresses in Soybean Production. Elsevier, Academic Press, USA. ISBN: 978-0-12-801730-2, p. 344.

Munns, R., Gilliham, M., 2015. Salinity tolerance of crops – what is the cost? New Phytologist 208, 668–673.

Munns, R., James, R., 2003. Screening methods for salinity tolerance: a case study with tetraploid wheat. Plant Soil 253, 201–218.

Munns, R., Tester, M., 2008. Mechanisms of salinity tolerance. Annual Review of Plant Biology 59, 651–681.

Munns, R., 2002. Comparative physiology of salt and water stress. Plant, Cell & Environment 25, 239–250.

Pellegrino, E., Öpik, M., Bonari, E., Ercoli, L., 2015. Responses of wheat to arbuscular mycorrhizal fungi: a meta-analysis of field studies from 1975 to 2013. Soil Biology and Biochemistry 84, 210–217.

Rabie, G., 2005. Influence of arbuscular mycorrhizal fungi and kinetin on the response of mungbean plants to irrigation with seawater. Mycorrhiza 15, 225–230.

Ristova, D., Busch, W., 2014. Natural variation of root traits: from development to nutrient uptake. Plant Physiology 166, 518–527.

Selvakumar, G., Kim, K., Hu, S., Sa, T., 2014. Effect of salinity on plants and the role of arbuscular mycorrhizal fungi and plant growth-promoting rhizobacteria in alleviation of salt stress. In: Ahmad, P., Wani, M. (Eds.), Physiological Mechanisms and Adaptation Strategies in Plants Under Changing Environment. Springer, New York. ISBN: 978-1-4614-8590-2, pp. 115–144.

Smith, S., Jakobsen, I., Grønlund, M., Smith, F., 2011. Roles of arbuscular mycorrhizas in plant phosphorus nutrition: interactions between pathways of phosphorus uptake in arbuscular mycorrhizal roots have important implications for understanding and manipulating plant phosphorus acquisition. Plant Physiology 156, 1050–1057.

Uehlein, N., Fileschi, K., Eckert, M., Bienert, G.P., Bertl, A., Kaldenhoff, R., 2007. Arbuscular mycorrhizal symbiosis and plant aquaporin expression. Phytochemistry 68, 122–129.

Wu, Q., Zou, Y., He, H., 2010. Contributions of arbuscular mycorrhizal fungi to growth, photosynthesis, root morphology and ionic balance of citrus seedlings under salt stress. Acta Physiologiae Plantarum 32, 297–304.

Wu, Q., Zou, Y., 2009. Mycorrhiza has a direct effect on reactive oxygen metabolism of drought-stressed citrus. Plant, Soil & Environment 55, 436–442.

Xia, T., Apse, M., Aharon, G., Blumwald, E., 2002. Identification and characterization of a NaCl-inducible vacuolar Na^+/H^+ antiporter in *Beta vulgaris*. Physiologiae Plantarum 116, 206–212.

Yang, H., Zhang, Q., Dai, Y., Liu, Q., Tang, J., Bian, X., Chen, X., 2015. Effects of arbuscular mycorrhizal fungi on plant growth depend on root system: a meta-analysis. Plant and Soil 389, 361–374.

Yamada, Y., Furusawa, S., Nagasaka, S., Shimomura, K., Yamaguchi, S., Umehara, M., 2014. Strigolactone signaling regulates rice leaf senescence in response to a phosphate deficiency. Planta 240, 399–408.

Zhu, X., Song, F., Xu, H., 2010. Influence of arbuscular mycorrhiza on lipid peroxidation and antioxidant enzyme activity of maize plants under temperature stress. Mycorrhiza 20, 325–332.

CHAPTER

16

Mycorrhizas, Drought, and Host-Plant Mortality

C.A. Gehring[1], *R.L. Swaty*[2], *R.J. Deckert*[1]

[1]Northern Arizona University, Flagstaff, AZ, United States; [2]The Nature Conservancy's LANDFIRE Team, Evanston, IL, United States

16.1 INTRODUCTION

Although there is debate about the role of global climate change in causing drought (Dai, 2011, 2013), there is agreement that warming temperatures will allow droughts to develop more quickly and to become more intense (Trenberth et al., 2013). Recent studies have documented intense drought (Ummenhofer et al., 2009; Allen et al., 2010), including 98% certainty that the 1998–2012 drought in the Levant region of the Mediterranean was the driest period in 500 years (Cook et al., 2016). The Amazon basin also has experienced intense drought during the past decade, with models indicating that the area of the Amazon affected by severe drought will triple by 2100 (Duffy et al., 2015). Dai (2013) used climate models to project increased aridity during the 21st century for most of Africa, the Americas, Australia, and Southeast Asia, along with portions of Europe and the Middle East.

More intense drought already has contributed to significant declines in the productivity and survival of plants. In 2002–03, trees perished as a result of drought and associated insect infestations in woodlands across millions of hectares of the southwestern United States (Breshears et al., 2005; Mueller et al., 2005a). In the same region of the United States, intense drought led to mortality and reduced productivity of highly drought-adapted shrub species in desert ecosystems (Hamerlynck and McAuliffe, 2008). The effects of drought will not be restricted to arid and semiarid regions. Choat et al. (2012) showed that more than two-thirds of 226 forest species from sites around the world were vulnerable to drought, given climate change projections. Importantly, both angiosperms and gymnosperms were vulnerable, as were species from tropical seasonal forest and tropical rain forest (Choat et al., 2012). Expectations for agriculture are also dire, because warming temperatures exacerbate water shortages in more arid areas (McKersie, 2015).

Drought-related plant mortality and changes in plant community composition are also likely to have significant impacts on soil nutrient dynamics, soil structure, and carbon (C)

Mycorrhizal Mediation of Soil
http://dx.doi.org/10.1016/B978-0-12-804312-7.00016-4

sequestration. Forests represent the largest component of the terrestrial C sink (Pan et al., 2011), a sink that has the potential to increase with climate change because warming temperatures, increasing carbon dioxide (CO_2), and increasing nutrient availability could promote tree growth (Brzostek et al., 2014). However, drying conditions could lead to reduced C sequestration because of tree mortality (Anderegg et al., 2013), but also because of reductions in photosynthesis that result from water limitation. In a 13-year study of a temperate forest, trees reduced annual accrual of C in woody biomass by 41% as a result of increasing water stress, despite an extra 42 days of wood production (Brzostek et al., 2014).

Efforts are underway to better predict the responses of plants to the intense droughts projected for the future. Choat et al. (2012) assessed vulnerability based on risk of hydraulic failure, and other studies have categorized plants based on their drought avoidance (isohydric) versus drought tolerance (anisohydric) strategies (e.g., McDowell et al., 2008; Attia et al., 2015). The role of root traits has been highlighted as an overlooked but important factor influencing plant response to drought (Brunner et al., 2015).

Mycorrhizal fungi are well known to improve the ability of their hosts to resist, tolerate, and recover from drought (Augé, 2001; Lehto and Zwiazek, 2011). Recent reviews have described both the extent of these effects as well as the mechanisms by which they occur. However, mycorrhizal fungi are also affected by drought, potentially in ways that influence their ability to promote the drought tolerance of their hosts. In this chapter, we briefly review the patterns and mechanisms of improved host drought tolerance for arbuscular mycorrhizal (AM) and ectomycorrhizal (EcM) fungi along with studies examining how mycorrhizal fungi are influenced by drought. We focus our review on the effects of mycorrhizal colonization or inoculation on measurements of individual plants because the importance of mycorrhizal networks and mycorrhizal fungal contributions to soil water retention and hydraulic lift are considered in Chapters 17 and 18. After the literature review, we explore the effects of recent droughts in the United States on trees and their associated mycorrhizas, describing the potential consequences for fungi, plants, and soils. While this exploration is restricted geographically, what we learn from it may be broadly applicable given the projections for increased drought described previously.

16.2 MYCORRHIZAS, PLANTS, AND DROUGHT

The relationships among drought, mycorrhizal fungi, and plants have been viewed from two perspectives, but rarely in the same study. Several studies have examined whether mycorrhizal abundance in roots or soil, mycorrhizal fungal species diversity, or mycorrhizal fungal species composition are altered by drought (e.g., Swaty et al., 2004; Azul et al., 2010). A smaller number of studies also have examined the influence of drought on enzymatic activity, respiration, or growth in culture of mycorrhizal fungal species or isolates (e.g., Coleman et al., 1989; Jany et al., 2003; di Pietro et al., 2007). A larger number of studies have assessed whether inoculation with mycorrhizal fungi promotes drought tolerance in their host plants. Studies in both of these categories have been reviewed recently (Kivlin et al., 2013; Worchel et al., 2013; Jayne and Quigley, 2014; Karst et al., 2014; Mohan et al., 2014). We briefly discuss the results of those reviews and subsequent publications to create a broader synthesis and to highlight areas of future research. We emphasize AM and EcM fungi because they have been the focus of most studies of the influence of mycorrhizal fungi on drought and host-plant–water relationships. It is difficult to assess drought effects on ericoid and orchid mycorrhizas because of the paucity of studies on these types of mycorrhizas. Jeliazkova and Percival (2003) saw no differences in commercial blueberry ericoid mycorrhizal colonization in water exclusion experiments, a result that may be inappropriate to extend to wild heathlands. Orchids often exhibit extreme symbiont fidelity, leading to high orchid mortality during drought, but orchid genotypes that can switch fungi show enhanced survival (McCormick et al., 2006). Thus the fate of orchids in the face of climate change may be strongly dependent on the tolerance of their fungal partners, which in turn are EcM with neighboring trees (McCormick et al., 2006) or are mycoheterotrophic symbionts (McCormick et al., 2009).

16.2.1 Effects of Drought on Mycorrhizas

Both AM and EcM fungi typically respond in some way to drought conditions. Of the 32 studies reviewed by Mohan et al. (2014), measures of mycorrhizal abundance, root length colonized, or activity (e.g., respiration) were commonly altered by drought, although the direction of the changes were variable. The criteria used by Mohan et al. (2014) limited the number of studies included, but their results were similar to those of Karst et al. (2014) who reported changes in EcM fungal abundance, biomass, community composition, and activity with drought among the studies of pines they reviewed. In an earlier review, Augé (2001) reported that root colonization by AM fungi was often affected by drought, with increases more common than decreases.

A few examples illustrate the variety of approaches used to assess the effects of drought on mycorrhizal fungi and their results. Several studies of AM fungi measured changes in abundance, diversity, or species composition in the field with drought or along soil moisture gradients. For example, path analysis showed that AM fungal hyphal length was directly correlated with precipitation in alpine grassland whereas AM fungal species richness was indirectly affected (Zhang et al., 2016). The abundance of AM fungal spores was negatively correlated with soil moisture and available phosphorus (P) in association with a leguminous shrub (*Caragana korshinskii*) used for restoration in China (Liu et al., 2009). Similar kinds of studies have been conducted with EcM fungi, primarily focused on associates of pines and oaks. EcM fungal community composition in cork oak (*Quercus suber*) across 15 sites was affected more by land use than site precipitation, despite more than threefold higher precipitation at one end of the gradient than the other (Azul et al., 2010). In contrast, field studies of drought effects on pine EcM fungi generally observed some change in abundance or community traits (Karst et al., 2014). In the largest Scots pine (*Pinus sylvestris*) forest in Spain, changes in forest mushroom abundance, including those of EcM species, were significantly negatively correlated with summer drought, warmer temperatures, and declining forest growth (Buentgen et al., 2015).

Other studies measured the effects of water limitation on mycorrhizal fungi in more controlled settings. For example, imposed drought in strawberries (*Fragaria × ananassa*) resulted in higher AM fungal colonization measured microscopically but lower fungal DNA measured using quantitative polymerase chain reaction (Boyer et al., 2015). Furthermore, *Funneliformis mosseae* had higher relative abundance in strawberry roots than *Funneliformis geosporus* under water-limited conditions (Boyer et al., 2015). Similar differences among species of EcM fungi in response to drought have also been observed in pure culture, root tip, and inoculation studies (e.g., Coleman et al., 1989; di Pietro et al., 2007; Kennedy and Peay, 2007). Root tips of beech (*Fagus sylvatica*) colonized by a widely distributed species of EcM fungus, *Cenococcum geophilum*, lost less volume and fewer electrolytes during drying than those of *Lactarius subdulcis* (di Pietro et al., 2007), confirming the greater drought tolerance of *C. geophilum* observed using other methods (e.g., Piggot, 1982). However, although *C. geophilum* tolerated both experimental drought and warming when associated with three species of oak, its relative abundance declined with drought, increased with warming, and was similar to the control in a combined drought by warming treatment (Herzog et al., 2012). Enzyme activity was also affected by drought and warming, but not in the same way as abundance (Herzog et al., 2012).

This variability in response within a single widespread fungal species raises several questions, including the importance of intraspecific variation in response to drought conditions. Despite its classification as an asexual fungus, *C. geophilum* has significant genetic diversity both within and among stands (e.g., Jany et al., 2002), and isolates are known to vary in drought tolerance (Coleman et al., 1989). Intraspecific variation in physiological traits appears to be widespread among fungi (Johnson et al., 2012), though often difficult to quantify, particularly in the obligately biotrophic AM fungi where studies of isolates are more challenging. At least within some AM fungal species, regional intraspecific genetic variation results in a spectrum of effects exerted on host plants and cosymbionts, which are related to drought tolerance and are more pronounced in arid environments (Stahl and Smith, 1984; Bethlenfalvay et al., 1989). The effects of drought on fungi also may vary with drought severity, although this is often difficult to compare across studies. For example, root colonization of *Pinus muricata* by two species of *Rhizopogon* declined significantly at soil moistures of 7% relative to 13%, with consequences for the benefits provided to host plants as well (Kennedy and Peay, 2007).

Most studies of drought effects on mycorrhizal fungi have focused on either AM fungi or EcM fungi, but a few have assessed both types of symbionts. Studies of dually colonized plant species such as members of the Salicaceae and Fagaceae families indicate that EcM fungi may be more sensitive to drought. For example, EcM colonization of oaks was reduced by drought, whereas AM colonization was not (Querejeta et al., 2009), a result similar to that observed in members of the genus *Populus*, in which higher soil moisture favored colonization by EcM fungi (Lodge, 1989; Gehring et al., 2006). In a global study in which site level intensity of root colonization by AM and EcM fungi was modeled against a suite of environmental variables, variability in precipitation was negatively associated with colonization by EcM fungi, whereas colonization by AM fungi was more responsive to temperature, particularly the temperature of the warmest month (Soudzilovskaia et al., 2015). The intensity of colonization of both types of fungi was responsive to varying soil characteristics as well, including pH for EcM fungi and nitrogen (N) availability for AM fungi. Arbuscular mycorrhizal associations predominate in more arid areas (Allen et al., 1995; Swaty et al., 2016), supporting the idea that these fungi might be better able to tolerate drought. However, variation in host plant ecology is challenging to factor out of these larger scale analyses.

Taken together, these studies suggest that drought often affects the abundance, species composition, and activity of both AM and EcM fungi. These studies also show that both species and genotypes of fungi can respond differently to changes in soil moisture even when associated with the same host plant species. Furthermore, different measures of the same trait such as fungal abundance can provide varying pictures of fungal response as observed in the study of AM fungi and strawberries described earlier (Boyer et al., 2015). The intriguing possibility that AM and EcM fungi might differ fundamentally in their drought tolerance also has some support, but the studies in which host plant species is held constant, allowing a more definitive test, are still limited.

16.2.2 Mycorrhizas and Host-Plant Drought Tolerance

Kivlin et al. (2013) and Mohan et al. (2014) examined the relationships between mycorrhizal fungi and the tolerance of plants to several global change factors, including drought, and reached similar conclusions. Kivlin et al. (2013) used a meta-analysis to assess whether plant biomass was affected by drought and whether inoculation with EcM or AM fungi altered that effect. Kivlin et al. (2013) also examined the effect of inoculation by leaf endophytes and dark septate root endophytes (DSEs). Experimentally imposed drought decreased plant biomass, but inoculation with DSEs, AM fungi, and EcM fungi significantly ameliorated that negative effect (Kivlin et al., 2013). The largest amelioration was observed with DSEs, with similar effect sizes for AM and EcM fungi. Mohan et al. (2014) used a vote-counting approach and a more restricted set of literature selection criteria (nonagricultural plant species studied since 1990 with a required suite of measurements) but reached similar conclusions. In more than 90% of 32 studies, plant productivity/growth under drought conditions was improved by inoculation with mycorrhizal fungi.

Two more focused meta-analyses of plant responses to inoculation with AM fungi under water-stressed and well-watered conditions found broadly similar results, but also showed that responses varied among plant hosts (Worchel et al., 2013; Jayne and Quigley, 2014) although not among taxa of AM fungi (Jayne and Quigley, 2014). Jayne and Quigley (2014) reviewed 54 papers, finding that inoculation with AM fungi improved above-ground, below-ground, and reproductive measurements of plant growth, although the effect sizes were generally smaller for below-ground growth. They also observed that the benefits of AM fungal inoculation were greater for perennial than annual plants, but that the seven most studied species of AM fungi had similar effects on plant growth (Jayne and Quigley, 2014). Interestingly, a mix of AM fungal species had the strongest positive effect on measures of plant reproduction (Jayne and Quigley, 2014). Grasses using the C3 photosynthetic pathway benefited more from AM fungi than C4 grasses under both well-watered and drought conditions (Worchel et al., 2013). Jayne and Quigley (2014) also concluded that inoculation with AM fungi had similar effects on plant growth in well-watered and water-stressed plants, but the effect size was more than 20% greater for water-stressed plants, a marginally significant difference ($p=0.053$; Jayne and Quigley, 2014). More pronounced benefits of inoculation by AM fungi as water limitation increased were also observed among the studies of grasses reviewed by Worchel et al. (2013).

Kivlin et al. (2013) and Mohan et al. (2014) pointed out that the studies they reviewed were biased in a number of ways and we expand upon this issue using the same studies. A majority of the studies were conducted in the greenhouse, and many of the studies of AM fungi involved crop-plant species (Kivlin et al., 2013; Mohan et al., 2014). Most of the nonagricultural studies reviewed by Mohan et al. (2014) were of short duration and focused on temperate ecosystems in North America and Europe. We tabulated data on plant and fungal species from these two reviews and observed that studies to date included only a small number of plant and fungal taxa. Among plants, most studies of drought tolerance/AM fungal relationships focused on crops, with nearly 30% focused on wheat and corn. Similarly, studies of drought/EcM fungal relationships were dominated by members of the Pinaceae. Representation among fungi was even more restricted. Of the 27 genera of AM fungi, only seven have been examined, with the genus *Rhizophagus* accounting for nearly 40% of studies using identified fungi (some studies used a mixture of species, soil inoculum, or unidentified field isolates). Among the highly diverse EcM fungi, only 12 of the 216 genera described as EcM by Tedersoo et al. (2010) have been studied, with an emphasis on basidiomycetes (10 of 12 genera).

The mechanisms by which both AM and EcM fungi influence plant–water relationships, and thus potentially drought tolerance, are varied and there are excellent reviews describing them (Augé, 2001; Lehto and Zwiazek, 2011; Augé et al., 2015). Augé (2001) reviewed more than 200 papers on AM fungal–plant–water relationships, finding that inoculation with AM fungi improved growth during drought nearly 80% of the time. The mechanisms for this effect included improved P uptake from dry soil (Augé, 2001), but there were also significant direct effects on measures of plant–water balance. More recently, Augé et al. (2015) reviewed the influence of inoculation with AM fungi on stomatal conductance to water vapor using a meta-analysis of 460 studies. Increased stomatal conductance of water vapor reflects transpiration rate and the inverse process, CO_2 influx, and therefore is an indicator of increased potential for photosynthesis by the plant. AM fungi generally increased stomatal conductance relative to nonmycorrhizal controls, even in well-watered conditions, but increases were substantially larger during moderate (two-fold) and severe (four-fold) drought. The largest promotion of stomatal conductance was observed after inoculation with native AM fungi and in mineral rather than organic soils (Augé et al., 2015). Interestingly, the patterns observed by Worchel et al. (2013) for growth promotion by AM fungi under drought by C3 versus C4 grasses were supported by studies of stomatal conductance, which were increased more in C3 than C4 plants (Augé et al., 2015). Amount of root colonization also emerged as an important predictor of increased stomatal conductance, as did plant size and P content, supporting the idea that it can be difficult to disentangle the direct and indirect effects of AM fungi on plant–water relationships (Smith et al., 2010). Although it is challenging to convincingly demonstrate its importance to plant drought tolerance (Smith et al., 2010), there is evidence that AM fungal hyphae can directly take up water from the soil during drought (Marulanda et al., 2003; Khalvati et al., 2005).

Although it has been demonstrated that EcM fungal hyphae can take up water from the soil in sufficient quantities to influence seedling performance (Boyd et al., 1986), Lehto & Zwiazek (2011) concluded that there was little evidence that this uptake is sufficient to influence the performance of mature trees. Instead they suggest that other influences of EcM fungi on host-plant–water relations were better substantiated, such as increased stomatal conductance to water vapor, altered hydraulic conductance of mycorrhizal roots, and osmotic adjustments facilitated by EcM fungi (Lehto and Zwiazek, 2011). However, these authors also point out that the mechanisms with the most support are those that are indirect, mediated through altered host-plant nutrition (Lehto and Zwiazek, 2011). In addition, the physiological responses of plants to inoculation with EcM fungi were highly variable and dependent upon the species and even genotype of fungus, as highlighted previously for studies of drought effects on mycorrhizal fungi. Lehto and Zwiazek (2011) suggest that future research should emphasize field studies using innovative approaches.

Although conducted in a greenhouse using field soil, a study of beech (*F. sylvatica*) and its ectomycorrhizas illustrates how manipulative experiments can be accompanied by isotopic labeling and molecular identification of fungi to yield new insights (Pena and Polle, 2014). Field soil inoculated and nonmycorrhizal beech seedlings were exposed to fully factorial treatments of light intensity (shade and full light) and soil moisture (normal and drought) and irrigated with ^{15}N-labeled water to quantify N uptake by the most common species of EcM fungi relative to nonmycorrhizal root tips. N uptake was enhanced with EcM fungi, but only under stressful conditions. Furthermore, the dominant species of EcM fungi colonizing beech roots varied in their responses to drought and shading treatments (Pena and Polle, 2014), demonstrating significant functional diversity, even within a single genus (*Tuber*).

16.2.3 Conclusions and Suggested Directions for Future Research

Overall, there is strong evidence that colonization by both AM and EcM fungi promotes drought tolerance in host plants and there are intriguing suggestions that mycorrhizal fungi have larger positive effects on plant–water relationships as drought severity increases. This result is encouraging, given the projections for increased drought in many areas of the world. However, there is much to learn, because studies are geographically (Mohan et al., 2014) and taxonomically limited for both plants and fungi. Studies of mycorrhizal fungal responses to drought stress and mycorrhizal fungal influences on host-plant–water relations are often done independently from one another, limiting our understanding of the effects of drought on the symbiosis. How well do we understand the function of the species of mycorrhizal fungi that increase the most during drought? Intraspecific variation in drought response is also important in both plants and fungi, but is most commonly considered in agricultural and plantation settings and less commonly in natural populations. Intriguing differences exist among plant functional groups, such as C3 and C4 grasses, but there is less evidence for similar differences among lineages of EcM and AM fungi where members of the same genus can respond very differently (e.g., Pena and Polle, 2014; Tedersoo et al., 2012). The mechanisms by which mycorrhizal fungi influence plant–water relationships are diverse and variable, but the consistent increases in AM colonized versus nonmycorrhizal plants in stomatal conductance to water vapor (Augé et al., 2015) suggest that this mechanism could be of widespread importance.

Based on these conclusions, we suggest several avenues of research. First, studies of more species of fungi and plants in more areas of the world are necessary. Intraspecific variation in plants and fungi also should be considered, although this may seem daunting, given how little we know about the drought tolerance of plants and fungi at the species level. An emphasis on functional traits as described across guilds of root infecting fungi by Aguilar-Trigueros et al. (2014) might be helpful. EcM fungi have been categorized by their hyphal exploration types in the soil (e.g., short distance, long distance; Tedersoo and Smith, 2013), a trait that might influence drought tolerance. Mohan et al. (2014) also point out the importance of studying multiple global change factors simultaneously; experimentally manipulating increased temperature with changes in moisture would be the most relevant with regard to drought given climate change predictions. We also suggest an emphasis on studies that combine measurements of both fungal and plant performance during drought, particularly in settings involving multiple fungal species or communities (e.g., Kennedy and Peay, 2007; Pena and Polle, 2014) and potentially also considering plants at a community scale, which rarely has been done but could be key to understanding changes in natural ecosystems with drought. For EcM systems in particular, we echo Lehto and Zwiazek's (2011) call for longer term studies that include analysis of the importance of EcM fungi to plant recovery from drought.

16.3 DROUGHT-RELATED HOST MORTALITY AND CONSEQUENCES FOR MYCORRHIZAS

The previous section of this chapter focuses on how the relationships among host plants and mycorrhizal fungi could be influenced by drought, generally focusing on situations in which the host and the fungus both survive. In many parts of the world severe drought has resulted in extensive host-plant mortality (McDowell et al., 2008). In this section of the chapter, we explore the consequences of host-plant mortality for mycorrhizal associations, emphasizing impacts on fungi, host-plant regeneration, and ecosystem processes.

Drought has caused plant mortality across six continents (McDowell et al., 2008) and diverse ecosystems, including tropical moist forests (Williamson et al., 2000), temperate forest and woodland (Suarez et al., 2004; Breshears et al., 2005), grassland (Moran et al., 2014), and desert (Hamerlynck and McAuliffe, 2008). In many cases, plant mortality has been focused on foundation plant species that have profound effects on associated communities and ecosystem properties (Breshears et al., 2005). Mortality events are widespread in forested ecosystems and appear to disproportionately affect larger trees, which play key roles in ecosystem C storage (Bennett et al., 2015). In some ecosystems, differential drought-related plant mortality could alter mycorrhizal associations with potentially long-term consequences for nutrient cycling and C storage. We use the pinyon-juniper woodlands of the southwestern United States as a case study to explore the consequences of differential host mortality on mycorrhizas and soils.

16.3.1 Pinyon Pine Mortality and Mycorrhizas

Pinyon-juniper woodlands, one of the largest vegetation types in the United States (West, 1984), have experienced extensive mortality from drought, particularly in the early part of the 21st century (Breshears et al., 2005; Mueller et al., 2005). Aerial surveys conducted by the United States Forest Service revealed significant tree mortality over 12,000 km^2 of the southwestern United States (Breshears et al., 2005). Mortality rates of pinyon and juniper were starkly different, with rates as high as 97% for some pinyon (*Pinus edulis*) populations but a far lower maximum mortality (25%) for juniper (*Juniperus* spp.) (Breshears et al., 2005; Mueller et al., 2005a). As observed in many other tree species (Bennett et al., 2015), pinyon mortality was higher in mature trees than juveniles (Mueller et al., 2005a). Pinyon and juniper also differ markedly in their hydraulic responses to drought; pinyon is isohydric, whereas juniper is anisohydric. Isohydric plants regulate stomatal conductance to maintain minimum water potential within a narrow range, reducing the risk of xylem cavitation but decreasing uptake of C, which can lead to C starvation (McDowell et al., 2008). Anisohydric species allow leaf-water potentials to decrease during drought, potentially incurring a great risk of xylem cavitation (McDowell et al., 2008). Several studies have predicted that ongoing climate change will lead to a dramatic change in the distribution of *P. edulis*, including its extirpation in the state of Arizona and large portions of other southwestern states (Rehfeldt et al., 2006; Cole et al., 2008).

Pinyon and juniper also differ in mycorrhizal associations, because pinyon forms EcM associations, whereas juniper and many of the other dominant plants of pinyon-juniper woodlands form AM associations (Haskins and Gehring, 2005). Although little is known about juniper mycorrhizas, pinyon has been comparatively well studied. The EcM communities of *P. edulis* are low in diversity and commonly dominated by ascomycete fungi, although this dominance varies among sites (Gehring et al., 1998), with climate (Gordon and Gehring, 2011; Gehring et al., 2014a), with the abundance of biotic stressors (Gehring et al., 2014b), and with host genetics (Sthultz et al., 2009a). Interestingly, the putatively drought tolerant ascomycete, *C. geophilum*, has not been commonly observed in studies of *P. edulis*, with ascomycetes in the genera *Geopora* and *Tuber* observed more often. Pinyon EcM fungi were more sensitive to N fertilization under drought conditions than the AM fungi of juniper (Allen et al., 2010). Comparisons of sites with high and low *P. edulis* mortality showed reduced abundance and species richness of EcM fungi on surviving trees at high mortality sites, with abundance shifts highly correlated with trunk growth (Swaty et al., 2004).

These studies indicate that pinyon is likely to have a severely contracted range as climate continues to change, whereas juniper will survive, leading to a vegetation shift from pinyon-juniper woodland to juniper grassland. We predict that a shift in mycorrhizal associations would be concomitant with this shift in dominant vegetation. We quantify this potential shift in mycorrhizas using pinyon-juniper woodland distribution data obtained from LANDFIRE (www.landfire.gov, version 1.2.0), a program that provides vegetation, fuel, disturbance, and fire regimes geospatial data products for the United States using peer-reviewed methods from multiple disciplines. Specifically, we utilized LANDFIRE's Biophysical Settings (BpS) descriptions and spatial dataset (www.landfire.gov, version 1.2.0) to estimate the land area of the western United States covered by pinyon-juniper woodland. The BpS in LANDFIRE represent the vegetation that would have been dominant on a particular site just before major European settlement (1800s) and includes comprehensive descriptions and linkage to disturbance models (www.landfire.gov; Rollins, 2009). We have combined data on mycorrhizal associations of dominant vegetation with BpS data previously to generate a map of the potential mycorrhizal associations of the conterminous United States (Swaty et al., 2016). We classified the landscape into one of 30 mycorrhizal categories based on the types of mycorrhizas formed by the dominant plants in that vegetation type. For example, the dominant vegetation of the BpS called *East Gulf Coastal Plain Northern Mesic Hardwood Slope Forest* consists of four species of trees that form EcM associations and two tree species that form AM associations and is categorized as EcM/am (Swaty et al., 2016). Here we focus only on data we tabulated from pinyon-juniper woodlands using our preestablished categories to characterize the mycorrhizal associations of pinyon-juniper woodlands with *P. edulis* present versus absent. Although *P. edulis* is not predicted to be lost from its entire current range due to climate change, dramatic reductions in its distribution (~80%) are predicted during this century (Rehfeldt et al., 2006; Cole et al., 2008). We removed *P. edulis* entirely as an heuristic exercise to explore the consequences of large-scale host mortality for fungi and soils.

Forests and woodlands dominated by *P. edulis* occupied an estimated 8.1 million hectares of the western United States before European settlement (Fig. 16.1). Extant vegetation just before massive *P. edulis* mortality in 2002 was similar to that estimated by LANDFIRE for the 1800s, because these particular ecosystems have not seen substantial conversion to agriculture or urban land uses (www.landfire.gov). These ecosystems occur across a variety of soil types and include a variable mix of other dominant vegetation, usually one or more species of juniper (e.g., *Juniperus monospera, Juniperus osteosperma*); other pines including *Pinus ponderosa* and a more arid-adapted species of pinyon, *Pinus monophylla*, a couple of species of oak (e.g., *Quercus gambelli*); various shrubs (e.g., *Artemisia tridentata*); and some grasses (e.g., *Bouteloua gracilis*). With *P. edulis* present, we identified three mycorrhizal categories based on this variation in dominant vegetation: EcM/am, in which EcM vegetation dominated, but AM vegetation was also important as in an EcM canopy and AM understory; EcM/AM, in which AM and EcM associations codominated; and AM/ecm, in which AM associations dominated, but EcM vegetation was also present, as in pinyon-juniper shrubland (Fig. 16.1). Most of the land area (84%) occupied by pinyon-juniper woodland was EcM/AM-codominated vegetation or AM-dominated (Fig. 16.1).

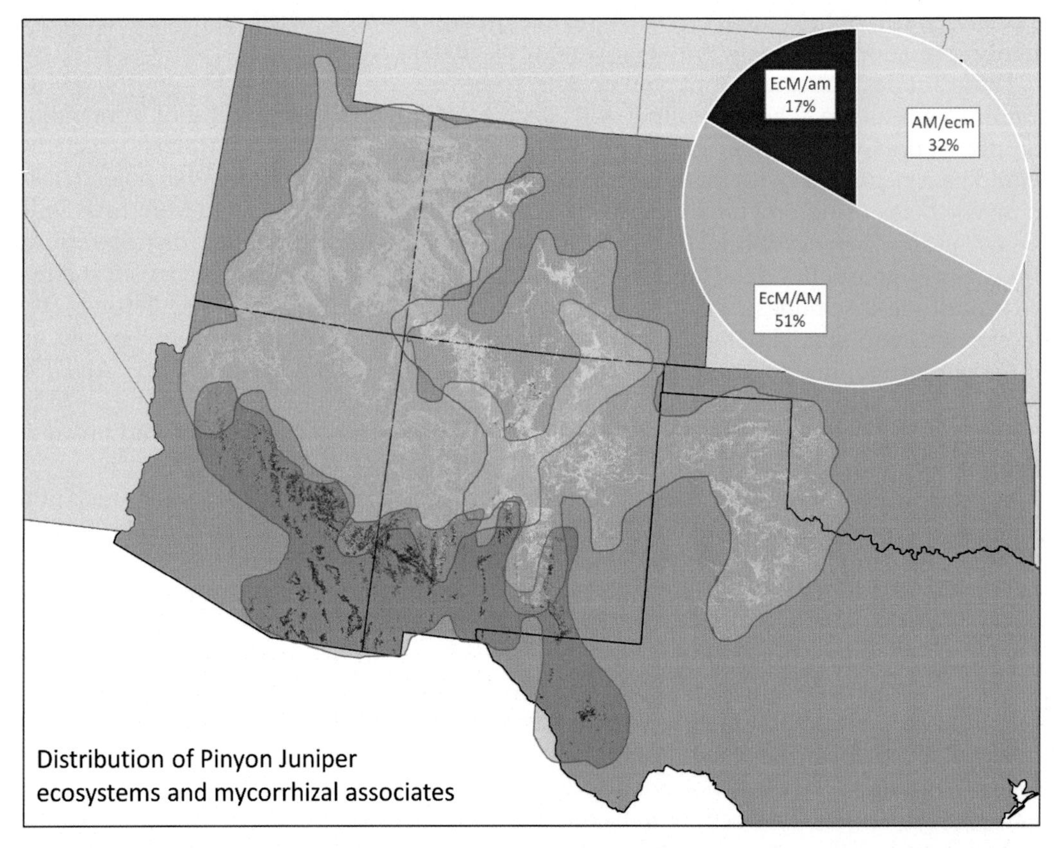

FIGURE 16.1 Map of the four corners states of Utah, Colorado, New Mexico, and Arizona (clockwise from upper left), and northern Texas showing the historical distribution of pinyon-juniper woodland ecosystems containing *Pinus edulis* as determined by the LANDFIRE program (www.landfire.gov). The types of mycorrhizal associations found in the regions are shown in color and their percentage area of the historical landscape in the inset pie chart. *Uppercase letters* indicate the dominant mycorrhizal type and *lowercase designations* indicate the associated less dominant type (e.g., *EcM/am*, ectomycorrhizal-dominant overstory with arbuscular mycorrhizal grass and forb understory). Polygons around the types of mycorrhizal association are included to aid visualization and do not indicate actual boundaries.

Large changes in mycorrhizal associations would occur with the loss of *P. edulis* from the landscape. AM associations, without EcM associations, would dominate (58% of land area) and EcM-dominated areas (EcM/am) would disappear (Fig. 16.2). Vegetation colonized by EcM fungi would occur across less than 45% of the land area (Fig. 16.2). Fig. 16.3 provides a slightly different perspective on these patterns, showing the land area where EcM-associated vegetation would still occur after *P. edulis* mortality. Approximately 4.7 million hectares of pinyon-juniper woodland would no longer contain significant numbers of EcM host plants after the demise of *P. edulis*, whereas the remaining land area (~3.4 million hectares) could retain EcM hosts including other species of pine and oaks. Although surrounding vegetation types are not shown, large areas of land in Utah and Arizona would no longer have dominant species that form EcM associations. Loss of tree species in the southwestern United States

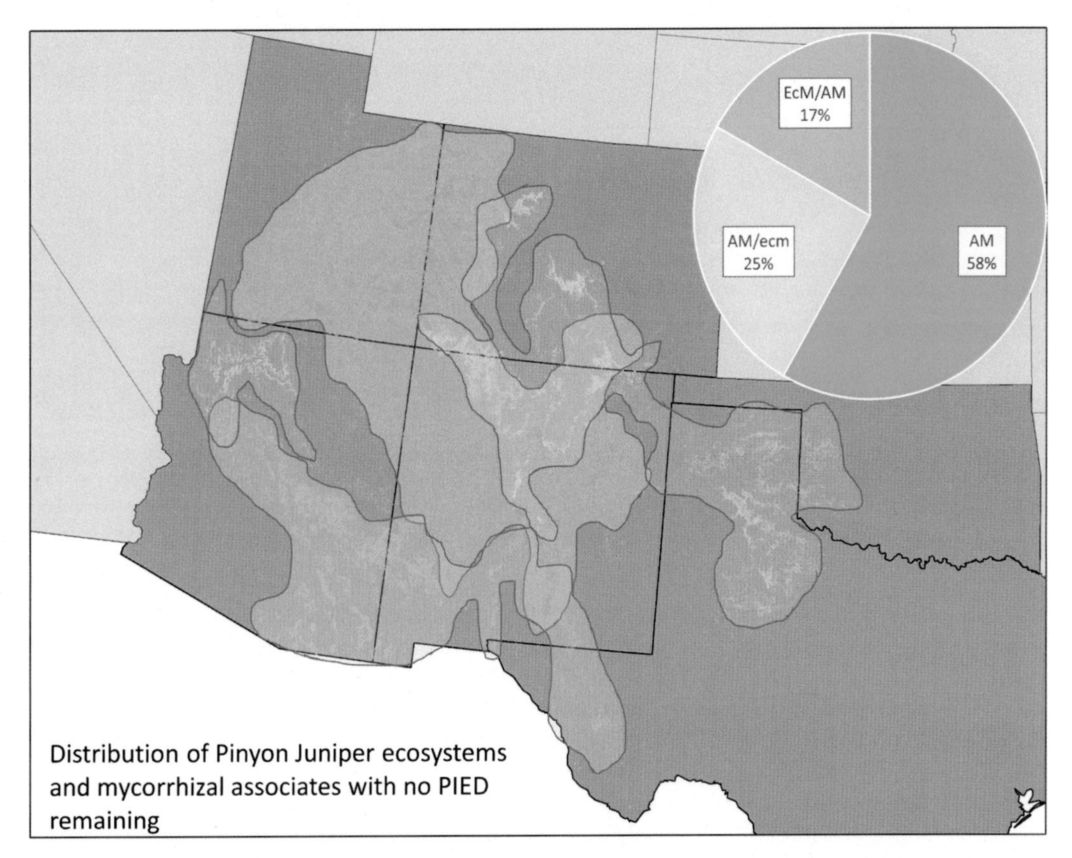

FIGURE 16.2 Map of the four corners states of Utah, Colorado, New Mexico, and Arizona (clockwise from upper left), and northern Texas showing how the types of mycorrhizal associations of formerly *Pinus edulis (PIED)*–dominated ecosystems would change after the anticipated demise of *P. edulis*. The percentage of the landscape occupied by the types of mycorrhizal association is shown in the inset pie chart. *Uppercase letters* indicate the dominant mycorrhizal type and *lowercase designations* indicate the associated less dominant type (e.g., *EcM/am*, ectomycorrhizal-dominant overstory with arbuscular mycorrhizal grass and forb understory). Polygons around the types of mycorrhizal association are included to aid visualization and do not indicate actual boundaries.

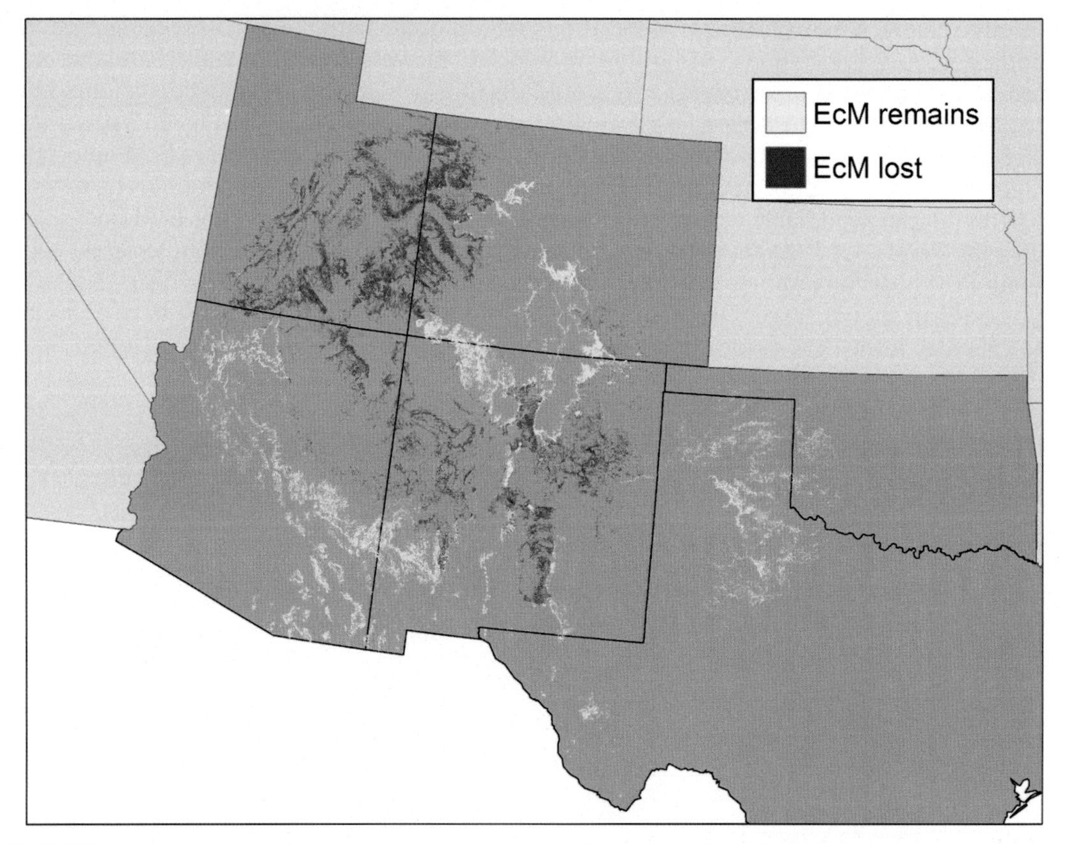

FIGURE 16.3 Map of the four corners states of Utah, Colorado, New Mexico, and Arizona (clockwise from upper left), and northern Texas highlighting the areas with *Pinus edulis*–dominated ecosystems and showing the persistence or loss of ectomycorrhizal associations upon the demise of *P. edulis*. Loss of ectomycorrhizal associations would occur primarily in areas where *P. edulis* was the sole ectomycorrhizal host. *EcM*, ectomycorrhizal.

may extend beyond *P. edulis*, with one study predicting the demise of all needle-leaved evergreen trees by 2100 (McDowell et al., 2016). The result would be extensive loss of EcM fungi, because the landscape would be dominated by AM-associated grasses and shrubs.

The consequences of these vegetation changes for EcM fungi could be extensive. The species richness of EcM fungi in these landscapes appears to be low and includes some species with widespread distributions such as *Lactarius deliciosus* and *Rhizopogon roseolus* (Gehring et al., 2014a). However, several species, particularly members of the division Ascomycota appear to be undescribed. Flores-Renteria et al. (2014) documented a new species of *Geopora* in the pinyon-juniper woodlands of northern Arizona and the phylogeny they developed included several other novel members of the genus from their study sites. Correlative field studies show that pinyon growth is positively correlated with the abundance of *Geopora* within a root-tip community (Gehring et al., 2014b), yet we know little about the functional traits of members of this genus. Apparently novel members of the genus *Tuber* also have been observed (Gehring, personal observation), suggesting that the loss of pinyon across much of its range could also mean the loss of significant undescribed fungal diversity. The fate of EcM fungi will depend, in part, on how long they are able to survive in the absence of a host. The ability of EcM fungi to acquire sufficient C through saprotrophic activity to sustain fungal growth is discussed in detail by Kuyper (Chapter 20), who finds little support for this "alternate C source hypothesis." Fungi could also survive as resistant propagules. Although few studies on the survival of the propagules of EcM fungi have been conducted, Bruns et al. (2009) and Nguyen et al. (2012) showed that survival varies among species and can be as long as 6 years for EcM fungi of *P. muricata* in California.

Given that *P. edulis* may suffer extirpation from a part of its range and EcM fungal propagule persistence may be limited, what are the prospects for maintenance of a pine root symbiont spore bank in the soil by immigration from neighboring areas? Mycorrhizal fungal spores are commonly airborne, but animals are also important vectors for spore dispersal, particularly for AM fungi and the sequestrate EcM fungi (Allen, 1991). Peay et al. (2012) observed that for *P. muricata*, EcM fungal spore dispersal exhibited patterns predicted by island biogeography theory (MacArthur and Wilson, 1967). Dispersal dropped off quickly with distance from propagule source, as did species richness. The long-distance dispersers were those species most abundant and prolific in spore production (Peay et al., 2012). Besides dispersal ability, successful mycorrhizal establishment was also influenced by the competitive ability of the fungus. Long-distance dispersers were not as successful at short distances, where they faced competition from the short-distance dispersers. In addition, several species collected in spore traps were not found on roots, indicating other factors at play (Peay et al., 2012). The dispersal ability of pinyon EcM fungi has not been assessed but is likely to differ markedly from that seen in the primarily airborne inocula of *P. muricata* (Peay et al., 2012). In contrast, pinyon EcM fungi are primarily composed of sequestrate or semisequestrate species that sporulate rarely (Gehring et al., 2014a). Our hypothetical redistribution of EcM symbioses based on a predicted future *P. edulis* distribution (Figs. 16.2 and 16.3) could create huge swathes of landscape extremely isolated from sources of EcM inoculum.

Loss of EcM hosts and fungi could have other consequences, including altering C storage and nutrient cycling. In Chapter 26 of this volume, Brzostek and colleagues briefly review studies both across ecosystems and within temperate forests, showing that shifts from AM to EcM vegetation are accompanied by decreasing litter quality, as reflected by higher C/N ratios, increasing N rather than P limitation, and increasing reliance on mycorrhizal symbionts to mobilize organic nutrients. EcM fungi are also thought to require more C investment by plants and to have slower turnover rates than AM fungi (Zhu and Miller, 2003; Ekblad et al., 2013). However, some of these trends do not apply in all ecosystems, such as EcM-dominated dipterocarp forests or AM-dominated podocarp forests (Chapters 19 and 25). Whether loss of *P. edulis* and its associated EcM fungi from millions of hectares of the western United States would alter soil and ecosystem properties (aside from loss of the living trees themselves), likely depends on how well *P. edulis* and species of *Juniperus* follow the AM/EcM trends identified for litter decomposition and other traits in other temperate ecosystems. In a comparison of leaf/needle litter decomposition of *P. edulis* and *J. monosperma*, *J. monosperma* lost 40% of its original mass in 26 weeks, whereas *P. edulis* lost only 10% (Gallo et al., 2006), a finding qualitatively similar to Murphy et al. (1998) and Phillips et al. (2013). As mentioned, comparatively little is known about the functional traits of the EcM fungi associated with *P. edulis*, either in terms of enzyme activity or rates of hyphal turnover, although EcM colonized *P. edulis* roots decomposed significantly more slowly than nonmycorrhizal *P. edulis* roots (Langley and Hungate, 2003). The data available suggest that shifts in ecosystem dynamics with loss of *P. edulis* are likely when considering differences in their mycorrhizal associates, but further study is necessary. Roman et al. (2015) use oaks as an example to propose that ecosystem-scale C dynamics differ significantly between isohydric and anisohydric plants during drought, with isohydric species like *P. edulis* significantly reducing gas exchange during drought, in contrast to oak species, which are anisohydric and maintain constant rates of gas exchange. Soil and ecosystem processes are thus likely to shift in pinyon-juniper woodlands as drought increases, even if *P. edulis* does not experience the extensive mortality that several models project.

Extensive *P. edulis* mortality, even if spatially variable and concentrated on adult trees, is likely to have consequences for *P. edulis* regeneration. Seed-cone production in *P. edulis* was 40% lower in the decade after extensive mortality than in a previous, wetter decade in New Mexico and Oklahoma (Redmond et al., 2012), whereas seedling recruitment was low across sites in northern Arizona (Redmond et al., 2015). In many stands, juvenile *P. edulis* survived the 2002–04 mortality event, but their postdrought survival was significantly lower than that of juniper and positively associated with soil available-water capacity and the presence of certain species of nurse trees or shrubs (Redmond et al., 2015). EcM fungal inoculum can be low in juniper-dominated ecosystems (Haskins and Gehring, 2005), compounding these other reestablishment difficulties. However, as shown in Fig. 16.3, this problem could be overcome in the portions of these woodlands that contain other EcM vegetation. Also, some genotypes of mature *P. edulis* survived drought more than three times better than other genotypes, with offspring showing the same patterns as their mothers (Sthultz et al., 2009b). Drought-tolerant and drought-intolerant *P. edulis* had consistent differences in EcM communities, with *Geopora* dominance associated with low mortality during extreme drought and high growth during ongoing but less extreme drought conditions (Gehring et al., 2014a). Although the outlook for *P. edulis* and its associated EcM fungi appears dire, drought-tolerant genotypes and their associated fungi offer some hope for survival.

16.3.2 Implications for Other Ecosystems

The *P. edulis* example highlights the consequences of the mortality of a plant species that forms a type of mycorrhizal association not commonly found in other associated vegetation. This may not be a common occurrence, but drought-induced mortality is predicted to fall heavily on needle-leaved evergreen trees throughout the western United States (McDowell et al., 2016), potentially also leaving behind a landscape dominated by AM-associated vegetation with accompanying changes in C and nutrient dynamics in the soil. However, oaks (*Quercus*) may represent an exception in some areas. In the western United States, arid climates have selected for drought-tolerant oak species. In a review of forest mortality induced by drought as a result of global climate change, California has already experienced forest mortality, primarily in montane mixed-conifer forests (Allen et al., 2010). Historical data from 1929–36, when compared with data from 2001–10 (McIntyre et al., 2015), show increasing oak dominance at the expense of pine in California. Patterns in the western United States contrast with those in the eastern, southeastern and midwestern states, where oaks are experiencing drought-associated mortality (Allen et al., 2010). However, a potential problem for the future of oak forests in California is recruitment.

Even in areas where mature oaks can tolerate fairly intense levels of drought once they have connected to permanent ground water, the seedling-to-sapling age cohort is a chokepoint where mortality is particularly high because oaks at this life stage rely on ephemeral water supplies (Mahall et al., 2009). Estimates based on adult tolerance of drought will be overly optimistic of a species' ability to persist in a particular area (McLaughlin and Zavaleta, 2012). Juvenile tolerance of increased temperature in *Quercus lobata* is lower by 3°C (summer maximum temperature) compared to mature trees, and rather than moving higher in latitude and elevation, the species will likely constrict around water bodies (McLaughlin and Zavaleta, 2012). Constriction of oaks around refugia that fall within the limits of seedling tolerance present an additional problem: reproductive and genetic isolation. Although a wind-pollinated genus, the number of oak fathers is declining as a result of geographic isolation for at least one species studied and is expected to decline further, resulting in increased selfing and loss of fitness (Sork et al., 2002).

Predicting the consequences of changes in oak distribution for mycorrhizal fungi is more complicated than it is for pines because of variability in the mycorrhizal associations they form. Most studies of oak mycorrhizas focus on EcM fungal associates and treat oaks as primarily EcM (e.g., Brundrett, 2009). Several studies have observed AM-colonized roots, particularly in red oaks, *Quercus* section Lobatae (see Dickie et al., 2001) but these have been mostly mesic forest species with few examples from arid lands (Querejeta et al., 2003). By way of contrast, there has been virtually no examination of white oaks, *Quercus* section *Quercus*. Watson et al. (1990) did not detect AM fungi in *Quercus alba* but they have been documented in a western scrub oak, *Quercus turbinella* (Mueller et al., 2005b).

These studies of *P. edulis* and oaks suggest that drought-induced host-plant mortality could have significant consequences for mycorrhizal fungi and the soil and ecosystem processes they influence. In the temperate zone of the northern hemisphere, recent tree-mortality events were concentrated on hosts of EcM fungi (e.g., *Pinus, Picea, Abies*). Members of the genera *Eucalyptus* and *Nothofagus*, which also form EcM associations, experienced extensive drought-related mortality in Australia and New Zealand. However, like oaks, members of the genus *Eucalpytus* can form associations with both AM and EcM fungi (Brundrett, 2009). Given that so little is known about the mycorrhizal interactions of dual host plants, this would be a productive area for future research, particularly given the potentially differential consequences of loss of EcM versus AM fungi for soil C and N dynamics. In addition, there is a lack of general knowledge about mycorrhizas and mycorrhizal fungi of arid land foundation-tree species, many of which are unidentified and undescribed. Key questions to examine would include the cost and benefit of arid land mycorrhizas (including the relation to host age cohort), community shifts with increasing drought, effects of a changing climate on mycorrhizal diversity at a variety of scales, and the ability of plant and fungal symbioses to remain coupled in the face of changing climate, differential dispersal of symbionts, and the extirpation of interacting species.

Acknowledgments

We thank NSF EF-1340852 for support of CAG and RJD. We acknowledge Haley Michael and The Nature Conservancy's LANDFIRE team (Dr. Jim Smith, Kori Blankenship, Sarah Hagen, Kim Hall and Jeannie Patton) for database and/or GIS support.

References

Aguilar-Trigueros, C.A., Powell, J.R., Anderson, I.C., Antonovics, J., Rillig, M.C., 2014. Ecological understanding of root-infecting fungi using trait-based approaches. Trends in Plant Science 19, 432–437.

Allen, E.B., Allen, M.F., Helm, D.J., Trappe, J.M., Molina, R., Rincon, E., 1995. Patterns and regulation of mycorrhizal plant and fungal diversity. Plant and Soil 170, 47–62.

Allen, C.D., Macalady, A.K., Chenchouni, H., Bachelet, D., McDowell, N., Vennetier, M., Kitzberger, T., Rigling, A., Breshears, D.D., (Ted) Hogg, E.H., et al., 2010. A global overview of drought and heat-induced tree mortality reveals emerging climate change risks for forests. Forest Ecology and Management 259, 660–684.

Allen, M.F., 1991. The Ecology of Mycorrhizas. Cambridge University Press, UK.

Anderegg, W.R.L., Kane, J.M., Anderegg, L.D.L., 2013. Consequences of widespread tree mortality triggered by drought and temperature stress. Nature Climate Change 3, 30–36.

Attia, Z., Domec, J.C., Oren, R., Way, D.A., Moshelion, M., 2015. Growth and physiological responses of isohydric and anisohydric poplars to drought. Journal of Experimental Botany 66, 4373–4381.

Augé, R.M., Toler, H.D., Saxton, A.M., 2015. Arbuscular mycorrhizal symbiosis alters stomatal conductance of host plants more under drought than under amply watered conditions: a meta-analysis. Mycorrhiza 25, 13–24.

Augé, R.M., 2001. Water relations, drought and vesicular-arbuscular mycorrhizal symbiosis. Mycorrhiza 11, 3–42.

Azul, A.M., Sousa, J.P., Agerer, R., Martín, M.P., Freitas, H., 2010. Land use practices and ectomycorrhizal fungal communities from oak woodlands dominated by *Quercus suber* L. considering drought scenarios. Mycorrhiza 20, 73–88.

Bennett, A.C., McDowell, N.G., Allen, C.D., Anderson-Teixeira, K.J., 2015. Larger trees suffer most during drought in forests worldwide. Nature Plants 1, 15139.

Bethlenfalvay, G.J., Franson, R.L., Brown, M.S., Mihara, K.L., 1989. The *Glycine-Glomus-Bradyrhizobium* symbiosis. IX. Nutritional, morphological and physiological responses of nodulated soybean to geographic isolates of the mycorrhizal fungus *Glomus mosseae*. Physiologia Plantarum 76, 226–232.

Boyd, R., Furbank, R., Read, D., 1986. Ectomycorrhiza and the water relations of trees. In: Gianinazzi-Pearson, V., Gianinazzi, S. (Eds.), Physiological and Genetic Aspects of Mycorrhizas. INRA, Paris, pp. 689–694.

Boyer, L.R., Brain, P., Xu, X.M., Jeffries, P., 2015. Inoculation of drought-stressed strawberry with a mixed inoculum of two arbuscular mycorrhizal fungi: effects on population dynamics of fungal species in roots and consequential plant tolerance to water deficiency. Mycorrhiza 25, 215–227.

Breshears, D.D., Cobb, N.S., Rich, P.M., Price, K.P., Allen, C.D., Balice, R.G., Romme, W.H., Kastens, J.H., Floyd, M.L., Belnap, J., et al., 2005. Regional vegetation die-off in response to global-change-type drought. Proceedings of the National Academy of Sciences of the United States of America 102, 15144–15148.

Brundrett, M.C., 2009. Mycorrhizal associations and other means of nutrition of vascular plants: understanding the global diversity of host plants by resolving conflicting information and developing reliable means of diagnosis. Plant and Soil 320, 37–77.

Brunner, I., Herzog, C., Dawes, M.A., Arend, M., Sperisen, C., 2015. How tree roots respond to drought. Frontiers in Plant Science 6, 547.

Bruns, T.D., Peay, K.G., Boynton, P.J., Grubisha, L.C., Hynson, N.A., Nguyen, N.H., Rosenstock, N.P., 2009. Inoculum potential of *Rhizopogon* spores increases with time over the first 4 yr of a 99-yr spore burial experiment. New Phytologist 181, 463–470.

Brzostek, E.R., Dragoni, D., Schmid, H.P., Rahman, A.F., Sims, D., Wayson, C.A., Johnson, D.J., Phillips, R.P., 2014. Chronic water stress reduces tree growth and the carbon sink of deciduous hardwood forests. Global Change Biology 20, 2531–2539.

Buentgen, U., Egli, S., Galvan, J.D., Diez, J.M., Aldea, J., Latorre, J., Martinez-Pena, F., 2015. Drought-induced changes in the phenology, productivity and diversity of Spanish fungi. Fungal Ecology 16, 6–18.

Choat, B., Jansen, S., Brodribb, T.J., Cochard, H., Delzon, S., Bhaskar, R., Bucci, S.J., Feild, T.S., Gleason, S.M., Hacke, U.G., et al., 2012. Global convergence in the vulnerability of forests to drought. Nature 491, 752–755.

Cole, K.L., Fisher, J., Arundel, S.T., Cannella, J., Swift, S., 2008. Geographical and climatic limits of needle types of one- and two-needled pinyon pines. Journal of Biogeography 35, 257–269.

Coleman, M.D., Bledsoe, C.S., Lopushinsky, W., 1989. Pure culture response of ectomycorrhizal fungi to imposed water stress. Canadian Journal of Botany 67, 29–39.

Cook, B.I., Anchukaitis, K.J., Touchan, R., Meko, D.M., Cook, E.R., 2016. Spatiotemporal drought variability in the Mediterranean over the last 900 years. Journal of Geophysical Research: Atmospheres 121, 2060–2074.

Dai, A., 2011. Drought Under Global Warming: A Review. Wiley Interdisciplinary Reviews: Climate Change, vol. 2, pp. 45–65.
Dai, A.G., 2013. Increasing drought under global warming in observations and models. Nature Climate Change 3, 52–58.
Di Pietro, M., Churin, J.L., Garbaye, J., 2007. Differential ability of ectomycorrhizas to survive drying. Mycorrhiza 17, 547–550.
Dickie, I.A., Koide, R.T., Fayish, A.C., 2001. Vesicular-arbuscular mycorrhizal infection of *Quercus rubra* seedlings. New Phytologist 151, 257–264.
Duffy, P.B., Brando, P., Asner, G.P., Field, C.B., 2015. Projections of future meteorological drought and wet periods in the Amazon. Proceedings of the National Academy of Sciences 112, 13172–13177.
Ekblad, A., Wallander, H., Godbold, D.L., Cruz, C., Johnson, D., Baldrian, P., Björk, R.G., Epron, D., Kieliszewska-Rokicka, B., Kjøller, R., et al., 2013. The production and turnover of extramatrical mycelium of ectomycorrhizal fungi in forest soils: role in carbon cycling. Plant and Soil 366, 1–27.
Flores-Renteria, L., Lau, M.K., Lamit, L.J., Gehring, C., 2014. An elusive ectomycorrhizal fungus reveals itself: a new species of *Geopora* (Pyronemataceae) associated with *Pinus edulis*. Mycologia 106, 553–563.
Gallo, M.E., Sinsabaugh, R.L., Cabaniss, S.E., 2006. The role of ultraviolet radiation in litter decomposition in arid ecosystems. Applied Soil Ecology 34, 82–91.
Gehring, C.A., Theimer, T.C., Whitham, T.G., Keim, P., 1998. Ectomycorrhizal fungal community structure of pinyon pines growing in two environmental extremes. Ecology 79, 1562–1572.
Gehring, C.A., Mueller, R.C., Whitham, T.G., 2006. Environmental and genetic effects on the formation of ectomycorrhizal and arbuscular mycorrhizal associations in cottonwoods. Oecologia 149, 158–164.
Gehring, C., Flores-Rentería, D., Sthultz, C.M., Leonard, T.M., Flores-Rentería, L., Whipple, A.V., Whitham, T.G., 2014a. Plant genetics and interspecific competitive interactions determine ectomycorrhizal fungal community responses to climate change. Molecular Ecology 23, 1379–1391.
Gehring, C.A., Mueller, R.C., Haskins, K.E., Rubow, T.K., Whitham, T.G., 2014b. Convergence in mycorrhizal fungal communities due to drought, plant competition, parasitism, and susceptibility to herbivory: consequences for fungi and host plants. Frontiers in Microbiology 5, 1–9.
Gordon, G., Gehring, C., 2011. Molecular characterization of pezizalean ectomycorrhizas associated with pinyon pine during drought. Mycorrhiza 21 (5), 431–441.
Hamerlynck, E.P., McAuliffe, J.R., 2008. Soil-dependent canopy die-back and plant mortality in two Mojave Desert shrubs. Journal of Arid Environments 72, 1793–1802.
Haskins, K.E., Gehring, C.A., 2005. Evidence for mutualist limitation: the impacts of conspecific density on the mycorrhizal inoculum potential of woodland soils. Oecologia 145, 123–131.
Herzog, C., Peter, M., Pritsch, K., Günthardt-Goerg, M.S., Egli, S., 2012. Drought and air warming affects abundance and exoenzyme profiles of Cenococcum geophilum associated with *Quercus robur*, *Q. petraea* and *Q. pubescens*. Plant Biology 15, 230–237.
Jany, J.L., Garbaye, J., Martin, F., 2002. *Cenococcum geophilum* populations show a high degree of genetic diversity in beech forests. New Phytologist 154, 651–659.
Jany, J.-L., Martin, F., Garbaye, J., 2003. Respiration activity of ectomycorrhizas from *Cenococcum geophilum* and *Lactarius* sp. in relation to soil water potential in five beech forests. Plant and Soil 255, 487–494.
Jayne, B., Quigley, M., 2014. Influence of arbuscular mycorrhiza on growth and reproductive response of plants under water deficit: a meta-analysis. Mycorrhiza 24, 109–119.
Jeliazkova, E., Percival, D., 2003. Effect of drought on ericoid mycorrhizas in wild blueberry (*Vaccinium angustifolium* Ait.). Canadian Journal of Plant Science 83, 583–586.
Johnson, D., Martin, F., Cairney, J.W.G., Anderson, I.C., 2012. The importance of individuals: intraspecific diversity of mycorrhizal plants and fungi in ecosystems. The New Phytologist 194, 614–628.
Karst, J., Randall, M.J., Gehring, C.A., 2014. Consequences for ectomycorrhizal fungi of the selective loss or gain of pine across landscapes. Botany 92, 855–865.
Kennedy, P.G., Peay, K.G., 2007. Different soil moisture conditions change the outcome of the ectomycorrhizal symbiosis between *Rhizopogon* species and *Pinus muricata*. Plant and Soil 291, 155–165.
Khalvati, M.A., Hu, Y., Mozafar, A., Schmidhalter, U., 2005. Quantification of water uptake by arbuscular mycorrhizal hyphae and its significance for leaf growth, water relations, and gas exchange of barley subjected to drought stress. Plant Biology 7, 706–712.
Kivlin, S.N., Emery, S.M., Rudgers, J.A., 2013. Fungal symbionts alter plant responses to global change. American Journal of Botany 100, 1445–1457.

Langley, J.A., Hungate, B.A., 2003. Mycorrhizal controls on belowground litter quality. Ecology 84, 2302–2312.

Lehto, T., Zwiazek, J.J., 2011. Ectomycorrhizas and water relations of trees: a review. Mycorrhiza 21, 71–90.

Liu, Y., He, L., An, L., Helgason, T., Feng, H., 2009. Arbuscular mycorrhizal dynamics in a chronosequence of *Caragana korshinskii* plantations. FEMS Microbiology Ecology 67, 81–92.

Lodge, D.J., 1989. The influence of soil-moisture and flooding on formation of VA-endomycorrhizas and ectomycorrhizas in *Populus* and *Salix*. Plant and Soil 117, 243–253.

MacArthur, R.H., Wilson, E.O., 1967. The Theory of Island Biogeography. Princeton University Press, Princeton, New Jersey.

Mahall, B.E., Tyler, C.M., Cole, E.S., Mata, C., 2009. A comparative study of oak (*Quercus*, Fagaceae) seedling physiology during summer drought in southern California. American Journal of Botany 96, 751–761.

Marulanda, A., Azcón, R., Ruiz-Lozano, J.M., 2003. Contribution of six arbuscular mycorrhizal fungal isolates to water uptake by *Lactuca sativa* plants under drought stress. Physiologia Plantarum 119, 526–533.

McCormick, M.K., Whigham, D.F., Sloan, D., O'Malley, K., Hodkinson, B., 2006. Orchid-fungus fidelity: a marriage meant to last? Ecology 87, 903–911.

McCormick, M.K., Whigham, D.F., O'Neill, J.P., Becker, J.J., Sarah, W., Rasmussen, H.N., Bruns, T.D., Taylor, D.L., 2009. Abundance and distribution of *Corallorhiza odontorhiza* reflect variations in climate and ectomycorrhizas. Ecological Monographs 79, 619–635.

McDowell, N., Pockman, W.T., Allen, C.D., Breshears, D.D., Cobb, N., Kolb, T., Plaut, J., Sperry, J., West, A., Williams, D.G., et al., 2008. Mechanisms of plant survival and mortality during drought: why do some plants survive while others succumb to drought? New Phytologist 178, 719–739.

McDowell, N.G., Williams, A.P., Xu, C., Pockman, W.T., Dickman, L.T., Sevanto, S., Pangle, R.E., Limousin, J.-M., Plaut, J.A., Mackay, D.S., et al., 2016. Multi-scale predictions of massive conifer mortality due to chronic temperature rise. Nature Climate Change 6, 295–300.

McIntyre, P.J., Thorne, J.H., Dolanc, C.R., Flint, A.L., Flint, L.E., Kelly, M., Ackerly, D.D., 2015. Twentieth-century shifts in forest structure in California: denser forests, smaller trees, and increased dominance of oaks. Proceedings of the National Academy of Sciences of the United States of America 112, 1458–1463.

McKersie, B., 2015. Planning for food security in a changing climate. Journal of Experimental Botany 66, 3435–3450.

McLaughlin, B., Zavaleta, E., 2012. Predicting species responses to climate change: demography and climate microrefugia in California valley oak (*Quercus lobata*). Global Change Biology 18 (7).

Mohan, J.E., Cowden, C.C., Baas, P., Dawadi, A., Frankson, P.T., Helmick, K., Hughes, E., Khan, S., Lang, A., Machmuller, M., et al., 2014. Mycorrhizal fungi mediation of terrestrial ecosystem responses to global change: mini-review. Fungal Ecology 10, 3–19.

Moran, M.S., Ponce-Campos, G.E., Huete, A., McClaran, M.P., Zhang, Y., Hamerlynck, E.P., Augustine, D.J., Gunter, S.A., Kitchen, S.G., Peters, D.P., Starks, P.J., Hernandez, M., 2014. Functional response of U.S. grasslands to the early 21st-century drought. Ecology 95 (8), 2121–2133. http://dx.doi.org/10.1890/13-1687.1.

Mueller, R.C., Scudder, C.M., Porter, M.E., Talbot Trotter, R., Gehring, C.A., Whitham, T.G., 2005a. Differential tree mortality in response to severe drought: evidence for long-term vegetation shifts. Journal of Ecology 93, 1085–1093.

Mueller, R.C., Sthultz, C.M., Martinez, T., Gehring, C.A., Whitham, T.G., 2005b. The relationship between stem-galling wasps and mycorrhizal colonization of *Quercus turbinella*. Canadian Journal of Botany-Revue Canadienne De Botanique 83, 1349–1353.

Murphy, K.L., Klopatek, J.M., Klopatek, C.C., 1998. The effects of litter quality and climate on decomposition along an elevational gradient. Ecological Applications 8, 1061–1071.

Nguyen, N.H., Hynson, N.A., Bruns, T.D., 2012. Stayin' alive: survival of mycorrhizal fungal propagules from 6-yr-old forest soil. Fungal Ecology 5, 741–746.

Pan, Y., Birdsey, R.A., Fang, J., Houghton, R., Kauppi, P.E., Kurz, W.A., Phillips, O.L., Shvidenko, A., Lewis, S.L., Canadell, J.G., et al., 2011. A large and persistent carbon sink in the world's forests. Science (New York, N.Y.) 333, 988–993.

Peay, K.G., Schubert, M.G., Nguyen, N.H., Bruns, T.D., 2012. Measuring ectomycorrhizal fungal dispersal: macroecological patterns driven by microscopic propagules. Molecular Ecology 21, 4122–4136.

Pena, R., Polle, A., 2014. Attributing functions to ectomycorrhizal fungal identities in assemblages for nitrogen acquisition under stress. The ISME Journal 8, 1–10.

Phillips, R.P., Brzostek, E., Midgley, M.G., 2013. The mycorrhizal-associated nutrient economy: a new framework for predicting carbon-nutrient couplings in temperate forests. The New Phytologist 199, 41–51.

Piggot, C.D., 1982. Survival of mycorrhiza formed by *Cenococcum geophilum* Fr. In dry soils. New Phytologist 92, 513–517.

Querejeta, J.I., Egerton-Warburton, L.M., Allen, M.F., 2003. Direct nocturnal water transfer from oaks to their mycorrhizal symbionts during severe soil drying. Oecologia 134, 55–64.

Querejeta, J.I., Egerton-Warburton, L.M., Allen, M.F., 2009. Topographic position modulates the mycorrhizal response of oak trees to interannual rainfall variability. Ecology 90, 649–662.

Redmond, M.D., Forcella, F., Barger, N.N., 2012. Declines in pinyon pine cone production associated with regional warming. Ecosphere 3 art120.

Redmond, M.D., Cobb, N.S., Clifford, M.J., Barger, N.N., 2015. Woodland recovery following drought-induced tree mortality across an environmental stress gradient. Global Change Biology 21, 3685–3695.

Rehfeldt, G.E., Crookston, N.L., Warwell, M.V., Evans, J.S., 2006. Empirical analyses of plant-climate relationships for the western United States. International Journal of Plant Sciences 167, 1123–1150.

Rollins, M.G., 2009. LANDFIRE: a nationally consistent vegetation, wildland fire, and fuel assessment. International Journal of Wildland Fire 18, 235.

Roman, D.T., Novick, K.A., Brzostek, E.R., Dragoni, D., Rahman, F., Phillips, R.P., 2015. The role of isohydric and anisohydric species in determining ecosystem-scale response to severe drought. Oecologia 179, 641–654.

Smith, S.E., Facelli, E., Pope, S., Smith, F.A., 2010. Plant performance in stressful environments: interpreting new and established knowledge of the roles of arbuscular mycorrhizas. Plant and Soil 326, 3–20.

Sork, V.L., Davis, F.W., Smouse, P.E., Apsit, V.J., Dyer, R.J., Fernandez-M, J.F., Kuhn, B., 2002. Pollen movement in declining populations of California Valley oak, *Quercus lobata*: where have all the fathers gone? Molecular Ecology 11, 1657–1668.

Soudzilovskaia, N.A., Douma, J.C., Akhmetzhanova, A.A., van Bodegom, P.M., Cornwell, W.K., Moens, E.J., Treseder, K.K., Tibbett, M., Wang, Y.P., Cornelissen, J.H.C., 2015. Global patterns of plant root colonization intensity by mycorrhizal fungi explained by climate and soil chemistry. Global Ecology and Biogeography 24, 371–382.

Stahl, P.D., Smith, W.K., 1984. Effects of different geographic isolates of *Glomus* on the water relations of *Agropyron smithii*. Mycologia 76, 261–267.

Sthultz, C.M., Whitham, T.G., Kennedy, K., Deckert, R., Gehring, C.A., 2009a. Genetically based susceptibility to herbivory influences the ectomycorrhizal fungal communities of a foundation tree species. New Phytologist 184, 657–667.

Sthultz, C.M., Gehring, C.A., Whitham, T.G., 2009b. Deadly combination of genes and drought: increased mortality of herbivore-resistant trees in a foundation species. Global Change Biology 15, 1949–1961.

Suarez, M.L., Ghermandi, L., Kitzberger, T., 2004. Factors predisposing episodic drought induced tree mortality in *Nothofagus*–site, climatic sensitivity and growth trends. Journal of Ecology 92, 954–966.

Swaty, R.L., Deckert, R.J., Whitham, T.G., Gehring, C.A., 2004. Ectomycorrhizal abundance and community composition shifts with drought: predictions from tree rings. Ecology 85, 1072–1084.

Swaty, R., Michael, H.M., Deckert, R., Gehring, C.A., 2016. Mapping the potential mycorrhizal associations of the conterminous United States of America. Fungal Ecology 1–9.

Tedersoo, L., Smith, M.E., 2013. Lineages of ectomycorrhizal fungi revisited: foraging strategies and novel lineages revealed by sequences from belowground. Fungal Biology Reviews 27, 83–99.

Tedersoo, L., May, T.W., Smith, M.E., 2010. Ectomycorrhizal lifestyle in fungi: global diversity, distribution, and evolution of phylogenetic lineages. Mycorrhiza 20, 217–263.

Tedersoo, L., Bahram, M., Toots, M., Diedhiou, A.G., Henkel, T.W., Kjoller, R., Morris, M.H., Nara, K., Nouhra, E., Peay, K.G., et al., 2012. Towards global patterns in the diversity and community structure of ectomycorrhizal fungi. Molecular Ecology 21, 4160–4170.

Trenberth, K.E., Dai, A., van der Schrier, G., Jones, P.D., Barichivich, J., Briffa, K.R., Sheffield, J., 2013. Global warming and changes in drought. Nature Climate Change 4, 17–22.

Ummenhofer, C.C., England, M.H., McIntosh, P.C., Meyers, G.A., Pook, M.J., Risbey, J.S., Gupta, A.S., Taschetto, A.S., 2009. What causes southeast Australia's worst droughts? Geophysical Research Letters 36, L04706.

Watson, G.W., von der Heide Spravka, K.G., Howe, V.K., 1990. Ecological significance of endo/ectomicorrhizae in the oak subgenus *Erythrobalanus*. Arboricultural Journal 107–116.

West, N.E., 1984. Successional Patterns and Productivity Potentials of Pinyon-Juniper Ecosystem Developing Strategies for Rangeland Management: A Report Prepared by the Committee on Developing Strategies for Rangeland Management. Westview Press, Boulder, Colorado, pp. 1301–1332.

Williamson, G.B., Laurance, W.F., Oliveira, A.A., Delamonica, P., Gascon, C., Lovejoy, T.E., Pohl, L., 2000. Amazonian tree mortality during the 1997 El Nino drought. Conservation Biology 14, 1538–1542.

Worchel, E.R., Giauque, H.E., Kivlin, S.N., 2013. Fungal symbionts alter plant drought response. Microbial Ecology 65, 671–678.

Zhang, J., Wang, F., Che, R., Wang, P., Liu, H., Ji, B., Cui, X., 2016. Precipitation shapes communities of arbuscular mycorrhizal fungi in Tibetan alpine steppe. Scientific Reports 6, 23488.

Zhu, Y., Miller, R.M., 2003. Carbon cycling by arbuscular mycorrhizal fungi in soil – plant systems. Trends in Plant Science 8, 2002–2004.

CHAPTER

17

Soil Water Retention and Availability as Influenced by Mycorrhizal Symbiosis: Consequences for Individual Plants, Communities, and Ecosystems

J.I. Querejeta

Spanish Research Council (CEBAS-CSIC), Murcia, Spain

17.1 INTRODUCTION

Soil water-retention capacity and related hydraulic properties (hydraulic conductivity, permeability) are primarily determined by texture (sand, silt, clay contents), structure (bulk density and porosity), and organic matter content (Hudson, 1994; Schaap et al., 1998; Wösten et al., 2001; Rawls et al., 2003; Lado et al., 2004a,b; Saxton and Rawls, 2006; Tóth et al., 2015). Mycorrhizal fungi are known to influence soil structure and organic matter content both directly and indirectly through their multiple effects on vegetation and soil. Together with plant roots, mycorrhizal extraradical hyphae grow into the soil matrix to create the three-dimensional structural framework that holds primary soil particles together through physical enmeshment and entanglement, a "sticky-string-bag" mechanism that contributes to soil aggregate formation and stabilization (Tisdall and Oades, 1982; Oades and Waters, 1991; Miller and Jastrow, 1990, 2000). Mycorrhizal mycelial networks modify soil structure in the mycorrhizosphere through the physical and chemical binding of soil particles by extraradical hyphae (and their exudates and degradation by-products), thus leading to changes in soil aggregation, pore size distribution, and bulk density (Tisdall, 1991; Caravaca et al., 2003; Rillig and Mummey, 2006; Wilson et al., 2009; Chapter 14, this volume). Mycorrhizal hyphal networks favor the formation of soil microaggregates and their stabilization into macroaggregate structures in a complex hierarchical process, thus altering total pore space

Mycorrhizal Mediation of Soil
http://dx.doi.org/10.1016/B978-0-12-804312-7.00017-6

and pore-size distribution in soil (Bearden and Petersen, 2000). Moreover, mycorrhizal fungi make large, direct contributions to soil organic matter accumulation through the production of fungal-derived carbon compounds, and promote soil organic matter storage through aggregate stabilization and protection of labile carbon pools inside aggregates (Rillig et al., 2001; Godbold et al., 2006; Wilson et al., 2009). Mycorrhizal fungi are also important indirect contributors to soil carbon sequestration and storage through their influence on plant physiology and growth, and plant community composition and productivity (van der Heijden et al., 2008). Mycorrhizal-induced changes in soil organic matter content and soil structure are expected to significantly affect the relationship of water to soil (Augé et al., 2001; Augé, 2004), including key parameters such as soil water retention capacity, hydraulic conductivity, and infiltration capacity (Franzluebbers, 2002; Saxton and Rawls, 2006). However, the potentially important role of mycorrhizal mycelial networks in shaping soil hydraulic properties has received surprisingly little research attention, likely because of the difficulty of disentangling the distinct effects of mycorrhizal fungi from those of plant roots.

17.2 INFLUENCE OF VEGETATION ON SOIL HYDRAULIC PROPERTIES

Vegetation enhances water infiltration rates, soil water retention, and soil hydraulic conductivity as a result of soil organic carbon accumulation and improved soil physical properties (Joffre and Rambal, 1988; Bonell et al., 2010; Germer et al., 2010; Asbjorssen et al., 2011). Plant roots increase and stabilize soil structural porosity through aggregate enmeshment and coalescence (Six et al., 2004), and enhance heterogeneity of the soil pore system and saturated hydraulic conductivity through the formation of highly conductive macropores (Angers and Caron, 1998; Scholl et al., 2014). Vegetation removal often leads to rapid deterioration of soil physical and hydraulic properties, with sharp decreases in aggregate stability and infiltration capacity and increases in bulk density linked to fast declines in soil organic matter content, which ultimately increases soil water erodibility (Castillo et al., 1997). Carminati et al. (2010) and Moradi et al. (2011) showed that soil water content is generally higher in rhizosphere soil compared with bulk soil over a wide range of soil water potentials, even during active plant transpiration, which indicates an altered soil water retention curve that leads to an enhanced soil water-holding capacity in the rhizosphere. Carminati et al. (2010) reported that the rhizosphere of several plant species held more water than the surrounding bulk soil during a soil drying cycle, although this effect varied with time depending on the drying/wetting history of the soil. The rhizosphere, with its higher soil water-holding capacity, avoids water depletion next to roots and slows down the decline in water potential as the soil dries, which favors soil–root hydraulic conductivity and thus water availability to plants during drought (Carminati et al., 2011; Moradi et al., 2011). Rhizosphere soil has physical and hydraulic properties that are different from those of the surrounding bulk soil as a result of several mechanisms, including root and soil shrinking and swelling during drying/wetting cycles, soil compaction by root penetration, interaction of mucilage exuded by roots with soil particles, mucilage shrinking/swelling, and mucilage biodegradation (Carminati and Vetterlein, 2013). Other additional mechanisms involving mycorrhizal fungi may also be at play, because this is still an open research area that requires further investigation. Whereas many studies have shown that plant roots significantly alter soil physical and hydraulic properties in the rhizosphere (Johnson-Maynard et al., 2002; Hallett et al., 2003; Kahle et al., 2005; Feeney et al., 2006; Hinsinger et al., 2009; Carminati et al., 2010), fewer studies have specifically examined the distinct contribution of symbiotic mycorrhizal fungi to the modification of soil hydraulic properties in the mycorrhizosphere.

17.3 MYCORRHIZAL FUNGAL INFLUENCE ON SOIL HYDRAULIC PROPERTIES: REVIEW OF PUBLISHED EVIDENCE

17.3.1 Arbuscular Mycorrhizal Fungi

In an early study, Thomas et al. (1986) compared the influence of mycorrhizal and nonmycorrhizal onion plants (*Allium cepa*) on the structure and permeability of potted soil. Plants inoculated with the arbuscular mycorrhizal (AM) fungus *Glomus macrocarpum*, or left uninoculated, were grown in pots filled with a calcareous silty clay loam soil for 230 days. At the end of the study, soil from the mycorrhizal treatment was significantly better aggregated, more porous, and had higher saturated hydraulic conductivity than the soil from the nonmycorrhizal treatment. Root dry mass and AM fungal soil hyphal density were both significantly positively correlated with the relative abundance of water-stable soil macroaggregates in the mycorrhizal treatment. However, given that the total dry mass of mycorrhizal plants was 5–6 times larger than that of nonmycorrhizal plants, it was not possible to disentangle the relative contributions of roots and AM fungal hyphae to the observed changes in soil structural and hydraulic properties in the mycorrhizal treatment. Thomas et al. (1986) concluded that the observed increases in soil aggregation, porosity, and hydraulic conductivity in the mycorrhizal treatment were mainly attributable to direct root effects, and largely mediated through plant growth promotion by the mycorrhizal fungal symbiont.

The pioneering work of Augé et al. (2001) first showed that soil colonization by AM roots and extraradical hyphae had a subtle but significant direct effect on the moisture characteristic curve of a fine sandy loam soil. The soil moisture characteristic (also termed the water-retention curve or moisture release plot) describes the relationship between soil water content and soil matric potential, and is used to predict the water holding capacity of a soil and the ease with which soils release water as they dry. The amount of water held by soil at relatively high soil matric potentials (and water contents) depends on the pore-size distribution and capillary properties of soil, and is thus particularly strongly affected by soil structure (Saxton and Rawls, 2006). By comparing the water retention properties of rhizosphere soil with pot-grown mycorrhizal and nonmycorrhizal plants (*Vigna unguiculata*) of similar size, Augé et al. demonstrated that soil colonization by the AM fungus *Rhizophagus intraradices* (formerly *Glomus intraradices*) over 7 months enhanced soil aggregation, and slightly but significantly increased the amount of plant-available water contained in soil during a soil drying cycle. Soil colonization by AM fungal hyphae shifted the soil moisture retention curve toward a higher soil water potential at any given soil water content, relative to nonmycorrhizal soil. As soil water potential began to decline during a drying cycle, mycorrhizal soil had a tendency to dry more (i.e., to give up a larger proportion of its water content) before reaching the plant permanent wilting point (−1.5 MPa), which means that slightly more water was available to roots in mycorrhizal soil than in nonmycorrhizal soil. Importantly, these improvements in the soil–water relationship were distinctly attributable to soil colonization by AM fungal hyphae and linked to increases in soil hyphal density, given that rooting patterns and characteristics (including root mass and length density and specific root length) were very similar between mycorrhizal and nonmycorrhizal soils. In other words, the potentially confounding influence of differential root growth between mycorrhizal and nonmycorrhizal plants was successfully factored out in this study. Augé et al. (2001) speculated that the relatively small extra amount of plant-available water contained in mycorrhizal soil in their experiment could be sufficient to postpone the onset of lethal drought stress during a drought episode (thus potentially extending the drought survival time of mycorrhizal plants relative to nonmycorrhizal plants), but would probably be insufficient to significantly contribute to plant transpiration during more than a few extra hours. Augé (2004) hypothesized that mycorrhizal fungal improvement of soil aggregation, structure, and water-retention capacity could be cumulative and thus become more substantial over a longer time (>7 months), potentially leading to much greater increases in soil water retention and availability to plants than those reported in their study.

In a follow-up study, Augé et al. (2004) used pot-grown *Phaseolus vulgaris* plants to develop mycorrhizal soils inoculated with the AM fungi *R. intraradices* or *Gigaspora margarita*. They found that the soil moisture characteristics of mycorrhizal and nonmycorrhizal soils were similar and did not differ significantly in soil water contents at soil water potential values of −0.2 MPa (very moderate drought) or −1.5 MPa (permanent wilting point), despite the relatively long time allowed for AM fungal extraradical hyphae to grow into the soil matrix and alter its structural and hydraulic properties (12 months). However, it was noted that mycorrhizal soils lost slightly but significantly more water than nonmycorrhizal soils to reach the threshold at which soil water potential first started to decline during the early stages of soil drying (−0.02 MPa), which coincided with the point of stomatal closure for the plants in this experiment (Augé, 2004). Moreover, they found that soil water potentials were slightly but significantly higher in mycorrhizal than in nonmycorrhizal soils under well-water conditions near field capacity. Soils colonized by AM fungi thus needed to dry more (i.e., give off more water to roots) than nonmycorrhizal soils before their water potential values began to decline from their well-watered state during the early stages of soil drying. In another elegant experiment, Augé et al. (2004) used wild type plants and nonmycorrhizal mutants of *P. vulgaris* grown into previously produced mycorrhizal or nonmycorrhizal soils in order to partition the AM fungal influence on stomatal conductance between the soil and root components. This approach allowed them to uncouple the influence of AM fungal root colonization from AM fungal-induced changes in the hydraulic properties of the soil in which plants where grown. Interestingly, they found that merely growing in mycorrhizal soil led to a significant enhancement of stomatal conductance in nonmycorrhizal mutant plants under both well-watered and drought conditions, indicating that the mycorrhizal soil altered the stomatal behavior of plants growing in it, regardless of whether the roots were mycorrhizal or not. It was concluded that about half of the substantial promotion of stomatal conductance in *P. vulgaris* by AM fungi was directly attributable to soil colonization by AM fungi (possibly, but not necessarily exclusively, through AM fungal-induced changes in soil hydraulic properties), whereas the other half was attributable to root colonization by AM fungi (Augé, 2001, 2004).

van der Heijden et al. (2006) evaluated the impact of AM fungi on several model grassland characteristics, including plant productivity and soil structure. They established experimental grassland mesocosms by planting a mixture of 11 plant species inoculated with four species of AM fungi. The soil in the mesocosms was a mix of field-collected soil with quartz sand (2:1 v/v). After 20 months, they found that the presence of AM fungi improved soil aggregate stability and permeability (measured as water percolation), and also reduced erosion and soil loss from the mesocosms during watering, despite equal soil organic matter contents in mycorrhizal and nonmycorrhizal mesocosms. Interestingly, the mycorrhizal mesocosms showed lower plant biomass (and root length) than the nonmycorrhizal mesocosms in this experiment, so the improvements in soil aggregation, permeability, and resistance to erosion found in mycorrhizal mesocosms could not be attributed to plant growth enhancement by AM fungi. However, the authors noted that it was not possible to identify the precise mechanisms responsible for the positive effects of AM fungi on soil structural and hydraulic properties. These effects could be at least in part independent of the direct effects of the AM extraradical hyphal network on soil structure, and partly related to indirect AM fungal effects (such as mycorrhizal-induced changes in plant community composition, given that different plant species may have distinctive effects on soil properties).

Hallett et al. (2009) investigated the relative importance of roots and AM fungi in altering soil physical properties in the mycorrhizosphere. They cultivated a mycorrhizal wild-type tomato (*Solanum lycopersicum*) inoculated with *Funneliformis mosseae* and *R. intraradices*, and a mycorrhiza-defective tomato mutant. They used a split-pot system with a root-excluding nylon mesh screen to separate soil colonized by both roots and AM fungal hyphae from soil colonized only by AM fungal extraradical hyphae (hyphal compartment). At the end of the plant growth period (84 days), plant shoot and root biomass were similar between mycorrhizal wild plants and nonmycorrhizal mutant plants. Soil aggregate stability, soil water sorptivity (at both −200 kPa and air-dry conditions) and soil water contents were generally higher in the rhizosphere of mycorrhizal plants (both roots and AM fungi present) than in the rhizosphere of nonmycorrhizal plants (no AM fungi present). These results suggest that the presence of AM fungi significantly enhanced the structural stability, unsaturated hydraulic conductivity, water infiltration capacity, and water-retention properties of rhizosphere soil, in spite of similar root biomass between mycorrhizal and nonmycorrhizal plants. Moreover, Hallett et al. (2009) reported that the root-free hyphal compartments colonized by AM fungi maintained 4–5% higher volumetric soil water content than the fungal compartments with no presence of AM fungi throughout the whole growing period (despite the same irrigation regimes in both treatments), which suggests that soil colonization by AM fungal extraradical hyphae by itself increased the water retention properties (i.e., water holding capacity) of root-free soil. It was concluded that the presence of AM fungi had a positive impact on soil aggregation in the mycorrhizosphere, although this increase in soil stability could not be explained from measured changes in soil porosity or water repellency. They hypothesized that the higher transpiration rates and larger fluctuations in rhizosphere soil water content (wetting/drying cycles) observed in mycorrhizal plants could be a significant mechanism of soil stabilization, either through enhanced soil shrinking and swelling or through enhanced production of root (and AM fungal hyphal) exudates. It is well established that changes in soil structure are stimulated by the production of exudates by roots and microbes in the rhizosphere, as well as by intense wetting and drying cycles caused by evapotranspiration (Czarnes et al., 2000).

Milleret et al. (2009) evaluated the effects of plant inoculation with the AM fungus *R. intraradices* on the development of soil structure and on soil hydraulic properties. They cultivated mycorrhizal and nonmycorrhizal leek plants (*Allium porrum*) in soil mesocosms filled with a carbonated loam soil for 35 weeks. They found that the presence of both roots and AM fungi had a synergistic decompacting and stabilizing effect on soil structure, and that the structural pore volume generated by root and AM fungal growth was several orders of magnitude larger than the volume of roots and AM fungal hyphae themselves. Moreover, the diameters of the newly generated structural pores were not the same as those of roots or hyphae, so they were not attributable to the mechanical intrusion of roots and AMF hyphae into the soil matrix. The authors proposed that root and hyphal exudates probably served as carbon sources for other soil microorganisms present in the mycorrhizosphere, which in turned enhanced soil aggregation and porosity and thus altered the soil moisture retention curve. These changes led to more porous and stable soils with greater plant-available water induced by the combined presence of roots and AM fungal hyphae. As a result, soil in microcosms with mycorrhizal plants had 12% higher total plant-available water, and up to 32% higher easily available water, than soil in microcosms with nonmycorrhizal plants. However, it was noted that the AM fungal-inoculated plants had significantly larger shoot and root biomass than the nonmycorrhizal plants in this study, so the observed AM fungal-induced improvement of soil physical and hydraulic properties was probably at least partly mediated through plant growth enhancement (i.e., it was not possible to disentangle the direct and indirect effects of AM fungi on soil properties).

Daynes et al. (2013) examined the relative contributions of soil organic matter, plant roots and AM fungi to the self-organization of mine spoil material in a pot experiment. They found that all three factors, organic matter, living plant roots, and AM fungi were needed for the development of a stable soil structure. In the presence of sufficient soil organic matter (which was essential for the formation of water-stable aggregates by roots and AM fungi), plant roots were the key contributors to the development of soil structure, which was further stabilized by AM fungi. A key result in this study was that the presence of AM fungi increased the accretion of organic carbon in the macroaggregate fraction of soil (>710 μm), even though mycorrhizal and nonmycorrhizal plants had similar biomass and presumably led to similar carbon inputs to soil over the 6-month experimental period. Total soil pore volume and soil bulk density were not significantly affected by the presence of AM fungi, but mesocosms with AM fungi had soil aggregates with higher mean weight diameter and 50% more soil in the greater than 2000 μm soil aggregate fraction than mesocosms without AM fungi. Moreover, plants with AM fungi significantly enhanced the retention of plant-available water in soil (in the −1 to −55 kPa matric potential range) relative to nonmycorrhizal plants. Interestingly, the plant-unavailable water content of soil (−3 to −30 MPa) was also significantly increased by the presence of AM fungi. In this experiment, the presence of plant roots without AM fungi initiated the development of soil structure, but the presence of roots with AM fungi caused a further improvement of the water-retention capacity of soil by creating microaggregates and macroaggregates and by maximizing the presence of finer pores (<300 μm). The authors concluded that whereas plant roots were essential for the development of soil structure, their role was largely indirect and mediated through the provision of habitat and carbon resources for mycorrhizal fungi, whose presence was necessary for stabilizing the newly formed structure into water stable macroaggregates.

In a recent study, Akhzari et al. (2015) evaluated the simultaneous effects of compost amendments and inoculation with (unidentified) AM fungi on plant growth and soil properties in a pot experiment. After 3 months of cultivation, they found that inoculation of *Medicago polymorpha* plants with AM fungi significantly decreased soil bulk density and increased soil water holding capacity across a wide range of compost application rates, relative to nonmycorrhizal mesocosms. Whereas soil water-holding capacity ranged from 39.6% to 52.7% in nonmycorrhizal pots (depending on compost application rates), it increased to 55.3–70.3% in the presence of AM fungi. However, given that plant biomass was consistently greater in AM fungal–inoculated than in noninoculated plants across compost treatments, this study does not permit disentangling the direct and indirect effects of mycorrhizal fungi on soil structure and hydraulic properties (i.e., direct soil modification by AM fungal extraradical hyphal networks and their products vs. plant growth enhancement by AM fungi). In another study, Samanei et al. (2015) cultivated potted mycorrhizal and nonmycorrhizal barley plants (*Hordeum vulgare*) in a coarse-textured (sandy loam) soil of low organic matter content and low water-retention capacity in order to evaluate the effects of two different AM fungi (*R. intraradices* and *Claroideoglomus* (formerly *Glomus*) *claroideum*) on soil physical properties. After 4 months, they found that AM fungal inoculation significantly increased the mean weight diameter of soil aggregates by 113–202%, total soil porosity by 2.2–2.6%, and plant-available soil water-holding capacity by 13–27%, relative to nonmycorrhizal soil. Plant inoculation with AM fungi altered the pore size distribution of this sandy soil, increasing micropores and mesopores but decreasing macropores, which led to a 69–88% decrease in the saturated hydraulic conductivity of soil in the AM fungal treatment (owing to decreased macroporosity) compared with the nonmycorrhizal control treatment. In this study, plant inoculation with *Claroideoglomus* (formerly *Glomus*) *etunicatum* caused significantly greater increases in mean weight diameter of aggregates and in plant-available soil water capacity than inoculation with *R. intraradices* (paralleling the greater plant nutrient uptake and growth stimulation by *C. etunicatum*), thus illustrating the differential effects of different AM fungi on soil hydraulic properties. Once again, the data presented in this study do not allow the separation of direct and indirect (plant-mediated) effects of AM fungi on soil structural and hydraulic properties.

Several other studies that were not specifically aimed at investigating the influence of mycorrhizal fungi on the water-retention properties of soil nonetheless provide data that are consistent with (and supportive of) mycorrhizal enhancement of soil water-holding capacity. In a pot study investigating the influence of AM fungi on soil–root hydraulic conductance in *Agrostis stolonifera*, González-Dugo (2010) found that a coarse textured soil (sand–silt mixture) colonized by the AM fungus *R. intraradices* maintained significantly higher water content than nonmycorrhizal soil under well-watered conditions (~20% vs. ~15%, respectively), despite similar plant biomass and transpiration rates in mycorrhizal and nonmycorrhizal plants. In another study, Ruth et al. (2011) used mycorrhizal and nonmycorrhizal barley plants (*H. vulgare*, inoculated or not with *R. intraradices*) grown in two-compartment mesocosms to quantify AM fungal hyphal water uptake. The two-compartment mesocosms were composed of a plant compartment (roots and AM fungi present) and a root-free hyphal compartment separated by root-excluding mesh screen. The silt-loam soil in the root-free hyphal compartments had approximately 3% higher water content in mycorrhizal than in nonmycorrhizal mesocosms after irrigation, which suggests that soil colonization by the extraradical hyphae of *R. intraradices* by itself enhanced the soil water-retention properties. Moreover, soil in the plant compartments also maintained 1–5% higher water content in the mycorrhizal-inoculated mesocosms (roots and AM fungi present) than in the nonmycorrhizal mesocosms (only roots present) under well-watered conditions and throughout successive soil drying cycles, despite the larger size (50% greater dry mass) of mycorrhizal plants. Overall, these data suggest that the presence of AM fungi by itself enhanced the water retention properties of soil in this experiment, regardless of the presence or absence of roots.

17.3.2 Ectomycorrhizal Fungi

The potential influence of ectomycorrhizal (EcM) fungi on the water-retention properties of soil has been examined by very few studies. Abundant extraradical mycelium production by many EcM fungi (as well as abundant production of hyphal exudates and decay compounds) suggests a high potential for altering the structure, porosity, and hydraulic properties of soil through mechanisms such as hyphal binding and enmeshment of soil particles into aggregates, hyphal exudate production, or enhanced soil carbon accretion, among others (Leake et al., 2004). Bogeat-Triboulot et al. (2004) reported that pot-grown maritime pine (*Pinus pinaster*) seedlings cultivated on a sandy loam soil and inoculated with the EcM fungus *Hebeloma cylindrosporum* showed greater soil adherence to roots than seedlings colonized by other EcM fungal species (*Thelephora terrestris* or *Laccaria bicolor*). The comparatively greater ability of *H. cylindrosporum* to aggregate sandy soil particles and to develop a rhizosheath was associated with a reduced root hydraulic resistance and a higher root water uptake capacity under both well-watered and drought conditions. In another mesocosm experiment, Querejeta et al. (2012) used fungicide to reduce mycorrhizal (EcM and AM) root colonization and soil hyphal density in coast live oak (*Quercus agrifolia*) saplings grown on a sandy loam soil. EcM colonization of roots and soil hyphal density were both reduced by the fungicide relative to the control treatment (from 74% to 44%, and from 71 to 33 m/g, respectively). Five days after irrigation to field capacity, soil in the fungicide-treated mesocosms had significantly lower water potential than soil in the control mesocosms (−0.7 vs. −0.4 MPa, respectively), despite similar size of plants in both treatments. Moreover, soil water potential correlated positively with viable soil hyphal length across treatments, suggesting that the decreases in mycorrhizal hyphal abundance with fungicide application may have also reduced the water-holding capacity of this sandy soil. In a follow-up study, Prieto et al. (2016) used fungicide to reduce EcM root colonization and soil hyphal density in pot-grown Aleppo pine (*Pinus halepensis*) saplings. It was found that the water content of rhizosphere soil (measured gravimetrically 9 days after irrigation to field capacity) was significantly decreased by fungicide application, despite similar plant sizes in fungicide-treated and control treatments. Moreover, soil water content at the end of the drying cycle correlated positively with soil hyphal length across treatments, which led the authors to suggest that fungicide addition may have decreased the water-retention capacity of the soil (2:1 mix of sand with field-collected soil) through reductions in soil EcM fungal hyphal density. However, it should be noted that the studies mentioned above were not specifically designed to test the influence of EcM fungi on the water-retention capacity of soils, so the evidence of an EcM fungal role in enhancing soil water-holding capacity remains largely correlative. More targeted experimental testing of EcM fungal influences on soil hydraulic properties (by studies specifically designed to address this question) is thus needed before any definitive conclusions can be drawn.

17.3.3 Long-Term Field Studies

The results of several field studies also suggest that mycorrhizal fungi may play an important role in modifying and improving the hydraulic properties of soils in the mid to long term. For example, Riestra et al. (2012) compared the water retention properties of a sandy loam soil of two neighboring *P. halepensis* plantations of the same age (40 years) in the Argentinian La Pampa (an AM fungal-dominated ecosystem with no native EcM fungi). Pine seedlings were inoculated with (unidentified) EcM fungi at the time of planting in one of the plantations, and left uninoculated in the other. Forty years after plantation establishment, it was found that soil water content at field capacity in the 6–12-cm depth interval was significantly higher under EcM fungal-inoculated trees than under noninoculated trees (15.4% vs. 11.4%, respectively), indicating a significant EcM fungal-induced improvement in the water-retention capacity of soil that was associated with greater accretion of soil organic carbon under EcM fungal–inoculated trees. In another field experiment, Kumar et al. (2016) evaluated the effects of AM fungal inoculation on soil physical properties in an okra–pea cropping system on an acid alfisol in the Himalayas. They found that plants inoculated with AM fungi (*F. mosseae*) increased the mean weight diameter of soil aggregates by 7% and improved soil water holding capacity by 6% after 2 years relative to noninoculated plants. In another field study, Wuest et al. (2005) reported that the ponded infiltration rate of a silt loam soil correlated positively with the soil content in easily extractable glomalin (an insoluble glycoprotein with gluelike properties that is abundantly produced by AM fungi; Rillig and Mummey, 2006), and this correlation held across soils subjected to a wide range of agricultural management practices. However, these field studies can only provide correlative (rather than causal) evidence of mycorrhizal fungal effects on soil hydraulic properties. Moreover, it is impossible to disentangle the relative contributions of direct and indirect long-term effects of mycorrhizal fungi in these field studies; in most cases, it appears likely that indirect mycorrhizal effects (i.e., through mycorrhizal enhancement of plant productivity) may have played a major role in the observed changes in soil hydraulic properties.

17.4 MYCORRHIZAL FUNGAL ROLE IN HYDRAULIC REDISTRIBUTION AND HYDRAULIC CONNECTIVITY IN THE VADOSE ZONE

Several mesocosm studies have shown that mycorrhizal fungi greatly enhance the unsaturated hydraulic conductivity of the soil matrix by providing pathways of preferential water movement along their extraradical hyphae (Cairney, 1992; Ruiz-Lozano and Azcón, 1995; Marulanda et al., 2003; Plamboeck et al., 2007; Ruth et al., 2011). Early studies by Duddrige et al. (1980) and Faber et al. (1991) demonstrated that the extraradical hyphae of both EcM and AM fungi can take up soil water and facilitate its transport to their host plants over ecologically significant distances. In particular, EcM fungal rhizomorphs and mycelial strands with central vessellike structures provide a pathway for the transport of physiologically relevant quantities of soil water to their host plants (Brownlee et al., 1983). Moreover, the external mycelia of mycorrhizal fungi enhance root–soil hydraulic conductivity and connectivity, particularly when root and soil shrinking during drying compromise root–soil contact and thus root water uptake (Augé, 2001; Lehto and Zwiazek, 2011). The extraradical mycelia of EcM and AM fungi also play a role in the redistribution of water released to soil by plant roots through hydraulic lift (and of water transferred directly from host roots to mycorrhizal fungi; Querejeta et al., 2003). Hydraulic lift (also termed *hydraulic redistribution*) is the process by which plant root systems passively transfer water from wet (usually deep) to drier (usually shallow) layers of the soil profile along a water potential gradient (Caldwell et al., 1998). The extraradical mycelia of mycorrhizal fungi increase the volume of soil affected by this process through hyphal transport and exudation of hydraulically redistributed water beyond the reach of roots, and also facilitate the transfer of water between neighboring plants via common mycorrhizal networks (Egerton-Warburton et al., 2007, 2008; Warren et al., 2008; Booth and Hoeksema, 2010; Querejeta et al., 2012; Prieto et al., 2016). Bidirectional water flows through the soil–fungal–plant mycorrhizal continuum and fungal uptake and translocation of water within the (often extensive) mycorrhizal external mycelia, thus significantly increasing three-dimensional unsaturated hydraulic conductivity and connectivity throughout the soil profile (Allen, 2007, 2009; Lilleskov et al., 2009). Mycorrhizal fungal hyphae might also significantly expand the volume of soil exploited for water by accessing moisture stored in the underlying weathered bedrock regolith. Water-filled micropores within the bedrock are inaccessible to plant roots but are accessible to AM and EcM fungal extraradical hyphae capable of colonizing the bedrock matrix, thanks to their comparatively smaller diameters, which might increase hydraulic conductivity and connectivity between these two compartments of the vadose zone (soil and underlying bedrock regolith) (Egerton-Warburton et al., 2003; Bornyasz et al., 2005).

17.5 MYCORRHIZAL FUNGAL ROLE IN REDUCING SOIL EROSION

Recent research has demonstrated the key role of mycorrhizal fungi in improving soil resistance to wind and water erosion. In a laboratory wind tunnel experiment, Burri et al. (2013) found that wind-induced particle loss in sandy soil decreased significantly with increasing percentage of root colonization by AM fungi in plants (*Anthyllis vulneraria*) inoculated with a mixture of three species of AM fungus. Soil loss by wind erosion under nonmycorrhizal seedlings was more than twice the amount of that under mycorrhizal seedlings, despite lack of any AM fungal–induced enhancement of plant growth in this study. Total root length was actually smaller in mycorrhizal plants than in nonmycorrhizal plants, so the AM fungi themselves must have exerted a direct stabilizing effect on soil that overcompensated for lower root growth in the former. In another greenhouse experiment, Mardhiah et al. (2016) evaluated the role of AM fungi in enhancing soil resistance to surface water flow erosion, using *Achillea millefolium* plants inoculated with *R. intraradices* or left uninoculated. Although mycorrhizal and nonmycorrhizal plants had similar size, the presence of AM fungi significantly reduced cumulative soil loss by surface water flow in a sandy loam alluvial soil. The ability of AM fungi to decrease soil detachment by concentrated water flow was primarily the result of their ability to produce extraradical hyphae, because the amount of soil loss by water erosion correlated negatively with AMF soil hyphal length.

17.6 CONSEQUENCES FOR INDIVIDUAL PLANTS, COMMUNITIES, AND ECOSYSTEMS, AND IMPLICATIONS FOR TERRESTRIAL ECOSYSTEMS RESPONSE TO GLOBAL CHANGE

Overall, the review of the published literature indicates that mycorrhizal fungi play a subtle but important role in shaping the water-retention and transmission capacity of soils through multiple direct and indirect (plant-mediated) mechanisms related to mycorrhizal-induced changes in soil structure and organic matter content. A majority of studies report that the presence and/or abundance of mycorrhizal mycelial networks correlates positively with soil water-retention capacity, saturated and unsaturated hydraulic conductivity, and water infiltration capacity, and negatively with soil water erodibility, although further experimental research in this area is clearly warranted. Direct mechanisms include mycorrhizal enhancement of soil aggregation and porosity (through physical enmeshment and chemical binding of soil particles by hyphal networks and their exudates) (Rillig and Mummey, 2006), enhanced soil organic matter accumulation and storage caused by fungal carbon inputs to soil and protection of labile soil carbon pools inside aggregates (Wilson et al., 2009), mycorrhizal inhibition of soil organic matter decomposition through competitive interactions with other soil microbial guilds (Gadgil effect) (Fernandez and Kennedy, 2016), or mycorrhizal regulation of soil water repellency (Hallett, 2007). At the ecosystem scale, indirect (plant-mediated) mechanisms may be equally or even more important than direct mechanisms, given that mycorrhizal enhancement of individual plant growth and plant community diversity and productivity (van der Heijden et al., 2008) in turn stimulate soil organic matter accumulation and soil aggregation, porosity, and water-retention capacity across soil textural classes (Hudson, 1994; Saxton and Rawls, 2006; Carminati et al., 2010).

The recognition that mycorrhizal fungi significantly influence soil hydraulic properties opens the question of whether mycorrhizal fungal responses to key global change drivers (Compant et al., 2010; Mohan et al., 2014) might lead to changes in the soil–water relationship, with feedback consequences for ecosystem primary production. Global change drivers that alter the abundance of mycorrhizal mycelial networks and their products in soil may lead to long-term changes in soil water retention and transmission capacity, which could be an overlooked mechanism of mycorrhizal mediation of ecosystem responses to global change.

As a tentative example, a conceptual model is outlined to illustrate the potential role of mycorrhizal-mediated changes in soil hydraulic properties during the long-term response of dryland ecosystems to climate aridification (Fig. 17.1). In drylands, plant photosynthate carbon availability for allocation to mycorrhizal fungi decreases with increasing plant drought stress, whereas severe soil desiccation has direct detrimental impacts on EcM and AM fungal extraradical hyphal growth and survival. Mesocosm and field studies have shown that severe soil drying causes large declines in EcM and AM fungal extraradical mycelial production and hyphal density in soil (Hunt and Fogel, 1983; Trent et al., 1994; Clark et al., 2009; Lutgen et al., 2003; Querejeta et al., 2007, 2009; Neumann et al., 2009; Staddon et al., 2003; Allen and Kitajima, 2013; Zhang et al., 2016). Large, persistent declines in the abundance of mycorrhizal mycelia and their products in soil would translate into substantial decreases in soil aggregation and organic matter storage in the mid to long terms (Rillig and Mummey, 2006; Wilson et al., 2009), which would eventually reduce the water retention and transmission capacities of dryland soils and would enhance soil erodibility (Hudson, 1994; Lado et al., 2004a,b). Any reductions in the infiltration and plant-available water capacities of dryland

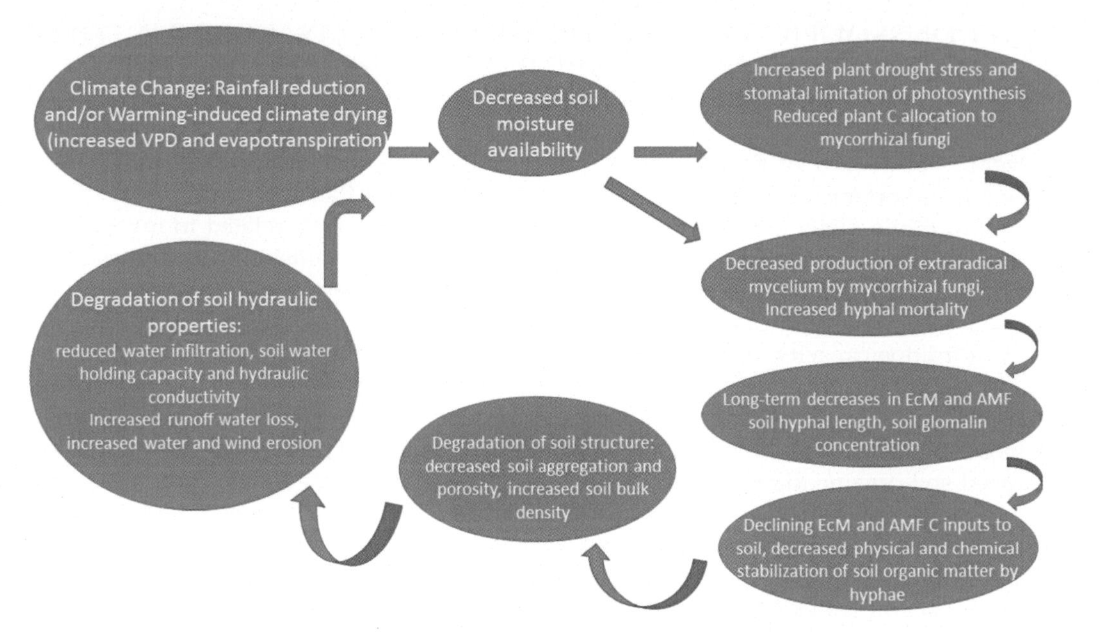

FIGURE 17.1 The potential role of mycorrhizal-mediated changes in soil hydraulic properties during the long-term response of dryland ecosystems to climate aridification. AM, arbuscular mycorrhizal; C, carbon; EcM, ectomycorrhizal; *VPD*, vapor pressure deficit.

soils, even if subtle, would increase vegetation vulnerability to rainfall variability and recurrent drought, with negative consequences for ecosystem primary productivity.

Conversely, other key global change drivers such as elevated atmospheric carbon dioxide (CO_2) enhance AM and EcM fungal extraradical mycelial production and soil hyphal length (Treseder and Allen, 2000; Treseder et al., 2003; Pritchard et al., 2008; Compant et al., 2010; Antoninka et al., 2011; Mohan et al., 2014), and might thereby improve the water-retention and transmission properties of soil through increases in soil aggregation, porosity, and organic matter accumulation. Together with concurrent decreases in plant transpiration and increases in water-use efficiency induced by elevated CO_2 (Wullschleger et al., 2002), enhanced extraradical hyphal production could lead to substantial mycorrhizal-mediated increases in soil water storage and availability to plants, with feedback consequences for ecosystem productivity.

17.7 KNOWLEDGE GAPS, RESEARCH NEEDS, AND FUTURE RESEARCH DIRECTIONS

Much more experimental research is needed to fully disentangle the direct and indirect (plant-mediated) effects of mycorrhizal fungi on soil hydraulic properties. Mesocosm studies using root-excluding hyphal compartments or fungicides may prove useful for this purpose (Ruth et al., 2011; Bingham and Simard, 2011a; Prieto et al., 2016). Long-term field studies in natural and agricultural ecosystems using intact versus periodically rotated hyphal ingrowth mesh cores (Johnson et al., 2001; Bingham and Simard, 2011b; Warren et al., 2008; Booth and Hoeksema, 2010; Leifheit et al., 2014a) could help characterize potential changes in soil hydraulic properties (soil water retention capacity, soil hydraulic conductivity, soil infiltration capacity) associated with the presence or absence of mycorrhizal extraradical mycelia. As an example, the use of rotated versus intact mesh cores in long-term field experiments would be useful to evaluate whether more efficient mycorrhizal hyphal exploitation of soil water (Ruiz-Lozano and Azcón, 1995; Duddrige et al., 1980) combined with water efflux by extraradical hyphae (Unestam and Sun, 1995; Sun et al., 1999; Egerton-Warburton et al., 2008) leads to more extreme wet/dry cycles, which are known to enhance soil aggregation and could thus improve soil hydraulic properties (Rillig and Mummey, 2006; Carminati et al., 2010).

Mycorrhizal suppression or inhibition of soil organic matter decomposition by saprotrophic fungi and bacteria (Gadgil effect) (Fernández and Kennedy, 2016; Leifheit et al., 2014b; Verbruggen et al., 2016) might be a pervasive mechanism by which the presence of EcM and AM fungi improves the hydraulic properties of soils across biomes, given that the water-retention properties of soil and its plant-available water capacity correlate strongly with organic matter content across soil textural classes (Hudson, 1994). Mycorrhizal modulation of soil organic matter turnover and storage processes and rates through their competitive interactions with other soil microbial guilds (saprotrophs) could thus lead to strong cumulative effects on soil organic carbon content and hydraulic properties in the long term (Averill et al., 2014; but see also Cheng et al., 2012). Using intact versus periodically rotated hyphal ingrowth mesh cores in multiyear field studies could help detect potential changes in soil hydraulic properties linked to changes in organic matter dynamics with or without the presence of mycorrhizal extraradical mycelia.

Given the key role of root-exuded mucilage in altering the hydraulic properties of rhizosphere soil relative to bulk soil (Carminati et al., 2010), further research is needed to advance understanding of mycorrhizal fungal influences on root exudate production and rhizodeposition (e.g., putative mycorrhizal-induced changes in mucilage amount or chemical composition) (Jones et al., 2004). The potential direct role of mycorrhizal hyphal exudates in sustaining or stimulating soil bacterial activity and in promoting soil aggregation, organic matter accretion, and soil water-retention capacity in the mycorrhizal hyphosphere also warrants further research.

The effects of mycorrhizal mycelium hydrophobicity and mycorrhizal-induced soil water repellency on soil hydraulic properties also merits further research attention. The potentially negative impact of hydrophobic EcM and AM fungal mycelia, hyphal exudates, and glomalin on soil infiltration and water retention has been mentioned by several studies, but this question is far from being resolved (Unestam & Sun, 1995; Feeney et al., 2004; Rillig et al., 2010; Zheng et al., 2014; Prieto et al., 2016). Fungal exudates are commonly amphiphilic in nature, so they tend to be strongly hydrophilic when wet, but below a critical moisture threshold the hydrophilic surfaces bond strongly with each other and with soil particles, leaving an exposed hydrophobic surface (Morales et al., 2010). Therefore if mycorrhizal soil dries beyond this critical water-content threshold, the soil behavior can shift abruptly from wettable to water-repellent, although prolonged wetting and field saturation during rainy periods usually allows soils to regain wettability. As suggested by Hallett (2007) and Claridge et al. (2009), seasonal water repellency in the topsoil layer could actually be a mechanism of soil water conservation in dryland ecosystems, because mycorrhizal-induced soil hydrophobicity may uncouple water stored in subsoil layers from surface evaporation during prolonged drought periods.

Virtually nothing is known about the potential importance of mycorrhizal diversity and community composition for shaping soil hydraulic properties. The putative relevance of mycorrhizal functional diversity for soil hydraulic properties merits research attention, with a focus on fungal functional traits that are most likely to be relevant to soil physical properties and water relations (e.g., extraradical hyphal production, hyphal longevity and turnover rates, hyphal tensile strength, mycelial architecture and hyphal exploration type, hydrophilic or hydrophobic properties of extraradical mycelia) (Unestam and Sun, 1995; Agerer, 2001; Rillig et al., 2015). Using hyphal ingrowth cores in field studies could help elucidate whether shifts in the identity and relative abundance of EcM and AM fungal taxa with contrasting functional traits (as measured by high throughput DNA sequencing) are associated with discernible changes in soil physical and hydraulic properties.

According to soil physics theory, the role of mycorrhizal fungi in modifying and improving the water retention properties of soil (and related hydraulic properties) should be particularly important for coarse-textured (sandy) soils of low or medium organic matter contents. Conversely, the influence of mycorrhizal fungi on the hydraulic properties of soil could be comparatively less relevant for fine-textured soils (clay) with high organic matter content, although this is just a tentative hypothesis that requires further experimental testing. Over long timescales (centennial to millennial), mycorrhizal fungi might even alter soil texture, particle size distribution, and clay mineralogy through their role in enhancing the weathering of mineral soil particles (Landeweert et al., 2001; Taylor et al., 2009; Arocena et al., 2012; Koele et al., 2014), which could eventually alter the hydraulic properties of soils. However, knowledge about mycorrhizal effects on soil texture is still in its infancy.

Finally, manipulative global change experiments using periodically rotated versus intact hyphal ingrowth mesh cores would contribute to advance an understanding of how (and whether) EcM and AM fungal responses to major global change drivers and their interactions (Compant et al., 2010; Mohan et al., 2014) could eventually alter the physical and hydraulic properties of soil in the mid to long term, which would be particularly consequential for the ecohydrology and primary productivity of water-limited ecosystems.

References

Agerer, R., 2001. Exploration types of ectomycorrhizae. Mycorrhiza 11, 107–114.

Akhzari, D., Attaeian, B., Arami, A., Mahmoodi, F., Aslani, F., 2015. Effects of vermicompost and arbuscular mycorrhizal fungi on soil properties and growth of *Medicago polymorpha* L. Compost Science and Utilization 23, 142–153.

Allen, M.F., Kitajima, K., 2013. *In situ* high-frequency observations of mycorrhizas. New Phytologist 200, 222–228.

Allen, M.F., 2007. Mycorrhizal fungi: highways for water and nutrients in arid soils. Vadose Zone Journal 6, 291–297.

Allen, M.F., 2009. Bidirectional water flows through the soil–fungal–plant mycorrhizal continuum. New Phytologist 182, 290–293.

Angers, D.A., Caron, J., 1998. Plant-induced changes in soil structure: processes and feedbacks. Biogeochemistry 42, 55–72.

Antoninka, A., Reich, P.B., Johnson, N.C., 2011. Seven years of carbon dioxide enrichment, nitrogen fertilization and plant diversity influence arbuscular mycorrhizal fungi in a grassland ecosystem. New Phytologist 192, 200–214.

Arocena, J.M., Velde, B., Robertson, S.J., 2012. Weathering of biotite in the presence of arbuscular mycorrhizae in selected agricultural crops. Applied Clay Science 64, 12–17.

Asbjornsen, H., Goldsmith, G.R., Alvarado-Barrientos, M.S., Rebel, K., van Osch, F.P., Rietkerk, M., Chen, J., Gotsch, S., Tobón, C., Geissert, D.R., et al., 2011. Ecohydrological advances and applications in plant-water relations research: a review. Journal of Plant Ecology 4, 3–22.

Augé, R.M., Stodola, A.J.W., Tims, J.E., Saxton, A.M., 2001. Moisture retention properties of a mycorrhizal soil. Plant and Soil 230, 87–97.

Augé, R.M., Sylvia, D.M., Park, S., Buttery, B.R., Saxton, A.M., Moore, J.L., Cho, K., 2004. Partitioning mycorrhizal influence on water relations of *Phaseolus vulgaris* into soil and plant components. Canadian Journal of Botany 82, 503–514.

Augé, R.M., 2001. Water relations, drought and vesicular-arbuscular mycorrhizal symbiosis. Mycorrhiza 11, 3–42.

Augé, R.M., 2004. Arbuscular mycorrhizae and soil/plant water relations. Canadian Journal of Soil Science 84, 373–381.

Averill, C., Turner, B.L., Finzi, A.C., 2014. Mycorrhiza-mediated competition between plants and decomposers drives soil carbon storage. Nature 505, 543–545.

Bearden, B.N., Petersen, L., 2000. Influence of arbuscular mycorrhizal fungi on soil structure and aggregate stability of a vertisol. Plant and Soil 218, 173–183.

Bingham, M.A., Simard, S.W., 2011a. Do mycorrhizal network benefits to survival and growth of interior Douglas-fir seedlings increase with soil moisture stress? Ecology and Evolution 1, 306–316.

Bingham, M.A., Simard, S., 2011b. Ectomycorrhizal networks of *Pseudotsuga menziesii* var. *glauca* trees facilitate establishment of conspecific seedlings under drought. Ecosystems 15, 188–199.

Bogeat-Triboulot, M.B., Bartoli, F., Garbaye, J., Marmeisse, R., Tagu, D., 2004. Fungal ectomycorrhizal community affects root hydraulic properties and soil adherence to roots of *Pinus pinaster* seedlings. Plant and Soil 267, 213–223.

Bonell, M., Purandara, B., Venkatesh, B., Krishnaswamy, J., Acharya, H.A.K., Singh, U.V., Jayakumar, R., Chappell, N., 2010. The impact of forest use and reforestation on soil hydraulic conductivity in the Western Ghats of India: implications for surface and sub-surface hydrology. Journal of Hydrology 391, 49–64.

Booth, M.G., Hoeksema, J.D., 2010. Mycorrhizal networks counteract competitive effects of canopy trees on seedling survival. Ecology 91, 2294–2302.

Bornyasz, M.A., Graham, R.C., Allen, M.F., 2005. Ectomycorrhizae in a soil-weathered granitic bedrock regolith: linking matrix resources to plants. Geoderma 126, 141–160.

Brownlee, C., Duddridge, J.A., Malibari, A., Read, D.J., 1983. The structure and function of mycelial systems of ectomycorrhizal roots with special reference to their role in forming inter-plant connections and providing pathways for assimilate and water transport. Plant and Soil 71, 433–443.

Burri, K., Gromke, C., Graf, F., 2013. Mycorrhizal fungi protect the soil from wind erosion: a wind tunnel study. Land Degradation & Development 24, 385–392.

Cairney, J.W.G., 1992. Translocation of solutes in ectomycorrhizal and saprotrophic rhizomorphs. Mycological Research 96, 135–141.

Caldwell, M.M., Dawson, T.E., Richards, J.H., 1998. Hydraulic lift: consequences of water efflux from the roots of plants. Oecologia 113, 151–161.

Caravaca, F., Alguacil, M.M., Figueroa, D., Barea, J.M., Roldán, A., 2003. Re-establishment of *Retama sphaerocarpa* as a target species for reclamation of soil physical and biological properties in a semi-arid Mediterranean area. Forest Ecology and Management 182, 49–58.

Carminati, A., Vetterlein, D., 2013. Plasticity of rhizosphere hydraulic properties as a key for efficient utilization of scarce resources. Annals of Botany 112, 277–290.

Carminati, A., Moradi, A.B., Vetterlein, D., Vontobel, P., Lehmann, E., Weller, U., Vogel, H.J., Oswald, S.E., 2010. Dynamics of soil water content in the rhizosphere. Plant and Soil 332, 163–176.

Carminati, A., Schneider, C.L., Moradi, A.B., Zarebanadkouki, M., Vetterlein, D., Vogel, H.J., Hildebrandt, A., Weller, U., Schüler, L., Oswald, S.E., 2011. How the rhizosphere may favor water availability to roots. Vadose Zone Journal 10, 988–998.

Castillo, V.M., Martinez-Mena, M., Albaladejo, J., 1997. Runoff and soil loss response to vegetation removal in a semi-arid environment. Soil Science Society of America Journal 61, 1116–1121.

Cheng, L., Booker, F.L., Tu, C., Burkey, K.O., Zhou, L.S., Shew, H.D., Rufty, T.W., Hu, S.J., 2012. Arbuscular mycorrhizal fungi increase organic carbon decomposition under elevated CO_2. Science 337, 1084–1087.

Claridge, A.W., Trappe, J.M., Hansen, K., 2009. Do fungi have a role as soil stabilizers and remediators after forest fire? Forest Ecology Management 257, 1063–1069.

Clark, N.M., Rillig, M.C., Nowak, R.S., 2009. Arbuscular mycorrhizal fungal abundance in the Mojave Desert: seasonal dynamics and impacts of elevated CO_2. Journal of Arid Environments 73, 834–843.

Compant, S., van der Heijden, M.G.A., Sessitsch, A., 2010. Climate change effects on beneficial plant–microorganism interactions. FEMS Microbiology Ecology 73, 197–214.

Czarnes, S., Hallett, P.D., Bengough, A.G., Young, I.M., 2000. Root- and microbial-derived mucilages affect soil structure and water transport. European Journal of Soil Science 51, 435–443.

Daynes, C.N., Field, D.J., Saleeba, J.A., Cole, M.A., McGee, P.A., 2013. Development and stabilisation of soil structure via interactions between organic matter, arbuscular mycorrhizal fungi and plant roots. Soil Biology and Biochemistry 57, 683–694.

Duddridge, J.A., Malibari, A., Read, D.J., 1980. Structure and function of mycorrhizal rhizomorphs with special reference to their role in water transport. Nature 287, 834–836.

Egerton-Warburton, L.M., Graham, R.C., Hubbert, K.R., 2003. Spatial variability in mycorrhizal hyphae and nutrient and water availability in a soil-weathered bedrock profile. Plant and Soil 249, 331–342.

Egerton-Warburton, L.M., Querejeta, J.I., Allen, M.F., 2007. Common mycorrhizal networks provide a potential pathway for the transfer of hydraulically lifted water between plants. Journal of Experimental Botany 58, 1473–1483.

Egerton-Warburton, L.M., Querejeta, J.I., Allen, M.F., 2008. Efflux of hydraulically lifted water from mycorrhizal fungal hyphae during imposed drought. Plant Signaling & Behavior 3, 68–71.

Faber, B.A., Zasoski, R.J., Munns, D.N., 1991. A method for measuring hyphal nutrient and water uptake in mycorrhizal plants. Canadian Journal Botany 69, 87–94.

Feeney, D.S., Daniell, T., Hallett, P.D., Illian, J., Ritz, K., Young, I.M., 2004. Does the presence of glomalin relate to reduced water infiltration through hydrophobicity? Canadian Journal of Soil Science 84, 365–372.

Feeney, D.S., Crawford, J.W., Daniell, T., Hallett, P.D., Nunan, N., Ritz, K., Rivers, M., Young, I.M., 2006. Three-dimensional microorganisation of the soil-root-microbe system. Microbial Ecology 52, 151–158.

Fernandez, C.W., Kennedy, P.G., 2016. Revisiting the 'Gadgil effect': do interguild fungal interactions control carbon cycling in forest soils? New Phytologist 209, 1382–1394.

Franzluebbers, A.J., 2002. Water infiltration and soil structure related to organic matter and its stratification with depth. Soil and Tillage Research 66, 197–205.

Germer, S., Neill, C., Krusche, A.V., Elsenbeer, H., 2010. Influence of land-use change on near-surface hydrological processes: undisturbed forest to pasture. Journal of Hydrology 380, 473–480.

Godbold, D.L., Hoosbeek, M.R., Lukac, M., Cotrufo, M.F., Janssens, I.A., Ceulemans, R., Polle, A., Velthorst, E.J., Scarascia-Mugnozza, G., DeAngelis, P., et al., 2006. Mycorrhizal hyphal turnover as a dominant process for carbon input into soil organic matter. Plant and Soil 281, 15–24.

Gonzalez-Dugo, V., 2010. The influence of arbuscular mycorrhizal colonization on soil-root hydraulic conductance in *Agrostis stolonifera* L. under two water regimes. Mycorrhiza 20, 365–373.

Hallet, P.D., Feeney, D.S., Bengough, A.G., Rillig, M.C., Scrimgeour, C.M., Young, I.M., 2009. Disentangling the impact of AM fungi versus roots on soil structure and water transport. Plant and Soil 314, 183–196.

Hallett, P.D., Gordon, D.C., Bengough, A.G., 2003. Plant influence on rhizosphere hydraulic properties: direct measurements using a miniaturized infiltrometer. New Phytologist 157, 597–603.

Hallett, P.D., 2007. Soil water repellency is wonderful-the positive environmental implications. In: Abstracts of the Bouyoucos Conference on the Origin of Water Repellency in Soils, Sanibel Island, Florida.

Hinsinger, P., Bengough, A.G., Vetterlein, D., Young, I.M., 2009. Rhizosphere: biophysics, biogeochemistry and ecological relevance. Plant and Soil 321, 117–152.

Hudson, B.D., 1994. Soil organic matter and available water capacity. Journal of Soil and Water Conservation 49, 189–194.

Hunt, G.A., Fogel, R., 1983. Fungal hyphal dynamics in a Western Oregon Douglas-fir stand. Soil Biology and Biochemistry 15, 641–649.

Joffre, R., Rambal, S., 1988. Soil water improvement by trees in the rangelands of southern Spain. Acta Oecologica 9, 405–422.

Johnson, D., Leake, J.R., Read, D.J., 2001. Novel in-growth core system enables functional studies of grassland mycorrhizal mycelial networks. New Phytologist 152, 555–562.

Johnson-Maynard, J.L., Graham, R.C., Wu, L., Shouse, P.J., 2002. Modification of soil structural and hydraulic properties after 50 years of imposed chaparral and pine vegetation. Geoderma 110, 227–240.

Jones, D.L., Hodge, A., Kuzyakov, Y., 2004. Plant and mycorrhizal regulation of rhizodeposition. New Phytologist 163, 459–480.

Kahle, P., Baum, C., Boelcke, B., 2005. Effect of afforestation on soil properties and mycorrhizal formation. Pedosphere 15, 754–760.

Koele, N., Dickie, I.A., Blum, J.D., Gleason, J.D., de Graaf, L., 2014. Ecological significance of mineral weathering in ectomycorrhizal and arbuscular mycorrhizal ecosystems from a field-based comparison. Soil Biology and Biochemistry 69, 63–70.

Kumar, A., Choudhary, A.K., Suri, V.K., 2016. Influence of AM fungi, inorganic phosphorus and irrigation regimes on plant water relations and soil physical properties in okra (*Abelmoschus esculentus* L.)–pea (*Pisum sativum* L.) cropping system in Himalayan acid alfisol. Journal of Plant Nutrition 39 (5), 666–682 .

Lado, M., Paz, A., Ben-Hur, M., 2004a. Organic matter and aggregate size interactions in infiltration, seal formation, and soil loss. Soil Science Society of America Journal 68, 935–942.

Lado, M., Paz, A., Ben-Hur, M., 2004b. Organic matter and aggregate-size interactions in saturated hydraulic conductivity. Soil Science Society of America Journal 68, 234–242.

Landeweert, R., Hoffland, E., Finlay, R.D., Kuyper, T.W., Van Breemen, N., 2001. Linking plants to rocks: ectomycorrhizal fungi mobilize nutrients from minerals. Trends in Ecology and Evolution 16, 248–254.

Leake, J.R., Johnson, D., Donnelly, D., Muckle, G., Boddy, L., Read, D.J., 2004. Networks of power and influence: the role of mycorrhizal mycelium in controlling plant communities and agroecosystem functioning. Canadian Journal of Botany 82, 1016–1045.

Lehto, T., Zwiazek, J., 2011. Ectomycorrhizas and water relations of trees: a review. Mycorrhiza 21, 71–90.

Leifheit, E.F., Verbruggen, E., Rillig, M.C., 2014a. Rotation of hyphal in-growth cores has no confounding effects on soil abiotic properties. Soil Biology and Biochemistry 79, 78–80.

Leifheit, E.F., Verbruggen, E., Rillig, M.C., 2014b. Arbuscular mycorrhizal fungi reduce decomposition of woody plant litter while increasing soil aggregation. Soil Biology and Biochemistry 81, 323–328.

Lilleskov, E.A., Bruns, T.D., Dawson, T.E., Camacho, F.J., 2009. Water sources and controls on water-loss rates of epigeous ectomycorrhizal fungal sporocarps during summer drought. New Phytologist 182, 483–494.

Lutgen, E.R., Muir-Clairmont, D., Graham, J., Rillig, M.C., 2003. Seasonality of arbuscular mycorrhizal hyphae and glomalin in a western Montana grassland. Plant and Soil 257, 71–83.

Mardhiah, U., Caruso, T., Gurnell, A., Rillig, M.C., 2016. Arbuscular mycorrhizal fungal hyphae reduce soil erosion by surface water flow in a greenhouse experiment. Applied Soil Ecology 99, 137–140.

Marulanda, A., Azcón, R., Ruiz-Lozano, J.M., 2003. Contribution of six arbuscular mycorrhizal fungal isolates to water uptake by *Lactuca sativa* plants under drought stress. Physiologia Plantarum 119, 526–533.

Miller, R.M., Jastrow, J.D., 1990. Hierarchy of root and mycorrhizal fungal interactions with soil aggregation. Soil Biology and Biochemistry 22, 579–584.

Miller, R.M., Jastrow, J.D., 2000. Mycorrhizal fungi influence soil structure. In: Kapulnik, Y., Douds, D.D. (Eds.), Arbuscular Mycorrhizas: Molecular Biology and Physiology. Kluwer Academic, Dordrecht, The Netherlands, pp. 3–18.

Milleret, R., Le Bayon, R.C., Lamy, F., Gobat, J.M., Boivin, P., 2009. Impact of roots, mycorrhizas and earthworms on soil physical properties as assessed by shrinkage analysis. Journal of Hydrology 373, 499–507.

Mohan, E., Cowden, C.C., Baas, P., Dawadi, A., Frankson, P.T., Helmick, K., Hughes, E., Khan, S., Lang, A., Machmuller, M., et al., 2014. Mycorrhizal fungi mediation of terrestrial ecosystem responses to global change: mini-review. Fungal Ecology 10, 3–19.

Moradi, A.B., Carminati, A., Vetterlein, D., Vontobel, P., Lehmann, E., Weller, U., Hopmans, J.W., Vogel, H.J., Oswald, S.E., 2011. Three-dimensional visualization and quantification of water content in the rhizosphere. New Phytologist 192, 653–663.

Morales, V.L., Parlange, J.Y., Steenhuis, T.S., 2010. Are preferential flow paths perpetuated by microbial activity in the soil matrix? A review. Journal of Hydrology 393, 29–36.

Neumann, E., Schmid, B., Römheld, V., George, E., 2009. Extraradical development and contribution to plant performance of an arbuscular mycorrhizal symbiosis exposed to complete or partial rootzone drying. Mycorrhiza 20, 13–23.

Oades, J.M., Waters, A.G., 1991. Aggregate hierarchy in soils. Australian Journal of Soil Research 29, 815–828.

Plamboeck, A.H., Dawson, T.E., Egerton-Warburton, L.M., North, M., Bruns, T.D., Querejeta, J.I., 2007. Water transfer via ectomycorrhizal fungal hyphae to conifer seedlings. Mycorrhiza 17, 439–447.

Prieto, I., Roldán, A., Huygens, D., Alguacil, M.M., Navarro, J.A., Querejeta, J.I., 2016. Species-specific roles of ectomycorrhizal fungi in facilitating interplant transfer of hydraulically redistributed water between *Pinus halepensis* saplings and seedlings. Plant and Soil 406, 15–27.

Pritchard, S.G., Strand, A.E., McCormack, M.L., Davis, M.A., Oren, R., 2008. Mycorrhizal and rhizomorph dynamics in a loblolly pine forest during 5 years of free-air-CO_2-enrichment. Global Change Biology 14, 1252–1264.

Querejeta, J.I., Egerton-Warburton, L.M., Allen, M.F., 2003. Direct nocturnal water transfer from oaks to their mycorrhizal symbionts during severe soil drying. Oecologia 134, 55–64.

Querejeta, J.I., Egerton-Warburton, L.M., Allen, M.F., 2007. Hydraulic lift may buffer rhizosphere hyphae against the negative effects of severe soil drying in a California Oak savanna. Soil Biology and Biochemistry 39, 409–417.

Querejeta, J.I., Egerton-Warburton, L.M., Allen, M.F., 2009. Topographic position modulates the mycorrhizal response of oak trees to inter-annual rainfall variability. Ecology 90, 649–662.

Querejeta, J.I., Egerton-Warburton, L.M., Prieto, I., Vargas, R., Allen, M.F., 2012. Changes in soil hyphal abundance and viability can alter the patterns of hydraulic redistribution by plant roots. Plant and Soil 355, 63–73.

Rawls, W.J., Pachepsky, Y.A., Ritchie, J.C., Sobecki, T.M., Bloodworth, H., 2003. Effect of soil organic carbon on soil water retention. Geoderma 116, 61–76.

Riestra, D., Noellemeyer, E., Quiroga, A., 2012. Soil texture and forest species condition the effect of afforestation on soil quality parameters. Soil Science 177, 279–287.

Rillig, M.C., Mummey, D.L., 2006. Mycorrhizas and soil structure. New Phytologist 171, 41–53.

Rillig, M.C., Wright, S.F., Nichols, K.A., Schmidt, W.F., Torn, M.S., 2001. Large contribution of arbuscular mycorrhizal fungi to soil carbon pools in tropical forest soils. Plant and Soil 233, 167–177.

Rillig, M.C., Mardatin, N.F., Leifheit, E.F., Antunes, P.M., 2010. Mycelium of arbuscular mycorrhizal fungi increases soil water repellency and is sufficient to maintain water-stable soil aggregates. Soil Biology and Biochemistry 42, 1189–1191.

Rillig, M.C., Aguilar-Trigueros, C.A., Bergmann, J., Verbruggen, E., Veresoglou, S.D., Lehmann, A., 2015. Plant root and mycorrhizal fungal traits for understanding soil aggregation. New Phytologist 205, 1385–1388.

Ruiz-Lozano, J.M., Azcón, R., 1995. Hyphal contribution to water uptake in mycorrhizal plants as affected by the fungal species and water status. Physiologia Plantarum 95, 472–478.

Ruth, B., Khalvati, M., Schmidhalter, U., 2011. Quantification of mycorrhizal water uptake via high-resolution on-line water content sensors. Plant and Soil 342, 459–468.

Samanei, F., Asghari, S., Aliasgharzad, N., 2015. The effects of two arbuscular mycorrhizal fungi on some physical properties of a sandy loam soil and nutrients uptake by spring barley. Journal of Soil Environment 1, 1–9.

Saxton, K.E., Rawls, W.J., 2006. Soil water characteristic estimates by texture and organic matter for hydrologic solutions. Soil Science Society of America Journal 70, 1569–1578.

Schaap, M.G., Leij, F.J., van Genuchten, M.T., 1998. Neural network analysis for hierarchical prediction of soil hydraulic properties. Soil Science Society of America Journal 62, 847–855.

Scholl, P., Leitner, D., Kammerer, G., Loiskandl, W., Kaul, H.P., Bodner, G., 2014. Root induced changes of effective 1D hydraulic properties in a soil column. Plant and Soil 381, 193–213.

Six, J., Bossuyt, H., Degryze, S., Denef, K., 2004. A history of research on the link between (micro) aggregates, soil biota, and soil organic matter dynamics. Soil and Tillage Research 79, 7–31.

Staddon, P.L., Thompson, K., Jakobsen, I., Grime, J.P., Askew, A.P., Fitter, A.H., 2003. Mycorrhizal fungal abundance is affected by long-term climatic manipulations in the field. Global Change Biology 9, 186–194.

Sun, Y.P., Unestam, T., Lucas, S.D., Johanson, K.J., Kenne, L., Finlay, R., 1999. Exudation–reabsorption in a mycorrhizal fungus, the dynamic interface for interaction with soil and soil microorganisms. Mycorrhiza 9, 137–144.

Taylor, L.L., Leake, J.R., Quirk, J., Hardy, K., Banwart, S.A., Beerling, D.J., 2009. Biological weathering and the long-term carbon cycle: integrating mycorrhizal evolution and function into the current paradigm. Geobiology 7, 171–191.

Thomas, R.S., Dakessian, S., Ames, R.N., Brown, M.S., Bethlenfalvay, G.J., 1986. Aggregation of a silty clay loam oil by mycorrhizal onion roots. Soil Science Society of America Journal 50, 1494–1499.

Tisdall, J.M., Oades, J.M., 1982. Organic matter and water-stable aggregates in soils. Journal of Soil Science 33, 141–163.

Tisdall, J.M., 1991. Fungal hyphae and structural stability of soil. Australian Journal of Soil Research 29, 729–743.

Tóth, B., Weynants, M., Nemes, A., Makó, A., Bilas, G., Tóth, G., 2015. New generation of hydraulic pedotransfer functions for Europe. European Journal of Soil Science 66, 226–238.

Trent, J.D., Svejcar, T.J., Blank, R.R., 1994. Mycorrhizal colonization, hyphal lengths, and soil moisture associated with two *Artemisia tridentata* subspecies. Great Basin Naturalist 54, 291–300.

Treseder, K.K., Allen, M.F., 2000. Mycorrhizal fungi have a potential role in soil carbon storage under elevated CO_2 and nitrogen deposition. New Phytologist 147, 189–200.

Treseder, K.K., Egerton-Warburton, L.M., Allen, M.F., Chen, Y., Oechel, W., 2003. Alteration of soil carbon pools and communities of mycorrhizal fungi in chaparral exposed to elevated CO_2. Ecosystems 6, 786–796.

Unestam, T., Sun, Y.P., 1995. Extramatrical structures of hydrophobic and hydrophyilic ectomycorrhizal fungi. Mycorrhiza 5, 301–311.

van der Heijden, M.G.A., Streitwolf-Engel, R., Riedl, R., Siegrist, S., Neudecker, A., Ineichen, K., Boller, T., Wiemken, A., Sanders, I.R., 2006. The mycorrhizal contribution to plant productivity, plant nutrition and soil structure in experimental grassland. New Phytologist 172, 739–752.

van der Heijden, M.G.A., Bardgett, R.D., Van Straalen, N.M., 2008. The unseen majority: soil microbes as drivers of plant diversity and productivity in terrestrial ecosystems. Ecology Letters 11, 296–310.

Verbruggen, E., Jansa, J., Hammer, E.C., Rillig, M.C., 2016. Do arbuscular mycorrhizal fungi stabilize litter-derived carbon in soil? Journal of Ecology 104, 261–269.

Warren, J.M., Brooks, J.R., Meinzer, F.C., Eberhart, J.L., 2008. Hydraulic redistribution of water from *Pinus ponderosa* trees to seedlings: evidence for an ectomycorrhizal pathway. New Phytologist 178, 382–394.

Wilson, G.W.T., Rice, C.W., Rillig, M.C., Springer, A., Hartnett, D.C., 2009. Soil aggregation and carbon sequestration are tightly correlated with the abundance of arbuscular mycorrhizal fungi: results from long-term field experiments. Ecology Letters 12, 452–461.

Wösten, J.H.M., Pachepsky, Y.A., Rawls, W.J., 2001. Pedotransfer functions: bridging the gap between available basic soil data and missing soil hydraulic characteristics. Journal of Hydrology 251, 123–150.

Wuest, S.B., Caesar-TonThat, T.C., Wright, S.F., Williams, J.D., 2005. Organic matter addition, N, and residue burning effects on infiltration, biological, and physical properties of an intensively tilled silt-loam soil. Soil Tillage Research 84, 154–167.

Wullschleger, S.D., Tschaplinski, T.J., Norby, R.J., 2002. Plant water relations at elevated CO_2– implications for water-limited environments. Plant Cell and Environment 25, 319–331.

Zhang, J., Wang, F., Che, R., Wang, P., Liu, H., Ji, B., Cui, X., 2016. Precipitation shapes communities of arbuscular mycorrhizal fungi in Tibetan alpine steppe. Scientific Reports 6, 23488.

Zheng, W., Morris, E.K., Rillig, M.C., 2014. Ectomycorrhizal fungi in association with *Pinus sylvestris* seedlings promote soil aggregation and soil water repellency. Soil Biology and Biochemistry 78, 326–331.

CHAPTER

18

Mycorrhizal Networks and Forest Resilience to Drought

B.J. Pickles[1], *S.W. Simard*[2]

[1]University of Reading, Reading, England, United Kingdom; [2]University of British Columbia, Vancouver, BC, Canada

18.1 INTRODUCTION

The ecological impacts of global warming through climate change are already becoming evident in forest biomes around the globe (Allen et al., 2010; Williams et al., 2012; Anderegg et al., 2013, 2015). All climate models predict that further warming is inevitable (IPCC, 2013); however, the consequences of this will vary spatially and temporally, and often in unpredictable ways. The evidence points toward major impacts of climate change on terrestrial ecosystems being driven by temperature and precipitation, leading to increases in aridity in some areas (Sherwood and Fu, 2014; Huang et al., 2015), flooding in others (Piao et al., 2010; Davidson et al., 2012; Frank et al., 2015), and an increase in the occurrence of extreme and stochastic weather events in general (Williams and Jackson, 2007; IPCC, 2013). For example, droughts followed by forest fires are occurring with increasing regularity and intensity (Wotton et al., 2010). The impacts of these changes are likely to be costly, given the extent of land use and ecosystem modification by humans, although in some areas it may prove beneficial for certain species or land uses. Regardless, it is clear that forests will be affected in multiple interactive ways that vary depending upon their resilience to change (Millar et al., 2007; Aitken et al., 2008; Thompson et al., 2009; Blois et al., 2013; Urban et al., 2013).

It has long been recognized that mycorrhizal fungi play an important role in nutrient and water uptake and transport (Safir et al., 1971; Duddridge et al., 1980; Smith and Read, 2008). Fungal hyphae provide a means to access nutrients and water from soil and rock pores that are inaccessible to plant roots (Bornyasz et al., 2005), and possess a much higher absorptive and adsorptive capacity because of their greater surface area. Thus plants that associate with mycorrhizal fungi increase their access to limiting nutrients [nitrogen (N), phosphorus (P)] and water at the cost of supplying their fungal partners with photosynthetically-derived sugars. The discovery that certain species (Molina and Horton, 2015) of mycorrhizal fungi can

Mycorrhizal Mediation of Soil
http://dx.doi.org/10.1016/B978-0-12-804312-7.00018-8

form networks with the potential to create linkages between individual plants of the same species (Brownlee et al., 1983; Selosse et al., 2006; Beiler et al., 2010) or even widely different phylogenies (Simard et al., 1997a; Dickie et al., 2004) adds an entirely new level of complexity to forest ecology.

Here we review the role of these mycorrhizal networks (MNs) in forest resilience to drought, examine the role of drought in forest decline around the world, suggest how management of MNs can be incorporated into the mitigation of drought impacts on forests, and propose new research strategies for studying MNs and drought effects on forests. Although arbuscular mycorrhizal (AM) fungi and ectomycorrhizal (EcM) fungi are both capable of forming networks, we focus our review primarily on EcM networks and their interactions with drought resistance in forests.

18.2 FOREST RESILIENCE

The concept of forest resilience has been used in various ways with different emphases depending on the focus of the literature, so we begin here by defining our terminology. Throughout this chapter we use the term *forest resilience* to mean (adapted from Thompson et al., 2009) the capacity of a forest to maintain its species composition, structure, and ecological functions, in response to abiotic and/or biotic pressure. Resilience in forests in particular is intrinsically linked to the regenerative capacity of the original forest.

18.3 THE ROLE OF MYCORRHIZAS IN WATER UPTAKE

Early studies on the role of AM fungi in host–water relationships suggested an indirect effect resulting from improved nutrition caused by increased P uptake (Safir et al., 1971), which becomes less available to plants as soil dries. Thirty years later, Augé's (2001) review of the subsequent literature assimilated over 200 papers on AM fungi and water relationships, drought, and photosynthetic rates (90 host species from 69 genera; 5 AM fungal genera with 22 species then classified in the genus *Glomus*). The AM fungi were found to have a direct impact on the water balance of both healthy and drought-affected host plants. In practice, plant–water relationships are linked somewhat inevitably to mycorrhizal symbioses, plant phenology, nutrient acquisition [especially carbon (C), N, and P], and therefore to biomass, growth rates, and colonization by AM fungi (Augé, 2001; Smith et al., 2010; Zhang et al., 2014). Direct uptake of water by AM hyphae has been shown in cases with multiple plant species (Marulanda et al., 2003; Smith et al., 2010). Studies of AM effects on citrus seedling drought responses have indicated that mycorrhizas are involved in several different processes, from increased water potential and photosynthetic rate to increased water content (Wu and Xia, 2006). Oxidative resistance has also been observed, with the roots of AM orange (*Poncirus trifoliata*) and tangerine (*Citrus tangerine*) plants expressing increased production of antioxidants under water stressed conditions compared with nonmycorrhizal controls (Wu et al., 2006a,b). Drought resistance conferred by AM fungi in the form of increased survival, growth, or water flow under drought stress has also been observed in herbaceous agricultural crops [e.g., maize and tomato (Barzana et al., 2012) and spinach (Zuccarini and Savé, 2015)].

Ectomycorrhizal fungi have also been found to play important roles in host water uptake. Rhizomorphs of a *Suillus* species were observed to transfer tritium (^3H_2O) up to 27 cm from a source in soil to the needles of small *Pinus sylvestris* (Scots pine) seedlings (Duddridge et al., 1980). Furthermore, colonization by EcM fungi was found to increase the drought tolerance of *Pseudotsuga menziesii* (Douglas-fir) seedlings (Kazantseva et al., 2010) by reducing the physiological effects on the host, evidenced by higher photosynthetic rate, C fixation and growth, and subsequently by enabling faster postdrought recovery (Parke et al., 1983). EcM fungi can also provide their hosts with similar drought resistance benefits to those mentioned for AM hosts. For example, in a laboratory experiment, seedlings of *Nothofagus dombeyi* colonized by *Pisolithus tinctorius* and *Descolea antarctica* displayed higher relative water content, lower oxidative stress, and higher antioxidant enzyme activity than nonmycorrhizal plants (Alvarez et al., 2009). The majority of the work on water relationships between EcM tree hosts and their fungi has been focused on dry temperate forests (British Columbia and Spain) and Mediterranean forests (Spain and California) (Table 18.1).

Both AM and EcM fungi increase the absorptive surface of plants below ground, enabling access to water sources that would otherwise be out of reach to a nonmycorrhizal plant. If colonization by mycorrhizal fungi stimulates a higher total surface area of roots and fungi as compared with roots alone, then root conductance is increased above levels that the plant could achieve by itself (Sands et al., 1982). The extensive length and biomass of mycorrhizal mycelium in many soils shows the potential significance of mycorrhizas in water uptake (Godbold et al., 2006; Joergensen and Wichern, 2008; Wallander et al., 2013). For example, a mean hyphal length of 275 m/g soil (0–10 cm depth) was observed in a Californian oak stand (Querejeta et al., 2007), compared with 365 m/g soil (0–10 cm depth) in a Douglas-fir stand in Oregon, with the biomass of EcM extramatrical mycelium estimated at 660 g/m^2 (Fogel and Hunt, 1979; Godbold et al., 2006).

Complicating the picture of fungus–plant–water relationships is the observation of significant variation in water use strategies and uptake abilities between genera and species of both plants and fungi. For example, AM fungal species differ in their provision of soil water to hosts under drought stress (e.g., *Glomus* spp.) (Marulanda et al., 2003). Fungal species differences in water uptake are likely to be a function of the production and structure of their hyphae (Allen, 2007). In EcM fungi, exploration types (Agerer, 2001) appear to be a useful guide to the ability of a species to translocate water, with long and medium range morphologies (especially those that produce rhizomorphs, which are unique to EcM fungi) showing evidence of being more conducive to transport (Lilleskov et al., 2009; Bingham and Simard, 2011), and short distance types better at absorption. In addition to direct pathways of water uptake and translocation through hyphae and rhizomorphs, mycorrhizal fungi may also indirectly improve the host–water relationship through "leaky hyphae" that exude water into soil, which is then reabsorbed by hyphae and transferred to other plants (Sun et al., 1999; Allen, 2007). This suggests that some plants may act as "water donors" (Sekija et al., 2011; Pang et al., 2013; Sun et al., 2014), with the hydraulic redistribution of water providing a benefit to other individuals that are in close proximity to their roots, potentially via hyphal uptake of lifted, redistributed, and exuded water [Andrea Nardini and José Ignacio (Nacho) Querejeta, pers. comm., April 2016]. In seasonally dry African woodlands, different AM hosts express different water-use strategies with the benefits of AM fungal association varying between species such that they occur at either high or low water availability (Birhane et al.,

TABLE 18.1 Field Studies Examining the Water Relationships of Ectomycorrhizal Hosts and Fungi

Biome/Climate (Location)	Host spp.	Fungal Identity	Parent Material	Soil Depth (cm)	Max. Depth Sampled (cm)	Reference
Mediterranean (California)	*Quercus agrifolia*	–	Granitic	24–100	400	Bornyasz et al. (2005)
Mediterranean (California)	*Pinus jeffreyi* *Arctostaphylos patula*	–	Granitic	75	–	Rose et al. (2003)
Mediterranean (California)	*P. jeffreyi* *Pinus contorta*	–	Granitic	–	80	Lilleskov et al. (2009)
Mediterranean (California)	*Pinus ponderosa*	–	Granitic	20–300	600	Witty et al. (2003)
Mediterranean (Spain)	*Quercus ilex* *Quercus faginea*	*Tuber melanosporum*	–	30–200+	–	Domínguez Núñez et al. (2006)
Mediterranean (California)	*Q. agrifolia*	*Amanita, Boletus, Cenococcum, Cortinarius, Elasmomyces, Entoloma, Hebeloma, Hydrotrya, Inocybe, Laccaria, Lactarius, Macowanites, Melanogaster, Pisolithus, Russula, Thelephoraceae, Tricholoma, Truncocolumella, Tuber, Zelleromyces*	Granitic	30–200+	30	Querejeta et al. (2009)
Dry hot to moist warm temperate (British Columbia)	*Pseudotsuga menzeisii* var. *glauca*	–	–	–	32	Bingham and Simard (2013)
Mediterranean (California)	*Adenostoma fasciculatum* *Arctostaphylos glandulosa* *Ceanothus greggi* *Salvia* spp. *Eriogonum* spp.	*Cenococcum, Pisolithus, Rhizopogon, Wilcoxina*	Granitic	35–220	220	Egerton-Warburton et al. (2003)
Mediterranean (Spain)	*Helianthemum almeriense*	*Terfezia claveryi*	–	–	20	Morte et al. (2010)

TABLE 18.1 Field Studies Examining the Water Relationships of Ectomycorrhizal Hosts and Fungi—cont'd

Biome/Climate (Location)	Host spp.	Fungal Identity	Parent Material	Soil Depth (cm)	Max. Depth Sampled (cm)	Reference
Dry temperate (Oregon)	*P. ponderosa*	Cortinariaceae, *Rhizopogon* spp., *Lactarius* spp., Suillaceae, *Wilcoxina* spp.	–	–	20	Warren et al. (2008)
Dry moorland (France)	*Pinus pinaster*	*Cenococcum geophilum*, *Rhizopogon* spp., *Scleroderma* spp., Thelephoraceae sp., *Lactarius*-like	Aeolian	120–300 (rooting)	80 (120)[a]	Bakker et al. (2006)
Humid moorland (France)	*P. pinaster*	*C.*, *Rhizopogon* spp., *Scleroderma* spp., Thelephoraceae spp., *Lactarius*-like	Aeolian	80–100 (rooting)	80	Bakker et al. (2006)
Semiarid temperate (British Columbia)	*P. menzeisii* var. *glauca*	–	–	–	35	Schoonmaker et al. (2007)
Mediterranean (California)	*Q. agrifolia*	–	Granitic	200	200	Querejeta et al. (2007)
Mediterranean (California)	*P. contorta*	*Boletus edulis*, *Amanita muscaria*, *Tricholoma* spp., *T. saponaceum*, *Suillus* spp., *Russula albonigra*	–	–	80	Lilleskov et al. (2009)

[a] *Figures show ectomycorrhizal (EcM) fungi to 80 cm depth and fine roots to 120 cm depth, but state that all fine roots were EcM.*

2015). There is a possibility that these types of interactions are more widely applicable to different dry and seasonally dry systems, but there is a need for more information on water-use strategies across different ecosystems and a broader range of species.

The overall picture that emerges from these studies is that mycorrhizal fungi appear to have the potential to contribute substantially to host drought avoidance.

18.4 MYCORRHIZAL NETWORKS AND THEIR ROLE IN HYDRAULIC REDISTRIBUTION AND DROUGHT RESPONSES

Mycorrhizal networks are emergent structures that arise when an individual fungus forms functioning mycorrhizas on multiple plants linked by its hyphae (Newman, 1988; Fitter, 2001; Simard and Durall, 2004; Selosse et al., 2006). It is often assumed that MNs play a facilitative role in plant establishment, but positive, neutral, and negative responses have been observed in different systems, with AM systems tending toward more neutral and negative outcomes and EcM systems toward more neutral or positive outcomes for plant hosts (van der Heijden and Horton, 2009). Observations of the presence of EcM "mycelial networks" and their involvement in interplant nutrient transfer were first made in axenic culture using loblolly pine (*Pinus taeda*) seedlings and *P. tinctorius* or *Thelephora terrestris* EcM fungi in an experiment tracing the movement of ^{14}C isotopes (Reid and Woods, 1968). Later, in laboratory experiments using seedlings grown in soils, Brownlee et al. (1983) noted that the EcM root tips of *P. sylvestris* seedlings were connected to each other by the hyphae and rhizomorphs of *Suillus bovinus*, and that the water potential of seedling needles dropped rapidly when mycelia connecting a water source were severed. The presence and activity of MNs in the field at larger scales has subsequently been noted in different ecosystems (Simard et al., 1997b; Horton and Bruns, 1998; Horton et al., 1999; Booth, 2004; Plamboeck et al., 2007; Warren et al., 2008; van der Heijden and Horton, 2009; Beiler et al., 2010), with MNs forming among mature trees and with seedlings (e.g., interior Douglas-fir, ponderosa pine, savanna oak) and between plants of varied phylogenetic history (e.g., *Arctostaphylos* and *Pinus* or *Pseudotsuga*). The majority of the research on EcM MNs has involved evergreen tree species that typically experience periods of sustained aridity in at least part of their range (Table 18.1). Well-studied host species include interior Douglas-fir (*Pseudotsuga menziesii* var. *glauca*) and ponderosa pine (*Pinus ponderosa*) in hot, dry areas of western North America (Simard et al., 1997a; Kennedy et al., 2003; Plamboeck et al., 2007; Warren et al., 2008; Beiler et al., 2010; Bingham and Simard, 2012a, 2013), and oaks in the hot, dry Mediterranean ecosystems of Portugal, Spain, and California (e.g., coast live oak, *Quercus agrifolia*) (Kurz-Besson et al., 2006; Egerton-Warburton et al., 2007, 2008; Querejeta et al., 2009).

Although the evidence shows that EcM MNs can affect seedling water budgets and enhance their survival and growth, the nature and mechanisms are not always clear (Warren et al., 2008; Bingham and Simard, 2011). For example, positive effects of MNs on seedling drought resistance have been observed in the laboratory, and this has been associated with transport of water through fungal linkages from replete to water-stressed seedlings (Bingham and Simard, 2011). In the field, however, resistance to drought was more strongly influenced by host provenance genetic history and local adaptation than access to MNs (Bingham and Simard, 2012a, 2013). Provenance trials across a wider range of sites, soils, and seed origins have revealed that local adaptation of conifer seedlings is at least partly driven by soil biology and soil climatic history, particularly moisture and drought regime (Pickles et al., 2015). However, severe weather events can overwhelm the positive effects of soil factors, at least in the case of planted seedlings (Pickles, unpublished data).

In dry forests, trees and shrubs are known to transfer water from deep to shallow soil layers via hydraulic lift (Moreira et al., 2003; Kurz-Besson et al., 2006; Egerton-Warburton et al., 2007, 2008; Bauerle et al., 2008; Scholz et al., 2008; Neumann and Gardon, 2012; Prieto et al., 2012; Quijano et al., 2013). During the day, water is absorbed by roots both in shallow and deep soil layers, then transported via the xylem to the foliage, where it is then transpired to the atmosphere. During the night, water is taken up by the root system from deep soil layers, where water potential is higher, to shallow layers, where water potential is generally lower because of daytime water uptake and evaporation, in order to sustain daytime transpiration. Importantly this translocation of water, "hydraulic redistribution," can occur from deep to shallow and shallow to deep soil horizons as well as horizontally (Burgess et al., 1998). An illustrated example of hydraulic lift and hydraulic redistribution is presented in Fig. 18.1.

Hydraulic redistribution of water by host plants has been shown in several studies, and is associated with enhanced plant nutrient uptake via maintenance of fine roots and their associated bacteria (e.g., root nodules) and fungi (e.g., mycorrhizas), or by stimulating microbial activity through the release of water into dry shallow soils (Querejeta et al., 2003; Austin et al., 2004; Kurz-Besson et al., 2006; Egerton-Warburton et al., 2007; Snyder et al., 2008; Warren et al., 2008; Aanderud and Richards, 2009; Lehto and Zwiazek, 2011). Because roots of trees and plants in nature are predominantly mycorrhizal, hydraulic uplift and redistribution inevitably involves the mycelium (Fig. 18.1). Although it is often reported that host plants are maintaining their symbionts in this manner (Querejeta et al., 2003; Schoonmaker et al.,

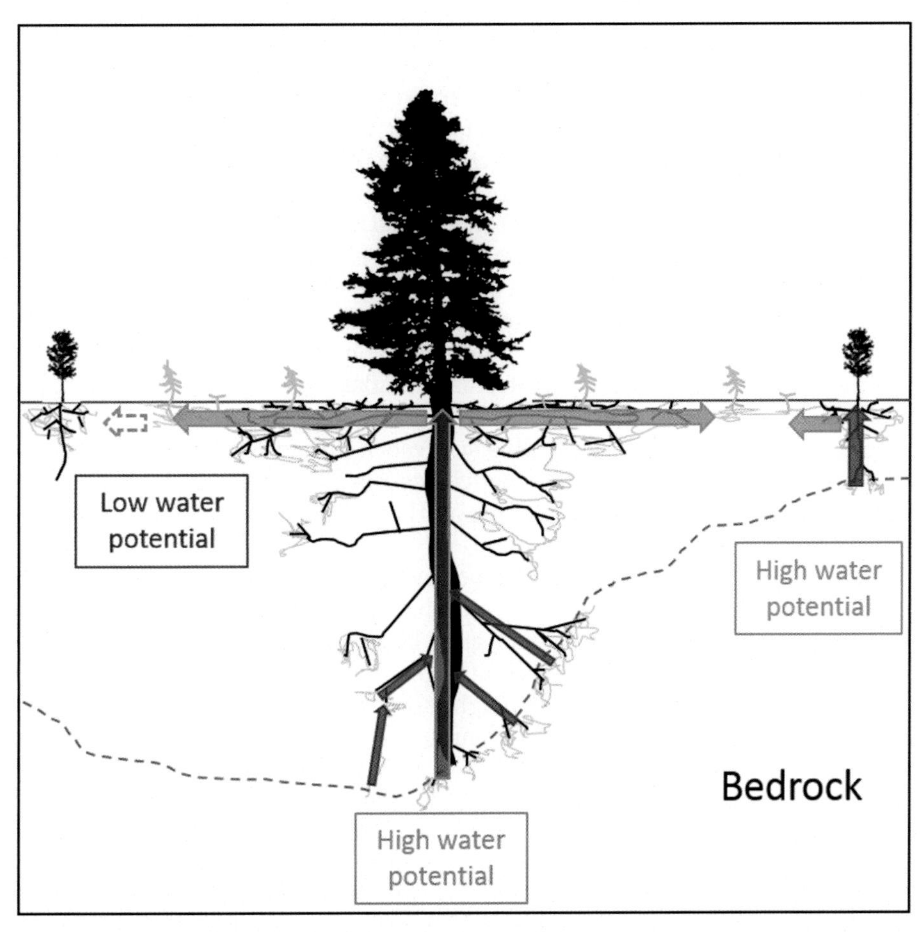

FIGURE 18.1 Illustration showing hypothetical water translocation pathways via roots (black) and mycorrhizal hyphae (orange) in a drought-prone ecosystem. *Vertical (dark blue) arrows* indicate hydraulic lift, *horizontal (light blue) arrows* indicate hydraulic redistribution, and a *dashed arrow* indicates possible water pathway resulting from "leaky" hyphae. *Black silhouettes* represent mature trees or shrubs; *green silhouettes* represent seedlings.

2007; Muñoz et al., 2008), there is also evidence showing that EcM fungi play a role in providing their hosts with access to deep water supplies during drought periods (Bornyasz et al., 2005) and in translocating hydraulically lifted water between hosts through MNs (Brooks et al., 2006; Egerton-Warburton et al., 2007, 2008; Plamboeck et al., 2007; Warren et al., 2008). According to Egerton-Warburton et al. (2007), the extent of hydraulic redistribution through AM fungal networks may be greater than through EcM networks because of the potential for large interconnected hyphal networks to form via anastomosis, although certain EcM fungal species with rhizomorphs are known to transport large amounts of water. It is thought that water moves through these MNs via mass flow along water potential gradients on the hyphal surfaces or inside the MN via cytoplasmic streaming (Plamboeck et al., 2007; Warren et al., 2008; Bingham and Simard, 2011). The extensive use of deep water supplies by *Boletus edulis* and *Suillus* species during fruiting (Lilleskov et al., 2009) points either to translocation from the host or to direct access through extensive vertical distribution of mycelium. Observations of *Pisolithus* species, *Rhizopogon* species (Egerton-Warburton et al., 2003), and other unidentified EcM fine roots (Bornyasz et al., 2005) in contact with bedrock show the potential for rhizomorph-forming EcM fungi to access deep water directly (Allen, 2009), although this would likely operate in conjunction with host redistribution.

Old-growth interior Douglas-fir and ponderosa pine trees are able to minimize drought stress by keeping soil water potential high in shallow soils during drought via hydraulic redistribution (Domec et al., 2004; Schoonmaker et al., 2007). Stomatal conductance and root embolism collectively aid in keeping shallow roots functioning and preventing stomatal closure. In a field study, EcM MNs were shown to provide a direct pathway for hydraulic redistribution of deuterium ($^{2}H_2O$) from old stumps to nearby ponderosa pine seedlings (Warren et al., 2008).

Little is known about the potential for competition in water access between AM and EcM MNs. In drought-prone chaparral in California, Douglas-fir seedlings establish in patches of EcM *Arctostaphylos* but not in AM patches of *Adenostoma*, and share a large proportion of their EcM community (Horton et al., 1999). This suggests a link to water access in these seasonally drought-stressed ecosystems with beneficial effects on Douglas-fir regeneration through either direct water supply via EcM fungi, transport of water through MNs, or hyphal leakage of water by *Arctostaphylos* roots or EcM fungi.

Access to EcM MNs of old trees has also been shown as crucial to seedling establishment in arid forests of southern interior British Columbia (Bingham and Simard, 2012a). Seedlings establishing within the MN of mature trees grew larger and had greater water use efficiency (i.e., higher $\delta^{13}C$ values) than those without access to the networks. The EcM community of these seedlings was similar to that of neighboring mature hosts, suggesting a significant contribution of mature tree MNs to EcM community formation on seedlings (Bingham and Simard, 2012b). Community similarity increased under drought stress, and seedling communities were dominated by long-range (*Rhizopogon*/*Suillus*) and medium-range (*Amphinema*, *Boletus*, *Cortinarius*, *Tomentella*, *Tricholoma*) species, with species richness decreasing under increasing drought strength (presumably to a community mainly consisting of EcM with long-range exploration types). It appears that linkage into the network of mature trees before the onset of summer drought was necessary for benefits to accrue to the seedlings, as was found by Horton et al. (1999). These results suggests that local provenances, which are adapted to drought stress through earlier germination, emergence, and mycorrhizal colonization, could perform better than populations from less drought-prone sites because of longer and more extensive linkage to existing MNs.

18.5 ROOTING DEPTH

The rooting depths of plants are often much greater than is typically studied (Stone and Kalisz, 1991; Canadell et al., 1996; Schenk and Jackson, 2002a), and this is particularly true with reference to their mycorrhizas (Pickles and Pither, 2014), with mean depths on the order of several meters and certain species observed to root as deeply as 60 m (*Eucalyptus* spp. in Australia) (Stone and Kalisz, 1991). Rooting depths vary within and among species (Stone and Kalisz, 1991; Canadell et al., 1996) and within and among ecosystems, with dry and drought-stressed systems tending to produce deeper rooting systems (Schenk and Jackson, 2002b). A global study has also revealed that mycorrhizal colonization patterns vary among ecosystems, with significant effects of climate and soil chemistry (Soudzilovskaia et al., 2015). EcM fungal community composition varies with depth (Dickie et al., 2002; Rosling et al., 2003; Genney et al., 2006; Taylor et al., 2014) and MN-forming species have been observed to segregate themselves vertically in the upper organic horizons (Beiler et al., 2012; Mujic et al., 2016), as well as at significant depths (4 m) extending into weathered bedrock (Bornyasz et al., 2005). Mesquite roots colonized by AM fungi have been recovered from 4.5–4.8 m depth in the Sonoran desert (Virginia et al., 1986) and viable AM spores associated with tree roots (*Faidherbia albida*) have been recovered from soils at the water table (34 m depth) in semiarid ecosystems of Senegal (Dalpé et al., 2000).

Parent material and soil type are also significant factors in plant and ecosystem water relationships (Jenny, 1961; Richter and Billings, 2015). Soils in mountainous regions are typically thin and unable to store large volumes of water, but the bedrock beneath them may be highly porous (Sternberg et al., 1996; Witty et al., 2003). Especially in granitic bedrock and chaparral systems, the ability to tap into bedrock sources of water provides a means to bypass drought effects, and as described previously, EcM fungi appear to provide a key functional role in this process through the ability of hyphae to infiltrate rock pores and access stored water therein (Fig. 18.1). Shrubs and trees in these regions have been shown to extract water from bedrock (Sternberg et al., 1996; Hubbert et al., 2001; Witty et al., 2003), and EcM fungi were found to be key players in the extraction of ground water by oak trees (*Q. agrifolia*) from granitic bedrock (Bornyasz et al., 2005). Ponderosa pine seedlings (*P. ponderosa*) have also been observed to extract water from weathered granitic bedrock through their EcM fungi (Witty et al., 2003). Lilleskov et al. (2009) found that hydraulic redistribution through mycorrhizal pathways was great enough to push sporocarps through crusts, rock, and cement fractures during extremely dry seasons. In a comparison of *Pinus pinaster* rooting depth and EcM associations in humid and dry sites in France, Bakker et al. (2006) found that humid sites displayed shallower rooting depth (<80 cm) compared with dry sites (>120 cm). They also observed a shift in the composition of exploration types of EcM fungi, from hydrophilic contact types in their humid site to short- and long-range exploration types (Agerer, 2001) in their dry site.

The overall picture from studies that assess rooting depth is that plants vary both interspecifically and intraspecifically, and tend to root more deeply in the soil profile than is typically considered, although we note that this may largely be because of practical difficulties associated with the cost (time and money) of access (Harrison et al., 2011). It is often suggested in the mycorrhizal literature that the majority of fine roots, and hence EcM fungi, are located in shallow organic soils (down to ≈20 cm) (Koide and Dickie, 2002). However, this has not been systematically studied and could represent a bias toward sampling in locations or environments that favor shallower rooting. Indeed, one of the first examinations of EcM fungal hyphae at depth found evidence of niche partitioning, with EcM species identified in all soil horizons that were assessed (Dickie et al., 2002). It is also speculated by some authors that their own work did not sample deeply enough to determine the maximum extent of EcM fine roots, especially in dry environments (Bakker et al., 2006). Several approaches may prove useful for estimating the importance of mycorrhizas at depth, including ground-penetrating radar (GPR) (Sucre et al., 2011). Deep rooting is particularly important to plants in drought-stressed areas. Furthermore, the observation of mycorrhizal symbioses and nodule-forming bacteria several meters below ground emphasizes the fact that these organisms play a role in ecosystem activity that extends from the soil surface all the way into bedrock (Richter and Billings, 2015).

18.6 THE ROLE OF DROUGHT IN GLOBAL FOREST DECLINE

Climate change is anticipated to increase the rate of severe forest drought events caused by increasing aridity (Sherwood and Fu, 2014) resulting from increased temperatures, decreased precipitation, and altered weather patterns (Allen et al., 2010; Williams et al., 2012; Anderegg et al., 2013, 2015). Dryland biomes are predicted to expand globally (Huang et al., 2015) through continued desertification and the recruitment of drought-tolerant plant species. An increase in the occurrence of extreme weather events will contribute to forest die-back and shifts in community replacement resulting from: (1) greater susceptibility of seedlings to extreme events (most mortality occurs at the seedling stage), and (2) enhancement of existing pressures on forests in marginal environments. Furthermore, increasing drought stress is associated with increasing susceptibility to pathogens and an inability to access or effectively redistribute water (e.g., drought increases host susceptibility to fungal pathogens, often delivered by insect pests) (Thompson et al., 2009). In interior British Columbia, increased incidence and severity of insect and pathogen infections of planted *Pinus contorta* was related to changing precipitation patterns and increasing summer drought (Heineman et al., 2010; Mather et al., 2010).

Comparative studies of dry-forest tree species (e.g., *Acacia* spp. and *Boswellia papyrifera* in Senegal) reveal that coexisting species can show very different responses to drought, and very different allocation strategies, with AM fungal colonization leading to differences in the benefits to the host (Birhane et al., 2015); *Acacia* performed better under higher water availability, *Boswellia* under low water availability. Scholz et al. (2008) explored the function of root morphologies in deciduous versus evergreen trees in Brazilian dry ecosystems. Significant differences in hydraulic lift were found between deciduous and evergreen plants, with the latter tending not to exhibit water transfer to upper roots and soil, and instead maintaining a constant upward movement of water toward leaves. This suggest that deeper rooting evergreens use a deep water supply year round, whereas deciduous trees make use of upper soil water in the rainy season and may supplement roots with water extracted at depth in order to maintain fine root function and symbionts.

Increases in the incidence and strength of drought events have been noted as a cause for concern in regard to forest recruitment (Hamann and Wang, 2006; Spittlehouse, 2008), as well as a driver of the observed and predicted increases in forest fire size and occurrence (Gillett et al., 2004; Nitschke and Innes, 2008; Wotton et al., 2010). Severe drought events often lead to mortality of planted seedlings, with potentially significant impacts on forest productivity, particularly in regions already subject to drought stress (Mather et al., 2010). In the southern interior of British Columbia, climatic drought has been a significant factor contributing to the 25% failure rate of commercial lodgepole pine (*P. contorta*) plantations (Roach et al., 2015). In the Assisted Migration Adaptation Trials throughout western North America (O'Neill et al., 2013), extreme drought events caused 95–100% mortality of planted seedlings in two of the most xeric sites (M. Carlson, pers. comm.; B.J. Pickles, pers. obs.). Although locally sourced provenances were similarly affected by drought-inflicted mortality, it was noted at multiple field locations that whereas planted seedlings were experiencing high mortality, naturally regenerating seedlings were surviving in situ (Mather et al., 2010).

The vulnerability of trees to climatically induced drought, especially higher temperatures, lowered precipitation, and weather extremes, is linked to a multiplicity of factors such as drought severity and rooting behavior. Rooting depth itself is linked to parent material, depth to water table, and depth to bedrock (Stone and Kalisz, 1991; Canadell et al., 1996; Schenk and Jackson, 2002a; Sucre et al., 2011). Planted seedlings are used extensively in reforestation efforts, but have significantly more restricted root systems than naturally regenerated seedlings in their first few years, leaving them more vulnerable to extreme drought. Studies examining the contribution of root behavior to recent tree mortality in forest droughts are lacking but are likely to prove revealing, given the importance of root branching patterns, diameter, density, and distribution to absorptive capacity (McCormack et al., 2015).

18.7 CLIMATE CHANGE PROJECTIONS FOR DROUGHT EFFECTS ON FORESTS AND THE DOMINO EFFECT

Global climate models predict that increasing aridity will generate stronger, more frequent, or more prolonged drought effects in already stressed zones and initiate seasonal drought conditions in zones where this has typically been absent (Bonan, 2008; Sherwood and Fu, 2014).

Once tree die-back has been initiated, it tends to produce cascading effects in forests as a result of the structural consequences of tree fall. This has been described as the "domino effect" (Bauhus et al., 2010), in which tree fall can be divided up into "initiator" trees and "domino" trees. Initiators are the trees that die and subsequently fall owing to their own mortality, whereas dominos are those trees that are felled by falling initiators. Once a certain critical environmental threshold is reached to cause mature tree mortality, initiator deaths result in wider mortality or damage of otherwise healthy trees, which in turn can increase their susceptibility to the original or secondary sources of mortality. Indeed, the concept can be extended to any triggering event in which an initial wave of mortality increases the chances of further mortality. In forests, trees already under stress, such as those on dry sites or in subordinate crown positions, are more vulnerable to drought. These trees are unable to access sufficient water to meet transpiration demands because of low soil water potential or inadequate root and mycorrhizal systems for accessing scarce water. In dry conditions, tree fall often results in root upheaval and soil disturbance, and hence MNs are likely disturbed by these events.

Human land-use practices can interact with climate to enhance or mitigate drought effects. In temperate and subboreal forests of North America, conifer mortality from bark beetles has been amplified by fire suppression and climate change, including warmer winters and drier summers (Kurtz et al., 2015). Combined with land-use change, extreme summer drought and heat waves like those of 2003 and 2012 have caused substantial yield losses, as well as widespread tree mortality events worldwide (Allen et al., 2015). The spring 2016 megafire in northern Canada is such an example, where unseasonably hot temperatures and drought conditions extending from the preceding fall because of El Niño combined with high winds to cause massive forest losses. Conversion of complex primary forest to simplified plantations has further increased vulnerability of temperate forests to drought. In tropical forest ecosystems, and particularly peat forests (Yule, 2010; Rose et al., 2011), increasing aridity often arises as a result of the application of irrigation techniques for agricultural purposes. This has the potential to cause cascading effects on forest resilience when amplified by climate change and El Niño drought events (Marlier et al., 2012). Thus the lowering of water tables through extraction for human use and associated conversion of forests to agricultural land is likely to add additional stress on any ecosystems in which deep rooting and hydraulic extraction of ground water is utilized as a strategy for drought resistance.

18.8 INCORPORATING MYCORRHIZAL NETWORKS IN FOREST MANAGEMENT

There are several steps that can be taken to incorporate the preservation of MN function into forest management. As an initial step associated with operational planning, it would be beneficial to identify forested areas that are anticipated to experience increasing drought severity by using the best available climate projections (Hamann and Wang, 2006; Coops et al., 2011; Hamann et al., 2011; Mathys et al., 2014; Anderegg et al., 2015). Identification and monitoring of the mycorrhizal fungal species in forested areas has been rather limited in scope, and there is a need for better landscape-scale information on the distribution of different species and functional types of fungi (Suz et al., 2015). Characterization of soils and parent materials should also be integrated, along with any information on rooting depth of mature trees. During harvesting operations, mature trees and deep-rooted shrubs should be retained for the purpose of maintaining water relationships via connection to water table for the provision of deep water to surface soils via hydraulic lift. Large, old trees have been shown to serve as hubs in MNs, providing linkages to seedlings, and thus serving key roles in facilitating regeneration via access to deep water and nutrient redistribution as well as mycorrhizal colonization. During replanting it may be beneficial to plant seedlings close to these old trees so as to maximize their chances of tapping into MNs. Tree species with deep root systems and mycorrhizal fungal communities with high networking potential should be selected for planting. A diversity of tree and shrub species, including genotypes migrated from warmer or drier climates (Pickles et al., 2015), should be managed for the resilience they confer on the ecosystem (Aitken and Bemmels, 2016). Protection of naturally regenerated trees should also be considered a priority because of their superior performance compared with planted trees when under stress (Mather et al., 2010). Attempting to minimize soil compaction around retained trees during harvesting may also help preserve MNs. In systems where the drought intensity is predicted to increase, it is likely that it will also be important to consider replanting in multiple years after harvest to attempt to maximize the chances of successful regeneration. This is because a severe drought is liable to inflict high mortality on seedlings, which is likely to be a consistent problem across forests of similar climatic and soil types such as dry and seasonally dry EcM host woodlands (coniferous, and also dry oak woodlands/savannas). Forest management recommendations for incorporating MNs into climate change mitigation and resilience are detailed elsewhere (Simard et al., 2013).

18.9 KNOWLEDGE GAPS AND FUTURE RESEARCH DIRECTIONS

Much of the existing research on MNs and water relationships has examined a particular subset of tree species in a particular subset of climatic zones (e.g., oaks and savanna/chaparral, oaks and conifers in Mediterranean ecosystems, interior Douglas-fir and ponderosa pine in dry temperate forests, thin soils over weathered granitic bedrock; Table 18.1). This has resulted in a tendency to extrapolate patterns derived from a few systems across the breadth of mycorrhizal associations. To clarify whether these patterns are in fact general, we need to extend our knowledge to additional species, soils, and ecosystems. Furthermore, there is a distinct need to expand our investigations beyond individual plants and examine collective impacts on forest stands. Comparative studies between EcM and AM MN–forming trees in the same and different environments are needed, as well as studies of water relationships in those species that form dual associations. No-analogue climates (Williams and Jackson, 2007) are predicted as a result of climate change, in which forests will be exposed to abiotic conditions that they have not previously experienced. Accounting for these impacts experimentally is difficult outside of growth chambers, but the effects on seedlings are expected to differ substantially from the effects on mature trees. Deep root associations also need to be studied more widely to get a better understanding of the depths to which functional mycorrhizas/MNs are typically formed in different ecosystems (Pickles and Pither, 2014). How these symbioses relate to rooting depth, soil moisture content, depth to bedrock, parent material, and drought stress will provide important information on the capabilities and function of mycorrhizas in nature.

Based on our review of the literature on water translocation by mycorrhizas and MN, we have identified the following future research areas that are likely to prove productive.

1. Exploiting climate and forest management data: a large body of research exists on the past, current, and predicted future impacts of drought stress in forests. This information could be used to identify a suitable gradient of field sites under different levels of drought stress with which to assess the relative contributions of MN to forest resilience. Additionally the data could be used to explore whether MNs change the predicted outcomes and what interactions occur with soil depth, depth to water table, and parent material.
2. Assessing a greater diversity of plant species at each site (Ishida et al., 2007; Dulmer et al., 2014): the potential for plants from widely different phylogenetic origins to be linked by a single MN and access water at depth greatly expands the range of potential water transport opportunities, but most research tends to focus on tree hosts alone.

3. Assessing a greater diversity of plant species across sites: most research on MNs has been restricted to a handful of host species, typically in Mediterranean or seasonally dry temperate forests. Thus there are opportunities to expand our knowledge of the importance of MNs for host–water relationships in a wider array of ecosystems and host species. Coupled with the predictive approaches suggested earlier, we can begin to select sites for field investigations by identifying locations that are experiencing drought stress and locations that are drought-free but predicted to experience drought in the future (e.g., using detailed climate models) (Wang et al., 2012). By visiting these sites, researchers can explore whether and where MN are being formed, and which fungal species are forming them. Field-based, stable-isotope probing experiments or other means of labeling the transfer of dissolved nutrients or compounds (e.g., dyes) may prove to be the most effective way of assessing the networking potential of fungal species. These types of study may help identify whether a wider array of network-forming fungi exists, given that most networking studies have focused on a small number of fungal and host species, typically *Suillus* and *Rhizopogon* associated with conifers.
4. Identifying promising "water donor trees," or nurse tree and plant species: to improve water status of crop species in plantations. Promising species would be deep rooted, have hydraulic lift potential, and have the ability to form MNs with crop trees in areas where drought is an issue. Likewise, identification of mycorrhizal fungi with high networking (Molina and Horton, 2015) and water transport (Allen, 2007) capabilities should be identified concurrently with tree species selection. The possible exploitation of such species for bioirrigation to facilitate establishment of young saplings requires careful experimental testing. These research activities may improve forest and agroforestry productivity in arid areas while limiting water use for irrigation and increasing the overall sustainability and resilience of forested systems to ongoing climate changes. This is especially true for areas where surface water is scarce but deep soil water is available.
5. Quantifying the amount of water hydraulically lifted and redistributed by different plant and fungal species, in different environments, and across seasons.
6. Quantifying the spatial proximity of planted seedlings to water donor trees, including the minimum and maximum distance for planting to maximize exploitation of hydraulic lift and improve resilience of crop species to drought.
7. Multiple source labeling approaches (Whitman and Lehmann, 2015): natural differences in the isotopic ratios of water from different sources have been used to successfully partition the proportion of uptake by source and tissue type. Experimental manipulation and partitioning of water sources using $H_2{}^{18}O$ and/or heavy water (^2H_2O), coupled with ^{15}N and $_2{}^{13}C$, may be an effective way of simultaneously examining multiple pathways of translocation and nutrient uptake.[13]
8. Deep sampling of host plant roots in drought-stressed environments: we hypothesize that drought stressed environments are more likely to promote deep rooting of hosts and hence the formation of deep symbioses. Similarly, the presence of porous parent material may promote the formation of deep symbioses by encouraging access to rock water and minerals (which may then promote drought resistance after a change in

climate). The distribution of and identity of fungi in deep soils is still rarely studied owing to practical difficulties, and so the ecological importance of these sorts of associations outside of ecosystems on weathered granite is relatively unknown. Remote sensing techniques such as GPR could be used to help assess tree rooting depths compared with depth to bedrock. We also note that much of the early data on tree rooting depth came about in relation to the mining industry (Canadell et al., 1996; Stone & Kalisz, 1991), and this may prove to be a useful industrial partnership for gathering such data. Other industrial activities that involve deep coring or excavation through soil layers would also represent a useful means of extracting data on the potential for mycorrhizal activity in deep soils (Nogué et al., 2013).

9. Examining potential for competition between MNs: (1) EcM versus AM network competition (e.g., pinegrass vs. Douglas-fir and/or ponderosa pine), (2) EcM versus EcM, or AM versus AM network competition (e.g., pinegrass network 1 vs. pinegrass network 2, Douglas-fir network 1 vs. Douglas-fir network 2, or Douglas-fir network vs. ponderosa pine network). Competitive effects may help explain vertical differentiation between mycorrhizal communities. We predict better competitive success for EcM fungi with water-transporting functions as drought stress increases. It is also possible that resource competition arises when morphologies (exploration types) are more similar.
10. Understanding the fungal component of ecosystem switches: If EcM networks are closely linked to host drought resistance, then by studying their changes in response to aridity and land-use change, it may be possible to identify the processes that are driving the switches from EcM forests to AM grasslands under drought stress (e.g., British Columbia) as compared with the reverse (e.g., California). For example, although EcM networks are hypothesized to provide a means by which Douglas-fir seedlings could resist drought, in practice little regeneration is occurring in drought-stressed areas. Is this in fact a cumulative effect of grazing in conjunction with drought, rather than the absence of an MN effect?

18.10 CONCLUSIONS

Managing forests for resilience to account for increasing aridity will require consideration of many biotic and geological components that have not been widely utilized. MNs provide a mechanism for the transport and release of water accessed through hydraulic lift, both via direct transport to connected plants and through leakage from hyphae into upper soil horizons. Increasing intensity and occurrence of drought in forest ecosystems represent a challenge to future forest management, placing increased stress on forests globally. Management of dry ecosystems in particular should consider the potential importance of water donor trees and their symbionts; take steps to maintain a sufficient stock of mature, deep-rooting trees; and consider the spatial distribution of replanting so as to maximize the potential for tapping in to MNs. Although the current state of knowledge indicates that MN are important for host water relationships in both AM and EcM symbioses, there is a need to assess a greater number of species and ecosystems.

References

Aanderud, Z.T., Richards, J.H., 2009. Hydraulic redistribution may stimulate decomposition. Biogeochemistry 95, 323–333.

Agerer, R., 2001. Exploration types of ectomycorrhizae: a proposal to classify ectomycorrhizal mycelial systems according to their patterns of differentiation and putative ecological importance. Mycorrhiza 11, 107–114.

Allen, M.F., 2007. Mycorrhizal fungi: highways for water and nutrients in arid soils. Vadose Zone Journal 6, 291–297.

Aitken, S.N., Bemmels, J.B., 2016. Time to get moving: assisted gene flow of forest trees. Evolutionary Applications 9, 271–290.

Aitken, S.N., Yeaman, S., Holliday, J.A., Wang, T., Curtis-McLane, S., 2008. Adaptation, migration or extirpation: climate change outcomes for tree populations. Evolutionary Applications 1, 95–111.

Allen, C.D., Macalady, A.K., Chenchouni, H., Bachelet, D., McDowell, N.G., Vennetier, M., Kitzberger, T., Rigling, A., Breshears, D.D., Hogg, E.H.T., et al., 2010. A global overview of drought and heat induced tree mortality reveals emerging climate change risk for forests. Forest Ecology and Management 259, 660–684.

Allen, C.D., Breshears, D.D., McDowell, N.G., 2015. On underestimation of global vulnerability to tree mortality and forest die-off from hotter drought in the Anthropocene. Ecosphere 6, 129.

Allen, M., 2009. Bidirectional water flows through the soil–fungal– plant mycorrhizal continuum. New Phytologist 182, 290–293.

Alvarez, M., Huygens, D., Fernandez, C., Gacita, Y., Olivares, E., Saavedra, I., Alberdi, M., Valenzuela, E., 2009. Effect of ectomycorrhizal colonization and drought on reactive oxygen species metabolism of Nothofagus dombeyi roots. Tree Physiology 29, 1047–1057.

Anderegg, W.R.L., Kane, J.M., Anderegg, L.D.L., 2013. Consequences of widespread tree mortality triggered by drought and temperature stress. Nature Climate Change 3, 30–36.

Anderegg, W., Schwalm, C., Biondi, F., Camarero, J., Koch, G., Litvak, M., Ogle, K., Shaw, J., Shevliakova, E., Williams, A.P., et al., 2015. Pervasive drought legacies in forest ecosystems and their implications for carbon cycle models. Science 349, 528–532.

Augé, R.M., 2001. Water relations, drought and vesicular-arbuscular mycorrhizal symbiosis. Mycorrhiza 11, 3–42.

Austin, A.T., Yahdjian, L., Stark, J.M., Belnap, J., Porporato, A., Norton, U., Ravetta, D.A., Schaeffer, S.M., 2004. Water pulses and biogeochemical cycles in arid and semiarid ecosystems. Oecologia 141, 221–235.

Bakker, M.R., Augusto, L., Achat, D.L., 2006. Fine root distribution of trees and understory in mature stands of maritime pine (*Pinus pinaster*) on dry and humid sites. Plant and Soil 286, 37–51.

Barzana, G., Aroca, R., Paz, J.A., Chaumont, F., Martinez-Ballesta, M.C., Carvajal, M., Ruiz-Lozano, J.M., 2012. Arbuscular mycorrhizal symbiosis increases relative apoplastic water flow in roots of the host plant under both well-watered and drought stress conditions. Annals of Botany 109, 1009–1017.

Bauerle, T.L., Richards, J.H., Smart, D.R., Eissenstat, D.M., 2008. Importance of internal hydraulic redistribution for prolonging the lifespan of roots in dry soil. Plant, Cell and Environment 31, 177–186.

Bauhus, J., Van der Meer, P., Kanninen, M., 2010. Ecosystem Goods and Services from Plantation Forests. Earthscan, London.

Beiler, K.J., Durall, D.M., Simard, S.W., Maxwell, S.A., Kretzer, A.M., 2010. Architecture of the wood-wide web: *Rhizopogon* spp. genets link multiple Douglas-fir cohorts. New Phytologist 185, 543–553.

Beiler, K.J., Simard, S.W., Lemay, V., Durall, D.M., 2012. Vertical partitioning between sister species of *Rhizopogon* fungi on mesic and xeric sites in an interior Douglas-fir forest. Molecular Ecology 21, 6163–6174.

Bingham, M.A., Simard, S.W., 2011. Do mycorrhizal network benefits to survival and growth of interior Douglas-fir seedlings increase with soil moisture stress? Ecology and Evolution 1, 306–316.

Bingham, M.A., Simard, S.W., 2012a. Ectomycorrhizal networks of *Pseudotsuga menziesii* var. *glauca* trees facilitate establishment of conspecific seedlings under drought. Ecosystems 15, 188–199.

Bingham, M.A., Simard, S.W., 2012b. Mycorrhizal networks affect ectomycorrhizal fungal community similarity between conspecific trees and seedlings. Mycorrhiza 22, 317–326.

Bingham, M.A., Simard, S.W., 2013. Seedling genetics and life history outweigh mycorrhizal network potential to improve conifer regeneration under drought. Forest Ecology and Management 287, 132–139.

Birhane, E., Kuyper, T.W., Sterck, F.J., Gebrehiwot, K., Bongers, F., 2015. Arbuscular mycorrhiza and water and nutrient supply differently impact seedling performance of dry woodland species with different acquisition strategies. Plant Ecology & Diversity 8, 387–399.

Blois, J.L., Zarnetske, P.L., Fitzpatrick, M.C., Finnegan, S., 2013. Climate change and the past, present, and future of biotic interactions. Science 341, 499–504.

Bonan, G.B., 2008. Forests and climate change: forcings, feedbacks, and the climate benefits of forests. Science 320, 1444–1449.

Booth, M.G., 2004. Mycorrhizal networks mediate overstorey-understorey competition in a temperate forest. Ecology Letters 7, 538–546.

Bornyasz, M.A., Graham, R.C., Allen, M.F., 2005. Ectomycorrhizae in a soil-weathered granitic bedrock regolith: linking matrix resources to plants. Geoderma 126, 141–160.

Brooks, J.R., Meinzer, F.C., Warren, J.M., Domec, J.C., Coulombe, R., 2006. Hydraulic redistribution in a Douglas-fir forest: lessons from system manipulations. Plant, Cell and Environment 29, 138–150.

Brownlee, C., Duddridge, J.A., Malibari, A., Read, D.J., 1983. The structure and function of mycelial systems of ectomycorrhizal roots with special reference to their role in forming inter-plant connections and providing pathways for assimilate and water transport. Plant and Soil 71, 433–443.

Burgess, S.S.O., Turner, N.C., Ong, C.K., 1998. The redistribution of soil water by tree root systems. Oecologia 115, 306–311.

Canadell, J., Jackson, R.B., Ehleringer, J.R., Mooney, H.A., Sala, O.E., Schulze, E.-D., 1996. Maximum rooting depth of vegetation types at the global scale. Oecologia 108, 583–595.

Coops, N.C., Waring, R.H., Beier, C., Roy-Jauvin, R., Wang, T., 2011. Modeling the occurrence of 15 coniferous tree species throughout the Pacific Northwest of North America using a hybrid approach of a generic process-based growth model and decision tree analysis. Applied Vegetation Science 14, 402–414.

Dalpé, Y., Diop, T.A., Plenchette, C., Gueye, M., 2000. Glomales species associated with surface and deep rhizosphere of *Faidherbia albida* in Senegal. Mycorrhiza 10, 125–129.

Davidson, E.A., de Araújo, A.C., Artaxo, P., Balch, J.K., Brown, I.F., Bustamante, M.M.C., Coe, M.T., DeFries, R.S., Keller, M., Longo, M., et al., 2012. The Amazon basin in transition. Nature 481, 321–328.

Dickie, I.A., Xu, B., Koide, R.T., 2002. Vertical niche differentiation of ectomycorrhizal hyphae in soil as shown by T-RFLP analysis. New Phytologist 156, 527–535.

Dickie, I.A., Guza, R.C., Krazewski, S.E., Reich, P.B., 2004. Shared perennial ectomycorrhizal fungi between and a herbaceous perennial (*Helianthemum bicknellii*) and oak (Quercus) seedlings. New Phytologist 164, 375–382.

Domec, J.C., Warren, J.M., Meinzer, F.C., Brooks, J.R., Coulombe, R., 2004. Native root xylem embolism and stomatal closure in stands of Douglas-fir and ponderosa pine: mitigation by hydraulic redistribution. Oecologia 141, 7–16.

Domínguez Núñez, J.A., Serrano, J.S., Barreal, J.A.R., González JAS de, O., 2006. The influence of mycorrhization with *Tuber melanosporum* in the afforestation of a Mediterranean site with *Quercus ilex* and *Quercus faginea*. Forest Ecology and Management 231, 226–233.

Duddridge, J.A., Malibari, A., Read, D.J., 1980. Structure and function of mycorrhizal rhizomorphs with special reference to their role in water transport. Nature 287, 834–836.

Dulmer, K.M., Leduc, S.D., Horton, T.R., 2014. Ectomycorrhizal inoculum potential of northeastern US forest soils for American chestnut restoration: results from field and laboratory bioassays. Mycorrhiza 24 (1), 65–74.

Egerton-Warburton, L.M., Graham, R.C., Hubbert, K.R., 2003. Spatial variability in mycorrhizal hyphae and nutrient and water availability in a soil-weathered bedrock profile. Plant and Soil 249, 331–342.

Egerton-Warburton, L.M., Querejeta, J.I., Allen, M.F., 2007. Common mycorrhizal networks provide a potential pathway for the transfer of hydraulically lifted water between plants. Journal of Experimental Botany 58, 1473–1483.

Egerton-Warburton, L.M., Querejeta, J.I., Allen, M.F., 2008. Efflux of hydraulically lifted water from mycorrhizal fungal hyphae during imposed drought. Plant Signaling & Behavior 3, 68–71.

Fitter, A.H., 2001. Specificity, links and networks in the control of diversity in plant and microbial communities. In: Press, M.C., Huntly, N.J., Levin, S. (Eds.), Ecology: Achievement and Challenge. 41st Symposium of the British Ecological Society. Blackwell Scientific, Oxford, UK, pp. 95–114.

Fogel, R., Hunt, G., 1979. Fungal and arboreal biomass in a western Oregon Douglas-fir ecosystem: distribution patterns and turnover. Canadian Journal of Forest Research 9, 245–256.

Frank, D., Reichstein, M., Bahn, M., Thonicke, K., Frank, D., Mahecha, M.D., Smith, P., van der Velde, M., Vicca, S., Babst, F., et al., 2015. Effects of climate extremes on the terrestrial carbon cycle: concepts, processes and potential future impacts. Global Change Biology 21, 2861–2880.

Genney, D.R., Anderson, I.C., Alexander, I.J., 2006. Fine-scale distribution of pine ectomycorrhizas and their extramatrical mycelium. New Phytologist 170, 381–390.

Gillett, N.P., Weaver, A.J., Zwiers, F.W., Flannigan, M.D., 2004. Detecting the effect of climate change on Canadian forest fires. Geophysical Research Letters 31.

Godbold, D.L., Hoosbeek, M.R., Lukac, M., Cotrufo, M.F., Janssens, I.A., Ceulemans, R., Polle, A., Velthorst, E.J., Scarascia-Mugnozza, G., De Angelis, P., et al., 2006. Mycorrhizal hyphal turnover as a dominant process for carbon input into soil organic matter. Plant and Soil 281, 15–24.

Hamann, A., Wang, T., 2006. Potential effects of climate change on ecosystem. Ecological Society of America 87, 2773–2786.

Hamann, A., Gylander, T., Chen, P.Y., 2011. Developing seed zones and transfer guidelines with multivariate regression trees. Tree Genetics and Genomes 7, 399–408.

Harrison, R.B., Footen, P.W., Strahm, B.D., 2011. Deep soil horizons: contribution and importance to soil carbon pools and in assessing whole-ecosystem response to management and global change. Forest Science 57, 67–76.

Heineman, J.L., Sachs, D.L., Mather, W.J., Simard, S.W., 2010. Investigating the influence of climate, site, location, and treatment factors on damage to young lodgepole pine in southern British Columbia. Candian Journal of Forest Research 40, 1109–1127.

Horton, T.R., Bruns, T.D., 1998. Multiple-host fungi are the most frequent and abundant ectomycorrhizal types in a mixed stand of Douglas fir (*Pseudostuga menziesii*) and bishop pine (*Pinus muricata*). New Phytologist 139, 331–339.

Horton, T.R., Bruns, T.D., Parker, V.T., 1999. Ectomycorrhizal fungi associated with *Arctostaphylos* contribute to *Pseudotsuga menziesii* establishment. Canadian Journal of Botany 77, 93–102.

Huang, J., Yu, H., Guan, X., Wang, G., Guo, R., 2015. Accelerated dryland expansion under climate change. Nature Climate Change 1–22.

Hubbert, K.R., Beyers, J.L., Graham, R.C., 2001. Roles of weathered bedrock and soil in seasonal water relations of *Pinus jeffreyi* and *Arctostaphylos patula*. Canadian Journal of Forest Research-Revue Canadienne de Recherche Forestiere 31, 1947–1957.

IPCC, 2013. In: Stocker, T., Qin, D., Plattner, G.-K., Tignor, M., Allen, S., Boschung, J., Nauels, A., Xia, Y., Bex, V., Midgley, P. (Eds.), Climate Change 2013: The Physical Science Basis. Cambridge University Press, Cambridge, UK/New York, NY, USA.

Ishida, T.A., Nara, K., Hogetsu, T., 2007. Host effects on ectomycorrhizal fungal communities: insight from eight host species in mixed conifer-broadleaf forests. New Phytologist 174, 430–440.

Jenny, H., 1961. Derivation of soil factor equations of soils and ecosystems. Proceedings of the Soil Science Society of America 25, 385–388.

Joergensen, R.G., Wichern, F., 2008. Quantitative assessment of the fungal contribution to microbial tissue in soil. Soil Biology and Biochemistry 40, 2977–2991.

Kazantseva, O., Bingham, M., Simard, S.W., Berch, S.M., 2010. Effects of growth medium, nutrients, water, and aeration on mycorrhization and biomass allocation of greenhouse-grown interior Douglas-fir seedlings. Mycorrhiza 20, 51–66.

Kennedy, P.G., Izzo, A.D., Bruns, T.D., 2003. There is high potential for the formation of common mycorrhizal networks between understorey and canopy trees in a mixed evergreen forest. Journal of Ecology 91, 1071–1080.

Koide, R.T., Dickie, I.A., 2002. Effects of mycorrhizal fungi on plant populations. Plant and Soil 244, 307–317.

Kurtz, Z.D., Müller, C.L., Miraldi, E.R., Littman, D.R., Blaser, M.J., Bonneau, R.A., 2015. Sparse and compositionally robust inference of microbial ecological networks. In: von Mering, C. (Ed.), PLoS Computational Biology, vol. 11, p. e1004226.

Kurz-Besson, C., Otieno, D., Lobo Do Vale, R., Siegwolf, R.T.W., Schmidt, M., Herd, A., Nogueira, C., David, T.S., David, J.S., Tenhunen, J., et al., 2006. Hydraulic lift in cork oak trees in a savannah-type Mediterranean ecosystem and its contribution to the local water balance. Plant and Soil 282, 361–378.

Lehto, T., Zwiazek, J.J., 2011. Ectomycorrhizas and water relations of trees: a review. Mycorrhiza 21, 71–90.

Lilleskov, E.A., Bruns, T.D., Dawson, T.E., Camacho, F.J., 2009. Water sources and controls on water-loss rates of epigeous ectomycorrhizal fungal sporocarps during summer drought. New Phytologist 182, 483–494.

Marlier, M.E., DeFries, R.S., Voulgarakis, A., Kinney, P.L., Randerson, J.T., Shindell, D.T., Chen, Y., Faluvegi, G., 2012. El Niño and health risks from landscape fire emissions in southeast Asia. Nature Climate Change 2, 1–6.

Marulanda, A., Azcón, R., Ruiz-Lozano, J.M., 2003. Contribution of six arbuscular mycorrhizal fungal isolates to water uptake by *Lactuca sativa* plants under drought stress. Physiologia Plantarum 119, 526–533.

Mather, W.J., Simard, S.W., Heineman, J.L., Sachs, D.L., 2010. Decline of planted lodgepole pine in the southern interior of British Columbia. The Forestry Chronicle 86, 484–497.

Mathys, A., Coops, N.C., Waring, R.H., 2014. Soil water availability effects on the distribution of 20 tree species in western North America. Forest Ecology and Management 313, 144–152.

McCormack, M.L., Dickie, I.A., Eissenstat, D.M., Fahey, T.J., Fernandez, C.W., Guo, D., Helmisaari, H.-S., Hobbie, E.A., Iversen, C.M., Jackson, R.B., et al., 2015. Redefining fine roots improves understanding of below-ground contributions to terrestrial biosphere processes. New Phytologist 207, 505–518.

Millar, C.I., Stephenson, N.L., Stephens, S.L., 2007. Climate change and forest of the future: managing in the face of uncertanity. Ecological Applications 17, 2145–2151.

Molina, R., Horton, T.R., 2015. Mycorrhiza specificity: its role in the development and function of common mycelial networks. In: Horton, T.R. (Ed.), Mycorrhizal Networks. Springer, Dordrecht.

Moreira, M.Z., Scholz, F.G., Bucci, S.J., Sternberg, L.S., Goldstein, G., Meinzer, F.C., Franco, A.C., 2003. Hydraulic lift in a neotropical savanna. Functional Ecology 17, 573–581.

Morte, A., Navarro-Ródenas, A., Nicolás, E., 2010. Physiological parameters of desert truffle mycorrhizal *Helianthemun almeriense* plants cultivated in orchards under water deficit conditions. Symbiosis 52, 133–139.

Mujic, A.B., Durall, D.M., Spatafora, J.W., Kennedy, P.G., 2016. Competitive avoidance not edaphic specialization drives vertical niche partitioning among sister species of ectomycorrhizal fungi. New Phytologist 209, 1174–1183.

Muñoz, M.R., Squeo, F.A., León, M.F., Tracol, Y., Gutiérrez, J.R., 2008. Hydraulic lift in three shrub species from the Chilean coastal desert. Journal of Arid Environments 72, 624–632.

Neumann, R.B., Gardon, Z.G., 2012. The magnituede of hydraulic redistribution by plant roots: a review and synthesis of empirical and modeling studies. New Phytologist 194, 337–352.

Newman, E.I., 1988. Mycorrhizal links between plants: their functioning and ecological significnce. Advances in Ecological Research 18, 243–270.

Nitschke, C.R., Innes, J.L., 2008. Climatic change and fire potential in South-Central British Columbia, Canada. Global Change Biology 14, 841–855.

Nogué, S., de Nascimento, L., Fernández-Palacios, J.M., Whittaker, R.J., Willis, K.J., 2013. The ancient forests of La Gomera, Canary Islands, and their sensitivity to environmental change. Journal of Ecology 101, 368–377.

O'Neill, G.A., Carlson, M., Berger, V., Ukrainetz, N., 2013. Assisted Migration Adaptation Trial Workplan. BC Ministry of Forests, Land and Natural Resource Operations, Research Branch. [WWW document] URL http://www.for.gov.bc.ca/hre/forgen/interior/AMAT.htm.

Pang, J., Wang, Y., Lambers, H., Tibbett, M., Siddique, K.H.M., Ryan, M.H., 2013. Commensalism in an agroecosystem: hydraulic redistribution by a deep-rooted legume improves survival of a droughted shallow-rooted legume companion. Physiologia Plantarum 149, 79–90.

Parke, J.L., Linderman, R.G., Black, C.H., 1983. The role of ectomycorrhizas in drought tolerance of Douglas-fir seedlings. New Phytologist 95, 83–95.

Piao, S., Ciais, P., Huang, Y., Shen, Z., Peng, S., Li, J., Zhou, L., Liu, H., Ma, Y., Ding, Y., et al., 2010. The impacts of climate change on water resources and agriculture in China. Nature 467, 43–51.

Pickles, B.J., Pither, J., 2014. Still scratching the surface: how much of the 'black box' of soil ectomycorrhizal communities remains in the dark? New Phytologist 201, 1101–1105.

Pickles, B.J., Twieg, B.D., O'Neill, G.A., Mohn, W.W., Simard, S.W., 2015. Local adaptation in migrated interior Douglas-fir seedlings is mediated by ectomycorrhizas and other soil factors. New Phytologist 207, 858–871.

Plamboeck, A.H., Dawson, T.E., Egerton-Warburton, L.M., North, M., Bruns, T.D., Querejeta, J.I., 2007. Water transfer via ectomycorrhizal fungal hyphae to conifer seedlings. Mycorrhiza 17, 439–447.

Prieto, I., Armas, C., Pugnaire, F.I., 2012. Water release through plant roots: new insights into its consequences at the plant and ecosystem level. New Phytologist 193, 830–841.

Querejeta, J.I., Egerton-Warburton, L.M., Allen, M., 2003. Direct nocturnal water transfer from oaks to their mycorrhizal symbionts during severe soil drying. Oecologia 134, 55–64.

Querejeta, J.I., Egerton-Warburton, L.M., Allen, M.F., 2007. Hydraulic lift may buffer rhizosphere hyphae against the negative effects of severe soil drying in a California Oak savanna. Soil Biology and Biochemistry 39, 409–417.

Querejeta, J.I., Egerton-Warburton, L.M., Allen, M.F., 2009. Topographic position modulates the mycorrhizal response of oak trees to interannual rainfall variability. Ecology 90, 649–662.

Quijano, J.C., Kumar, P., Drewry, D.T., 2013. Passive regulation of soil biogeochemical cycling by root water transport. Water Resources Research 49, 3729–3746.

Reid, C.P.P., Woods, F.W., 1968. Translocation of C14-labeled compounds in mycorrhizae and its implications in interplant nutrient cycling. Ecology 50, 179–187.

Richter, D.d.B., Billings, S.A., 2015. 'One physical system': Tansley's ecosystem as Earth's critical zone. New Phytologist 206, 900–912.

Roach, W.J., Simard, S.W., Sachs, D.L., 2015. Evidence against planting lodgepole pine monocultures in the cedar-hemlock forests of southeastern British Columbia. Forestry 88, 345–358.

Rose, K.L., Graham, R.C., Parker, D.R., 2003. Water source utilization by *Pinus jeffreyi* and *Arctostaphylos patula* on thin soils over bedrock. Oecologia 134, 46–54.

Rose, M., Posa, C., Wijedasa, L.S., Corlett, R.T., 2011. Biodiversity and conservation of tropical peat swamp forests. BioScience 61, 49–57.

Rosling, A., Landeweert, R., Lindahl, B.D., Larsson, K.H., Kuyper, T.W., Taylor, A.F.S., Finlay, R.D., 2003. Vertical distribution of ectomycorrhizal fungal taxa in a podzol soil profile. New Phytologist 159, 775–783.

Safir, G.R., Boyer, J.S., Gerdemann, J.W., 1971. Mycorrhizal enhancement of water transport in soybean. Science 172, 581–583.

Sands, R., Fiscus, E.L., Reid, C.P.P., 1982. Hydraulic properties of pine and bean roots with varying degrees of suberization, vascular differentiation and mycorrhizal infection. Functional Plant Biology 9, 559–569.

Schenk, H.J., Jackson, R.B., 2002a. Rooting depths, lateral root spreads and belowground aboveground allometries of plants in water limited ecosystems. Journal of Ecology 480–494.

Schenk, H.J., Jackson, R.B., 2002b. The global biogeography of roots. Ecological Monographs 72, 311–328.

Scholz, F.G., Bucci, S.J., Goldstein, G., Moreira, M.Z., Meinzer, F.C., Domec, J.C., Villalobos-Vega, R., Franco, A.C., Miralles-Wilhelm, F., 2008. Biophysical and life-history determinants of hydraulic lift in Neotropical savanna trees. Functional Ecology 22, 773–786.

Schoonmaker, A.L., Teste, F.P., Simard, S.W., Guy, R.D., 2007. Tree proximity, soil pathways and common mycorrhizal networks: their influence on the utilization of redistributed water by understory seedlings. Oecologia 154, 455–466.

Sekija, N., Araki, H., Yano, K., 2011. Applying hydraulic lift in an agroecosystem: forage plants with shoots removed supply water to neighboring vegetable crops. Plant and Soil 341, 39–50.

Selosse, M.A., Richard, F., He, X., Simard, S.W., 2006. Mycorrhizal networks: des liaisons dangereuses? Trends in Ecology and Evolution 21, 621–628.

Sherwood, S., Fu, Q., 2014. A drier future? Science 343, 737–739.

Simard, S.W., Durall, D.M., 2004. Mycorrhizal networks: a review of their extent, function, and importance. Canadian Journal of Botany 82, 1140–1165.

Simard, S.W., Jones, M.D., Durall, D.M., Perry, D.A., David, D., Molina, R., 1997a. Reciprocal transfer of carbon isotopes between ectomycorrhizal *Betula papyfera* and *Pseudotsuga menziesii*. New Phytologist 137, 529–542.

Simard, S.W., Perry, D.A., Jones, M.D., Myrold, D.D., Durall, D.M., Molina, R., 1997b. Net transfer of carbon between ectomycorrhizal tree species in the field. Nature 388, 579–582.

Simard, S.W., Martin, K., Vyse, A., Larson, B., 2013. In: Messier, C., Klaus, K., Coates, D. (Eds.), Managing Forests as Complex Adaptive Systems. Taylor & Francis, New York, NY.

Smith, S.E., Read, D., 2008. Mycorrhizal Symbiosis. Academic Press, New York, NY, USA.

Smith, S.E., Facelli, E., Pope, S., Smith, F.A., 2010. Plant performance in stressful environments: interpreting new and established knowledge of the roles of arbuscular mycorrhizas. Plant and Soil 326, 3–20.

Snyder, K.A., James, J.J., Richards, J.H., Donovan, L.A., 2008. Does hydraulic lift or nighttime transpiration facilitate nitrogen acquisition? Plant and Soil 306, 159–166.

Soudzilovskaia, N.A., Douma, J.C., Akhmetzhanova, A.A., van Bodegom, P.M., Cornwell, W.K., Moens, E.J., Treseder, K.K., Tibbett, M., Wang, Y.P., Cornelissen, J.H.C., 2015. Global patterns of plant root colonization intensity by mycorrhizal fungi explained by climate and soil chemistry. Global Ecology and Biogeography 24, 371–382.

Spittlehouse, D.L., 2008. Climate Change, Impacts, and Adaptation Scenarios: Climate Change and Forest and Range Management in British Columbia. B.C. Min. For. Range, Res. Br., Victoria, BC, Canada. Tech. Rep. 045 http://www.for.gov.bc.ca/hfd/pubs/Docs/Tr/Tr045.htm.

Sternberg, P.D., Anderson, M.A., Graham, R.C., Beyers, J.L., Tice, K.R., 1996. Root distribution and seasonal water status in weathered granitic bedrock under chaparral. Geoderma 72, 89–98.

Stone, E.L., Kalisz, P.J., 1991. On the maximum extent of tree roots. Forest Ecology and Management 46, 59–102.

Sucre, E.B., Tuttle, J.W., Fox, T.R., 2011. Depth in rocky forest soils. Water 57, 59–66.

Sun, Y.P., Unestam, T., Lucas, S.D., Johanson, K.J., Kenne, L., Finlay, R., 1999. Exudation-reabsorption in a mycorrhizal fungus, the dynamic interface for interaction with soil and soil microorganisms. Mycorrhiza 9, 137–144.

Sun, S.J., Meng, P., Zhang, J.S., Wan, X., 2014. Hydraulic lift by *Juglans regia* relates to nutrient status in the intercropped shallow-root crop plant. Plant and Soil 374, 629–641.

Suz, L.M., Barsoum, N., Benham, S., Cheffings, C., Cox, F., Hackett, L., Jones, A.G., Mueller, G.M., Orme, D., Seidling, W., et al., 2015. Monitoring ectomycorrhizal fungi at large scales for science, forest management, fungal conservation and environmental policy. Annals of Forest Science 72, 877–885.

Taylor, D.L., Hollingsworth, T.N., McFarland, J.W., Lennon, N.J., Nusbaum, C., Ruess, R.W., 2014. A first comprehensive census of fungi in soil reveals both hyperdiversity and fine-scale niche partitioning. Ecological Monographs 84, 3–20.

Thompson, I., Mackey, B., McNulty, S., Mosseler, A., 2009. Forest Resilience, Biodiversity, and Climate Change. A Synthesis of the Biodiversity/Resilience/Stability Relationship in Forest Ecosystems. Technical Series 43. Secretariat of the Convention on Biological Diversity, Montreal, p. 67.

Urban, M.C., Zarnetske, P.L., Skelly, D.K., 2013. Moving forward: dispersal and species interactions determine biotic responses to climate change. Annals of the New York Academy of Sciences 1297, 44–60.

van der Heijden, M.G.A., Horton, T.R., 2009. Socialism in soil? The importance of mycorrhizal fungal networks for facilitation in natural ecosystems. Journal of Ecology 97, 1139–1150.

Virginia, R.A., Jenkins, M.B., Jarrell, W.M., 1986. Depth of root symbiont occurrence in soil. Biology and Fertility of Soils 2, 127–130.

Wallander, H., Ekblad, A., Godbold, D.L., Johnson, D., Bahr, A., Baldrian, P., Björk, R.G., Kieliszewska-Rokicka, B., Kjøller, R., Kraigher, H., et al., 2013. Evaluation of methods to estimate production, biomass and turnover of ectomycorrhizal mycelium in forests soils – a review. Soil Biology and Biochemistry 57, 1034–1047.

Wang, T., Campbell, E.M., O'Neill, G.A., Aitken, S.N., 2012. Projecting future distributions of ecosystem climate niches: uncertainties and management applications. Forest Ecology and Management 279, 128–140.

Warren, J.M., Brooks, J.R., Meinzer, F.C., Eberhart, J.L., 2008. Hydraulic redistribution of water from *Pinus ponderosa* trees to seedlings: evidence for an ectomycorrhizal pathway. New Phytologist 178, 382–394.

Whitman, T., Lehmann, J., 2015. A dual-isotope approach to allow conclusive partitioning between three sources. Nature Communications 6, 8708.

Williams, J.W., Jackson, S.T., 2007. Novel climates, no-analog communities, and ecological surprises. Frontiers in Ecology and the Environment 5 (9), 475–482.

Williams, A.P., Allen, C.D., Macalady, A.K., Griffin, D., Woodhouse, C.A., Meko, D.M., Swetnam, T.W., Rauscher, S.A., Seager, R., Grissino-Mayer, H.D., et al., 2012. Temperature as a potent driver of regional forest drought stress and tree mortality. Nature Climate Change 3, 292–297.

Witty, J.H., Graham, R.C., Hubbert, K.R., Doolittle, J.A., Wald, J.A., 2003. Contributions of water supply from the weathered bedrock zone to forest soil quality. Geoderma 114, 389–400.

Wotton, B.M., Nock, C.A., Flannigan, M.D., 2010. Forest fire occurence and climate change in Canada. International Journal of Wildland Fire 19, 253–271.

Wu, Q.-S., Xia, R.-X., 2006. Arbuscular mycorrhizal fungi influence growth, osmotic adjustment and photosynthesis of citrus under well-watered and water stress conditions. Journal of Plant Physiology 163, 417–425.

Wu, Q.-S., Xia, R.-X., Zou, Y.N., 2006a. Reactive oxygen metabolism in mycorrhizal and non-mycorrhizal citrus (*Poncirus trifoliata*) seedlings subjected to water stress. Journal of Plant Physiology 163, 1101–1110.

Wu, Q.-S., Zou, Y.N., Xia, R.-X., 2006b. Effects of water stress and arbuscular mycorrhizal fungi on reactive oxygen metabolism and antioxidant production by citrus (*Citrus tangerine*) roots. European Journal of Soil Biology 42, 166–172.

Yule, C.M., 2010. Loss of biodiversity and ecosystem functioning in Indo-Malayan peat swamp forests. Biodiversity and Conservation 19, 393–409.

Zhang, Y., Yao, Q., Li, J., Hu, Y., Chen, J., 2014. Growth response and nutrient uptake of *Eriobotrya japonica* plants inoculated with three isolates of arbuscular mycorrhizal fungi under water stress condition. Journal of Plant Nutrition 37, 690–703.

Zuccarini, P., Savé, R., 2015. Three species of arbuscular mycorrhizal fungi confer different levels of resistance to water stress in *Spinacia oleracea* L. Plant Biosystems - An International Journal Dealing With All Aspects of Plant Biology 3504, 1–4.

SECTION IV

MYCORRHIZAL MEDIATION OF ECOSYSTEM CARBON FLUXES AND SOIL CARBON STORAGE

Jan Jansa, Lead Editor

CHAPTER

19

Introduction: Mycorrhizas and the Carbon Cycle

J. Jansa[1], *K.K. Treseder*[2]

[1]Academy of Sciences of the Czech Republic, Prague, Czech Republic; [2]University of California, Irvine, CA, United States

19.1 THE CARBON CYCLE

Carbon (C) is the fourth most abundant element in the current universe, yet much less abundant in Earth's crust (webelements.com). It is the chemical basis of all life as we know it (Bergin et al., 2014). It serves as the main building block of all living organisms on Earth and the main energy carrier for metabolic processes. Its concentration in living organisms fluctuates around 50% on a dry mass basis (Fierer et al., 2009). Because of the importance of C for life as we know it, boundary conditions for habitability of extrasolar planets have been defined based on the potential for C and water cycling (von Bloh et al., 2005; Unterborn et al., 2014). With only few exceptions, photosynthesis is the main mechanism responsible for the entrance of C into the biosphere, with virtually all organic compounds on Earth being dependent on this very process, including fossil fuels (natural gas, crude oil, coal) and materials derived from it such as plastics and tars, timber, all sorts of agricultural products, living biomass of microbes, plants and animals (including humans), organic residues, composts, peats, biochars, and the soil organic matter (SOM).

The terrestrial C cycle is the biogeochemical cycle in which C is exchanged among its reservoirs in atmosphere, biosphere, hydrosphere, and lithosphere (excluding pedosphere and geosphere from this list because of their overlaps with the previously used terms). It has slow and fast components. The slow C cycle is dominated by fully oxidized C (carbonate or CO_2), moving at geological time scales among oceans, rocks, and atmosphere, and including processes such as volcanic CO_2 emissions (see also Chapter 2). It moves between 10^{13} and 10^{14} g C each year (earthobservatory.nasa.gov). The fast C cycle is dominated by the SOM pool, which is larger than the biosphere and atmosphere C pools combined, and in terms of size it is only surpassed by calcareous rocks and ocean carbonate pools. The fast C cycle moves between 10^{16} and 10^{17} g C each year (earthobservatory.nasa.gov), is mostly biologically dominated, and includes oxidation–reduction reactions such as photosynthesis and respiration. The two components of

Mycorrhizal Mediation of Soil
http://dx.doi.org/10.1016/B978-0-12-804312-7.00019-X

the C cycle overlap in several reservoirs (e.g. hydrosphere and lithosphere), but the position of the atmosphere is somewhat special—it is truly the mixing point, where nearly all of the C is present in the same form (CO_2) and the pool is being fed from the slow and the fast C cycles. In addition, it is largely affected by human-induced CO_2 emissions in a range close to the natural fast C cycle (i.e. 10^{15}–10^{16} g C per year; Kammen and Marino, 1993; Ver et al., 1999).

19.2 THE KEY ROLE OF THE SOM IN SOIL PROCESSES

The SOM is not only by far the largest pool involved in the fast C cycle, but it is also the most heterogeneous pool in terms of chemical variability, spatial heterogeneity, and diversity of processes involved in its turnover (Gaillard et al., 1999; Schaumann and Thiele-Bruhn, 2011; Yuan et al., 2013; Angst et al., 2016). By definition, it consists of all C bound in organic molecules in soil, which are not parts of living organisms. This C pool is continuously built up through inputs such as plant (both aboveground and root) litter, exudates, wood and charcoal particles, dung, animal remains, and dead microbes. It is recycled through myriads of soil macro- and microorganisms (Fierer et al., 2009) and encompasses particularly labile C compounds such as simple organic acids with a half-life counting in minutes to hours (Watt et al., 2006; Fujii et al., 2013) through more recalcitrant compounds such as starch, cellulose, and chitin degrading over weeks to months (German et al., 2011; Zeglin et al., 2013; Tischer et al., 2015), through lignin and black C (metabolically carbonized organic matter or charred organic materials through exposure to high temperatures upon limited oxygen availability; Waggoner et al., 2015), which could resist decomposition over years to millenia (Valášková et al., 2007; Foereid et al., 2011; Singh et al., 2012). Black C, especially the biochar, which is the intentionally created black C through charring of organic residues, has attracted particular attention in the recent decades for environmentally friendly waste management, as efficient soil conditioner, and a measure to offset climatic changes, allowing to sequester large amounts of C belowground over long periods of time (Novotny et al., 2015). Current interest in implementing biochar application in agricultural practices, especially in infertile tropical soils, require taking its effects and interactions into account with soil microbes in general and mycorrhizas in particular (Quilliam et al., 2013; Hammer et al., 2014; for more information see also Chapter 25).

19.3 POSITION OF MYCORRHIZAL FUNGI WITHIN THE SOIL FOOD WEBS

Similar to all fungi, mycorrhizal fungi are heterotrophs; thus they are completely dependent on reduced C as their only energy source for all of their metabolism and growth. In contrast to many other soil organisms, mycorrhizal fungi derive most or all of their energy directly from their living host plants, often becoming literally part of their plant hosts (Fig. 19.1; see also Chapters 20 and 21 for more details). Their mycelia either enwrap the roots by dense hyphal mats in case of ectomycorrhizas (EcM) or penetrate deep inside of the host plant roots in the other types of mycorrhizal symbioses. Through the intimate physical contact, mycorrhizal fungi obtain exclusive access to simple organic molecules (mainly sugars) in exchange for symbiotic benefits conferred to the plants, mainly in form of improved nutrient acquisition from

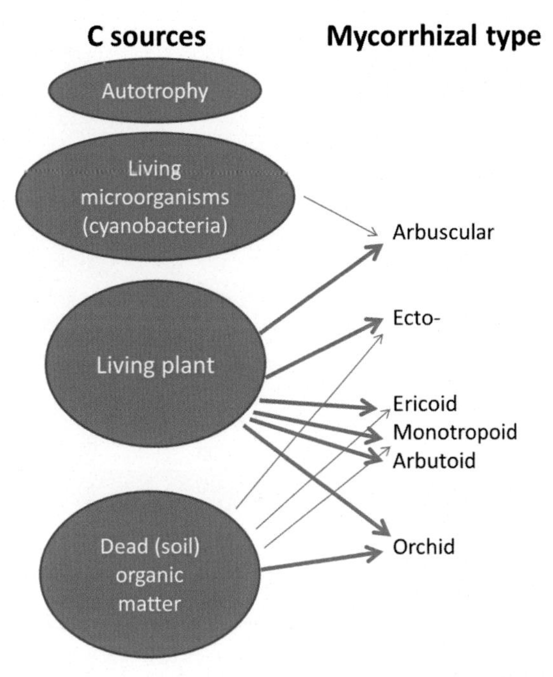

FIGURE 19.1 Direct dependency of the different mycorrhizal types (recruiting from different taxonomic group of fungi; see Table 1.1 for details) on different C sources, illustrating the central role of a living plant for all mycorhizal fungi and the quantitatively less important role of the soil organic matter, with a complete absence of autotrophy and a very special case of symbiosis of *Geosiphon* (Glomeromycota) with phototrophic prokaryotes (Schüßler, 2002).

the soil beyond the reach of roots (van der Heijden et al., 2015). Once in the fungal hyphae, the C obtained from plants is quickly redistributed within the mycorrhizosphere, which is the soil zone colonized by the extraradical (or extramatrical) mycelium (Schrey et al., 2015). The mycelium grows and/or produces extracellular materials (i.e., exudates, secretions, exoenzymes) that could potentially be consumed by other soil organisms (Fig. 19.2; see also Chapter 9). The hyphae themselves are either eaten by biotrophic organisms (fungal pathogens such as *Trichoderma*; Rousseau et al., 1996) while they are alive or processed by soil saprophytes after they have died (e.g. Fernandez et al., 2016). The organisms dependent on the C shuffled through mycorrhizal hyphae (i.e., the mycorrhiza-dependent food chains) recruit from various taxonomic and ecological groups, including bacteria, collembolans and mites (Scheu and Folger, 2004; Schneider et al., 2005), but there is also an intriguing opportunity that some EcM and possibly also other mycorrhizal fungi could actually recycle C from dead mycorrhizal hyphae in the soil (see Chapter 23 for more details). Ecologically interesting yet still poorly documented is the possibility of specific association of mycorrhizal hyphae with soil microbes fulfilling complementary functions to the hyphae themselves. For example, microbes capable of degrading complex organic molecules can associate with the surfaces of arbuscular mycorrhizal (AM) hyphae (Jansa et al., 2013), which have in turn very limited capacity to degrade such substrates alone (Joner and Johansen, 2000). In addition, there can be functional dependencies between soil saprotrophs and EcM hyphae in accessing recalcitrant protein–tannin complexes (Wu et al., 2003). Recent developments in molecular techniques such as ^{13}C-stable isotope probing allow

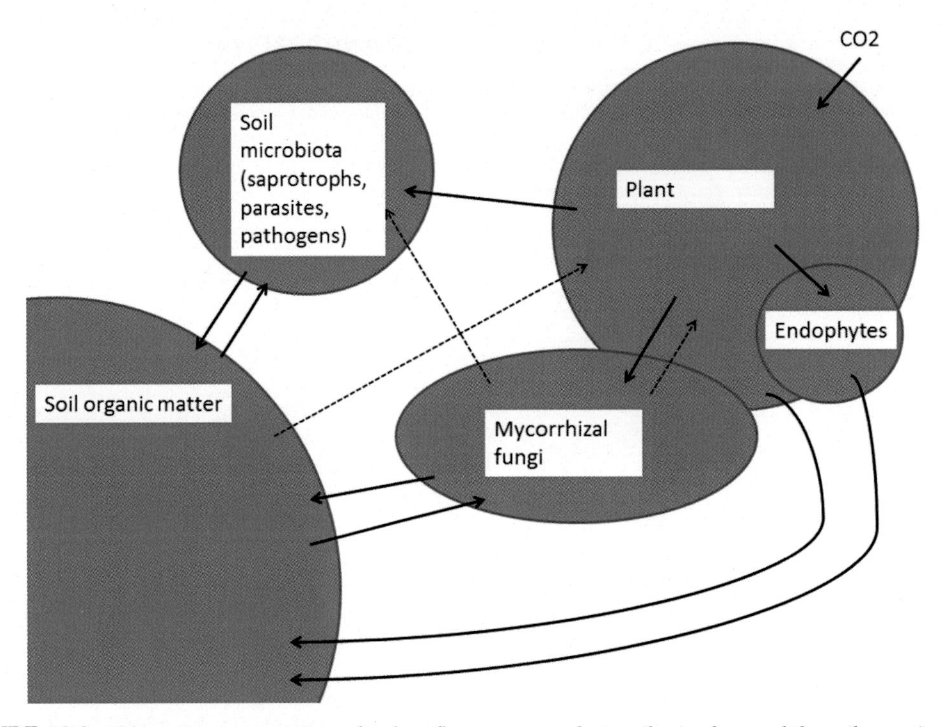

FIGURE 19.2 Schematic representation of carbon fluxes among plant, soil microbes, and the soil organic matter. *Thick lines* represent common and quantitatively important pathways, and *dashed lines* represent less common/ less important pathways. Overlaps between circles representing different pools stand for tissues composed of two entities (e.g., root and rhizobia in the nodules or mycorrhizal hyphae inside roots). Pools are not exactly to scale.

unprecedented insights into the identity and dynamics of the interorganismal linkages in complex environments such as soils. Several studies have now used this approach to address the pathways and players active in C channeling through the soil via mycorrhizal fungi (Leake et al., 2006; Drigo et al., 2010; Lekberg et al., 2013; see Chapter 22 for further details).

19.4 MYCORRHIZAL SYMBIOSIS AND THE SOIL C CYCLING

Mycorrhizal fungi contribute to the C cycle through redistribution of recently fixed C through the soil (Jakobsen and Rosendahl, 1990; Drigo et al., 2010; Nottingham et al., 2013; Fernandez et al., 2016), feeding (and potentially also priming) organic matter mineralization pathways (Staddon et al., 2003; Balasooriya et al., 2013; Lindahl and Tunlid, 2015; Fernandez et al., 2016) and immobilizing/stabilizing the C in highly recalcitrant organic compounds (Sousa et al., 2012). These pathways are further discussed in Chapters 21, 23 and 24. However, there is another key mechanism by which mycorrhizal fungi are involved in soil C cycling, and this is related to their important contribution to the mineral nutrition of their plants hosts (see Section II of this book). Soil colonized by mycorrhizal hyphae could quickly be depleted of easily available P and/or N, and the mineralization of the SOM may be slowed as a consequence of starvation for mineral nutrients of the (saprotrophic) microbes involved in the SOM mineralization. This so-called Gadgil effect has been described for ectomycorrhizal pines in

New Zealand (Gadgil and Gadgil, 1971), but similar results have also been reported recently for AM systems (Verbruggen et al., 2016). Although the outlined mechanism of the Gadgil effect could further be confounded by other factors such as water availability (Koide and Wu, 2003) and/or mycorrhiza-mediated increase of primary productivity (Orwin et al., 2011), it seems that the differential exploitation (high and low, respectively) of nutrients and the C/energy locked up in the SOM (Lindahl and Tunlid, 2015) is the most important mechanism behind the observed stabilization of SOM due to mycorrhizal activity (see also Chapters 20 and 24 for more details). On the other hand, there is at least one case where mycorrhizal fungi do indeed mineralize SOM to obtain C/energy. This C is then used by the fungi and transferred to the host plant. Such a lifestyle is the rule in the orchids and other plants with dust seeds (Eriksson and Kainulainen, 2011) during seed germination and sometimes during the entire life span of achlorophyllous species (Zettler et al., 2005; Keel et al., 2011; Rasmussen and Rasmussen, 2014; Stockel et al., 2014; see also Chapter 21 for more details). However, in some orchid and also other types of mycorrhizas, where the plant gains (or at least is thought to gain) the C from the fungal partner, in the so-called mycoheterotrophic plant species, it is more likely that the mycorrhizal fungus obtains C from a neighboring green plant rather than from the SOM (Taylor et al., 2004; Barrett et al., 2010; Courty et al., 2011). In some of these cases, particularly those involving achlorophyllous plants associating with the AM fungi (e.g., Bidartondo et al., 2002), the mechanism of C transfer from the fungus to the plant is remaining completely unknown and actually contradicts earlier experimental evidence of absence of C transfer from AM fungal hyphae to the plant tissues (Fitter et al., 1998; Pfeffer et al., 2004).

19.5 FUNCTIONAL DIVERSITY IN MYCORRHIZAL SYMBIOSES WITH RESPECT TO C CYCLING

Different types of mycorrhizal fungi (e.g. EcM, ericoid, orchid, and AM fungi) have been traditionally viewed as fulfilling different ecosystem roles in terms of utilization of soil nutrients, supporting plant nutritional demands and access to the different C sources apart from colonizing phylogenetically (mostly) disjunctive groups of plant taxa (Table 1.1; Read and Perez-Moreno, 2003; Gartner et al., 2012; van der Heijden et al., 2015). However, recent research blurs some of the previously established wisdoms and tends to include a broader range of fungal taxa among the mycorrhizal (or mycorrhiza-like) fungi such as Mucoromycotina, *Sebacina*, *Colletotrichum*, and the elusive group of dark septate endophytes (Jumpponen, 2001; Weiss et al., 2011; Field et al., 2015; Hiruma et al., 2016). In addition, recent findings have demonstrated unusual partner associations such as *Russula* with some mixotrophic or mycotrophic orchids (Girlanda et al., 2006; Ogura-Tsujita et al., 2012; Kong et al., 2015). Moreover, we now have evidence for complex interactions and their functional consequences between various soil microorganisms (saprotrophs, chemolithotrophs) and mycorrhizal fungi (Lindahl et al., 1999; Herman et al., 2012; Bukovská et al., 2016). In addition, recent studies have detected some typical mycorrhizal fungi (e.g., *Tuber* sp.) in roots of many nonhost neighboring plants (Gryndler et al., 2014), which suggests either much broader specificity of (some) mycorrhizal associations or a possible shortcut in C cycling from a nonhost plant to mycorrhizal fungi, effectively bypassing SOM formation. Next we specifically elaborate on the traditional and novel aspects of functioning and mutual interactions between the different mycorrhizal types, with a particular attention to C/energy fluxes, some of which are addressed in more detail in subsequent chapters.

19.5.1 Arbuscular Mycorrhiza

In the AM symbiosis, a few hundred fungal species establish the symbiosis with tens of thousands of unrelated plant species (van der Heijden et al., 2008; Krüger et al., 2012; see also Table 1.1). This means that a single AM fungus can possibly colonize several to many plant species (individuals) at once, forming so-called common mycorrhizal networks (Egerton-Warburton et al., 2007; Bever et al., 2010; Lekberg et al., 2010). It has been shown that under such situations one plant partner could feed the AM hyphal network with the C whereas another plant preferentially derives the symbiotic benefits (mainly in terms of improved mineral nutrition) without giving much C in return (Walder et al., 2012; Walder and van der Heijden, 2015). How common this is in nature and how consistent is it with biological market theory (Fellbaum et al., 2014; Werner and Kiers, 2015) is currently a subject of much controversy. Previous experimental evidence also showed that the C flow is always unidirectional (i.e., from plant to the AM fungus, ascribed to a very effective trehalose/lipid "valve" effectively preventing a reverse flow of C from the fungus to the plant; Pfeffer et al., 2004). However, how would the presumed mycoheterotrophic AM hosts (Bidartondo et al., 2002) acquire their C then if not from its mycorrhizal partner? With the experimental proof still missing of the C transfer from a neighboring green plant to the achlorophyllous host via AM hyphae, this remains one of the major and long-standing research challenges. It is also one of the great hopes of mycorrhizal physiology with respect to elucidating the molecular mechanisms and their regulation of C transfer from plant host to the AM fungus by using the exceptional case where it is possibly not working the ordinary way.

19.5.2 Ectomycorrhiza

Similarly as in the AM symbiosis, the fungal partner in EcM symbiosis is fed mainly by plant sugars, although many of the EcM fungi (unlike the AM fungi) are capable of degrading complex organic molecules (Lindahl and Tunlid, 2015). In contrast to AM fungi that have not yet been grown in the absence of host roots, many EcM fungi can easily be grown on axenic nutrient media in the absence of a host plant (Langdale and Read, 1990; Nehls et al., 2007; Larsen et al., 2011). Compared with the AM fungi, there is also much greater progress with respect to identification of missing key genes/pathways of C transfer from the plant to the EcM fungi because of available genome sequences of several of the EcM fungi and their plant hosts (Larsen et al., 2011; Ceccaroli et al., 2015; Kohler et al., 2015). By using combinations of isotopic and molecular approaches, previous assertions on the partially saprotrophic lifestyle of some truffles and other EcM fungi under field conditions (e.g., Hobbie et al., 2013) seem to have been largely refuted now (Le Tacon et al., 2015; Lindahl and Tunlid, 2015). However, there are some cases in which EcM fungi obviously feed mycoheterotrophic (achlorophyllous) plants with the C obtained from a neighboring green plant host. These mycoheterotrophic plants recruit from various plant groups, often not typically EcM ones, such as a liverwort *Cryptothallus* sp. (Wickett and Goffinet, 2008); all monotropes (Cullings et al., 1996) and some orchids (Selosse et al., 2004; Girlanda et al., 2006; Barrett and Freudenstein, 2008); parasitizing hyphal networks formed by truffles (*Tuber* sp.); and *Tulasnella*, *Russula*, and other typically EcM fungi. The molecular mechanisms and their regulation of this atypical C transfer from the EcM fungus to the plant remain unknown thus far.

19.5.3 Ericoid Mycorrhiza

The typical ericoid mycorrhizal (ErM) fungi belong to the most efficient degraders of the complex organic materials in soil (Bending and Read, 1996, 1997; Kohler et al., 2015). These fungi are also capable of axenic cultivation on simple and complex sugars (Varma and Bonfante, 1994; Hughes and Mitchell, 1995; Midgley et al., 2004). Nevertheless, as with the EcM symbiosis, it remains unclear to what extent they utilize SOM to cover for their energy demand under natural conditions or just to derive mineral nutrients (above all the N) from it. Very little work has been done so far on the molecular basis of C exchanges between the ErM fungi and their plant hosts as well as on deciphering the stoichiometry of trades of mineral nutrients for C between the symbiotic partners as depends on environmental conditions such as nutrient and/or light availability (Michelsen et al., 1996; Hofland-Zijlstra and Berendse, 2009). One notable exception is the work by Grelet et al. (2009), directly demonstrating reciprocal exchange of C and N between symbiotic partners in ErM symbiosis—although the fungus used in that experiment has originally been isolated from EcM root tips of a pine. Thus these and other (e.g., Vrålstad, 2004) results suggest a possible fungal partner overlap between EcM and ErM symbioses. The arbutoid and monotropoid mycorrhizas established between phylogenetically close relative plant taxa to typical ErM plants and typical EcM fungi could then be regarded rather as one extremity of a gradient than as a true exception.

19.5.4 Orchid Mycorrhiza

Orchid mycorrhizas represent one of the evolutionarily youngest types of mycorrhizal symbiosis, involving a great number of extant orchid species (Brundrett, 2002; see also Table 1.1 for more details). From the C cycling perspective, it is clearly distinct from the other types of mycorrhizal symbiosis because the C flow is typically bidirectional—the fungi often supply significant amounts of C to the germinating dust seeds of orchids (Selosse et al., 2011) whereas at later stages of plant ontogeny the net C flux is typically reversed, dominated by the C flow from the green plant to the fungus (Cameron et al., 2006, 2008). Among the orchids, there is also a high number of mixotrophic and mycoheterotrophic species, supposedly feeding on their fungal partner (Selosse and Roy, 2009), yet the direct experimental evidence for the C exchange and C sources (another plant or the SOM) in such relationship remains mostly unclear. See Chapter 21 for more details.

19.6 OPEN QUESTIONS, EXPERIMENTAL CHALLENGES

Carbon fluxes between the plant and fungal partners in mycorrhizal symbiosis still remain largely enigmatic with respect to their mechanistic basis (genes and proteins), their magnitude, temporal variation, and interactions with environmental conditions such as light intensity. For example, mycorrhizal C costs may well be, on average, significantly lower than the 20% net photosynthetic production previously measured in young cucumber plants (Jakobsen and Rosendahl, 1990); indeed, the measured figures rarely exceed 10% of the plant's C budget (Paul and Kucey, 1981; Grimoldi et al., 2006; Lendenmann et al., 2011). With the technical issues of high-throughput screening of whole (meta)transcriptomes and use of ^{13}C for tracing the C fluxes in plant-mycorrhiza-soil systems being solved or nearly so (Slavíková et al., 2016), we just need carefully designed experiments to answer many of the remaining open questions. Among those, particular attention should be paid to mycoheterotrophic plants, and especially those associating with AM fungi. Our future experiments should also pay due attention to the shared mycorrhizal networks interconnecting several different plant individuals belonging to the same or to different species. Ecological significance of recycling C from the SOM by mycorrhizal fungi (Hobbie et al., 2013) and transfer of this "dead C" from the fungus to associated plants should be quantified and compared to the "recent C" fixed by a living mycorrhizal host plant. Interactions among plant hosts, mycorrhizal fungi, and other soil microbes should be studied mainly from the perspective of metabolic/trophic dependencies and competition for resources. In this regard, direct interactions between mycorrhizal fungi belonging to different functional types (e.g., AM and EcM) should be studied in relevant ecosystem settings where their hyphospheres intermingle, such as in some mixed deciduous forests or tropical rainforests (see also Table 1.1). And the consequences of these interactions for ecosystem processes (Fig. 19.3) should finally be established and fed into the

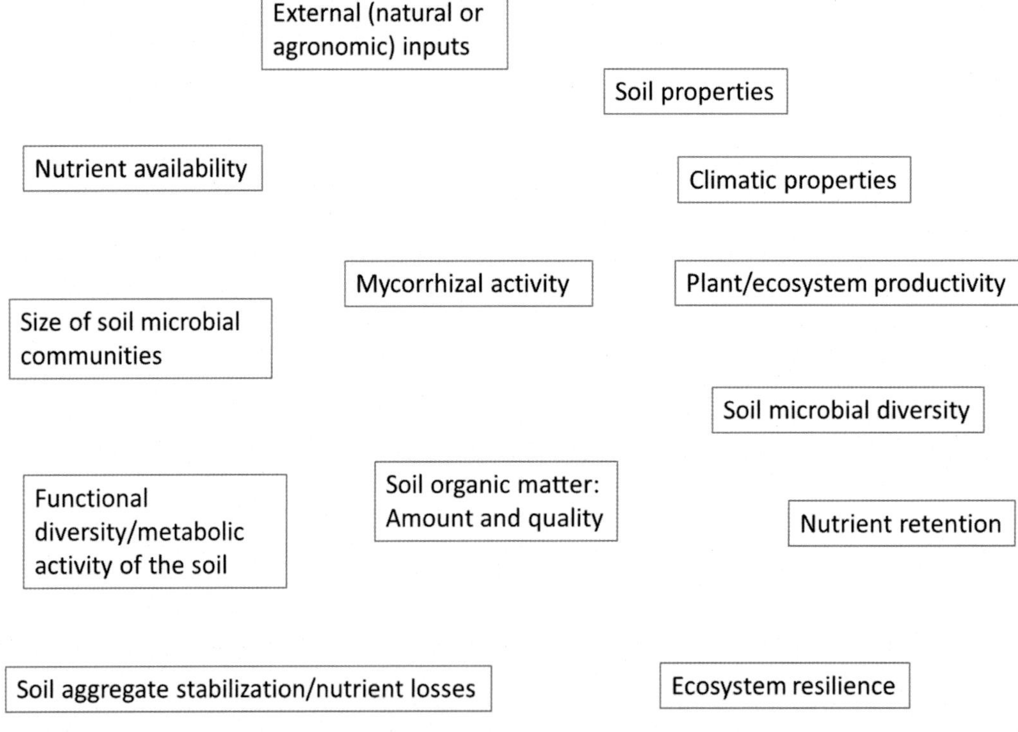

FIGURE 19.3 Various ecosystem features affecting and feeding back on the soil organic matter, causing changes in its pool size, turnover rates, and/or quality. Although multiple pieces of evidence documented several direct and indirect interactions between the different ecosystem features displayed here under specific environmental situations, the *arrows* representing direct and indirect linkages are intentionally left out of this picture. This is to illustrate the as yet unmet challenge to assign broadly valid, mutually comparable, quantitative, and dynamically changing values to such arrows in this heavily interlinked multidimensional space to allow for realistic predictions of interactions between mycorrhizal communities and soil carbon pools and their dynamics.

global C cycling models (see Chapter 26 for more details). Mycorrhizas are mostly missing from C cycling models despite large amounts of C being globally fluxed through their hyphal networks (Talbot et al., 2008). However, with experimental evidence mounting on the role of mycorrhizas in both soil C loss and soil C stabilization as well as plant nutrition and performance in nearly all terrestrial ecosystems, they emerge as important ecosystem engineers, strongly modulating ecosystem productivity, functioning, and resilience as well as SOM turnover (Jastrow et al., 2007; Bever, 2015; Fernandez and Kennedy, 2015; Sochorová et al., 2016; Verbruggen et al., 2016).

Acknowledgment

J.J. was supported by the Czech Science Foundation (project 14-19191S); the Ministry of Education, Youth, and Sports of the Czech Republic (LK11224); Fellowship J. E. Purkyně provided by the Czech Academy of Sciences; and the long-term developmental program (RVO61388971). K.K.T. was supported by funding from the US National Science Foundation (DEB-1457160). The authors declare no competing interests.

References

Angst, G., Kogel-Knabner, I., Kirfel, K., Hertel, D., Mueller, C.W., 2016. Spatial distribution and chemical composition of soil organic matter fractions in rhizosphere and non-rhizosphere soil under European beech (*Fagus sylvatica* L.). Geoderma 264, 179–187.

Balasooriya, W.K., Huygens, D., Denef, K., Roobroeck, D., Verhoest, N.E.C., Boeckx, P., 2013. Temporal variation of rhizo deposit-C assimilating microbial communities in a natural wetland. Biology and Fertility of Soils 49, 333–341.

Barrett, C.F., Freudenstein, J.V., 2008. Molecular evolution of *rbc*L in the mycoheterotrophic coralroot orchids (*Corallorhiza* Gagnebin, Orchidaceae). Molecular Phylogenetics and Evolution 47, 665–679.

Barrett, C.F., Freudenstein, J.V., Taylor, D.L., Koljalg, U., 2010. Rangewide analysis of fungal associations in the fully mycoheterotrophic *Corallorhiza striata* complex (Orchidaceae) reveals extreme specificity on ectomycorrhizal *Tomentella* (Thelephoraceae) across North America. American Journal of Botany 97, 628–643.

Bending, G.D., Read, D.J., 1996. Nitrogen mobilization from protein-polyphenol complex by ericoid and ectomycorrhizal fungi. Soil Biology and Biochemistry 28, 1603–1612.

Bending, G.D., Read, D.J., 1997. Lignin and soluble phenolic degradation by ectomycorrhizal and ericoid mycorrhizal fungi. Mycological Research 101, 1348–1354.

Bergin, E., Cleeves, L.I., Crockett, N., Blake, G., 2014. Exploring the origins of carbon in terrestrial worlds. Faraday Discussions 168, 61–79.

Bever, J.D., Dickie, I.A., Facelli, E., Facelli, J.M., Klironomos, J., Moora, M., et al., 2010. Rooting theories of plant community ecology in microbial interactions. Trends in Ecology and Evolution 25, 468–478.

Bever, J.D., 2015. Preferential allocation, physio-evolutionary feedbacks, and the stability and environmental patterns of mutualism between plants and their root symbionts. New Phytologist 205, 1503–1514.

Bidartondo, M.I., Redecker, D., Hijri, I., Wiemken, A., Bruns, T.D., Dominguez, L., et al., 2002. Epiparasitic plants specialized on arbuscular mycorrhizal fungi. Nature 419, 389–392.

Brundrett, M.C., 2002. Coevolution of roots and mycorrhizas of land plants. New Phytologist 154, 275–304.

Bukovská, P., Gryndler, M., Gryndlerová, H., Püschel, D., Jansa, J., 2016. Organic nitrogen-driven stimulation of arbuscular mycorrhizal fungal hyphae correlates with abundance of ammonia oxidizers. Frontiers in Microbiology 7, 711.

Cameron, D.D., Leake, J.R., Read, D.J., 2006. Mutualistic mycorrhiza in orchids: evidence from plant-fungus carbon and nitrogen transfers in the green-leaved terrestrial orchid *Goodyera repens*. New Phytologist 171, 405–416.

Cameron, D.D., Johnson, I., Read, D.J., Leake, J.R., 2008. Giving and receiving: measuring the carbon cost of mycorrhizas in the green orchid, *Goodyera repens*. New Phytologist 180, 176–184.

Ceccaroli, P., Saltarelli, R., Polidori, E., Barbieri, E., Guescini, M., Ciacci, C., et al., 2015. Sugar transporters in the black truffle *Tuber melanosporum*: from gene prediction to functional characterization. Fungal Genetics and Biology 81, 52–61.

Courty, P.E., Walder, F., Boller, T., Ineichen, K., Wiemken, A., Rousteau, A., et al., 2011. Carbon and nitrogen metabolism in mycorrhizal networks and mycoheterotrophic plants of tropical forests: a stable isotope analysis. Plant Physiology 156, 952–961.

Cullings, K.W., Szaro, T.M., Bruns, T.D., 1996. Evolution of extreme specialization within a lineage of ectomycorrhizal epiparasites. Nature 379, 63–66.

Drigo, B., Pijl, A.S., Duyts, H., Kielak, A., Gamper, H.A., Houtekamer, M.J., et al., 2010. Shifting carbon flow from roots into associated microbial communities in response to elevated atmospheric CO_2. Proceedings of the National Academy of Sciences of the United States of America 107, 10938–10942.

earthobservatory.nasa.gov (accessed on 03.03.2016.).

Egerton-Warburton, L.M., Querejeta, J.I., Allen, M.F., 2007. Common mycorrhizal networks provide a potential pathway for the transfer of hydraulically lifted water between plants. Journal of Experimental Botany 58, 1473–1483.

Eriksson, O., Kainulainen, K., 2011. The evolutionary ecology of dust seeds. Perspectives in Plant Ecology Evolution and Systematics 13, 73–87.

Fellbaum, C.R., Mensah, J.A., Cloos, A.J., Strahan, G.E., Pfeffer, P.E., Kiers, E.T., et al., 2014. Fungal nutrient allocation in common mycorrhizal networks is regulated by the carbon source strength of individual host plants. New Phytologist 203, 646–656.

Fernandez, C.W., Kennedy, P.G., 2015. Moving beyond the black-box: fungal traits, community structure, and carbon sequestration in forest soils. New Phytologist 205, 1378–1380.

Fernandez, C.W., Langley, J.A., Chapman, S., McCormack, M.L., Koide, R.T., 2016. The decomposition of ectomycorrhizal fungal necromass. Soil Biology and Biochemistry 93, 38–49.

Field, K.J., Rimington, W.R., Bidartondo, M.I., Allinson, K.E., Beerling, D.J., Cameron, D.D., et al., 2015. First evidence of mutualism between ancient plant lineages (*Haplomitriopsida* liverworts) and Mucoromycotina fungi and its response to simulated Palaeozoic changes in atmospheric CO_2. New Phytologist 205, 743–756.

Fierer, N., Strickland, M.S., Liptzin, D., Bradford, M.A., Cleveland, C.C., 2009. Global patterns in belowground communities. Ecology Letters 12, 1238–1249.

Fitter, A.H., Graves, J.D., Watkins, N.K., Robinson, D., Scrimgeour, C., 1998. Carbon transfer between plants and its control in networks of arbuscular mycorrhizas. Functional Ecology 12, 406–412.

Foereid, B., Lehmann, J., Major, J., 2011. Modeling black carbon degradation and movement in soil. Plant and Soil 345, 223–236.

Fujii, K., Morioka, K., Hangs, R., Funakawa, S., Kosaki, T., Anderson, D.W., 2013. Rapid turnover of organic acids in a Dystric Brunisol under a spruce-lichen forest in northern Saskatchewan, Canada. Canadian Journal of Soil Science 93, 295–304.

Gadgil, R.L., Gadgil, P.D., 1971. Mycorrhiza and litter decomposition. Nature 233, 133.

Gaillard, V., Chenu, C., Recous, S., Richard, G., 1999. Carbon, nitrogen and microbial gradients induced by plant residues decomposing in soil. European Journal of Soil Science 50, 567–578.

Gartner, T.B., Treseder, K.K., Malcolm, G.M., Sinsabaugh, R.L., 2012. Extracellular enzyme activity in the mycorrhizospheres of a boreal fire chronosequence. Pedobiologia 55, 121–127.

German, D.P., Chacon, S.S., Allison, S.D., 2011. Substrate concentration and enzyme allocation can affect rates of microbial decomposition. Ecology 92, 1471–1480.

Girlanda, M., Selosse, M.A., Cafasso, D., Brilli, F., Delfine, S., Fabbian, R., et al., 2006. Inefficient photosynthesis in the Mediterranean orchid *Limodorum abortivum* is mirrored by specific association to ectomycorrhizal Russulaceae. Molecular Ecology 15, 491–504.

Grelet, G.A., Johnson, D., Paterson, E., Anderson, I.C., Alexander, I.J., 2009. Reciprocal carbon and nitrogen transfer between an ericaceous dwarf shrub and fungi isolated from *Piceirhiza bicolorata* ectomycorrhizas. New Phytologist 182, 359–366.

Grimoldi, A.A., Kavanova, M., Lattanzi, F.A., Schaufele, R., Schnyder, H., 2006. Arbuscular mycorrhizal colonization on carbon economy in perennial ryegrass: quantification by $^{13}CO_2/^{12}CO_2$ steady-state labelling and gas exchange. New Phytologist 172, 544–553.

Gryndler, M., Černá, L., Bukovská, P., Hršelová, H., Jansa, J., 2014. *Tuber aestivum* association with non-host roots. Mycorrhiza 24, 603–610.

Hammer, E.C., Balogh-Brunstad, Z., Jakobsen, I., Olsson, P.A., Stipp, S.L.S., Rillig, M.C., 2014. A mycorrhizal fungus grows on biochar and captures phosphorus from its surfaces. Soil Biology and Biochemistry 77, 252–260.

Herman, D.J., Firestone, M.K., Nuccio, E., Hodge, A., 2012. Interactions between an arbuscular mycorrhizal fungus and a soil microbial community mediating litter decomposition. Fems Microbiology Ecology 80, 236–247.

Hiruma, K., Gerlach, N., Sacristán, S., Nakano, R.T., Hacquard, S., Kracher, B., et al., 2016. Root endophyte *Colletotrichum tofieldiae* confers plant fitness benefits that are phosphate status dependent. Cell 65, 464–474.

Hobbie, E.A., Ouimette, A.P., Schuur, E.A.G., Kierstead, D., Trappe, J.M., Bendiksen, K., Ohenoja, E., 2013. Radiocarbon evidence for the mining of organic nitrogen from soil by mycorrhizal fungi. Biogeochemistry 114, 381–389.

Hofland-Zijlstra, J.D., Berendse, F., 2009. The effect of nutrient supply and light intensity on tannins and mycorrhizal colonisation in Dutch heathland ecosystems. Herbaceous Plant Ecology: Recent Advances in Plant Ecology 297–311.

Hughes, E., Mitchell, D.T., 1995. Utilization of sucrose by *Hymenoscyphus ericae* (an ericoid endomycorrhizal fungus) and ectomycorrhizal fungi. Mycological Research 99, 1233–1238.

Jakobsen, I., Rosendahl, L., 1990. Carbon flow into soil and external hyphae from roots of mycorrhizal cucumber plants. New Phytologist 115, 77–83.

Jansa, J., Bukovská, P., Gryndler, M., 2013. Mycorrhizal hyphae as ecological niche for highly specialized hypersymbionts—or just soil free-riders? Frontiers in Plant Science 4, 134.

Jastrow, J.D., Amonette, J.E., Bailey, V.L., 2007. Mechanisms controlling soil carbon turnover and their potential application for enhancing carbon sequestration. Climatic Change 80, 5–23.

Joner, E.J., Johansen, A., 2000. Phosphatase activity of external hyphae of two arbuscular mycorrhizal fungi. Mycological Research 104, 81–86.

Jumpponen, A., 2001. Dark septate endophytes—are they mycorrhizal? Mycorrhiza 11 (4), 207–211.

Kammen, D.M., Marino, B.D., 1993. On the origin and magnitude of preindustrial anthropogenic CO_2 and CH_4 emissions. Chemosphere 26, 69–86.

Keel, B.G., Zettler, L.W., Kaplin, B.A., 2011. Seed germination of *Habenaria repens* (Orchidaceae) in situ beyond its range, and its potential for assisted migration imposed by climate change. Castanea 76, 43–54.

Kohler, A., Kuo, A., Nagy, L.G., Morin, E., Barry, K.W., Buscot, F., et al., 2015. Convergent losses of decay mechanisms and rapid turnover of symbiosis genes in mycorrhizal mutualists. Nature Genetics 47, 410–415.

Koide, R.T., Wu, T., 2003. Ectomycorrhizas and retarded decomposition in a *Pinus resinosa* plantation. New Phytologist 158, 401–407.

Kong, A., Cifuentes, J., Estrada-Torres, A., Guzman-Davalos, L., Garibay-Orijel, R., Buyck, B., 2015. Russulaceae associated with mycoheterotroph *Monotropa uniflora* (Ericaceae) in Tlaxcala, Mexico: a phylogenetic approach. Cryptogamie Mycologie 36, 479–512.

Krüger, M., Krüger, C., Walker, C., Stockinger, H., Schüßler, A., 2012. Phylogenetic reference data for systematics and phylotaxonomy of arbuscular mycorrhizal fungi from phylum to species level. New Phytologist 193, 970–984.

Langdale, A.R., Read, D.J., 1990. Substrate decomposition and product release by ericoid and ectomycorrhizal fungi grown on protein. Agriculture Ecosystems and Environment 28, 285–291.

Larsen, P.E., Sreedasyam, A., Trivedi, G., Podila, G.K., Cseke, L.J., Collart, F.R., 2011. Using next generation transcriptome sequencing to predict an ectomycorrhizal metabolome. BMC Systems Biology 5, 70.

Le Tacon, F., Zeller, B., Plain, C., Hossann, C., Brechet, C., Martin, F., et al., 2015. Study of nitrogen and carbon transfer from soil organic matter to *Tuber melanosporum* mycorrhizas and ascocarps using ^{15}N and ^{13}C soil labelling and whole-genome oligoarrays. Plant and Soil 395, 351–373.

Leake, J.R., Ostle, N.J., Rangel-Castro, J.I., Johnson, D., 2006. Carbon fluxes from plants through soil organisms determined by field $^{13}CO_2$ pulse-labelling in an upland grassland. Applied Soil Ecology 33, 152–175.

Lekberg, Y., Hammer, E.C., Olsson, P.A., 2010. Plants as resource islands and storage units - adopting the mycocentric view of arbuscular mycorrhizal networks. Fems Microbiology Ecology 74, 336–345.

Lekberg, Y., Rosendahl, S., Michelsen, A., Olsson, P.A., 2013. Seasonal carbon allocation to arbuscular mycorrhizal fungi assessed by microscopic examination, stable isotope probing and fatty acid analysis. Plant and Soil 368, 547–555.

Lendenmann, M., Thonar, C., Barnard, R.L., Salmon, Y., Werner, R.A., Frossard, E., Jansa, J., 2011. Symbiont identity matters: carbon and phosphorus fluxes between *Medicago truncatula* and different arbuscular mycorrhizal fungi. Mycorrhiza 21, 689–702.

Lindahl, B.D., Tunlid, A., 2015. Ectomycorrhizal fungi—potential organic matter decomposers, yet not saprotrophs. New Phytologist 205, 1443–1447.

Lindahl, B., Stenlid, J., Olsson, S., Finlay, R., 1999. Translocation of ^{32}P between interacting mycelia of a wood-decomposing fungus and ectomycorrhizal fungi in microcosm systems. New Phytologist 144, 183–193.

Michelsen, A., Jonasson, S., Sleep, D., Havstrom, M., Callaghan, T.V., 1996. Shoot biomass, delta ^{13}C, nitrogen and chlorophyll responses of two arctic dwarf shrubs to in situ shading, nutrient application and warming simulating climatic change. Oecologia 105, 1–12.

Midgley, D.J., Chambers, S.M., Cairney, J.W.G., 2004. Utilisation of carbon substrates by multiple genotypes of ericoid mycorrhizal fungal endophytes from eastern Australian Ericaceae. Mycorrhiza 14, 245–251.

Nehls, U., Grunze, N., Willmann, M., Reich, M., Kuster, H., 2007. Sugar for my honey: Carbohydrate partitioning in ectomycorrhizal symbiosis. Phytochemistry 68, 82–91.

Nottingham, A.T., Turner, B.L., Winter, K., Chamberlain, P.M., Stott, A., Tanner, E.V.J., 2013. Root and arbuscular mycorrhizal mycelial interactions with soil microorganisms in lowland tropical forest. Fems Microbiology Ecology 85, 37–50.

Novotny, E.H., Maia, C.M.B.D., Carvalho, M.T.D., Madari, B.E., 2015. Biochar: Pyrogenic carbon for agricultural use—a critical review. Revista Brasileira De Ciencia Do Solo 39, 321–344.

Ogura-Tsujita, Y., Yokoyama, J., Miyoshi, K., Yukawa, T., 2012. Shifts in mycorrhizal fungi during the evolution of autotrophy to mycoheterotrophy in *Cymbidium* (Orchidaceae). American Journal of Botany 99, 1158–1176.

Orwin, K.H., Kirschbaum, M.U.F., St John, M.G., Dickie, I.A., 2011. Organic nutrient uptake by mycorrhizal fungi enhances ecosystem carbon storage: a model-based assessment. Ecology Letters 14, 493–502.

Paul, E.A., Kucey, R.M.N., 1981. Carbon flow in plant microbial associations. Science 213 (4506), 473–474.

Pfeffer, P.E., Douds, D.D., Bucking, H., Schwartz, D.P., Shachar-Hill, Y., 2004. The fungus does not transfer carbon to or between roots in an arbuscular mycorrhizal symbiosis. New Phytologist 163, 617–627.

Quilliam, R.S., Glanville, H.C., Wade, S.C., Jones, D.L., 2013. Life in the 'charosphere'—does biochar in agricultural soil provide a significant habitat for microorganisms? Soil Biology and Biochemistry 65, 287–293.

Rasmussen, H.N., Rasmussen, F.N., 2014. Seedling mycorrhiza: a discussion of origin and evolution in Orchidaceae. Botanical Journal of the Linnean Society 175, 313–327.

Read, D.J., Perez-Moreno, J., 2003. Mycorrhizas and nutrient cycling in ecosystems - a journey towards relevance? New Phytologist 157, 475–492.

Rousseau, A., Benhamou, N., Chet, I., Piche, Y., 1996. Mycoparasitism of the extramatrical phase of Glomus intraradices by Trichoderma harzianum. Phytopathology 86, 434–443.

Schaumann, G.E., Thiele-Bruhn, S., 2011. Molecular modeling of soil organic matter: Squaring the circle? Geoderma 166, 1–14.

Scheu, S., Folger, M., 2004. Single and mixed diets in Collembola: effects on reproduction and stable isotope fractionation. Functional Ecology 18, 94–102.

Schneider, K., Renker, C., Maraun, M., 2005. Oribatid mite (Acari, Oribatida) feeding on ectomycorrhizal fungi. Mycorrhiza 16, 67–72.

Schrey, S.D., Hartmann, A., Hampp, R., 2015. Rhizosphere interactions. In: Krauss, G.J., Nies, D.H. (Eds.), Ecological Biochemistry: Environmental and Interspecies Interactions. Wiley, Weinheim, pp. 293–310.

Schüßler, A., 2002. Molecular phylogeny, taxonomy, and evolution of *Geosiphon pyriformis* and arbuscular mycorrhizal fungi. Plant and Soil 244, 75–83.

Selosse, M.A., Roy, M., 2009. Green plants that feed on fungi: facts and questions about mixotrophy. Trends in Plant Science 14, 64–70.

Selosse, M.A., Faccio, A., Scappaticci, G., Bonfante, P., 2004. Chlorophyllous and achlorophyllous specimens of *Epipactis microphylla* (Neottieae, Orchidaceae) are associated with ectomycorrhizal septomycetes, including truffles. Microbial Ecology 47, 416–426.

Selosse, M.A., Boullard, B., Richardson, D., 2011. Noel Bernard (1874-1911): orchids to symbiosis in a dozen years, one century ago. Symbiosis 54, 61–68.

Singh, N., Abiven, S., Torn, M.S., Schmidt, M.W.I., 2012. Fire-derived organic carbon in soil turns over on a centennial scale. Biogeosciences 9, 2847–2857.

Slavíková, R., Püschel, D., Janoušková, M., Hujslová, M., Konvalinková, T., Gryndlerová, H., Gryndler, M., Weiser, M., Jansa, J., 2016. Monitoring CO_2 emissions to gain a dynamic view of carbon allocation to arbuscular mycorrhizal fungi. Mycorrhiza, in press. http://dx.doi.org/10.1007/s00572-016-0731-2.

Sochorová, L., Jansa, J., Verbruggen, E., Hejcman, M., Schellberg, J., Kiers, E.T., et al., 2016. Long-term agricultural management maximizing hay production can significantly reduce belowground C storage. Agriculture Ecosystems and Environment 220, 104–114.

Sousa, C.D., Menezes, R.S.C., Sampaio, E.V.D.B., Lima, F.D., 2012. Glomalin: characteristics, production, limitations and contribution to soils. Semina-Ciencias Agrarias 33, 3033–3044.

Staddon, P.L., Ramsey, C.B., Ostle, N., Ineson, P., Fitter, A.H., 2003. Rapid turnover of hyphae of mycorrhizal fungi determined by AMS microanalysis of C-14. Science 300, 1138–1140.

Stockel, M., Tesitelova, T., Jersakova, J., Bidartondo, M.I., Gebauer, G., 2014. Carbon and nitrogen gain during the growth of orchid seedlings in nature. New Phytologist 202, 606–615.

Talbot, J.M., Allison, S.D., Treseder, K.K., 2008. Decomposers in disguise: mycorrhizal fungi as regulators of soil C dynamics in ecosystems under global change. Functional Ecology 22, 955–963.

Taylor, D.L., Bruns, T.D., Hodges, S.A., 2004. Evidence for mycorrhizal races in a cheating orchid. Proceedings of the Royal Society B-Biological Sciences 271, 35–43.

Tischer, A., Blagodatskaya, E., Hamer, U., 2015. Microbial community structure and resource availability drive the catalytic efficiency of soil enzymes under land-use change conditions. Soil Biology and Biochemistry 89, 226–237.

Unterborn, C.T., Kabbes, J.E., Pigott, J.S., Reaman, D.M., Panero, W.R., 2014. The role of carbon in extrasolar planetary geodynamics and habitability. Astrophysical Journal 793, 124.

Valášková, V., Šnajdr, J., Bittner, B., Cajthaml, T., Merhautová, V., Hoffichter, M., et al., 2007. Production of lignocellulose-degrading enzymes and degradation of leaf litter by saprotrophic basidiomycetes isolated from a *Quercus petraea* forest. Soil Biology and Biochemistry 39, 2651–2660.

van der Heijden, M.G.A., Bardgett, R.D., van Straalen, N.M., 2008. The unseen majority: soil microbes as drivers of plant diversity and productivity in terrestrial ecosystems. Ecology Letters 11, 296–310.

van der Heijden, M.G.A., Martin, F.M., Selosse, M.A., Sanders, I.R., 2015. Mycorrhizal ecology and evolution: the past, the present, and the future. New Phytologist 205, 1406–1423.

Varma, A., Bonfante, P., 1994. Utilization of cell-wall related carbohydrates by ericoid mycorrhizal endophytes. Symbiosis 16, 301–313.

Ver, L.M.B., Mackenzie, F.T., Lerman, A., 1999. Biogeochemical responses of the carbon cycle to natural and human perturbations: past, present, and future. American Journal of Science 299, 762–801.

Verbruggen, E., Jansa, J., Hammer, E.C., Rillig, M.C., 2016. Do arbuscular mycorrhizal fungi stabilize litter-derived carbon in soil? Journal of Ecology 104, 261–269.

von Bloh, W., Bounama, C., Franck, S., 2005. Dynamic habitability of extrasolar planetary systems. In: Dvorak, R., Ferraz-Mello, S. (Eds.), Comparison of the Dynamical Evolution of Planetary Systems - Proceedings of the 6th Alexander von Humboldt Colloquium on Celestial Mechanics. Springer, Bad Hofgastein, Austria, pp. 287–300.

Vrålstad, T., 2004. Are ericoid and ectomycorrhizal fungi part of a common guild? New Phytologist 164, 7–10.

Waggoner, D.C., Chen, H.M., Willoughby, A.S., Hatcher, P.G., 2015. Formation of black carbon-like and alicyclic aliphatic compounds by hydroxyl radical initiated degradation of lignin. Organic Geochemistry 82, 69–76.

Walder, F., van der Heijden, M.G.A., 2015. Regulation of resource exchange in the arbuscular mycorrhizal symbiosis. Nature Plants 1, 15159.

Walder, F., Niemann, H., Natarajan, M., Lehmann, M.F., Boller, T., Wiemken, A., 2012. Mycorrhizal networks: common goods of plants shared under unequal terms of trade. Plant Physiology 159, 789–797.

Watt, M., Kirkegaard, J.A., Passioura, J.B., 2006. Rhizosphere biology and crop productivity - a review. Australian Journal of Soil Research 44, 299–317.

webelements.com (accessed on 03.03.2016.).

Weiss, M., Sýkorová, Z., Garnica, S., Riess, K., Martos, F., Krause, C., et al., 2011. Sebacinales everywhere: previously overlooked ubiquitous fungal endophytes. Plos One 6, e16793.

Werner, G.D.A., Kiers, E.T., 2015. Partner selection in the mycorrhizal mutualism. New Phytologist 205, 1437–1442.

Wickett, N.J., Goffinet, B., 2008. Origin and relationships of the myco-heterotrophic liverwort *Cryptothallus mirabilis* Malmb. (Metzgeriales, Marchantiophyta). Botanical Journal of the Linnean Society 156, 1–12.

Wu, T.H., Sharda, J.N., Koide, R.T., 2003. Exploring interactions between saprotrophic microbes and ectomycorrhizal fungi using a protein-tannin complex as an N source by red pine (*Pinus resinosa*). New Phytologist 159, 131–139.

Yuan, Z.Q., Gazol, A., Lin, F., Ye, J., Shi, S., Wang, X.G., et al., 2013. Soil organic carbon in an old-growth temperate forest: spatial pattern, determinants and bias in its quantification. Geoderma 195, 48–55.

Zeglin, L.H., Kluber, L.A., Myrold, D.D., 2013. The importance of amino sugar turnover to C and N cycling in organic horizons of old-growth Douglas-fir forest soils colonized by ectomycorrhizal mats. Biogeochemistry 112, 679–693.

Zettler, L.W., Piskin, K.A., Stewart, S.L., Hartsock, J.J., Bowles, M.L., Bell, T.J., 2005. Protocorm mycobionts of the Federally threatened eastern prairie fringed orchid, *Platanthera leucophaea* (Nutt.) Lindley, and a technique to prompt leaf elongation in seedlings. Studies in Mycology 53, 163–171.

CHAPTER

20

Carbon and Energy Sources of Mycorrhizal Fungi: Obligate Symbionts or Latent Saprotrophs?

T.W. Kuyper

Wageningen University, Wageningen, The Netherlands

20.1 INTRODUCTION

Our simple problems often grew
to mysteries we fumbled over,
because of lines we nimbly drew
and later neatly stumbled over.

Piet Hein.

These words by the Danish poet indicate the problem of the human attempt of pigeonholing natural phenomena. Can scientific advances be best achieved by creating and defining separate categories, or should we see these categories as human constructs and is there a need to look much more at overlapping concepts because many of these created categories can be better considered as continua along a line? This problem has been around in the case of the (putative) saprotrophic capabilities of mycorrhizal fungi, particularly ectomycorrhizal fungi. Are concepts such as saprotrophy and biotrophy (a lifestyle to which the ectomycorrhizal lifestyle belongs) in fact continua, as suggested by Koide et al. (2008), and do ectomycorrhizal fungi exhibit blurred lifestyles (Plett and Martin, 2011)? Or can we better understand C and nutrient cycling if we consider these lifestyles as separate categories? However, in the latter case, how do we then indicate that ectomycorrhizal fungi do influence the cycles of soil organic matter and of the nutrients contained therein? Are they, in fact, decomposers in disguise (Talbot et al., 2008)? Are they decomposers or are they saprotrophs (Lindahl and Tundlid, 2015)? (And what is the difference between a decomposer and a saprotroph?)

This debate has ancient roots. Ideas on a potential role of ectomycorrhizal fungi as transformers of C, ranging from fresh litter to humified soil organic matter, are as old as the field of mycorrhizal research. The pioneer of mycorrhizal research, Frank, already suggested a

Mycorrhizal Mediation of Soil
http://dx.doi.org/10.1016/B978-0-12-804312-7.00020-6

major contribution to soil organic matter turnover and nutrient mineralization (especially N) as a consequence of their saprotrophic activity (Read, 1987; Lindahl and Tunlid, 2015). Field observations suggested saprotrophic capabilities of ectomycorrhizal fungi, especially of the species that were found on dead, often strongly decomposed wood. However, ectomycorrhizal fungi on and in decayed wood are similar to those in adjacent mineral soil, suggesting that there are no ectomycorrhizal fungal species that specialize in the use of dead woody debris (Walker et al., 2014) and suggesting that field observations were consistent with, but not a test for, saprotrophic functioning. Laboratory studies with synthetic complex C compounds such as lignin (Trojanowski et al., 1984) equally confirmed the saprotrophic potential of several species of ectomycorrhizal fungi. Laboratory (Colpaert and Van Tichelen, 1996) and field studies (Treseder et al., 2006) contradicted the use of fresh litter by ectomycorrhizal fungi. Read et al. (2004) equally concluded that ectomycorrhizal fungi have low capabilities to decompose soil organic matter, but they proposed that ericoid mycorrhizal fungi do a better job.

Focus on the degradation of organic matter as a mechanism to acquire nutrients that are stored in organic forms has been more prominent. Read (1987) proposed a major role of ectomycorrhizal (and ericoid mycorrhizal) fungi in the acquisition of organic N and P. His work lies at the basis for conceptual models that attribute a more organic nutrient economy to ectomycorrhizal plants and fungi and a mineral nutrient economy to arbuscular mycorrhizal plants and fungi (Phillips et al., 2013). Their mycorrhizal-associated nutrient economy (MANE) framework is an explicit and testable hypothesis of how differences between both mycorrhizal types drive and are driven by differences in soil properties and nutrient cycling. The putative differential effect of ectomycorrhizal and arbuscular mycorrhizal trees on the acquisition of nutrients in organic forms ultimately drives these soil–plant feedbacks.

In the debate most attention has been given to the question of whether ectomycorrhizal fungi have saprotrophic capabilities, whereas for arbuscular mycorrhizal fungi the answer has generally been that these fungi lack saprotrophic capabilities. In fact, they cannot even be grown in the laboratory on a Petri dish in the absence of a plant root that provides the C compounds for the fungus. On the other hand, almost all ectomycorrhizal fungi can be grown in Petri dishes as pure cultures, provided external sources of C and nutrients are provided. However, there is virtually no evidence that these fungi can complete their life cycle in the absence of a host.

To address that question of saprotrophic capabilities and saprotrophic functioning under field conditions of mycorrhizal fungi, researchers have used a large range of observations and methods, both experimental and modeling. As this review will show, many data have been published that are consistent with a hypothesis that ectomycorrhizal fungi exert saprotrophic lifestyles under certain ecological conditions. However, data that are consistent with a preferred theory do not necessarily constitute a critical test for that theory against competing theories. Furthermore, model outcomes should be evaluated in the framework of the underlying assumptions and mechanisms that have been built into the model.

In this Chapter I review the various arguments that have been put forward to claim differential effects of ectomycorrhizal and arbuscular mycorrhizal plants on C and nutrient cycling, especially the part of the nutrient cycle in which organic forms are involved. However, before engaging with these various lines of putative evidence, it is important to discuss the two major ways in which the term *saprotrophic capability* has been used.

20.2 TWO CONCEPTS OF SAPROTROPHY

This brief historical overview of the underlying question about the saprotrophic capabilities of ectomycorrhizal fungi indicates that two different perspectives played a role in claims about putative saprotrophic lifestyles of these fungi.

One concept assumes that the C obtained from dead organic material is used for the buildup of biomass. "True" saprotrophic fungi use C for making biomass—as ectomycorrhizal fungi do on Petri dishes, but not or hardly ever so under field conditions. The other concept looks at the transformations of C. In that concept the focus is on the role of organic nutrients (N, P—although almost all focus in the previously published literature has been on N) that become available through the transformation and degradation of organic material. The C may be completely respired or only function as a C skeleton of amino acids. In that case the fungus "mines" the organic material for nutrients. The term *mining* as used here for acquisition of organic nutrients while transforming organic matter seems different from the conceptual separation between nutrient scavengers (plants with arbuscular mycorrhizal fungi) and nutrient miners (plants that excrete organic anions or acids to enhance P acquisition from strongly adsorbed or fixed sources) as proposed by Lambers et al. (2008). The excretion of organic anions is likely a relevant trait in relation to the saprotrophic capabilities of ectomycorrhizal fungi.

However, in both cases the final result of ectomycorrhizal fungal activity may be similar (i.e. a decrease of soil organic matter). However, this outcome is not inevitable. Saprotrophic activity by mycorrhizal fungi could equally result in increases in soil organic matter. Furthermore, although arbuscular mycorrhizal fungi lack saprotrophic capabilities, their activity could increase or decrease soil organic matter through priming processes.

Talbot et al. (2008) proposed three models in which a mycorrhizal (specifically ectomycorrhizal) role in C cycling can be envisaged. They used these models to predict how global environmental change would affect the potential saprotrophic activity of mycorrhizal fungi. Their first hypothesized model, called the "plan B hypothesis" or "alternate C source hypothesis," proposes that ectomycorrhizal fungi under conditions of reduced C supply (partly) switch their lifestyle and obtain (part of) their C demand for biomass buildup through saprotrophy. In this review I will use the term *alternate C source hypothesis*. If that hypothesis were valid, then one would predict that elevated CO_2 levels (which increase C supply to the fungus from the host plant; Alberton et al., 2005) reduce that switch toward saprotrophic activity. On the other hand, elevated N availability, when the C costs for the tree of acquiring N in the nonmycorrhizal condition become less and the costs of maintaining the symbiosis become (too) high, creates conditions for the fungus to switch to plan B, the alternative C source. The second hypothesized model, called the "coincidental decomposer hypothesis," makes the opposite predictions. The hypothesis assumes that mycorrhizal fungi decompose soil organic matter as a byproduct of mining for nutrients, especially N. This model would then fit with a hypothesis that very little (if at all) of the C acquired is used for producing fungal biomass (except the C skeleton of the organic N molecules). This model predicts that saprotrophic activity would increase under elevated CO_2 and be reduced under conditions of elevated N availability. This mechanism is part of what I would describe as the priming hypothesis. Talbot et al. (2008) reserved the term *priming* for a third mechanism (the "priming effects hypothesis"), in which ectomycorrhizal fungi exhibit saprotrophic activity when C allocation to roots is high and the plant primes the growth and activity of mycorrhizal fungi. However, the third hypothesis makes exactly the same predictions as the second hypothesis with regard to changes of saprotrophic activity under conditions of elevated CO_2 and N availability. For that reason I will take both hypotheses together because we currently lack an explicit test to separately evaluate the second and third hypothesis. Talbot et al. (2008) did not indicate which hypothesis, on the evidence available at the time of writing their paper, they considered more likely.

The review by Lindahl and Tunlid (2015) uses the same categorization. They separate cases in which the fungus primarily uses the soil organic matter for the acquisition of C and in which part of the C acquired is incorporated in biomass (saprotrophs) from cases in which the fungi primarily acquire N and P bound in organic molecules and in which the C is respired rather than used for biomass formation (decomposers). The first case accords with the alternate C source hypothesis and the second with the coincidental decomposer and priming effects hypotheses. I do consider their terminology (the term *saprotroph* being used for fungi using the C from soil organic matter degradation for biomass growth and the term *decomposer* for fungi that mine soil organic matter for nutrients, thereby transforming it without using the C for biomass growth) ambiguous. For that reason I will refer to both theories as the alternate C source hypothesis and to the nutrient mining by priming hypothesis. The summary of the literature by Lindahl and Tundlid (2015), as indicated in the title of their paper, claims very little support for the first but very substantial support for the second hypothesis.

I prefer to use the term *nutrient mining by priming* to describe cases in which litter or soil organic matter is transformed by ectomycorrhizal fungi to acquire nutrients that are present in organic forms. One further benefit of that terminology is that it can provide an explicit link with two competing hypotheses in the literature about priming of soil organic matter degradation (Chen et al., 2014). There are currently two competing hypotheses on priming that make opposite predictions. The first hypothesis "stoichiometric decomposition," proposes that organic matter decomposition is maximal when inputs of C and mineral nutrients match with microbial demands. A corollary of that hypothesis is that priming of organic matter decomposition is highest under conditions of high N availability. This is then similar to the alternate C source hypothesis of Talbot et al. (2008), in which high N availability is hypothesized to increase mycorrhizal fungal saprotrophic activity, with the additional condition that the mycorrhizal fungus switches to an alternative or additional C source when the tree downregulates its C supply to the fungus. The second mechanism by Chen et al. (2014) is called "microbial N mining" and proposes that microbes, to acquire sufficient nutrients from recalcitrant organic sources, do require labile (easily available) C to increase the decomposition process. A corollary of that hypothesis is that priming decreases under conditions of high N availability, similar to what is proposed under the hypothesis of nutrient mining by priming by ectomycorrhizal fungi by Talbot et al. (2008). Linking the specific issue of saprotrophic capabilities of mycorrhizal fungi to the general microbial ecological theory of priming is advantageous because it can help understanding why arbuscular mycorrhizal fungi, which lack saprotrophic capabilities, can still affect rates of organic matter degradation and why saprotrophic activities by mycorrhizal fungi can result in decreases (positive priming) and increases (negative priming) in soil organic matter. A further advantage of this broader concept is that it can link abiotic priming (destabilization of organic matter through competitive desorption; Keiluweit et al., 2015) with a similar mechanism in ectomycorrhizal fungi that would suggest saprotrophic activity (Clarholm et al., 2015).

In the next sections I will review the various forms of evidence about the saprotrophic capabilities of ectomycorrhizal fungi. At the end I will discuss whether and how arbuscular mycorrhizal fungi can affect decomposition and soil organic matter dynamics. I will conclude with suggesting several questions that in my view demand an increased research effort.

20.3 PHYLOGENETIC EVIDENCE

Phylogenetic research indicates that the ectomycorrhizal Basidiomycetes and Ascomycetes evolved from saprotrophic ancestors, with general losses of the cell-wall–degrading machinery (Kohler et al., 2015). Within the ectomycorrhizal Basidiomycetes the mycorrhizal lifestyle repeatedly evolved from predecessors with varying saprotrophic lifestyles: brown-rotting ancestors (e.g. Boletales), white-rotting wood-degrading fungi (e.g. Russulales), and white-rotting litter-decaying fungi (e.g. *Amanita*; Wolfe et al., 2012). There is no evidence for the reverse evolutionary pathway (i.e. that saprotrophic fungi evolved secondarily out of ectomycorrhizal fungi). The only possible exception to that pattern is *Hydnomerulius pinastri*. Its phylogenetic position (Nuhn et al., 2013; Kohler et al., 2015) implies that either the ectomycorrhizal habit originated three times within the Boletales (if *Hydnomerulius* is a member of the Paxillaceae; Kohler et al., 2015) or twice (if *Hydnomerulius* is sister to the Paxillaceae plus Boletaceae; Nuhn et al., 2013). Independent evolution of the ectomycorrhizal habit with concomitant losses of GH6 and GH7 (two cellobiohydrolases) has to be evaluated against one origin with a reversal to a saprotrophic lifestyle with independent acquisition of these enzymes. Cullings and Courty (2009), who supported the hypothesis of plan B decomposers, noted in their critique to a paper by Baldrian (2009) that there are numerous reversals from ectomycorrhizal fungi to saprotrophs but did not specify the fungi involved.

In that sense the mycorrhizal habit, because of the loss of several crucial cell-wall–degrading enzymes, is a dead-end street. However, the possibility that some genes have been retained and are still functional does exist. This set of retained ancestral genes is different for different clades of ectomycorrhizal fungi (Kohler et al., 2015). Therefore the data leave open the possibility that mycorrhizal fungal species can combine both lifestyles (see Section 20.5 for mixotrophy).

20.4 ENZYMATIC EVIDENCE

Fruitbodies of certain ectomycorrhizal fungi produce enzymes that have been known to play a major role in C cycling (e.g. laccases). In addition, the mycelium and the ectomycorrhizal root tip can produce these enzymes. Such enzymes have been used as taxonomic markers, for instance in studies of ectomycorrhizal fungi when grown in culture conditions.

The last decade has seen a major interest in enzymatic assays of ectomycorrhizal root tips. Standardized methods for such enzyme assays have been previously published (e.g. Pritsch et al., 2011). The utility of such enzyme tests for the question about the ability of ectomycorrhizal fungi to transform organic matter has been reviewed by Pritsch and Garbaye (2011). Enzymatic tests include enzymes for C acquisition (both labile compounds such as cellulose and hemicelluloses and aromatic compounds such as lignin and humic compounds), N acquisition (from plant material as leucine aminopeptidase; from fungal material as *N*-acetylglucosaminidase), and P acquisition (phosphate monoesterases). Enzyme activities of ectomycorrhizal root tips are much higher than those of nonmycorrhizal root tips, implying a major fungal role in the production of those enzymes.

Comparison of the various studies has not yet indicated clear signals because there was often little consistency between the results for various fungal species. Neither phylogeny, nor mycorrhizal exploration type (Agerer, 2001), nor soil properties can be used to predict enzyme expression profiles for defined ectomycorrhizal fungal species. The method for assessment of enzyme activities is also not without problems. First, the method assesses potential activity (i.e. activity in a buffer solution rather than at the water potential of the soil). The method discounts the possibility of enzyme inactivation through the formation of protein–polyphenol complexes or through association with mineral surfaces or in aggregates. As in realistic field conditions, the rate-liming step for enzyme functioning is usually not the level of enzyme activity but substrate supply. Therefore the enzymatic methods may show a poor correlation with the actual processes. A further problem with applying enzymatic tests was noted by Talbot et al. (2013). They noted no correlations between enzymatic activities of ectomycorrhizal root tips and of bulk soil and explained the discrepancy by suggesting that most enzymatic activity might occur at the extraradical mycelium and not at the root tip. Novel approaches, such as described by Phillips et al. (2014) through the use of hyphal traps, would allow to assess enzymatic activities at the hyphal level and at the same time allow discrimination between enzymes produced by saprotrophic and ectomycorrhizal fungi. The authors observed high enzymatic activities of hydrolytic and oxidative enzymes by ectomycorrhizal fungi, often at levels higher than those of saprotrophic fungi. The fact that hydrolytic enzymes (β-1,4 glucosidase), involved in litter degradation, and oxidative enzymes (peroxidase and phenol oxidase, but not laccase), involved in humus transformation and release of organic N from forests humus, were high in ectomycorrhizal fungi could suggest that ectomycorrhizal fungi make a major contribution to organic C turnover in these forests. Consistent with this suggestion is the observation that leucine aminopeptidase was higher in trap bags that were colonized by ectomycorrhizal fungi than in trap bags that also allowed access to saprotrophic fungi. Leucine amine peptidase activity was highest in trap bags with *Cortinarius*, consistent with their hypothesized major role in the cycling of organic N (see Section 20.7). In a study in West African ectomycorrhizal rain forests, leucine aminopeptidase activity of ectomycorrhizas was not significantly different from that of nonmycorrhizal root tips, which fits with other data (^{15}N signatures), indicating that N nutrition was through mineral, not organic, forms (Tedersoo et al., 2012; Kuyper, 2012).

Talbot et al. (2015) compared 10 different enzymes of 48 model fungal species (45 Basidiomycetes, 3 Ascomycetes) belonging to ectomycorrhizal and saprotrophic species in vitro. Activities of carbohydrases (cellobiohydrolase) and of acid phosphatase were significantly lower in ectomycorrhizal fungi than in saprotrophic fungi. There were no significant differences between both guilds in activities of polyphenol oxidase and leucine aminopeptidase, contrary to the Phillips et al. (2014) study. On the basis of these data, I would conclude support for the nutrient mining by priming hypothesis (for N, but apparently not for P) but less so for the alternate C source hypothesis (because enzymes to degrade litter as a source of C were lower than for saprotrophic fungi).

20.5 CARBON SIGNATURES

It has been a common practice to derive plausible inferences about the lifestyle of unknown fungi based on the ^{13}C signature, sometimes accompanied by ^{15}N signature. The observation that ectomycorrhizal and saprotrophic fungi generally differ in these isotopic signatures forms the basis for the saprotrophic-ectomycorrhizal divide (Henn and Chapela, 2001). Admittedly, there is partial overlap, but the strong link with ^{13}C sources (recently fixed C versus older C) suggests that fungal fruitbodies closely track the source signal. If ectomycorrhizal fungi acquire N from organic sources and take up N and C in the form of amino acids, then some weak signature of saprotrophy could result. ^{14}C signatures could equally be used to infer the C sources of ectomycorrhizal fungi. Treseder et al. (2006) calculated that less than 2% of C in ectomycorrhizal root tips was directly derived from litter that was degraded by the ectomycorrhizal fungi. Data by Bréda et al. (2013) equally indicated that less than 1% of C of oak trees during spring reactivation was derived from litter, thereby contradicting an earlier study by Courty et al. (2007). Hobbie et al. (2013) separately assessed the ^{14}C of structural C and protein C in fruitbodies. They concluded that structural C was all very recent, indicating no ability of ectomycorrhizal fungi to acquire C through saprotrophic activity for biomass growth (alternate C source hypothesis). The ^{14}C signal of protein was different, with substantial variation indicating C sources for protein that were very recent to old C. Species of *Cortinarius* and *Leccinum* had distinct signatures that showed the uptake of organic N from old soil organic matter (nutrient mining through priming hypothesis). In general, these studies confirm that there is a relatively clear separation between fungi that use C from soil organic matter ("true" saprotrophic fungi) and fungi that use recently fixed C that is provided by the tree (ectomycorrhizal fungi). In fungi evidence for mixotrophy is very limited, contrary to that for plants, in which several cases have been reported that plants acquire C autotrophically and heterotrophically, usually through C transfer mediated by ectomycorrhizal fungi (Selosse and Roy, 2009).

Studies as the ones referred to herein average over feeding guilds and do not exclude the possibility that individual species do exhibit mixotrophic lifestyles. Only a few fungal species have been repeatedly mentioned. Truffle growers have suggested that truffles can become independent from their host after several months and can use soil organic matter or dead wood for the production of fruitbodies. Data by Hobbie et al. (2013) support the use of (very) old C by *Tuber* species, and approximately 10% of protein C originated from very old C from mineral, rather than from humus, layers. Studies by Zeller et al. (2008) and Le Tacon et al. (2015) have not yielded support for a mixotrophic lifestyle by *Tuber melanosporum*. However, C for the growth of fruitbodies can be mobilized after the trees have stopped C assimilation. Likely the assimilated C can be stored in wood and supplied to the fungus later. A case of mixotrophy was also suggested for *Lactarius quietus*, which was hypothesized to provide oak trees with saprotrophically acquired C from litter and soil organic matter before and during the period of bud break (Courty et al., 2007). Their conclusion was based on a significant correlation between tree phenology and upregulation of four enzymes involved in C acquisition. They also suggested that *L. quietus* during that period acted as a brown-rot fungus, which is curious considering that this fungus evolved out of wood-inhabiting white-rot fungi. The species is also extremely variable in its enzymatic profile. Rineau and Courty (2011) reported that, of the 10 clusters of enzymatic activity profiles that they recognized, *L. quietus* occurred in 8, and this very high plasticity currently does not allow generalization.

Mixotrophy has also been claimed for *Tricholoma matsutake*. Vaario et al. (2012) observed a positive correlation between its productivity and activities of hemicellulases in spots where the fungus abundantly occurred. Laboratory studies indicated that the species could grow exclusively on hemicelluloses. Unfortunately, field studies with natural substrates with hemicellulose have not yet been conducted.

20.6 ECTOMYCORRHIZAL FUNGI INVOLVED

In many studies statements that ectomycorrhizal fungi exhibit saprotrophic activity were not specified with regard to the fungal species involved. Such statements may then have resulted in overconfidence in claims about the saprotrophic ability of ectomycorrhizal fungi in general. Species most commonly involved in claims on saprotrophic capabilities (as support for the nutrient mining by priming hypothesis) often belong to the ectomycorrhizal Boletales and the genus *Cortinarius*.

The Boletales are nested within the brown-rot fungi. Depending on the exact position of the brown-rot fungus *H. pinastri*, the ectomycorrhizal habit evolved twice (*Suillus*, Boletineae) or thrice (*Suillus*, Paxillaceae, Boletaceae). The species *Paxillus involutus* has been extensively studied for saprotrophic activity. It has been shown to have a Fenton-like mechanism to oxidize soil organic matter, similar to the mechanism of brown-rot fungi, thereby releasing organic N forms (Rineau et al., 2012, 2013). The enzymes expressed by *P. involutus* are similar to the enzymes expressed by the related brown-rot fungi; however, *P. involutus* lacks transcripts and the genes to metabolize the carbohydrates released (Rineau et al., 2012; Kohler et al., 2015), especially GH6 and GH7.

The extent to which other boletes retain the ability to oxidize organic matter is less well known. For *Suillus luteus* there is indirect evidence that it has extensive saprotrophic capabilities. Chapela et al. (2001) studied the effects of pine invasion in the paramó grasslands of Ecuador. They noted large losses of soil organic matter in these plantations (up to 30% of soil organic C was lost in the 20 years after the plantation was established) that were almost exclusively colonized by the introduced ectomycorrhizal fungus *S. luteus*. The ^{13}C signature of fruitbodies was closer to that of soil samples than that of the pine tissues. Additional evidence came from the ^{14}C signature in fruitbodies that indicated that fruitbody C was fixed 2–5 years before sampling. That number may be based on recently fixed C and on C (as part of protein skeletons?) based on much older soil organic C. Such cases may be exceptional; pines and their associated mycorrhizal fungi are not native, and the plantations are very strongly dominated by just one species. Cullings et al. (2010) reported increased laccase activity by *Suillus granulatus* after litter addition and suggested that the species may act as a facultative saprotroph. However, their report of upregulated laccase activity is surprising, considering that Kohler et al. (2015) in their genomic analyses of a *Suillus* species did not report the presence of lignin-degrading enzymes.

One further case has been described in which the introduction of exotic ectomycorrhizal trees depleted old soil organic matter. In Hawaii the introduction of eucalypts resulted in substantial soil C losses, approximately 10% in 10 years, but these were offset by newly formed C in the plantation (Baskin and Binkley, 1998). No information on the ectomycorrhizal fungal species involved was provided, but it is known that species of the genus *Pisolithus* (Boletales) are common ectomycorrhizal associates of exotic eucalypts.

In the case of *P. involutus* the ability to oxidize soil organic matter was linked explicitly to N mining. Curiously, both this species and its sister species *Paxillus rubicundulus*, an exclusive associate of the N-fixing alders (*Alnus*), are tolerant of high N (and especially nitrate) availability (Huggins et al., 2014). Therefore the selective benefit of N mining for these species demands more attention. It may not be surprising that in studies with *P. involutus* the addition of mineral ammonium had very limited effects on the decomposition of litter (Rineau et al., 2013), an observation that seems in contrast with the hypothesis of nutrient mining through priming.

In other cases the link between N mining and ectomycorrhizal fungal saprotrophy has been more clearly established. This is especially evident for species of the genus *Cortinarius*. The occurrence of Class II peroxidases in *Cortinarius* has been described by Bödeker et al. (2009, 2014). *Cortinarius* species are very sensitive to N loading of forest ecosystems and likely have evolved a trophic lifestyle to acquire N in cases in which mineral N is present in very low concentrations. Class II peroxidases (including manganese-dependent peroxidases) play a major role in the degradation of litter, wood, and humus (Sinsabaugh, 2010). Furthermore, Bödeker et al. (2009) noted the frequent occurrence of these peroxidases in members of the genera *Lactarius* and *Russula*. It is interesting to note that the protein signature of these latter species did not include old C, derived from soil organic matter, whereas the protein signature of *Cortinarius* species did (Hobbie et al., 2013). Clearly the roles played by those enzymes (they are also involved in detoxification of phenolic compounds and as adaptation against manganese toxicity at low pH) need further study.

Parker et al. (2015) studied ectomycorrhizal tree (*Betula nana* and *Betula pubescens*) encroachment in the subarctic and noted that tree encroachment resulted in massive losses of soil organic C. They suggested that ectomycorrhizal mining for soil N drove the disappearance of soil C. Consistent with that hypothesis, Deslippe et al. (2016) provided evidence that birch encroachment was correlated with the increased abundance of a species of *Cortinarius*, which would confirm the link among nutrient mining by these species, increased growth of ectomycorrhizal trees, and reduced C stocks.

Not all species of ectomycorrhizal fungi may possess this ability to oxidize soil organic matter and acquire organic N through this process. Wu (2011) actually was of the opinion that this ability is relatively rare. A recent study by Shah et al. (2016) confirmed this mechanism of oxidation of soil organic matter in two members of the Boletales (*P. involutus*, *S. luteus*), two of the Agaricales (*Laccaria bicolor*, *Hebeloma cylindrosporum*), and one member of the Atheliales (*Piloderma croceum*). The authors suggested that the ability to oxidize soil organic matter was widespread, but that the ability to degrade fresh litter was extremely limited because most carbohydrase genes seemed to have been lost in the evolutionary transition from a saprotrophic to an ectomycorrhizal lifestyle. Again, the species with this capability include species that are very nitrophobic (*P. croceum*) to highly N tolerant (*L. bicolor*).

20.7 NONENZYMATIC NUTRIENT MINING BY ECTOMYCORRHIZAL FUNGI

Ectomycorrhizal fungal species could, next to the production of the enzymatic oxidative machinery, also contribute to enhanced C cycling and uptake of nutrients by exudation of organic anions (citrate, oxalate). Such anions (or acids, depending on soil pH) could destabilize mineral-protected C, making it accessible for degradation and N mining (Keiluweit et al., 2015). Clarholm et al. (2015) proposed a model, the unbutton model, in which C destabilization through citrate and oxalate would, similar to expression of ligninolytic enzymes, increase access to N from easily degradable but currently protected sources. The model by Clarholm et al. (2015) separates a first step, chemical priming, that increases access to such relatively labile but protected C sources (Keiluweit et al., 2015) from a second step, the proper priming. The authors suggested that citrate was especially beneficial in acid environments, and oxalate in calcareous, high-pH environments, and that the fungi involved would be ectomycorrhizal fungi with hydrophobic mycelium in acid environments and species with more hydrophilic mycelium in calcareous environments. Further research toward the fungi involved and their functional traits in relation to the various low-molecular–weight organic anions exuded seems a clear research challenge. In this nonenzymatic case the focus is more on mineral-protected C whereas the oxidative mechanism focuses on recalcitrant C. The relative importance of recalcitrance versus protection (through sorption to metal hydroxides or through incorporation in aggregates that are inaccessible for soil enzymes) is currently a hotly debated topic in organic matter studies, where a shift toward more focus on protection (and less on recalcitrance) is currently occurring (Schmidt et al., 2011; Lehmann and Kleber, 2015). Whereas both mechanisms for acquisition of organic N could work with comparable effectiveness, they could have large differences for the acquisition of organic P sources, with the nonenzymatic strategy being superior.

20.8 STOICHIOMETRIC CONSIDERATIONS

Although the evidence to date suggests very little or no support for the alternate C source hypothesis, various lines of evidence provide solid support for the nutrient mining through priming hypothesis. Quantification of that effect in terms of nutrient and C dynamics is still sparse. The C acquired through degradation of litter and soil organic matter that ends up, as part of the C skeleton of amino acids, in fungal mycelia and fruitbodies is often small, ranging from less than 2% (Treseder et al., 2006) to 10% (Hobbie et al., 2013). The amount of C that was lost is likely much higher, with numbers up to 30% of soil organic C lost in an exotic pine plantation in 20 years (Chapela et al., 2001). Such selective mining for N should be visible in changes in C:N stoichiometry in soil organic matter along a depth gradient. Lindahl et al. (2007) investigated changes in C:N ratios along a depth profile in a boreal pine forest. They reported an increase in soil organic matter C:N ratio from the fragmented litter (F) layer to the humus (H) layer from 41 to 51. If we assume that this change was caused by N mining (species of *Cortinarius* and *Piloderma* were abundant in these deeper layers, whereas there very few saprotrophic fungi in those layers), then we can calculate the contribution by ectomycorrhizal fungi to C cycling. For the calculation we assume that amino acids are taken up with a C:N ratio of 6, a number that accords very well with a study by Murphy et al. (2015), who reported a C:N of primed material of 5. In that case (and assuming very low decomposition rates in the F and H layers), approximately 20% of the C must have been lost. Lindahl et al. (2007) also provided data on ^{15}N abundance, with ^{15}N abundance (δ-values) from the F to the lower H layer increasing from −2.8‰ to +1.6‰. If this increase is (largely) due to the removal of proteinaceous material, then this material must be strongly depleted in ^{15}N (Hobbie and

Horton, 2007). Next to selective removal of ^{15}N-depleted proteinaceous material, a large contribution of ectomycorrhizal fungal necromass (which is usually enriched in ^{15}N, and fruitbodies and mycelium of *Cortinarius* species are known to have high ^{15}N values) can explain these changes in the N signature over soil depth. Therefore these data are indicative for the major contribution that ectomycorrhizal fungi make in C cycling in forests.

A further explanation of the increase in C:N in deeper soil layers may be N allocation by saprotrophic fungi from deeper organic layers to fresh litter layers with a much higher C:N ratio (Boberg et al., 2014). That hypothesis was further investigated by Bödeker et al. (2016). They observed N loss from organic matter when it was colonized by ectomycorrhizal fungi but translocation of N to more superficial layers when it was colonized by saprotrophic fungi. Relatively fresh organic material was decomposed efficiently by saprotrophic and ectomycorrhizal fungi (but the latter group was outcompeted by the first group), but old organic matter was poorly decomposed by both functional guilds. The low mass loss of old organic material by saprotrophs is consistent with the low energy content contained in this recalcitrant fraction; however, the low mass loss by ectomycorrhizal fungi is surprising in the light of the 20% that must have been decomposed in the study by Lindahl et al. (2007). It may be significant in this regard that in the study by Bödeker et al. (2016) species of *Cortinarius* were extremely rare (whereas they were dominant in the study by Lindahl et al. (2007)). Further research to understand the nature and N signature of the primed material and of the ectomycorrhizal fungi involved in N and C transformations is clearly warranted.

Stoichiometric considerations underlie the model by Averill et al. (2014). They argued that competition for N between ectomycorrhizal and saprotrophic fungi reduced N availability for saprotrophs, resulting in reduced decomposition rates and C accumulation. They noted that in ectomycorrhizal systems approximately 70% more C could be stored per unit N than in arbuscular mycorrhizal systems. However, although their data are consistent with a theory that links saprotrophic capabilities of ectomycorrhizal fungi to C storage (similar as in the model by Orwin et al. (2011)), they do not provide a crucial test for it. Their study included four biomes, of which two contained ectomycorrhizal and arbuscular mycorrhizal systems. For temperate forests the pattern holds (C:N ratios for soil organic matter in ectomycorrhizal and arbuscular mycorrhizal forests being 19.9 and 11.8, respectively), whereas there were no significant differences for (sub-)tropical forests (C:N ratios for soil organic matter in ectomycorrhizal and arbuscular mycorrhizal forests being 13.0 and 12.1, respectively). The large discrepancy in C storage per unit N (or conversely, the extent of N mining from soil organic matter) between temperate and tropical forests indicates that, for causes still unknown, nutrient mining by ectomycorrhizal fungi is not a global phenomenon. Data from an afrotropical forest (Tedersoo et al., 2012; Kuyper, 2012) confirm that in ectomycorrhizal and arbuscular mycorrhizal forests the N cycle is equally open and characterized by a predominance of mineral N. The MANE framework (Phillips et al., 2013) states explicitly that the organic nutrient cycle in ectomycorrhizal vegetation is typical for boreal and temperate ecosystems and is not necessarily a global phenomenon.

Whereas C:N stoichiometry has received attention in the framework of the debate on saprotrophic capabilities of ectomycorrhizal fungi, there seems to be an almost complete neglect for N:P stoichiometry. In theory, arguments about mining of organic nutrients could equally apply to N and P, but most focus has been directed toward N. There may be crucial differences between the organic cycles of N and P. Whereas organic N, upon oxidative release from phenolic and humus complexes, can be depolymerized into small organic molecules that can be directly taken up, organic P cannot be taken up as such. Organic P releases orthophosphate, due to enzymatic activity of phosphomonoesterases and phytases, in the external environment and the uptake is in the form of that orthophosphate anion. Mineral protection is more important in the case of organic P than for organic N. Therefore it is an unanswered issue whether the enzymatic oxidation of soil organic matter is equally important for N and P or whether increased availability of P depends more on mineral priming in accordance to the unbutton model (Clarholm et al., 2015; Keiluweit et al., 2015). In that case the oxidation of soil organic matter is a major source for organic N and the production of low-molecular-weight organic anions (citrate, oxalate) the major source for organic P. Theoretical arguments (Dijkstra et al., 2013) suggest that nutrient mining through priming may not be of major importance for P acquisition. Therefore it is important to study the consequences of ectomycorrhizal fungal saprotrophic behavior for N and P.

20.9 MODELING STUDIES

Two studies have been published in which the ability of ectomycorrhizal plants to exhibit saprotrophic activity has been incorporated. In these models the consequences for C cycling and C storage have been explored.

Moore et al. (2015a) modeled C decomposition in the presence of ectomycorrhiza. They compared a null model, in which ectomycorrhizal fungi obtain all C from the tree, with alternative models, in which part of the C is soil derived. In one of these alternative models the fraction of C delivered from the soil is constant (constant uptake); in the other alternative model the fraction of C acquired through saprotrophic activity increases with decreasing soil fertility, which reduces plant productivity and hence C flow to the fungus (variable uptake). Mycorrhizal fungi are either assumed to assimilate C at limiting availability as fast as "true" saprotrophs or to be more efficient (a lower K_M in uptake kinetics) at low substrate availability. Their two model scenarios bear resemblance to (but are not equal to) the nutrient mining by priming hypothesis (constant uptake) and the alternate C source hypothesis (variable uptake). C flux from the soil toward the mycorrhizal fungus in the variable-uptake model ranges between 20% and 25%, with higher values in less productive sites; the proportion of soil C that goes to ectomycorrhizal fungi rather than saprotrophic microbes ranges between 15% and 30%. In these models soil C pools are reduced by 15–40% compared with the null model of no saprotrophic activity by ectomycorrhizal fungi, depending on differences in substrate affinity and uptake rates between ectomycorrhizal and saprotrophic fungi. Their model contains a simple mathematical function to describe the switch from biotrophy to partial saprotrophy (mixotrophy); this switch occurs when C expenditure by the fungus becomes larger than the C supplied by the tree. However, the basis for the claim of a mycorrhizal switch and actually C acquisition by mycorrhizal fungi is very weak, and, possibly apart from *T. matsutake*, little evidence for such a switch has been reported. Their model does not provide a linkage between C decomposition and N acquisition by the mycorrhizal fungus and plant. In a subsequent study, Moore et al. (2015b) investigated and modeled the effect of (ectomycorrhizal) roots and extraradical mycelia of ectomycorrhizal fungi on soil C turnover. That model suggests a large effect on labile (particulate) and stable (mineral-bound) organic matter.

In the presence of mycorrhizal roots and mycorrhizal mycelia, the amount of mineral-protected C was found to be less than 50% than in cases without roots, irrespective of the amount of extraradical mycelia. For particulate organic matter, the presence of mycorrhizal mycelia had a large effect, reducing soil C by almost 75%. The authors noted that their simulation results are incomplete because they do not include the turnover of extraradical mycelia. If fungal tissues (mycelia, ectomycorrhizal root tips) are more recalcitrant than plant material (Langley et al., 2006), then the mycorrhizal symbiosis could still contribute to soil C storage. Their model does not provide a feedback with N availability; therefore changes in C inputs are not included in the model.

Deckmyn et al. (2011) modeled C cycles in forests. Their highly parameterized model includes mycorrhizal and nonmycorrhizal fungi that interact in the highly degraded, recalcitrant soil C pool. Their model does not specify whether the ectomycorrhizal fungi, while acquiring organic N from the degraded organic matter, can use the C for biomass buildup. Their model run, without trees, in which ectomycorrhizal biomass was reduced by 90%, suggests that the ectomycorrhizal fungi can switch to plan B and have access to an alternative C source. In a subsequent paper (Deckmyn et al., 2014) it was equally stated that ectomycorrhizal fungi have an ability to degrade organic matter, but at a rate that is much lower than that of saprotrophic fungi.

Other modeling studies have also included the possibility of saprotrophic behavior of mycorrhizal fungi. The trade model by Grman et al. (2012) is based on the generalist behavior of mycorrhizal fungi (i.e. the ability to acquire C independent of the host) based on the study by Vaario et al. (2012) on *T. matsutake*.

Orwin et al. (2011) proposed a different model, in which ectomycorrhizal fungi could access organically bound nutrients, on the basis of the nutrient mining by priming hypothesis. Their model indicated that ectomycorrhizal fungi changed soil organic C quantity and quality. Because of changes in organic matter quality, C:N ratios increased (consistent with observations by Lindahl et al. (2007)). Selective mining of the nutrient-rich part of soil organic matter by ectomycorrhizal fungi reduced decomposition rates by saprotrophs, resulting in increased C storage due to ectomycorrhizal saprotrophic activity. Therefore their model fits with observations that are in agreement with the so-called Gadgil effect (Fernandez and Kennedy, 2016). The outcome of enhanced C storage is opposite to the model outcome of Moore et al. (2015a). The model by Orwin et al. (2011) included a feedback to nutrient acquisition to ectomycorrhizal fungi and hence to plant productivity. Increased productivity and C inputs to the soil resulted in even larger soil C storage due to ectomycorrhizal activity. However, note that their model implicitly assumes that the acquired organically bound nutrients are transferred to the plants rather than immobilized in the extraradical mycelium, an assumption that may not always be valid (Franklin et al., 2014; Kuyper and Kiers, 2014). The effect of decomposition rates of mycelia was not included in their model.

20.10 ARBUSCULAR MYCORRHIZAL FUNGI

It is widely agreed that arbuscular mycorrhizal fungi do not possess the enzymes to make a substantial contribution to the degradation of litter and soil organic matter (Hodge, 2014; see also Chapter 8). They lack the enzymes to depolymerize proteins, although they can take up amino acids as part of their nutrition, and this could lead to minor enrichment of C that is derived from dead organic material. However, Talbot et al. (2008) suggested that there is also a role of arbuscular mycorrhizal fungi in the decomposition of organic matter; however, this is likely related to priming by other microbes (see also Chapter 21). Because the consequences of priming, with C and nutrient transformations being executed by saprotrophic microbes, could result in increases or decreases of soil organic matter due to the arbuscular mycorrhizal symbiosis, they need to be included in this review. The consequences of priming, through exudates released by the arbuscular mycorrhizal fungus, are enhanced access to organic and mineral N due to enhanced litter and humus degradation (Tu et al., 2006; Atul-Nayyar et al., 2009).

The current literature does not seem to allow for conclusive generalizations because positive priming (enhanced C loss and N release) and negative priming (enhanced C storage) have been observed (see Chapter 8). Carillo et al. (2016) proposed that priming by arbuscular mycorrhizal plants (and fungi) may enhance litter decomposition but stabilize soil organic matter and thereby promote soil C storage. Negative priming of fresh litter (root litter) was observed by Carillo et al. (2016) Verbruggen et al. (2016, see also Chapter 24). The authors hypothesized that the arbuscular mycorrhizal fungi effectively competed with and suppressed saprotrophic fungi and hence litter accumulated, a mechanism that is equivalent to the mechanism proposed by Orwin et al. (2011). Cheng et al. (2012) provided data that under conditions of elevated CO_2 positive priming by arbuscular mycorrhizal fungi became more important, thereby effectively reducing soil C storage.

20.11 SAPROTROPHIC CAPABILITIES OF ECTOMYCORRHIZAL FUNGI: THE WAY FORWARD

In this review I have evaluated the support for two theories that ascribe capabilities to transform litter and soil organic matter to ectomycorrhizal fungi. The theory that ectomycorrhizal fungi could acquire and use C for biomass growth (alternate C source hypothesis) was found to be almost without support. The theory that ectomycorrhizal fungi oxidize organic material and acquire organic nutrients (nutrient mining through priming hypothesis) was found to have substantial support, in agreement with the conclusions drawn by Lindahl and Tunlid (2015). The evidence would be in line with the current priming theory that describes how saprotrophic fungi mine old organic matter for mineral nutrients. However, several questions cannot be addressed currently, and I use this concluding section to mention these issues.

Nutrient mining through priming has almost exclusively been focused toward N, and P acquisition has been rather neglected. We need a more explicit stoichiometric approach (Johnson, 2010; see also Chapter 8), the more so because the specific mechanisms by which organic N and P can be acquired are likely different. Although both mechanisms (enzymatic oxidation, production of low-molecular–weight organic anions) likely result in decomposition of organic C, the actual quantities and the quality of the remaining material will differ. Quantitative balances on C lost and nutrients gained need to be refined and assessed more often, and changes in stable isotope signatures along a vertical gradient in which ectomycorrhizal fungal saprotrophic activity is occurring need to be combined to better characterize the nature (quality) of the primed material. Current models allow positive and negative effects of mycorrhizal fungal saprotrophic activity (or priming) on C stocks, depending on assumptions of how effectively mycorrhizal and saprotrophic fungi compete for nutrients (Bödeker et al., 2016; see also Chapter 24). Therefore we need to better link theories on saprotrophic activities of ectomycorrhizal fungi and the Gadgil effect. Next to direct changes through mycorrhizal (or mycorrhiza-induced) saprotrophic nutrient mining, we need to increase focus on further changes that could contribute to C sequestration or C losses. For instance, despite the nutrient mining activity by ectomycorrhizal fungi in boreal forests, it is likely that in the absence of fire ectomycorrhizal fungi and plants are replaced by ericoid mycorrhizal fungi and plants. Because these plants and the fungi produce litter of lower quality, C and nutrient turnover decrease, resulting in less productive vegetation growing on soils with ever-increasing levels of soil organic matter (Clemmensen et al., 2013, 2015). And finally, with an evident link between saprotrophic activity of ectomycorrhizal fungi and global environmental change (where mycorrhizal activity can generate positive and negative feedbacks to C cycling), we need to understand why our concepts apply only in boreal and temperate forests and not (or less so) in tropical rain forests. Does the lack of a mechanism that holds across biomes relate to the absence of specific clades of mycorrhizal fungi (species of *Cortinarius* are very rare in tropical Africa) or to specific traits of tropical and temperate tree species? Experiments that yielded data that were consistent with theory have been frequent; it is now important to come up with experiments that critically test competing theories.

Acknowledgments

The author is grateful to an anonymous reviewer for helpful suggestions.

References

Agerer, R., 2001. Exploration types of ectomycorrhizae. A proposal to classify ectomycorrhizal mycelial systems according to their patterns of differentiation and putative ecological importance. Mycorrhiza 11, 107–114.

Alberton, O., Kuyper, T.W., Gorissen, A., 2005. Taking mycocentrism seriously: mycorrhizal fungal and plant responses to elevated CO_2. New Phytologist 167, 859–868.

Atul-Nayyar, A., Hamel, C., Hanson, K., Germida, J., 2009. The arbuscular mycorrhizal symbiosis links N mineralization to plant demand. Mycorrhiza 19, 239–246.

Averill, C., Turner, B.L., Finzi, A.C., 2014. Mycorrhiza-mediated competition between plants and decomposers drives soil carbon storage. Nature 505, 543–545.

Baldrian, P., 2009. Ectomycorrhizal fungi and their enzymes in soils: is there enough evidence for their role of facultative soil saprotrophs? Oecologia 161, 657–660.

Baskin, M.A., Binkley, D., 1998. Changes in soil carbon following afforestation in Hawaii. Ecology 79, 828–833.

Boberg, J.B., Finlay, R.D., Stenlid, J., Ekblad, A., Lindahl, B.D., 2014. Nitrogen and carbon reallocation in fungal mycelia during decomposition of boreal forest litter. PLoS One 9, e92897.

Bödeker, I.T.M., Nygren, C.M.R., Taylor, A.F.S., Olson, Å., Lindahl, B.D., 2009. Class II peroxidase-encoding genes are present in a phylogenetically wide range of ectomycorrhizal fungi. ISME Journal 3, 1387–1395.

Bödeker, I.T.M., Clemmensen, K.E., de Boer, W., Martin, F., Olson, Å., Lindahl, B.D., 2014. Ectomycorrhizal *Cortinarius* species participate in enzymatic oxidation of humus in northern forest ecosystems. New Phytologist 203, 245–256.

Bödeker, I.T.M., Lindahl, B.D., Olson, Å., Clemmensen, K.E., 2016. Mycorrhizal and saprotrophic fungal guilds compete for the same organic substrates but affect decomposition differently. Functional Ecology. http://dx.doi.org/10.1111/1365-2435.12677 (in press).

Bréda, N., Maillard, P., Montpied, P., Bréchet, C., Garbaye, J., Courty, P.-E., 2013. Isotopic evidence in adult oak trees of a mixotrophic life style during spring reactivation. Soil Biology and Biochemistry 58, 136–139.

Carillo, Y., Dijkstra, F.A., LeCain, D., Pendall, E., 2016. Mediation of soil C decomposition by arbuscular mycorrhizal fungi in grass rhizospheres under elevated CO_2. Biogeochemistry 127, 45–55.

Chapela, I.H., Osher, L.J., Horton, T.R., Henn, M.R., 2001. Ectomycorrhizal fungi introduced with exotic pine plantations induce soil carbon depletion. Soil Biology and Biochemistry 33, 1733–1740.

Chen, R., Senbayram, M., Blagodatsky, S., Myachina, O., Dittert, K., Lin, X., Blagodatskaya, E., Kuzyakov, Y., 2014. Soil C and N availability determine the priming effect: microbial mining and stoichiometric decomposition theories. Global Change Biology 20, 2356–2367.

Cheng, L., Booker, F.L., Tu, C., Burkey, K.O., Zhou, L., Shaw, H.D., Rufty, T.W., Hu, S., 2012. Arbuscular mycorrhizal fungi increase organic carbon decomposition under elevated CO_2. Science 337, 1084–1087.

Clarholm, M., Skyllberg, U., Rosling, A., 2015. Organic acid induced release of nutrients from metal-stabilized soil organic matter – the unbutton model. Soil Biology and Biochemistry 84, 168–176.

Clemmensen, K.E., Bahr, A., Ovaskainen, O., Dahlberg, A., Ekblad, A., Wallander, H., Stenlid, J., Finlay, R.D., Wardle, D.A., Lindahl, B.D., 2013. Roots and associated fungi drive long-term carbon sequestration in boreal forests. Science 339, 1615–1618.

Clemmensen, K.E., Finlay, R.D., Dahlberg, A., Stenlid, J., Wardle, D.A., Lindahl, B.D., 2015. Carbon sequestration is related to mycorrhizal fungal community shifts during long-term succession in boreal forests. New Phytologist 205, 1525–1536.

Colpaert, J.V., Van Tichelen, K.K., 1996. Decomposition, nitrogen and phosphorus mineralization from beech leaf litter colonized by ectomycorrhizal or litter-decomposing basidiomycetes. New Phytologist 134, 123–132.

Courty, P.-E., Bréda, N., Garbaye, J., 2007. Relation between oak tree phenology and the secretion of organic matter degrading enzymes by *Lactarius quietus* ectomycorrhizas before and during bud break. Soil Biology and Biochemistry 39, 1655–1663.

Cullings, K., Courty, P.-E., 2009. Saprotrophic capabilities to study functional diversity and resilience of ectomycorrhizal community. Oecologia 161, 661–664.

Cullings, K., Ishkhanova, G., Ishkhanov, G., Henson, J., 2010. Induction of saprophytic behavior in the ectomycorrhizal fungus *Suillus granulatus* by litter addition in a *Pinus contorta* (Lodgepole pine) stand in Yellowstone. Soil Biology and Biochemistry 42, 1176–1178.

Deckmyn, G., Campioli, M., Muys, B., Kraigher, H., 2011. Simulating C cycls in forest soils: including the active role of micro-organisms in the ANAFORE forest model. Ecological Modelling 222, 1972–1985.

Deckmyn, G., Meyer, A., smits, M.M., Ekblad, A., Grebenc, T., Komarov, A., Kraigher, H., 2014. Simulating ectomycorrhizal fungi and their role in carbon and nitrogen cycling in forest ecosystems. Canadian Journal of Forest Research 44, 535–553.

Deslippe, J.R., Hartmann, M., Grayston, S.J., Simard, S.W., Mohn, W.W., 2016. Stable isotope probing implicates a species of *Cortinarius* in carbon transfer through ectomycorrhizal fungal mycelial networks in Arctic tundra. New Phytologist 210, 383–390.

Dijkstra, F.A., Carillo, Y., Pendall, E., Morgan, J.A., 2013. Rhizosphere priming: a nutrient perspective. Frontiers in Microbiology 4 article 216.

Fernandez, C.W., Kennedy, P.G., 2016. Revisiting the 'Gadgil effect': do interguild fungal interactions control carbon cycling in forest soils? New Phytologist 209, 1382–1394.

Franklin, O., Näsholm, T., Högberg, P., Högberg, M., 2014. Forests trapped in nitrogen limitation – an ecological market perspective on ectomycorrhizal symbiosis. New Phytologist 203, 657–666.

Grman, E., Robinson, T.M.P., Klausmeier, C.A., 2012. Ecological specialization and trade affect the outcome of negotiations in mutualism. American Naturalist 179, 567–581.

Henn, M.R., Chapela, I.H., 2001. Ecophysiology of ^{13}C and ^{15}N isotope fractionation in forest fungi and the roots of the saprotrophic-mycorrhizal divide. Oecologia 128, 480–487.

Hobbie, E.A., Horton, T.R., 2007. Evidence that saprotrophic fungi mobilise carbon and mycorrhizal fungi mobilise nitrogen during litter decomposition. New Phytologist 173, 447–449.

Hobbie, E.A., Ouimette, A.P., Schuur, E.A.G., Kierstead, D., Trappe, J.M., Bendiksen, K., Ohenoja, E., 2013. Radiocarbon evidence for the mining of organic nitrogen from soil by mycorrhizal fungi. Biogeochemistry 114, 381–389.

Hodge, A., 2014. Interactions between arbuscular mycorrhizal fungi and organic material substrates. Advanced in Applied Microbiology 89, 47–99.

Huggins, J.A., Talbot, J., Gardes, M., Kennedy, P.G., 2014. Unlocking environmental keys to host specificity: differential tolerance of acidity and nitrate by *Alnus*-associated ectomycorrhizal fungi. Fungal Ecology 12, 52–61.

Johnson, N.C., 2010. Resource stoichiometry elucidates the structure and function of arbuscular mycorrhizas across scales. New Phytologist 185, 631–647.

Keiluweit, M., Bougoure, J.J., Nico, P.S., Pett-Ridge, J., Weber, P.K., Kleber, M., 2015. Mineral protection of soil carbon counteracted by root exudates. Nature Climate Change 5, 588–595.

Kohler, A., Kuo, A., Nagy, L.G., Morin, E., Barry, K.W., Buscot, F., Cambäck, B., Choi, C., Clum, A., Colpaert, J., et al., 2015. Convergent losses of decay mechanisms and rapid turnover of symbiosis genes in mycorrhizal mutualists. Nature Genetics 47, 410–415.

Koide, R.T., Sharda, J.N., Herr, J.R., Malcolm, G.M., 2008. Ectomycorrhizal fungi and the biotrophy – saprotrophy continuum. New Phytologist 178, 230–233.

Kuyper, T.W., Kiers, E.T., 2014. The danger of mycorrhizal traps? New Phytologist 203, 352–354.

Kuyper, T.W., 2012. Ectomycorrhiza and the open nitrogen cycle in an afrotropical rainforest. New Phytologist 195, 728–729.

Lambers, H., Raven, J.A., Shaver, G.R., Smith, S.E., 2008. Plant nutrient-acquisition strategies change with soil age. Trends in Ecology and Evolution 23, 95–103.

Langley, J.A., Chapman, S.K., Hungate, B.A., 2006. Ectomycorrhizal colonization slows root decomposition: the post-mortem fungal legacy. Ecology Letters 9, 955–959.

Le Tacon, F., Zeller, B., Plain, C., Hossann, C., Brechet, C., Martin, F., Kohler, A., Villerd, J., Robin, C., 2015. Study of nitrogen and carbon transfer from soil organic matter to *Tuber melanosporum* mycorrhizas and ascocarps using ^{15}N and ^{13}C soil labelling and whole-genome oligo-arrays. Plant and Soil 395, 351–373.

Lehmann, J., Kleber, M., 2015. The contentious nature of soil organic matter. Nature 528, 60–68.

Lindahl, B.D., Tunlid, A., 2015. Ectomycorrhizal fungi – potential organic matter decomposers, yet not saprotrophs. New Phytologist 205, 1443–1447.

Lindahl, B.D., Ihrmark, K., Boberg, J., Trumbore, S.E., Högberg, P., Stenlid, J., Finlay, R.D., 2007. Spatial separation of litter decomposition and mycorrhizal nitrogen uptake in a boreal forest. New Phytologist 173, 611–620.

Moore, J.A.M., Jiang, J., Post, W.M., Classen, A.T., 2015a. Decomposition by ectomycorrhizal fungi alters soil carbon storage in a simulation model. Ecosphere 6 (3) article 29.

Moore, J.A.M., Jiang, J., Patterson, C.M., Mayes, M.A., Wang, G., Classen, A.T., 2015b. Interactions among roots, mycorrhizas and free-living microbial communities differentially impact soil carbon processes. Journal of Ecology 103, 1442–1453.

Murphy, C.J., Baggs, E.M., Morley, N., Wall, D.P., Paterson, E., 2015. Rhizosphere priming can promote mobilisation of N-rich compounds from soil organic matter. Soil Biology and Biochemistry 81, 236–243.

Nuhn, M.E., Binder, M., Taylor, A.F.S., Halling, R.E., Hibbett, D.S., 2013. Phylogenetic overview of the Boletineae. Fungal Biology 117, 479–511.

Orwin, K.H., Kirschbaum, M.U.F., StJohn, M.G., Dickie, I.A., 2011. Organic nutrient uptake by mycorrhizal fungi enhances ecosystem carbon storage: a model-based assessment. Ecology Letters 14, 493–502.

Parker, T.C., Subke, J.-A., Wookey, P.A., 2015. Rapid carbon turnover beneath shrub and tree vegetation is associated with low soil carbon stocks at a subarctic treeline. Global Change Biology 21, 2070–2081.

Phillips, R.P., Brzostek, E., Midgley, M.G., 2013. The mycorrhizal associated nutrient economy: a new framework for predicting carbon-nutrient couplings in temperate forests. New Phytologist 199, 41–51.

Phillips, L.A., Ward, V., Jones, M.D., 2014. Ectomycorrhizal fungi contribute to soil organic matter cycling in sub-boreal forests. ISME Journal 8, 699–713.

Plett, J.M., Martin, F., 2011. Blurred boundaries: lifestyle lessons from ectomycorrhizal fungal genomes. Trends in Genetics 27, 14–22.

Pritsch, K., Garbaye, J., 2011. Enzyme secretion by ECM fungi and exploitation of mineral nutrients from soil organic matter. Annals of Forest Science 68, 25–32.

Pritsch, K., Courty, P.-E., Churin, J.-L., Cloutier-Hurteau, B., Ali, M.A., Damon, C., Duchemin, M., Egli, S., Ernst, J., Fraissinet-Tachet, L., et al., 2011. Optimized assay and storage conditions for enzyme activity profiling of ectomycorrhizae. Mycorrhiza 21, 589–600.

Read, D.J., Leake, J.R., Perez-Moreno, J., 2004. Mycorrhizal fungi as drivers of ecosystem processes in heathland and boreal forest biomes. Canadian Journal of Botany 82, 1243–1263.

Read, D.J., 1987. In support of Frank's organic nitrogen theory. Angewandte Botanik 61, 25–37.

Rineau, F., Courty, P.-E., 2011. Secreted enzymatic activities of ectomycorrhizal fungi as a case study of functional diversity and functional redundancy. Annals of Forest Science 68, 69–80.

Rineau, F., Roth, D., Shah, F., Smits, M.M., Johansson, T., Canback, B., Olsen, P.B., Persson, P., Grell, M.N., Lindquist, E., Grigoriev, I.V., Lange, L., Tunlid, A., 2012. The ectomycorrhizal fungus *Paxillus involutus* converts organic matter in plant litter using a trimmed brown-rot mechanism involving Fenton chemistry. Environmental Microbiology 14, 1477–1487.

Rineau, F., Shah, F., Smits, M.M., Persson, P., Johansson, T., Carleer, R., Troein, C., Tunlid, A., 2013. Carbon availability triggers the decomposition of plant litter and assimilation of nitrogen by an ectomycorrhizal fungus. The ISME Journal 7, 2010–2022.

Schmidt, M.W.I., Torn, M.S., Abiven, S., Dittmar, T., Guggenberger, G., Janssens, I.A., Kleber, M., Kögel-Knabner, I., Lehmann, J., Manning, D.A.C., et al., 2011. Persistence of soil organic matter as an ecosystem property. Nature 478, 49–56.

Selosse, M.-A., Roy, M., 2009. Green plants that feed on fungi: facts and questions about mixotrophy. Trends in Plant Sciences 14, 64–70.

Shah, F., Nicolás, C., Bentzer, J., Ellström, M., Smits, M.M., Rineau, F., Cambäck, B., Floudas, D., Carleer, R., Lackner, G., et al., 2016. Ectomycorrhizal fungi decompose soil organic matter using oxidative mechanisms adapted from saprotrophic ancestors. New Phytologist 209, 1705–1719.

Sinsabaugh, R.L., 2010. Phenol oxidase, peroxidase and organic matter dynamics of soil. Soil Biology and Biochemistry 42, 391–404.

Talbot, J.M., Allison, S.D., Treseder, K.K., 2008. Decomposers in disguise: mycorrhizal fungi as regulators of soil dynamics in ecosystems under global change. Functional Ecology 22, 955–963.

Talbot, J.M., Bruns, T.D., Smith, D.P., Branco, S., Glassman, S.I., Erlandson, S., Vilgalys, R., Peay, K.G., 2013. Independent roles of ectomycorrhizal and saprophytic communities in soil organic matter decomposition. Soil Biology and Biochemistry 57, 282–291.

Talbot, J.M., Martin, F., Kohler, A., Henrissat, B., Peay, K.G., 2015. Functional guild classification predicts the enzymatic role of fungi in litter and soil biogeochemistry. Soil Biology and Biochemistry 88, 441–456.

Tedersoo, L., Naadel, T., Bahram, M., Pritschm, K., Buegger, F., Leal, M., Kõljalg, U., Põldmaa, K., 2012. Enzymatic activities and stable isotope patterns of ectomycorrhizal fungi in relation to phylogeny and exploration types in an afrotropical rain forest. New Phytologist 195, 832–843.

Treseder, K.K., Torn, M.S., Masiello, C.A., 2006. An ecosystem-scale radiocarbon tracer to test use of litter carbon by ectomycorrhizal fungi. Soil Biology and Biochemistry 38, 1077–1082.

Trojanowski, J., Haider, K., Hüttermann, A., 1984. Decomposition of ^{14}C-labelled lignin, holocellulose and lignocellulose by mycorrhizal fungi. Archives of Microbiology 139, 202–206.

Tu, C., Booker, F.L., Watson, D.M., Chen, X., Rufty, T.W., Shi, W., Hu, S., 2006. Mycorrhizal mediation of plant N acquisition and residue decomposition: impact of mineral N inputs. Global Change Biology 12, 793–803.

Vaario, L.-M., Heinonsalo, J., Spetz, P., Pennanen, T., Heinonen, J., Tervahauta, A., Fritze, H., 2012. The ectomycorrhizal fungus *Tricholoma matsutake* is a facultative saprotroph in vitro. Mycorrhiza 22, 409–418.

Verbruggen, E., Jansa, J., Hammer, E.C., Rillig, M.C., 2016. Do arbuscular mycorrhizal fungi stabilize litter-derived carbon in soil? Journal of Ecology 104, 261–269.

Walker, J.K.M., Phillips, L.A., Jones, M.D., 2014. Ectomycorrhizal fungal hyphae communities vary more along a pH and nitrogen gradient than between decayed wood and mineral soil microsites. Botany 92, 453–463.

Wolfe, B.E., Tulloss, R.E., Pringle, A., 2012. The irreversible loss of a decomposition pathway marks the single origin of an ectomycorrhizal symbiosis. PLoS One 7, e39597.

Wu, T., 2011. Can ectomycorrhizal fungi circumvent the nitrogen mineralization for plant nutrition in temperate forest ecosystems? Soil Biology and Biochemistry 43, 1109–1117.

Zeller, B., Bréchet, C., Maurice, C.-P., Le Tacon, F., 2008. Saprotrophic versus symbiotic strategy during truffle ascocarp development under holm oak. A response based on ^{13}C and ^{15}N natural abundance. Annals of Forest Science 65, 607 1–10.

CHAPTER

21

Magnitude, Dynamics, and Control of the Carbon Flow to Mycorrhizas

K.J. Field[1], *S.J. Davidson*[2], *S.A. Alghamdi*[2,3], *D.D. Cameron*[2]

[1]University of Leeds, Leeds, United Kingdom; [2]University of Sheffield, Sheffield, United Kingdom; [3]King Abudul Aziz University, Jeddah, Kingdom of Saudi Arabia

21.1 INTRODUCTION

Approximately 90–95% of terrestrial plants form symbioses with mycorrhizal fungi of one type or another (Smith and Read, 2008), allocating an estimated 5–30% of host net photosynthates to their mycorrhizal partners (Staddon and Fitter, 1998). The movement of organic C from plants to symbiotic soil fungi accounts for approximately 12% of all of the land-based C cycle (Smith and Read, 2008), with a central role in landscape-scale processes such as mineral weathering (see Chapters 2 and 3). For example, mycorrhizal C is a critical factor in determining plant community structure and function within a wide variety of ecosystems on Earth. The allocation of photosythates to mycorrhizas depends upon a diverse array of drivers such as plant and fungal taxonomy, physiology, and biotic (coassociated microbes, pathogen load, and herbivory) and abiotic factors.

The overwhelming majority of plants across the land plant phylogeny associate with aseptate filamentous fungi of the Glomeromycota, forming arbuscular mycorrhizas (AMs). Other fungal taxa also form mutualistic mycorrhizal associations, including members of the higher fungal phyla, Basidiomycota and Ascomycota (Smith and Read, 2008), as well as more basal linages from the order Endogonales (Mucoromycotina) (Bidartondo et al., 2011). Each of these types of mycorrhizal associations redistribute plant-fixed C to the incipient mycorrhizal fungal community, albeit in differing quantities and via distinct physiological mechanisms.

In this chapter we will investigate the biological, evolutionary, and environmental factors that come together to define the magnitude, dynamics, and control of C fluxes between plant and fungal partners. We will address three overarching questions: (1) How does the physiology and magnitude of plant-to-fungus C flow depend on mycorrhizal functional group? (2) How does C availability (i.e. variation in CO_2 concentration and

Mycorrhizal Mediation of Soil
http://dx.doi.org/10.1016/B978-0-12-804312-7.00021-8

light intensity) influence C flux between plants and mycorrhizal fungal communities? (3) To what extent is C flow between plant and symbiotic fungal partners regulated by reciprocal nutrient exchange?

21.2 HOW DOES THE PHYSIOLOGY AND MAGNITUDE OF PLANT-TO-FUNGUS C FLOW DEPEND ON MYCORRHIZAL FUNCTIONAL GROUP?

21.2.1 Mycorrhizal Functional Groups

Mycorrhiza-forming fungi are found throughout the so-called higher and lower fungal lineages and are often defined in terms of their central physiological, morphological, and functional characteristics. The simplest delineation made is between the endomycorrhizas, those that penetrate the host plant cell wall, and the ectomycorrhizas, those fungi that grow between host cells in the apoplastic space. Again, these functional groups contain phylogenetically distinct fungal taxa.

These different functional traits are inherently underpinned by different physiological mechanisms of bidirectional resource transfer. In this section we will review the evidence underlying the mechanisms of C transfer across spatial scales, from the cell through to the underpinning molecular processes facilitating C transport and partitioning. For an overview of C allocation to different mycorrhizal functional types, see Table 24.1 (Chapter 24).

21.2.2 Arbuscular Mycorrhizas

The arbuscular mycorrhizas (AMs) are widely considered to be mutualistic in nature, with the fungal partner extracting and assimilating mineral nutrients and exchanging them with their plant host for photosynthates. AM fungi are able to access soil minerals from far beyond the nutrient depletion zones (Smith and Read, 2008) of plant roots and from soil pores too small to be accessible by root hairs. As such, AM fungi increase their host plant's ability to forage for and assimilate soil nutrients. In exchange for this, AM fungi exploit the photosynthetic capabilities of their plant hosts by assimilating plant-fixed organic C compounds and using them for their own growth and proliferation (Smith and Read, 2008).

Plants and mycorrhizal fungi interact across several interfacial tissues, the identity of which appears to be highly taxon specific. In endomycorrhizas C transfer is thought to be largely localized to specialized fungal structures with a large surface area:volume ratio such as the highly branched arbuscules or tightly wound fungal coils; termed the Arum and Paris types, respectively (Smith and Read, 2008). It is undoubtedly true that these structures play a central role in facilitating nutrient transfer from mycorrhizal fungi to plant hosts, with P transporters having been localized to the periarbuscular membrane (Javot et al., 2007). However, the role of arbuscular fungal membrane-bound transporters in mediating the flux of C from plant to fungus remains unclear (Fitter, 2006; Parniske, 2008). This is largely because the mechanistic basis for interspecific C transport in mycorrhizal associations remains equivocal (see Section 21.3).

The current paradigm suggests that plant photosynthates are transported to the fungus across the periarbuscular space as sucrose before undergoing hydrolysis and being taken up into the fungus as monosaccharides. The transportation of the simple hexose sugars across the fungal membrane is active; therefore it is likely to involve membrane-bound transporters. Upon transfer to the fungus, the hexoses are converted into triacylglycerol lipid droplets and transported toward fungal sinks (Parniske, 2008). Although hexose transporter genes have been known to occur in the family Glomales for some time (Fitter, 2006), it is only recently that their expression in the fungal arbuscular membrane has been demonstrated (Helber et al., 2011). However, it is important to note that the expression of AM fungal monosaccharide transporter *MST2* is not restricted to the arbuscule but is expressed endemically throughout the intra- and extramatrical fungal hyphae (Helber et al., 2011). This suggests that inter- and intracellular hyphae have the capacity to contribute significantly to plant C assimilation by the fungus, as suggested by Gianinazzi-Pearson et al. (1991) and by Fitter (2006). This prediction now requires detailed experimental validation to understand the extent to which hyphal C transfer contributes to the net C flux from plant to fungus and thereon throughout the soil community.

21.2.3 Orchid Mycorrhizas and Life-Cycle–Dependent C Flux

The family Orchidaceae is the largest family of flowering plants, comprising approximately 32,000 species in approximately 800 genera with an average of 100 discoveries of new species made each year (Arditti, 1992). As well as being specious, orchids are ecologically very successful, being found in most terrestrial ecosystems from temperate grasslands to tropical rain forests (Moore, 1980; Arditti and Ghani, 2000), reflecting their extraordinary success in dispersing and adapting to a wide variety of habitats and environments. Two specialized groups have been recognized within the family: the terrestrial orchids, which germinate underground in soil and represent approximately 25% of species that are widely distributed in forest and grassland biomes, and the epiphytes, which use other plants in tropical forests for mechanical support, growing on branches within their canopies and constituting at least 70% of all known orchid species (Ackerman, 1983).

The function of mycorrhizas that form between plants of the family Orchidaceae and Basidiomycete fungi is less well understood than their AM counterparts. In large part, this significant knowledge gap is a function of bidirectional fluxes of C between plant and fungal partners in ecologically relevant quantities. This complexity is further compounded by the complexity of the orchid life cycle in relation to their mycorrhizal symbionts.

Orchids are critically reliant on the provision of C and nutrients for germination by mycorrhizal fungi (Leake, 1994). In most cases these are basidiomycete fungi that derive their organic C via saprotrophy. This reliance on resource provisioning by fungi is because orchids produce thousands of tiny seeds per capsule that contain no or limited energy reserves for germination and early growth in their vestigial endosperms (Arditti, 1979; Eriksson and Kainulainen, 2011; see Fig. 21.1). After initiation of the symbiosis, each embryo develops into a protocorm, a specialized organ that defines the first stage of germination (Fig. 21.1). Therefore, for most terrestrial orchid species, autotrophy occurs only after an expanded stage of heterotrophic growth (Leake and Cameron, 2010; Phillips et al., 2011).

Orchid species are unable to hydrolyze starch or cellulose; therefore they cannot directly access C saprotrophically as has been historically suggested (Leake, 2005). As a result, orchids are entirely dependent upon compatible fungal partners to supply the essential C and

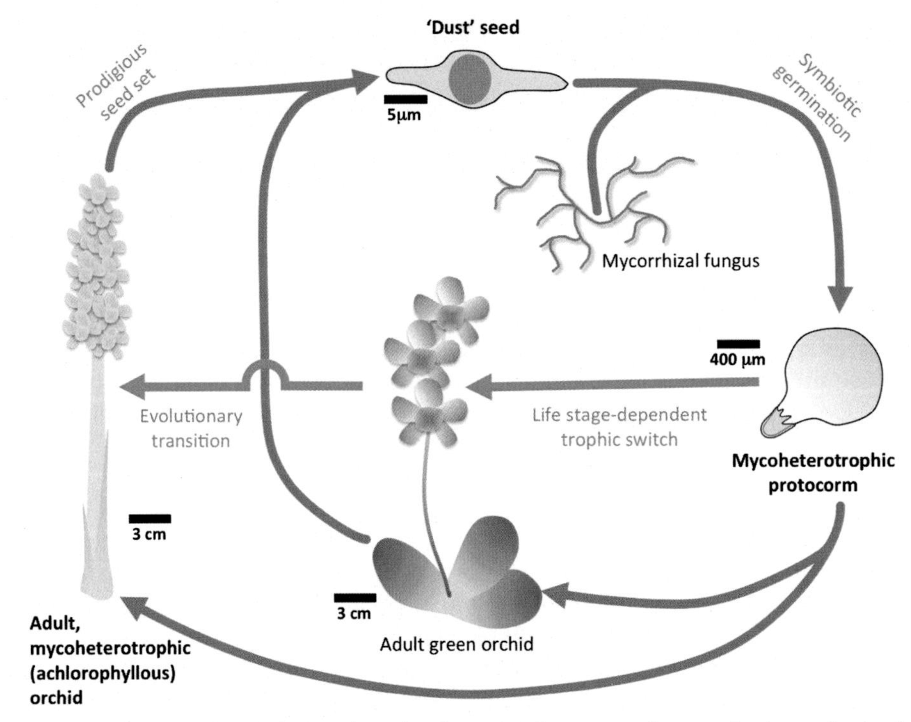

FIGURE 21.1 Lifecycle of orchids detailing the dynamic changes in degree of autotrophy in the different trophic strategies at both a life history and an evolutionary level.

nutrients required for successful establishment and germination in nature (Leake, 1994; Fig. 21.1). Thus orchid plants are described as mycoheterotrophic for part, or all, of their lives. The initially mycoheterotrophic nutrition in orchid seedlings normally changes to autotrophy in adult green orchid (life-stage–dependent trophic switch; Fig. 21.1); therefore any reliance on the association with fungi for C subsidies is reduced or lost (Cameron et al., 2008).

In terrestrial orchids mycorrhizal symbiosis is essential for the uptake of mineral nutrients such as P and N (Alexander et al., 1984; Cameron et al., 2006, 2007). Although transport of N (Cameron et al., 2006) and P (Cameron et al., 2007; Alexander et al., 1984) has been shown to occur between fungus and plant, our knowledge base is limited to a just few species, most notably *Goodyera repens*. In this species it has been shown that C continues to flow from fungus to the adult green plant while the adult simultaneously allocates photosynthates to the fungal partner (Cameron et al., 2006). The net polarity of this flux is from plant to fungus, with approximately 10 times more C being transferred to the fungus than is received from it (Cameron et al., 2008). The ultimate consequence of this complicated relationship is that orchids operate a mutualistic symbiosis based on the dynamic exchange of C for nutrients as adults, in line with our classical view of mycorrhizas. It is important to note that all other green orchids that associate with saprotrophic basidiomycete fungi that have been investigated to date have been shown to engage in an analogous mode of symbiosis (Lim et al., 2014), although this trophic strategy has not been investigated in any epiphytic species at the physiological level.

Further complexity arises in interpreting the dynamics of C flow to and from orchids via their mycorrhizal fungal symbionts because of cryptic adult heterotrophy (Cameron et al., 2009;

Gebauer and Meyer, 2003). While still green these plants are associated with fungi that can access a large and consistent pool of plant C stored within the mycorrhizal network. This may have underpinned the reliance of some green orchids associated with ectomycorrhizal (EcM)-forming basidomycetes on significant amounts of fungal C in the green adult life stage—a strategy referred to as *mixotrophy* (Selosse and Roy, 2009). This evidence comes from stable isotope studies that have demonstrated that some green orchids derive a significant amount of C from their mycorrhizal fungi rather than being fully reliant on autotrophy (Cameron et al., 2009; Gebauer and Meyer, 2003; Selosse and Roy, 2009; Merckx et al., 2010). A second evolutionary transition underpins the ultimate abandonment of autotrophy by some orchid species, becoming entirely reliant on mycoheterotrophic C acquisition throughout their lives (Fig. 21.1).

In contrast, the functional importance of mycorrhiza through the life cycles of epiphytic orchids are less well known and may be distinct from that in terrestrial orchids, especially during germination because epiphytic orchids develop in the canopy; thus they are exposed to relatively high light levels from an early stage of development (Leake and Cameron, 2010). Recent studies of the fungal associates of more than 150 tropical orchid species (Martos et al., 2012) have revealed a major ecological barrier between above- and belowground mycorrhizal fungal networks, with only approximately 10% of the fungal taxa partnering epiphytic and terrestrial orchids, raising the hypothesis that there are important functional differences in mycorrhiza between the two kinds of orchid related to their distinct habitats (Leake and Cameron, 2010). Light intensity during seedling germination may play a role in this because protocorms formed by epiphytic orchids are usually green; thus the extent of their reliance on heterotrophic C gain, even as seedlings, is equivocal. The optimal light requirement for growth varies between epiphytic and terrestrial orchid species (Heifetz et al., 2000). In some terrestrial orchid species such as *G. repens* even low levels of light inhibit germination, whereas in other species such as *Dactylorhiza purpurella* germination appears to be insensitive to light. Some studies have found that light can stimulate germination and early seedling establishment in the absence of exogenous C (Downie, 1941; Smith, 1973; Harvais, 1974).

21.2.4 Other Endomycorrhizas

Ericoid mycorrhizas (ErM) are another class of endomycorrhizas formed between fungi and the roots of hosts from the order Ericales. ErM fungi, recruiting from the ascomycetes and hyphomycete fungi, are the dominant fungal symbionts in heathland ecosystems and form strong symbioses with ericaceous dwarf shrubs such as *Empetrum*, *Vaccinium*, and *Calluna* (Read and Perez-Moreno, 2003). ErM fungi penetrate the plant through root hairs before forming endophytic structures (Harley and Smith, 1983; Schmidt and Stewart, 1999; Read, 1996). In contrast to ectomycorrhizal (EcM) fungi, the ErM fungi that form ErMs have evolved to produce many C-compound degrading enzymes with the capability of decomposing the complex organic material that makes up their host species (Smith and Read, 2008). These mycorrhizas again serve to free up C that would otherwise be locked within the rhizosphere in recalcitrant ericaceous leaf litter.

21.2.5 Ectomycorrhizas

In contrast to the endomycorrhiza, the key diagnostic feature of the ectomycorrhizas is the absence of intracellular fungal structures; instead, the bidirectional C-for-nutrient transfer occurs between intact plant cells and fungal hyphae within the apoplastic space—the Hartig net (Bücking et al., 2007). EcM-forming fungi fall within the families of Basidomycota and Ascomycota and colonize a diverse range of trees and woody shrubs. In addition, the same fungi are known to form endosymbiotic associations with some species of orchids, a small number of herbaceous plants of the family Ericaceae (e.g. *Monotropa hypopitys*; Leake, 2004), and a single species of mycoheterotrophic liverwort (*Aneura mirabilis*) (Bidartondo et al., 2003). Physiologically, ectomycorrhizas are distinct from their endomycorrhizal counterparts in that many species have retained the ability to hydrolyze complex carbohydrates and hence are not obligate biotrophs; however, obligate biotrophy still exists in a subset of species within this functional group of mycorrhiza. Ectomycorrhizas form with evergreen and deciduous species; thus the dynamics of their C nutrition is likely a function, at least in part, of their host's life history. Moreover, the C pool from which they are supplied is far greater in magnitude (on an individual host basis) than that of AMs, which form primarily in short-lived, herbaceous species, although there are exceptions to this generalization.

Several EcM-forming species of fungi, typical of arctic and boreal ecosystems, have been shown to produce various oxidative and hydrolytic enzymes capable of decomposing soil organic matter and freeing C within the rhizosphere (Talbot et al., 2008; Phillips et al., 2014; Brzostek et al., 2015). These fungi play a pivotal role in the circulation of C within their native ecosystems; for example, the boreal forest biome covers 11% of the Earth's land surface and contains approximately 16% of the C stocks sequestered within its soils (Clemmensen et al., 2015). The boreal zone hosts a distinctive plant community dominated by species forming ectomycorrhizas with basidiomycetes and ascomycetes (Smith and Read, 2008) along with ErMs formed by a range of higher fungi. More than half of the biological activity within boreal soils is driven by the movement of recent photosynthates to roots and their associated microbial communities (Högberg et al., 2001; Lindahl et al., 2007), resulting from plant demand for N (Read and Perez-Moreno, 2003).

The dominance of EcM fungi in well-decomposed litter supports the idea that mycorrhizas play a significant role in mobilizing N from organic matter in boreal soils (Lindahl et al., 2007). Climate change is driving an increase in productivity in the latitudes at which these ecosystems occur; in turn, this is resulting in a significant increase in soil C sequestration via increased litter fall (Todd-Brown et al., 2012). *Cortinarius*, a genus of EcM-forming fungus that commonly colonizes deciduous shrubs, retains the ability to use peroxidase enzymes, particularly when inorganic N is at a low concentration in the environment. The mechanism by which N is liberated from organic complexes for fungal uptake results in stimulated decomposition of organic matter in which C, which would otherwise be locked up within litter, becomes available for decomposition by the microbial community (Talbot et al., 2008). The extent to which this soil C contributes to the C budget of the mycorrhizal fungus itself requires further research.

21.3 HOW DOES C AVAILABILITY (CO_2 AND SHADING) INFLUENCE THE CARBON FLUX BETWEEN PLANT AND MYCORRHIZAL FUNGAL COMMUNITIES?

The movement and redistribution of C characterizes mycorrhizal networks and strongly influences (and is influenced by) above- and belowground community biodiversity and function. The amounts and distribution of photosynthates throughout plant communities are affected by a wide variety of interacting abiotic and biotic factors (Kaiser et al., 2015). These generate a cascade of ecologically relevant outcomes, some of which we will explore here.

21.3.1 Abiotic Influences on C Flux Between Plants and Mycorrhizal Fungi

Human activities are driving a rapid and unprecedented increase in atmospheric CO_2 concentrations, with the hourly concentration measured at Mauna Loa, Hawaii hitting the symbolic milestone of 400 ppm in May 2013 (reported by Richard Monastersky in *Nature*, May 2, 2013). Such levels were last reached in the Pliocene (~3 million years ago) when mean Earth temperatures are thought to have been approximately 2–3°C higher than today (Martínez-Botí et al., 2015). The most recent climate predictions suggest that unless drastic measures are taken to curb CO_2 emissions, concentrations of this potent greenhouse gas will continue to sharply increase, reaching up to 800 ppm by 2070 (Pachauri et al., 2014).

It has been known for a long time from studies using various experimental and/or computer modeling techniques that when plants grow under an elevated CO_2 atmosphere their growth and productivity usually increases (Ainsworth and Long, 2005). A large proportion of the additional photosynthates produced are believed to be transported belowground (Rillig and Allen, 1999). Depending on plant life history, this C may be stored as starch in specialized structures such as tubers, bulbs, or rhizomes, or it may be exuded directly into the soil as various low-molecular–weight compounds. Given that mycorrhizas form the critical interface between most plants and the soil, it is logical that they are directly affected by movement of increased amounts of C to the roots. It is important to note that mycorrhizal fungi occupy environments that are inherently of above-ambient CO_2 concentration (i.e. inside of the root and in the soil; Fitter et al., 2000); however, the movement of additional photosynthates belowground still translates into greater C allocation to, and assimilation by, mycorrhizal fungi (Drigo et al., 2010). It is well established that even moderate increases in atmospheric CO_2 concentrations lead to a mean 47% increase in mycorrhizal growth (36% increase in root colonization) across various land plants (see meta-analysis by Treseder, 2004). The clear trends for increased colonization and growth of AM and EcM fungi (Staddon and Fitter, 1998; Alberton et al., 2005) under elevated atmospheric CO_2 are likely facilitated by increased allocation of photosynthates by host plants, particularly in the case of obligately biotrophic AMs. However, it is also possible that these general observations have a more indirect source, being driven instead by faster plant growth (Staddon and Fitter, 1998) and/or substantial changes in the composition of mycorrhizal fungal communities themselves, with the ratio of Glomeraceae to Gigasporaceae having been shown to increase significantly under elevated CO_2 conditions (Cotton et al., 2015).

In addition to effects on mycorrhizal abundance and community composition, and contrary to the findings of Staddon and Fitter (1998), atmospheric CO_2 concentration has been shown to have dramatic effects on mycorrhizal functioning in terms of C-for-nutrient exchange between symbionts. The flux of C from plant to fungus has been measured directly using isotope tracers and measured indirectly using plant and fungal biomass as proxies for plant C investments. Such studies have been conducted across a range of plant clades across the land plant phylogeny and at different atmospheric CO_2 concentrations, ranging from 350 to 1500 ppm (see Treseder, 2004). The former approach, in which transfer of plant C to fungal symbionts is measured directly using isotope tracers (as in Field et al., 2012, 2015a, 2016; Zhang et al., 2015), has provided us with powerful new insights into how atmospheric CO_2 concentration affects the physiological functioning of the mycorrhizal symbiosis and the stoichiometry of the exchange between partners. This level of precision is simply not possible when using proxies such as plant and/or fungal biomass or hyphal densities. Isotope tracer studies have established that mycorrhizal fungi appear to benefit from increased allocation and assimilation of recent photosynthates when the plants are grown under a high-CO_2 atmosphere (Drigo et al., 2010; Field et al., 2012, 2015a, 2016), although it also appears that temperature may play a significant role in this (Gavito et al., 2003). It appears much lower when plants are exposed to a subambient CO_2 atmosphere (Zhang et al., 2015), although there are very few studies examining this. These observations are in accordance with the hypothesis that with increased CO_2 availability plants increase productivity and so are able to proportionally transfer more photosynthates to their fungal partners.

Root organ culture-based studies have shown that some fungi appear to respond to increases in plant C supply by increasing nutrient supply (in this case N and P) in a "reciprocal rewards" fashion (Kiers et al., 2011; Fellbaum et al., 2012; Hammer et al., 2011). This suggests that by adjusting C flow to the fungus, plants are able to regulate the symbiosis and thus ensure evolutionary stability (Kiers et al., 2011). However, this approach to quantifying C-for-nutrient exchange between mycorrhizal symbionts is not without constraints. The transformed root organ cultures that make up the plant part of the symbiosis are unable to synthesize their own C supply through photosynthesis. As such, the source element of the source–sink relationship between plants and fungi is removed. In addition, the fungus meanwhile faces little competition for nutrients; thus regulation of C-for-nutrient exchange between symbionts is very likely to deviate from that occurring in natural conditions. Finally, in root organ culture systems it is very difficult to distinguish whether the fungi within the root cells has released nutrient into the plant tissue or whether it is merely detected in roots tissues but is actually retained within fungal structures (Walder and van der Heijden, 2015). As such it is critical that careful consideration is given to interpreting the results from experiments in which these systems are used.

Shading of plants in experiments has been used as an alternative way in which to limit the amount of photosynthate generated by the plant (e.g. Fellbaum et al., 2014). Although such experimental designs do not separate roots from shoots and thereby allow plants some control over their C fixation, shaded plants are invariably stunted in growth (see supplementary plant biomass data in Fellbaum et al., 2014) with further inherent downstream effects on plant physiology (e.g. reduced respiration, transpiration, lower demand for nutrient, etc.). As such, it is imperative that appropriate controls are incorporated into the experimental design. Given that diffusion and mass flow of nutrients along hydrological gradients are the principal mechanisms for replenishing nutrient depletion zones within the soil generated by plant roots and mycorrhizal fungi (Hepworth et al., 2015), the reduced transpirational pull produced by shaded plants is less likely to influence this. By limiting fungal access to nutrients, shading studies artificially shift the achievable stoichiometry of C-for-nutrient exchange. From this collective evidence it is clear that data from experiments using shading as a treatment should be interpreted very carefully, taking into account the full ramifications of the shading treatment on plant mass flow and nutrient demand and how these factors may influence movement of C from plant to fungus and the surrounding rhizosphere.

In several studies in which atmospheric CO_2 concentration was directly manipulated to increase or decrease its availability to the plant, nutrient uptake is not always affected in the linear fashion described by root organ cultures or shading experiments (e.g. Kiers et al., 2011; Fellbaum et al., 2012, 2014; Hammer et al., 2011). Several land plants with symbiotic fungal partners have been shown to receive greater quantities of N and/or P under a lower atmospheric CO_2 concentration than under a higher one (Field et al., 2012, 2015a; 2016; Zhang et al., 2015). Although the underpinning physiological mechanisms for these observations remain unclear, an intriguing hypothesis is that the increased availability of organic C to symbiotic soil fungi serves to drive increased growth and proliferation of fungal hyphae. This is likely to lead to increased fungal demand for soil nutrients (Dumbrell et al., 2011) and thus result in reduced nutrient transfer to plant partners.

In all of the examples described here, plants and fungi are limited in their choice of symbiotic partners, with the majority having been placed together under controlled conditions. Future studies should look to make use of naturally occurring plant and fungal communities, complete with natural biodiversity and intraspecific genetic variation of plants and fungi, and consider the findings within a plant and soil community context.

21.3.2 Temporal Dynamics in C Allocation

In their now classic experiment, Högberg et al. (2001) demonstrated strong seasonality in soil respiration in northern coniferous pine forests, unsurprisingly with a peak in the summer months before declining in the autumn. Using an elegant girdling experiment restricting the flow of current assimilate, Högberg et al. (2001) were able to demonstrate that over half of this respiration is due to current assimilate feeding the EcM community. This coupled to the seasonality of respiratory fluxes strongly suggests strong seasonal regulation of C flow into EcM communities. The C required to meet metabolic C demands of the fungi over winter is presumably met by remobilization fungal C reserves or potentially supplemented by heterotrophic C assimilation in the winter months (see Section 21.2.5 for further details).

Although recent research has focused on the assumption that mycorrhizas are supplied exclusively with C from current photoassimilate, there is potential for host plants to support their mycorrhizas with stored C. However, this has been largely ignored, even in trees with large C stores and in tuberous and bulb-forming plants, which have significant capacity for C storage (Merryweather and Fitter, 1995). The role of stored C represents a significant and overlooked component of mycorrhizal C budgets, potentially leading to substantial underestimation of the C cost of mycorrhizas to their host plants. When considered at the global scale, an estimated 10–30% of primary productivity flows to mycorrhizal fungi (Taylor et al., 2009). This represents a substantial fraction of the planetary C budget. Moreover, the 10–30% figure could even be an underestimate because (1) such estimates of C allocation are derived on the assumption that only current assimilate is transferred to mycorrhizal fungi and (2) there is negligible transfer from C pools stored in plant tissues, both of which are equivocal.

21.3.3 Community Diversity

One of the distinctive features of mycorrhizal fungi is their ability to form vast underground networks of hyphae, colonizing multiple plant partners and sometimes fusing to link individual plants of multiple species together. By studying the C flux through these extensive common mycorrhizal networks (CMNs) we may begin to understand the wider effect of mycorrhizas on ecosystem structure and function.

Using the stable isotope ^{13}C as a tracer, Fitter et al. (1998) demonstrated that C may move between interconnected plants within an AM CMN, although in this case, the ^{13}C tracer remained within the plant root tissues. Likewise, Simard et al. (1997) demonstrated interplant C transfer in EcM tree seedlings. These findings suggest that a mechanism may be present by which mycorrhiza assimilate and mobilize recent photosynthates from roots containing younger fungal structures such as arbuscules and hyphae to roots containing older fungal structures, including storage bodies such as vesicles. However, given that plant-fixed C compounds would have to remain within fungal structures rather than having any measurable effect on plant fitness in this scenario, it is difficult to see how such interplant transfer of C would have any appreciable effect on wider plant community structure or function (Fitter et al., 1998). However, more recently it has been found that there is no detectable fungal-mediated transfer of C between neighboring plants at all (Zabinski et al., 2002). This supports the ecologically important hypothesis that mycorrhiza-enhanced plant growth is primarily the result of increased access to nutrients.

Different groups of plants that may coexist within a community have important distinctions in life histories. These are likely to have dramatic effects on C flow through CMNs, particularly where the plant is either fully or partially mycoheterotrophic in at least part of its life cycle (see Section 21.1). In cases in which young seedlings or gametophytes of nonseed plants are mycoheterotrophic, it has been shown that the symbiosis operates on a "take now, pay later" basis, in which fungal C invested in the plant during its mycoheterotrophic stage is repaid upon commencement of photosynthesis (Cameron et al., 2006; Field et al., 2015b; see Section 21.1). In such systems the flow of C from plants to the rhizosphere may become limited as a result of the high C sink strength of the mycoheterotrophic plants. Of course C sink strength varies not only between different plant species within communities but also between species and types of mycorrhizal fungi.

Soil microorganisms, other than mycorrhizal fungi, also have a major effect on C flow on multiple trophic levels (Cameron et al., 2013). A food web analysis of the movement of C via mycorrhizas and associated bacterial communities using ^{13}C analysis of fatty acids suggests that C flux through fungal and bacterial channels is essential to stabilizing tropical forest food webs (Pollierer et al., 2012). A more recent study has also demonstrated that AMs are an important conduit for the delivery of recent photoassimilate to soil bacteria (Kaiser et al., 2015), Various studies have shown that microarthropods such as mites and collembola graze on mycorrhizal fungal hyphae, disrupting C flow from plants into the soil (Johnson et al., 2005). However, studies also suggest that such grazing is low in frequency and intensity, with the microarthropods preferentially grazing on nonmycorrhizal fungal hyphae where possible (Klironomos and Kendrick, 1996; Klironomos and Ursic, 1998; Klironomos and Moutoglis, 1999). Together, these lines of research demonstrate how important a holistic view of the whole soil microbial community is essential to fully understand C fluxes between plants and soil on an ecosystem scale.

21.4 TO WHAT EXTENT IS THE CARBON FLOW BETWEEN PLANT AND SYMBIOTIC FUNGAL PARTNERS REGULATED BY RECIPROCAL NUTRIENT EXCHANGE?

21.4.1 Evolutionary History of Plant-to-Fungal C Flow

There are multiple lines of strong evidence that mycorrhizal symbiosis dates back more than 450 million years to around the time plants first moved onto the land in the early Devonian (Field et al., 2015c). Given their mutualistic nature, mycorrhiza-forming fungi likely played a key role in the colonization of the terrestrial environment by plants, providing the earliest, rootless plants with mineral nutrients extracted directly from primordial mineral soils in return for organic C supplied by the plant (Pirozynski and Malloch, 1975). This hypothesis is supported by numerous lines of evidence, with some of the most recent and most compelling being the conservation of the suite of genes (encoding the *SYM* signaling pathway) thought to be required to initiate and establish mycorrhizal symbioses throughout the land plant phylogeny (Delaux et al., 2013). This is strongly indicative of vertical inheritance from the earliest land plants.

The role of mycorrhizas in land plant terrestrialization is unlikely to be exclusive to AM fungi of the Glomeromycota. Multiple lines of evidence now challenge the long-held AM-focused paradigm, pointing instead toward a role for fungi of the partially saprotrophic Mucoromycotina lineage (Bidartondo et al., 2011). Quantification of C-for-nutrient exchange between AM and Mucoromycotina fungi and their nonvascular plant hosts has revealed that both symbioses are nutritional mutualisms (Field et al., 2012, 2015a, 2016). Together with molecular (Bidartondo et al., 2011; Rimington et al., 2015) and recent fossil (Strullu-Derrien et al., 2014) evidence, this lends weight to the idea that the earliest plant-fungal mutualisms were likely more transient in nature and of greater diversity than the current paradigm dictates (Field et al., 2015c). There is a distinct likelihood that the earliest plants were not limited in their choice of fungal symbionts; rather, there were various "symbiotic options" available to them either in exclusive single-fungus partnerships or in more catholic, multifungal partner situations (Field et al., 2015c, 2016; Rimington et al., 2015). Indeed, several early-diverging lineages of land plants today regularly associate with multiple fungi from various clades (Desirò et al., 2015; Rimington et al., 2015) in apparently mutualistic relationships, although the nutrient contribution and C cost of each fungal partner relative to one other and to single fungal associations remains to be fully elucidated.

21.4.2 Why Does C Flow Between Plants and Fungi Need to Be Regulated?

The evolution of symbioses, "the living together of unlike organisms," has been central to the evolution of complex life. Mutualisms, such as mycorrhizal symbiosis, are characterized by the reciprocal exchange of resources between partners to their mutual fitness benefit, thus ensuring their evolutionary stability (Raven and Allen, 2003). Inherently then, mutualistic symbioses are subject to strong evolutionary pressures to cheat. In an association, if one partner can provide fewer resources to the other while simultaneously maintaining the same levels of resource capture from their host, then the cheater will derive fitness benefits that will be selected for. One mechanism in which the host can prevent cheating is through the imposition of sanctions, withdrawing essential resources from the symbiont as a consequence of nonpayment (Kiers et al., 2011). Despite such complex mechanisms to apparently maintain symmetry in the reciprocal exchange of resources between symbiotic partners, the mycorrhizal symbiosis is rife with asymmetry when considered within a nutritional context alone (Leake and Cameron, 2010; Cameron et al., 2013). Walder et al. (2012) provide an elegant example of such asymmetry in C-for-nutrient exchange in a CMN. Specifically, they showed that flax plants (*Linum usitatissimum*) contribute little C to the CMN but gained up to 94% of the available N and P whereas cooccurring *Sorghum bicolor* plants invested substantially more C to the same CMN but received little in the way of nutritional benefit (Walder et al., 2012). Therefore understanding the regulation and dynamics of C-for-nutrient exchange is essential to understand the continued evolutionary success of the mycorrhizal habit and to highlight the extent of non-nutritional benefits underpinning stabilizing selection.

21.4.3 C-for-Nutrient Exchange: Initial Evidence Through to Current Hypotheses

Given their ancient evolutionary origins, it has been postulated that, for mutualistic interactions between plants and fungi to persist, they require stabilization in the form of "reciprocal rewards" (Kiers et al., 2011). According to biological market theory, the C-for-nutrient exchange between plants and fungi is regulated through these reciprocal rewards, in which a generous partner (one who supplies more nutritional benefit) is rewarded in some way by the other partner. In mycorrhizas, this hypothesis for stabilization must at some level involve detection by each party of variation in the amount of benefit received and so allow individuals to withhold nutrients until maximum reward for their investment is achieved (and vice versa; Kiers and van der Heijden, 2006; Walder and van der Heijden, 2015). This idea has been explored using various different methods, most of which involve some form of tracking of the movement of nutrients between partners. Such investigations have shown that plants may indeed favor generous fungal partners with a greater allocation of recent photosynthate. Experiments using transformed root organ cultures (Kiers et al., 2011; Fellbaum et al., 2012) and some whole plants (Fellbaum et al., 2014; Bever et al., 2009; Hammer et al., 2011) have shown plant hosts to favor fungi supplying the greatest complement of nutrients by supplying them with a greater share of the available photosynthates.

Much of the experimental evidence supporting a biological markets-style reciprocal-rewards–based stabilization of mycorrhizal symbiosis is drawn from single plants with limited options for fungal partners. In reality, mycorrhizas form large and complex mycelial networks between multiple plant partners of often different species (van der Heijden et al., 1998). In such complex networks in which different plant and fungal species have different source and sink strengths, it is easy to envisage that some species allocate relatively little to the network whereas others appear to invest more heavily and play a more central role in maintaining the CMN (Walder and van der Heijden, 2015). This is exemplified by Merrild et al. (2013), who demonstrate that larger plants (with presumably higher sink strengths and greater C resources for provisioning of mycorrhiza) receive more P from a CMN than coassociated seedlings. This dynamic likely makes it difficult for plants and fungi to actively make resource allocation decisions based on the expedience of their partners. The diverse nature of partners within a CMN adds further complications and difficulty in regulation of resource exchange in terms of functionality of each symbiont. As discussed previously, for

C-for-nutrient exchange to be tightly coupled, the location of transfer should be in close proximity. However, this is not necessarily the case because C transporters in *Arum*-type arbuscular mycorrhizal fungi (AMF) are located throughout the hyphae and not just in the arbuscules as AM-specific P transporters are (Helber et al., 2011).

Although there is little doubt that reciprocal exchange of photosynthates for nutrients can lend evolutionary stability to some plant-mycorrhizal symbioses, the model does not account for the persistence of all mycorrhizas. To be considered mutualistic by the de Bary (1879) definition, the reciprocal exchange of C for nutrients does not necessarily have to be symmetrical in terms of fitness benefit to either party. For example, Lendenmann et al. (2011) show that different AM fungi have different demands on plant C and in turn provide differing nutritional benefits. Specifically, *Gigaspora margarita* has been shown to have a high C demand but negligible benefits in terms of plant P uptake (Lendenmann et al., 2011). There are some further notable examples in which regulation of C-for-nutrient exchange between plants and mycorrhizal fungi simply cannot be regulated in a linear, reciprocal rewarding type strategy; orchids and their mycorrhizal partners provide an elegant reminder of this (Leake, 1994; see also Section 21.1). Therefore it is critical that we take a holistic view of the lifetime fitness benefit accrued by each partner.

It is also important to consider that stabilizing selection does not occur as a function of instantaneous bidirectional fluxes of resources between symbiotic partners but rather it is a function of the net effect of these processes on the fitness of each partner. Therefore it is necessary to understand the contribution of the non-nutritional benefits of (and to) each partner in the mycorrhizal symbiosis when accounting for apparent asymmetries in the magnitude of bidirectional resource flux, which appear difficult to reconcile at first glance. For example, the enhanced defense capacity of a mycorrhizal host plant and the role of the root as a refuge for the mycorrhizal fungus under unfavorable environmental conditions are important, non-nutritional factors contributing to the net fitness of each symbiont. These non-nutritional factors are often overlooked in a C-for-nutrient centric model such as latently applied in the biological market hypothesis.

A further non-nutritional source of evolutionary stability may lie in interspecific compatibility between plant and fungal partner. Many lycopods and ferns have a chlorophyll-free, subterranean gametophyte phase of their life cycle. Gametophytes of these species are fully reliant on symbiotic fungi for their C and nutrient uptake and can persist for several years (Winther and Friedman, 2007) until such a time that the sporophyte emerges aboveground and commences autotrophy. Recent research has shown that an Ophioglossoid fern becomes heterotrophic upon trophophore emergence and subsequently pays back the initial C investment made by the fungus in the mycoheterotrophic gametophyte (Field et al., 2015b). This "take now, pay later" mode of symbiosis requires us to estimate the fitness consequences of symbiosis for each partner over their entire lifespan to pinpoint the location of the symbiosis within the spectrum of mutualistic interactions. In these cases, and likely in the cases of many orchids in which such take now, pay later modes of C nutrition have been shown to be used (Cameron et al., 2008), intergenerational fungal fidelity and specificity are likely important for the evolutionary stability of the symbiosis (Leake et al., 2008; Field et al., 2015b, 2015c).

It is clear that coregulation of C-for-nutrient exchange between symbiotic partners represents one of the mechanisms stabilizing the mycorrhizal mutualism. However, any benefits conferred to the plant by the AM symbiosis are very much dependent on the context in which they are observed (Johnson et al., 1997). For example, under ample soil nutrient conditions, the C drain of mycorrhizal fungi can represent a fitness cost and over evolutionary timescales can lead to the mycorrhizal trait being selected against. Key evidence comes from the fact that the mycorrhizal symbiosis is not always beneficial in terms of enhancing plant growth and/or fitness, and in extreme situations some plants have abandoned it altogether (e.g. the Cruciferae) although they belong to wider clades with mycorrhizal members and arise from presumably mycorrhizal ancestors (Smith and Read, 2008). Several studies have shown that AMs in particular can actually cause retarded plant growth in some plant species (Hoeksema et al., 2010; Klironomos, 2003) while simultaneously promoting growth in heterospecifics. An elegant example of this is the AM fungus *Glomus macrocarpum*. *G. macrocarpum* is widely distributed throughout grassland, herbaceous, and woodland habitats, being especially abundant in association with *Acer* tree species (Klironomos and Kendrick, 1996). Although there are several reports of this fungal species increasing host plant biomass production and P uptake (Nandini and Tholkappian, 2012), it is also the causal agent of tobacco stunt disease (Guo et al., 1994). Although the underlying mechanisms are unknown, the large reproductive structures formed by *G. macrocarpum* and stunting of some host plants suggest that colonization by this fungus can place a high C demand on the host (Field et al., 2015b).

For example, the fitness benefit to the plant may be increased resilience to pathogens and/or environmental stress and/or a function of the interaction between host and fungal genotype; so-called G by G interactions (Rowntree et al., 2011). The examples again reinforce the case that the benefits of the symbiosis should be considered over the lifetime of each partner rather than at a single given time point.

21.5 CONCLUSIONS

The central recurring theme, in terms of understanding the magnitude, physiology, and regulation of C transfers from plant to fungal partner in the mycorrhizal symbiosis, is context dependency. It is clear that functional types of mycorrhiza differ in the magnitude of C they transfer, the processes facilitating these transfers, and the mechanisms through which those transfers are regulated. It is also clear that this variability is a function of the biotic and abiotic environment as well as the life history and evolutionary origins of the symbiosis. If there is one generality to be gleaned from our review of plant C allocation to fungus, it is that understanding the ecology of the partnership requires us to take a holistic viewpoint. This must integrate a diverse array of biotic and abiotic factors that converge to shape the dynamics of C exchanges between symbionts in the mycorrhizal symbiosis.

Acknowledgments

D.D.C. and K.J.F. are grateful to Professor Jonathan Leake, Professor Sir David Read FRS, Professor Marcel van der Heijden, and Dr Thorunn Helgason for insightful discussion. The authors gratefully acknowledge funding from the following: a BBSRC Translational Fellowship (BB/M026825/1) to K.J.F., a Royal Society University Research Fellowship to D.D.C., a doctoral training grant from the Saudi government to S.A.A., and a NERC PhD studentship to S.J.D. D.D.C. and K.J.F. thank the NERC for financial support (award numbers: NE/I024089/1, NE/F019033 and NE/N00941X/1). The authors thank the reviewers and editors for their helpful comments on the manuscript.

References

Ackerman, J.D., 1983. Specificity and mutual dependency of the orchid-euglossine bee interaction. Biological Journal of the Linnean Society 20 (3), 301–314.

Ainsworth, E.A., Long, S.P., 2005. What have we learned from 15 years of free-air CO_2 enrichment (FACE)? A meta-analytic review of the responses of photosynthesis, canopy properties and plant production to rising CO_2. New Phytologist 165 (2), 351–372.

Alberton, O., Kuyper, T.W., Gorissen, A., 2005. Taking mycocentrism seriously: mycorrhizal fungal and plant responses to elevated CO_2. New Phytologist 167 (3), 859–868.

Alexander, C., Alexander, I.J., Hadley, G., 1984. Phosphate-uptake by *Goodyera repens* in relation to mycorrhizal infection. New Phytologist 97, 401–411.

Arditti, J., Ghani, A.K.A., 2000. Tansley Review No. 110. Numerical and physical properties of orchid seeds and their biological implications. New Phytologist 145 (3), 367–421.

Arditti, J., 1979. Aspects of the Physiology of Orchids. Academic Press, London.

Arditti, J., 1992. Fundamentals of Orchid Biology. John Wiley & Sons.

Bever, J.D., Richardson, S.C., Lawrence, B.M., Holmes, J., Watson, M., 2009. Preferential allocation to beneficial symbiont with spatial structure maintains mycorrhizal mutualism. Ecology Letters 12 (1), 13–21.

Bidartondo, M.I., Bruns, T.D., Weiß, M., Sérgio, C., Read, D.J., 2003. Specialized cheating of the ectomycorrhizal symbiosis by an epiparasitic liverwort. Proceedings of the Royal Society of London B: Biological Sciences 270 (1517), 835–842.

Bidartondo, M.I., Read, D.J., Trappe, J.M., Merckx, V., Ligrone, R., Duckett, J.G., 2011. The dawn of symbiosis between plants and fungi. Biology Letters 7, 574–577.

Brzostek, E.R., Dragoni, D., Brown, Z.A., Phillips, R.P., 2015. Mycorrhizal type determines the magnitude and direction of root-induced changes in decomposition in a temperate forest. New Phytologist 206 (4), 1274–1282.

Bücking, H., Hans, R., Heyser, W., 2007. The apoplast of ectomycorrhizal roots–site of nutrient uptake and nutrient exchange between the symbiotic partners. In: The Apoplast of Higher Plants: Compartment of Storage, Transport and Reactions. Springer, Netherlands, pp. 97–108.

Cameron, D.D., Leake, J.R., Read, D.J., 2006. Mutualistic mycorrhiza in orchids: evidence from plant–fungus carbon and nitrogen transfers in the green-leaved terrestrial orchid *Goodyera repens*. New Phytologist 171 (2), 405–416.

Cameron, D.D., Johnson, I., Leake, J.R., Read, D.J., 2007. Mycorrhizal acquisition of inorganic phosphorus by the green-leaved terrestrial orchid *Goodyera repens*. Annals of Botany 99 (5), 831–834.

Cameron, D.D., Johnson, I., Read, D.J., Leake, J.R., 2008. Giving and receiving: measuring the carbon cost of mycorrhizas in the green orchid, *Goodyera repens*. New Phytologist 180 (1), 176–184.

Cameron, D.D., Preiss, K., Gebauer, G., Read, D.J., 2009. The chlorophyll-containing orchid *Goodyera repens* derives little carbon through photosynthesis. New Phytologist 183 (2), 358–364.

Cameron, D.D., Neal, A.L., van Wees, S.C., Ton, J., 2013. Mycorrhiza-induced resistance: more than the sum of its parts? Trends in Plant Science 18 (10), 539–545.

Clemmensen, K.E., Finlay, R.D., Dahlberg, A., Stenlid, J., Wardle, D.A., Lindahl, B.D., 2015. Carbon sequestration is related to mycorrhizal fungal community shifts during long-term succession in boreal forests. New Phytologist 205 (4), 1525–1536.

Cotton, T.E., Fitter, A.H., Miller, R.M., Dumbrell, A.J., Helgason, T., 2015. Fungi in the future: interannual variation and effects of atmospheric change on arbuscular mycorrhizal fungal communities. New Phytologist 205 (4), 1598–1607.

de Bary, A., 1879. Die Ercheinung der Symbiose. Karl J. Trübner, Strassburg, p. 30.

Delaux, P.M., Séjalon-Delmas, N., Bécard, G., Ané, J.M., 2013. Evolution of the plant–microbe symbiotic 'toolkit'. Trends in Plant Science 18 (6), 298–304.

Desirò, A., Faccio, A., Kaech, A., Bidartondo, M.I., Bonfante, P., 2015. *Endogone*, one of the oldest plant-associated fungi, host unique Mollicutes-related endobacteria. New Phytologist 205 (4), 1464–1472.

Downie, D.G., 1941. Notes on the germination of some British orchids. Transactions of the Botanical Society of Edinburgh 33 (2), 94–103 (Taylor & Francis Group).

Drigo, B., Pijl, A.S., Duyts, H., Kielak, A.M., Gamper, H.A., Houtekamer, M.J., Boschker, H.T., Bodelier, P.L., Whiteley, A.S., van Veen, J.A., Kowalchuk, G.A., 2010. Shifting carbon flow from roots into associated microbial communities in response to elevated atmospheric CO_2. Proceedings of the National Academy of Sciences 107 (24), 10938–10942.

Dumbrell, A.J., Ashton, P.D., Aziz, N., Feng, G., Nelson, M., Dytham, C., Fitter, A.H., Helgason, T., 2011. Distinct seasonal assemblages of arbuscular mycorrhizal fungi revealed by massively parallel pyrosequencing. New Phytologist 190 (3), 794–804.

Eriksson, O., Kainulainen, K., 2011. The evolutionary ecology of dust seeds. Perspectives in Plant Ecology, Evolution and Systematics 13 (2), 73–87.

Fellbaum, C.R., Gachomo, E.W., Beesetty, Y., Choudhari, S., Strahan, G.D., Pfeffer, P.E., Kiers, E.T., Bücking, H., 2012. Carbon availability triggers fungal nitrogen uptake and transport in arbuscular mycorrhizal symbiosis. Proceedings of the National Academy of Sciences 109 (7), 2666–2671.

Fellbaum, C.R., Mensah, J.A., Cloos, A.J., Strahan, G.E., Pfeffer, P.E., Kiers, E.T., Bücking, H., 2014. Fungal nutrient allocation in common mycorrhizal networks is regulated by the carbon source strength of individual host plants. New Phytologist 203 (2), 646–656.

Field, K.J., Cameron, D.D., Leake, J.R., Tille, S., Bidartondo, M.I., Beerling, D.J., 2012. Contrasting arbuscular mycorrhizal responses of vascular and non-vascular plants to a simulated Palaeozoic CO_2 decline. Nature Communications 3, 835.

Field, K.J., Rimington, W.R., Bidartondo, M.I., Allinson, K.E., Beerling, D.J., Cameron, D.D., Duckett, J.G., Leake, J.R., Pressel, S., 2015a. First evidence of mutualism between ancient plant lineages (Haplomitriopsida liverworts) and Mucoromycotina fungi and its response to simulated Palaeozoic changes in atmospheric CO_2. New Phytologist 205 (2), 743–756.

Field, K.J., Leake, J.R., Tille, S., Allinson, K.E., Rimington, W.R., Bidartondo, M.I., Beerling, D.J., Cameron, D.D., 2015b. From mycoheterotrophy to mutualism: mycorrhizal specificity and functioning in *Ophioglossum vulgatum* sporophytes. New Phytologist 205 (4), 1492–1502.

Field, K.J., Pressel, S., Duckett, J.G., Rimington, W.R., Bidartondo, M.I., 2015c. Symbiotic options for the conquest of land. Trends in Ecology & Evolution 30 (8), 477–486.

Field, K.J., Rimington, W.R., Bidartondo, M.I., Allinson, K.E., Beerling, D.J., Cameron, D.D., Duckett, J.G., Leake, J.R., Pressel, S., 2016. Functional analysis of liverworts in dual symbiosis with Glomeromycota and Mucoromycotina fungi under a simulated Palaeozoic CO_2 decline. The ISME Journal 10, 1514–1526.

Fitter, A.H., Graves, J.D., Watkins, N.K., Robinson, D., Scrimgeour, C., 1998. Carbon transfer between plants and its control in networks of arbuscular mycorrhizas. Functional Ecology 12 (3), 406–412.

Fitter, A.H., Heinemeyer, A., Staddon, P.L., 2000. The impact of elevated CO_2 and global climate change on arbuscular mycorrhizas: a mycocentric approach. New Phytologist 147 (1), 179–187.

Fitter, A.H., 2006. What is the link between carbon and phosphorus fluxes in arbuscular mycorrhizas? A null hypothesis for symbiotic function. New Phytologist 172, 3–6.

Gavito, M.E., Schweiger, P., Jakobsen, I., 2003. P uptake by arbuscular mycorrhizal hyphae: effect of soil temperature and atmospheric CO_2 enrichment. Global Change Biology 9, 106–116.

Gebauer, G., Meyer, M., 2003. ^{15}N and ^{13}C natural abundance of autotrophic and mycoheterotrophic orchids provides insight into nitrogen and carbon gain from fungal association. New Phytologist 160, 209–223.

Gianinazzi-Pearson, V., Smith, S.E., Gianinazzi, S., Smith, F.A., 1991. Enzymatic studies on the metabolism of vesicular-arbuscular mycorrhizas. New Phytologist 117, 61–74.

Guo, B.Z., An, Z.Q., Hendrix, J.W., 1994. A mycorrhizal pathogen (*Glomus macrocarpum* Tul. & Tul.) of tobacco: effects of long- and short-term cropping on the mycorrhizal fungal community and stunt disease. Applied Soil Ecology 1, 269–276.

Hammer, E.C., Pallon, J., Wallander, H., Olsson, P.A., 2011. Tit for tat? A mycorrhizal fungus accumulates phosphorus under low plant carbon availability. FEMS Microbiology Ecology 76 (2), 236–244.

Harley, J.L., Smith, S.E., 1983. Mycorrhizal Symbiosis. Academic Press, Inc.

Harvais, G., 1974. Notes on the biology of some native orchids of Thunder Bay, their endophytes and symbionts. Canadian Journal of Botany 52 (3), 451–460.

Heifetz, P.B., Förster, B., Osmond, C.B., Giles, L.J., Boynton, J.E., 2000. Effects of acetate on facultative autotrophy in *Chlamydomonas reinhardtii* assessed by photosynthetic measurements and stable isotope analyses. Plant Physiology 122, 1439–1445.

Helber, N., Wippel, K., Sauer, N., Schaarschmidt, S., Hause, B., Requena, N., 2011. A versatile monosaccharide transporter that operates in the arbuscular mycorrhizal fungus *Glomus* sp. Is crucial for the symbiotic relationship with plants. The Plant Cell 23, 3812–3823.

Hepworth, C., Doheny-Adams, T., Hunt, L., Cameron, D.D., Gray, J.E., 2015. Manipulating stomatal density enhances drought tolerance without deleterious effect on nutrient uptake. New Phytologist 208 (2), 336–341.

Hoeksema, J.D., Chaudhary, V., Gehring, C.A., Johnson, N.C., Karst, J., Koide, R.T., Pringle, A., Zabinski, C., Bever, J.D., Moore, J.C., Wilson, G.W., 2010. A meta-analysis of context-dependency in plant response to inoculation with mycorrhizal fungi. Ecology Letters 13 (3), 394–407.

Högberg, P., Nordgren, A., Buchmann, N., Taylor, A.F.S., Ekblad, A., Högberg, M.N., Nyberg, G., Ottosson-Löfvenius, M., Read, D.J., 2001. Large-scale forest girdling shows that current photosynthesis drives soil respiration. Nature 411, 789–792.

Javot, H., Penmetsa, R.V., Terzaghi, N., Cook, D.R., Harrison, M.J., 2007. A Medicago truncatula phosphate transporter indispensable for the arbuscular mycorrhizal symbiosis. Proceedings of the National Academy of Sciences 104 (5), 1720–1725.

Johnson, N.C., Graham, J.H., Smith, F.A., 1997. Functioning of mycorrhizal associations along the mutualism–parasitism continuum. New Phytologist 135 (4), 575–585.

Johnson, D., Krsek, M., Wellington, E.M., Stott, A.W., Cole, L., Bardgett, R.D., Read, D.J., Leake, J.R., 2005. Soil invertebrates disrupt carbon flow through fungal networks. Science 309 (5737), 1047.

Kaiser, C., Kilburn, M.R., Clode, P.L., Fuchslueger, L., Koranda, M., Cliff, J.B., Solaiman, Z.M., Murphey, D.V., 2015. Exploring the transfer of recent plant photosynthates to soil microbes: mycorrhizal pathway vs direct root exudation. New Phytologist 205, 1537–1551.

Kiers, E.T., van der Heijden, M.G., 2006. Mutualistic stability in the arbuscular mycorrhizal symbiosis: exploring hypotheses of evolutionary cooperation. Ecology 87 (7), 1627–1636.

Kiers, E.T., Duhamel, M., Beesetty, Y., Mensah, J.A., Franken, O., Verbruggen, E., Fellbaum, C.R., Kowalchuk, G.A., Hart, M.M., Bago, A., Palmer, T.M., 2011. Reciprocal rewards stabilize cooperation in the mycorrhizal symbiosis. Science 333 (6044), 880–882.

Klironomos, J.N., Kendrick, W.B., 1996. Palatability of microfungi to soil arthropods in relation to the functioning of arbuscular mycorrhizae. Biology and Fertility of Soils 21 (1–2), 43–52.

Klironomos, J.N., Moutoglis, P., 1999. Colonization of non-mycorrhizal plants by mycorrhizal neighbours as influenced by the collembolan, *Folsomia candida*. Biology and Fertility of Soils 29 (3), 277–281.

Klironomos, J.N., Ursic, M., 1998. Density-dependent grazing on the extraradical hyphal network of the arbuscular mycorrhizal fungus, *Glomus intraradices*, by the collembolan, *Folsomia candida*. Biology and Fertility of Soils 26 (3), 250–253.

Klironomos, J.N., 2003. Variation in plant response to native and exotic arbuscular mycorrhizal fungi. Ecology 84 (9), 2292–2301.

Leake, J.R., Cameron, D.D., 2010. Physiological ecology of mycoheterotrophy. New Phytologist 185 (3), 601–605.

Leake, J.R., Cameron, D.D., Beerling, D.J., 2008. Fungal fidelity in the myco-heterotroph-to-autotroph life cycle of Lycopodiaceae: a case of parental nurture? New Phytologist 177 (3), 572–576.

Leake, J.R., 1994. The biology of myco-heterotrophic ('saprophytic') plants. New Phytologist 127 (2), 171–216.

Leake, J.R., 2004. Myco-heterotroph/epiparasitic plant interactions with ectomycorrhizal and arbuscular mycorrhizal fungi. Current Opinion in Plant Biology 7 (4), 422–428.

Leake, J.R., 2005. Plants parasitic on fungi: unearthing the fungi in myco-heterotrophs and debunking the 'saprophytic' plant myth. Mycologist 19 (03), 113–122.

Lendenmann, M., Thonar, C., Barnard, R.L., Salmon, Y., Werner, R.A., Frossard, E., Jansa, J., 2011. Symbiont identity matters: carbon and phosphorus fluxes between *Medicago truncatula* and different arbuscular mycorrhizal fungi. Mycorrhiza 21 (8), 689–702.

Lim, W.-H., Newman, B., Dixon, K., Read, D.J., 2014. An improved understanding of orchid mycorrhizal ecology. In: Poster Presentation at the 33rd New Phytologist Symposium: Networks of Power & Influence: Ecology and Evolution of Symbioses between Plants and Mycorrhizal Fungi; 14–16 May. Agroscope, Zurich, Switzerland.

Lindahl, B.D., Ihrmark, K., Boberg, J., Trumbore, S.E., Högberg, P., Stenlid, J., Finlay, R.D., 2007. Spatial separation of litter decomposition and mycorrhizal nitrogen uptake in a boreal forest. New Phytologist 173 (3), 611–620.

Martínez-Botí, M.A., Marino, G., Foster, G.L., Ziveri, P., Henehan, M.J., Rae, J.W.B., Mortyn, P.G., Vance, D., 2015. Boron isotope evidence for oceanic carbon dioxide leakage during the last deglaciation. Nature 518 (7538), 219–222.

Martos, F., Munoz, F., Pailler, T., Kottke, I., Gonneau, C., Selosse, M.-A., 2012. The role of epiphytism in architecture and evolutionary constraint within mycorrhizal networks of tropical orchids. Molecular Ecology 21 (20), 5098–5109.

Merckx, V., Stöckel, M., Fleischmann, A., Bruns, T.D., Gebauer, G., 2010. ^{15}N and ^{13}C natural abundance of two mycoheterotrophic and a putative partially mycoheterotrophic species associated with arbuscular mycorrhizal fungi. New Phytologist 188, 590–596.

Merrild, M.P., Ambus, P., Rosendahl, S., Jakobsen, I., 2013. Common arbuscular mycorrhizal networks amplify competition for phosphorus between seedlings and established plants. New Phytologist 200 (1), 229–240.

Merryweather, J., Fitter, A., 1995. Phosphorus and carbon budgets: mycorrhizal contribution in *Hyacinthoides non-scripta* (L.) Chouard ex Rothm. under natural conditions. New Phytologist 129 (4), 619–627.

Monastersky, R., 2013. Global carbon dioxide levels near worrisome milestone. Nature 497 (7447), 13–14.

Moore, R.T., 1980. Taxonomic significance of septal ultrastructure in the genus *Onnia* Karsten (Polyporinae/Hymenochaetaceae). Botaniska Notiser 133, 169–175.

Nandini, R.T.S., Tholkappian, P., 2012. Co-inoculation of AM fungi *Glomus macrocarpum* and *Acaulospora laevis* and *Bacillus megaterium* var. *phosphaticum* on the growth and nutrient contents of *Solanum xanthocarpum* Schrad. and Wendl. International Journal of Pharmaceutical & Biological Archives 3, 1137–1141.

Pachauri, R.K., Allen, M.R., Barros, V.R., Broome, J., Cramer, W., Christ, R., Church, J.A., Clarke, L., Dahe, Q., Dasgupta, P., Dubash, N.K., 2014. Climate Change 2014: Synthesis Report. (Contribution of Working Groups I, II and III to the Fifth Assessment Report of the Intergovernmental Panel on Climate Change).

Parniske, M., 2008. Arbuscular mycorrhiza: the mother of plant root endosymbioses. Nature Reviews Microbiology 6 (10), 763–775.

Phillips, R.D., Brown, A.P., Dixon, K.W., Hopper, S.D., 2011. Orchid biogeography and factors associated with rarity in a biodiversity hotspot, the Southwest Australian Floristic Region. Journal of Biogeography 38 (3), 487–501.

Phillips, L.A., Ward, V., Jones, M.D., 2014. Ectomycorrhizal fungi contribute to soil organic matter cycling in sub-boreal forests. The ISME Journal 8 (3), 699–713.

Pirozynski, K.A., Malloch, D.W., 1975. The origin of land plants: a matter of mycotrophism. Biosystems 6 (3), 153–164.

Pollierer, M.M., Dyckmans, J., Scheu, S., Haubert, D., 2012. Carbon flux through fungi and bacteria into the forest soil animal food web as indicated by compound-specific ^{13}C fatty acid analysis. Functional Ecology 26 (4), 978–990.

Raven, J.A., Allen, J.F., 2003. Genomics and chloroplast evolution: what did cyanobacteria do for plants. Genome Biology 4 (3), 209.

Read, D.J., Perez-Moreno, J., 2003. Mycorrhizas and nutrient cycling in ecosystems – a journey towards relevance? New Phytologist 157 (3), 475–492.

Read, D.J., 1996. The structure and function of the ericoid mycorrhizal root. Annals of Botany 77 (4), 365–374.

Rillig, M.C., Allen, M.F., 1999. What is the role of arbuscular mycorrhizal fungi in plant-to-ecosystem responses to elevated atmospheric CO_2? Mycorrhiza 9 (1), 1–8.

Rimington, W.R., Pressel, S., Duckett, J.G., Bidartondo, M.I., 2015. Fungal associations of basal vascular plants: reopening a closed book? New Phytologist 205 (4), 1394–1398.

Rowntree, J.K., Cameron, D.D., Preziosi, R.F., 2011. Genetic variation changes the interactions between the parasitic plant-ecosystem engineer *Rhinanthus* and its hosts. Philosophical Transactions of the Royal Society of London B: Biological Sciences 366 (1569), 1380–1388.

Schmidt, S., Stewart, G.R., 1999. Glycine metabolism by plant roots and its occurrence in Australian plant communities. Functional Plant Biology 26 (3), 253–264.

Selosse, M.-A., Roy, M., 2009. Green plants that feed on fungi: facts and questions about mixotrophy. Trends in Plant Science 14 (2), 64–70.

Simard, S.W., Perry, D.A., Jones, M.D., Myrold, D.D., Durall, D.M., Molina, R., 1997. Net transfer of carbon between ectomycorrhizal tree species in the field. Nature 388 (6642), 579–582.

Smith, S.E., Read, D.J., 2008. Mycorrhizal Symbiosis, third ed. Academic Press, Cambridge, UK.

Smith, S.E., 1973. Asymbiotic germination of orchid seeds on carbohydrates of fungal origin. New Phytologist 72 (3), 497–499.

Staddon, P.L., Fitter, A.H., 1998. Does elevated atmospheric carbon dioxide affect arbuscular mycorrhizas? Trends in Ecology and Evolution 13 (11), 455–458.

Strullu-Derrien, C., Kenrick, P., Pressel, S., Duckett, J.G., Rioult, J.P., Strullu, D.G., 2014. Fungal associations in *Horneophyton ligneri* from the Rhynie Chert (c. 407 million year old) closely resemble those in extant lower land plants: novel insights into ancestral plant–fungus symbioses. New Phytologist 203 (3), 964–979.

Talbot, J.M., Allison, S.D., Treseder, K.K., 2008. Decomposers in disguise: mycorrhizal fungi as regulators of soil C dynamics in ecosystems under global change. Functional Ecology 22 (6), 955–963.

Taylor, L.L., Leake, J.R., Quirk, J., Hardy, K., Banwart, S.A., Beerling, D.J., 2009. Biological weathering and the long-term carbon cycle: integrating mycorrhizal evolution and function into the current paradigm. Geobiology 7 (2), 171–191.

Todd-Brown, K.E., Hopkins, F.M., Kivlin, S.N., Talbot, J.M., Allison, S.D., 2012. A framework for representing microbial decomposition in coupled climate models. Biogeochemistry 109 (1–3), 19–33.

Treseder, K.K., 2004. A meta-analysis of mycorrhizal responses to nitrogen, phosphorus, and atmospheric CO_2 in field studies. New Phytologist 164 (2), 347–355.

van der Heijden, M.G., Klironomos, J.N., Ursic, M., Moutoglis, P., Streitwolf-Engel, R., Boller, T., Wiemken, A., Sanders, I.R., 1998. Mycorrhizal fungal diversity determines plant biodiversity, ecosystem variability and productivity. Nature 396 (6706), 69–72.

Walder, F., van der Heijden, M.G., 2015. Regulation of resource exchange in the arbuscular mycorrhizal symbiosis. Nature Plants 1, 15159.

Walder, F., Niemann, H., Natarajan, M., Lehmann, M.F., Boller, T., Wiemken, A., 2012. Mycorrhizal networks: common goods of plants shared under unequal terms of trade. Plant Physiology 159 (2), 789–797.

Winther, J.L., Friedman, W.E., 2007. Arbuscular mycorrhizal symbionts in *Botrychium* (Ophioglossaceae). American Journal of Botany 94 (7), 1248–1255.

Zabinski, C.A., Quinn, L., Callaway, R.M., 2002. Phosphorus uptake, not carbon transfer, explains arbuscular mycorrhizal enhancement of *Centaurea maculosa* in the presence of native grassland species. Functional Ecology 16 (6), 758–765.

Zhang, H., Ziegler, W., Han, X., Trumbore, S., Hartmann, H., 2015. Plant carbon limitation does not reduce nitrogen transfer from arbuscular mycorrhizal fungi to *Plantago lanceolata*. Plant and Soil 396 (1–2), 369–380.

CHAPTER

22

Trading Carbon Between Arbuscular Mycorrhizal Fungi and Their Hyphae-Associated Microbes

B. Drigo[1,2], *S. Donn*[1]

[1]Western Sydney University, Penrith, NSW, Australia; [2]University of South Australia, Mawson Lakes, SA, Australia

22.1 MYCORRHIZAS AND HYPHAE-ASSOCIATED MICROBES

The most widespread symbiosis in the rhizosphere is the symbiosis between plant roots and mycorrhizal fungi (Smith and Read, 2008). Mycorrhizal fungi affect the exchange of nutrients at the plant root level (Brundrett, 2002), and have impacts on ecosystem functioning through their effects on plant productivity and community assembly (Bever et al., 2010). Mycorrhizas develop mutualistic interactions that allow the plants to exploit habitats that would be otherwise inaccessible to them, and boost their competitiveness over plants lacking these associations (Bever et al., 2010; Smith and Read, 2008).

The origin of mycorrhizal symbiosis, based on molecular clock dating, is estimated to have occurred around 600 million years ago (Redecker et al., 2000), supporting the hypothesis that those mycorrhizas were instrumental in the colonization of land by plants (Remy et al., 1994). The most ancient and widespread type of mycorrhiza is the arbuscular mycorrhiza, established by many plant species and a narrow taxonomic group of arbuscular mycorrhizal (AM) fungi (Chapter 1), followed by ericoid, orchid, or ectomycorrhizal (EcM) systems (Meharg and Cairney, 2000; Smith and Read, 2008; Schüßler et al., 2001). Nowadays, mycorrhizal fungi occur in the majority of existing plant families (Smith and Read, 1997) and have a direct influence on biogeochemical cycles and an indirect effect on ecosystem functions (van der Heijden et al., 2008; Veresoglou et al., 2012), structure, and productivity (van der Heijden et al., 1998; Jansa et al., 2008). Most plant roots are colonized by multiple mycorrhizal fungi and most mycorrhizal fungi are not host-specific, colonizing various host plants at the same time (van der Heijden et al., 2015). As a consequence, plants are usually interconnected by mycorrhizal mycelial networks in so-called "woodwide webs" (Simard et al., 1997).

Mycorrhizal Mediation of Soil
http://dx.doi.org/10.1016/B978-0-12-804312-7.00022-X

With the mycorrhizal hyphae on one side connected to plant roots, and from the rhizosphere extending to the soil, mycorrhizal fungi colonize and interconnect simultaneously two environments, specifically the roots of a host plant and the rhizosphere (Fig. 22.1) (Jansa et al., 2013; Drigo et al., 2013). Whereas the hyphae inside the roots are mainly surrounded by plant cells and thus exposed to a stable environment, the hyphae extending to the soil are exposed to great variations in abiotic characteristics and they constantly interact with other organisms in soil such as fungi, macrofauna, microfauna, and bacteria (Fig. 22.1) (Drigo et al., 2013; Kaiser et al., 2015). The total soil volume under the influence of mycorrhizal plants either through roots or through mycorrhizal hyphae is referred to as the *mycorrhizosphere* and includes the combined effects of the soil microbial communities in the rhizosphere and in the hyphosphere (Fig. 22.1) (Timonen and Marschner, 2006). Hence the mycorrhizosphere might be considered the crossroad of the root–soil habitat, where complex fine-scale gradients of substrate availability, water potential, and redox state modify the root–soil environment and consequently the composition, activity, and colonization ability of the surrounding beneficial, pathogenic, and commensal microbial communities (Fig. 22.1) (Timonen and Marschner, 2006). In past studies, using soil compartments where only hyphae could enter, Andrade et al. (1998) found that the presence of hyphae caused an increase in fluorescent pseudomonads and an *Alcaligenes eutrophus* strain, whereas Ravnskov et al. (1999) reported a decrease in *Pseudomonas fluorescens* strain DF57, and *Burkholderia* and *Bradyrhizobium* were present on

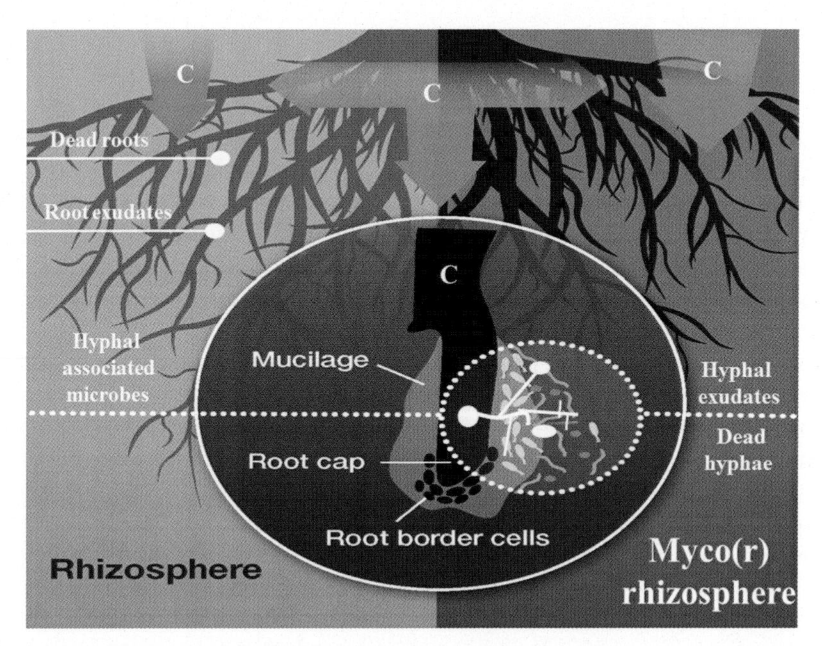

FIGURE 22.1 The mycorrhizosphere might be considered the crossroad of the root–soil habitat, where complex fine-scale gradients of substrate availability, water potential, and redox state modify the root–soil environment and consequently the composition, activity, and colonization ability of the surrounding beneficial, pathogenic, and commensal microbial community. The pathways of photosynthetically fixed carbon (C) in the below-ground compartment of the plant–fungi–soil system are described in this figure. Thickness of lines represents approximate volume/rate of fluxes. Respiration losses are not shown.

the EcM hyphae associated with pine trees (Timonen and Hurek, 2006). Several other studies observed that the presence of mycorrhizal hyphae can modify the bacterial community composition and growth (Scheublin et al., 2010; Toljander et al., 2007; Mansfeld-Giese et al., 2002; Marschner and Baumann, 2003; Rillig et al., 2006; Filion et al., 1999; Cheeke et al., 2015; Bender and van der Heijden, 2015; van der Heijden et al., 2015). There is little evidence for the association of mycorrhizal hyphae with eukaryotes (i.e., yeasts) and viruses, and the reasons behind recruiting a specific microflora in the mycorrhizosphere remain mostly unclear (Bonfante and Anca, 2009; Jansa et al., 2013; Bonfante and Genre, 2015).

In this chapter we provide an overview of the available scientific knowledge on the identity and putative roles of hyphae-associated microbes with respect to the mycorrhizal fungi and also to the mycorrhizal host plants. We explore the dynamics of these associations under fluctuating environmental conditions and the evolving insights to understand hyphae-associated microbes. More specifically, we analyze the potential involvement of the microbes in nutrient cycling and carbon (C) transformation in the hyphosphere. We focus mainly on the AM fungi, because they represent the most widespread mycorrhizal fungal group, forming a symbiosis with the vast majority of the land plants (Chapter 1), although examples of some EcM systems, another widely abundant mycorrhizal symbiosis (Chapter 1), are also presented.

22.2 CARBON ALLOCATION FROM MYCORRHIZAL FUNGI TO THE HYPHAE-ASSOCIATED MICROBES IN THE HYPHOSPHERE

Between 4% and (apparently up to) 30% of the net plant photosynthetic production is transferred to mycorrhizal fungi and subsequently released to soil microbes (Fig. 22.1) (Paul and Kucey, 1981; Jakobsen and Rosendahl, 1990; Drigo et al., 2010; Lendenmann et al., 2011; Calderon et al., 2012; Kaiser et al., 2015). This release can be via direct exudation from the mycorrhizal hyphae or by immobilization in the extraradical mycelium of the mycorrhizal fungi (Fig. 22.1) (Drigo et al., 2012; Chapter 23). Both root exudation and transfer of C accumulated via photosynthetic activity to mycorrhizal fungi occur within a few hours in grasses and up to a few days in trees (Drigo et al., 2010, 2013; Kaiser et al., 2015). Direct evidence for the transfer of plant-derived C from mycorrhizal hyphae to their associated microbes is still lacking; however, previous studies have shown that AM fungal hyphae act as a major hub for translocating fresh plant C to soil microbes and release C gradually to certain hyphae-associated microbes such as *Burkholderia* spp. and *Pseudomonas* spp. (Figs. 22.2 and 22.3) (Johansson et al., 2004; Toljander et al., 2007; Drigo et al., 2010, 2013; Kaiser et al., 2015). This has the indirect ability to shift the hyphosphere active microbial community, which might reward AM fungi with increased nutrient availability through stimulation of soil organic matter (SOM) depolymerization (Hodge and Storer, 2015; Jansa et al., 2013).

Another pathway for the photosynthates to move from the plant to the hyphae-associated microbes is through decay of mycorrhizal mycelium (Fig. 22.1). We know little about the fate of C contained in hyphal biomass once it dies (Drigo et al., 2012; Chapter 23). This is particularly important because when detached from the roots, external mycorrhizal mycelia become a large resource of both labile and recalcitrant C that can fuel the activity of hyphosphere microorganisms and contribute to the formation of SOM. For example, in a poplar plantation, EcM mycelium was the dominant pathway (62%) through which the C entered the SOM pool,

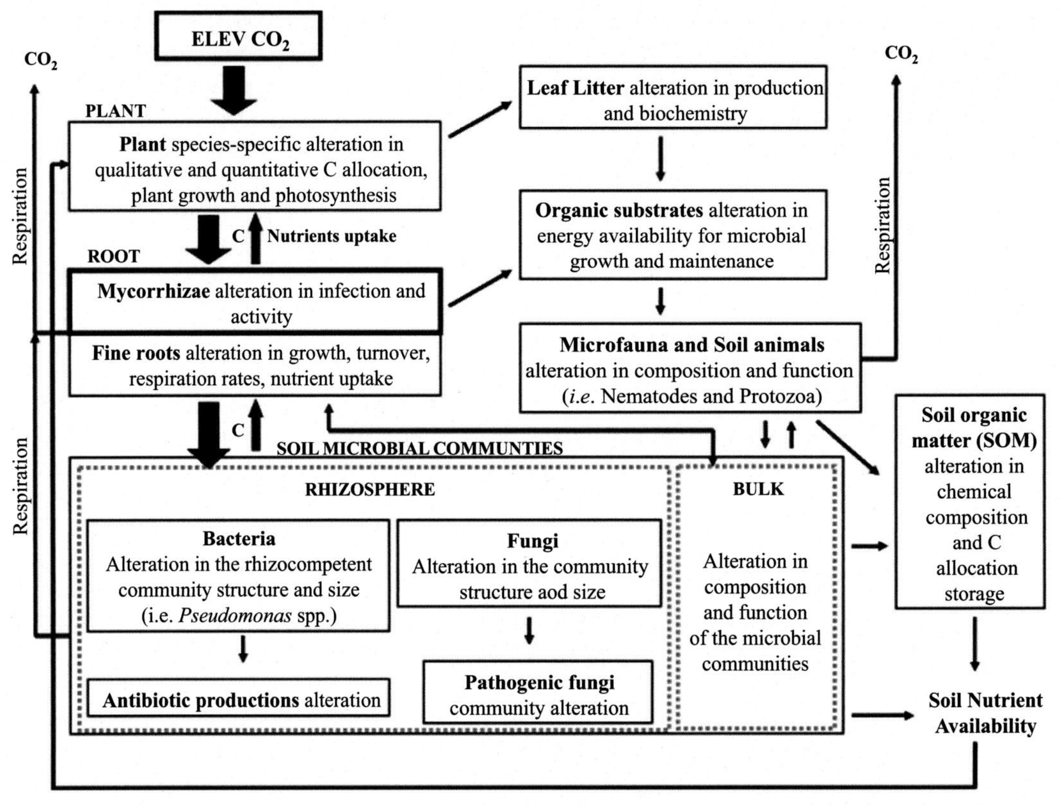

FIGURE 22.2 Conceptual model, which describes the response of a dominant plant species to elevated atmospheric carbon dioxide (ELEV CO_2). The increased photosynthetically fixed carbon (C) allocation is initially directed mainly to mycorrhizas and root tissues. Mycorrhizas are translocating the C into the soil microbial communities, thereby changing the structure, size, and activity of the rhizosphere microbial communities (bacteria and fungi) to a larger extent than the communities of the bulk soil. Soil microbial communities subsequently affect food web interactions and mediate the ecosystem feedback systems that regulate the cycling of both C and mineral nutrients such as nitrogen. *Adapted from Drigo, B., Kowalchuk, G.A., van Veen, J.A., 2008. Climate change goes underground: effects of elevated atmospheric CO^2 on microbial community structure and activities in the rhizosphere. Biology and Fertility of Soils 44, 667–679.*

exceeding the input via leaf litter and fine root turnover (Godbold et al., 2006). Use of mesh in-growth bags suggested that EcM mycelium had a mean residence time of about 10 years (Wallander and Thelin 2008), but this is likely to differ based on a number of factors including fungal species, morphology, and edaphic conditions. Both the speed at which dead mycelium decays and the factors controlling this in soils are poorly understood and the evidence is sometimes contradictory. For example, Koide and Malcolm (2009) analyses suggest a key role of the nitrogen (N) content of hyphae. Wilkinson et al. (2011) found that hyphal necromass rapidly stimulated microbial activity and that this was exaggerated by the effects of species richness of the dead fungi. This finding suggested that some nutrient resources in hyphae were being degraded preferentially by saprotrophic microbes. A similar situation has been found after additions of dissolved organic C (Cleveland et al., 2007). Lindahl et al. (2010) has also shown that saprotrophic fungi can respond rapidly to inputs of EcM mycelium. These

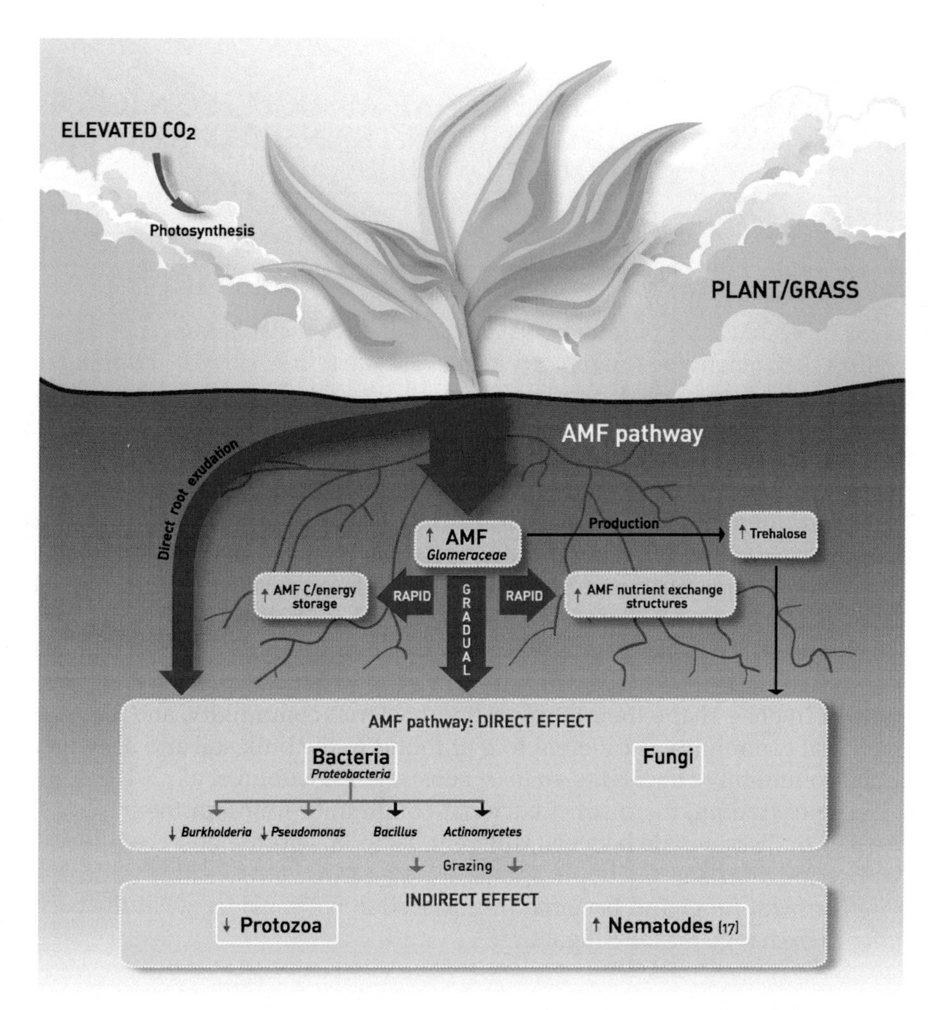

FIGURE 22.3 Conceptual model of carbon (C) flow in mycorrhizal plant–soil systems summarizing the observed effects of elevated carbon dioxide (CO_2) atmospheric concentrations on soil microbial communities. *Brown arrows* indicate increases and decreases in the respective community sizes, as determined by real-time polymerase chain reaction and lipid analyses in Drigo et al. (2007) and Drigo et al. (2010), as well as changes in community structure and C flow. Absence of an arrow indicates no significant change in the community size or structure. *Red arrows* indicate no effect of increased C availability because of elevated CO_2 on the *Actinomycetes* spp. and *Bacillus* spp. communities. The mechanism and magnitude of the C flow along the soil food web are indicated by the *green arrows*. *AM*, Arbuscular mycorrhizal. *Effects on nematodes are based on Drigo, B., Kowalchuk, G.A., Yergeau, E., Bezemer, T.M., Boschker, H.T.S., 2007. Impact of elevated C dioxide on the rhizosphere communities of* Carex arenaria *and* Festuca rubra. *Global Change Biology 13, 2396–2410, whereas the other data are based on Drigo, B., van Veen, J., Pijl, A.S., Kielak, A.M., Gamper, H., et al., 2010. Shifting C flow from roots into associated microbial communities in response to elevated atmospheric* CO^2. *Proceedings of the National Academy of Sciences United States of America 107, 10938–10942.*

studies indicate the need to better understand the link between C inputs from mycorrhizal hyphae and the soil biodiversity and microbial activity.

22.3 INVOLVEMENT OF THE HYPHAE-ASSOCIATED MICROBES IN NUTRIENT CYCLING AND CARBON TRANSFORMATION IN THE HYPHOSPHERE

Trading C implies a mutually beneficial symbiosis and whereas tracer experiments provide evidence for the flow of C from roots to bacteria via a mycorrhizal pathway (Drigo et al., 2010, 2013; Kaiser et al., 2015), the payment in the opposite direction, if any, remains largely unknown. Jansa et al. (2013) pose the question of whether hyphae-associated microbes perform functions that benefit the fungus, or whether they are "free-riders," mopping up hyphal exudates. To establish a trading relationship with beneficial microbes, mycorrhizal fungi must first be able to recognize and interact differently with potential partners they encounter in order to avoid supplying/releasing C to free-riders or parasites. In an in vitro study of interactions between an EcM fungus (*Laccaria bicolor* S238 N) and three soil bacteria, the fungus responds differently to each bacterial strain 14 days after inoculation (Deveau et al., 2015). The largest shifts in *L. bicolor* transcriptome were in response to the antagonistic bacteria, *Collimonas fungivorans*, but beneficial and neutral *P. fluorescens* strains also induced unique transcriptional responses. All three bacteria responded to the presence of the fungus with changes in transcripts related to nutrient acquisition and metabolism, but whether this indicates trading or competition with the fungus for nutrients is unknown.

Mycorrhizal hyphae shape their surrounding bacterial community, and the mycorrhizosphere microbial community is different to that found in the bulk soil and also distinct from the microbial community attached to an inert substrate (Scheublin et al., 2010). Although not direct evidence of trading, the distinct bacterial community found in the hyphosphere, the ability of fungi to discriminate between different bacteria, and the observed flow of C from fungal hyphae to bacteria in the hyphosphere set the scene for a possible trade of resources between mycorrhizal fungi and bacteria. In this section we consider what the bacteria may have to offer in exchange for the fungal C.

22.3.1 Mycorrhiza Helper Bacteria

One of the most well studied interactions between mycorrhizal fungi and bacteria are those of mycorrhiza helper bacteria (MHB) (Garbaye, 1994). Garbaye describes MHB mechanisms in terms of aiding establishment of the plant–fungus symbiosis and we follow this convention, considering postinfection fungal–bacterial interactions in the latter sections.

Examples of MHB in both the AM and EcM literature are plentiful and have been reviewed for example by Frey-Klett et al. (2007). Mechanisms of bacteria helping the mycorrhizal fungus to establish the symbiosis with a plant include improving germination of fungal spores and enhancement of hyphal growth preinfection, altering signaling between plant and fungus, and modifying rhizosphere soil (Garbaye, 1994), as well as modifying the plant root to improve cell permeability or increase root branching (Artursson et al., 2006). Direct evidence of transfer of C in return for these services is lacking and any transfer of plant C via mycorrhizas would be separated in time from the benefits provided by the bacteria.

22.3.2 Endocellular Bacteria

A more intimate relationship exists between mycorrhizas and endocellular bacteria (Bonfante and Anca, 2009). In the case of *Candidatus Glomeribacter gigasporarum*, an obligate endosymbiont of the AM fungus *Gigaspora margarita* (Bianciotto et al., 2001; Bonfante and Anca, 2009), the loss of metabolic pathways indicates a dependence on the host fungi for energy and nutrients and thus evidence for a transfer of C from the fungi to the bacteria (Ghignone et al., 2012). In comparative transcriptomic and proteomic studies of wild-type and cured lines of *G. margarita* spores, the presence of the endosymbiont increased the antioxidant capacity of the fungus and affected metabolism (Salvioli et al., 2016; Vannini et al., 2016), and the wild-type line also had a higher growth rate than the cured fungal isolate (Lumini et al., 2007). Although these studies focus on the establishment stages of mycorrhizas, endosymbiont activity was greatest in the extraradical mycelia during plant–fungus symbiosis (Anca et al., 2009), with the functional significance of the endosymbiont occurrence being not entirely understood.

22.3.3 Nutrient Uptake

22.3.3.1 Phosphorus

The greatest progress toward identifying a mutually beneficial trade has been made in the interactions between AM fungi and phosphorus (P)-solubilizing bacteria (PSBs). Improving P nutrition of plants is cited as a major benefit of the plant–AM fungi symbiosis. Plant P availability is limited by transformation of insoluble or moderately soluble forms of P to water-soluble orthophosphate and the rate of P diffusion through soil water (Richardson and Simpson, 2011). The extended reach of hyphae beyond the rhizosphere zone accounts for the ability of AM fungi to improve plant P status (Bolan, 1991), providing a direct route for the plant to obtain P from the bulk soil rather than relying on diffusion through the soil solution where the microbial community could interfere through competition for readily available P. However, the ability of AM fungi, or at least of *Rhizophagus irregularis*, to mobilize adsorbed P from solid surfaces or from organic forms is limited (Antunes et al., 2007; Tisserant et al., 2013) and the benefits of AM fungi to plant P nutrition are likely dependent on bacteria first mobilizing P from the SOM or from the solid surfaces.

PSBs release P in plant available forms by producing organic acids that solubilize mineral phosphates, and acid phosphatases that release P from organically bound P compounds (Rodriguez and Fraga, 1999). PSBs are common in soil and can constitute over 40% of all the culturable rhizosphere community (Jorquera et al., 2008). The capacity to mobilize P is shared among phylogenetically diverse taxa including Gammaproteobacteria [e.g., *Pseudomonas, Enterobacter* and *Pantoea* (Jorquera et al., 2008)] and Bacilli (Rodriguez and Fraga, 1999). PSBs are active in the rhizosphere, resulting in P depletion around the root (Richardson and Simpson, 2011); mycorrhizal hyphae extend the plants' access to P beyond this depletion zone. In addition, the hyphosphere bacterial community may have enhanced P solubilizing capacity either through an increased proportion of PSBs (Frey-Klett et al., 2007) or other bacteria improving P solubility (Taktek et al., 2015).

In vitro studies have demonstrated the ability of PSBs to grow with fungal hyphae as a sole source of C. In a split plate experiment, bacteria were separated from mycorrhizal chicory roots on a minimal medium, whereas the *R. irregularis* extraradical mycelia were allowed to grow between the two compartments (Taktek et al., 2015). Populations of four PSBs, including *Rhizobium*, *Burkholderia*, and *Rahnella* spp., grew and exhibited phosphate solubilizing activity with AM hyphae as a C source; however, the C may have been obtained through turnover of fungal hyphae. Zhang et al. (2016) collected hyphal exudates consisting of sugars (galactose, glucose, and trehalose) and carboxylates (aconitate, citrate, and, in higher P concentration media, succinate) from *R. irregularis*. Growth and phosphatase activity of a population of the PSBs, *Rahnella aquatilis*, was demonstrated in the presence of these *R. irregularis* exudates. These experiments show that the PSBs tightly associated with hyphae are able to grow on the AM fungi hyphal exudates as a sole nutrient source, but whether the bacteria offer anything in return to the fungi is less clear. It is also unclear whether the release of C compounds from hyphae is actively regulated by the fungi or just passively leaked.

Synergistic effects of inoculation of PSBs and AM fungi on plant biomass and P content have been observed through ^{32}P-tracer experiments (Azcón-Aguilar and Barea, 2015). For example, when a form of P not readily available for plant uptake, rock phosphate, was added to microcosms, onion plants co-inoculated with both AM fungi (*R. irregularis*) and a PSB strain (*Bacillus subtilis*) had greater shoot biomass, P, and N contents than noninoculated controls or plants inoculated with either AM fungi or PSB alone (Toro et al., 1997). *Bacillus subtilis* acted as an MHB, with AM colonization of the onion root increasing in its presence; in addition, co-inoculated plants had access to P that was otherwise unavailable.

Conversely, in maize soil microcosms, co-inoculation of *R. irregularis* and a mixture of three PSBs (*Pseudomonas alcaligenes*, *Bacillus megaterium*, and *R. aquatilis*) did not improve plant performance (Wang et al., 2016). Plant roots and PSBs were spatially separated in soil compartments linked by fungal hyphae. Whereas inoculation of *R. irregularis* increased shoot biomass and P content of the plants, addition of PSBs did not further improve plant performance. Instead, in the presence of PSB, hyphal length density was reduced and, compared with PSB inoculation alone, microbial biomass P increased in the presence of the fungi. Using ^{13}C-labeling, Wang et al. (2016) demonstrated transfer of plant C to PSBs via mycorrhizas, yet bacteria out-competed the AM fungus for P, providing no apparent benefit to the AM fungus in exchange for the increased C availability.

Zhang et al. (2014a) further compared the interaction between the AM fungus *R. irregularis* and PSB *R. aquatilis* at different concentrations of organic P (phytate) in the soil. *R. irregularis* increased shoot P content of *Medicago* spp. at high P (75 mg kg^{-1}), and acted synergistically with *R. aquatilis* to further increase shoot P. However, at low P, there was no further increase in shoot P in the presence of the PSB and instead the microbial biomass P increased in the presence of the AM fungus. The interactions between *R. irregularis* and the PSB *R. aquatilis* were dependent on soil conditions, with bacteria only providing services under favorable nutrient conditions, whereas under P limiting conditions, the interaction was competitive rather than mutualistic. Further studies are needed with different microbial strains/species and more complex communities to determine whether this is a widely occurring phenomenon.

In contrast to AM fungi, EcM fungi are able to mobilize P by their own phosphatase activities and organic acid excretion. Still, PSB were enriched in the hyphosphere of *L. bicolor* associated with Douglas fir (*Pseudostuga menziesii*) (Frey-Klett et al., 2005), suggesting a possible trade of P for C between the EcM fungi and their associated bacterial communities.

22.3.3.2 Other Nutrients

In addition to P, mineral weathering by EcM fungi can mobilize nutrients including potassium, calcium, and magnesium (van Breemen et al., 2000; contrasting views in Chapters 2 and 3). This too may be the result of complex interactions between the fungi and associated bacteria. The population density of a weathering-competent bacterial strain of *Burkholderia* significantly increased in the presence of EcM fungi on Scots pine (*Pinus sylvestris*) seedlings (Koele et al., 2009) and this bacterial strain worked synergistically with the EcM fungus *Scleroderma citrinum* to increase plant uptake of magnesium. Profiling of mycorrhizosphere bacterial communities suggests other potential exchanges of nutrients for C, such as trades including sulfur (Gahan and Schmalenberger, 2015) and iron (Frey-Klett et al., 2005).

Many of the bacteria found to be associated with EcM fungi at root tips have N-fixing capabilities, and though several studies provided evidence that bacterial N fixation did happen inside EcM root tips, this was only up to a very limited extent (Nguyen and Bruns, 2015). The supply of N to plants partnered by EcM fungi is obviously more dependent on the capabilities of the fungus itself to obtain N from the soil (Chalot and Brun, 1998).

There is a debate regarding the potential role of AM fungi in supplying N to plants (Veresoglou et al., 2012; Hodge and Storer, 2015; Chapter 8). Fungal N requirements are high compared with those of the host plant (Johnson et al., 2015) and whereas AM fungi are efficient competitors for ammonium (NH_4^+) (Hodge and Storer, 2015), their ability to mobilize N from other sources is unclear. This high demand for N and limited ability to access it provides a potential niche for microbial symbionts. AM fungi alter the composition of soil decomposer communities (Nuccio et al., 2013), and release of C from hyphae coupled with uptake of NH_4^+ can increase decomposition by soil bacteria and mineralization of N in the hyphosphere (Hodge and Storer, 2015). The N mineralized by soil communities is then potentially transported by hyphae to the plant (Nuccio et al., 2013), but whether this can be considered trade or competition remains to be resolved.

In terms of nutrient mobilization, the enzymatic activities of the mycorrhizal hyphae-associated bacteria are to a large extent shared with the rhizosphere bacteria. However, the larger surface area of mycorrhizal hyphae compared with that of roots and the larger soil volume explored by hyphal networks as compared with the root systems extend the potential contribution of these microbial traits beyond the rhizosphere depletion zone (Bolan, 1991). The common challenge remaining for all is to provide a direct evidence of uptake of nutrients by mycorrhizal hyphae released by bacteria associated with the hyphae.

22.3.4 Induced Protection of Plants Against Pathogens

Several mechanisms have been proposed for AM fungi–induced protection of plants against pathogens, including competition with fungal pathogens for space and nutrients, enhanced nutrient status of the plant, and priming of plant defenses. *Mycorrhizal induced resistance (MIR)* refers to the priming of plant defenses after colonization, including salicylic acid (SA) and jasmonic acid (JA) pathways (Pozo and Azcón-Aguilar, 2007). Cameron et al. (2013) propose that MIR relies not only on colonization of the plant cells by the fungus but is a combination of fungi priming SA defenses and bacteria in the mycorrhizosphere priming JA defense pathways. The mycorrhizosphere provides an environment where bacteria can achieve high cell densities (Andrade et al., 1998), enabling quorum sensing to occur. Quorum sensing molecules are important in eliciting systemic responses in the plant and could be taken up by the plants directly either at the root surface or through mycorrhizal hyphae (Barto et al., 2012; Cameron et al., 2013).

Association of AM fungi with biocontrol bacteria is an additional mechanism potentially providing protection against plant and fungal antagonists. *Paenibacillus* sp. B2, derived from sporocarps of *Glomus mosseae*, inhibited growth of several pathogenic fungi including *Phytopthora nicotianae* and reduced root damage to tomato plants (Budi et al., 1999). This isolate also acted as an MHB, increasing colonization of roots by *G. mosseae*, and the AM fungi and bacteria worked synergistically, reducing root necrosis caused by *P. nicotianae* more when added together than either of the organism separately (Budi et al., 1999). Other organisms with biocontrol potential such as *P. fluorescens* (Toljander et al., 2006; Viollet et al., 2011) and *Streptomyces* sp. (Scheublin et al., 2010) are reportedly enriched in the mycorrhizosphere. Although biocontrol bacteria are often effective against fungal pathogens and can also inhibit mycorrhizal colonization levels of the roots, AM fungi are not always adversely affected (Paulitz and Linderman, 1991). The nature of these interactions may be genotype or strain specific.

AM fungi can be subject to mycoparasitism, such as that of *Glomus intraradices* by *Trichoderma harzianum* T-203 (Rousseau et al., 1996). Whether biocontrol bacteria can protect the mycorrhizal hyphae themselves remains an area yet to be explored experimentally.

22.3.5 Other Hyphae-Associated Microbial Benefits

Horizontal transfer of bacterial genes involved in carbohydrate metabolism has been detected in the EcM species of the genus *Amanita* (Chaib De Mares et al., 2015). This genetic resource provides opportunities for the fungus to utilize new alternative substrates and thus colonize new niches. Although there is evidence of numerous transfers of bacterial genes that enhance the SOM degradation capacity to soil fungi, *Amanita* is thus far the only mycorrhizal example described of horizontal gene transfer from bacteria (Zhang et al., 2014b).

The distinct bacterial communities associated with mycorrhizas are not necessarily involved in C trading. Hyphae can aid dispersal of bacteria, enabling them to cross air-filled soil pores (Kohlmeier et al., 2005; Simon et al., 2015). Other bacteria that do receive C might do so to the detriment of the fungus; for example, oxalic acid released by EcM fungi may attract the mycophagous bacteria *Collimonas* (Rudnick et al., 2015). Still, the C does flow from mycorrhizas to the bacteria and there are a number of services bacteria potentially provide in return. Whether the interactions between mycorrhizas and bacteria are mutually beneficial trades, competition or parasitism likely depends on the species or strains present. Interactions between the same species may have different outcomes, and whether this is dependent on the particular strains in question or the experimental design requires further investigation because the nature of the interactions between mycorrhizal fungi and associated bacteria is sensitive to the context, such as nutrient background in the soil (Ding et al., 2014; Zhang et al., 2016).

22.4 DYNAMICS OF THE MYCORRHIZOSPHERE ASSOCIATIONS UNDER FLUCTUATING ENVIRONMENTAL CONDITIONS

Evidence that the climate is changing is overwhelming. Regardless of mitigation progress, atmospheric carbon dioxide (CO_2) levels will probably surpass the 500 ppm threshold by 2050, held by many as indicative of "dangerous" interference (Monastersky, 2013). Thus it is unavoidable that terrestrial ecosystems will be confronted with higher atmospheric CO_2 in the future (Berrang-Ford et al., 2011).

Under the condition of elevated CO_2, AM fungi usually respond by increasing colonization of plant roots (Figs. 22.2 and 22.3) (Johnson and Gehring, 2007; Drigo et al., 2008, 2010), rapidly acquiring recent photosynthates (Johnson et al., 2003; Carillo et al., 2016) and significantly contributing to both fast and slow pools of the soil organic C through the retention of C in the mycelium (Olsson and Johnson, 2005; Johnson et al., 2013; Meier et al., 2015; Chapters 23 and 24). This significantly influences the hyphae-associated microbial community dynamics in the mycorrhizosphere (Fig. 22.3) (Drigo et al., 2010, 2013; Cotton et al., 2015).

A shift from bacterial- to more fungal (including AM fungi)-dominated soil food webs has been observed a number of times and may prove to be a general response to elevated CO_2 (Figs. 22.2 and 22.3) (Rillig and Allen, 1999; Olsson and Johnson, 2005; Drigo et al., 2007, 2010; Pritchard, 2011; Drigo et al., 2013; Treseder, 2016). This stimulation of AM fungal activity could be beneficial for the ecosystem functioning because AM fungi may play a role in maintaining soil structure (Rillig et al., 2002), soil C stabilization (Treseder and Allen, 2000; Drigo et al., 2010), and nutrient immobilization through hyphal translocation (Frey et al., 2000). However, the extent to which mycorrhizas benefit from atmospheric CO_2 enrichment will probably depend heavily on the nutrient status of the plant and the fungus, and the AM fungi species present in the ecosystem (Treseder, 2016).

Increases in atmospheric CO_2 are thought to be largely responsible for increases in global mean surface temperatures, which are projected to increment by another 1.4–5°C by 2100 (IPCC, 2014). Such increases in temperature would further contribute to more than a 20% reduction in mean annual precipitation globally and altered patterns of rainfall, including more extreme events. Thus in the future, plants will likely experience increases in acute heat and drought stress, which can impact soil microbial function, dynamics, and biodiversity (Thomas et al., 2004; Evans and Wallenstein, 2014). Indeed, rapid rhizosphere microbial responses to sudden moisture availability, such as those resulting from rewetting events after extended periods of drought, often result in immediate mineralization of C- and N-containing compounds that accumulate during drought periods (Barnard et al., 2013). There is evidence that rhizosphere microbial communities may become more tolerant to drought when previously exposed to variable rainfall patterns through changes in microbial life history traits (Evans and Wallenstein, 2014). However, this may also influence the functional traits of the local microbial communities with potential implications for ecosystem functioning (Evans and Wallenstein, 2014). Tolerating moisture stress implies resource and nutrient investment in exopolysaccharide and spore production, and the accumulation of compatible solutes (reviewed in Schimel et al., 2007), which may become restricted as the water-film thickness is diminished by desiccation or by an abrupt flush upon rewetting dry soils (Barnard et al., 2013; Vries et al., 2012). AM fungi can help plants and soil microbes deal with drought and extreme rainfall patterns by acting, directly or indirectly, on plant functionality at the above- and below-ground levels (Chapter 17). In the plant leaves and roots, the osmotic stress caused by drought is balanced by mycorrhiza through biochemical changes that mostly imply increased biosynthesis of metabolites (i.e., proline and sugars) that act as osmolytes (Brunner et al., 2015). These metabolites help decrease the plant osmotic potential and allow the plant to sustain the cell physiological activity (Brunner et al., 2015). The AM plants withstand drought-induced oxidative stress by the increased production of antioxidant enzymes (Rapparini and Penuelas, 2014) and the creation of a highly functional root system for nutrient/water uptake (Hodge and Storer, 2015). At the same time, AM fungal hyphae in the rhizosphere constitute an efficient pathway for nutrient/water uptake and transport, allowing a more efficient exploitation of the water and nutrient reservoirs in the soil pores where only fungal hyphae can grow, thereby bypassing the zones of water and nutrient depletion around the roots and the mycorrhizosphere microbial communities (Rapparini and Penuelas, 2014). Molecular mechanisms initiated by AM fungi to respond to drought events include gene activation for production of functional proteins such as the membrane transporter aquaporin, ion, and sugar transporters (Li et al., 2012a). In addition, AM fungi might reinforce the resistance of associated microbes to drought by improving soil structural stability that in turn increases the retention of soil water (Medina and Azcon, 2010).

The response of mycorrhiza and associated soil microbial communities to climate change under field conditions, however, remains difficult to measure and is not predictable from single-factor experiments (Nielsen and Ball, 2015). Therefore multifactorial experiments are urgently needed to define ecologically relevant taxonomic and functional units for microorganisms and, consequently, improve the relevant models' reliability (Albert et al., 2011; Chapter 26).

22.5 UNRESOLVED QUESTIONS ON TRADING CARBON AND NUTRIENT BETWEEN MYCORRHIZAS AND HYPHAE-ASSOCIATED MICROBES

Mycorrhizas are in the front line of plant-soil–microbial interactions. Mycorrhizas and their associated microbes' growth, nutrient uptake, and release of solutes vary both in space and time, and interact with heterogeneous soil microenvironments that provide habitats for associated microbes at various scales. Despite tremendous progress in method development in the past decades, finding a suitable experimental set-up to research processes occurring at the dynamic conjunction of the mycorrhizosphere still represent a major challenge. Recent methodological developments in rhizosphere research, involving visualization of two- or three-dimensional rhizosphere processes via chemical imaging, microbial imaging, and non-invasive imaging has a significant potential in visualizing the mycorrhizosphere processes across a range of scale from centimeter to submicrometers.

Another recent methodological development is the use of isotopically labeled compounds with stable isotopes, such as ^{2}H [(the stable hydrogen isotope deuterium), ^{13}C, ^{15}N, and oxygen (^{18}O), in combination with complementary techniques such as mass spectrometry, high-throughput sequencing, nanoscale secondary ion mass spectrometry (NanoSIMS), and neutron radiography to trace the movement of water and C- and/or N-containing compounds through the mycorrhizosphere. These approaches have great potential to unravel mechanisms of uptake, storage, and translocation of C, N, and other nutrients at the mycorrhizosphere level. In particular, NanoSIMS combines unprecedented spatial resolution, sensitivity, and mass specificity to quantitatively visualize fine-scale differences in metabolism and cell–cell interactions (Pett-Ridge & Weber, 2011). The combination of DNA and RNA stable-isotope probing techniques with NanoSIMS and fluorescent in situ hybridization (FISH) will contribute to map the mycorrhizosphere microbial associations and reveal their function in trading nutrients. Indeed, these will allow tracking metabolic activities of single mycorrhizal hyphae-associated microbial cells by imaging natural isotopic/elemental composition or isotope distribution after stable-isotope probing. NanoSIMS with FISH is a particularly valuable combination for expanding our understanding of microbial relationships and nutrient flows in environmental microbial assemblages and characterizes microbial-mediated biogeochemical processes. NanoSIMS can further be combined with immunolabeling or quantitative molecular imaging to identify microbial communities and chemical exchanges within microbial energy webs.

The use of single-cell Raman spectra might aid to detect intrinsic chemical composition of living organisms in the mycorrhizosphere as signatures or "fingerprints" (Li et al., 2012b). The combination of stable-isotope probing techniques with Raman spectroscopy has been proven to be a valuable novel approach in microbiology, microbial ecology, and clinical research. The use of stable-isotope probing–Raman spectroscopy will allow tracking C, N, and/or oxygen flows in the complex microfauna and macrofauna of the mycorrhizosphere in a quantitative and nondestructive manner, providing crucial information on the biochemical pathways that lead to the synthesis of the basic biomolecules.

Although each of the presented techniques on their own provide important new insights, we believe that particularly the combined use of different methodological approaches will be promising in shedding more light onto the great number of yet unrevealed processes at the root-mycorrhiza–soil interface.

References

Albert, C.H., Grassein, F., Schurrd, F.M., Vieilledent, G., Violle, C., 2011. When and how should intraspecific trait variability be considered in plant ecology? Perspect. Plant Ecol. Evol. 13, 217–225.

Anca, I.A., Lumini, E., Ghignone, S., Salvioli, A., Bianciotto, V., Bonfante, P., 2009. The *ftsZ* gene of the endocellular bacterium '*Candidatus* Glomeribacter gigasporarum' is preferentially expressed during the symbiotic phases of its host mycorrhizal fungus. Molecular Plant-Microbe Interactions 22, 302–310. http://dx.doi.org/10.1094/Mpmi-22-3-0302.

Andrade, G., Linderman, R.G., Bethlenfalvay, G.J., 1998. Bacterial associations with the mycorrhizosphere and hyphosphere of the arbuscular mycorrhizal fungus *Glomus mosseae*. Plant and Soil 202, 79–87.

Andrade, G., Mihara, K., Linderman, R., et al., 1998. Plant and Soil 202, 89.

Antunes, P.M., Schneiderb, K., Hillisc, D., Klironomos, J.N., 2007. Can the arbuscular mycorrhizal fungus *Glomus intraradices* actively mobilize P from rock phosphates? Pedobiologia 51, 281–286.

Artursson, V., Finlay, R.D., Jansson, J., 2006. Interactions between arbuscular mycorrhiza fungi and bacteria and their potential for stimulating plant growth. Environmental Microbiology 8, 1–10.

Azcón-Aguilar, C., Barea, J.M., 2015. Nutrient cycling in the mycorrhizosphere. Journal of Soil Science and Plant Nutrition 15, 372–396.

Barnard, R.L., Osborne, C.A., Firestone, M.K., 2013. Responses of soil bacterial and fungal communities to extreme desiccation and rewetting. ISME Journal 7, 2229–2241.

Barto, K.E., Weidenhamer, J.D., Cipollini, D., Rillig, M.C., 2012. Fungal superhighways: do common mycorrhizal networks enhance below ground communication? Trends in Plant Science 17, 633–637.

Bender, S.F., van der Heijden, M.G.A., 2015. Soil biota enhance agricultural sustainability by improving crop yield, nutrient uptake and reducing nitrogen leaching losses. Journal of Applied Ecology 52, 228–239.

Berrang-Ford, L., Ford, J.D., Paterson, J., 2011. Are we adapting to climate change? Global Environmental Change 21, 25–33.

Bever, J.D., Dickie, I.A., Facelli, E., Facelli, J.M., Klironomos, J., Moora, M., et al., 2010. Rooting theories of plant community ecology in microbial interactions. Trends in Ecology and Evolution 25, 468–478.

Bianciotto, V., Andreotti, S., Balestrini, R., Bonfante, P., Perotto, S., 2001. Mucoid mutants of the biocontrol strain *Pseudomonas fluorescens* CHA0 show increased ability in biofilm formation on mycorrhizal and nonmycorrhizal carrot roots. Molecular Plant-Microbe Interactions 14, 255–260.

Bolan, N.S., 1991. A critical review on the role of mycorrhizal fungi in the uptake of phosphorus by plants. Plant and Soil 134, 189–207.

Bonfante, P., Anca, I.A., 2009. Plants, mycorrhizal fungi and bacteria: a network of interactions. Annual Review of Microbiology 63, 363–383.

Bonfante, P., Genre, A., 2015. Arbuscular mycorrhizal dialogues: do you speak 'plantish' or 'fungish'? Trends Plant Science 20, 150–154.

Brundrett, M.C., 2002. Coevolution of roots and mycorrhizas of land plants. New Phytologist 154, 275–304.

Brunner, I., Herzog, C., Dawes, M.A., Arend, M., Sperisen, C., 2015. How tree roots respond to drought. Frontiers in Plant Science 6, 547.

Budi, S.W., van Tuinen, D., Martinotti, G., Gianinazzi, S., 1999. Isolation from the *Sorghum bicolor* mycorrhizosphere of a bacterium compatible with arbuscular mycorrhiza development and antagonistic towards soilborne fungal pathogens. Applied and Environmental Microbiology 65, 5148–5150.

Calderon, F.J., Schultz, D.J., Paul, E.A., 2012. Carbon allocation, belowground transfers, and lipid turnover in a plant-microbial association. Soil Science Society of America Journal 76, 1614–1623. http://dx.doi.org/10.2136/sssaj2011.0440.

Cameron, D.D., Neal, A.L., van Wees, S.C.M., Ton, J., 2013. Mycorrhiza-induced resistance: more than the sum of its parts? Trends in Plant Science 18, 539–545.

Carillo, Y., Dijkstra, F.A., LeCain, D., Pendall, E., 2016. Mediation of soil C decomposition by arbuscular mycorrizhal fungi in grass rhizospheres under elevated CO_2. Biochemistery 127(1), 45–55.

Chalot, M., Brun, A., 1998. Physiology of organic nitrogen acquisition by ectomycorrhizal fungi and ectomycorrhizas. FEMS Microbiol Rev. 22 (1), 21–44.

Chaib De Mares, M., Hess, J., Floudas, D., Lipzen, A., Choi, C., Kennedy, M., Grigoriev, IV, Pringle, A., Horizontal transfer of carbohydrate metabolism genes into ectomycorrhizal *Amanita. New Phytologist* 205, 1552–1564.

Cheeke, T.E., Schütte, U.M., Hemmerich, C.M., Cruzan, M.B., Rosenstiel, T.N., Bever, J.D., 2015. Spatial soil heterogeneity has a greater effect on symbiotic arbuscular mycorrhizal fungal communities and plant growth than genetic modification with *Bacillus thuringiensis* toxin genes. Molecular Ecology 24, 2580–2593.

Cleveland, C.C., Nemergut, D.R., Schmidt, S.K., Townsend, A.R., 2007. Increases in soil respiration following labile carbon additions linked to rapid shifts in soil microbial community composition. Biogeochemistry 82, 229–240. http://dx.doi.org/10.1007/s10533-006-9065-z.

Cotton, T.E.A., Fitter, A.H., Miller, R.M., Dumbrell, A.J., Helgason, T., 2015. Fungi in the future: interannual variation and effects of atmospheric change on arbuscular mycorrhizal fungal communities. New Phytologist 205, 1598–1607.

Deveau, A., Barret, M., Diedhiou, A.G., Leveau, J., de Boer, W., Martin, F., et al., 2015. Pairwise transcriptomic analysis of the interactions between the EM fungus *Laccaria bicolor* S238N and three beneficial, neutral and antagonistic soil bacteria. Microbial Ecology 69, 146–159.

Ding, X., Zhang, L., Zhang, S., Feng, G., 2014. Phytate utilization of maize mediated by different nitrogen forms in a plant-arbuscular mycorrhizal fungus- phosphate—solubilizing bacterium system. Journal of Plant Interactions 9, 514–520.

Drigo, B., Kowalchuk, G.A., Yergeau, E., Bezemer, T.M., Boschker, H.T.S., 2007. Impact of elevated C dioxide on the rhizosphere communities of *Carex arenaria* and *Festuca rubra*. Global Change Biology 13, 2396–2410.

Drigo, B., Kowalchuk, G.A., van Veen, J.A., 2008. Climate change goes underground: effects of elevated atmospheric CO_2 on microbial community structure and activities in the rhizosphere. Biology and Fertility of Soils 44, 667–679.

Drigo, B., van Veen, J., Pijl, A.S., Kielak, A.M., Gamper, H., et al., 2010. Shifting C flow from roots into associated microbial communities in response to elevated atmospheric CO_2. Proceedings of the National Academy of Sciences United States of America 107, 10938–10942.

Drigo, B., Anderson, I.C., Kannangara, G.S.K., Cairney, J.W.G., Johnson, D., 2012. Rapid incorporation of carbon from ectomycorrhizal mycelial necromass into soil fungal communities. Soil Biology and Biochemistry 49, 4–10.

Drigo, B., Kowalchuk, G.A., Knaap, B.M., Pijl, A.S., Boschker, T.S., Veen, J.A., 2013. Impacts of three years of elevated atmospheric CO_2 on rhizosphere carbon flow and microbial community dynamics. Global Change Biology 19 (2), 621–636.

Evans, S.E., Wallenstein, M.D., 2014. Climate change alters ecological responses of soil bacteria. Ecology Letters 17, 155–164.

Filion, M., St-Arnaud, M., Fortin, J.A., 1999. Direct interaction between the arbuscular mycorrhizal fungus Glomus intraradices and different rhizosphere microorganisms. New Phytol. 141, 525–533.

Frey, S.D., Elliott, E.T., Paustian, K., Peterson, G.A., 2000. Fungal translocation as a mechanism for soil nitrogen inputs to surface residue decomposition in a no-tillage agroecosystem. Soil Biology and Biochemistry 32, 689–698.

Frey-Klett, P., Chavatte, M., Clausse, M., Courrier, S., Le Roux, C., Raaijmakers, J., et al., 2005. EM symbiosis affects functional diversity of rhizosphere fluorescent pseudomonads. New Phytologist 165, 317–328.

Frey-Klett, P., Garbaye, J., Tarkka, M., 2007. The mycorrhiza helper bacteria revisited. New Phytologist 176, 22–36.

Gahan, J., Schmalenberger, A., 2015. Arbuscular mycorrhizal hyphae in grassland select for a diverse and abundant hyphospheric bacterial community involved in sulfonate desulfurization. Applied Soil Ecology 89, 113–121.

Garbaye, J., 1994. Helper bacteria: a new dimension to the mycorrhizal symbiosis. New Phytologist 128, 197–210.

Ghignone, S., Salvioli, A., Anca, I., Lumini, E., Ortu, G., Petiti, L., et al., 2012. The genome of the obligate endobacterium of an AM fungus reveals an interphylum network of nutritional interactions. The ISME Journal 6, 136–145.

Godbold, D.L., Hoosbeek, M.R., Lukac, M., Cotrufo, M.F., Janssens, I.A., Ceulemans, R., Polle, A., Velthorst, E.J., Scarascia-Mugnozza, G., De Angelis, P., Miglietta, F., Peressotti, A., 2006. Mycorrhizal hyphal turnover as a dominant process for carbon input into soil organic matter. Plant and Soil 281, 15–24.

Hodge, A., Storer, K., 2015. Arbuscular mycorrhiza and nitrogen: implication for individual plants through to ecosystems. Plant and Soil 386, 1–19.

IPCC, 2014. Summary for Policymakers. In: Field, CB, Barros, VR, Dokken, DJ, Mach, KJ, Mastrandrea, MD, et al. (Eds.), Climate Change 2014: Impacts, Adaptation, and Vulnerability. Part A: Global and Sectoral Aspects. Contribution of Working Group II to the Fifth Assessment Report of the Intergovernmental Panel on Climate Change. Cambridge University Press, Cambridge, United Kingdom and New York, NY, USA, pp. 1–32.

Jakobsen, I., Rosendahl, L., 1990. Carbon flow into soil and external hyphae from roots of mycorrhizal cucumber plants. New Phytol. 115, 77–83.

Jansa, J., Smith, F.A., Smith, S.E., 2008. Are there benefits of simultaneous root colonization by different arbuscular mycorrhizal fungi? New Phytologist 177, 779–789.

Jansa, J., Bukovská, P., Gryndler, M., 2013. Mycorrhizal hyphae as ecological niche for highly specialized hypersymbionts - or just soil free-riders. Frontiers in Plant Science 4 Article 134.

Johansson, J.F., Paul, L.R., Finlay, R.D., 2004. Microbial interactions in the mycorrhizosphere and their significance for sustainable agriculture. FEMS Microbiology Ecology 48, 1–13.

Johnson, N.C., Gehring, C.A., 2007. Mycorrhizas: symbiotic mediators of rhizosphere and ecosystem processes. In: Cardon, Z.G., Whitbeck, J.L. (Eds.), The Rhizosphere: An Ecological Perspective. Elsevier Academic Press, London, UK, pp. 31–56.

Johnson, N.C., Wolf, J., Koch, G.W., 2003. Interactions among mycorrhizas, atmospheric CO_2 and soil N impact plant community composition. Ecology Letters 6, 532–540.

Johnson, N.C., Angelard, C., Sanders, I.R., Kiers, E.T., 2013. Predicting community and ecosystem outcomes of mycorrhizal responses to global change. Ecology Letters 16, 140–153.

Johnson, N.C., Wilson, G.W.T., Wilson, J.A., Miller, R.M., Bowker, M.A., 2015. Mycorrhizal phenotypes and the law of the minimum. New Phytologist 205, 1473–1484.

Jorquera, M.A., Hernández, M.T., Rengel, Z., Marschner, P., Luz Mora, M., 2008. Isolation of culturable phosphobacteria with both phytate-mineralization and phosphate-solubilization activity from the rhizosphere of plants grown in a volcanic soil. Biol Fertil Soils 44, 1025–1034.

Kaiser, C., Kilburn, M.R., Clode, P.L., Fuchslueger, L., Koranda, M., Cliff, J.B., et al., 2015. Exploring the transfer of recent plant photosynthates to soil microbes: mycorrhizal pathway vs direct root exudation. The New Phytologist 205, 1537–1551. http://dx.doi.org/10.1111/nph.13138.

Koele, N., Turpault, M.P., Hildebrand, E.E., Uroz, S., Frey-Klett, P., 2009. Interactions between mycorrhizal fungi and mycorrhizosphere bacteria during mineral weathering: budget analysis and bacterial quantification. Soil Biology and Biochemistry 41, 1935–1942.

Kohlmeier, S., Smits, T.H.M., Ford, R.M., Keel, C., Harms, H., Wick, L.Y., 2005. Taking the fungal highway: mobilization of pollutant-degrading bacteria by fungi. Environmental Science and Technology 39, 4640–4646. http://dx.doi.org/10.1021/es047979z.

Koide, R.T., Malcolm, G.M., 2009. N concentration controls decomposition rates of different strains of ectomycorrhizal fungi. Fungal Ecology 2, 197–202. http://dx.doi.org/10.1016/j.funeco.2009.06.001.

Lendenmann, M., Thonar, C., Barnard, R.L., Salmon, Y., Werner, R.A., Frossard, E., et al., 2011. Symbiont identity matters: carbon and phosphorus fluxes between *Medicago truncatula* and different arbuscular mycorrhizal fungi. Mycorrhiza 21, 689–702.

Li, M., Feng, F., Cheng, L., 2012a. Expression patterns of genes involved in sugar metabolism and accumulation during apple fruit development. PLoS ONE 7: e33055. http//:dx.doi.org.10.1371/journal.pone.0033055.

Li, M., Xu, J., Romero-Gonzalez, M., Banwart, S.A., et al., 2012b. Single cell Raman spectroscopy for cell sorting and imaging. Curr. Opin. Biotechnol. 23, 56–63.

Lindahl, B.D., de Boer, W., Finlay, R.D., 2010. Disruption of root carbon transport into forest humus stimulates fungal opportunists at the expense of mycorrhizal fungi. ISME Journal 4, 872–881. http://dx.doi.org/10.1038/ismej.2010.19.

Lumini, E., Bianciotto, V., Jargeat, P., Novero, M., Salvioli, A., Faccio, A., et al., 2007. Presymbiotic growth and sporal morphology are affected in the arbuscular mycorrhizal fungus *Gigaspora margarita* cured of its endobacteria. Cellular Microbiology 9, 1716–1729.

Mansfeld-Giese, K., Larsen, J., Bodker, L., 2002. Bacterial populations associated with mycelium of the arbuscular mycorrhizal fungus Glomus intraradices. FEMS Microbiol. Ecol. 41, 133–140.

Marschner, P., Baumann, K., 2003. Changes in bacterial community structure induced by mycorrhizal colonisation in split-root maize. Plant and Soil 251, 279–289.

Meier, M.S., Stoessel, F., Jungbluth, N., Juraske, R., Schader, C., Stolze, M., 2015. Environmental impacts of organic and conventional agricultural products – are the differences captured by life cycle assessment? J. Environ. Manag. 149, 193–208.

Meharg, A.A., Cairney, J.W.G., 2000. Ectomycorrhizas extending the capabilities of rhizosphere remediation? Soil Biology and Biochemistry 32 (11–12), 1475–1484.

Medina, A., Azcón, R., 2010. Effectiveness of the application of arbuscular mycorrhiza fungi and organic amendments to improve soil quality and plant performance under stress conditions. J Soil Sci Plant Nutr 10, 354–372.

Monastersky, R., 2013. Global carbon dioxide levels near worrisome. Nature 497 (7447)13–14.

Nielsen, U.N., Ball, B.A., 2015. Impacts of altered precipitation regimes on soil communities and biogeochemistry in arid and semi-arid ecosystems. Global Change Biology 21, 1407–1421.

Nguyen, N.H., Bruns, T.D., 2015. The microbiome of *Pinus muricata* ectomycorrhizas: community assembleges, fungal species effects, and *Burkholderia* as important bacteria in multipartnered symbioses. Microbial Ecology 69, 914–921.

Nuccio, E.E., Hodge, A., Pett-Ridge, J., Herman, D.J., Weber, P.K., Firestone, M.K., 2013. An arbuscular mycorrhizal fungus significantly modifies the soil bacterial community and nitrogen cycling during litter decomposition. Environ. Microbiol 15, 1870–1881.

Olsson, P.A., Johnson, N.C., 2005. Tracking C from the atmosphere to the rhizosphere. Ecology Letters 8, 1264–1270.

Paul, E.A., Kucey, R.M.N., 1981. Carbon flow in plant-microbial associations. Science 213, 473–474.

Paulitz, T.C., Linderman, R.G., 1991. Lack of antagonism between the biocontrol agent *Gliocladium virens* and vesicular arbuscular mycorrhizal fungi. New Phytologist 117, 303–308.

Pett-Ridge, J., Weber, P.K., 2011. NanoSIP: NanoSIMS applications for microbial biology. In: Navid, A. (Ed.), Microbial Systems Biology: Methods and Protocols. Humana Press.

Pozo, M., Azcón-Aguilar, C., 2007. Unraveling mycorrhiza-induced resistance. Current Opinion in Plant Biology 10, 393–398.

Pritchard, S.G., 2011. Soil organisms and global climate change. Plant Pathology 60, 82–99.

Rapparini, F., Penuelas, J., 2014. Mycorrhizal fungi to alleviate drought stress on plant growth. In: Miransari, M. (Ed.), Use of Microbes for the Alleviation of Soil Stresses. Springer Science and Business Media, New York, FL, pp. 21–42.

Ravnskov, S., Nybroe, O., Jakobsen, I., 1999. Influence of an arbuscular mycorrhizal fungus on *Pseudomonas fluorescens* DF57 in rhizosphere and hyphosphere soil. New Phytologist 142, 113–122.

Redecker, D., Morton, J.B., Bruns, T.D., 2000. Ancestral lineages of arbuscular mycorrhizal fungi (Glomales). Molecular Phylogenetics and Evolution 2, 276–284.

Remy, W., Taylor, T.N., Hass, H., Kerp, H., 1994. Four hundred-million-year-old vesicular arbuscular mycorrhizas. Proc. Natl. Acad. Sci. USA 91, 11841–11843.

Richardson, A.E., Simpson, R.J., 2011. Soil microorganisms mediating phosphorus availability. Plant Physiology 156, 989–996.

Rillig, M.C., Allen, M.F., 1999. What is the role of arbuscular mycorrhizal fungi in plant-to-ecosystem responses to elevated atmospheric CO_2? Mycorrhiza 9, 1–8.

Rillig, M.C., Wright, S.F., Shaw, M.R., Field, C.B., 2002. Artificial climate warming positively affects arbuscular mycorrhizas but decreases soil aggregate water stability in an annual grassland. Oikos 97, 52–58.

Rillig, M.C., Mummey, D.L., Ramsey, P.W., Klironomos, J.N., Gannon, J.E., 2006. Phylogeny of arbuscular mycorrhizal fungi predicts community composition of symbiosis-associated bacteria. FEMS Microbiology Ecology 57, 389–395.

Rodriguez, H., Fraga, R., 1999. Phosphate solubilizing bacteria and their role in plant growth promotion. Biotechnology Advances 17, 319–339.

Rousseau, A., Benhamou, N., Chet, I., Piche, Y., 1996. Mycoparasitism of the extramatrical phase of Glomus intraradices by Trichoderma harzianum. Phytopathology 86, 434–443.

Rudnick, M.B., van Veen, J.A., de Boer, W., 2015. Baiting of rhizosphere bacteria with hyphae of common soil fungi reveals a diverse group of potentially mycophagous secondary consumers. Soil Biology and Biochemistry 88, 73–82.

Vannini, A., Ghignone, S., Novero, M., Navazio, L., Venice, F., Bagnaresi, P., et al., 2016. Symbiosis with an endobacterium increases the fitness of a mycorrhizal fungus, raising its bioenergetic potential. The ISME Journal 10, 130–144.

Scheublin, T.R., Sanders, I.R., Keel, C., Roelof van der Meer, J., 2010. Characterisation of microbial communities colonising the hyphal surfaces of arbuscular mycorrhizal fungi. The ISME Journal 4, 752–763.

Schimel, J.P., Balser, T.C., Wallenstein, M., 2007. Microbial stress-response physiology and its implications for ecosystem function. Ecology 88, 1386–1394.

Schüßler, A., Schwarzott, D., Walker, C., 2001. A new fungal phylum, the Glomeromycota: phylogeny and evolution. Mycol. *Res* 105 (12), 1413–1421.

Simon, A., Bindschedler, S., Job, D., Wick, L.Y., Filippidou, S., Kooli, W.M., et al., 2015. Exploiting the fungal highway: development of a novel tool for the in situ isolation of bacteria migrating along fungal mycelium. FEMS Microbiology Ecology 91 fiv116.

Simard, S.W., Perry, D.A., Jones, M.D., Myrold, D.M.D., Molina, R., 1997. Net transfer of carbon between ectomycorrhizal tree species in the field. Nature 388, 579–582.

Smith, S.E., Read, D.J., 1997. Mycorrhizal Symbiosis. Academic Press, San Diego, CA.

Smith, S.E., Read, D.J., 2008. Mycorrhizal Symbiosis, third ed. Academic Press, London.

Taktek, S., Trèpanier, M., Servin, P.M., St-Arnaud, M., Pichè, Y., Fortin, J.-A., et al., 2015. Trapping of phosphate solubilizing bacteria on hyphae of the arbuscular mycorrhizal fungus *Rhizophagus irregularis* DAOM 197198. Soil Biology and Biochemistry 90, 1–9.

Thomas, C.D., Cameron, A., Green, R.E., Bakkenes, M., Beaumont, L.J., Collingham, Y.C., Erasmus, B.F.N., de Siqueira, M.F., Grainger, A., Hannah, L., Hughes, L., Huntley, B., van Jaarsveld, A.S., Midgley, G.F., Miles, L., Ortega-Huerta, M.A., Peterson, A.T., Phillips, O.L., Williams, S.E., 2004. Extinction risk from climate change. Nature 427, 145–148.

Timonen, S., Hurek, T., 2006. Characterization of culturable bacterial populations associating with *Pinus sylvestris-Suillus bovinus* mycorrhizospheres. Can. J. Microbiol 52, 769–778.

Timonen, S., Marschner, P., 2006. Mycorrhizosphere concept. In: Mukerji, K.G., Manoharachary, C., Singh, J. (Eds.). Mukerji, K.G., Manoharachary, C., Singh, J. (Eds.), Soil Biology, Microbial Activity in the Rhizosphere, vol. 7. Springer-Verlag, Berlin Heidelberg, pp. 155–172.

Tisserant, E., Malbreil, M., Kuo, A., Kohler, A., Symeonidid, A., Balestrini, R., et al., 2013. Genome of an arbuscular mycorrhizal fungus provides insight into the oldest plant symbiosis. Proceedings of the National Academy of the Sciences of the United States of America 110, 20117–20122.

Toljander, J.F., Artursson, V., Paul, L.R., Jansson, J.K., Finlay, R.D., 2006. Attachment of different soil bacteria to arbuscular mycorrhizal fungal extraradical hyphae is determined by hyphal vitality and fungal species. FEMS Microbiology Ecology 254, 34–40.

Toljander, J.F., Lindahl, B.D., Paul, L.R., Elfstrand, M., Finlay, R.D., 2007. Influence of arbuscular mycorrhizal mycelial exudates on soil bacterial growth and community structure. FEMS Microbiology Ecology 61, 295–304.

Toro, M., Azcón, R., Barea, J.M., 1997. Improvement of arbuscular mycorrhiza development by inoculation of soil with phosphate-solubilizing rhizobacteria to improve rock phosphate bioavailabilty (^{32}P) and nutrient cycling. Applied and Environmental Microbiology 63, 4408–4412.

Treseder, K.K., Allen, M.F., 2000. Mycorrhizal fungi have a potential role in soil C storage under elevated CO_2 and nitrogen deposition. New Phytologist 147, 189–200.

Treseder, K.K., 2016. A meta-analysis of mycorrhizal responses to nitrogen, phosphorus, and atmospheric CO_2 in field studies. New Phytologist 164, 347–355.

van Breemen, N., Finlay, R., Lundström, U., Jongmans, A.G., Giesler, R., Olsson, M., 2000. Mycorrhizal weathering: a true case of mineral plant nutrition? Biogeochemistry 49, 53–67.

van der Heijden, M.G.A., Klironomos, J.N., Ursic, M., Moutoglis, P., Streitwolf-Engel, R., et al., 1998. Mycorrhizal fungal diversity determines plant biodiversity, ecosystem variability and productivity. Nature 396, 69–72.

van der Heijden, M.G.A., Bardgett, R.D., van Straalen, N.M., 2008. The unseen majority: soil microbes as drivers of plant diversity and productivity in terrestrial ecosystems. Ecology Letters 11, 296–310.

van der Heijden, M.G.A., Martin, F.M., Selosse, M.A., Sanders, I.R., 2015. Mycorrhizal ecology and evolution: the past, the present, and the future. New Phytologist 205, 1406–1423.

Vannini, C., Carpentieri, A., Salvioli, A., Novero, M., Marsoni, M., Testa, L., del Pinto, M.C., Amoresano, A., Ortolani, F., Bracale, M., Bonfante, P., 2016. An interdomain network: the endobacterium of a mycorrhizal fungus promotes antioxidative responses in both fungal and plant hosts. New Phytologist 211, 265–275.

Veresoglou, S.D., Chen, B., Rillig, M.C., 2012. Arbuscular mycorrhiza and soil nitrogen cycling. Soil Biology and Biochemistry 46, 53–62.

Viollet, A., Corberand, T., Mougel, C., Robin, A., Lemanceau, P., Mazurier, S., 2011. Fluorescent pseudomonads harboring type III secretion genes are enriched in the mycorrhizosphere of *Medicago truncatula*. FEMS Microbiology Ecology 75, 457–467.

Vries, F.T., Manning, P., Tallowin, J.R.B., Mortimer, S.R., Pilgrim, E.S., et al., 2012. Abiotic drivers and plant traits explain landscape-scale patterns in soil microbial communities. Ecology Letters 15, 1230–1239.

Wallander, H., Thelin, G., 2008. The stimulating effect of apatite on ectomycorrhizal growth diminishes after PK fertilization. Soil Biology and Biochemistry 40, 2517–2522.

Wang, F., Shi, N., Jiang, R.F., Zhang, F.S., Feng, G., 2016. In situ stable isotope probing of phosphate-solubilizing bacteria in the hyphosphere. Journal of Experimental Botany 67, 1689–1701. http://dx.doi.org/10.1093/jxb/erv561.

Wilkinson, A., Alexander, I.J., Johnson, D., 2011. Species richness of ectomycorrhizal hyphal necromass increases soil CO_2 efflux under laboratory conditions. Soil Biology and Biochemistry 43, 1350–1355. http://dx.doi.org/10.1016/j.soilbio.2011.03.009.

Zhang, L., Fan, J., Ding, X., He, X., Zhang, F., Feng, G., 2014a. Hyphosphere interactions between an arbuscular mycorrhizal fungus and a phosphate solubilizing bacteria promote phytate mineralization in soil. Soil Biology and Biochemistry 74, 177–183.

Zhang, M., Pereira E Silva, Mde C., Chaib de Mares, M., van Elsas, J.D., 2014b. The mycosphere constitutes an arena for horizontal gene transfer with strong evolutionary implications for bacterial-fungal interactions. FEMS Microbiology Ecology 89, 516–526.

Zhang, L., Xu, M., Liu, Y., Zhang, F., Hodge, A., Feng, G., 2016. Carbon and phosphorus exchange may enable cooperation between an arbuscular mycorrhizal fungus and a phosphate-solubilizing bacterium. New Phytologist. http://dx.doi.org/10.1111/nph. 13838.

CHAPTER

23

Immobilization of Carbon in Mycorrhizal Mycelial Biomass and Secretions

R.D. Finlay, K.E. Clemmensen
Swedish University of Agricultural Sciences, Uppsala, Sweden

23.1 INTRODUCTION

The flow of carbon through plant roots (Jones et al., 2009) and their associated mycorrhizal mycelia (Finlay, 2008) is of fundamental importance for many biogeochemical processes. Allocation of C to mycorrhizal fungal mycelia and its subsequent partitioning into structural and nonstructural biomass components, secretions, respiratory losses, and various forms of necromass or secretions with different degrees of incorporation into organic material and secondary minerals is a complex process involving many interactions taking place on different temporal and spatial scales. It is impossible to discuss the flow of C through these pools without reference to the types of interaction that take place and the way that C transfer is regulated. Many of these interactions are dealt with in more detail in separate chapters, including the role of mycorrhizal mycelia in pedogenesis (Chapter 2), mineral weathering (Chapter 3), the mobilization of nutrients from organic material (Chapter 8), the effects of mycorrhizal mycelia and mycelial secretions on soil aggregate formation (Chapter 14), C and energy supply from organic substrates (Chapter 20) and plants (Chapter 21), interactions with hyphae-associated microbes (Chapter 22), decomposer fungi (Chapter 24), or animals. In each of these areas different types of interaction, occurring on different time scales, may result in sequestration or release of C. These processes are summarized in Fig. 23.1 and discussed here in so far as they influence the balance between C release and C sequestration.

In this chapter we first discuss C allocation in relation to the growth dynamics, distribution, overall production, and standing biomass of different types of mycorrhizal mycelia as well as different factors influencing these parameters. Secondly we identify some of the components of the fungal secretome that play an important role in interactions with mineral and organic substrates. Thirdly, we discuss fungal necromass properties and decomposition, and

Mycorrhizal Mediation of Soil
http://dx.doi.org/10.1016/B978-0-12-804312-7.00023-1

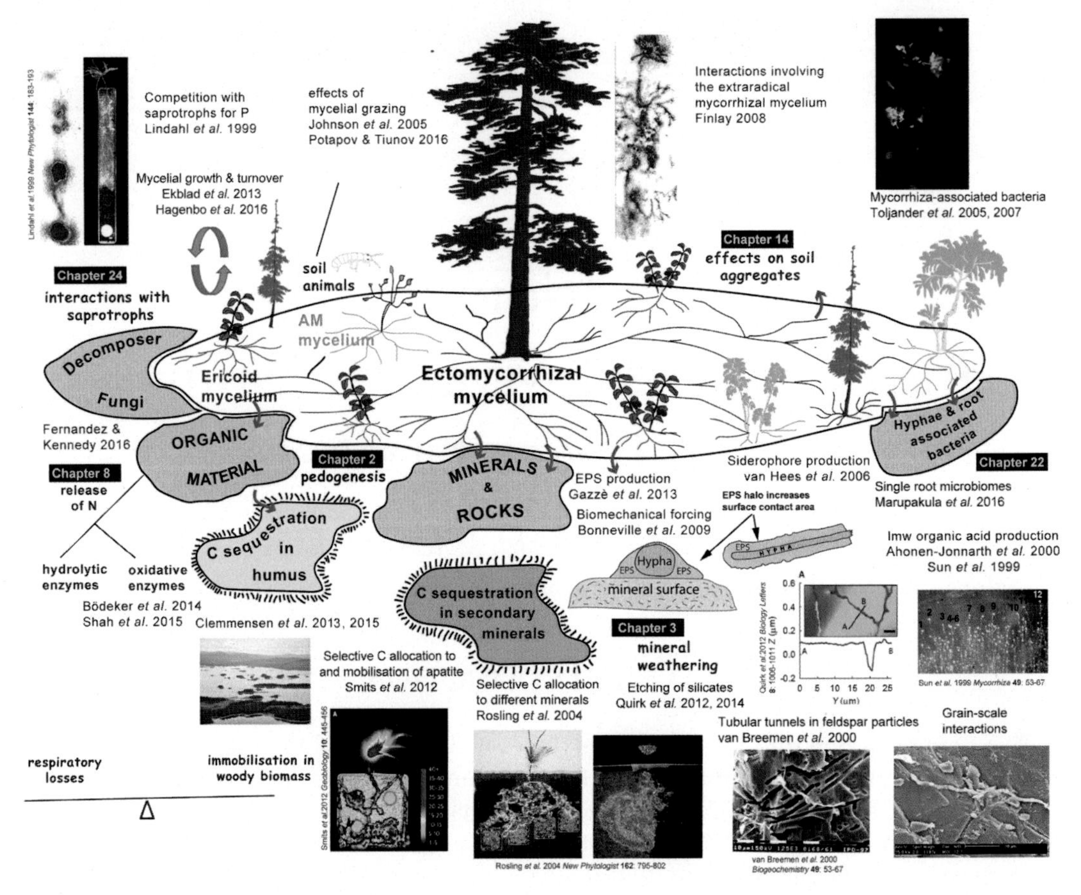

FIGURE 23.1 Summary of the different interactions of mycorrhizal mycelia involving transport of carbon.

finally we discuss the possible long-term immobilization of C into secondary minerals and stable soil organic matter (SOM).

Mycorrhizal fungal biomass in soils represents a large sink for current assimilates in many ecosystem types. Through their dependence on sugars, derived directly from their plant hosts, the mycorrhizal fungi represent a unique functional guild, with a C allocation pathway into the soil, that contrasts with the other major fungal guild—the saprotrophic fungi—that depend on dead organic matter. Once in mycorrhizal fungal biomass, C is allocated into either mycelial growth or metabolic activities, the latter leading to either respiratory losses or incorporation of C into secreted substances that interact with organic or mineral substrates, resulting either in further C losses through respiration or leaching or to incorporation of C into recalcitrant organic substrates or secondary minerals. The internal allocation of C within the mycorrhizal mycelium affects the C residence time in the living biomass as well as in various soil pools derived from the mycelial biomass. Residence times of the C allocated to the mycorrhizal mycelium probably vary from a few hours to centuries at the ecosystem level. The focus of our discussion in this chapter is on amounts and residence times rather than functions because these are treated in detail in other chapters.

23.2 MYCELIAL BIOMASS PRODUCTION AND TURNOVER

Mycorrhizal mycelial biomass in complex substrates such as soils has traditionally been quantified either as part of the total microbial biomass or as part of the total fungal biomass. No one biomarker gives a perfect picture of fungal biomass (Baldrian et al., 2013); rather, the measurements of different markers represent a continuum from active biomass to decomposing necromass (Ekblad et al., 1998; Yuan et al., 2008).

Likewise, mycorrhizal fungal activities have been integrated into measurements of total respiratory fluxes from soils and often interpreted as part of the autotrophic (i.e. plant), belowground respiration. Therefore mycorrhizal fungal C allocation, even between the major pathways—biomass production and metabolic processes is still not well understood. It has proved difficult to find any biochemical marker for mycorrhizal fungi that can be directly related to their biomass and/or activity. Ectomycorrhizal (EcM) and ericoid mycorrhizal (ErM) associations are formed mainly by members of the Ascomycota and Basidiomycota. These phyla also include most other fungal guilds, including the saprotrophic fungi, and the main chemical constituents of the fungal biomass are relatively similar across guilds—apart from the Glomeromycota. Thus, for biomass and activity estimates of the fungal partner, the approach has depended on the system and mycorrhizal type of interest.

During recent years the use of ingrowth cores (Wallander et al., 2001, 2013; Johnson et al., 2002a,b), coupled with the advance of new large-scale DNA- and RNA-based molecular tools (Lindahl et al., 2013), have made more specific measurements of mycorrhizal fungal biomass and activity possible (e.g. Ekblad et al., 2013, 2016; Wallander et al., 2010). Some estimates of mycorrhizal biomass in different systems are presented in Box 23.1 and are discussed in a comprehensive review by Leake et al. (2004). Production and turnover are less well studied than biomass, but recent estimates are also presented in Box 23.1 and discussed here.

23.2.1 Arbuscular Mycorrhizal Fungi

Estimates of the extraradical mycelial biomass of arbuscular mycorrhizal (AM) fungi are typically expressed in terms of hyphal length (m)/g soil or hyphal length (m)/m colonized root length. Typical estimates range between 3 and 30 m/g soil and lengths of extraradical hyphae are typically one to two orders of magnitude larger than the colonized root length (see Leake et al., 2004). Data are available from pot and hyphal compartment experiments, as well as field experiments, although field experiments are sometimes difficult to interpret because of the presence of saprotrophic fungal mycelium, which may be difficult to distinguish from the AM hyphae for the untrained eye. C flow to AM mycelia has been studied in field soils using cores that exclude roots but allow ingrowth of mycorrhizal hyphae (Johnson et al., 2002a,b). Using ^{13}C feeding to host plants, these experiments have shown rapid transport of labeled plant assimilates to the extraradical mycelium that, together with the estimated respiratory fluxes after 72 h, accounted for 9% of net C fixation.

23.2.2 Ectomycorrhizal Fungi

The biomass of EcM fungi is partitioned among three main components: mycorrhizal root tips, mycorrhizal extramatrical mycelium, and spore-forming carpophores. Depending on the species involved, the EcM fungal mycelium typically extends between just a few millimeters

BOX 23.1 ESTIMATES OF MYCORRHIZAL FUNGAL STANDING BIOMASS, PRODUCTION, AND TURNOVER

Mycorrhizal Type	System	*n*	Biomass (kg DM/ha) Unless Otherwise Indicated	Biomass Production (kg DM/ha/year) Unless Otherwise Indicated	Turnover (Times / year)	References
Arbuscular mycorrhiza	Pot system with *Plantago lanceolata* and *Glomus* spp.	15	3–30 m/g soil		60	Staddon et al. (2003)
	Hyphal compartment and pot studies					Leake et al. (2004)
Ectomycorrhiza	Boreal and temperate forest	140	100–900 (400)*	100–600 (200)*	0.5–1**	Ekblad et al. (2013), this chapter
	Boreal forest	7	40–300	90–220	0.3–7	Hagenbo et al. (2016)
	Subtropical forest	1	300	3000	10	Hendricks et al. (2016),
		1			13	Ekblad et al. (2016)
Ericoid mycorrhiza	Boreal forest	35	480–1500		1***	Clemmensen et al. (2013), Sterkenburg et al. (2015) Persson (1980)
	Temperate heath	1			0.5–1***	Aerts et al. (1992)

DM, dry mass. *, Amounts in parentheses represent mean values based on data in Ekblad et al. (2013) and estimates presented in this chapter, all excludes mineral soil; **, biomass and turnover includes biomass in mantles and mycelium; ***, turnover estimates are for ericaceous fine roots.

to several decimeters from the EcM root tips and forms efficient conduits for C into the soil matrix. Some species have been found to form extensive mycelial networks covering up to several tens of square meters (Dahlberg, 1997). Thus in forested ecosystems dominated by EcM trees the standing biomass and the annual production of EcM mycelium may be large and constitute a significant fraction of total belowground biomass and flow of C into the soil.

Using a field experiment in which 50-year-old pine trees were girdled to terminate belowground transport of C, Högberg and Högberg (2002) estimated EcM fungal biomass to constitute at least one third of the total microbial biomass in the soil (excluding roots) and to correspond to approximately 145 kg dry mass per hectare. However, this estimate may be an underestimate because it was based on decreases in total microbial biomass using the fumigation-extraction method 1–3 months after the girdling, at which time other microbial communities benefitting from the newly dead mycelium could have increased their biomass (Högberg and Högberg, 2002). In the same experiment, living EcM roots and mycelium were estimated to contribute at least 50% of total soil respiration (Högberg et al., 2001) and 40–50% of the dissolved organic carbon (DOC) pool in the soil (Högberg and Högberg, 2002).

Both chitin, one of the major structural components of fungal cell walls, and ergosterol, the fungal analogue to the membrane-bound cholesterol in animals, have been used as more direct markers for fungal biomass in soil and plant tissues. However, chitin—and to some extent ergosterol—may remain after the fungus dies (Ekblad et al., 1998; Yuan et al., 2008), and both markers indifferently target mycorrhizal and saprotrophic fungi. When combined with manipulations to physically separate root-associated and saprotrophic fungi or with DNA-based community quantification, these markers have proved particularly valuable.

Using a root-severing experiment in a coniferous forest, causing a decline in ergosterol and fungal phospholipid fatty acids supposedly due to necrosis of EcM mycelia, Wallander et al. (2001) estimated the annual production of EcM mycelium to be 125–200 kg/ha and the mycelial biomass (including mantles) to be 700–900 kg per hectare across three temperate coniferous forests. In boreal forests, ergosterol concentration in soils including fine roots is typically close to 50 μg/g (Wallander et al., 2010; Clemmensen et al., 2013; Sterkenburg et al., 2015). This corresponds to a total standing fungal biomass of 1600–2500 kg dry mass per hectare for organic soils with the typical 10–15 kg organic matter per square meter in the mor layer and assuming 3 μg ergosterol per mg fungal mass. With an approximately 1:1 contribution by saprotrophic and mycorrhizal fungi to fungal communities in these systems (based on DNA marker abundances), standing mycorrhizal biomass would then be 800–1250 kg dry mass per hectare, consistent with the estimate by Wallander et al. (2001). However, other root-associated fungi may contribute to biomass, particularly those associated with the ericoid understory, and based on ITS abundances EcM fungi have been estimated to contribute only 10–20% of total fungal biomass in the northern and the most nutrient-deprived forests (Clemmensen et al., 2013; Sterkenburg et al., 2015), suggesting 160–500 kg EcM fungal biomass per hectare (Box 23.1). In a humid subtropical pine forest, EcM mycelial biomass has been estimated to be 300 kg/ha in the upper 10 cm on the basis of difference in ergosterol before and after 5 months of laboratory incubations to eliminate EcM fungi deprived of their energy source (Hendricks et al., 2016).

The use of sand-filled ingrowth cores has made more specific measurements of mycorrhizal mycelial production and activities possible because the lack of an organic C source is reasoned to decrease the saprotrophic background in these cores (Wallander et al., 2001).

However, estimates based on long incubation times may underestimate production due to simultaneous turnover in the bags (Wallander et al., 2001, 2013), and the artificial nature of the sand substrate may also lead to possible underestimates (Hendricks et al., 2006). Using this method, across the 140 temperate and boreal forest stands reviewed by Ekblad et al. (2013), annual EcM mycelial production in the upper 10 cm of soil ranges between 100 and 600 kg dry mass ha^{-1} (Box 23.1). In the subtropical pine forest studied by Hendricks et al. (2016), mycelial production was estimated to reach 3000 kg/ha year using short-term incubated ingrowth cores filled with native soil. Significant amounts of mycelium are also produced in mineral soil layers (Wallander et al., 2004), and Ekblad et al. (2013) suggested a general production estimate in the upper 10 cm of soil in 40-year-old Swedish *Picea abies* forest to be approximately 200 and at least 400 kg/ha year for the whole soil profile. This corresponds to approximately 5% of net primary production in this forest type (Ekblad et al., 2013).

Thus growth and biomass of EcM mycelium vary widely among systems and likely depend on climatic, edaphic, and biotic factors. Stand age was found to affect mycelial production along a gradient of forty, 1- to 130-year-old managed spruce forests, with the highest production measured around canopy closure in the 10- to 30-year-old stands (Wallander et al., 2010). Mycelial production is also highly sensitive to forest fertilization, although responses have been found to be highly site specific (e.g. Wallander et al., 2011; Kjøller et al., 2012). Available biomass and production estimates are highly biased toward coniferous stands in Scandinavia, where most studies have been conducted so far (Ekblad et al., 2013). Future studies should elucidate whether EcM-dominated ecosystems found outside of this region—such as in the arctic and tropical zones—show similar biomass and production ranges.

23.2.3 Ericoid Mycorrhizal Fungi

ErM dwarf shrubs, belonging to the family Ericaceae, dominate natural arctic-alpine tundra, cultural heathlands in the temperate zone, and the understory vegetation in forests and mire vegetation of the boreal zone. However, the extent to which ErM fungi contribute to soil fungal biomass in these systems is largely unknown (Read and Perez-Moreno, 2003). The ErM fungi grow mainly as hyphal coils inside of the epidermal cells of the hair roots of the ericoid plant. Because of the short range of their extramatrical mycelium, their production cannot be estimated using ingrowth bags in the same way as for EcM systems. Furthermore, the abundance and growth dynamics of ErM fungi are little studied in the field because few of the involved species form conspicuous fruit bodies.

Early work with isolation of ericoid endophytes, verified as mycorrhizal endophytes by back inoculation, revealed that isolated fungi across all continents were predominantly slow growing, asexual, and dark colored (Smith and Read, 2008; and references therein). These were later identified, on the basis of fruit body formation in culture, as *Rhizoscyphus ericae*, a member of the Ascomycete order Helotiales. With increasing amounts of DNA sequence data associated with ErM isolates, it has since become evident that *R. ericae* is part of the larger *Hymenoscyphus* aggregate, also containing several species verified as capable of forming EcM and ErM (Vrålstad et al., 2002; Grelet et al., 2009).

A few other confirmed ErM endophytes have been identified as belonging to other Ascomycete clades within Helotiales or Chaetothyriales, or the Basidiomycete order Sebacinales (Selosse et al., 2007). However, based on molecular identification of fungi in surface-sterilized ericoid roots it has become increasingly evident that the diversity of endophytic fungi is much larger than known from culturing studies (e.g. Cairney and Meharg, 2003; Bougoure and Cairney, 2005; Bougoure et al., 2007; Kjøller et al., 2010; Walker et al., 2011).

In a study of soil fungal communities in organic soil profiles of a 100- to 5000-year northern boreal forest chronosequence, 60–95% of the fungal amplicons in the rooting zone belonged to Ascomycota, particularly Helotiales and Chaetothyriales (Clemmensen et al., 2013, 2015). To interpret this result in a functional manner, an approach was adopted in which DNA-based fungal species (i.e. species hypotheses *sensu*; Kõljalg et al., 2013) were classified as ErM based on close sequence similarity to either sequences derived from verified ErM fungi or sequences derived from surface-sterilized roots of ericoid plants. Assuming that such root-associated fungi were ErM, approximately 60% of the fungal sequences in the rooting zone (40% across all soil horizons) were ErM in the boreal forest chronosequence and approximately 30% across the 25 pristine hemiboreal forests studied by Sterkenburg et al. (2015). This corresponds to an estimated standing biomass of 480–1500 kg/ha, based on the same calculation as above for EcM (Box 23.1), and recognizing that fungal biomass estimated from amplicon proportions is merely a rough estimate (Baldrian et al., 2013).

23.2.4 Biomass Turnover

The rate of mycorrhizal fungal biomass turnover (i.e. the rate with which the biomass is replaced per unit time), together with the size of the standing biomass pool, is the process that ultimately determines how much mycelial necromass is deposited into the SOM pool. This process is still not very well understood because good methods are not available (Ekblad et al., 2013).

Estimates of turnover of AM fungal hyphae have been obtained in pot systems containing *Plantago lanceolata* plants colonized by different *Glomus* species. To quantify C turnover in hyphae, plants were exposed to fossil "^{14}C-dead" carbon dioxide for 5 h, then returned to a normal ambient CO_2 environment, and the ^{14}C content in hyphae was determined using accelerator mass spectrometry (Staddon et al., 2003). The results from this study suggest that the extraradical hyphae of AM fungi live on average 5–6 days, implying an annual turnover of at least 60 times per year (Box 23.1). Turnover of intraradical hyphae and spores may be slower than this, but the results highlight the importance of the mycorrhizal mycelium as a rapid pathway for C input to soil.

Estimates of standing mycelial biomass and mycelial production for EcM systems presented above are roughly in the same range as each other, suggesting biomass turnover of approximately 0.5–2 times per year. This is comparable to estimates of fine root turnover of 0.4–1.3 times per year in temperate and boreal forests (Brunner et al., 2012). However, these estimates of mycelial turnover may be too low because of a combination of underestimation of production, due to the use of sand as a substrate and to long-term incubations, and overestimation of biomass, which may include saprotrophic and ErM fungi as well as mycorrhizal root tips.

The use of short-term, sequentially incubated mycelial ingrowth bags has recently provided more specific estimates of biomass dynamics of the extramatrical EcM mycelium. In subtropical forest, Hendricks et al. (2016) estimated mycelial turnover by summing production in short-term (2-month) incubated ingrowth cores and dividing this annual production by the average standing biomass in the soil, similar to the sequential coring technique used for root production estimates. On this basis, the mycelium was estimated to turn over 10 times per year (i.e. have a mean residence time of 36 days). Likewise, in another subtropical pine forest, Ekblad et al. (2016) estimated mycelial biomass to turn over 13 times per year, based on a combination of ingrowth bags with overlapping and sequential incubation schemes and a mathematical model. Using a similar approach, Hagenbo et al. (2016) found increasing standing mycelial biomass despite decreasing mycelial production with forest age over a managed boreal pine stand chronosequence. However, whereas production decreased by a factor of 3, biomass turnover decreased by a factor of about 20, from 7 to 0.3 times per year across the 10- to 100-year-old stands. Thus variation in biomass turnover, rather than variation in production, controlled variation in standing biomass stocks over the gradient. Although mycelial turnover seemed rather stable over different seasons in these studies, seasonality in mycelial dynamics is evident from the seasonal occurrence of EcM sporocarps, and EcM mycelial growth has often been estimated to be highest at the end of the growing season, where tree C allocation belowground reaches a maximum (Wallander et al., 2001; Högberg and Högberg, 2002).

Interannual biomass dynamics are not well understood. It is likely that factors such as varying weather conditions, climatic changes, gap dynamics, and successional development affect competitive balance among mycorrhizal individuals in the soil, leading to biomass and species turnover at work at the decennial time scale (Pritchard et al., 2008). An interesting idea is that fruit body production in EcM fungal species may depend on a massive death of mycelium to make enough N available to produce the observed stocks of carpophores (Ekblad et al., 2016). Conversely, this would imply that the extent to which mycelial biomass builds up in the soil is intimately linked with the fruit body formation strategy of individual fungal species in a system and the factors triggering and affecting fruit body production.

Mycelial turnover estimates suggest lifespan to be 5 to 580 days for mycelial biomass, but different chemical and structural components clearly differ in longevity. Individual observable structures, such as mycorrhizal root tips, mycelial cords, and sclerotia, likely remain viable much longer than undifferentiated diffuse soil mycelium, which may turn over within days (e.g. Finlay and Read, 1986; Cairney, 2012). Mycelial cords consist of differentiated hyphae arranged in parallel to form tube-like structures used for long-distance transport between the mycorrhizal roots and soil resource patches. Most cord-forming fungi belong to the Basidiomycota and include saprotrophic and EcM fungi. Lifespans of individual cords of EcM fungi have been estimated to average 11 months using minirhizotron imaging in a *Pinus edulis–Juniperus monosperma* woodland (Treseder et al., 2005).

Sclerotia serve as resting structures that, analogous to the spore bank of some EcM fungi, remain dormant in the soil for several years until a disturbance triggers their growth and colonization of roots (Miller et al., 1994). Sclerotial structures attributed to the common EcM ascomycete *Cenococcum geophilum* amounted to as much as 440 kg per hectare in a temperate, old-growth *P. abies* forest, although not all sclerotia may have been viable (Dahlberg et al., 1997). Using ^{14}C dating, the age of the highly melanized sclerotial remains of *Cenococcum* in soils have been estimated to reach several thousand years (Benedict, 2011). Mycorrhizal roots colonized by *C. geophilum* were found to persist 4-10 times longer (821 vs. 129 days) than other EcM fungi as observed through minirhizotron imaging (Fernandez et al., 2013). However because of sturdy, heavily melanized cell walls, *Cenococcum*-colonized roots remained seemingly intact, even after death, suggesting that they either had a longer lifespan or a higher resistance to decomposition after death or both. Thus the balance between biomass components with different turnover times within a mycorrhizal fungal community is likely to be important for the mean turnover times observed in EcM systems.

Little is known about the rate of biomass turnover of ErM fungi. However, being root endophytes it may be hypothesized that their turnover is at least the same rate as the roots they colonize. For example, in a temperate heathland, the root biomass turnover rate of *Calluna vulgaris* was estimated to be 0.64 times year^{-1} based on sequential root harvesting, corresponding to one- to two-thirds of the root turnover in two cooccurring grasses (Aerts et al., 1992). This suggests lifespans of ericoid roots of roughly 1.5 years. Root biomass of the ErM understory of a 20-year-old southern boreal *Pinus sylvestris* forest also showed annual turnover (Persson, 1980). However, the lifespan of the finest roots—the hair roots—inside of which the ErM fungi have most of their biomass, is possibly faster. On the other hand, not much is known about the possible persistence of the often highly melanized ErM hyphal coils in the soil, where they may persist as survival and dispersal propagules. Another interesting aspect is whether all root-associated ascomycetes, such as ErM fungi, are fully dependent on recent assimilates via a symbiotic relationship or whether some have saprotrophic capacities and may take up significant amounts of C from the SOM pool. Evidence from sequenced genomes suggests that at least the ErM *Oidiodendron maius* (Ascomycota) has retained more of the genetic potential to decompose energy-rich C substrates, such as cellulose, than the studied EcM fungi (Kohler et al., 2015). However, this is still an open question, with the prospect of being resolved by ongoing mycorrhizal fungal genome sequencing efforts.

23.2.5 Biomass Recycling and Decomposition by Mycorrhizal Fungi

Agerer (2006) categorized EcM fungal species into short- and long-distance exploration types on the basis of the amount and differentiation of their extramatrical mycelium. The mycelium of short-distance types is composed of simple hyphae that explore soil closer to the roots whereas the mycelium of long-distance types differentiates into hydrophobic hyphal cords that connect mycorrhizal root tips with a more distant exploratory mycelium.

Cord-forming types have been proposed to be the most C demanding because of their extensive growth (Hobbie, 2006) and could potentially contribute large amounts of mycelial necromass to the SOM pool. However, cord-forming EcM fungi have been hypothesized to be particularly efficient at internal recycling of resources fixed in their own biomass, potentially leading to short residence times of C in mycelial biomass and necromass, compared with other EcM and ErM fungi (Clemmensen et al., 2015). The main structural element of the fungal biomass is the cell wall, which represents approximately 20–50% of fungal biomass (Ruiz-Herrera, 2012). With N-containing chitin making up a substantial proportion, the N content of the mycorrhizal mycelium is typically 2.5%.

On the basis of meta-barcoding of DNA markers, it has become evident that fungal communities are highly stratified, at least in EcM- and ErM-dominated ecosystems in boreal, temperate, and tropical biomes. Linked to their main C source, free-living saprotrophic fungi dominate in recently shed plant litter on the soil surface whereas mycorrhizal fungi dominate in older organic layers (Lindahl et al., 2007; Baldrian et al., 2012; Clemmensen et al., 2013; McGuire et al., 2013) as well as in mineral soils (Rosling et al., 2003).

This almost complete dominance of mycorrhizal fungi in energy-deprived older organic layers implies that they also have primary access to their own necromass. The high N content makes recently dead mycelium an important N source, suggesting that a closed nutrient cycle in which mycorrhizal fungi mobilize N directly from their own necromass would be beneficial (Kerley and Read, 1998). A range of ErM and EcM fungi have been found to be able to take up and survive solely on organic N sources, including those in fungal necromass (Leake and Read, 1990; Kerley and Read, 1998; Smith and Read, 2008). However, the cord-forming strategy may be an adaptation to particularly efficient colonization and exploitation of nutrient patches in environments where N is mostly locked up in solid organic substrates (Boddy, 1999; Falconer et al., 2007). Younger mycelia formed by such species consist of diffuse exploratory hyphal fans, whereas older mycelial systems persist as linear cords interlinking EcM tips and resources (Donnelly et al., 2004). The initial fans are produced and turned over within weeks through autolytic processes (Finlay and Read, 1986; Bending and Read, 1995), limiting "losses" of N to the SOM pool while maximizing nutrient delivery to the plant host. Furthermore, many cord-forming EcM fungi are equipped with a suite of particularly efficient extracellular enzymes that may contribute to efficient internal biomass recycling and degradation of organic complexes in the soil (Hobbie and Agerer, 2010; Hobbie et al., 2013; Bödeker et al., 2014; Lindahl and Tunlid, 2015). Thus the growth strategy of certain members of mycorrhizal fungal communities may be of utmost importance to the extent to which mycorrhiza mediate C immobilization in the soil.

23.3 SECRETIONS OF MYCORRHIZAL MYCELIA

The secretome of mycorrhizal mycelia includes a wide range of molecules, including low-molecular–weight (lmw) organic acids, amino acids, polyols, peptides, siderophores, glycoproteins, and a diverse range of enzymes such as proteases, phosphatases, peroxidases, and laccases. The production of these substances is highly variable within and between different types of mycorrhizal fungi and influenced by different environmental conditions.

A schematic representation of the flow of C through the mycorrhizal mycelium and its secretome is shown in Fig. 23.2, together with the longer term immobilization processes that result in storage of stable C in organic and mineral pools. Many of the molecules secreted by the mycelium are labile and subject to rapid decomposition; however, they play a collective role in mobilization of nutrients that can lead to a longer term sequestration of C in recalcitrant substrates. The rapid decreases in soil respiration demonstrated after girdling of forest trees in experiments by Högberg et al. (2001) suggest that the flux of current assimilates to mycorrhizal roots is directly connected to the supply and respiration of C in soil. In a 50-year-old boreal forest, extractable DOC was 45% lower in girdled plots than in control plots (Högberg and Högberg, 2002), suggesting a large contribution by roots and associated fungi to soluble C pools, although the contribution of these two components could not be separately determined. Field experiments using $^{13}CO_2$ tracer in 15-year-old pine stands (Högberg et al., 2008) have estimated the mean residence time of labile C in EcM tips to be approximately 4 days and that of total soil microbial cytoplasmic C to be 17 days, suggesting fast C through-flow in fungal biomass. Similar rapid transport of C through AM systems has been shown in soil from upland grassland ecosystems by Johnson et al. (2002b).

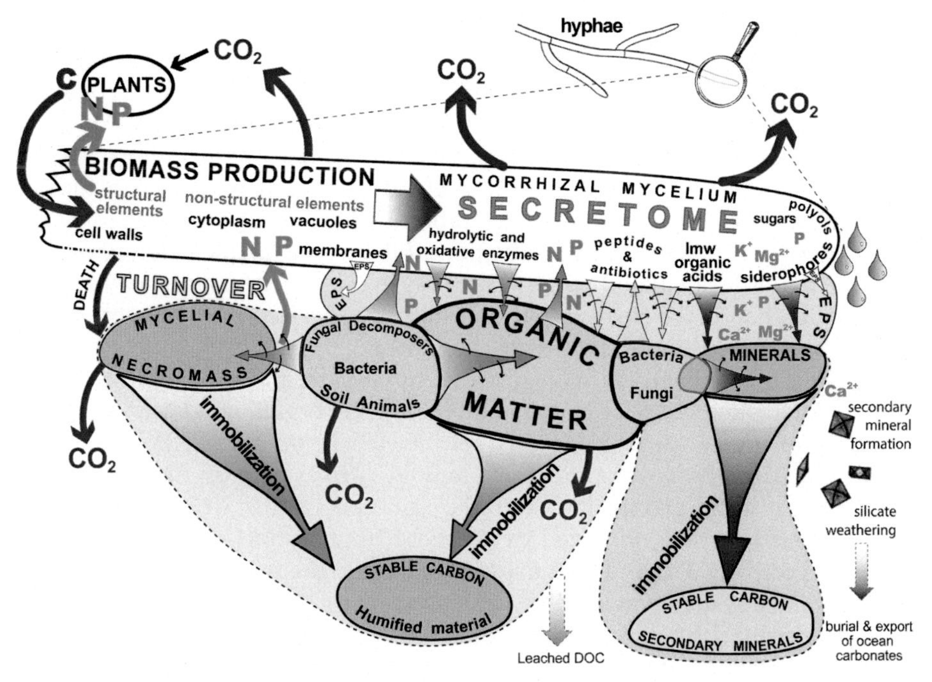

FIGURE 23.2 Schematic representation of C flow through the mycorrhizal mycelium and its secretome, showing immobilization in recalcitrant humified material and secondary mineral, as well as losses through respiration and leaching. *DOC*, dissolved organic carbon.

Conflicting evidence exists on the possible interactions between grazing soil animals and the C flow through the mycorrhizal mycelium. Johnson et al. (2005) demonstrated that grazing by Collembola disrupted the C flow through AM fungal networks (Johnson et al., 2002a,b). Remén et al. (2008) demonstrated reductions of 82–92% in the abundance of the oribatid mite *Oppiella nova*, which they attributed to the reduction of EcM hyphae as a food source after girdling of spruce trees (Högberg et al., 2001). On the other hand, stable isotope data published by Potapov and Tiunov (2016) suggest that soil collembolans do not use mycorrhizal fungi as the main food source and support the emerging view that the extramatrical mycorrhizal mycelium can be retained in the soil to serve as a progenitor of stabilized SOM.

The role of mycorrhizal fungi in stabilizing soil aggregates has been discussed intensively in the literature, but it is also the subject of Chapter 14 and will not be discussed here. The lmw organic acids are important components of the exudates produced by mycorrhizal fungi and have been identified in a range of studies. Simple carboxylic acids are present in most soil solutions and can act as weathering agents, promoting dissolution of primary minerals, and have been implicated in pedogenic processes. Their sorption characteristics were studied by van Hees et al. (2003), who found adsorbed-to-solution ratios as high as 3100. The organic acids are readily adsorbed to the solid phase, and sorption provides an important buffering role in maintaining soil solution concentrations at low organic acid concentrations, inhibiting microbial degradation of the acids. The lmw compounds are maintained at low concentrations (<50 μM) in soil solution, but the flux through this pool is extremely rapid, with mean residence times of 1–10h due to microbial mineralization to CO_2 (van Hees et al., 2005). Therefore these soluble, transitory compounds may contribute substantially to the total CO_2 efflux from soil. Direct measurements of oxalate exudation from the hyphal tips of the EcM fungus *Hebeloma crustuliniforme* have been made by Van Hees et al. (2006), who calculated exudation rates of 19 ± 3 fmol oxalate per hyphal tip h^{-1}. These authors calculated that this could theoretically result in concentrations of 30mM oxalate within 1h inside of feldspar tunnels occupied by fungal hyphae, a concentration 10,000 times higher than in the surrounding soil solution. In the same study, production of the hydroxamate siderophore ferricrocin was also detected and calculated to be able to reach 1.5μM concentrations, approximately 1000 times higher than in the surrounding soil solution. It is interesting to note that Reichard et al. (2005) demonstrated that steady-state dissolution of goethite by 2′-deoxymugineic acid phytosiderophores was synergistically enhanced by oxalate, and it is possible that there may be synergistic interactions between other combinations of organic acids and siderophores.

Direct study of organic acid production by intact EcM mycelia of *P. sylvestris* seedlings was made possible by the development of a new in vitro system by Ahonen-Jonnarth et al. (2000). In this study the production of oxalic acid (per g root dry weight) by seedlings exposed to elevated Al and colonized by the EcM fungi *Suillus variegatus* and *Rhizopogon roseolus* was respectively 39.5 and 26 times higher than in the nonmycorrhizal control plants. The same laboratory system was used by Johansson et al. (2008, 2009) to investigate the effect of different mycorrhizal fungi on production of lmw organic acids, amino acids, and DOC. In these experiments the identifiable lmw organic acids constituted only a small proportion (3–5%) of the DOC fraction, but DOC production was increased in mycorrhizal treatments relative to the nonmycorrhizal controls.

Other studies of mycorrhizal hyphal exudates have been conducted using nuclear magnetic resonance spectroscopy. Sun et al. (1999) studied exudation of fluid droplets by the EcM fungus *Suillus bovinus* and found that sugars and polyols comprised 32% and peptides 14% of the exudate mass. Oxalic acids and acetic acid were also found. Polyols such as mannitol and arabitol are thought to be important for retaining turgor in fungal hyphae during C translocation along hydrostatic pressure gradients. The exudation of droplets at hyphal tips may be an important mechanism for conditioning the immediate environment of hyphal tips and facilitating interactions with substrates and associated microorganisms, even in drier soils. Similar observations have been made by Querejeta et al. (2003), who demonstrated that water obtained by *Quercus agrifolia* plants, using hydraulic lift, can be transferred to associated AM and EcM fungi to maintain their integrity and activity during drought, even when the fertile upper soil is dry.

Exudation of C compounds at hyphal tips is undoubtedly important for interactions between mycorrhizal fungi and bacteria (Marupakula et al., 2016; see also Chapter 22). These probably involve priming of bacterial activity through supply of exudates from vital hyphae (Toljander et al., 2007), but they may also include recycling of C from damaged or senescing hyphae (Toljander et al., 2006). Carbon supply from mycorrhizal hyphae can also provide energy for associated bacteria to solubilize phosphate (Zhang et al., 2014, 2016).

Fungi secrete a vast number of different extracellular enzymes involved in hyphal growth and differentiation, nutrient and C mobilization, and antagonistic interactions with other biota (Baldrian, 2008). There are two main classes of enzymes—the hydrolytic and the oxidative enzymes. Among hydrolytic enzymes are the proteases and phosphatases, important for release of N and P from organic and mineral substrates, as well as a suite of enzymes attacking the C-rich organic polymers in cell walls of plants and microorganisms, particularly the cellulases and chitinases. Although mycorrhizal fungi—depending on mycorrhizal type and species (see Chapters 8 and 20)—have retained a well-developed capacity to produce most hydrolytic enzymes important for mobilization of nutrients (including chitinases), their genetic potential for hydrolytic decomposition of cellulose, in particular, and other cell wall components has contracted in comparison with their saprotrophic ancestors (Kohler et al., 2015).

Among oxidative enzymes, the peroxidases and laccases are important. Mn peroxidases are extracellular enzymes that evolved uniquely within the Agaricomycetes (Floudas et al., 2012), and certain EcM fungal taxa have retained the genetic potential to produce these enzymes (Bödeker et al., 2009; Shah et al., 2016). Mn peroxidases can catalyze complete mineralization of organic matter and play a pivotal role in decomposition of lignin and other phenol-rich macromolecules (Sinsabaugh, 2010). Laccases are found in both Ascomycetes and Basidiomycetes, and their roles can be various (Baldrian, 2008). Apart from their role in oxidative depolymerization of organic matter, they may have an important role in the build-up of organic matter because they are also involved in synthesis of humic substances, e.g., when melanin is synthesized from plant and fungal derived monomers (Stevenson, 1994; Butler and Day, 1998). The production of these enzymes depends strongly on fungal species. In field studies, enzymes and fungal communities are highly patchy, and strong colocalization of species and enzymes has been found both across small-scale vertical profiles and across horizontal transects on the meter scale, as well as across seasons (Baldrian, 2014; Bödeker et al., 2014).

Very little is known about the actual amounts of exoenzymes secreted and their turnover times. As judged from high variation in soil enzyme activity over time and rapid responses in enzyme activities to experimental manipulation (Bödeker et al., 2014), enzymes are probably more important in terms of their effects on SOM turnover processes than in terms of their own direct contribution to stable SOM build-up.

23.4 NECROMASS PROPERTIES AND DECOMPOSITION

23.4.1 Mycorrhizal Fungal Necromass Decomposition

It is clear that a recently dead mycelium contributes to the SOM pool, but the extent to which mycorrhizal fungal remains contribute to longer term SOM accumulation depends on the balance between necromass inputs (via biomass turnover) and necromass decomposition processes on longer time scales. If the necromass input rate is greater than its decomposition rate, then the fungal remains will contribute directly to the build-up of SOM stocks. However, decomposition products may also be further transformed and/or incorporated into decomposer biomass/necromass.

As for other organic materials, mycorrhizal fungal necromass decomposition depends on intrinsic characteristics and external factors. In parallel with plant litters, an intrinsic factor of importance is the chemical quality and structure of the necromass, which likely vary widely among mycorrhizal fungi and different biomass fractions. In this chapter we focus mainly on the role of necromass quality because this has been most intensively studied, although we acknowledge that a range of physicochemical factors, such as temperature, moisture, and pH, as well as the specific soil matrix (e.g. clay, sand, organic) in which the necromass is deposited, may be equally important. However, it is important to acknowledge that decomposition is a biologically mediated process and that these external factors act mainly through their effects on the presence, abundance, activity, and efficiency of decomposer organisms. Thus, once mycorrhizal fungal biomass components die and are released to the environment, they constitute a potential resource pool of labile and complexed C and nutrients that can support the activity and growth of various soil organisms.

The fraction of necromass, which is either inaccessible or not energetically beneficial for the specific decomposer organisms present in the system, can then contribute to the SOM, which can eventually be further transformed and stabilize in a longer temporal perspective.

23.4.2 Effects of Mycorrhizal Tissue Quality

Fungal hyphae behind the actively growing hyphal tips are often highly vacuolated because cytoplasm is continuously translocated to metabolically active regions. Most of the nonstructural cell contents, such as the membranes and cytoplasm, exhibit relatively high turnover rates and are rapidly incorporated into soil microbiota (Miltner et al., 2012; Drigo et al., 2012). Thus, senescing hyphae consist mainly of cell walls, and cell wall components have been suggested to exert important controls on necromass decomposition rates (Ekblad et al., 1998; Fernandez et al., 2016).

The quantitatively most important component of fungal cell walls is the polysaccharides, which typically account for 80–90% of the dry mass (Baldrian et al., 2013). Glucans and chitin are the major structural polysaccharides in fungi, except in Oomycetes and certain ascomycete yeasts. The cell wall consists of a skeleton of elastic microfibrils made of cross-linked β-glucans and chitin suspended in a matrix of glycoproteins and amorphous α-glucans (Feofilova, 2010). Chitin is a hard polymer synthesized from *N*-acetyl glucosamine giving physical support for the wall (Wessels, 1993). The chitin–glucan complex also interacts with other cell wall components, such as amino acids and melanin, affecting its properties and functions as well as its chemical recalcitrance (Feofilova, 2010). Such chitin–glucan complexes are not present in the Zygomycota. On the cell wall surface, some fungi form a mucous polysaccharide material consisting of β-glucans. On the outer surface special glucans, proteins termed *hydrophobins* and the phenolic pigment melanin also occur.

The glucans are polymers exclusively of glucose, but chitin and protein represent sources of C and N to decomposer organisms. Across different ecosystem types, C:N ratios of the EcM mycelium have been found to be close to 20 (Wallander et al., 2003; Clemmensen et al., 2006), which is much lower than that of plant litters, which typically have C:N ratios of 25–170 (Berg, 2000; Hobbie, 2005). The high content of N and relatively simple polysaccharides should make mycorrhizal necromass a high-quality nutrient and C source for soil microbiota.

In line with this, in laboratory incubations it has been found that chitin is not recalcitrant relative to other compounds in fungal tissues and that its concentration is positively related to the decomposition rate of fungal tissues (Drigo et al., 2012; Fernandez and Koide, 2012). Likewise, in an experiment in which chitin, *N*-acetyl glucosamine, or cell wall material were added to EcM mycelial mat-colonized or nonmat soil organic horizons in an old-growth Douglas fir forest, both substrates stimulated respiration, N mineralization, and fungal biomass production (Zeglin et al., 2013; Zeglin and Myrold, 2013). Mycelial necromass with high initial N content has also been found to decompose more rapidly than necromass with lower N content (Fernandez and Koide, 2014), paralleling observations of the initial stages of plant litter decomposition.

The highly ephemeral nature of EcM fungal carpophores supports the idea that fungal necromass is potentially easily decomposed, although the soil environment may be different and soil mycelium and carpophore qualities may differ. In a study following decomposition of fleshy fungal sporocarps (*Tylopilus felleus*) buried in mycelial ingrowth bags in organic layers of an oak forest, approximately 80% of dry weight was lost after 3 weeks (Brabcová et al., 2016). A rapid incorporation of C from EcM mycelial necromass into soil fungal communities has also been observed (Drigo et al., 2012; Miltner et al., 2012). Thus initial rates of mycelial decomposition seem much faster than those of plant litters. For various plant litters, 20–50% mass loss is typically recorded during the first year of decomposition (Berg, 2000; Hobbie, 2005). However, even under artificial experimental conditions in the mentioned studies, a fraction of the mass remained in its initial form after a few weeks or was incorporated into microbial pools derived from the mycorrhizal mycelium.

The few existing studies of mycelium decomposition mainly cover the shorter term fate of mycelial necromass. Only a few fungal species have been studied, and drivers of long-term immobilization in soil pools are not easily inferred. One idea, supported by shorter term decomposition experiments and by longer term mycelial necromass inputs to stable soil pools, is that the phenolic pigment melanin retards decomposition in a manner similar to the effect of lignin in plant materials (Fernandez et al., 2016). This means that mycelium with melanized cell walls, such as many mycorrhizal (and other root-associated) ascomycetes (Robinson, 2001; Smith and Read, 2008), may be more resistant to decomposition, leading to a higher proportion of their necromass being preserved in long-term humus stores (Coelho et al., 1997; Koide et al., 2013; Fernandez and Koide, 2014). However, it is still unknown whether high necromass N content retards later stages of decomposition in a manner similar to that observed in plant litter (Berg, 2000)—a question of relevance when predicting effects of N fertilization or deposition on soil C sequestration. An experiment in which roots were left to decompose in nylon mesh bags in forest soil did indeed show that despite higher N concentrations, roots colonized by EcM fungal mantles decomposed much more slowly than nonmycorrhizal roots over a 2-year period (Langley et al., 2006). This suggests that mycorrhizal fungal tissues may be preferentially preserved during decomposition, although the mechanism is not clear.

23.5 INCORPORATION INTO STABLE CARBON

Extensive literature exists on how partially degraded plant remains form SOM (Paul, 2007), and frameworks have been suggested on how to link short-term litter decomposition dynamics with long-term stabilization in the soil matrix (Cotrufo et al., 2013). The decomposition and stabilization dynamics of mycorrhizal remains may to a large extent be driven by similar mechanisms as those of plant remains (see earlier about tissue quality). However, by their directed allocation of a separate root-derived C flow into the soil matrix, mycorrhizal fungi can grow and deposit necromass in soil volumes that are otherwise not extensively colonized by soil organisms and perform energy-demanding, qualitatively different activities that affect long-term C stabilization in mineral and organic complexes.

23.5.1 Sequestration of Carbon Into Secondary Minerals

Fungi exert a significant influence on biogeochemical processes, especially in soil, rock and mineral surfaces, and the plant root–soil interface, where, as mycorrhizal fungi, they are responsible for major mineral transformations, redistribution of inorganic nutrients, and flow of C. Fungi are important components in rock-inhabiting communities with roles in mineral dissolution and secondary mineral formation (Gadd, 2007, 2010, 2013). Their role in weathering of rocks and minerals through biomechanical and biochemical attack has been extensively studied. Proton-promoted dissolution is supplemented by ligand-promoted dissolution of minerals by strong chelators such as oxalic and citric acid that may act synergistically with siderophores. Secondary minerals may be deposited as carbonates, oxalates, or other mycogenic minerals, and the role of "rock-building fungi" has been discussed in addition to the role of "rock-eating fungi" (Fomina et al., 2010). Fungi, including many EcM fungi, are prolific producers of oxalate, and oxalotrophic bacteria are capable of oxidizing calcium oxalate to calcium carbonate. Because the oxalate is organic in origin, and half of its C is transformed into mineral C with a much longer residence time than organic C, this process represents a potential major sink for sequestration of C from the atmosphere (Verrecchia et al., 2006). Precipitation of carbonate minerals by microorganisms in connection with silicate weathering has also been discussed by Ferris et al. (1994) in relation to its potential role as a sink for atmospheric CO_2.

The possible roles of mycorrhizal fungi in pedogenesis (Chapter 2) and weathering of minerals (Chapter 3) are discussed elsewhere in this book and will not be discussed in detail here. However, certain processes will be briefly reviewed with respect to their role in C allocation and immobilization. The possible role of mycorrhizal fungi in relation to mineral weathering has been discussed in relation to the discovery of numerous tubular pores, 3–10 μm in diameter, in weatherable minerals in podzol surface soils and shallow granitic rock under European coniferous forests (van Breemen et al., 2000; Jongmans et al., 1997; Landeweert et al., 2001). Fungal hyphae were found occupying some of these pores, and it was speculated that they could be formed by the weathering action of hyphae releasing organic acids and siderophores. However, the etiology of pore formation has been questioned, and their quantitative contribution to total weathering rates has been calculated to be negligible (Smits et al., 2005). Different arguments concerning the possible role of EcM fungi as agents of biogenic weathering are discussed by Finlay et al. (2009). Differential allocation of plant-derived C to patches of primary minerals such as quartz and potassium feldspar (Rosling et al., 2004) and to apatite and quartz (Smits et al., 2012) suggest tightly coupled plant-fungal interactions underlying weathering. In the experiment by Smits et al. (2012), when P was limiting, 17 times more ^{14}C was allocated to wells containing apatite than to those containing only quartz, and fungal colonization of the substrate increased the release of P by a factor of almost three. Selective allocation of biomass to grains of different minerals by *Paxillus involutus* has also been demonstrated (Leake et al., 2008; Smits et al., 2008), suggesting grain-scale "biosensing."

Carbon allocation in the form of sugars and polyols (Sun et al., 1999) may be important in generating turgor pressure in hyphae and have consequences for weathering of minerals with lattice structure. Ultramicroscopic and spectroscopic observations of fungus–biotite interfaces during weathering of biotite flakes by *P. involutus* (Bonneville et al., 2009) revealed evidence of biomechanical forcing and altered interlayer spacing, and it appears that physical distortion of the lattice structure takes place before chemical alteration through dissolution and oxidation. High internal pressures in hyphae are thought to be an evolutionary adaptation to facilitate penetration of plant tissues and rock surfaces (Jongmans et al., 1997). Carbon allocation is also necessary for the production of the extracellular polymeric substances (EPS) that many fungi produce. These consist primarily of proteins and polysaccharides and can contain lipids and humic substances. Studies of EPS produced by *P. involutus* colonizing biotite using atomic force microscopy (Gazzè et al., 2013) suggest that EPS halos increase the surface area of contact between mineral surfaces and hyphae, which would have consequences for delivery (and possibly protection) of organic acids and siderophores to mineral surfaces, as suggested by Finlay et al. (2009).

Studies of "trenching" of silicate mineral surfaces (basalt) buried under different tree species growing in an arboretum using vertical scanning interferometry suggest that trenching and hyphal colonization increased with evolutionary progression from AM fungi to EcM fungi and with progression from gymnosperm to angiosperm host plants (Quirk et al., 2012). The authors concluded that this evolutionary progression resulted in a release of calcium from the basalt by EcM gymnosperms and angiosperms at twice the rate achieved by AM gymnosperms, and that forested ecosystems have become major engines of continental silicate weathering, regulating global CO_2 concentrations by driving calcium export into ocean carbonates (Quirk et al., 2012). Additional studies of the same tree species using different CO_2 environments suggest that weathering intensified during evolutionary progression from AM fungal symbionts to EcM symbionts and that calcium dissolution rates were related to photosynthate energy fluxes and were higher during simulated past CO_2 atmosphere (1500 ppm) under which the EcM fungi evolved (Quirk et al., 2014). The extent of C sequestration associated with global silicate weathering and associated export of carbonates to the ocean floor is argued to be sufficient to reduce global CO_2 levels. Artificial acceleration of this process through distribution of pulverized silicate rocks across tropical terrestrial landscapes has been proposed as a method to offset anthropogenic CO_2 emissions (Taylor et al., 2016) with a view to reducing global CO_2 concentrations and ameliorating ocean acidification by 2100. Recent studies on carbonate weathering by EcM fungi-colonizing tree roots (Thorley et al., 2015) suggest that EcM tree species weather calcite-containing rock grains more rapidly than AM trees because of greater acidification by the EcM trees. Weathering and corresponding alkalinity export to oceans may increase with rising atmospheric CO_2 and associated climate change, slowing rates of ocean acidification.

23.5.2 Sequestration of Carbon Into Soil Organic Matter

It has become increasingly clear that microbial residues, including fungal residues, play an important role in the formation of stable SOM (Gleixner, 1999; Ehleringer et al., 2000; Miltner et al., 2012; Wallander et al., 2011). However, the importance of the contribution by C derived from mycorrhizal root allocation versus C initially derived from aboveground plant litter has been elusive and the belowground inputs largely neglected in C cycling models. Ingrowth cores containing C4-crop soil incubated in detritus from C3 trees enabled new (tree-derived) soil C to be distinguished from the old soil C on the basis of the different ^{13}C signatures (Godbold et al., 2006). On the basis of this, the EcM mycelium was concluded to be the dominant pathway (62%) through which C entered the SOM pool, exceeding the input via leaf litter and fine root turnover. A model quantifying above- and belowground C inputs and losses in boreal forest soil and parameterized by bomb ^{14}C dating of organic matter estimated that 50–70% of C had entered the soil via roots and associated fungi rather than aboveground plant litter (Clemmensen et al., 2013). Variation in total soil C stocks across the studied 100- to 5000-year boreal forest chronosequence was shown to depend primarily on dynamics of the root-derived inputs. Furthermore, the accumulation of chitin, the stable isotope patterns, and the distribution of fungal guilds in the soil profiles suggested that a major fraction of the organic matter was of mycorrhizal fungal origin (Clemmensen et al., 2013). These findings highlighted the quantitative importance of belowground inputs to accumulating humus layers (i.e. that boreal forest organic soils grow from below).

Mechanisms for stabilization of mycorrhizal fungal remains are still largely unresolved and are likely to vary between systems. As discussed earlier, it is evident that the resistance of mycelial necromass to decomposition varies among fungal species as a consequence of differences in tissue qualities. Further studies are needed of the fate of chitin, melanin, and glomalin during decomposition and their possible role as precursors of stable soil C. The growth form and decomposer capacity of the active species, particularly whether the cord-forming strategy is dominant, also seems important for the necromass production and preservation. Furthermore, there may be spatial separation between deposited necromass and decomposers (e.g. across horizons or through mineral–organic interactions; Schmidt et al., 2011), suggesting that spatial organization of the soil system should not be neglected in future studies. In systems where an organic layer accumulates without contact with the mineral subsoil, a "humification" process (c.f. Stevenson, 1994) involving random oxidation of organic remains may be important for stabilization of old organic matter (Lindahl and Tunlid, 2015). Environmental conditions, such as waterlogging and permafrost, certainly are important controls of decomposition processes in some systems.

Historically, the large C stocks in mor layers of heathlands have been interpreted as effects of low plant litter quality and unfavorable soil environment (low pH) and ErM fungi as generally better equipped to access organically bound nutrients than EcM fungi (Read, 1991; Read et al., 2004). If the mor layer, to a large extent, is built from mycorrhizal roots and mycelium, then this would suggest that ErM communities, instead—on balance—produce more SOM than they decompose, which does not contradict their dependency on organic N sources for their nutrition. Conversely, in systems with lower rates of SOM accumulation, the more efficient EcM decomposers may be found.

Strong environmental or successional gradients have great value as model systems for disentangling relations between mycorrhizal fungal communities, their drivers, and their effects on long-term ecosystem processes. Ecotones between forests, dominated by EcM trees, and heathlands, dominated by ErM dwarf shrubs, are common in natural and cultural landscapes. Such ecotones are naturally found in transitions toward arctic and alpine zones, and ericaceous shrubs comprise an increasing share of primary production in aging boreal forests. The chronosequence of boreal forested islands (Clemmensen et al., 2013) indeed provided

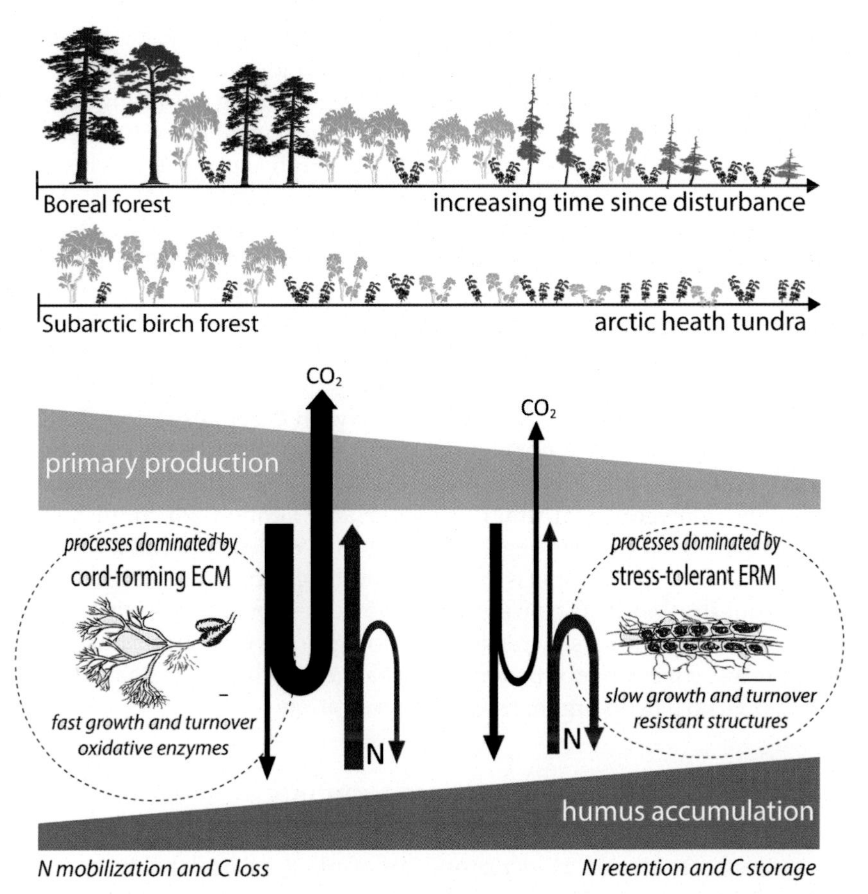

FIGURE 23.3 Conceptual framework depicting how shifts in mycorrhizal symbionts in a 5000-year boreal forest succession after wildfire or along the subarctic-alpine forest–heath ecotone affect N and C dynamics. The scale bars within the dashed ellipses represent 0.1 mm. *Modified from Clemmensen, K.E., Finlay, R.D., Dahlberg, A., Stenlid, J., Wardle, D.A., Lindahl, B.D., 2015. Carbon sequestration is related to mycorrhizal fungal community shifts during long-term succession in boresl forests. New Phytologist 205, 1525–1536.*

evidence that belowground C sequestration was lowest when cord-forming EcM fungi dominated soil processes and increased when ErM fungi increased in dominance in old-growth boreal forests. On the basis of these results, a conceptual framework to describe a possible mechanistic relationship—involving mycorrhizal fungal successions—was developed (Fig. 23.3; Clemmensen et al., 2015). In this succession aboveground primary production by trees and ericoid dwarf shrubs decreases whereas belowground C sequestration increases with time since fire. At earlier successional stages, C allocation (black arrows) to mycorrhizal fungi is presumed to be high, and cord-forming EcM basidiomycetes (mainly *Suillus*, *Cortinarius* and *Piloderma*) dominate root zone processes. Through rapid growth and turnover of their exploratory mycelium they facilitate N mobilization (red arrows) to the host plants but restrict the amount of C and N transferred to long-term humus pools. At later successional stages less C is allocated to mycorrhizal symbionts, and stress-tolerant, root-associated ErM ascomycetes dominate C and N dynamics in the root zone. By building biomass structures resistant to decomposition they facilitate N retention and long-term C storage in the humus. This suggests that the higher abundance of certain cord-forming EcM fungi at earlier successional stages promotes efficient recycling of N and C from fungal mycelium and older humus while stress-adapted, root-associated ascomycetes generally promote biochemical stabilization of C and N in organic matter derived from mycelium in the later stages. Thus long-term ecosystem succession involves impairment of mycorrhizal N recirculation and, consequently, progressive nutrient limitation and vegetation changes with time since fire (Alberton et al., 2007; Wardle et al., 2012), a cycle that would naturally be reset by wildfire disturbance.

Functionally similar mycorrhizal fungal community shifts were found across a subarctic-to-alpine ecotone from mountain birch forest to heath tundra. Here a strong positive coupling between tree production and abundance of EcM fungi (in this case *Cortinarius* and *Leccinum*) was found, both of which were negatively coupled with C sequestration along the ecotone (Fig. 23.3; Clemmensen et al. unpublished). These findings support the existence and importance of an evolutionary tradeoff between conservative/stress-tolerance versus exploitative resource-use traits in mycorrhizal fungi (Treseder and Lennon, 2015), in parallel to those known from plant resource-use strategies (Chapin III, 1980). These traits seem to positively link biomass turnover and necromass turnover, both of which are negatively linked with organic soil C sequestration. This conceptual framework provides a mechanistic explanation for plant production being inversely related to belowground C sequestration, and it suggests that increasing forest cover or production with changes in climate or management practices could have negative effects on organic soil C sequestration.

23.6 CONCLUSIONS

The more or less continuous, direct provision of plant assimilates to mycelia of mycorrhizal fungi enables them to interact with recalcitrant organic substrates and mineral substrates in ways that are not possible for fungi in other functional guilds. Immobilization of C may take place on different temporal and spatial scales, and the mycelia of different mycorrhizal fungi vary greatly in their biomass and turnover, depending upon the plant hosts, fungal species composition, and types of substrate involved. The fungal secretome consists of diverse, labile molecules that may have inherently short half-lives and contribute to short-term C loss. However, these molecules may be protected from decomposition if secreted into matrices of extracellular polysaccharides also containing antibiotics and may attain sufficient concentrations to modify mineral and organic substrates. Long-term stabilization of C in organic and mineral substrates may take place through interactions involving glycoproteins, melanin, EPS, and formation of secondary minerals. Relationships between above- and belowground C storage require further study across ecotones and along other environmental gradients, and further studies of the ways in which mycorrhizal fungi and saprotrophic fungi interact (Fernandez and Kennedy, 2016) will improve our understanding of C cycling. Effects of fungal species composition may be important for amounts and patterns of C allocation via mycorrhizal mycelia to the soil, and for the ultimate recalcitrance of the mycelial necromass, and further research is needed to determine whether there are tradeoffs between immobilization and mobilization-related traits at the community level in different systems. The further development of trait theory for mycorrhizal fungi (Crowther et al., 2014; Fernandez and Kennedy, 2015) and the use of functional traits to analyze functional aspects of fungal communities without a priori binning of functional guilds have great potential for resolving community-level effects on soil C immobilization as well as predicting responses of important community characteristics with environmental changes.

There is still disagreement about the quantitative contribution of biogenic weathering to overall weathering (see Chapters 2 and 3), and problems of upscaling from hyphal etching at the microscopic scale to catchment and ecosystem assessments of weathering necessitate better integration across these different spatial scales. Important trends associated with the evolution of vascular plants have become apparent. Much of the success of plants and fungi as biogeochemical engineers is due to their ability to form symbiotic mergers. The role of fungi as phototrophic symbionts in lichens and mycorrhizas is well known, and the ubiquity and significance of lichens as pioneer organisms in the early stages of mineral soil formation, and as a model for understanding weathering in a wider context, are also well understood (Banfield et al., 1999). Successive increases in size of plant hosts and the extent of substrate colonization by their fungal symbionts (Quirk et al., 2015) have enabled them to have larger effects as biogeochemical engineers, affecting the cycling of nutrients and C at an ecosystem and global level. Improved understanding of the role of mycorrhizal fungi in C storage (Averill et al., 2014) requires a more holistic view of these symbiotic relationships, but further detailed knowledge about the flow of C between different pools will also require advances in our understanding of the biology of individual fungi and the effects of changes in community structure. Mycelial transport and allocation of C in labile forms is rapid and dynamic but leads to different forms of immobilization in stable organic material and biogenic minerals that take place on time scales of hundreds, thousands, or even millions of years. These issues all deserve particular attention in the future (Box 23.2).

BOX 23.2 FUTURE ISSUES/PERSPECTIVES

- What is the relative importance of long-term C sequestration into stable organic substrates and C in minerals, and over what time scales do these processes operate?
- What are the key drivers of mycorrhizal community structure and function in different systems? What are the roles of plant community structure, carbon supply from hosts, and environmental factors such as nutrient availability, pH, and herbivory?
- Is there a general tradeoff between immobilization and mobilization-related traits within mycorrhizal fungal communities, and is this important for belowground C sequestration?
- What is the relative importance of biotic interactions involving priming or competition between mycorrhizal fungi and saprotrophs?
- Do general negative relationships exist between aboveground and belowground C storage (ecotones/environmental gradients)?

Acknowledgments

The authors acknowledge financial support from the Swedish Research Council for Environment, Agricultural Sciences, and Spatial Planning (FORMAS); the Swedish Research Council (VR); the European Commission; and the Swedish University of Agricultural Sciences (SLU).

References

Aerts, R., Bakker, C., Decaluwe, H., 1992. Root turnover as a determinant of the cycling of C, N and P in a dry heathland ecosystem. Biogeochemistry 15, 175–190.

Agerer, R., 2006. Fungal relationships and structural identity of their ectomycorrhizae. Mycological Progress 5, 67–107.

Ahonen-Jonnarth, U., Van Hees, P.A.W., Lundström, U.S., Finlay, R.D., 2000. Production of organic acids by mycorrhizal and non-mycorrhizal *Pinus sylvestris* L. seedlings exposed to elevated concentrations of aluminium and heavy metals. New Phytologist 146, 557–567.

Alberton, O., Kuyper, T.W., Gorissen, A., 2007. Competition for nitrogen between *Pinus sylvestris* and ectomycorrhizal fungi generates potential for negative feedback under elevated CO_2. Plant and Soil 296, 159–172.

Averill, C., Turner, B.L., Finzi, A.C., 2014. Mycorrhiza-mediated competition between plants and decomposers drives soil carbon storage. Nature 505, 543–545.

Baldrian, P., Kolařík, M., Štursová, M., Kopecký, J., Valášková, V., Větrovský, T., Žifčáková, L., Šnajdr, J., Rídl, J., Vlček, C., et al., 2012. Active and total microbial communities in forest soil are largely different and highly stratified during decomposition. ISME Journal 6, 248–258.

Baldrian, P., Větrovský, T., Cajthaml, T., Dobiášová, P., Petranková, M., Snajdr, J., Eichlerova, I., 2013. Estimation of fungal biomass in forest litter and soil. Fungal Ecology 6, 1–11.

Baldrian, P., 2008. Enzymes of saprotrophic basidiomycetes. In: Boddy, L., Frankland, J.C., van West, P. (Eds.), Ecology of Saprotrophic Basidiomycetes. Elsevier Ltd., London, UK, pp. 19–41.

Baldrian, P., 2014. Distribution of extracellular enzymes in soils: spatial heterogeneity and determining factors at various scales. Soil Science Society of America Journal 78, 11–18.

Banfield, J.F., Barker, W.W., Welch, S.A., Taunton, A., 1999. Biological impact on mineral dissolution: application of the lichen model to understanding mineral weathering in the rhizosphere. Proceedings of the National Academy of Sciences of the United States of America 96, 3404–3411.

Bending, G.D., Read, D.J., 1995. The structure and function of the vegetative mycelium of ectomycorrhizal plants V. Foraging behaviour and translocation of nutrients from exploited litter. New Phytologist 130, 401–409.

Benedict, J.B., 2011. Sclerotia as indicators of mid-Holocene tree-limit altitude, Colorada Front Range, USA. The Holocene 21, 1021–1023.

Berg, B., 2000. Litter decomposition and organic matter turnover in northern forest soils. Forest Ecology and Management 133, 13–22.

Boddy, L., 1999. Saprotrophic cord-forming fungi: meeting the challenge of heterogeneous environments. Mycologia 91, 13–32.

Bödeker, I.T.M., Nygren, C.M.R., Taylor, A.F.S., Olson, A., Lindahl, B.D., 2009. ClassII peroxidase-encoding genes are present in a phylogenetically wide range of ectomycorrhizal fungi. ISME Journal 3, 1387–1395.

Bödeker, I.T.M., Clemmensen, K.E., de Boer, W., Martin, F., Olson, A., Lindahl, B.D., 2014. Ectomycorrhizal *Cortinarius* species participate in enzymatic oxidation of humus in northern forest ecosystems. New Phytologist 203, 245–256.

Bonneville, S., Smits, M.M., Brown, A., Harrington, J., Leake, J.R., Brydson, R., Benning, L.G., 2009. Plant-driven fungal weathering: early stages of mineral alteration at the nanometer scale. Geology 37, 615–618.

Bougoure, D.S., Cairney, J.W.G., 2005. Assemblages of ericoid mycorrhizal and other root-associated fungi from *Epacris pulchella* (Ericaceae) as determined by culturing and direct DNA extraction from roots. Environmental Microbiology 7, 819–827.

Bougoure, D.S., Parkin, P.I., Cairney, J.W.G., Alexander, I.J., Anderson, I.C., 2007. Diversity of fungi in hair roots of Ericaceae varies along a vegetation gradient. Molecular Ecology 16, 4624–4636.

Brabcová, V., Nováková, A.D., Davidová, A., Baldrian, P., 2016. Dead fungal mycelium in forest soil represents a decomposition hotspot and a habitat for specific microbial community. New Phytologist 210, 1369–1381.

Brunner, I., Bakker, M., Björk, R., Hirano, Y., Lukac, M., Aranda, X., Børja, I., Eldhuset, T., Helmisaari, H., Jourdan, C., et al., 2012. Fine-root turnover rates of European forests revisited: an analysis of data from sequential coring and ingrowth cores. Plant and Soil 362, 357–372.

Butler, M.J., Day, A.W., 1998. Fungal melanins: a review. Canadian Journal of Microbiology 44, 1115–1136.

Cairney, J.W.G., Meharg, A.A., 2003. Ericoid mycorrhiza: a partnership that exploits harsh edaphic conditions. European Journal of Soil Science 54, 735–740.

Cairney, J.W.G., 2012. Extramatrical mycelia of ectomycorrhizal fungi as moderators of carbon dynamics in forest soil. Soil Biology & Biochemistry 47, 198–208.

Chapin III, F.S., 1980. The mineral nutrition of wild plants. Annual Review of Ecology and Systematics 11, 233–260.

Clemmensen, K.E., Michelsen, A., Jonasson, S., Shaver, G.R., 2006. Increased ectomycorrhizal fungal abundance after long-term fertilization and warming of two arctic tundra ecosystems. New Phytologist 171, 391–404.

Clemmensen, K.E., Bahr, A., Ovaskainen, O., Dahlberg, A., Ekblad, A., Wallander, H., Stenlid, J., Finlay, R.D., Wardle, D.A., Lindahl, B.D., 2013. Roots and associated fungi drive long-term carbon sequestration in boreal forest. Science 339, 1615–1618.

Clemmensen, K.E., Finlay, R.D., Dahlberg, A., Stenlid, J., Wardle, D.A., Lindahl, B.D., 2015. Carbon sequestration is related to mycorrhizal fungal community shifts during long-term succession in boresl forests. New Phytologist 205, 1525–1536.

Coelho, R.R.R., Sacramento, D.R., Linhares, L.F., 1997. Amino sugars in fungal melanins and soil humic acids. European Journal of Soil Science 48, 425–429.

Cotrufo, M.F., Wallenstein, M.D., Boot, C.M., Denef, K., Paul, E., 2013. The Microbial Efficiency-Matrix Stabilization (MEMS) framework integrates plant litter decomposition with soil organic matter stabilization: do labile plant inputs form stable soil organic matter? Global Change Biology 19, 988–995.

Crowther, T.W., Maynard, D.S., Crowther, T.R., Peccia, J., Smith, J.R., 2014. Untangling the fungal niche: trait-based approach. Frontiers in Microbiology 5:Article 579.

Dahlberg, A., Jonsson, L., Nylund, J.E., 1997. Species diversity and distribution of biomass above and below ground among ectomycorrhizal fungi in an old-growth Norway spruce forest in south Sweden. Canadian Journal of Botany 75, 1323–1335.

Dahlberg, A., 1997. Population ecology of *Suillus variegatus* in old Swedish Scots pine forests. Mycological Research 101, 47–54.

Donnelly, D.P., Boddy, L., Leake, J.R., 2004. Development, persistence and regeneration of foraging ectomycorrhizal mycelial systems in soil microcosms. Mycorrhiza 14, 37–45.

Drigo, B., Anderson, I.C., Kannangara, G.S.K., Cairney, J.W.G., Johnson, D., 2012. Rapid incorporation of carbon from ectomycorrhizal mycelial necromass into soil fungal communities. Soil Biology & Biochemistry 49, 4–10.

Ehleringer, J.R., Buchmann, N., Flanagan, L.B., 2000. Carbon isotope ratios in belowground carbon cycle processes. Ecological Applications 10, 412–422.

Ekblad, A., Wallander, H., Näsholm, T., 1998. Chitin and ergosterol combined to measure total and living fungal biomass in ectomycorrhizas. New Phytologist 138, 143–149.

Ekblad, A., Wallander, H., Godbold, D.L., Cruz, C., Johnson, D., Baldrian, P., Bjork, R.G., Epron, D., Kieliszewska-Rokicka, B., Kjøller, R., et al., 2013. The production and turnover of extramatrical mycelium of ectomycorrhizal fungi in forest soils: role in carbon cycling. Plant and Soil 366, 1–27.

Ekblad, A., Mikusinska, A., Ågren, G.I., Menichetti, L., Wallander, H., Vilgalys, R., Bahr, A., Eriksson, U., 2016. Production and turnover of ectomycorrhizal extramatrical mycelial biomass and necromass under elevated CO_2 and nitrogen fertilization. New Phytologist 211, 874–885.

Falconer, R.E., Bown, J.L., White, N.A., Crawford, J.W., 2007. Biomass recycling: a key to efficient foraging by fungal colonies. Oikos 116, 1558–1568.

Feofilova, E.P., 2010. The fungal cell wall: modern concepts of its composition and biological function. Microbiology 79, 711–720.

Fernandez, C.W., Kennedy, P.G., 2015. Moving beyond the black-box: fungal traits, community structure, and carbon sequestration in forest soils. New Phytologist 205, 1378–1380.

Fernandez, C.W., Kennedy, P.G., 2016. Revisiting the 'Gadgil effect': do interguild fungal interactions control carbon cycling in forest soils? New Phytologist 209, 1382–1394.

Fernandez, C.W., Koide, R.T., 2012. The role of chitin in the decomposition of ectomycorrhizal fungal litter. Ecology 93, 24–28.

Fernandez, C.W., Koide, R.T., 2014. Initial melanin and nitrogen concentrations control the decomposition of ectomycorrhizal fungal litter. Soil Biology & Biochemistry 77, 150–157.

Fernandez, C.W., McCormack, M.L., Hill, J.M., Pritchard, S.G., Koide, R.T., 2013. On the persistence of *Cenococcum geophilum* ectomycorrhizas and its implications for forest carbon and nutrient cycles. Soil Biology & Biochemistry 65, 141–143.

Fernandez, C.W., Langley, J.A., Chapman, S., McCormack, M.L., Koide, R.T., 2016. The decomposition of ectomycorrhizal fungal necromass. Soil Biology and Biochemistry 93, 38–49.

Ferris, F.G., Wiese, R.G., Fyfe, W.S., 1994. Precipitation of carbonate minerals by microorganisms: implications for silicate weathering and the global carbon dioxide budget. Geomicrobiology Journal 12, 1–13.

Finlay, R.D., Read, D.J., 1986. The structure and function of the vegetative mycelium of ectomycorrhizal plants. I. Translocation of ^{14}C-labelled carbon between plants interconnected by a common mycelium. New Phytologist 103, 143–156.

Finlay, R.D., Wallander, H., Smits, M., Holmström, S., van Hees, P.A.W., Lian, B., Rosling, A., 2009. The role of fungi in biogenic weathering in forest soils. Fungal Biology Reviews 23, 101–106.

Finlay, R.D., 2008. Ecological aspects of mycorrhizal symbiosis with special emphasis on the functional diversity of interactions involving the extraradical mycelium. Journal of Experimental Botany 59, 1115–1126.

Floudas, D., Binder, M., Riley, R., Barry, K., Blanchette, R.A., Henrissat, B., Martinez, A.T., Otillar, R., Spatafora, J.W., Yadav, J.S., et al., 2012. The Paleozoic origin of enzymatic lignin decomposition reconstructed from 31 fungal genomes. Science 336 (6089), 1715–1719.

Fomina, M., Burford, E.P., Hillier, S., Kierans, M., Gadd, G.M., 2010. Rock-building fungi. Geomicrobiology Journal 27, 624–629.

Gadd, G.M., 2007. Geomycology: biogeochemical transformations of rocks, minerals, metals and radionuclides by fungi, bioweathering and bioremediation. Mycological Research 111, 3–49.

Gadd, G.M., 2010. Metals, minerals and microbes: geomicrobiology and bioremediation. Microbiology 156, 609–643.

Gadd, G.M., 2013. Geomycology: fungi as agents of biogeochemical change. Biology and Environment-Proceedings of the Royal Irish Academy 113b (2), 139–153.

Gazzè, S.A., Saccone, L., Smits, M.M., Duran, A.L., Leake, J.R., Banwart, S.A., Ragnarsdottir, K.V., McMaster, T.J., 2013. Nanoscale observations of extracellular polymeric substances deposition on phyllosilicates by an ectomycorrhizal fungus. Geomicrobiology Journal 30, 721–730.

Gleixner, G., 1999. Molecular insights into soil carbon turnover. Rapid Communication in Mass Spectrometry 13, 1278–1283.

Godbold, D., Hoosbeek, M., Lukac, M., Cotrufo, M., Janssens, I., Ceulemans, R., Polle, A., Velthorst, E., Scarascia-Mugnozza, G., De Angelis, P., et al., 2006. Mycorrhizal hyphal turnover as a dominant process for carbon input into soil organic matter. Plant and Soil 281, 15–24.

Grelet, G.A., Johnson, D., Paterson, E., Anderson, I.C., Alexander, I.J., 2009. Reciprocal carbon and nitrogen transfer between an ericaceous dwarf shrub and fungi isolated from *Piceirhiza bicolorata* ectomycorrhizas. New Phytologist 182, 359–366.

Hagenbo, A., Clemmensen, K.E., Finlay, R.D., Kyaschenko, J., Lindahl, B.D., Fransson, P., Ekblad, A., 2016. Changes in Turnover Rather than Production Regulate Ectomycorrhizal Biomass Across a *Pinus sylvestris* Chronosequence. (submitted).

Hendricks, J.J., Mitchell, R.J., Kuehn, K.A., Pecot, S.D., Sims, S.E., 2006. Measuring external mycelia production of ectomycorrhizal fungi in the field: the soil matrix matters. New Phytologist 171, 179–186.

Hendricks, J.J., Mitchell, R.J., Kuehn, K.A., Pecot, S.D., 2016. Ectomycorrhizal fungal mycelia turnover in a longleaf pine forest. New Phytologist 209, 1693–1704.

Hobbie, E.A., Agerer, R., 2010. Nitrogen isotopes in ectomycorrhizal sporocarps correspond to belowground exploration types. Plant and Soil 327, 71–83.

Hobbie, E.A., Ouimette, A.P., Schuur, E.A.G., Kierstead, D., Trappe, J.M., Bendiksen, K., Ohenoja, E., 2013. Radiocarbon evidence for the mining of organic nitrogen from soil by mycorrhizal fungi. Biogeochemistry 114, 381–389.

Hobbie, S.E., 2005. Contrasting effects of substrate and fertilizer nitrogen on the early stages of litter decomposition. Ecosystems 8, 644–656.

Hobbie, E.A., 2006. Carbon allocation to ectomycorrhizal fungi correlates with belowground allocation in culture studies. Ecology 87, 563–569.

Högberg, M.N., Högberg, P., 2002. Extramatrical ectomycorrhizal mycelium contributes one-third of microbial biomass and produces, together with associated roots, half the dissolved organic carbon in a forest soil. New Phytologist 154, 791–795.

Högberg, P., Nordgren, A., Buchmann, N., Taylor, A.F.S., Ekblad, A., Högberg, M.N., Nyberg, G., Ottosson-Lofvenius, M., Read, D.J., 2001. Large-scale forest girdling shows that current photosynthesis drives soil respiration. Nature 411, 789–792.

Högberg, P., Högberg, M.N., Gottlicher, S.G., Betson, N.R., Keel, S.G., Metcalfe, D.B., Campbell, C., Schindlbacher, A., Hurry, V., Lundmark, T., et al., 2008. High temporal resolution tracing of photosynthate carbon from the tree canopy to forest soil microorganisms. New Phytologist 177, 220–228.

Johansson, E., Fransson, P.M.A., Finlay, R.D., van Hees, P.A.W., 2008. Quantitative analysis of root and ectomycorrhizal exudates as a response to heavy metal (Pb, Cd and As) stress. Plant and Soil 313, 39–54.

Johansson, E., Fransson, P.M.A., Finlay, R.D., van Hees, P.A.W., 2009. Quantitative analysis of soluble exudates produced by ectomycorrhizal roots as a response to ambient and elevated CO_2. Soil Biology & Biochemistry 41, 1111–1116.

Johnson, D., Leake, J.R., Ostle, N., Ineson, P., Read, D.J., 2002a. In situ $^{13}CO_2$ pulse-labelling of upland grassland demonstrates that a rapid pathway of carbon flux from arbuscular mycorrhizal mycelia to the soil. New Phytologist 153, 327–334.

Johnson, D., Leake, J.R., Read, D.J., 2002b. Transfer of recent photosynthate into mycorrhizal mycelium of an upland grassland: short-term respiratory losses and accumulation of ^{14}C. Soil Biology & Biochemistry 34, 1521–1524.

Johnson, D., Krsek, M., Wellington, E.M.H., Stott, A.W., Cole, L., Bardgett, R.D., Read, D.J., Leake, J.R., 2005. Soil invertebrates disrupt carbon flow through fungal networks. Science 309, 1047.

Jones, D.L., Nguyen, C., Finlay, R.D., 2009. Carbon flow in the rhizosphere: carbon trading at the soil-root interface. Plant and Soil 321, 5–33.

Jongmans, A.G., van Breemen, N., Lundström, U., Finlay, R.D., van Hees, P.A.W., Giesler, R., Melkerud, P.-A., Olsson, M., Srinivasan, M., Unestam, T., 1997. Rock-eating fungi: a true case of mineral plant nutrition? Nature 389, 682–683.

Kerley, S.J., Read, D.J., 1998. The biology of mycorrhiza in the Ericaceae XX. Plant and mycorrhizal necromass as nitrogenous substrates for the ericoid mycorrhizal fungus *Hymenscyphus ericae* and its host. New Phytologist 139, 353–360.

Kjøller, R., Olsrud, M., Michelsen, A., 2010. Co-existing ericaceous plant species in a subarctic mire community share fungal root endophytes. Fungal Ecology 3, 205–214.

Kjøller, R., Nilsson, L.-O., Hansen, K., Schmidt, I.K., Vesterdal, L., Gundersen, P., 2012. Dramatic changes in ectomycorrhizal community composition, root tip abundance and mycelial production along a stand-scale nitrogen deposition gradient. New Phytologist 194, 278–286.

Kohler, A., Kuo, A., Nagy, L.G., Morin, E., Barry, K.W., Buscot, F., Canback, B., Choi, C., Cichocki, N., Clum, A., et al., 2015. Convergent losses of decay mechanisms and rapid turnover of symbiosis genes in mycorrhizal mutualists. Nature Genetics 47, 410 U176.

Koide, R.T., Fernandez, C., Malcolm, G., 2013. Determining place and process: functional traits of ectomycorrhizal fungi that affect both community structure and ecosystem function. New Phytologist 201, 433–439.

Kõljalg, U., Nilsson, R.H., Abarenkov, K., Tedersoo, L., Taylor, A.F.S., Bahram, M., Bates, S.T., Bruns, T.D., Bengtsson-Palme, J., Callaghan, T.M., et al., 2013. Towards a unified paradigm for sequence-based identification of fungi. Molecular Ecology 22, 5271–5277.

Landeweert, R., Hofflund, E., Finlay, R.D., van Breemen, N., 2001. Linking plants to rocks: ectomycorrhizal fungi mobilize nutrients from minerals. Trends in Ecology & Evolution 16, 248–254.

Langley, J.A., Chapman, S.K., Hungate, B.A., 2006. Ectomycorrhizal colonization slows root decomposition: the post-mortem fungal legacy. Ecology Letters 9, 955–959.

Leake, J.R., Read, D.J., 1990. Chitin as a nitrogen source for mycorrhizal fungi. Mycological Research 94, 993–995.

Leake, J.R., Johnson, D., Donnelly, D., Muckle, G.E., Boddy, L., Read, D.J., 2004. Networks of power and influence: the role of mycorrhizal mycelium in controlling plant communities and agro-ecosystem functioning. Canadian Journal of Botany 82, 1016–1045.

Leake, J.R., Duran, A.L., Hardy, K.E., Johnson, I., Beerling, D.J., Banwart, S.A., Smits, M.M., 2008. Biological weathering in soil: the role of symbiotic root-associated fungi biosensing minerals and directing photosynthate-energy into grain-scale mineral weathering. Mineralogical Magazine 72, 85–89.

Lindahl, B.D., Tunlid, A., 2015. Ectomycorrhizal fungi – potential organic matter decomposers, yet not saprotrophs. New Phytologist 205, 1443–1447.

Lindahl, B.D., Ihrmark, K., Boberg, J., Trumbore, S.E., Högberg, P., Stenlid, J., Finlay, R.D., 2007. Spatial separation of litter decomposition and mycorrhizal nitrogen uptake in a boreal forest. New Phytologist 173, 611–620.

Lindahl, B.D., Nilsson, R.H., Tedersoo, L., Abarenkov, K., Carlsen, T., Kjøller, R., Koljalg, U., Pennanen, T., Rosendahl, S., Stenlid, J., et al., 2013. Fungal community analysis by high-throughput sequencing of amplified markers – a user's guide. New Phytologist 199, 288–299.

Marupakula, S., Mahmood, S., Finlay, R.D., 2016. Analysis of single root-tip microbiomes suggests that distinctive bacterial communities are selected by *Pinus sylvestris* roots colonised by different ectomycorrhizal fungi. Environmental Microbiology 18, 1470–1483.

McGuire, K.L., Allison, S.D., Fierer, N., Treseder, K.K., 2013. Ectomycorrhizal-dominated boreal and tropical forests have distinct fungal communities, but analogous spatial patterns across soil horizons. PLoS One 8, e68278.

Miller, S.L., Torres, P., McClean, T.M., 1994. Persistence of basidiospored and sclerotia of ectomycorrhizal fungi and *Morchella* in soil. Mycologia 86, 89–95.

Miltner, A., Bombach, P., Schmidt-Brücken, B., Kästner, M., 2012. SOM genesis: microbial biomass as a significant source. Biogeochemistry 111, 41–55.

Paul, E.A., 2007. Soil Microbiology, Ecology and Biochemistry. Academic Press, San Diego, CA, USA.

Persson, H.A., 1980. Spatial distribution of fine root growth, mortality and decomposition in a young Scots pine stand in Central Sweden. Oikos 34, 77–87.

Potapov, A.M., Tiunov, A.V., 2016. Stable isotope composition of mycophagous collembolans versus mycotrophic plants: do soil invertebrates feed on mycorrhizal fungi? Soil Biology & Biochemistry 93, 115–118.

Pritchard, S.G., Strand, A.E., McCormack, M.L., Davis, M.A., Oren, R., 2008. Mycorrhizal and rhizomorph dynamics in a loblolly pine forest during 5 years of free-air-CO_2-enrichment. Global Change Biology 14, 1252–1264.

Querejeta, J.I., Egerton-Warburton, L.M., Allen, M.F., 2003. Direct nocturnal water transfer from oaks to their mycorrhizal symbionts during severe soil drying. Oecologia 134, 55–64.

Quirk, J., Beerling, D.J., Banwart, S.A., Kakonyi, G., Romero-Gonzalez, M.E., Leake, J.R., 2012. Evolution of trees and mycorrhizal fungi intensifies silicate mineral weathering. Biology Letters 8, 1006–1011.

Quirk, J., Andrews, M.Y., Leake, J.R., Banwart, S.A., Beerling, D.J., 2014. Ectomycorrhizal fungi and past high CO_2 atmospheres enhance mineral weathering through increased below-ground carbon-energy fluxes. Biology Letters 10.

Quirk, J., Leake, J.R., Johnson, D.A., Taylor, L.L., Saccone, L., Beerling, D.J., 2015. Constraining the role of early land plants in Palaeozoic weathering and global cooling. Proceedings of the Royal Society of London B 282 20151115.

Read, D.J., Perez-Moreno, J., 2003. Mycorrhizas and nutrient cycling in ecosystems – a journey towards relevance? New Phytologist 157, 475–492.

Read, D.J., Leake, J.R., Perez-Moreno, J., 2004. Mycorrhizal fungi as drivers of ecosystem processes in heathland and boreal forest biomes. Canadian Journal of Botany-Revue Canadienne de Botanique 82, 1243–1263.

Read, D.J., 1991. Mycorrhizas in ecosystems. Experimenta 47, 376–391.

Reichard, P.U., Kraemer, S.M., Frazier, S.W., Kretzschmar, R., 2005. Goethite dissolution in the presence of phytosiderophores: rates, mechanisms, and the synergistic effect of oxalate. Plant and Soil 276, 115–132.

Remén, C., Persson, T., Finlay, R.D., Ahlström, K., 2008. Responses of oribatid mites to tree girdling and nutrient addition in boreal coniferous forests. Soil Biology & Biochemistry 40, 2881–2890.

Robinson, C.H., 2001. Cold adaptation in arctic and Antarctic fungi. New Phytologist 151, 341–353.

Rosling, A., Landeweert, R., Lindahl, B.D., Larsson, K.H., Kuyper, T.W., Taylor, A.F.S., Finlay, R.D., 2003. Vertical distribution of ectomycorrhizal fungal taxa in a podzol soil profile. New Phytologist 159, 775–783.

Rosling, A., Lindahl, B.D., Finlay, R.D., 2004. Carbon allocation in intact mycorrhizal systems of *Pinus sylvestris* L. seedlings colonizing different mineral substrates. New Phytologist 162, 795–802.

Ruiz-Herrera, J., 2012. Fungal Cell Wall: Structure, Synthesis, and Assembly. CRC Press, Boca Raton, FL, USA.

Schmidt, M.W.I., Torn, M.S., Abiven, S., Dittmar, T., Guggenberger, G., Janssens, I.A., Kleber, M., Kogel-Knabner, I., Lehmann, J., Manning, D.A.C., et al., 2011. Persistence of soil organic matter as an ecosystem property. Nature 478 (7367), 49–56.

Selosse, M.-A., Setaro, S., Glatard, F., Richard, F., Urcelay, C., Weiss, M., 2007. Sebacinales are common mycorrhizal associates of Ericaceae. New Phytologist 174, 864–878.

Shah, F., Nicolas, C., Bentzer, J., Ellstrom, M., Smits, M., Rineau, F., Canback, B., Floudas, D., Carleer, R., Lackner, G., et al., 2016. Ectomycorrhizal fungi decompose soil organic matter using oxidative mechanisms adapted from saprotrophic ancestors. New Phytologist 209, 1705–1719.

Sinsabaugh, R.L., 2010. Phenol oxidase, peroxidase and organic matter dynamics of soil. Soil Biology & Biochemistry 42, 391–404.

Smith, S.E., Read, D.J., 2008. Mycorrhizal Symbiosis. Academic Press, Cambridge, UK.
Smits, M.M., Hoffland, E., Jongmans, A.G., van Breemen, N., 2005. Contribution of mineral tunneling to total feldspar weathering. Geoderma 125, 59–69.
Smits, M.M., Bonneville, S., Haward, S., Leake, J.R., 2008. Ectomycorrhizal weathering, a matter of scale? Mineralogical Magazine 72, 131–134.
Smits, M.M., Bonneville, S., Benning, L.G., Banwart, S.A., Leake, J.R., 2012. Plant-driven weathering of apatite – the role of an ectomycorrhizal fungus. Geobiology 10, 445–456.
Staddon, P.L., Ramsey, C.B., Ostle, N., Ineson, P., Fitter, A.H., 2003. Rapid turnover of hyphae of mycorrhizal fungi determined by AMS microanalysis of ^{14}C. Science 300, 1138–1140.
Sterkenburg, E., Bahr, A., Durling, M.B., Clemmensen, K.E., Lindahl, B.D., 2015. Changes in fungal communities along a boreal forest soil fertility gradient. New Phytologist 207, 1145–1158.
Stevenson, F.J., 1994. Humus Chemistry: Genesis, Composition, Reactions. John Wiley & Son, Hoboken, NJ, USA.
Sun, Y.-P., Unestam, T., Lucas, S.D., Johanson, K.J., Kenne, L., Finlay, R.D., 1999. Exudation–reabsorption in mycorrhizal fungi, the dynamic interface for interaction with soil and other microorganisms. Mycorrhiza 9, 137–144.
Taylor, L.L., Quirk, J., Thorley, R.M.S., Kharecha, P.A., Hansen, J., Ridgwell, A., Lomas, M.R., Banwart, S.A., Beerling, D.J., 2016. Enhanced weathering strategies for stabilizing climate and averting ocean acidification. Nature Climate Change 6, 402–406.
Thorley, R.M.S., Taylor, L.L., Banwart, S.A., Leake, J.R., Beerling, D.J., 2015. The role of forest trees and their mycorrhizal fungi in carbonate rock weathering and its significance for global carbon cycling. Plant Cell and Environment 38, 1947–1961.
Toljander, J.F., Artursson, V., Paul, L.R., Jansson, J.K., Finlay, R.D., 2006. Attachment of different soil bacteria to arbuscular mycorrhizal fungi is determined by hyphal vitality and fungal species. FEMS Microbiology Letters 254, 34–40.
Toljander, J.F., Paul, L., Lindahl, B.D., Elfstrand, M., Finlay, R.D., 2007. Influence of AM fungal exudates on bacterial community structure. FEMS Microbiology Ecology 61, 295–304.
Treseder, K.K., Lennon, J.T., 2015. Fungal traits that drive ecosystem dynamics on land. Microbiology and Molecular Biology Reviews 79, 243–262.
Treseder, K.K., Allen, M.F., Ruess, R.W., Pregitzer, K.S., Hendrick, R.L., 2005. Lifespans of fungal rhizomorphs under nitrogen fertilization in a pinyon-juniper woodland. Plant and Soil 270, 249–255.
van Breemen, N., Finlay, R., Lundström, U., Jongmans, A.G., Giesler, R., Olsson, M., 2000. Mycorrhizal weathering: a true case of mineral plant nutrition? Biogeochemistry 49, 53–67.
van Hees, P.A.W., Vinogradoff, S.I., Edwards, A.C., Godbold, D.L., Jones, D.L., 2003. Low molecular weight organic acid adsorption in forest soils: effects on soil solution concentrations and biodegradation rates. Soil Biology & Biochemistry 35, 1015–1026.
van Hees, P.A.W., Jones, D.L., Finlay, R.D., Godbold, D.L., Lundström, U.S., 2005. The carbon we do not see: do low molecular weight compounds have a significant impact on carbon dynamics and respiration in forest soils? Soil Biology & Biochemistry 37, 1–13.
Van Hees, P.A.W., Rosling, R., Essén, S., Godbold, D.L., Jones, D.L., Finlay, R.D., 2006. Oxalate and ferricrocin exudation by the extramatrical mycelium of an ectomycorrhizal fungus in symbiosis with *Pinus sylvestris*. New Phytologist 169, 367–378.
Verrecchia, E.P., Braissant, O., Cailleau, G., 2006. The oxalate-carbonate pathway in soil carbon storage: the role of fungi and oxalotrophic bacteria. In: Gadd, G.M. (Ed.), Fungi in Biogeochemical Cycles. Cambridge University Press, Cambridge, pp. 289–310.
Vrålstad, T., Schumacher, T., Taylor, A.F.S., 2002. Mycorrhizal synthesis between fungal strains of the *Hymenoscyphus ericae* aggregate and potential ectomycorrhizal and ericoid hosts. New Phytologist 153, 143–152.
Walker, J.F., Aldrich-Wolfe, L., Riffel, A., Barbare, H., Simpson, N.B., Trowbridge, J., Jumpponen, A., 2011. Diverse Helotiales associated with the roots of three species of Arctic Ericaceae provide no evidence for host specificity. New Phytologist 19, 515–527.
Wallander, H., Nilsson, L.O., Hagerberg, D., Baath, E., 2001. Estimation of the biomass and seasonal growth of external mycelium of ectomycorrhizal fungi in the field. New Phytologist 151, 753–760.
Wallander, H., Nilsson, L.O., Hagerberg, D., Rosengren, U., 2003. Direct estimates of C:N ratios of ectomycorrhizal mycelia collected from Norway spruce forest soils. Soil Biology & Biochemistry 35, 997–999.

Wallander, H., Goransson, H., Rosengren, U., 2004. Production, standing biomass and natural abundance of ^{15}N and ^{13}C in ectomycorrhizal mycelia collected at different soil depths in two forest types. Oecologia 139, 89–97.

Wallander, H., Johansson, U., Sterkenburg, E., Durling, M.B., Lindahl, B.D., 2010. Production of ectomycorrhizal mycelium peaks during canopy closure in Norway spruce forests. New Phytologist 187, 1124–1134.

Wallander, H., Ekblad, A., Bergh, J., 2011. Growth and carbon sequestration by ectomycorrhizal fungi in intensively fertilized Norway spruce forests. Forest Ecology and Management 262, 999–1007.

Wallander, H., Ekblad, A., Godbold, D.L., Johnson, D., Bahr, A., Baldrian, P., Bjork, R.G., Kieliszewska-Rokicka, B., Kjøller, R., Kraigher, H., et al., 2013. Evaluation of methods to estimate production, biomass and turnover of ectomycorrhizal mycelium in forests soils – a review. Soil Biology & Biochemistry 57, 1034–1047.

Wardle, D.A., Jonsson, M., Bansal, S., Bardgett, R.D., Gundale, M.J., Metcalfe, D.B., 2012. Linking vegetation change, carbon sequestration and biodiversity: insights from island ecosystems in a long-term natural experiment. Journal of Ecology 100, 16–30.

Wessels, J.G.H., 1993. Wall growth, protein excretion and morphogenesis in fungi. New Phytologist 123, 397–413.

Yuan, J.P., Kuang, H.C., Wang, J.H., Liu, X., 2008. Evaluation of ergosterol and its esters in the pileus, gill, and stipe tissues of agaric fungi and their relative changes in the comminuted fungal tissues. Applied Microbiology and Biotechnology 80 (3), 459–465.

Zeglin, L.H., Myrold, D.D., 2013. Fate of decomposed fungal cell wall material in organic horizons of old-growth Douglas-fir forest soils. Soil Science Society of America Journal 77, 489–500.

Zeglin, L.H., Kluber, L.A., Myrold, D.D., 2013. The importance ofamino sugar turnover to C and N cycling in organic horizons of old-growth Douglas-fir forest soils colonized by ectomycorrhizal mats. Biogeochemistry 112, 679–693.

Zhang, L., Fan, J., Ding, X., He, X., Zhang, F., Feng, G., 2014. Hyphosphere interactions between an arbuscular mycorrhizal fungus and a phosphate solubilizing bacterium promote phytate mineralization in soil. Soil Biology & Biochemistry 74, 177–183.

Zhang, L., Xu, M., Liu, Y., Zhang, F., Hodge, A., Feng, G., 2016. Carbon and phosphorus exchange may enable cooperation between an arbuscular mycorrhizal fungus and a phosphate-solubilizing bacterium. New Phytologist 210, 1022–1032.

CHAPTER

24

Mycorrhizal Interactions With Saprotrophs and Impact on Soil Carbon Storage

E. Verbruggen[1], *R. Pena*[2], *C.W. Fernandez*[3], *J.L. Soong*[1]

[1]University of Antwerp, Wilrijk, Belgium; [2]University of Goettingen, Goettingen, Germany; [3]University of Minnesota, St. Paul, MN, United States

24.1 INTRODUCTION

As we have seen in previous chapters a unifying feature of all mycorrhizal fungi is that they prolifically forage the rhizosphere and often the soil beyond for nutrients. In so doing, they encounter a myriad of bacteria, saprotrophic fungi, protists, and soil fauna with which they engage in neutral, positive, or negative interactions and thereby contribute to the structure and functioning of the soil food web. The soil food web is in turn responsible for cycling of nutrients from soil organic matter (SOM), which ultimately determines how much C entering soil is stored or respired back to the atmosphere as CO_2. Therefore the interactions between mycorrhizal fungi and soil saprotrophs are of primary interest in understanding how much C is stored in different ecosystems and under different environmental regimes. This understanding has gained additional importance in the past decades as a crucial unknown in predicting the global climate response to increasing levels of atmospheric CO_2. In this chapter we will discuss the different ways by which mycorrhizal fungi interact with saprotrophs as a potential source of nutrition and as a competitor for nutrients and space. Furthermore, we will explain how these different interactions can affect the amount of C stored in soil.

24.1.1 How Different Are Roots and Mycorrhizal Fungi in Relation to Saprotrophs?

When considering the effect of mycorrhizal fungi on the soil food web and soil C, it is important to ask how C allocated to fungal hyphae differs from C invested in roots. The reason is that variation in plant allocation between roots and mycorrhizal fungi is a very

Mycorrhizal Mediation of Soil
http://dx.doi.org/10.1016/B978-0-12-804312-7.00024-3

commonly observed response to natural environmental gradients (e.g. Soudzilovskaia et al., 2015); thus it may be a particularly potent source of variation in soil C storage. These environmental drivers include land use change (Gosling et al., 2006) and global change factors such as elevated atmospheric CO_2 (Sanders et al., 1998), temperature (Heinemeyer et al., 2004), a combination of both (Vicca et al., 2009), nutrient availability (Liu et al., 2015), and precipitation (Antoninka et al., 2015). In Table 24.1 we have summarized a list of mycorrhizal and fine root traits that may affect soil-dwelling saprotrophs. Indeed, we see there are large general differences in anatomy, physiology, and composition between these different biological entities, such as turnover time, C allocation, uptake capacities, and anatomical traits. Fine roots are thought to interact intensively with saprotrophs by releasing easily degradable carbohydrates, which "prime" saprotrophic activity in the rhizosphere, releasing nutrients held in the SOM (Kuzyakov et al., 2000). It is less clear to what extent mycorrhizal fungi are involved in similar interactions, although many compounds have been isolated from the outside of hyphae (e.g. Toljander et al., 2007; Fransson et al., 2016), suggesting that they do share this capacity with roots. A rough estimation of the typical fraction of net fixed C allocated to rhizodeposits is 11% (Jones et al., 2009), which includes as primary constituents mucilage, root exudates, and mycorrhizal fungi because these latter are not formally distinguished in this

TABLE 24.1 General Traits of Plant Roots and Mycorrhizal Fungi That Partly Determine Their Interactions With Saprotrophs

	Plant Roots (Fine)	AM Fungi	EcM Fungi	ErM Fungi
C allocation	Average 22% (Jackson et al., 1997; Jones et al., 2009; McCormack et al., 2015)	3–20% (Johnson et al., 2002; Leake et al., 2004)	1–22% (Leake et al., 2004; Hobbie, 2006)	3–40% (of total hair roots) (Olsrud and Christensen, 2004; Olsrud et al., 2004; Hobbie and Hobbie, 2008)
Median lifespan	20 days to 1 year (Mccormack et al., 2012)	5–6 days (Staddon et al., 2003)	33–48 days (Hendricks et al., 2006, 2015; Allen and Kitajima, 2013)	?
Organic nutrient use/uptake	Limited; sugars, amino acids, organic P	Limited; all before, potentially chitin (Whiteside et al., 2012)	Substantial, but highly variable; all before, also more complex, humified organic pools	High; all before (Bending and Read, 1996; Burke and Cairney, 2002)
Main exudates/deposits	Sugars, amino acids, organic acids (Jones et al., 2004)	Sugars, organic acids (Hooker et al., 2007; Toljander et al., 2007)	Sugars, amino acids, organic acids (Johansson et al., 2009)	?
Diameter	100–2000 μm	1–10 μm	2–10 μm (Mccormack et al., 2015)	1 μm in the hyphal sheath and up to 3 μm inside of the cells (Setaro et al., 2006)
C:N:P ratio	450:11:1 (Jackson et al., 1997)	20:2:1 (Allen et al., 2003)	20:2:1 (Allen et al., 2003)	20:2:1 (Allen et al., 2003)

Except for C allocation, traits refer to extraradical absorptive hyphae in case of mycorrhizal fungi. Mycorrhizal fungi are subdivided in the broad functional groups of arbuscular mycorrhizal (AM), ectomycorrhizal (EcM), and ericoid mycorrhizal (ErM) fungi.

terminology. These are responsible for the "rhizosphere effect", named after the observation that the biological activity within a few millimeters around roots, and beyond in the case of mycorrhiza, is orders of magnitude greater than in the "bulk soil."

24.1.2 Mycorrhizal Effects on Soil Organic Matter

According to current understanding of SOM dynamics (Box 24.1), mycorrhizal fungi can affect formation and preservation of SOM in multiple ways. Allocation of C by plants to

BOX 24.1 A CHANGING UNDERSTANDING OF SOIL ORGANIC MATTER ORIGINS

Over the past decade a new understanding of the processes underlying long-term soil C sequestration have emerged. It was previously assumed that the selective preservation of recalcitrant soil organic matter (SOM) compounds was the main source of soil C persistence, as would be apparent from short-term litter decomposition studies. Such SOM is a main component of soil C stocks in organic soils of peatlands and high-latitude forests by virtue of its pool size and residence time. However, for most mineral soils over longer time scales research has shown that recalcitrant plant-derived compounds are of relatively minor importance in contributing to the SOM pool (Von Lützow et al., 2006; Marschner et al., 2008). Advancements in chemical and molecular techniques have revealed how degradable organic compounds can persist in the soil for much longer than predicted because of chemical and physical protection within the soil physical matrix (Schmidt et al., 2011; Lehmann and Kleber, 2015). The buildup of SOM that contributes to soil C storage is a result of inputs of plant-derived material to the soil, the preservation of which is constrained by microbial ecology, enzyme kinetics, environmental drivers, and matrix protection (Kleber, 2010). Particulate organic matter contributes to soil C storage and can be protected from mineralization through incorporation into soil aggregates or eventually decomposed. Microbial residues and products are mostly produced via the decomposition of more labile organic residues such as dissolved organic carbon (DOC; Cotrufo et al., 2013, 2015; Soong et al., 2015). It is these microbial residues and products that have also been found to contribute to mineral-associated SOM fractions, which have the longest mean residence times in the soil (Ehleringer, 2000; Grandy and Neff, 2008). The direct flow of C from plants to mycorrhizal symbionts makes mycorrhizal residues particularly significant to soil C storage. In an experiment on a poplar plantation, mycorrhizal hyphae turnover was found to be the dominant pathway of plant inputs of C to SOM, exceeding inputs from fine root turnover and leaf litter (Godbold et al., 2006). This high rate of C input, along with the potential stabilization of mycorrhizal residues on mineral surfaces, gives mycorrhizal inputs a high potential to contribute to soil C storage. In this framework, interactions with saprotrophs will determine which compounds are deposited (mycorrhizal fungal or saprotrophic?) and the extent to which SOM is used as a C source through their specific enzymes. Thus mycorrhizal biomass residues are a source of C inputs to the soil while also stimulating soil aggregation through glomalin production and hyphal networks (see also Chapter 14 for further reading).

mycorrhizal fungi could increase the amount of C from photosynthate to belowground biomass. Belowground inputs can be a potent source of soil C (Godbold et al., 2006; Clemmensen et al., 2013); root litter has been found to contribute more to soil C storage than aboveground litter because of the vulnerability of leaf litter to high rates of saprotroph activities in the upper soil and the proximity of root litter within the soil matrix (Rasse et al., 2005; Hatton et al., 2015). Similarly, mycorrhizal hyphal litter may contribute more to soil C storage than C allocation to aboveground biomass that eventually falls to the soil as surface litter because of reduced microbial biomass and activity with soil depth after root senescence. Whether mycorrhizal fungi actively decompose available SOM or compete with other saprotrophs would also affect SOM decomposition rates. Talbot et al. (2008) describe three hypotheses for why ectomycorrhizal (EcM) and ericoid mycorrhizal (ErM) fungi metabolize soil C, which vary in support from empirical evidence. The first hypothesis is that they use soil C as an alternative energy source when photosynthate supply is low (e.g. on cloudy days or in early spring; Courty et al., 2007). Support for this hypothesis is poor for EcM (Lindahl and Tunlid, 2015), but it is tentatively possible in ErM that have a larger suite of saprotrophic genes at their disposal (Kohler et al., 2015). The second is that soil C is mineralized as a byproduct of SOM decomposition to retrieve nutrients. The use of extracellular enzymes by mycorrhizal fungi to release nutrients from organic matter would also release C from SOM. The third hypothesis is that mycorrhizal activity is stimulated by the availability of photosynthate C so that when plant supply of C is high, mycorrhizal fungi have the energy needed to break down more recalcitrant SOM. This "priming effect" leads to increased decomposition of SOM and loss of soil C when provisioning of C from plants is highest (Kuzyakov et al., 2000). The third hypothesis can also apply to arbuscular mycorrhizal (AM) fungi but then intermediated by stimulation of soil saprotrophs, more in line with the traditional definition of soil priming (Kuzyakov et al., 2000). Hypotheses 2 and 3 are similar in cause and consequence, with the distinction that under the third hypothesis the emphasis is on higher mineralization rates in the presence of mycorrhizal fungi than would be energetically feasible in their absence.

24.2 MYCORRHIZAL FUNGI AS A SOURCE OF C IN SOIL

The importance of DOC to soil C storage is now gaining attention due to its dynamic production and consumption in the soil, driving the need for explicit representation in soil C models (Campbell et al., 2014; Ahrens et al., 2015; see Box 24.1). Climate, soil type, and vegetation type are known factors that control DOC concentrations and composition in the soils; however, the magnitude of the influence of each factor varies across ecosystems (Camino-Serrano et al., 2014). Leaching of DOC during surface litter decomposition is an important flux of low-molecular–weight organic matter into the soil, where it can be immediately taken up by soil microbes, adsorb directly to mineral surfaces, or leach further down into the soil profile (Kaiser and Kalbitz, 2012; Soong et al., 2015). DOC also enters the soil via root and mycorrhizal exudation of organic acids and carbohydrates. In a study by Kluber et al. (2010), oxalate concentrations were found to be 40 times higher under EcM hydromorphic mats than nonmat mineral soils, indicating that some EcM fungi may play a significant role in the exudation of oxalate into the soil. Oxalate is the salt of oxalic acid, the exudation of which could help EcM fungi acquire P and other nutrients through mineral weathering while simultaneously contributing to increased C inputs to the soil (Kluber et al., 2010). Although DOC inputs may contribute to soil C accumulation via stimulation of microbial biomass that is later stabilized by mineral adsorption, labile C substrates also provide energy to stimulate microbial biomass and activity, leading to increased decomposition or "priming" of older SOM (Kuzyakov et al., 2000; Gunina and Kuzyakov, 2015). Throughout this chapter we will invoke this mechanism when referring to priming, in contrast to increased SOM decomposition by the mycorrhizal fungi in response to increasing plant photosynthate supply, which is sometimes also used in literature (see Box 24.1).

Differences among EcM and AM fungal physiology, biochemistry, and morphology likely alter their fate as decomposition residues in the soil (Fernandez et al., 2016). EcM fungi have thicker cell walls and more pigmentation and septa between cells, making them more recalcitrant than AM fungal residues, contributing to their longer mean residence time in the soil (Langley and Hungate, 2003; see Table 24.1). Although long, thin AM fungal hyphae may be more easily degradable than EcM hyphae, the production of glomalin-related soil proteins and enmeshment of soil by AM fungi increases soil aggregate formation and physical protection of AM fungal residues from decomposition (see also Chapter 14).

24.2.1 Provision of Labile C to Microbial Saprotrophs

Saprotrophs in mineral soil are nearly always (co-)limited by C because of the rapid depletion of available labile C-rich substrates in the soil (e.g. Griffiths et al., 2012; Hopkins et al., 2014). The addition of energy-rich, labile substrates such as sugars provides the energy needed for microbial growth and the production of extracellular enzymes involved in mineralization and acquisition of other nutrients, such as N, P, K, and other (micro-)nutrients. AM fungi are known to exude multiple sugars (Hooker et al., 2007); thus it is highly plausible that they fuel microbial activity (see also Chapters 8, 22, and 23) that can lead to priming of SOM decomposition. Studies tracking the belowground fate of photosynthates have found rapid transfers of C to AM fungi (Hannula et al., 2012) followed by release into free-living microbial biomass (Drigo et al., 2010; Kaiser et al., 2015). Most experimental studies report increased abundances or activity of free-living heterotrophic fungi and bacteria in response to the presence of mycorrhizal fungi, especially in cases in which decomposition rates are stimulated by AM fungi (Table 24.2). Therefore priming is often considered as one of the principal mechanisms responsible for increased SOM decomposition by mycorrhizal fungi (see Table 24.2). However, there are several studies that highlight some uncertainty around the role of AM fungi in SOM priming. For example, Shahzad et al. (2015) evaluated the independent effects of root exudates and turnover versus AM fungi on the rhizosphere priming effect. They report that nearly all (up to 98%) of priming was caused by roots (exudates and turnover) rather than by mycorrhizal fungi. This suggests that either the combined effects of exudation and turnover of AM fungi have minor effects on priming compared with those of plant roots by virtue of their low quantity or that they differ in the extent to which they specifically prime the microbial community to mineralize SOM.

The latter could be related to different microbial community responses to AM than non-AM rhizodeposition. Toljander et al. (2007), Hannula et al. (2012), and Kaiser et al. (2015) report significant changes in microbial communities in the vicinity of AM fungi and/or a significant microbial incorporation of AM fungi-derived C (see also Table 24.2). Kaiser et al. (2015) further

TABLE 24.2 Selected Experiments in Which Mycorrhizal Abundance was Manipulated and a Saprotrophic Response (Primarily Microbial) and an Aspect of Decomposition was Measured

Study	Mycorrhizal Type	Decomposers	Effect Direction	Decomposition Measured	Effect Direction	Time	Substrate	Setting
Hodge et al. (2001)	AM; *Glomus hoi*	PLFA; composition unchanged	-	N release/uptake litter C loss	↑	6 weeks	Milled grass shoots	Laboratory
Leifheit et al. (2014)	AM; *Rhizophagus irregularis*	Non-AM fungal hyphal length reduced	↓	Litter mass loss	↓	22 weeks	Wooden stick	Laboratory
Verbruggen et al. (2016)	AM; community	PLFA; fungi reduced	↓	Litter C loss	↓	26 weeks	Intact wheat roots	Laboratory
Koller et al. (2013)	AM; *R. intraradices*	Increased microbial respiration	↑	N release/uptake	↑	6 weeks	Milled grass shoots	Laboratory
Atul-Nayyar et al. (2009)	AM; *Claroideoglomus claroideum* *R. intraradices* *G. clarum*	PLFA; composition fungi and bacteria changed, abundance unchanged	↑	N release/uptake Reduced C:N	↑	24 weeks	Milled wheat shoot and root	Laboratory
Albertsen et al. (2006)	AM; *R. intraradices*	PLFA; increases of fungi and bacteria	↑	nd	nd	8 weeks	Ground barley leaves	Laboratory
Welc et al. (2010)	AM; *R. intraradices*, *Funneliformis mosseae*	PLFA; decreases of fungi and bacteria	↓	nd	nd	6 weeks	Cellulose	Laboratory
Cheng et al. (2012)	AM; *Acaulospora morrowiae*, *Gigaspora margarita*, *G. clarum*	nd	-	Litter C loss under elevated CO_2	↑	Up to 15 weeks	Fragments grass shoots	Laboratory and field

Nuccio et al. (2013)	AM; *G. hoi*	Micro-array; Positive and negative on different bacterial groups	↓↑	N release/uptake Increase litter-derived C:N	↑	10 weeks	*Plantago lanceolata* roots	Laboratory
Nottingham et al. (2013)	AM; Community	PLFA: Decreases of fungi and bacteria	↓	Reduced microbial SOM uptake	↓	26 weeks	Soil-derived C	Field; tropical forest
Moore et al. (2015)	EcM, ±AM; Community	Enzyme activities	-↑	Reduced respiration (only in presence roots)	↓	22 weeks (resp. Immediate)	Starch	Field; temperate forest
Brzostek et al. (2015)	EcM AM; Community	Enzyme activities	↑-	Respiration/C mineralization	↑- ↓	117 weeks	Soil-derived C. litter	Field; temperate forest
Carrillo et al. (2016)	AM; Community	PLFA; Changed community; some gram negative bacteria and protists increased	↑↓	CO_2 induced priming	↓-	9 weeks	Soil-derived C	Laboratory
Bödeker et al. (2016)	EcM; community	Next generation sequencing; reduced colonization of fresh litter by saprotrophic fungi	↓	Reduced mass loss	↓	4 months (after 16 months effects were less pronounced)	Litter	Field; boreal forest

AM, arbuscular mycorrhizal; *EcM*, ectomycorrhizal; *ErM*, ericoid mycorrhizal; *nd*, not determined; *PFLA*, phospholipid fatty acid; *SOM*, soil organic matter.

suggest that this pathway differs from the normal root exudation pathway in that it is more directed and active as opposed to a typically more passive C release by plants. This explains why C moved faster in AM fungi and appeared to be more responsive to localized nutrient pulses. As in Shahzad et al. (2015), AM fungi were a quantitatively less important source of microbial C than roots (Kaiser et al., 2015). A less constitutive and more directed C flow from AM fungi than roots could provide an explanation for the greater stimulation of microbial decomposition by AM fungi in high-nutrient patches (Hodge et al., 2001; Cheng et al., 2012; Nuccio et al., 2013) than in nutrient-poor SOM (Nottingham et al., 2013; Shahzad et al., 2015; Carrillo et al., 2016; Table 24.2). A recent microcosm study also suggests that whether or not AM fungi would stimulate microbial saprotroph activity may depend on nutrient levels (Zhang et al., 2016b). Here, an AM fungus did not stimulate the activity of a phosphate solubilizing bacterium when P levels were low and effectively competed with them for organic P (phytate). When P levels were elevated the effect was reversed, suggesting that stimulation of these bacteria was contingent upon whether a base demand of phosphorus had been met (Zhang et al., 2016b). Together these data suggest that AM fungi are involved in the priming of SOM (Table 24.2), but that the effect can depend strongly on context, such as nutritional demand, CO_2 levels, and substrate (fresh litter vs. SOM).

If AM fungi will indeed induce weaker positive priming effects on SOM than roots, then it is important to recognize that this may not mean that a shift from roots to AM fungi will positively affect soil organic carbon (SOC) stocks. For instance, Nottingham et al. (2013) report that in a lowland tropical forest system AM fungi significantly affected microbial communities measured by phospholipid fatty acid analysis. This was accompanied with a higher incorporation of plant-derived C into these microbes, but it suppressed the incorporation of SOM-derived C compared with a treatment in which plant roots also interacted with these microbes (Nottingham et al., 2013). Therefore the priming effect induced by plant roots was absent when only mycorrhizal fungi had access to soil. However, in an earlier study in a similar system it was found that AM fungi significantly contribute to respiration of autotrophically derived C, accounting for 12–40% of C respired by roots and AM fungi together (Nottingham et al., 2010). As such, although AM fungi may not contribute to decomposition of SOM in that study, the high proportion of C respired by AM fungi compared with roots may still mean that C allocated toward AM fungi leads to higher CO_2 fluxes from the soil. This is in line with studies in which ^{13}C or ^{14}C gas labeling was adopted to estimate C flux through mycorrhizal fungi, where very fast and abundant C return to the atmosphere appears to be the norm (Olsson and Johnson, 2005), and is faster than typical estimates for roots (Table 24.1).

24.2.2 Importance in Different Mycorrhizal Fungi and Effect on Soil Organic Carbon

Whether mycorrhizal fungi affect saprotrophs, for instance through exudation of labile C and turnover, will also depend on their encounter frequency. The highest densities of saprotrophs will commonly be found in young litter, before easily degradable compounds have been diminished (Clemmensen et al., 2015). In many studies AM fungi have been reported to actively grow toward and colonize litter patches (Hodge et al., 2001). For instance, Hodge and Fitter (2010) report an average of three times higher hyphal density in an experimental greenhouse system where AM fungi could colonize compartment with grass leaves as compared with the one without leaves. In a montane tropical forest, AM fungi were found to extensively colonize the surfaces and insides of decomposing leaves, amounting to 29% of total AM fungal length found at a given area (Camenzind and Rillig, 2013). Therefore AM fungi are likely to heavily interact with the litter decomposition process. In EcM fungi this behavior is less apparent. Encounters between EcM fungi and litter decomposing fungi may be less prevalent because some forests have highly stratified soils, where these fungi inhabit spatially distinct layers: EcM fungi tend to be dominant in deeper layers where partially decomposed litter with low C:N ratios is found (Lindahl et al., 2007). In an experiment manipulating the presence of EcM fungi in forest soil plots via trenching, Lindahl et al. (2010) found increased abundance of saprotrophic fungi lower in the soil profile, suggesting that the scarcity of saprotrophic fungi in these deeper soil horizons is a result of competitive exclusion by EcM fungi rather than vertical niche specialization. However, part of this effect may have been caused by an immediate flux of labile substrate from dying EcM hyphae and roots into deeper soil layers, making it difficult to distinguish between these two mechanisms (Lindahl et al., 2010). Still, in a study where organic substrates from both soil horizons was reciprocally placed in either horizon and fungi colonizing these substrates followed, both depth and litter type explained fungal communities found to colonize it to a similar extent (Bödeker et al., 2016). This indicates that fungi dominant in either soil profile (EcM vs. saprotrophs) can suppress the other type regardless of substrate type. This may be caused by interference competition, which is documented to be particularly strong among some basidiomycetes (Boddy, 2000). Regardless of the cause, the resulting highly stratified soils typical of EcM-dominated ecosystems may cause EcM fungi to interact less intensively with saprotrophs than AM fungi and may partly explain why AM fungi are more commonly found to stimulate saprotrophs in fresh litter than EcM.

24.3 COMPETITION FOR NUTRIENTS AND HABITAT

Saprotrophs and mycorrhizal fungi partly depend on the same nutrient sources and share habitats, leading to the potential for exploitative competition and antagonistic interference. In the case of direct competition for resources, preemption of a resource will diminish the growth of a competitor and thereby affect its population size and potential functioning—in this case decomposition. Indeed, there is some evidence that suppression of other soil biota by mycorrhizal fungi can lead to reduced mineralization of SOC. For instance, field studies suppressing AM fungi using the fungicide benomyl have found that soil respiration was increased and SOC stocks decreased, suggesting an inhibitive effect of AM fungi on the activity of other soil microbes (Wilson et al., 2009; Zhang et al., 2016a). Suppression of saprotrophs was also suggested to be responsible for the well-known Gadgil effect (Gadgil and Gadgil, 1971), in which trenching of forest soil plots eliminated recently fixed C supply into the plots, thereby reducing EcM fungi, which then led to an increased mass loss of the litter layer by more than 50%. However, this effect is far from universal (see Fernandez & Kennedy, 2016, for an in-depth review on the topic). Unlike some EcM fungi that are able to break down relatively complex organic substrates for resources, AM fungi likely compete with saprotrophs for NH_4^+ and inorganic P, given their ability to efficiently scavenge for both (Hodge et al., 2010; Jansa et al., 2011). Although reductions in nutrient availability may have negative effects on saprotrophs and thus decrease SOC loss, the opposite can also occur. Many studies have reported that a reduction of NH_4^+ can cause reductions in SOM (e.g. Janssens et al., 2010), which may be explained by saprotrophs mining SOM for N (Craine et al., 2007; Ramirez et al., 2012). In this case competition for N between mycorrhizal fungi and saprotrophs could also increase their SOC utilization. This is exemplified in a study by Cheng et al. (2012) in which AM fungi were found to reduce soil NH_4^+ pools but had no negative effect on NO_3^- and even increased the pool size of the latter. The authors reported a sharp increase of litter decomposition when AM fungi were present and suggested that the combination of mycorrhiza-mediated C supply (turnover and exudation) to surrounding saprotrophs and the removal of NH_4^+ strongly stimulated saprotroph activity.

In the case of P there are strong indications that under P-limiting conditions P additions will increase SOM decomposition, likely by relieving saprotrophs from a (co-)limitation by P and C of their growth (Craine et al., 2007; Nottingham et al., 2015). Moreover, data from multiple field experiments suggest that P fertilization can reduce SOC stocks despite increased primary productivity (Poeplau et al., 2015; Sochorová et al., 2016). Therefore if mycorrhizal fungi can reduce P availability to a sufficient extent to induce P (co-)limitation on saprotrophs, then this could be a significant contributor to SOC retention. AM fungi are common in ecosystems with low P and are generally stimulated by low levels of available P in the soils (e.g. Xiang et al., 2014), which makes it plausible that there is a threshold below which AM fungi may cause and even exacerbate P limitation of other microbes. EcM fungi generally occur in less P-limited ecosystems such as temperate boreal forests and shrublands, making it less likely for them to impose P limitation on saprotrophs. However, many of them have highly efficient P immobilization capabilities (Plassard and Dell, 2010), and in P-limited situations such as lowland tropical rainforests and eucalypt forests in Australia (Jones et al., 1998) EcM activity could plausibly lead to P limitation of saprotrophs.

24.3.1 Organic Nutrient Uptake

EcM fungi in particular have the capacity to release and take up organic nutrients (Table 24.1, Box 24.2) making them strong resource competitors with saprotrophs. Although AM fungi can take up organic low-molecular–weight compounds (Whiteside et al., 2012), the recently sequenced genome of an AM fungus suggests a limited repertoire of depolymerizing enzymes (Tisserant et al., 2013). In contrast, approximately 87% of the EcM fungi investigated in laboratory studies produce proteases, and they are able to grow with protein as the only source of N (reviewed in Talbot and Treseder, 2010). Some of them have the genetic potential to secrete enzymes that hydrolyze other organic compounds such as chitin (Lindahl and Taylor, 2004; Frettinger et al., 2006), which is a primary source of organic N in soil (Gooday, 1990; Clemmensen et al., 2013). Because EcM fungi do not depend on SOM as a C source, they are thought to be more important in decomposing the nonhydrolyzable organic N pools, which contain lignin, condensed tannins, and other phenolic compounds, whereas saprotrophic fungi dominate hydrolytic decomposition in early stages of litter decomposition (Colpaert and Van Tichelen, 1996; Lindahl and Tunlid, 2015). Still, a large-scale genome comparison of mycorrhizal and saprotrophic fungi reveals that all EcM fungal species as well as orchid and ericoid mycorrhiza have a somewhat restricted suite of degrading enzymes (Kohler et al., 2015). The analysis included gene families of carbohydrate-active enzymes involved in

BOX 24.2 SHORT-CIRCUITING NUTRIENT CYCLES

Organic nutrient uptake, likely limited in AM fungi but prevalent in EcM and ErM fungi, can be thought of as shortcutting the N cycle, because typically N release from organic residues and its further uptake by plants upon the death of an organism is intermediated by dedicated decomposers. Some EcM fungi seem to be able to immobilize springtails (a main soil fungivore) while still alive and effectively extract N contained in them, thereby bypassing the typical N mineralization step (Klironomos and Hart, 2001). The high effectivity of EcM fungi in removing organic N from litter and rendering it unsuitable to saprotrophs as a result is predicted to lead to SOC buildup and widening of C:N ratio of the SOM (Orwin et al., 2011; Averill et al., 2014). Alternatively (under certain conditions and for some EcM species), organic N use might be associated with complete mineralization of C and a loss of SOC (Clemmensen et al., 2015). In a recent study that sought to explain why some EcM trees are found in monodominant patches in an otherwise highly diverse tropical forest, this shortcut was found to be the most likely candidate for this EcM tree's success (Corrales et al., 2016). These lines of evidence are a powerful indication that mycorrhizas can efficiently produce shortcuts in the soil food web and effectively alter its structure and functioning.

cellulose and hemicellulose degradation, and the Class II peroxidase (POD), which degrade lignin. In general, EcM fungi have been found to lack the ability to degrade lignocellulose, with most species examined possessing no POD genes. However, several taxa do possess the genes coding for POD enzymes, notably *Cortinarius glaucopus*, but also *Piloderma croceum*, *Laccaria bicolor*, and *Hebeloma cylindrosporum* (Bödeker et al., 2014; Kohler et al., 2015). Soil POD activity has been found to positively correlate with EcM fungal species richness (Talbot et al., 2013) and relative abundance (Phillips et al., 2013; Bödeker et al., 2014). Soil POD activity has also been shown to decline in response to ammonium addition in forest soils (Sinsabaugh, 2010; Bödeker et al., 2014), which could be the result of a reduction in allocation to these costly enzymes (Talbot and Treseder, 2010) or a shift in dominance toward EcM fungi possessing different enzymatic suites (Lilleskov et al., 2011). Rineau et al. (2012) showed that *Paxillus involutus* may degrade SOM using a glucose-dependent (Rineau et al., 2013) Fenton reaction mechanism, similar to saprotrophic so-called brown rot fungi.

Despite (or because) of these capabilities, mycorrhizal fungi do not depend on SOM as their main source of C, which may cause them to transform the SOM into a state less suitable for saprotrophs, thereby preserving C contained in it (Lindahl and Tunlid, 2015; see also Box 24.2). However, although competition for nutrients between mycorrhizal fungi and saprotrophs is likely to occur, it is hard to prove with certainty in the field (Fernandez and Kennedy, 2016). It may be relatively easy to show negative effects of mycorrhizal fungi on saprotrophs and their activity, and this is indeed fairly commonly reported (e.g. Table 24.2), but it is very challenging to separate competition for resources from direct antagonism such as production of compounds that inhibit saprotrophs or reducing water availability (Koide and Wu, 2003). Likewise, saprotrophic biomass may go unaffected by competitive interactions with mycorrhizal fungi, but significant changes in phenotype or community structure may ultimately cause functional shifts. For instance, as we see in Table 24.2, shifts in saprotrophic communities in response to mycorrhizal fungi can be associated with changes in decomposition without a measurable effect on microbial abundance.

24.4 INTERACTIONS AMONG MYCORRHIZAL FUNGI, SOIL FAUNA, AND SOIL ORGANIC CARBON

Mycorrhizal fungi are a quantitatively important source of belowground C and are an integral component of the belowground soil food web. In a study partitioning aboveground versus belowground plant inputs to different functional groups of decomposers in forest soils (Pollierer et al., 2012), root-channeled C was found to be a major component of soil faunal diet. Further partitioning this into root or fungal C suggested that mycorrhizal fungi were a particularly important dietary resource to soil fauna (Pollierer et al., 2012). Although aboveground litter C is commonly assumed to be the main input for the soil food web, recent studies have highlighted the role of root-derived C in fueling the soil food web (Pollierer et al., 2007; Bonkowski et al., 2009; Pausch et al., 2016). Grazing on mycorrhizal fungi is typically performed by micro- (Ruess et al., 2000) and mesofauna (Ngosong et al., 2014; Kanters et al., 2015) whereas macrofauna typically contract root-derived C supply through a bacteria-based diet (Eisenhauer et al., 2008; Gilbert et al., 2014). In doing so macrofauna can affect mycelial turnover by imposing massive physical damage to fungal external mycelia (Bonkowski et al., 2009). Both of these processes, ingestion and disruption, can significantly contribute to SOM formation if it increases microfaunal (Soong et al., 2016) or macrofaunal (such as earthworm-intermediated) incorporation of C into soil (de Vries et al., 2013).

Although soil fauna can have a great effect on mycorrhizal performance, the opposite is also common. Some mycorrhizal fungi have low palatability, supposedly by producing deterring compounds or even channeling secondary metabolites produced by plant hosts (Duhamel et al., 2013). Although most soil fauna are generalists (Setälä et al., 2005; Anslan et al., 2016), they may exhibit grazing preferences for the more nutritious and palatable fungal species (Maraun et al., 2003). This preferential feeding reduces biomass or even eliminates certain taxa from the community at the benefit of the less-preferred or the fast-growing hyphal taxa (Kanters et al., 2015). For example, springtails, which serve as a main component of the SOM decomposition chain (Davidson, 1993), prefer feeding on dark pigmented over hyaline hyphae (Maraun et al., 2003). Indeed, when given a choice springtails appear not to feed on hyaline AM fungi, and ingestion of AM hyphae can even reduce their fitness (Klironomos and Ursic, 1998). This preference is surprising because melanin is a recalcitrant compound to digest. The same preference was reported for nematodes, which in a multiple choice experiment largely fed on *Cenococcum geophilum* and *Rhizoscyphus ericae* mycelia. These preferences are expected to result in alteration of mycorrhizal community composition and structure (Kanters et al., 2015; Yang et al., 2015) that may further affect mycorrhizal activity and contribution to soil C storage (Clemmensen et al., 2015). Consumption of melanized hyphae, which are characterized by long life span and slow decomposition (Fernandez and Koide, 2014; Fernandez et al., 2016), could favor EcM species that produce hyaline hyphae, which as a side effect could increase mycorrhizal mycelial turnover and alter contributions of the different fungal species to SOM buildup or loss.

Fungal biomass C storage, turnover (Hobbie and Agerer, 2010), and necromass decomposition (Fernandez et al., 2016) also vary with fungal mycelial exploration type (Agerer, 2001). Thus any induced shift in mycorrhizal or saprotrophic fungal community composition may alter the rates of these processes. For example, cord-forming hydrophobic hyphal structures, specific to medium-distance fringe and long-distance exploration types, tend to persist in soils longer compared with other types (Fernandez et al., 2016). In a field experiment Kanters et al. (2015) reported a massive decline of *Amphinema byssoides* and *Cortinarius* sp. through the actions of collembola, and these EcM species are characterized by the above-mentioned mycelial type. This may lead to lower decomposition rates of fungal necromass in collembola-shaped mycorrhizal communities. Such effects could be further amplified by the reduction of plant contributions to soil C via mycorrhizal hyphal networks in the presence of soil mesofauna due to hyphal feeding or disruption (Johnson, 2005). The outcome of interactions between mycorrhizal fungi and soil saprotrophs may vary with animal density (Steinaker and Wilson, 2008) and environmental conditions (Yang et al., 2015). In addition, competitive interactions of mycorrhizal and saprotrophic fungi might be modulated by animal decomposers via preferential feeding of the two guilds (Gange, 2000; Tiunov and Scheu, 2005; Endlweber and Scheu, 2007; Jonas et al., 2007; Crowther et al., 2011). Although new research has significantly advanced our understanding of the effects of soil fauna and mycorrhizal fungi on soil C stocks, the complex interactive effects between mycorrhizal fungi and soil fauna warrant further investigation (Van der Wal et al., 2013).

24.5 CONCLUSION

In this chapter we have discussed the multiple ways by which mycorrhizal fungi affect saprotrophs and soil C storage. There is evidence that the two most important but opposing mechanisms by which this may occur, priming and competitive interactions with saprotrophs, are commonly thought to be performed by EcM and AM fungi. We have also seen that there are general differences among roots, mycorrhizal fungal types, and mycorrhizal fungal species in the ways in which they interact with the SOM, detritivores, and saprotrophs. Although we have focused on direct effects of mycorrhizas on saprotrophs and their activities, we acknowledge that there are strong indirect effects of a similar magnitude in affecting nutrient cycling and soil C. For instance, mycorrhizal fungi may have strong effects on soil biogeochemical processes through changes in plant nutrition, productivity, and being part of a trait "syndrome" that causes nutritional feedbacks (McGuire et al., 2010; Phillips et al., 2013; Midgley et al., 2015; but see Koele et al., 2012). Still, even limiting the scope to direct interactions with saprotrophs, although it is a very active field of research and our understanding has increased substantially in recent years, it is still incomplete. This is particularly true for the importance of behavioral and morphological differences within mycorrhizal types and the ways in which nutrient context (such as C, N, and P limitation) modulates interactions with saprotrophs. Moreover, in part driven by the fast-evolving understanding of SOM dynamics, research should distinguish how mycorrhizal fungi affect new versus old SOM because the response of the former may be a poor predictor of the latter (Verbruggen et al., 2013). This knowledge will be crucial for a true understanding of ecosystem responses to alterations at the plant–soil interface as imposed by anthropogenic activities, including the global change.

Acknowledgments

The authors thank Björn Lindahl for his thoughtful suggestions on an earlier version of this chapter. Support from a grant of the German Research Community (DFG, Priority Program Biodiversity Exploratorien, PE 2256/1-1) to R.P. is gratefully acknowledged.

References

Agerer, R., 2001. Exploration types of ectomycorrhizae: a proposal to classify ectomycorrhizal mycelial systems according to their patterns of differentiation and putative ecological importance. Mycorrhiza 11, 107–114.

Ahrens, B., Braakhekke, M.C., Guggenberger, G., Schrumpf, M., Reichstein, M., 2015. Contribution of sorption, DOC transport and microbial interactions to the ^{14}C age of a soil organic carbon profile: insights from a calibrated process model. Soil Biology and Biochemistry 88, 390–402.

Albertsen, A., Ravnskov, S., Green, H., Jensen, D.F., Larsen, J., 2006. Interactions between the external mycelium of the mycorrhizal fungus Glomus intraradices and other soil microorganisms as affected by organic matter. Soil Biology and Biochemistry 38, 1008–1014.

Allen, M., Kitajima, K., 2013. In situ high frequency observations of mycorrhizas. New Phytologist 200, 222–228.

Allen, M.F., Swenson, W., Querejeta, J.I., Egerton-Warburton, L.M., Treseder, K.K., 2003. Ecology of mycorrhizae: a conceptual framework for complex interactions among plants and fungi. Annual Review of Phytopathology 41, 271–303.

Anslan, S., Bahram, M., Tedersoo, L., 2016. Temporal changes in fungal communities associated with guts and appendages of Collembola as based on culturing and high-throughput sequencing. Soil Biology and Biochemistry 96, 152–159.

Antoninka, A.J., Ritchie, M.E., Johnson, N.C., 2015. The hidden Serengeti-Mycorrhizal fungi respond to environmental gradients. Pedobiologia 58, 165–176.

Atul-Nayyar, A., Hamel, C., Hanson, K., Germida, J., 2009. The arbuscular mycorrhizal symbiosis links N mineralization to plant demand. Mycorrhiza 19, 239–246.

Averill, C., Turner, B.L., Finzi, A.C., 2014. Mycorrhiza-mediated competition between plants and decomposers drives soil carbon storage. Nature 505, 543–545.

Bending, G.D., Read, D.J., 1996. Nitrogen mobilization from protein-polyphenol complex by ericoid and ectomycorrhizal fungi. Soil Biology and Biochemistry 28, 1603–1612.

Boddy, L., 2000. Interspecific combative interactions between wood-decaying basidiomycetes. FEMS Microbiology Ecology 31, 185–194.

Bödeker, I.T.M., Clemmensen, K.E., de Boer, W., Martin, F., Olson, Å., Lindahl, B.D., 2014. Ectomycorrhizal Cortinarius species participate in enzymatic oxidation of humus in northern forest ecosystems. New Phytologist 203, 245–256.

Bödeker, I.T., Lindahl, B.D., Olson, Å., Clemmensen, K.E., 2016. Mycorrhizal and saprotrophic fungal guilds compete for the same organic substrates but affect decomposition differently. Functional Ecology. http://dx.doi.org/10.1111/1365-2435.12677 in press.

Bonkowski, M., Villenave, C., Griffiths, B., 2009. Rhizosphere fauna: the functional and structural diversity of intimate interactions of soil fauna with plant roots. Plant and Soil 321, 213–233.

Brzostek, E.R., Dragoni, D., Brown, Z.A., Phillips, R.P., 2015. Mycorrhizal type determines the magnitude and direction of root-induced changes in decomposition in a temperate forest. New Phytologist 206, 1274–1282.

Burke, R., Cairney, J.W.G., 2002. Laccases and other polyphenol oxidases in ecto-and ericoid mycorrhizal fungi. Mycorrhiza 12, 105–116.

Camenzind, T., Rillig, M.C., 2013. Extraradical arbuscular mycorrhizal fungal hyphae in an organic tropical montane forest soil. Soil Biology and Biochemistry 64, 96–102.

Camino-Serrano, M., Gielen, B., Luyssaert, S., Ciais, P., Vicca, S., Guenet, B., De, V.B., Cools, N., Ahrens, B., Altaf Arain, M., et al., 2014. Linking variability in soil solution dissolved organic carbon to climate, soil type, and vegetation type. Global Biogeochemical Cycles 28, 497–509.

Campbell, E., Parton, W., Soong, J., Cotrufo, M., Paustian, K., 2014. Litter decomposition and leaching (LIDEL) model: modeling plant litter decomposition to CO_2, dissolved organic matter and microbial products through nitrogen and lignin controls on microbial carbon use efficiency. Soil Biology and Biochemistry.

Carrillo, Y., Dijkstra, F.A., LeCain, D., Pendall, E., 2016. Mediation of soil C decomposition by arbuscular mycorrizhal fungi in grass rhizospheres under elevated CO_2. Biogeochemistry 127, 45–55.

Cheng, L., Booker, F.L., Tu, C., Burkey, K.O., Zhou, L., Shew, H.D., Rufty, T.W., Hu, S., 2012. Arbuscular mycorrhizal fungi increase organic carbon decomposition under elevated CO_2. Science 337, 1084–1087.

Clemmensen, K.E., Bahr, A., Ovaskainen, O., Dahlberg, A., Ekblad, A., Wallander, H., Stenlid, J., Finlay, R.D., Wardle, D.A., Lindahl, B.D., 2013. Roots and associated fungi drive long-term carbon sequestration in boreal forest. Science 339, 1615–1618.

Clemmensen, K.E., Finlay, R.D., Dahlberg, A., Stenlid, J., Wardle, D.A., Lindahl, B.D., 2015. Carbon sequestration is related to mycorrhizal fungal community shifts during long-term succession in boreal forests. New Phytologist 205, 1525–1536.

Colpaert, J.V., Van Tichelen, K.K., 1996. Decomposition, nitrogen and phosphorus mineralization from beech leaf litter colonized by ectomycorrhizal or litter-decomposing basidiomycetes. New Phytologist 134, 123–132.

Corrales, A., Mangan, S.A., Turner, B.L., Dalling, J.W., 2016. An ectomycorrhizal nitrogen economy facilitates monodominance in a neotropical forest. Ecology Letters 19, 383–392.

Cotrufo, M.F., Soong, J.L., Horton, A.J., Campbell, E.E., Haddix, M.L., Wall, D.H., Parton, W.J., 2015. Formation of soil organic matter via biochemical and physical pathways of litter mass loss. Nature Geoscience.

Cotrufo, M.F., Wallenstein, M.D., Boot, C.M., Denef, K., Paul, E., 2013. The microbial efficiency-matrix stabilization (MEMS) framework integrates plant litter decomposition with soil organic matter stabilization: do labile plant inputs form stable soil organic matter? Global Change Biology 19, 988–995.

Courty, P.E., Bréda, N., Garbaye, J., 2007. Relation between oak tree phenology and the secretion of organic matter degrading enzymes by *Lactarius quietus* ectomycorrhizas before and during bud break. Soil Biology and Biochemistry 39, 1655–1663.

Craine, J.M., Morrow, C., Fierer, N., 2007. Microbial nitrogen limitation increases decomposition. Ecology 88, 2105–2113.

Crowther, T.W., Boddy, L., Jones, T.H., 2011. Outcomes of fungal interactions are determined by soil invertebrate grazers. Ecology Letters 14, 1134–1142.

Davidson, D., 1993. The effects of herbivory and granivory on terrestrial plant succession. Oikos 68, 23–35.

Drigo, B., Pijl, A.S., Duyts, H., Kielak, A., Gamper, H.A., Houtekamer, M.J., Boschker, H.T.S., Bodelier, P.L.E., Whiteley, A.S., van Veen, J.A., et al., 2010. Shifting carbon flow from roots into associated microbial communities in response to elevated atmospheric CO_2. Proceedings of the National Academy of Sciences of the United States of America 107, 10938–10942.

Duhamel, M., Pel, R., Ooms, A., Bucking, H., Jansa, J., Ellers, J., Van Straalen, N.M., Wouda, T., Vandenkoornhuyse, P., Kiers, E.T., 2013. Do fungivores trigger the transfer of protective metabolites from host plants to arbuscular mycorrhizal hyphae? Ecology 94, 2019–2029.

Ehleringer, J.R., 2000. Carbon isotope ratios in belowground carbon cycle processes. Ecological Applications 10, 412–422.

Eisenhauer, N., Marhan, S., Scheu, S., 2008. Assessment of anecic behavior in selected earthworm species: effects on wheat seed burial, seedling establishment, wheat growth and litter incorporation. Applied Soil Ecology 38, 79–82.

Endlweber, K., Scheu, S., 2007. Interactions between mycorrhizal fungi and Collembola: effects on root structure of competing plant species. Biology and Fertility of Soils 43, 741–749.

Fernandez, C.W., Kennedy, P.G., 2016. Revisiting the 'Gadgil effect': do interguild fungal interactions control carbon cycling in forest soils? New Phytologist 209, 1382–1394.

Fernandez, C.W., Koide, R.T., 2014. Initial melanin and nitrogen concentrations control the decomposition of ectomycorrhizal fungal litter. Soil Biology and Biochemistry 77, 150–157.

Fernandez, C.W., Langley, J.A., Chapman, S., McCormack, M.L., Koide, R.T., 2016. The decomposition of ectomycorrhizal fungal necromass. Soil Biology and Biochemistry 93, 38–49.

Fransson, P., Andersson, A., Norström, S., Bylund, D., Bent, E., 2016. Ectomycorrhizal exudates and pre-exposure to elevated CO_2 affects soil bacterial growth and community structure. Fungal Ecology 20, 211–224.

Frettinger, P., Herrmann, S., Lapeyrie, F., Oelmüller, R., Buscot, F., 2006. Differential expression of two class III chitinases in two types of roots of *Quercus robur* during pre-mycorrhizal interactions with *Piloderma croceum*. Mycorrhiza 16, 219–223.

Gadgil, R.L., Gadgil, P., 1971. Mycorrhiza and litter decomposition. Nature 233, 133.

Gange, A., 2000. Arbuscular mycorrhizal fungi, Collembola and plant growth. Trends in Ecology and Evolution 15, 369–372.

Gilbert, K.J., Fahey, T.J., Maerz, J.C., Sherman, R.E., Bohlen, P., Dombroskie, J.J., Groffman, P.M., Yavitt, J.B., 2014. Exploring carbon flow through the root channel in a temperate forest soil food web. Soil Biology and Biochemistry 76, 45–52.

Godbold, D.L., Hoosbeek, M.R., Lukac, M., Cotrufo, M.F., Janssens, I.A., Ceulemans, R., Polle, A., Velthorst, E.J., Scarascia-Mugnozza, G., De Angelis, P., et al., 2006. Mycorrhizal hyphal turnover as a dominant process for carbon input into soil organic matter. Plant and Soil 281, 15–24.

Gooday, G.W., 1990. Physiology of microbial degradation of chitin and chitosan. Biodegradation 1, 177–190.

Gosling, P., Hodge, A., Goodlass, G., Bending, G.D., 2006. Arbuscular mycorrhizal fungi and organic farming. Agriculture, Ecosystems & Environment 113, 17–35.

Grandy, A.S., Neff, J.C., 2008. Molecular C dynamics downstream: the biochemical decomposition sequence and its impact on soil organic matter structure and function. Science of the Total Environment 404, 297–307.

Griffiths, B.S., Spilles, A., Bonkowski, M., 2012. C:N:P stoichiometry and nutrient limitation of the soil microbial biomass in a grazed grassland site under experimental P limitation or excess. Ecological Processes 1, 6.

Gunina, A., Kuzyakov, Y., 2015. Sugars in soil and sweets for microorganisms: review of origin, content, composition and fate. Soil Biology and Biochemistry 90, 87–100.

Hannula, S.E., Boschker, H.T.S., de Boer, W., van Veen, J.A., 2012. ^{13}C pulse-labeling assessment of the community structure of active fungi in the rhizosphere of a genetically starch-modified potato (*Solanum tuberosum*) cultivar and its parental isoline. The New Phytologist 194, 784–799.

Hatton, P.J., Castanha, C., Torn, M.S., Bird, J.A., 2015. Litter type control on soil C and N stabilization dynamics in a temperate forest. Global Change Biology 21, 1358–1367.

Heinemeyer, A., Ridgway, K.P., Edwards, E.J., Benham, D.G., Young, J.P.W., Fitter, A.H., 2004. Impact of soil warming and shading on colonization and community structure of arbuscular mycorrhizal fungi in roots of a native grassland community. Global Change Biology 10, 52–64.

Hendricks, J.J., Hendrick, R.L., Wilson, C.A., Mitchell, R.J., Pecot, S.D., Guo, D., 2006. Assessing the patterns and controls of fine root dynamics: an empirical test and methodological review. Journal of Ecology 94, 40–57.

Hendricks, J.J., Mitchell, R.J., Kuehn, K.A., Pecot, S.D., 2015. Ectomycorrhizal fungal mycelia turnover in a longleaf pine forest. New Phytologist 1693–1704.

Hobbie, E.A., 2006. Carbon allocation to ectomycorrhizal fungi correlates with belowground allocation in culture studies. Ecology 87, 563–569.

Hobbie, E.A., Agerer, R., 2010. Nitrogen isotopes in ectomycorrhizal sporocarps correspond to belowground exploration types. Plant and Soil 327, 71–83.

Hobbie, E.A., Hobbie, J.E., 2008. Natural abundance of ^{15}N in nitrogen-limited forests and tundra can estimate nitrogen cycling through mycorrhizal fungi: a review. Ecosystems 11, 815–830.

Hodge, A., Campbell, C.D., Fitter, A.H., 2001. An arbuscular mycorrhizal fungus accelerates decomposition and acquires nitrogen directly from organic material. Nature 413, 297–299.

Hodge, A., Fitter, A.H., 2010. Substantial nitrogen acquisition by arbuscular mycorrhizal fungi from organic material has implications for N cycling. Proceedings of the National Academy of Sciences of the United States of America 107, 13754–13759.

Hodge, A., Helgason, T., Fitter, A.H., 2010. Nutritional ecology of arbuscular mycorrhizal fungi. Fungal Ecology 3, 267–273.

Hooker, J.E., Piatti, P., Cheshire, M.V., Watson, C.A., 2007. Polysaccharides and monosaccharides in the hyphosphere of the arbuscular mycorrhizal fungi Glomus E3 and Glomus tenue. Soil Biology and Biochemistry 39, 680–683.

Hopkins, F.M., Filley, T.R., Gleixner, G., Lange, M., Top, S.M., Trumbore, S.E., 2014. Increased belowground carbon inputs and warming promote loss of soil organic carbon through complementary microbial responses. Soil Biology and Biochemistry 76, 57–69.

Jackson, R.B., Mooney, H.A., Schulze, E.D., 1997. A global budget for fine root biomass, surface area, and nutrient contents. Proceedings of the National Academy of Sciences of the United States of America 94, 7362–7366.

Jansa, J., Finlay, R.D., Wallander, H., Smith, F.A., Smith, S.E., 2011. Role of mycorrhizal symbioses in phosphorus cycling. In: Bünemann, E.K., Oberson, A., Frossard, E. (Eds.), Phosphorus in Action. Springer, Heidelberg, pp. 137–168.

Janssens, I.A., Dieleman, W., Luyssaert, S., Subke, J.-A., Reichstein, M., Ceulemans, R., Ciais, P., Dolman, A.J., Grace, J., Matteucci, G., et al., 2010. Reduction of forest soil respiration in response to nitrogen deposition. Nature Geoscience 3, 315–322.

Johansson, E.M., Fransson, P.M.A., Finlay, R.D., van Hees, P.A.W., 2009. Quantitative analysis of soluble exudates produced by ectomycorrhizal roots as a response to ambient and elevated CO_2. Soil Biology and Biochemistry 41, 1111–1116.

Johnson, D., 2005. Soil invertebrates disrupt carbon flow through fungal networks. Science 309, 1047.

Johnson, D., Leake, J.R., Ostle, N., Ineson, P., Read, D.J., 2002. In situ $^{13}CO_2$ pulse-labelling of upland grassland demonstrates a rapid pathway of carbon flux from arbuscular mycorrhizal mycelia to the soil. New Phytologist 153, 327–334.

Jonas, J.L., Wilson, G.W.T., White, P.M., Joern, A., 2007. Consumption of mycorrhizal and saprophytic fungi by Collembola in grassland soils. Soil Biology and Biochemistry 39, 2594–2602.

Jones, M.D., Durall, D.M., Tinker, P.B., 1998. A comparison of arbuscular and ectomycorrhizal *Eucalyptus coccifera*: growth response, phosphorus uptake efficiency and external hyphal production. New Phytologist 140, 125–134.

Jones, D.L., Hodge, A., Kuzyakov, Y., 2004. Plant and mycorrhizal regulation of rhizodeposition. New Phytologist 163, 459–480.

Jones, D.L., Nguyen, C., Finlay, R.D., 2009. Carbon flow in the rhizosphere: carbon trading at the soil–root interface. Plant and Soil 321, 5–33.

Kaiser, K., Kalbitz, K., 2012. Cycling downwards - dissolved organic matter in soils. Soil Biology and Biochemistry 52, 29–32.

Kaiser, C., Kilburn, M.R., Clode, P.L., Fuchslueger, L., Koranda, M., Cliff, J.B., Solaiman, Z.M., Murphy, D.V., 2015. Exploring the transfer of recent plant photosynthates to soil microbes: mycorrhizal pathway vs direct root exudation. New Phytologist 205, 1537–1551.

Kanters, C., Anderson, I.C., Johnson, D., 2015. Chewing up the wood-wide web: selective grazing on ectomycorrhizal fungi by collembola. Forests 6, 2560–2570.

Kleber, M., 2010. What is recalcitrant soil organic matter? Environmental Chemistry 7, 320–332.

Klironomos, J.N., Hart, M.M., 2001. Animal nitrogen swap for plant carbon. Nature 410, 651–652.

Klironomos, J.N., Ursic, M., 1998. Density-dependent grazing on the extraradical hyphal network of the arbuscular mycorrhizal fungus, Glomus intraradices, by the collembolan, *Folsomia candida*. Biology and Fertility of Soils 26, 250–253.

Kluber, L.A., Tinnesand, K.M., Caldwell, B.A., Dunham, S.M., Yarwood, R.R., Bottomley, P.J., Myrold, D.D., 2010. Ectomycorrhizal mats alter forest soil biogeochemistry. Soil Biology and Biochemistry 42, 1607–1613.

Koele, N., Dickie, I.A., Oleksyn, J., Richardson, S.J., Reich, P.B., 2012. No globally consistent effect of ectomycorrhizal status on foliar traits. New Phytologist 196, 845–852.

Kohler, A., Kuo, A., Nagy, L.G., Morin, E., Barry, K.W., Buscot, F., Canbäck, B., Choi, C., Cichocki, N., Clum, A., et al., 2015. Convergent losses of decay mechanisms and rapid turnover of symbiosis genes in mycorrhizal mutualists. Nature Genetics 47, 410–415.

Koide, R.T., Wu, T., 2003. Ectomycorrhizas and retarded decomposition in a *Pinus resinosa* plantation. New Phytologist 158, 401–407.

Koller, R., Rodriguez, A., Robin, C., Scheu, S., Bonkowski, M., 2013. Protozoa enhance foraging efficiency of arbuscular mycorrhizal fungi for mineral nitrogen from organic matter in soil to the benefit of host plants. New Phytologist 199, 203–211.

Kuzyakov, Y., Friedel, J.K., Stahr, K., 2000. Review of mechanisms and quantification of priming effects. Soil Biology and Biochemistry 32, 1485–1498.

Langley, J.A., Hungate, B.A., 2003. Mycorrhizal controls on belowground litter quality. Ecology 84, 2302–2312.

Leake, J.R., Johnson, D., Donnelly, D.P., Muckle, G.E., Boddy, L., Read, D.J., 2004. Networks of power and influence: the role of mycorrhizal mycelium in controlling plant communities and agroecosystem functioning. Canadian Journal of Botany-Revue Canadienne De Botanique 82, 1016–1045.

Lehmann, J., Kleber, M., 2015. The contentious nature of soil organic matter. Nature 1–9.

Leifheit, E.F., Verbruggen, E., Rillig, M.C., 2014. Arbuscular mycorrhizal fungi reduce decomposition of woody plant litter while increasing soil aggregation. Soil Biology and Biochemistry 81, 323–328.

Lilleskov, E.A., Hobbie, E.A., Horton, T.R., 2011. Conservation of ectomycorrhizal fungi: exploring the linkages between functional and taxonomic responses to anthropogenic N deposition. Fungal Ecology 4, 174–183.

Lindahl, B.D., de Boer, W., Finlay, R.D., 2010. Disruption of root carbon transport into forest humus stimulates fungal opportunists at the expense of mycorrhizal fungi. The ISME Journal 4, 872–881.

Lindahl, B.D., Ihrmark, K., Boberg, J., Trumbore, S.E., Hogberg, P., Stenlid, J., Finlay, R.D., 2007. Spatial separation of litter decomposition and mycorrhizal nitrogen uptake in a boreal forest. New Phytologist 173, 611–620.

Lindahl, B.D., Taylor, A.F.S., 2004. Occurrence of N -acetylhexosaminidase-encoding genes in ectomycorrhizal basidiomycetes. New Phytologist 193–199.

Lindahl, B.D., Tunlid, A., 2015. Ectomycorrhizal fungi – potential organic matter decomposers, yet not saprotrophs. New Phytologist 205, 1443–1447.

Liu, B., Li, H., Zhu, B., Koide, R.T., Eissenstat, D.M., Guo, D., 2015. Complementarity in nutrient foraging strategies of absorptive fine roots and arbuscular mycorrhizal fungi across 14 coexisting subtropical tree species. New Phytologist.

Von Lützow, M.V., Kögel-Knabner, I., Ekschmitt, K., Matzner, E., Guggenberger, G., Marschner, B., Flessa, H., 2006. Stabilization of organic matter in temperate soils: mechanisms and their relevance under different soil conditions - a review. European Journal of Soil Science 57, 426–445.

Maraun, M., Martens, H., Migge, S., Theenhaus, A., Scheu, S., 2003. Adding to 'the enigma of soil animal diversity': fungal feeders and saprophagous soil invertebrates prefer similar food substrates. European Journal of Soil Biology 39, 85–95.

Marschner, B., Brodowski, S., Dreves, A., Gleixner, G., Gude, A., Grootes, P.M., Hamer, U., Heim, A., Jandl, G., Ji, R., et al., 2008. How relevant is recalcitrance for the stabilization of organic matter in soils? Journal of Plant Nutrition and Soil Science 171, 91–110.

Mccormack, M.L., Adams, T.S., Smithwick, E.A.H., Eissenstat, D.M., 2012. Predicting fine root lifespan from plant functional traits in temperate trees. New Phytologist 195, 823–831.

Mccormack, M.L., Dickie, I.A., Eissenstat, D.M., Fahey, T.J., Fernandez, C.W., Guo, D., Erik, A., Iversen, C.M., Jackson, R.B., 2015. Redefining fine roots improves understanding of below-ground contributions to terrestrial biosphere processes. New Phytologist 207, 505–518.

McGuire, K.L., Zak, D.R., Edwards, I.P., Blackwood, C.B., Upchurch, R., 2010. Slowed decomposition is biotically mediated in an ectomycorrhizal, tropical rain forest. Oecologia 164, 785–795.

Midgley, M.G., Brzostek, E., Phillips, R.P., 2015. Decay rates of leaf litters from arbuscular mycorrhizal trees are more sensitive to soil effects than litters from ectomycorrhizal trees. Journal of Ecology 103, 1454–1463.

Moore, J.A.M., Jiang, J., Patterson, C.M., Mayes, M.A., Wang, G., Classen, A.T., 2015. Interactions among roots, mycorrhizas and free-living microbial communities differentially impact soil carbon processes. Journal of Ecology 103, 1442–1453.

Ngosong, C., Gabriel, E., Ruess, L., 2014. Collembola grazing on arbuscular mycorrhiza fungi modulates nutrient allocation in plants. Pedobiologia 57, 171–179.

Nottingham, A.T., Turner, B.L., Stott, A.W., Tanner, E.V.J., 2015. Nitrogen and phosphorus constrain labile and stable carbon turnover in lowland tropical forest soils. Soil Biology and Biochemistry 80, 26–33.

Nottingham, A.T., Turner, B.L., Winter, K., Chamberlain, P.M., Stott, A., Tanner, E.V.J., 2013. Root and arbuscular mycorrhizal mycelial interactions with soil microorganisms in lowland tropical forest. FEMS Microbiology Ecology 85, 37–50.

Nottingham, A.T., Turner, B.L., Winter, K., van der Heijden, M.G.a, Tanner, E.V.J., 2010. Arbuscular mycorrhizal mycelial respiration in a moist tropical forest. New Phytologist 186, 957–967.

Nuccio, E.E., Hodge, A., Pett-Ridge, J., Herman, D.J., Weber, P.K., Firestone, M.K., 2013. An arbuscular mycorrhizal fungus significantly modifies the soil bacterial community and nitrogen cycling during litter decomposition. Environmental Microbiology 15, 1870–1881.

Olsrud, M., Christensen, T.R., 2004. Carbon cycling in subarctic tundra; seasonal variation in ecosystem partitioning based on in situ ^{14}C pulse-labelling. Soil Biology and Biochemistry 36, 245–253.

Olsrud, M., Melillo, J.M., Christensen, T.R., Michelsen, A., Wallander, H., Olsson, P.A., 2004. Response of ericoid mycorrhizal colonization and functioning to global change factors. New Phytologist 162, 459–469.

Olsson, P.A., Johnson, N.C., 2005. Tracking carbon from the atmosphere to the rhizosphere. Ecology Letters 8, 1264–1270.

Orwin, K.H., Kirschbaum, M.U.F., St John, M.G., Dickie, I.A., 2011. Organic nutrient uptake by mycorrhizal fungi enhances ecosystem carbon storage: a model-based assessment. Ecology Letters 14, 493–502.

Pausch, J., Kramer, S., Scharroba, A., Scheunemann, N., Butenschoen, O., Kandeler, E., Marhan, S., Riederer, M., Scheu, S., Kuzyakov, Y., et al., 2016. Small but active – pool size does not matter for carbon incorporation in below-ground food webs. Functional Ecology 30, 479–489.

Phillips, R.P., Brzostek, E., Midgley, M.G., 2013. The mycorrhizal-associated nutrient economy: a new framework for predicting carbon-nutrient couplings in temperate forests. The New Phytologist 199, 41–51.

Plassard, C., Dell, B., 2010. Phosphorus nutrition of mycorrhizal trees. Tree Physiology 30, 1129–1139.

Poeplau, C., Bolinder, M.A., Kirchmann, H., Kätterer, T., 2015. Phosphorus fertilisation under nitrogen limitation can deplete soil carbon stocks – evidence from Swedish meta-replicated long-term field experiments. Biogeosciences Discussions 12, 16527–16551.

Pollierer, M.M., Dyckmans, J., Scheu, S., Haubert, D., 2012. Carbon flux through fungi and bacteria into the forest soil animal food web as indicated by compound-specific ^{13}C fatty acid analysis. Functional Ecology 26, 978–990.

Pollierer, M.M., Langel, R., Körner, C., Maraun, M., Scheu, S., 2007. The underestimated importance of belowground carbon input for forest soil animal food webs. Ecology Letters 10, 729–736.

Ramirez, K.S., Craine, J.M., Fierer, N., 2012. Consistent effects of nitrogen amendments on soil microbial communities and processes across biomes. Global Change Biology 18, 1918–1927.

Rasse, D.P., Rumpel, C., Dignac, M.F., 2005. Is soil carbon mostly root carbon? Mechanisms for a specific stabilisation. Plant and Soil 269, 341–356.

Rineau, F., Roth, D., Shah, F., Smits, M., Johansson, T., Canbäck, B., Olsen, P.B., Persson, P., Grell, M.N., Lindquist, E., et al., 2012. The ectomycorrhizal fungus *Paxillus involutus* converts organic matter in plant litter using a trimmed brown-rot mechanism involving Fenton chemistry. Environmental Microbiology 14, 1477–1487.

Rineau, F., Shah, F., Smits, M.M., Persson, P., Johansson, T., Carleer, R., Troein, C., Tunlid, A., 2013. Carbon availability triggers the decomposition of plant litter and assimilation of nitrogen by an ectomycorrhizal fungus. The ISME Journal 7, 2010–2022.

Ruess, L., Garcia Zapata, E., Dighton, J., 2000. Food preferences of a fungal-feeding *Aphelenchoides* species. Nematology 2, 223–230.

Sanders, I.R., Streitwolf-Engel, R., van der Heijden, M.G.A., Boller, T., Wiemken, A., 1998. Increased allocation to external hyphae of arbuscular mycorrhizal fungi under CO_2 enrichment. Oecologia 117, 496–503.

Schmidt, M.W.I., Torn, M.S., Abiven, S., Dittmar, T., Guggenberger, G., Janssens, I.A., Kleber, M., Kögel-Knabner, I., Lehmann, J., Manning, D.A.C., et al., 2011. Persistence of soil organic matter as an ecosystem property. Nature 478, 49–56.

Setälä, H.M., Berg, M.P., Jones, T., 2005. Trophic structure and functional redundancy in soil communities. In: Bardgett, R.D., Usher, M.B., Hopkins, D. (Eds.), Biological Diversity and Function in Soils, pp. 236–249.

Setaro, S., Kottke, I., Oberwinkler, F., 2006. Anatomy and ultrastructure of mycorrhizal associations of neotropical Ericaceae. Mycological Progress 5, 243–254.

Shahzad, T., Chenu, C., Genet, P., Barot, S., Perveen, N., Mougin, C., Fontaine, S., 2015. Contribution of exudates, arbuscular mycorrhizal fungi and litter depositions to the rhizosphere priming effect induced by grassland species. Soil Biology and Biochemistry 80, 146–155.

Sinsabaugh, R.L., 2010. Phenol oxidase, peroxidase and organic matter dynamics of soil. Soil Biology and Biochemistry 42, 391–404.

Sochorová, L., Jansa, J., Verbruggen, E., Hejcman, M., Schellberg, J., Kiers, E.T., Collins, N., 2016. Long-term agricultural management maximizing hay production can significantly reduce belowground C storage. Agriculture, Ecosystems and Environment 220, 104–114.

Soong, J.L., Parton, W.J., Calderon, F., Campbell, E.E., Cotrufo, M.F., 2015. A new conceptual model on the fate and controls of fresh and pyrolized plant litter decomposition. Biogeochemistry 124, 27–44.

Soong, J.L., Vandegehuchte, M.L., Horton, A.J., Nielsen, U.N., Denef, K., Shaw, E.A., de Tomasel, C.M., Parton, W., Wall, D.H., Cotrufo, M.F., 2016. Soil microarthropods support ecosystem productivity and soil C accrual: evidence from a litter decomposition study in the tallgrass prairie. Soil Biology and Biochemistry 92, 230–238.

Soudzilovskaia, N.A., Douma, J.C., Akhmetzhanova, A.A., van Bodegom, P.M., Cornwell, W.K., Moens, E.J., Treseder, K.K., Tibbett, M., Wang, Y.-P., Cornelissen, J.H.C., 2015. Global patterns of plant root colonization intensity by mycorrhizal fungi explained by climate and soil chemistry. Global Ecology and Biogeography 24, 371–382.

Staddon, P.L., Ramsey, C.B., Ostle, N., Ineson, P., Fitter, A.H., 2003. Rapid turnover of hyphae of mycorrhizal fungi determined by AMS microanalysis of ^{14}C. Science 300, 1138–1140.

Steinaker, D.F., Wilson, S.D., 2008. Scale and density dependent relationships among roots, mycorrhizal fungi and collembola in grassland and forest. Oikos 117, 703–710.

Talbot, J.M., Allison, S.D., Treseder, K.K., 2008. Decomposers in disguise: mycorrhizal fungi as regulators of soil C dynamics in ecosystems under global change. Functional Ecology 22, 955–963.

Talbot, J.M., Bruns, T.D., Smith, D.P., Branco, S., Glassman, S.I., Erlandson, S., Vilgalys, R., Peay, K.G., 2013. Independent roles of ectomycorrhizal and saprotrophic communities in soil organic matter decomposition. Soil Biology and Biochemistry 57, 282–291.

Talbot, J.M., Treseder, K.K., 2010. Controls over mycorrhizal uptake of organic nitrogen. Pedobiologia 53, 169–179.

Tisserant, E., Malbreil, M., Kuo, A., Kohler, A., Symeonidi, A., Balestrini, R., Charron, P., Duensing, N., Frei dit Frey, N., Gianinazzi-Pearson, V., et al., 2013. Genome of an arbuscular mycorrhizal fungus provides insight into the oldest plant symbiosis. Proceedings of the National Academy of Sciences 1–6.

Tiunov, A.V., Scheu, S., 2005. Facilitative interactions rather than resource partitioning drive diversity-functioning relationships in laboratory fungal communities. Ecology Letters 8, 618–625.

Toljander, J.F., Lindahl, B.D., Paul, L.R., Elfstrand, M., Finlay, R.D., 2007. Influence of arbuscular mycorrhizal mycelial exudates on soil bacterial growth and community structure. FEMS Microbiology Ecology 61, 295–304.

Verbruggen, E., Jansa, J., Hammer, E.C., Rillig, M.C., 2016. Do arbuscular mycorrhizal fungi stabilize litter-derived carbon in soil? Journal of Ecology 104, 261–269.

Verbruggen, E., Veresoglou, S.D., Anderson, I.C., Caruso, T., Hammer, E.C., Kohler, J., Rillig, M.C., 2013. Arbuscular mycorrhizal fungi – short-term liability but long-term benefits for soil carbon storage? New Phytologist 197, 366–368.

Vicca, S., Zavalloni, C., Fu, Y.S.H., Voets, L., Dupré de Boulois, H., Declerck, S., Ceulemans, R., Nijs, I., Janssens, I.A., 2009. Arbuscular mycorrhizal fungi may mitigate the influence of a joint rise of temperature and atmospheric CO_2 on soil respiration in grasslands. International Journal of Ecology Article ID: 1–10.

de Vries, F.T., Thebault, E., Liiri, M., Birkhofer, K., Tsiafouli, M.A., Bjornlund, L., Bracht Jorgensen, H., Brady, M.V., Christensen, S., de Ruiter, P.C., et al., 2013. Soil food web properties explain ecosystem services across European land use systems. Proceedings of the National Academy of Sciences 110, 14296–14301.

Van der Wal, A., Geydan, T.D., Kuyper, T.W., De Boer, W., 2013. A thready affair: linking fungal diversity and community dynamics to terrestrial decomposition processes. FEMS Microbiology Reviews 37, 477–494.

Welc, M., Ravnskov, S., Kieliszewska-Rokicka, B., Larsen, J., 2010. Suppression of other soil microorganisms by mycelium of arbuscular mycorrhizal fungi in root-free soil. Soil Biology and Biochemistry 42, 1534–1540.

Whiteside, M.D., Digman, M.A., Gratton, E., Treseder, K.K., 2012. Organic nitrogen uptake by arbuscular mycorrhizal fungi in a boreal forest. Soil Biology and Biochemistry 55, 7–13.

Wilson, G.W.T., Rice, C.W., Rillig, M.C., Springer, A., Hartnett, D.C., 2009. Soil aggregation and carbon sequestration are tightly correlated with the abundance of arbuscular mycorrhizal fungi: results from long-term field experiments. Ecology Letters 12, 452–461.

Xiang, D., Verbruggen, E., Hu, Y., Veresoglou, S.D., Rillig, M.C., Zhou, W., Xu, T., Li, H., Hao, Z., Chen, Y., et al., 2014. Land use influences arbuscular mycorrhizal fungal communities in the farming-pastoral ecotone of northern China. New Phytologist 204, 968–978.

Yang, N., Schuetzenmeister, K., Grubert, D., Jungkunst, H.F., Gansert, D., Scheu, S., Polle, A., Pena, R., 2015. Impacts of earthworms on nitrogen acquisition from leaf litter by arbuscular mycorrhizal ash and ectomycorrhizal beech trees. Environmental and Experimental Botany 120, 1–7.

Zhang, B., Li, S., Chen, S., Ren, T., Yang, Z., Zhao, H., Liang, Y., Han, X., 2016a. Arbuscular mycorrhizal fungi regulate soil respiration and its response to precipitation change in a semiarid steppe. Scientific Reports 6, 19990.

Zhang, L., Xu, M., Liu, Y., Zhang, F., Hodge, A., Feng, G., 2016b. Carbon and phosphorus exchange may enable cooperation between an arbuscular mycorrhizal fungus and a phosphate-solubilizing bacterium. New Phytologist. http://dx.doi.org/10.1111/nph.13838.

CHAPTER

25

Biochar—Arbuscular Mycorrhiza Interaction in Temperate Soils

R.T. Koide

Brigham Young University, Provo, UT, United States

25.1 INTRODUCTION

The discovery of unexpectedly fertile soils containing unusually high concentrations of charred organic materials in the Amazon (Woods and Deneven, 2009) and elsewhere in the world (Wiedner et al., 2015) has resulted in a great deal of enthusiasm for using biochar (i.e. charcoal used as a soil amendment; Lehmann and Joseph, 2015) to improve agricultural productivity and sustainability in temperate climates (Lehmann, 2007; Laird, 2008). Biochars, produced via pyrolysis (Kambo and Dutta, 2015), are fundamentally different from hydrochars, which will not be covered in this review. For various reasons the enthusiasm for using biochar has not yet been translated into its enthusiastic use, possibly because some of the zeal for biochar is based more on hope than on science. Indeed, if Mark Twain were alive today he might be inclined to say that interest in biochar has been fueled mainly by parties who have had the largest opportunities to know the least about it.

In the temperate regions of the world, large concentrations of biochar-type materials occur primarily in forested areas, where large accumulations of biomass are available for charring, either due to natural or manmade fire (Tryon, 1948), and where ectomycorrhizas are often the most important type of mycorrhiza. However, much of the current interest in biochar is as a soil amendment in temperate, agronomic settings, where wood is scarce and arbuscular mycorrhizas (AMs) are the most important. It is not often that either lightning or farmers set fire to fields of corn or soybean to make useful pyrogenic byproducts. This is not to say that small amounts of biochar made from biomass grown on the farm cannot or should not be applied to the soil in relatively small quantities year by year. However, under those conditions many of the reported benefits from biochar may require many years to achieve. Alternatively, the movement of large quantities of biochar produced from off-farm locations (e.g. forests) might achieve results more rapidly but, of course, would require incurring the additional financial and environmental costs of transportation. For that reason, regions in

Mycorrhizal Mediation of Soil
http://dx.doi.org/10.1016/B978-0-12-804312-7.00025-5

which relatively small farms are dispersed in a matrix of forest, such as much of Pennsylvania (Leslie et al., 2014), may be the most practical places in which to promote the use of biochar.

The addition of significant quantities of biochar to soil may result in positive effects on plant growth stemming from improvement of soil quality such as has been frequently demonstrated in highly weathered, acidic, tropical soils (Glaser et al., 2002). It seems doubtful that the addition of biochar to soils in most temperate regions of the world, where soils possess more neutral pH, using practical rates of application and in the absence of supplemental inputs will result in the increased crop yields demonstrated in the terra preta and terra mulata soils of the Amazon (Lehmann, 2009; Woods and Deneven, 2009; Birk et al., 2009) or even in some Japanese soils (Ogawa and Okimori, 2010). In addition to biochar, those soils were simultaneously amended with all manner of other organic materials, including excrement from humans and livestock (Factura et al., 2010; Ogawa and Okimori, 2010; Wiedner et al., 2015). There is understandably now a growing interest in mixing biochar with other materials to increase its efficacy in agriculture (Joseph et al., 2015), but it may not be realistic to expect farmers to add biochar and other materials, such as manure, to their soils to achieve the yield increases about which so much has been written. The simultaneous amendment of soil with biochar and manure might be relatively easily accomplished for at least the 5% of the cropland soils that currently receive manure (MacDonald et al., 2009). However, the physical separation between plant and animal agriculture in many locations represents a significant barrier to the use of manure; more than half of the crop acreage in the United States occurs on farms with no animal production (MacDonald et al., 2009). Therefore we must again consider the costs (financial and environmental) of transporting large quantities of biochar and manure to fields where they can be co-incorporated. Moreover, it is not always clear how long biochar must be weathered before it possesses desirable properties (Quilliam et al., 2012a). It is possible that the current fertility of terra preta soils is at least a partial function of the age of the biochar (Lehmann, 2007).

Despite the many impediments to widespread adoption of biochar in temperate agriculture, amendment of soils with biochar may yet prove to be a good investment simply because of its ability to sequester C (Lehmann, 2007). Because biochars have long half-lives, even in perennially moist soils, they may be used to sequester large amounts of C in the soil (Fang et al., 2014; Nguyen et al., 2014). Unfortunately, C sequestration by itself may not be a large enough consideration to motivate widespread use of biochar by most farmers. However, biochar could be used in place of other soil amendments. For example, a tremendous amount of agricultural lime is used in North American agriculture, and most C in that lime eventually leaves the soil and travels to the atmosphere in the form of CO_2 (West and McBride, 2005). Depending on the feedstock and pyrolysis conditions, biochar can also have significant capacity to increase soil pH (Glaser et al., 2002). Indeed, a large amount of the benefit from biochar in acidic, tropical soils may simply be due to its alkalizing effect and ability to reduce exchangeable Al ions (Ogawa and Okimori, 2010; Van Zwieten et al., 2010). The replacement of even a fraction of the lime currently used with biochar to increase soil pH could sequester significant amounts of C. Moreover, because biochar may absorb more water than coarse-textured soils, it may be used to improve plant access to water (Koide et al., 2015).

If the use of biochar becomes common, then it will be important to know how it interacts with other components of the soil that are important to crop production. Because biochar influences various chemical and physical properties of soils, it will undoubtedly influence soil dwellers of all kinds, including roots and mycorrhizal fungi. Therefore the purpose of this short review is to highlight some of those potential interactions. Much more than roots, mycorrhizal fungi, and biochar is obviously involved in growing a crop. Nonetheless, this review is restricted to the extremely narrow topic of the interaction between biochar and the AM symbiosis.

25.2 BIOCHAR AND MYCORRHIZAS

The effect of biochar on mycorrhizal symbiosis is difficult to review in a conventional way for at least four reasons. First, biochar has been used in widely varying amounts. In field experiments researchers have used as little as 0.3 t/ha (Blackwell et al., 2015) and as much as 15 t/ha (Saito, 1990). In greenhouse studies researchers have used from 5% (Hammer et al., 2015) to 70% biochar by volume (Conversa et al., 2015). This extreme variation in the proportion of biochar to soil has undoubtedly contributed to the large variation in the outcomes of its use.

Second, even when concentrations are reported, it is still difficult to predict the effect of the biochar because biochars vary so widely in density. The particle densities of biochars range from 1.3 to 1.4 g/cm^3 for switchgrass and bamboo biochars, respectively (Koide et al., unpublished; Hernandez-Mena et al., 2014), to 2.2 g/cm^3 for municipal biosolids biochar (Khanmohammadi et al., 2015). Bulk densities of biochars range from 0.08 to 0.07 g/cm^3 for switchgrass and coir biochars, respectively (Koide et al., 2015; Kaudal et al., 2016), to 0.31 for pine bark and biosolids biochars (Kaudal et al., 2016). Therefore, depending on the biochar type, a given weight of biochar may occupy very different volumes. As root systems explore a given volume of soil, for a given weight of biochar the roots will differ greatly in their exposure to it depending on its density.

Third, irrespective of the weight or volume of biochar used in an experiment, its chemical and physical nature are remarkably variable depending on the type of feedstock and the conditions of its pyrolysis (Atkinson et al., 2010). Thus, in addition to the variation in the concentration of biochar used in any particular experiment, the outcome of mixing biochar with soil will be determined by the nature of the biochar in relation to the nature of the soil. For example, some biochars possess high surface area and high cation exchange capacity (Liang et al., 2006). Therefore many have pointed to these desirable properties of biochar, thus suggesting that its addition to soil should improve it. However, many naturally abundant components of soil, including the colloidal materials (humus), have high surface area and high cation exchange capacity (Brady, 1974); therefore the addition of biochar may be superfluous or even detrimental with respect to these functions, depending on the characteristics of the biochar and the soil.

Fourth, most of the studies documenting effects of biochar on AM fungi have thus far been performed in pots under conditions that may not be relevant in the field. It is hardly worth mentioning that laboratory and greenhouse studies are easier to perform than field studies; therefore it is no wonder that there are probably 10 or 20 greenhouse pot studies for every study performed in the field. However, in pot studies it is easy to produce high concentrations of biochar that would never be practical in the field on a large scale. The spatial distribution of biochar in a pot is likely to be far more homogeneous than in the field. Finally, pot studies are often of relatively short duration, but the ultimate effects of biochar under field conditions may not be realized for a long period of time, possibly years or decades (Lehmann, 2007). Therefore, although in this review I do refer to many pot studies, it is well to remember that we can only glean from them some of the potential (and possibly transient) effects of biochar on soil properties or on the symbiosis.

Because of the variation in the quantity of biochar applied, the chemical and physical qualities of the biochar used, the homogeneity of biochar–soil mixtures, and the duration of the experiments, it seems pointless, at this point, to document the many context-specific cases in which biochar has had this or that effect on mycorrhizal symbiosis. All of the possible interactions of application rate, soil type, crop type, and biochar type have yet to be captured. Thus one can easily find positive (Nishio and Okano, 1991; Ishii and Kadoya, 1994; Ezawa et al., 2002; Matsubara et al., 2002; Yamato et al., 2006; Blackwell et al., 2015; LeCroy et al., 2013; Solaiman et al., 2010; Warnock et al., 2010; Güereña et al., 2015; Vanek and Lehmann, 2015) and neutral or negative (Ezawa et al., 2002; Habte and Antal, 2010; George et al., 2012; Nzanza et al., 2012; Quilliam et al., 2012b; Biederman and Harpole, 2013; LeCroy et al., 2013; Hu et al., 2014; Akhter et al., 2015; Hall and Bell, 2015) effects of biochar on mycorrhizal colonization or mycorrhiza function, but it is difficult to discern the reasons for these effects.

Instead, it seems more useful to take another approach altogether. In Section 25.3, I document some of the possible effects of biochar on soil properties (green arrows, Fig. 25.1) and then show how these altered soil properties influence the amount of growth mycorrhizal

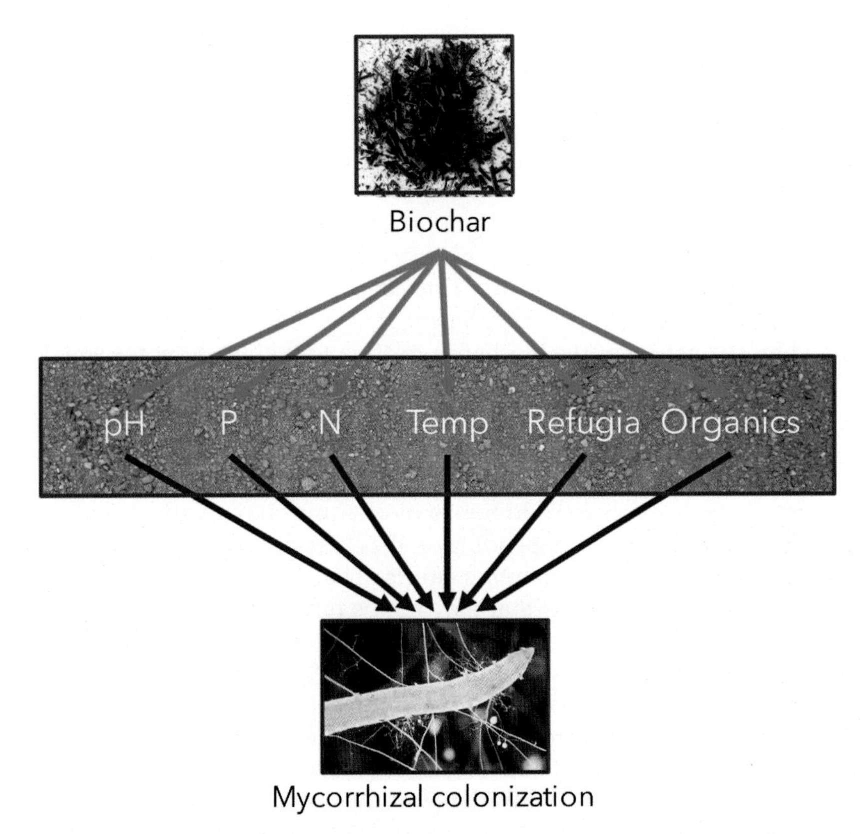

FIGURE 25.1 Biochar may influence mycorrhizal colonization via its effects on various soil properties (as elaborated in Section 25.3). *Green arrows* represent effects of biochar on various soil properties. *Blue arrows* then represent effects of those soil properties on mycorrhizal colonization. *Temp*, temperature.

hyphae make in the soil, the structure of mycorrhizal fungal communities, and the extent to which mycorrhizal fungi colonize roots (blue arrows, Fig. 25.1).

Many want to know how biochar can improve plant growth by invigorating the mycorrhizal symbiosis and thus increasing the contribution that the fungi make toward plant vigor (Warnock et al., 2007). Indeed, biochar may increase the contribution made by the fungi toward plant vigor. In Section 25.4 I discuss some situations in which the addition of biochar to soil increases plant responsiveness to mycorrhizal colonization (the difference between mycorrhizal and nonmycorrhizal plants). However, it is also possible for biochar to reduce the contribution of mycorrhizal fungi. Mycorrhizal fungi improve plant growth primarily under conditions in which the plant is somehow stressed, either because of insufficiency of nutrients (Koide, 1991) or because of pressure from pathogens (Newsham et al., 1995). When biochar itself alleviates either of these stresses, under some circumstances one might actually expect biochar to reduce the effectiveness of the symbiosis, not enhance it. This is also covered in Section 25.4. These effects of biochar on plant response to mycorrhizal colonization, mediated by level of nutrient stress or disease, are summarized in Fig. 25.2.

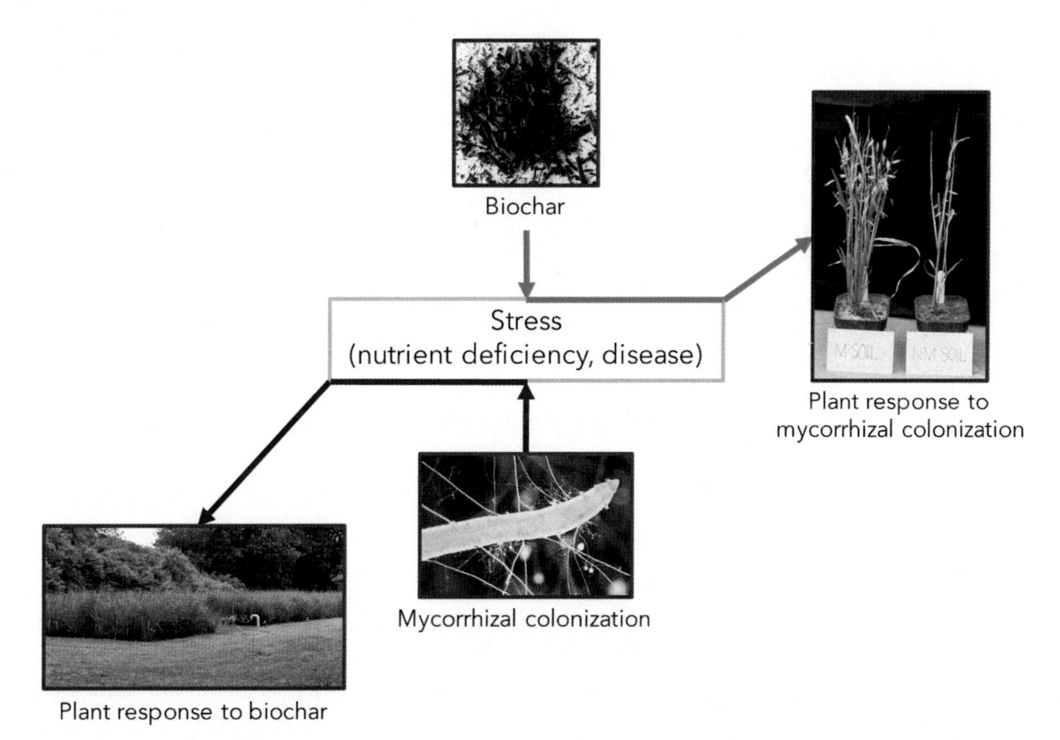

FIGURE 25.2 Biochar may alleviate nutrient stress and disease experienced by a plant, the same stresses alleviated by mycorrhizal colonization. Depending on the mechanism by which biochar influences stress, biochar may reduce, increase, or have no effect on plant response to mycorrhizal colonization (*green arrows*), and mycorrhizal colonization may reduce, increase, or have no effect on plant response to biochar (*blue arrows*). I discuss the former in Section 25.4.

25.3 BIOCHAR INFLUENCES MYCORRHIZAL COLONIZATION VIA ITS EFFECTS ON SOIL PROPERTIES

The extent to which soil properties are influenced by the addition of biochar will of course be specific to the context. Among other factors, this will depend on the amount of biochar applied, the physical and chemical characteristics of the biochar (Lehmann et al., 2011), the physical and chemical characteristics of the soil (its pH, water content, nutrient availability, etc.), the ability of the soil to buffer these variables, and environmental variables such as air temperature and rainfall. Therefore the addition of biochar to soil will influence its chemical and physical properties to various extents. Nevertheless, when biochar influences either physical or chemical properties of soil, it has the potential to influence the many organisms dwelling in the soil (Atucha and Litus, 2015; Elad et al., 2011; Grossman et al., 2010; Kim et al., 2007; Lehmann et al., 2011; O'Neill et al., 2009; Song et al., 2014), including AM fungi.

It is likely that for all of the various ways in which biochar influences soil properties, their effects on mycorrhizal fungi will vary depending on the species of mycorrhizal fungus. For example, P availability (Hayman et al., 1975), N availability (Hayman, 1970) and pH (Lekberg et al., 2007) influence the relative abundance of the component species in mycorrhizal fungal communities. Thus, although this is seldom reported, the addition of biochar to soil is likely to influence the structure of mycorrhizal fungal communities, which is known to influence plant performance (van der Heijden et al., 1998).

Mycorrhizal colonization of roots is obviously of great importance to the functioning of the symbiosis, specifically to the capture of carbohydrate by the fungi. Nevertheless, uptake of phosphate and other nutrients from the soil may be determined more by exploration of the soil by external hyphae than by the level of colonization of the root system (Hart and Reader, 2002). Thus, although most researchers report only the level of mycorrhizal colonization of roots, this value is not necessarily related to the benefit derived from colonization (Smith et al., 2004; Lekberg and Koide, 2005). The effects of soil properties on extraradical hyphal growth have usually gone unreported for the obvious reason that it is more tedious to assess despite its clear importance. One notable exception is the work by Warnock et al. (2010). Thus, in general, we have as yet an incomplete picture of the effect of soil characteristics on the functioning of the mycorrhizal symbiosis.

25.3.1 Soil pH

Many biochars derived from pyrolysis are alkaline in nature; therefore amending soil with them frequently increases soil pH (Tryon, 1948; Van Zwieten et al., 2010; Quilliam et al., 2012b; Biederman and Harpole, 2013; Curaqueo et al., 2014; Gul et al., 2015; Conversa et al., 2015; Güereña et al., 2015). The opposite may be true for acid-treated hydrochars (George et al., 2012). The soil pH may influence, among other things, the degree of mycorrhization (El-Kherbawy et al., 1989) and the composition of AM fungal communities through environmental filtering (Bainard et al., 2014; Jansa et al., 2014; Kohout et al., 2015).

25.3.2 P Availability

In acidic soils, much of the P may be bound by Al or Fe (White and Taylor 1977; Naidu et al., 1990). Increasing pH due to biochar application can increase the availability of P from these metal complexes (Yamato et al., 2006; Biederman and Harpole, 2013; Zhu et al., 2014). In addition to its influence on P availability via its effect on pH, many biochars possess anionic surface functional groups, which contribute to their cation exchange capacity. These anionic functional groups may compete for phosphate adsorption sites on other soil surfaces, increasing phosphate availability by reducing phosphate adsorption (Jiang et al., 2015). Some biochars contain significant concentrations of P, resulting in an increase in P availability upon addition to soil (Revell et al., 2012). In neutral soils, much of the P may be bound by Ca, and slight increases in pH associated with biochar addition may result in a reduction in the solubility of P (Brady, 1974). Indeed, Warnock et al. (2010) found that in some cases biochar addition to soil reduced soil P availability. Some biochars may remove phosphate from the soil solution via adsorption (Yao et al., 2011). The effect of P availability on the mycorrhizal symbiosis is well known. At very low P availability and, again, at very high P availability colonization of roots by mycorrhizal fungi may be inhibited (Bolan et al., 1984; Koide and Li, 1990). The negative effects of high P availability on mycorrhizal colonization are especially well known (Mosse, 1973).

25.3.3 N Availability

The addition of biochar to soil may help to retain N (Liu et al., 2014; Bai et al., 2015), resulting in increased concentrations of ammonium (Spokas et al., 2011; Taghizadeh-Toosi et al., 2012). Thus, under some circumstances, biochar may improve the N fertility of soils (Atkinson et al., 2010). However, in some cases biochar addition to soil may temporarily immobilize N (Bruun et al., 2012). The alteration in N availability by biochar may influence the mycorrhizal symbiosis. Addition of N fertilizer to *Trifolium subterraneum* (Chambers et al., 1980) or wheat (Hayman, 1970) reduced mycorrhizal colonization. Blanke et al. (2005) hypothesized that N deficiency stimulates mycorrhizal colonization as a mechanism to improve N uptake. However, Hepper (1983) found that N addition increased mycorrhizal colonization in lettuce, especially when plants were P deficient.

25.3.4 Soil Temperature

Biochar is black. Thus, when mixed with soil in substantial proportions, the soil may become darker and more absorptive of infrared radiation (Ventura et al., 2013). Nevertheless, this will not be an issue when plants cover the soil and, even when they do not, the addition of biochar to soil may reduce its density and thermal conductivity (Zhang et al., 2002), reducing the effect of biochar-induced heating below the first few centimeters of soil (Nelissen et al., 2015). However, early in the season before the crop is sufficiently developed to reduce absorption of solar radiation by the soil surface it is possible that increased soil surface temperatures will influence the mycorrhizal symbiosis. In the first place, seed germination may be possible earlier in the season, giving rise to living roots that can then be colonized by mycorrhizal fungi. Furthermore, temperature may affect the rate at which mycorrhizal fungi colonize roots. In the temperature range of 10–20°C, an increase in temperature may positively affect colonization (Carvalho et al., 2015).

25.3.5 Mycorrhizal Fungal Refugia

Biochars made from plant materials are porous, largely because of the presence of tracheary elements that conduct water. The hyphae of AM fungi may grow into biochar pores (Hammer et al., 2014). Thus some have hypothesized that biochar, particularly biochar from plant materials, may serve as a refugium for AM fungi (Saito, 1990; Saito and Marumoto, 2002; Douds et al., 2014). It is logical to hypothesize that biochar provides a refuge for nonfilamentous microbes such as bacteria or even protozoa. However, the case for fungi is less clear. Although hyphae inside biochar pores may be protected from soil animals larger than the pores, the hyphae nonetheless need to find their way into the pores from the outside. The hyphae outside of the pores are obviously still susceptible to attack. Thus it is not clear that the refuge provided by biochar particles is particularly useful in maintaining nutrient transport to the root if the hyphae are severed between the root and the biochar particle, which seems likely.

However, hyphae or spores may function as sources of inoculum. Furthermore, when they are found within the pores of a biochar particle, they exist in a protected environment. Thus addition to soils may substantially increase the inoculum potential of the soil in subsequent years. The phenomenon may be similar to the use of other porous materials containing mycorrhizal fungal propagules previously proposed as sources of inoculum (Dehne and Backhaus, 1986).

The pores in plant-derived biochars vary in diameter among species by at least one order of magnitude, and within a given ring-porous species by half that (Woodcock, 1989). Thus the distribution of pore sizes may be determined by the type of biochar used. Moreover, overall soil porosity may be altered by the addition of biochar due to the interaction between biochar particles and soil aggregates (Curaqueo et al., 2014). This variation in pore diameter may affect the kinds of mycorrhizal fungi in the community. For example, the size of the spores produced by AM fungi varies by more than an order of magnitude, largely corresponding to family (Trappe, 1982). Thus the species composition of AM fungal spores found within biochar particles might depend on the distribution of soil pore sizes, which is a function of the biochar feedstock species.

It must be noted that ground biochars, with no or limited capacity to provide refugia for microorganisms, may also significantly stimulate mycorrhizal colonization (Nishio and Okano, 1991; Ezawa et al., 2002). Thus biochar must be capable of stimulating mycorrhizal fungi in at least one other way.

25.3.6 Organic Inhibitors and Signals

In addition to the persistent C (largely fused aromatic structures) and interspersed ash inclusions, biochar contains a variable amount of easily mineralizable C that is sometimes referred to as volatile and that can have significant effects on the soil biota (Lehmann et al., 2011). Some of these volatile compounds are organic (Cross and Sohi, 2011) and are capable of inhibiting roots and microorganisms, including mycorrhizal fungi. These include pyrolysis oil, wood vinegar, tar, pitch, and essential oils (Tiilikkala et al., 2010). The size of this fraction depends on, among other things, the temperature of pyrolysis, being less abundant in biochars produced at higher temperatures (Luo et al., 2013). Warnock et al. (2010) found that biochars produced at higher temperatures (430°C), which presumably contain fewer volatiles, had less of an inhibitory effect on mycorrhizal fungi than biochar produced at a lower temperature (360°C).

In the same way that activated charcoal can adsorb organic compounds, biochars can adsorb organic compounds produced elsewhere in the soil. Root systems may produce substances that inhibit mycorrhizal colonization (Koide and Li, 1991). To the extent that organic inhibitors of mycorrhizal fungi or mycorrhizal colonization exist in the soil, their adsorption to biochar could release mycorrhizal fungi from inhibition (Elmer and Pignatello, 2011; Sosa-Rodriguez et al., 2014). On the other hand, organic signaling molecules between roots and mycorrhizal fungi (Akiyama et al., 2005) could also be adsorbed by biochar, possibly interfering with the colonization process.

25.4 BIOCHAR INFLUENCES PLANT RESPONSE TO MYCORRHIZAL COLONIZATION VIA ITS IMPACT ON THE LEVEL OF PLANT STRESS

There is little doubt that under certain circumstances addition of biochar to soil reduces nutrient and disease stress, the very benefits plants frequently derive from mycorrhizal colonization. On the one hand, if biochar reduces stress by a mechanism that is different from the one used by mycorrhizal fungi, then biochar may actually increase the effectiveness of mycorrhizal fungi (synergistic effects) or have no effect (additive effects). As an example of synergism, when biochar increases soil solution phosphate concentrations, mycorrhizal colonization could benefit the plant more than it would without biochar by taking up this greater concentration of phosphate. As an example of additive effects, biochar may suppress root pathogenesis by changing the pH of the rhizosphere whereas mycorrhizal fungi may suppress root pathogenesis using a different mechanism, such as by competing with pathogens that colonize the root.

On the other hand, when biochar and mycorrhizal fungi influence plants in the very same way, the benefit from mycorrhizal colonization would be smaller than it would be in the absence of biochar. For example, if biochar were to stimulate saprotrophic fungal growth in the soil and on roots, causing roots to maintain better contact with soil aggregates and increasing the capacity of the roots to take up water, then the mycorrhizal fungi would be largely superfluous in this context.

25.4.1 Biochar Decreases Nutrient Stress in Plants

Alkaline biochars may increase the concentration of phosphate in the soil solution, particularly for acid soils. For example, an increase of phosphate concentration may occur when the pH is increased in acid soils that contain substantial quantities of Al or Fe phosphates (Yamato et al., 2006; Biederman and Harpole, 2013; Zhu et al., 2014). In that case, mycorrhizal fungi could be particularly useful in capturing the additional free phosphate, resulting in an additive or even synergistic effect on plant phosphate uptake.

The same additive or synergistic interaction between biochar and AM fungi may occur for N uptake to the extent that AM fungi increase N uptake, which may be significant for ammonium (Tanaka and Yano, 2005). Biochar has been shown to improve N availability in some soils by reducing ammonium leaching and denitrification (Atkinson et al., 2010), by increasing N fixation (Mia et al., 2014; Guereña et al., 2015), or by increasing soil aggregation (see below), which may act to increase C and N retention by soils (Liu et al., 2014).

Patches of pure switchgrass biochar inhibited growth of switchgrass roots (Koide et al., unpublished). Mycorrhizal colonization may be more beneficial to plants with reduced root system development than plants with better developed root systems (Bryla and Koide, 1990). In contrast, amendment of soil with more uniformly dispersed biochar may increase soil aeration and reduce soil tensile strength and thus permit greater root elongation (Lehmann et al., 2011; references therein, Biederman and Harpole, 2013; Atucha and Litus, 2015), increasing nutrient uptake by roots and reducing response to mycorrhizal colonization (Bryla and Koide, 1990).

Biochar may improve phosphate availability by increasing soil water content and thus increasing the diffusivity of phosphate (Bhadoria et al., 1991). Biochar may increase soil water content by improving soil aggregation (George et al., 2012; Curaqueo et al., 2014; Gul et al., 2015 and references therein), which may increase water retention (Tisdall and Oades, 1982). Biochar itself, in the absence of effects on soil aggregation, may also directly improve soil water retention, particularly in coarse-textured soils (Koide et al., 2015). Because mycorrhizal fungi have their largest effects on phosphate uptake when soils are dry (Nelsen and Safir, 1982), addition of biochar to soil, resulting in greater water retention, could decrease the response to mycorrhizal colonization.

Biochar may increase soil aggregation by increasing bacterial biomass and the availability of organic materials (Gul et al., 2015) that bind primary soil particles together into small aggregates. If mycorrhizal fungi bind small aggregates into larger aggregates (Miller and Jastrow, 1990), biochar and mycorrhizal fungi could have additive or synergistic effects. On the other hand, biochar could increase soil aggregation primarily by stimulating growth of AM fungi. If that were the case, then the biochar effect and the AM fungal effect are one and the same.

As mentioned previously, amending soil with biochar frequently increases soil pH (Tryon, 1948; Van Zwieten et al., 2010; Quilliam et al., 2012b; Biederman and Harpole, 2013; Curaqueo et al., 2014; Gul et al., 2015; Conversa et al., 2015; Güereña et al., 2015), which may influence the effectiveness of AM fungi on plant growth (Graw, 1979; Ouzounidou et al., 2015). For example, increasing pH may decrease the availability of Zn and Mn (El-Kherbawy et al., 1989; Jansen et al., 2003; Van Zwieten et al., 2010), which may influence plant response to mycorrhizal colonization; mycorrhizal colonization may enhance plant growth when soil Zn concentrations are low (Watts-Williams et al., 2015) and when Mn concentrations are high (Alho et al., 2015).

25.4.2 Biochar Reduces Disease Severity in Plants

Biochar can reduce disease severity (Elmer and Pignatello, 2011; Lehmann et al., 2011 and references therein), possibly by altering the abundance of certain bacteria, by releasing specific compounds that induce systemic acquired resistance in plants (Elad et al., 2011), by improving the overall vigor of the plant (Lehmann et al., 2011), or by stimulating AM fungal colonization. Colonization by AM fungi has been shown to positively affect plants by reducing the incidence or severity of disease (Newsham et al., 1995), possibly via competition (Wacker et al., 1990; Matsubara et al., 2002; Elmer and Pignatello, 2011; Atucha and Litus, 2015). If biochar reduces disease by stimulating mycorrhizal fungi, then the biochar effect and the AM fungal effect are one and the same. However, if biochar and AM fungi reduce disease by different mechanisms, then they could act additively or synergistically.

25.5 CONCLUSIONS

The large-scale adoption of biochar in temperate agriculture is impeded in many regions by several factors, including the costs of transporting large quantities of biochar from areas of production to those used in row-crop agriculture and the need to mix biochar with other organic materials such as manure or with inorganic nutrients to achieve maximum yield benefits. Nevertheless, one can point to some clear advantages to amending temperate soils with biochar, which include its capacity to sequester C in soil for long periods, replace lime as a more C-stable means of increasing soil pH, and improve soil water retention in coarse-textured soils. Thus we may see the day when biochar is more widely used in temperate agriculture.

What effect will biochar additions to soil have on the mycorrhizal symbiosis and plant responsiveness to mycorrhizal colonization? One can safely say (with complete dissatisfaction) "it all depends on the context." The amount of biochar used, the feedstock, the pyrolysis conditions, the nature of the soil and the climate, etc. will all determine how biochar influences the symbiosis. However, all is not lost. Research on biochar has shown it to influence certain soil physical and chemical properties, including pH, P availability, N availability, inhibitor concentration, temperature, and physical refugia. Because of other research relating these soil physical and chemical properties to mycorrhizal colonization and mycorrhizal fungal community structure, one may be able to (cautiously) predict the effect of biochar on mycorrhizal colonization and mycorrhizal fungal community structure if one first understands how the biochar will affect the soil. Furthermore, because we know in some cases how biochar influences nutrients stress and disease in plants and, through other studies, how these variables influence response to mycorrhizal colonization by plants, we can indirectly predict how biochar might influence plant response to mycorrhizal colonization.

Should one be concerned whether biochar affects mycorrhizal colonization or plant response to mycorrhizal colonization? In any agricultural system, modification of the one component will have many direct and indirect effects on other components of the system. Therefore it should never be surprising to see that by altering the soil, other parts, including the plants and the soil microbial communities, are also altered in some fashion. If biochar is used to sequester C and actually improves yield while otherwise having no adverse effects, should one care whether it influences mycorrhizal colonization, the community of mycorrhizal fungi, or plant response to colonization simply because it does influence these things? All agricultural practices in the developed world are motivated by economics, either short term or long term. Short-term economic considerations exist without regard to issues of sustainability such as soil erosion, greenhouse gas emissions, leaching, energy efficiency, etc. Long-term economic considerations include issues of sustainability. The fact is that we do not know how changes in mycorrhizal colonization, etc. that are due to the use of biochar will influence our appreciation of either short-term or long-term economic considerations. Moreover, the mycorrhizal symbiosis is just one of many aspects of agricultural systems that will be affected by biochar. Whole chapters might just as well have been devoted to the effects of biochar on soil erosion, energy use, C sequestration, greenhouse gas emissions, soil bacteria, other soil fungi, soil arthropods, weeds, etc.

Biochar should be considered in as many contexts as necessary. However, it seems unlikely that mycorrhizal fungi should be the first consideration. In some situations, the most important context may be soil erosion, in which case the development of hyphae of mycorrhizal fungi may be important. In others it may be greenhouse gas emissions, in which case mycorrhizal fungi may be unimportant. As stated previously, there are more considerations in growing a crop than biochar and mycorrhizal fungi. Thus for some I am sure that this chapter is entirely unsatisfying, for I have not been able to make any sweeping conclusions or broad generalizations about biochar and mycorrhizas. However, as Mark Twain once remarked, "All generalizations are false, including this one."

Acknowledgments

This research was supported by USDA-NIFA-AFRI 2011-67009-20072 with additional support from Brigham Young University, including the Roger and Victoria Sant Educational Endowment for a Sustainable Environment. The author is grateful for the expert comments made on a previous version of this chapter by Drs. J. Lehmann and J. Jansa.

References

Akhter, A., Hage-Ahmed, K., Soja, G., Steinkellner, S., 2015. Compost and biochar alter mycorrhization, tomato root exudation, and development of *Fusarium oxysporum* f. sp. *lycopersici*. Frontiers in Plant Science 6, 1–13.

Akiyama, K., Matsuzaki, K., Hayashi, H., 2005. Plant sesquiterpenes induce hyphal branching in arbuscular mycorrhizal fungi. Nature 435, 824–827.

Alho, L., Carvalho, M., Brito, I., Goss, M.J., 2015. The effect of arbuscular mycorrhiza fungal propagules on the growth of subterranean clover (*Trifolium subterraneum* L.) under Mn toxicity in ex situ experiments. Soil Use and Management 31, 337–344.

Atkinson, C.J., Fitzgerald, J.D., Hipps, N.A., 2010. Potential mechanisms for achieving agricultural benefits from biochar application to temperate soils: a review. Plant and Soil 337, 1–18.

Atucha, A., Litus, G., 2015. Effect of biochar amendments on peach replant disease. HortScience 50, 863–868.

Bai, S.H., Reverchon, F., Xu, C.Y., Xu, Z., Blumfield, T.J., Zhao, H., Van Zwieten, L., Wallace, H.M., 2015. Wood biochar increases nitrogen retention in field settings mainly through abiotic processes. Soil Biology and Biochemistry 90, 232–240.

Bainard, L.D., Bainard, J.D., Hamel, C., Gan, Y., 2014. Spatial and temporal structuring of arbuscular mycorrhizal communities is differentially influenced by abiotic factors and host crop in a semi-arid prairie agroecosystem. FEMS Microbiology Ecology 88, 333–344.

Bhadoria, P., Claassen, N., Jungk, A., 1991. Phosphate diffusion coefficients in soil as affected by bulk density and water content. Zeitschrift für Pflanzenernährung und Bodenkunde 154, 53–57.

Biederman, L.A., Harpole, W.S., 2013. Biochar and its effects on plant productivity and nutrient cycling: a meta-analysis. GCB Bioenergy 5, 202–214.

Birk, J.J., Steiner, C., Teixiera, W.C., Zech, W., Glaser, B., 2009. Microbial response to charcoal amendments and fertilization of a highly weathered tropical soil. In: Woods, W.I., Teixeira, W.G., Lehmann, J., Steiner, C., WinklerPrins, A.M.G.A., Rebellato, L. (Eds.), Amazonian Dark Earths: Wim Sombroek's Vision. Springer, pp. 309–324.

Blackwell, P., Joseph, S., Munroe, P., Anawar, H.M., Storer, P., Gilkes, R.J., Solaiman, Z.M., 2015. Influences of biochar and biochar-mineral complex on mycorrhizal colonisation and nutrition of wheat and sorghum. Pedosphere 25, 686–695.

Blanke, V., Renker, C., Wagner, M., Füllner, K., Held, M., Kuhn, A.J., Buscot, F., 2005. Nitrogen supply affects arbuscular mycorrhizal colonization of Artemisia vulgaris in a phosphate-polluted field site. New Phytologist 166, 981–992.

Bolan, N., Robson, A., Barrow, N., 1984. Increasing phosphorus supply can increase the infection of plant roots by vesicular-arbuscular mycorrhizal fungi. Soil Biology and Biochemistry 16, 419–420.

Brady, N.C., 1974. The Nature and Properties of Soils, eighth ed. MacMillan, New York.

Bruun, E.W., Ambus, P., Egsgaard, H., Hauggaard-Nielsen, H., 2012. Effects of slow and fast pyrolysis biochar on soil C and N turnover dynamics. Soil Biology and Biochemistry 46, 73–79.

Bryla, D.R., Koide, R.T., 1990. Regulation of reproduction in wild and cultivated *Lycopersicon esculentum* Mill. by vesicular-arbuscular mycorrhizal infection. Oecologia 84, 74–81.

Carvalho, M., Brito, I., Alho, L., Goss, M.J., 2015. Assessing the progress of colonization by arbuscular mycorrhiza of four plant species under different temperature regimes. Journal of Plant Nutrition and Soil Science 515–522.

Chambers, C., Smith, S., Smith, F., 1980. Effects of ammonium and nitrate ions on mycorrhizal infection, nodulation and growth of *Trifolium subterraneum*. New Phytologist 85, 47–62.

Conversa, G., Bonasia, A., Lazzizera, C., Elia, A., June 16, 2015. Influence of biochar, mycorrhizal inoculation, and fertilizer rate on growth and flowering of Pelargonium (*Pelargonium zonale* L.) plants. Frontiers in Plant Science. http://dx.doi.org/10.3389/fpls.2015.00429.

Cross, A., Sohi, S.P., 2011. The priming potential of biochar products in relation to labile carbon contents and soil organic matter status. Soil Biology and Biochemistry 43, 2127–2134.

Curaqueo, G., Meier, S., Khan, N., Cea, M., Navia, R., 2014. Use of biochar on two volcanic soils: effects on soil properties and barley yield. Journal of Soil Science and Plant Nutrition 14, 911–924.

Dehne, H.W., Backhaus, G.F., 1986. The use of vesicular-arbuscular mycorrhizal fungi in plant production. 1. Inoculum production. Zeitschrift fur Pflanzenkrankheiten und Pflanzenschutz 93, 415–424.

Douds, D.D., Lee, J., Uknalis, J., Boateng, A.A., Ziegler-Ulsh, C., 2014. Pelletized biochar as a carrier for AM fungi in the on-farm system of inoculum production in compost and vermiculite mixtures. Compost Science & Utilization 22, 253–262.

El-Kherbawy, M., Angle, J.S., Heggo, A., Chaney, R.L., 1989. Soils effects on growth and heavy metal uptake of alfalfa (*Medicago sativa* L.). Biology and Fertility of Soils 8, 61–65.

Elad, Y., Cytryn, E., Meller Harel, Y., Lew, B., Graber, E.R., 2011. The biochar effect: plant resistance to biotic stress. Phytopathologia Mediterranea 50, 335–349.

Elmer, W.H., Pignatello, J.J., 2011. Effect of biochar amendments on mycorrhizal associations and Fusarium crown and root rot of asparagus in replant soils. Plant Disease 95, 960–966.

Ezawa, T., Yamamoto, K., Yoshida, S., 2002. Enhancement of the effectiveness of indigenous arbuscular mycorrhizal fungi by inorganic soil amendments. Soil Science and Plant Nutrition 48, 897–900.

Factura, H., Bettendorf, T., Buzie, C., Pieplow, H., Reckin, J., Otterpohl, R., 2010. Terra Preta sanitation: re-discovered from an ancient Amazonian civilisation – integrating sanitation, bio-waste management and agriculture. Water Science and Technology 61, 2673–2679.

Fang, Y., Singh, B., Singh, B.P., Krull, E., 2014. Biochar carbon stability in four contrasting soils. European Journal of Soil Science 65, 60–71.

George, C., Wagner, M., Kücke, M., Rillig, M.C., 2012. Divergent consequences of hydrochar in the plant-soil system: arbuscular mycorrhiza, nodulation, plant growth and soil aggregation effects. Applied Soil Ecology 59, 68–72.

Glaser, B., Lehmann, J., Zech, W., 2002. Ameliorating physical and chemical properties of highly weathered soils in the tropics with charcoal – a review. Biology and Fertility of Soils 35, 219–230.

Graw, D., 1979. The influence of soil pH on the efficiency of vesicular-arbuscular mycorrhiza. New Phytologist 82, 687–695.

Grossman, J., O'Neill, B., Tsai, S., Liang, B., Neves, E., Lehmann, J., Thies, J., 2010. Amazonian anthrosols support similar microbial communities that differ distinctly from those extant in adjacent, unmodified soils of the same mineralogy. Microbial Ecology 60, 192–205.

Güereña, D.T., Lehmann, J., Thies, J.E., Enders, A., Karanja, N., Neufeldt, H., 2015. Partitioning the contributions of biochar properties to enhanced biological nitrogen fixation in common bean (*Phaseolus vulgaris*). Biology and Fertility of Soils 51, 479–491.

Gul, S., Whalen, J.K., Thomas, B.W., Sachdeva, V., Deng, H., 2015. Physico-chemical properties and microbial responses in biochar-amended soils: mechanisms and future directions. Agriculture. Ecosystems & Environment 206, 46–59.

Habte, M., Antal Jr., M.J., 2010. Reaction of mycorrhizal and nonmycorrhizal *Leucaena leucocephala* to charcoal amendment of mansand and soil. Communications in Soil Science and Plant Analysis 41, 540–552.

Hall, D.J.M., Bell, R.W., 2015. Biochar and compost increase crop yields but the effect is short term on sandplain soils of Western Australia. Pedosphere 25, 720–728.

Hammer, E.C., Balogh-Brunstad, Z., Jakobsen, I., Olsson, P.A., Stipp, S.L.S., Rillig, M.C., 2014. A mycorrhizal fungus grows on biochar and captures phosphorus from its surfaces. Soil Biology and Biochemistry 77, 252–260.

Hammer, E.C., Forstreuter, M., Rillig, M.C., Kohler, J., 2015. Biochar increases arbuscular mycorrhizal plant growth enhancement and ameliorates salinity stress. Applied Soil Ecology 96, 114–121.

Hart, M., Reader, R., 2002. Host plant benefit from association with arbuscular mycorrhizal fungi: variation due to differences in size of mycelium. Biology and Fertility of Soils 36, 357–366.

Hayman, D.S., 1970. Endogone spore numbers in soil and vesicular-arbuscular mycorrhiza in wheat as influenced by season and soil treatment. Transactions of the British Mycological Society 54, 53–63.

Hayman, D.S., Johnson, A.M., Ruddlesdin, I., 1975. The influence of phosphate and crop species on endogone spores and vesicular-arbuscular mycorrhizal under field conditions. Plant and Soil 43, 489–495.

Hepper, C., 1983. The effect of nitrate and phosphate on the vesicular-arbuscular mycorrhizal infection of lettuce. New Phytologist 92, 389–399.

Hernandez-Mena, L.E., Pecora, A.A.B., Beraldo, A.L., 2014. Slow pyrolysis of bamboo biomass: analysis of biochar properties. Chemical Engineering Transactions 37, 115–120.

Hu, J., Wu, F., Wu, S., Lam, C.L., Lin, X., Wong, M.H., 2014. Biochar and *Glomus caledonium* influence Cd accumulation of upland kangkong (*Ipomoea aquatica* Forsk.) intercropped with Alfred stonecrop (*Sedum alfredii* Hance). Scientific Reports 4, 4671.

Ishii, T., Kadoya, K., 1994. Effects of charcoal as a soil conditioner on citrus growth and vesicular-arbuscular mycorrhizal development. Journal of the Japanese Society for Horticultural Science (Japan) 63, 529–535.

Jansa, J., Erb, A., Oberholzer, H.R., Šmilauer, P., Egli, S., 2014. Soil and geography are more important determinants of indigenous arbuscular mycorrhizal communities than management practices in Swiss agricultural soils. Molecular Ecology 23, 2118–2135.

Jansen, B., Nierop, K.G.J., Verstraten, J.M., 2003. Mobility of Fe(II), Fe(III) and Al in acidic forest soils mediated by dissolved organic matter: influence of solution pH and metal/organic carbon ratios. Geoderma 113, 323–340.

Jiang, J., Yuan, M., Xu, R., Bish, D.L., 2015. Mobilization of phosphate in variable-charge soils amended with biochars derived from crop straws. Soil and Tillage Research 146, 139–147.

Joseph, S., Anawar, H.M., Storer, P., Blackwell, P., Chia, C., Lin, Y., Munroe, P., Donne, S., Horvat, J., Wang, J., et al., 2015. Effects of enriched biochars containing magnetic iron nanoparticles on mycorrhizal colonisation, plant growth, nutrient uptake and soil quality improvement. Pedosphere 25, 749–760.

Kambo, H.S., Dutta, A., 2015. A comparative review of biochar and hydrochar in terms of production, physico-chemical properties and applications. Renewable and Sustainable Energy Reviews 45, 359–378.

Kaudal, B.B., Chen, D., Madhavan, D.B., Downie, A., Weatherley, A., 2016. An examination of physical and chemical properties of urban biochar for use as growing media substrate. Biomass and Bioenergy 84, 49–58.

Khanmohammadi, Z., Afyuni, M., Mosaddeghi, M.R., 2015. Effect of pyrolysis temperature on chemical and physical properties of sewage sludge biochar. Waste Management and Research 33, 275–283.

Kim, J.-S., Sparovek, G., Longo, R.M., De Melo, W.J., Crowley, D., 2007. Bacterial diversity of terra preta and pristine forest soil from the Western Amazon. Soil Biology and Biochemistry 39, 684–690.

Kohout, P., Doubková, P., Bahram, M., Suda, J., Tedersoo, L., Vorisková, J., Sudová, R., 2015. Niche partitioning in arbuscular mycorrhizal communities in temperate grasslands: a lesson from adjacent serpentine and nonserpentine habitats. Molecular Ecology 24, 1831–1843.

Koide, R.T.R., 1991. Nutrient supply, nutrient demand and plant response to mycorrhizal infection. New Phytologist 117, 365–386.

Koide, R., Li, M., 1990. On host regulation of the vesicular-arbuscular mycorrhizal symbiosis. New Phytologist 114, 59–64.

Koide, R., Li, M., 1991. Mycorrhizal fungi and the nutrient ecology of three oldfield annual plant species. Oecologia 85, 403–412.

Koide, R., Nguyen, B.T., Skinner, R.H., Dell, C.J., Peoples, M.S.M., Adler, P.P.R., Drohan, P.P.J., 2015. Biochar amendment of soil improves resilience to climate change. Global Change Biology Bioenergy 7, 1084–1091.

Laird, D.A., 2008. The charcoal vision: a win-win-win scenario for simultaneously producing bioenergy, permanently sequestering carbon, while improving soil and water quality. Agronomy Journal 100, 178–181.

LeCroy, C., Masiello, C.A., Rudgers, J.A., Hockaday, W.C., Silberg, J.J., 2013. Nitrogen, biochar, and mycorrhizae: alteration of the symbiosis and oxidation of the char surface. Soil Biology and Biochemistry 58, 248–254.

Lehmann, J., 2007. A handful of carbon. Nature 447, 143–144.

Lehmann, J., 2009. Terra preta nova – where to from here? In: Woods, W.I., Teixeira, W.G., Lehmann, J., Steiner, C., WinklerPrins, A., Rebellato, L. (Eds.), Amazonian Dark Earths: Wim Sombroek's Vision. Springer, Berlin, pp. 473–486.

Lehmann, J., Joseph, S., 2015. Biochar for environmental management: an introduction. In: Lehman, J., Joseph, S. (Eds.), Biochar for Environmental Management. Science and Technology. Earthscan, Sterling, VA, USA, pp. 1–12.

Lehmann, J., Rillig, M.C., Thies, J., Masiello, C.A., Hockaday, W.C., Crowley, D., 2011. Biochar effects on soil biota – a review. Soil Biology and Biochemistry 43, 1812–1836.

Lekberg, Y., Koide, R.T., 2005. Is plant performance limited by abundance of arbuscular mycorrhizal fungi? A meta-analysis of studies published between 1988 and 2003. New Phytologist 168, 189–204.

Lekberg, Y., Koide, R., Rohr, J., Aldrich-Wolfe, L., Morton, J., 2007. Role of niche restrictions and dispersal in the composition of arbuscular mycorrhizal fungal communities. Journal of Ecology 95, 95–105.

Leslie, T.W., Biddinger, D.J., Rohr, J.R., Hulting, A.G., Mortensen, D.A., Fleischer, S.J., 2014. Examining shifts in Carabidae assemblages across a forest-agriculture ecotone. Environmental Entomology 43, 18–28.

Liang, B., Lehmann, J., Solomon, D., Kinyangi, J., Grossman, J., O'Neill, B., Skjemstad, J.O., Thies, J., Luizão, F.J., Petersen, J., et al., 2006. Black carbon increases cation exchange capacity in soils. Soil Science Society of America Journal 70, 1719–1730.

Liu, Z., Chen, X., Jing, Y., Li, Q., Zhang, J., Huang, Q., 2014. Effects of biochar amendment on rapeseed and sweet potato yields and water stable aggregate in upland red soil. Catena 123, 45–51.

Luo, Y., Durenkamp, M., De Nobili, M., Lin, Q., Devonshire, B.J., Brookes, P.C., 2013. Microbial biomass growth, following incorporation of biochars produced at 350°C or 700°C, in a silty-clay loam soil of high and low pH. Soil Biology and Biochemistry 57, 513–523.

MacDonald, J.M., Ribaudo, M., Livingston, M., Beckman, J., Huang, W.-Y., 2009. Manure Use for Fertilizer and for Energy: Report to Congress. Administrative Publication No. AP-037. USDA, Economic Research Service, p. 53.

Matsubara, Y., Hasegawa, N., Fukui, H., 2002. Incidence of Fusarium root rot in asparagus seedlings infected with arbuscular mycorrhizal fungus as affected by several soil amendments. Journal of the Japanese Society of Horticultural Science 71, 370–374.

Mia, S., van Groenigen, J.W., van de Voorde, T.F.J., Oram, N.J., Bezemer, T.M., Mommer, L., Jeffery, S., 2014. Biochar application rate affects biological nitrogen fixation in red clover conditional on potassium availability. Agriculture. Ecosystems and Environment 191, 83–91.

Miller, R.M., Jastrow, J.D., 1990. Hierarchy of root and mycorrhizal fungal interactions with soil aggregation. Soil Biology and Biochemistry 22, 579–584.

Mosse, B., 1973. Plant growth responses to vesicular-arbuscular mycorrhiza IV. In soil given additional phosphate. New Phytologist 72, 127–136.

Naidu, R., Syersy, J.K., Tillman, R.W., Kirkman, J.H., 1990. Effect of liming on phosphate sorption by acid soils. Journal of Soil Science 41, 165–175.

Nelissen, V., Ruysschaert, G., Manka'Abusi, D., D'Hose, T., De Beuf, K., Al-Barri, B., Cornelis, W., Boeckx, P., 2015. Impact of a woody biochar on properties of a sandy loam soil and spring barley during a two-year field experiment. European Journal of Agronomy 62, 65–78.

Nelsen, C.E., Safir, G.R., 1982. Increased drought tolerance of mycorrhizal onion plants caused by improved phosphorus nutrition. Planta 154, 407–413.

Newsham, K., Fitter, A., Watkinson, A., 1995. Arbuscular mycorrhiza protect an annual grass from root pathogenic fungi in the field. Journal of Ecology 83, 991–1000.

Nguyen, B.T., Koide, R.T., Dell, C., Drohan, P., Skinner, H., Adler, P.R., Nord, A., 2014. Turnover of soil carbon following addition of switchgrass-derived biochar to four soils. Soil Science Society of America Journal 78, 531–537.

Nishio, M., Okano, S., 1991. Stimulation of the growth of alfalfa and infection of roots with indigenous vesicular-arbuscular mycorrhizal fungi by the application of charcoal. Bulletin of the National Grasslands Research Institute 45, 61–71.

Nzanza, B., Marais, D., Soundy, P., 2012. Effect of arbuscular mycorrhizal fungal inoculation and biochar amendment on growth and yield of tomato. International Journal of Agriculture and Biology 14, 965–969.

Ogawa, M., Okimori, Y., 2010. Pioneering works in biochar research, Japan. Australian Journal of Soil Research 48, 489–500.

O'Neill, B., Grossman, J., Tsai, M.T., Gomes, J.E., Lehmann, J., Peterson, J., Neves, E., Thies, J.E., 2009. Bacterial community composition in Brazilian anthrosols and adjacent soils characterized using culturing and molecular identification. Microbial Ecology 58, 23–35.

Ouzounidou, G., Skiada, V., Papadopoulou, K.K., Stamatis, N., Kavvadias, V., Eleftheriadis, E., Gaitis, F., 2015. Effects of soil pH and arbuscular mycorrhiza (AM) inoculation on growth and chemical composition of chia (*Salvia hispanica* L.) leaves. Brazilian Journal of Botany 38, 487–495.

Quilliam, R.S., DeLuca, T.H., Jones, D.L., 2012a. Biochar application reduces nodulation but increases nitrogenase activity in clover. Plant and Soil 366, 83–92.

Quilliam, R.S., Marsden, K.A., Gertler, C., Rousk, J., DeLuca, T.H., Jones, D.L., 2012b. Nutrient dynamics, microbial growth and weed emergence in biochar amended soil are influenced by time since application and reapplication rate. Agriculture. Ecosystems & Environment 158, 192–199.

Revell, K.T., Maguire, R.O., Agblevor, F.A., 2012. Influence of poultry litter biochar on soil properties and plant growth. Soil Science 177, 402–408.

Saito, M., 1990. Charcoal as a micro-habitat for VA mycorrhizal fungi, and its practical implication. Agriculture. Ecosystems & Environment 29, 341–344.

Saito, M., Marumoto, T., 2002. Inoculation with arbuscular mycorrhizal fungi: the status quo in Japan and the future prospects. Diversity and Integration in Mycorrhizas 244, 273–279.

Smith, S.E., Smith, F.A., Jakobsen, I., 2004. Functional diversity in arbuscular mycorrhizal (AM) symbioses: the contribution of the mycorrhizal P uptake pathway is not correlated with mycorrhizal responses in growth or total P uptake. New Phytologist 162, 511–524.

Solaiman, Z.M., Blackwell, P., Abbott, L.K., Storer, P., 2010. Direct and residual effect of biochar application on mycorrhizal root colonisation, growth and nutrition of wheat. Australian Journal of Soil Research 48, 546.

Song, Y., Zhang, X., Ma, B., Chang, S.X., Gong, J., 2014. Biochar addition affected the dynamics of ammonia oxidizers and nitrification in microcosms of a coastal alkaline soil. Biology and Fertility of Soils 50, 321–332.

Sosa-Rodriguez, T., Dupré de Boulois, H., Granet, F., Gaurel, S., Declerck, S., 2014. Effect of activated charcoal and pruning of the taproot on the in vitro mycorrhization of *Hevea brasiliensis* Müll. Arg. In: In Vitro Cellular & Developmental Biology – Plant, vol. 50, pp. 317–325.

Spokas, K.A., Novak, J.M., Venterea, R.T., 2011. Biochar's role as an alternative N-fertilizer: ammonia capture. Plant and Soil 350, 35–42.

Taghizadeh-Toosi, A., Clough, T.J., Sherlock, R.R., Condron, L.M., 2012. Biochar adsorbed ammonia is bioavailable. Plant and Soil 350, 57–69.

Tanaka, Y., Yano, K., 2005. Nitrogen delivery to maize via mycorrhizal hyphae depends on the form of N supplied. Plant. Cell & Environment 28, 1247–1254.

Tiilikkala, K., Fagernas, L., Tiilikkala, J., 2010. History and use of wood pyrolysis liquids as biocide and plant protection product. Open Agriculture Journal 4, 111–118.

Tisdall, J.M., Oades, J.M., 1982. Organic matter and water-stable aggregates in soils. Journal of Soil Science 33, 141–163.

Trappe, J.M., 1982. Synoptic keys to the genera and species of zygomycetous mycorrhizal fungi. Phytopathology 72, 1102–1108.

Tryon, E.H., 1948. Effect of charcoal on certain physical, chemical, and biological properties of forest soils. Ecological Monographs 18, 81–115.

van der Heijden, M., Klironomos, J., Ursic, M., Moutoglis, P., Streitwolf-Engel, R., Boller, T., Wiemken, A., Sanders, I., 1998. Mycorrhizal fungal diversity determines plant biodiversity, ecosystem variability and productivity. Nature 74, 69–72.

Van Zwieten, L., Kimber, S., Morris, S., Chan, K.Y., Downie, A., Rust, J., Joseph, S., Cowie, A., 2010. Effects of biochar from slow pyrolysis of papermill waste on agronomic performance and soil fertility. Plant and Soil 327, 235–246.

Vanek, S.J., Lehmann, J., 2015. Phosphorus availability to beans via interactions between mycorrhizas and biochar. Plant and Soil 395, 105–123.

Ventura, F., Salvatorelli, F., Piana, S., Pieri, L., Pisa, P.R., 2013. The effects of biochar on the physical properties of bare soil. Earth and Environmental Science Transactions of the Royal Society of Edinburgh 103, 5–11.

Wacker, T.L., Safir, G.R., Stephens, C.T., 1990. Mycorrhizal fungi in relation to asparagus growth and Fusarium wilt. Acta Horticulturae 271, 417–422.

Warnock, D.D., Lehmann, J., Kuyper, T.W., Rillig, M.C., 2007. Mycorrhizal responses to biochar in soil – concepts and mechanisms. Plant and Soil 300, 9–20.

Warnock, D.D., Mummey, D.L., McBride, B., Major, J., Lehmann, J., Rillig, M.C., 2010. Influences of non-herbaceous biochar on arbuscular mycorrhizal fungal abundances in roots and soils: results from growth-chamber and field experiments. Applied Soil Ecology 46, 450–456.

Watts-Williams, S.J., Smith, F.A., McLaughlin, M.J., Patti, A.F., Cavagnaro, T.R., 2015. How important is the mycorrhizal pathway for plant Zn uptake? Plant and Soil 390, 157–166.

West, T.O., McBride, A.C., 2005. The contribution of agricultural lime to carbon dioxide emissions in the United States: dissolution, transport, and net emissions. Agriculture. Ecosystems and Environment 108, 145–154.

White, R.E., Taylor, A.W., 1977. Effect of pH on phosphate adsorption and isotopic exchange in acid soils at low and high additions of soluble phosphate. Journal of Soil Science 28, 48–61.

Wiedner, K., Schneeweiß, J., Dippold, M.A., Glaser, B., 2015. Anthropogenic dark earth in northern Germany — the Nordic analogue to terra preta de Índio in Amazonia. Catena 132, 114–125.

Woodcock, D., 1989. Climate sensitivity of wood-anatomical features in a ring-porous oak (*Quercus macrocarpa*). Canadian Journal of Forest Research 19, 639–644.

Woods, W.I., Deneven, W.M., 2009. Amazonian dark earths: the first century of reports. In: Woods, W.I., Teixeira, W.G., Lehmann, J., Steiner, C., WinklerPrins, A., Rebellato, L. (Eds.), Amazonian Dark Earths: Wim Sombroek's Vision. Springer, Berlin, pp. 1–14.

Yamato, M., Okimorii, Y., Wibowo, I.F., Anshori, S., Ogawa, M., 2006. Effect of the application of charred bark of *Acacia mangium* on the yield of maize, cowpea and peanut, and soil chemical properties in South Sumatra, Indonesia. Soil Science and Plant Nutrition 52, 489–495.

Yao, Y., Gao, B., Inyang, M., Zimmerman, A.R., Cao, X., Pullammanappallil, P., Yang, L., 2011. Biochar derived from anaerobically digested sugar beet tailings: characterization and phosphate removal potential. Bioresource Technology 102, 6273–6278.

Zhang, W., Cai, Y., Tu, C., Ma, L.Q., 2002. Arsenic speciation and distribution in an arsenic hyperaccumulating plant. The Science of the Total Environment 300, 167–177.

Zhu, Q.-H., Peng, X.-H., Huang, T.-Q., Xie, Z.-B., Holden, N.M., 2014. Effect of biochar addition on maize growth and nitrogen use efficiency in acidic red soils. Pedosphere 24, 699–708.

CHAPTER

26

Integrating Mycorrhizas Into Global Scale Models: A Journey Toward Relevance in the Earth's Climate System

E.R. Brzostek[1], *K.T. Rebel*[2], *K.R. Smith*[1], *R.P. Phillips*[3]

[1]West Virginia University, Morgantown, WV, United States; [2]University of Utrecht, Utrecht, Netherlands; [3]Indiana University, Bloomington, IN, United States

26.1 INTRODUCTION

Mycorrhizal fungi are ubiquitous soil microbes that form symbiotic associations with nearly 95% of plants on earth (van der Heijden et al., 2015). In this evolutionarily ancient symbiosis, plants provide carbohydrates to the fungi in exchange for nutrients obtained from soil solution and soil organic matter (SOM) (Chapter 20). Estimates of this exchange indicate that up to 20% of net primary production (NPP) can be used to support mycorrhizal fungi (Hobbie, 2006; Högberg et al., 2008), and that up to 50% of plant demand for nitrogen (N) and phosphorus (P) can be met by mycorrhizal uptake (Hobbie, 2006; Hobbie and Hobbie, 2008; Högberg et al., 2008). In addition to their role as nutritional mutualists, mycorrhizal fungi also affect ecosystem processes indirectly by enhancing soil structure, stabilizing SOM, and conferring pathogen resistance and drought tolerance to plants (Rillig, 2004b; Pozo and Azcon-Aguilar, 2007; Rillig et al., 2015). For these reasons, mycorrhizal fungi are considered key mediators of community dynamics and ecosystem processes in all terrestrial ecosystems from the tropics to the arctic (Johnson et al., 2013).

Despite their importance to ecosystem functioning, mycorrhizal fungi are rarely represented explicitly in ecosystem/biogeochemistry models (Johnson et al., 2006), and are virtually absent from large-scale terrestrial biosphere models (TBMs). This omission stems primarily from two factors: data limitations and the lack of theoretical frameworks for accounting for variability in mycorrhizal form and function. Data limitations, particularly for

Mycorrhizal Mediation of Soil
http://dx.doi.org/10.1016/B978-0-12-804312-7.00026-7

climatically sensitive regions of the earth (e.g., tropical forests, ecotones), result from the fact that unlike leaf processes that can be measured nondestructively, mycorrhizal fungi occur in the opaque and heterogeneous medium of soils. As such, mycorrhizal-driven processes (e.g., nutrient uptake, extracellular enzyme release) are rarely measured in situ, and are often estimated using proxies (e.g., enzyme activities inside ingrowth cores) or by measuring the consequences of their removal via trenching or girdling (Fernandez and Kennedy, 2016). The second factor, the scarcity of conceptual frameworks for modeling mycorrhizal functioning, results in large part from the diversity of these fungi (Lindahl et al., 2007; Peay et al., 2008). Like plants, mycorrhizal fungi exhibit an extraordinary diversity of functional traits that are controlled by both phylogenetic constraints and environmental conditions. Consequently the innumerable interactions that can occur among plants, fungi, and their environment are viewed as a poor fit for first-order models that contain limited representations of biotic interactions and are run at relatively coarse spatial scales (e.g., from half a degree to 1-km grid for most TBMs).

Given that including such variation in TBMs would require "big data" (which does not exist for mycorrhizal fungi) and could lead to model "overparameterization," any justification for including mycorrhizal representations must outweigh the cost of increasing model complexity and potential sources of error. Two critical questions logically follow. If mycorrhizal dynamics are already implicitly parameterized in TBMs (which perform reasonably well), what are some potential benefits of adding explicit representations of these fungi? And why should mycorrhizal functions be prioritized over other processes that are known to be important but lack explicit representations in the current generation of TBMs (e.g., root orders, microbial food webs)?

One of the primary motivations for including mycorrhizal fungi in TBMs is that the omission of these dynamics can lead to incorrect projections of ecosystem responses to global change. A key finding of the free-air carbon dioxide (CO_2) enrichment experiments was that some forests sustained NPP over time by mining N from SOM, whereas others did not (Norby and Zak, 2011). None of the TBMs run at the site level could replicate these patterns, because the mechanisms that putatively drive these dynamics [e.g., increased plant allocation of carbon (C) to soil microbes to accelerate SOM decomposition] are absent from most models (Zaehle et al., 2014; Walker et al., 2015). This failure of the models presents a clear case for prioritizing the integration of more mechanistic model structures, including but not exclusive to mycorrhizal function, in the next generation of models.

A similar case can be made for the inclusion of mycorrhizal dynamics from global model runs. Recent model intercomparison experiments have shown wide disagreement among TBMs in projecting the magnitude of C uptake by terrestrial ecosystems (Arora et al., 2013; Friedlingstein et al., 2014), a discrepancy that relates to how TBMs represent nutrient limitation (Fisher et al., 2014). In one model, adding N constraints to NPP reduced C uptake by approximately 40% in N-limited ecosystems, whereas adding P constraints to NPP reduced C uptake by 20% in a P-limited ecosystem (Wang et al., 2010). These results highlight the consequences of incorrect representations of nutrient limitation in TBMs (Thomas et al., 2015) and underscore the need for modeling the actual processes that control nutrient acquisition (e.g., enzymatic activity of mycorrhizal fungi, plant stimulation of microbial priming).

The high degree of sensitivity of mycorrhizal fungi to global changes such as atmospheric CO_2 and N deposition (Treseder and Allen, 2000) presents a further justification for including explicit representations of mycorrhizal dynamics (Johnson et al., 2013; Mohan et al., 2014). However, explicit representations in TBMs may not always be necessary if mycorrhizal fungi function as "trait integrators" for a suite of biogeochemical processes. Studies have demonstrated that ecosystem responses to global changes may depend in part on the type of mycorrhizal fungi that are dominant in the ecosystem (Phillips et al., 2013). In a synthesis of nearly 100 studies from across the globe, Terrer et al. (2016) reported that only plants that associate with a single type of mycorrhizal fungi [ectomycorrhizal (EcM) fungi] are capable of acquiring the extra N needed to support a sustained growth response to elevated CO_2. In a separate synthesis, Averill et al. (2014) reported that ecosystems dominated by EcM fungi possess larger stocks of soil C than ecosystems dominated by arbuscular mycorrhizal (AM) fungi. This synthesis suggests that ecosystems dominated by EcM plants may be highly susceptible to losses of C owing to rising temperatures, a process that would reduce the magnitude of any C gains realized under elevated CO_2. Thus modeling mycorrhizal functioning indirectly, using mycorrhizal associations to define new plant functional types (PFTs), may represent a practical compromise for improving model behavior while minimizing model uncertainty (Shi et al., 2016).

Here we present a framework for including mycorrhizal fungi in large-scale models that balances some of the complexity of mycorrhizal ecology with the simplicity necessary to model mycorrhizal processes at regional and global scales. Although we are well aware of other priorities for TBM development (e.g., the inclusion of roots and microbial groups with dynamic functionality), we maintain that the inclusion of mycorrhizal dynamics into TBMs has the potential to link plant and microbial processes, in essence coupling C and nutrient cycles. First we describe the strengths of ecosystem/biogeochemistry models that include explicit representations of mycorrhizal dynamics. Second we highlight key ecosystem characteristics controlled by mycorrhizal fungi and make the case for which mycorrhizal functions and traits are critical to model at the global scale. Third we discuss how even coarse representations of mycorrhizal dynamics (characterizing ecosystems by their dominant mycorrhizal association) may improve current TBMs in their ability to predict ecosystem sensitivity to global changes. To conclude, we discuss key challenges moving forward, including data needs for model validation and benchmarking strategies that balance the logistics of empirical research with the data necessary to drive model development.

26.2 EXISTING MODEL FRAMEWORKS

There is a rich history of modeling mycorrhizal dynamics in evolutionary models, as well as in population and plant community models (Fitter, 1991; Johnson, 1997; Landis and Fraser, 2008; Bever, 2015; Johnson et al., 2006). Although many of these models take into account the role of nutrient availability in mediating plant–mycorrhizal interactions, most do so at the level of an individual plant and fungal partner rather than at the ecosystem scale (Schnepf and Roose, 2006; Schnepf et al., 2008; Gashaw Deressa and Schenk, 2008). Most of these models are also parameter-rich, which precludes their adoption and implementation into large-scale models.

The inclusion of mycorrhizal dynamics in ecosystem/biogeochemistry models is a relatively recent occurrence. These models fall into one of two categories: (1) parameter-based functional (i.e., process) models, and (2) optimization models. Process models include functional attributes of mycorrhizal fungi such as their uptake of N and P. Although these models are structured to capture complex dynamics occurring below ground (e.g., competition for nutrients between mycorrhizal and free-living fungi), a critical limitation is that they are highly specific to the ecosystem in which they were developed. Optimization models, in contrast, tend to be more generalizable across ecosystems because these models use economic principles to govern how processes such as the C cost of nutrient acquisition are determined in the model. Thus there is a trade-off between the two model types that relates to their specificity (process models) and generality (optimization models).

Of the process models that include mycorrhizal dynamics, there are notable differences in how the models represent C allocation and nutrient uptake (Table 26.1). In MYCOFON, the amount of C allocated to mycorrhizas is generally a fixed percentage of the amount of C allocated to foliage, and nutrients can only be acquired in inorganic forms (Meyer et al., 2010). In contrast to MYCOFON, C allocation in ANAFORE depends on the nutrient status of the plant (Deckmyn et al., 2011; Orwin et al., 2011) and ANAFORE is the only model that explicitly separates C allocated to mycorrhizas versus C released as root exudates (based on plant nutrient demand). ANAFORE differs from MYCOFON in that it permits organic N uptake, building on emerging evidence and theory that organic nutrient uptake occurs in most ecosystems (Näsholm et al., 2009) and is more energetically favorable than inorganic nutrient uptake (Schimel and Bennett, 2004). The third process model, MySCaN, dynamically allocates C to mycorrhizal fungi based on the nutritional status of the fungi, thus enabling a feedback between plant and mycorrhizal stoichiometry.

Unlike ANAFORE, MySCaN does not distinguish between C flux to mycorrhizal fungi versus C flux to root exudates (like MYCOFON), although it does permit uptake of both organic and inorganic nutrients (including organic P uptake). Notably, none of the three models allow C allocated to mycorrhizal hyphae to be released to soil as soluble exudates, a process that has been shown to fuel microbial activity, nutrient transformations, and nutrient release from minerals (Fransson and Johansson, 2010; Kaiser et al., 2015; Chapters 3 and 22).

Two optimization models that include explicit representations of mycorrhizal fungi are the model of Franklin et al. (2014) and the Fixation and Uptake of Nitrogen (FUN) 2.0 model (Fisher et al., 2010; Brzostek et al., 2014). Although both models use similar economic principles to control C allocation to mycorrhizas, there are notable differences in how the models are structured. The model of Franklin et al. (2014) uses a dynamic marketplace between multiple fungal symbionts and multiple plant partners, where the aggregated terms of trade (across all fungal and plant partners) are dictated by each partner maximizing its fitness. In contrast, FUN 2.0 models C allocation by first calculating the cost of mycorrhizal root N uptake as a function of inorganic N concentrations in soil and root biomass (i.e., based on availability and access). C allocation to mycorrhizal roots is then optimized to consider the cost of other plant N acquisition strategies such as retranslocation and N fixation. Whereas both models differentiate between root and mycorrhizal uptake, the mechanism by which roots and fungi compete for N is different. In the Franklin et al. (2014) model, mycorrhizas have a high affinity for N uptake that enables them to outcompete roots at low N concentrations. In contrast, FUN 2.0 includes nonmycorrhizal root uptake as a strategy that competes with mycorrhizal uptake and is advantageous in high N environments.

TABLE 26.1 Key Characteristics and Processes of Existing Model Frameworks

Model	MySCaN	ANAFORE	MYCOFON	Franklin	FUN 2.0
Reference	Orwin et al. (2011)	Deckmyn et al. (2011)	Meyer et al. (2010)	Franklin et al. (2014)	Brzostek et al. (2014)
Modeling approach	Functional	Functional	Functional	Optimality, market	Optimality
MF	No explicit difference	EcM	EcM	EcM	EcM & AM
Nutrient cycling	C, N, P	C, N (and detailed water)	C, N	C, N	C, N
Based on	CENTURY	CENTURY	DNDC	Franklin et al. (2012)	Fisher et al. (2010)
Temporal resolution	Daily	Daily	Canopy hourly, rest daily	Daily	Adaptable
Organic matter pools	Soil litter (metabolic and structural), SOM (slow and resistant)	Litter (surface, soil) is divided into woody and nonwoody and then separated into accessible, cellulose, and recalcitrant. Soil litter is divided into fine roots, coarse roots, and microbes	Litter (resistant, labile, very labile)	Not explicit	FUN is a subroutine designed to run coupled within larger TBMs. Relies on TBM decomposition output
Soil biota	MF, saprotrophic fungi, bacteria, below-ground grazers, predators	MF, saprotrophic fungi, bacteria, below-ground grazers, predators	MF, roots	MF	MF, roots
Soil biota functions	Growth, turnover, nutrient uptake	Fungi and bacteria can actively degrade organic compounds for energy and nutrients	Fungal turnover equals a fraction of fungal C and N, fungal respiration	N uptake, C transfers to mycorrhizas.	N uptake, C transfers to mycorrhizas and other rhizosphere microbes.
Plant growth	Regulated by most limiting nutrient (N, P), GPP divided over leaves and roots, senescence is at constant rate; decrease in leaf C/N ratio leads to higher growth		Detailed 40-layer canopy, GPP dependent on stomatal conductance, temperature, CO_2 in atmosphere, foliage N concentration and photosynthetically active radiation	GPP (function of N content, light, CO_2) minus C cost [resp (function of N content) + litter production] − C cost MF (function of N demand, fungal N return)	Calculates a C cost of N uptake to downregulate growth when N is limiting

(Continued)

TABLE 26.1 Key Characteristics and Processes of Existing Model Frameworks—cont'd

Model	MySCaN	ANAFORE	MYCOFON	Franklin	FUN 2.0
Root exudate	Not explicit	Increases when plant is nutrient deficient, as a % of total C allocated to roots	Not explicit	Not explicit	Not explicit
Plant C allocation to MF	Allocation is a function of the nutrient limitation of the plant and mycorrhizas and their stoichiometric constraints	Increases when plant is nutrient deficient as a % of total C allocated to roots	C allocation to MF is a fraction of C allocation to foliage; fraction is species specific	C allocation to MF is a tradeoff between C investment in N uptake (via EcM) and C-investment in growth	Yes, but not explicitly separated from C allocation to other rhizosphere microbes
MF C uptake soil	No	Yes, but less efficient that using plant C	No, C from plant is only C-source.	No	No
MF N allocation to plant	Maximum 40% per year; first MF gets C from plant, resulting in a nutrient imbalance. MF take up N from litter, SOM, and inorganic pools. Transfer dependent on plant and MF stoichiometry	Constant fraction of MF N uptake	There is N demand that is either met by roots or by MF (roots covered by MF cannot take up N)	N demand by growth MF leads to N uptake (following fixed N/C ratio); when N left, the plant can take it up	Mycorrhizal root uptake is one of five pathways by which the plant can acquire N
Exchange from plant perspective	High C/N ratio > nutrient limited > most C transfer and N transfer from MF high	Constant fraction	Fraction of C allocation to foliage, depending on phenology of plant. This is then adjusted for nutrient supply: soil N up, plant C supply to MF down	C allocation to MF = tradeoff between C investment in N uptake (via EcM) and C investment in growth	Plant optimally allocates C to a suite of N uptake pathways including MF, depending upon their relative cost
Exchange from MF perspective	When MF have a high C/N ratio, the MF are nutrient limited and less nutrients go to plant (immobilization), resulting in less C to MF	Constant fraction	Dependent on N demand of plant (difference between optimal and actual N content of all plant compartments). When plant N demand cannot be met by roots, it will be met by MF	Fungi optimize C allocated to reproduction vs. that allocated to N uptake	Varies as a function of both accessibility and availability of soil N

TABLE 26.1 Key Characteristics and Processes of Existing Model Frameworks—cont'd

Model	MySCaN	ANAFORE	MYCOFON	Franklin	FUN 2.0
Inorganic N uptake	First by MF, what is left over can be taken up by plant	Plants take up ammonium and nitrate; all biota take up mineral N in equal distribution before degrading organic N	Uptake inhibited when roots covered by hyphal mantle; N uptake determined by N demand of plant. Roots take up first, inorganic N left over can be taken up by MF	Yes	Yes
Organic N uptake	Only by MF	Only by MF	No	Not explicit	Not explicit
Inorganic P uptake	Available to plant dependent on root density & MF by density	No	No	No	No
Organic P uptake	No	No	No	No	No
Other key processes		Degradation of SOM leads to: (1) growth of organism, (2) more stable SOM, (3) respiration. N-rich material is degraded first		Plants and EcM fungi have multiple potential trading partners and both optimize fitness	Optimization of plant growth

AM, arbuscular mycorrhizal; *C*, Carbon; *CO_2*, carbon dioxide; *DNDC*, DeNitrification-DeComposition; *EcM*, ectomycorrhizal; *FUN*, Fixation and Uptake of Nitrogen; *GPP*, gross primary production; *MF*, mycorrhizal fungi; *N*, nitrogen; *P*, phosphorus; *SOM*, soil organic matter; *TBM*, terrestrial biosphere model.

A critical difference between all the models (process and optimization) is how they deal with different mycorrhizal strategies. Only FUN 2.0 explicitly partitions mycorrhizal types into AM and EcM fungi. This distinction is relevant, given differences in the costs and benefits of associating with AM versus EcM fungi. AM fungi benefit plants by increasing their access to soil nutrients, extending hyphae beyond nutrient depletion zones around roots. EcM fungi also acquire nutrients from beyond depletion zones, but additionally benefit plants by increasing the size of the available nutrient pool by degrading SOM (Talbot et al., 2008; Lindahl and Tunlid, 2015). These functions also impart different costs. AM fungi typically produce hyphae that are less costly to construct and maintain than EcM hyphae/mycelium, but that are also shorter-lived (Read and Perez-Moreno, 2003). Given differences between AM and EcM fungi, models that can accommodate both mycorrhizal types or predict which mycorrhizal associations should dominate based on known physiological tradeoffs and environmental constraints should lead to improved representations of mycorrhizal dynamics.

An additional feature of FUN 2.0 is that it can be run as a stand-alone model or as subroutine in a TBM. FUN 2.0 requires only nine inputs (e.g., NPP, root biomass, soil N), all common in TBMs, and has only 11 parameters that control the shape of the C cost functions for each nutrient uptake strategy (e.g., mycorrhizal roots, retranslocation, and N fixation). Consequently FUN is the only mycorrhizal model that has been run at the global scale (Shi et al., 2016).

The development and validation of FUN 2.0 may provide a roadmap for integrating mycorrhizal dynamics into TBMs (Fig. 26.1). First, model inputs and outputs (i.e., retranslocation, C allocation to mycorrhizal roots) were measured in forests that varied in their abundance of EcM and AM trees (Fig. 26.1A). Model estimates of the C cost to support AM versus EcM fungi were validated with measurements of hyphal production (ingrowth bags) and root exudation from published studies (Fig. 26.1B) (Wallander et al., 2013; Yin et al., 2014). After assuming

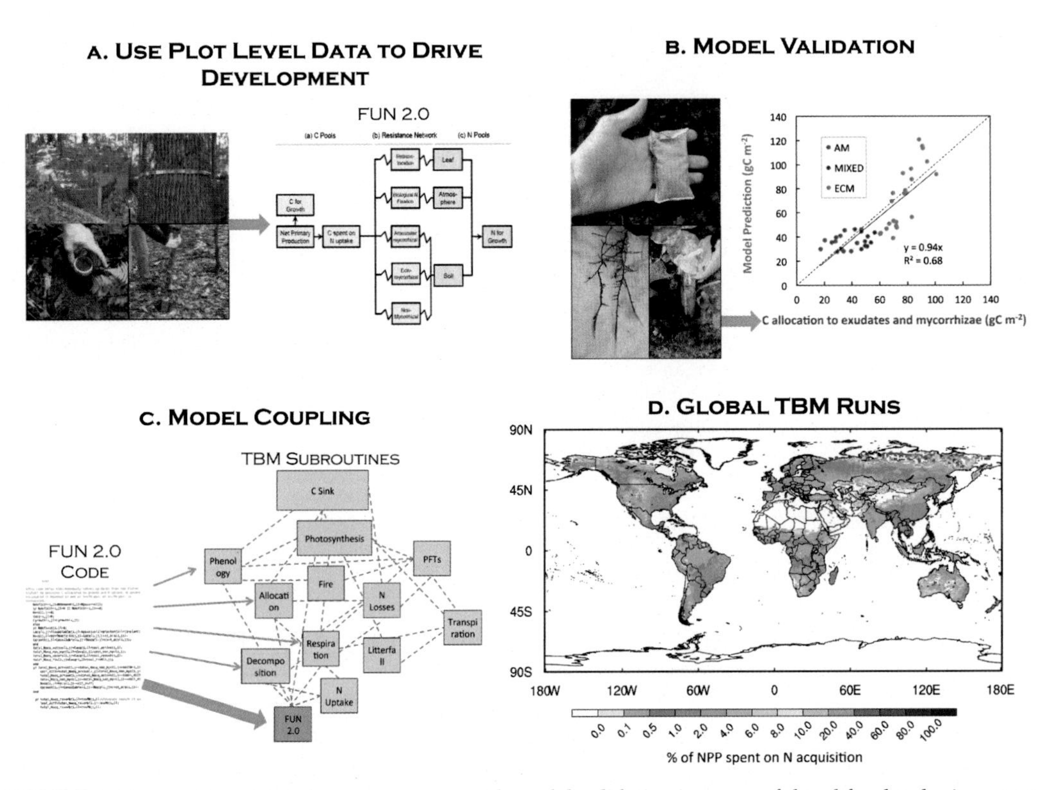

FIGURE 26.1 Linking plot scale measurements with model validation is a powerful tool for developing mycorrhizal subroutines for terrestrial biosphere models (TBMs). Mycorrhizal uptake was integrated into Fixation and Uptake of Nitrogen (FUN) 2.0 by targeted measurements of (A) model inputs and (B) outputs for validation across plots that varied in the abundance of ectomycorrhizal (EcM) and arbuscular mycorrhizal (AM) trees. (C) Fully coupling FUN 2.0 into a TBM required the addition of a novel subroutine as well as the modification of subroutines across the TBM code. (D) The end result is an estimate of the carbon (C) cost of nitrogen (N) acquisition that includes the impact of EcM and AM fungal traits. *NPP*, Net primary production; *PFT*, plant functional type. *(A) Adapted from Brzostek, E.R., Fisher, J.B., Phillips, R.P., 2014. Modeling the carbon cost of plant nitrogen acquisition: mycorrhizal trade-offs and multipath resistance uptake improve predictions of retranslocation. Journal of Geophysical Research-Biogeosciences 119 (8), 1684–1697. (D) Adapted from Shi, M., Fisher, J.B., Brzostek, E.R., Phillips, R.P., 2016. Carbon cost of plant nitrogen acquisition: global carbon cycle impact from an improved plant nitrogen cycle in the Community Land Model. Global Change Biology 22 (3), 1299–1314.*

that most biomes are either AM- or EcM-dominated (except temperate forests, which were modeled as 50% AM and 50% EcM), FUN 2.0 was coupled into version 4.0 of the Community Land Model (CLM) (Fig. 26.1C). This enabled FUN 2.0 to produce a global scale estimate of the C cost of N acquisition that includes both EcM and AM fungi (Fig. 26.1D). Coupling FUN into CLM 4.0 required not only the addition of the FUN subroutine but also the modification of other subroutines including decomposition, N uptake, and phenology.

The diversity of model representations (Table 26.1) provides a foundation from which to build a globally applicable mycorrhizal subroutine that can be integrated into TBMs. However, in the process of scaling these models from one biome to the globe, models need to include: (1) both EcM and AM fungi, (2) separate plant C allocation to exudates versus mycorrhizas, (3) the role of mycorrhizal fungi in P uptake, and (4) differences in mycorrhizal physiology between AM and EcM fungi. Further, none of the models include differences between EcM and AM fungi in C use efficiency (CUE). Given that modeling physiological differences between mycorrhizal types has the potential to substantially alter our ability to predict the future size of the land C sink (Wieder et al., 2013), mycorrhizal models need to include differences between mycorrhizas in their CUE and how they allocate plant C to various pathways to take up nutrients.

26.3 CRITICAL MYCORRHIZAL FUNCTIONS FOR TERRESTRIAL BIOSPHERE MODELS

TBMs represent the scientific community's best hypotheses about how energy, water, and CO_2 are exchanged between the land and atmosphere. As such, model structures in TBMs need to reflect our best understanding of which factors drive ecosystem processes such as NPP and decomposition. Cleveland et al. (2013) reported that internally recycled nutrients, and not externally-derived nutrients, are the primary constraint on NPP globally. If true, this suggests that nutrient limitation cannot be modeled merely by balancing nutrient inputs with outputs (Thomas et al., 2015); rather, nutrient limitation should be modeled as a function of plant and microbial stoichiometry, growth efficiencies, and turnover times. Despite this, most model structures include limited representations of both soil microbes (which control nutrient availability by decomposing plant detritus) and roots (which control primary productivity by acquiring nutrients and water from soil). Moreover, roots and microbes are often represented simply as C and N pools that have fixed stoichiometry and turnover rates but little interaction (Treseder et al., 2012). These below-ground parameterizations are in stark contrast to the detailed physiological representation of photosynthesis and respiration of leaves in these same models.

Although there have been several efforts to include more root and microbial functioning in models, the new model structures have taken parallel tracks that neglect the critical importance of root–microbe interactions in coupling C nutrient cycles. Recent "microbial models" include an active microbial biomass pool with enzyme activities and uptake rates based on Michaelis–Menten kinetics, and a turnover rate that controls SOM stabilization (Wieder et al., 2013; Sulman et al., 2014). However, only one of these models can simulate root exudation and its impacts on decomposition via priming effects (Sulman et al., 2014). Moreover, none of the current generation of TBMs have model structures to account for biome-scale differences in root traits or root order (Smithwick et al., 2014; Warren et al., 2015) despite the importance of these factors in mediating the magnitude and direction of priming (Cheng et al., 2014) and nutrient acquisition (McCormack et al., 2015).

Mycorrhizal fungi possess functions that incorporate processes that are traditionally separated into the domains of either roots or free-living soil microbes (Fig. 26.2). As such, modeling mycorrhizas as specialized roots or as a specialized microbial guild fails to capture the dynamic interactions that can occur. For example, fungal hyphae can forage into nutrient-rich patches (like roots) but can also produce enzymes (like microbes) and take up nutrients in organic forms (Read and Perez-Moreno, 2003). Further, both of these functions entail C investments by the mycorrhizas (i.e., biomass vs. enzyme production) that should be optimized to maximize nutrient uptake (Johnson et al., 2013). The trading of these nutrients for C between fungi and plants adds another layer of complexity (Walder and van der Heijden, 2015).

An essential feature of a mycorrhizal model is one that allocates C below-ground dynamically in response to shifts in the availability of limiting resources. Because C allocated for

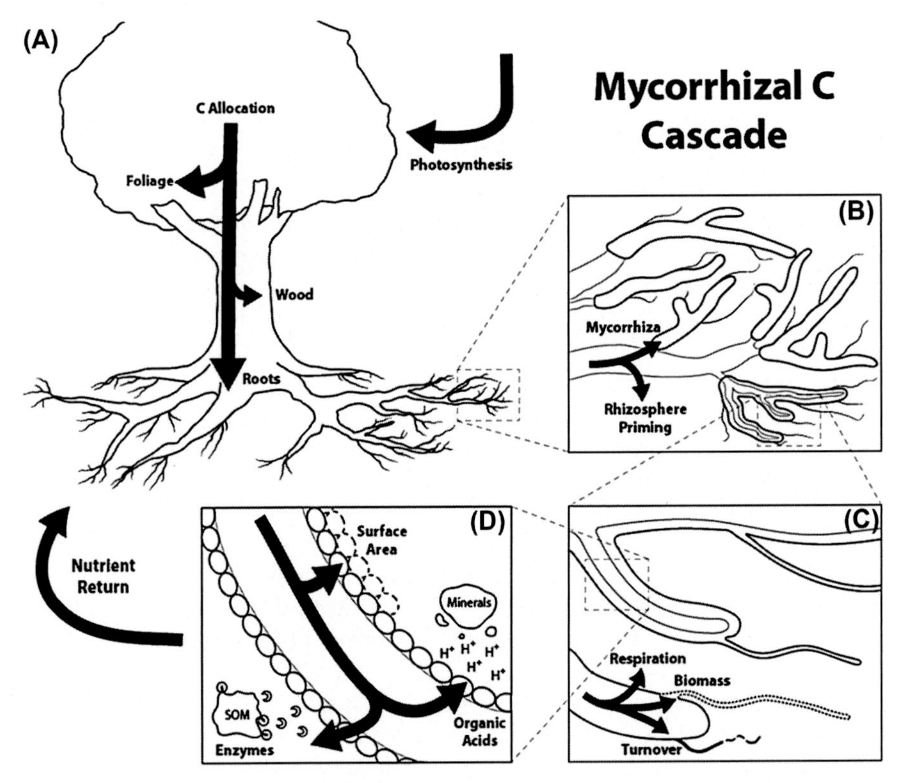

FIGURE 26.2 Models need to dynamically predict carbon (C) allocation to mycorrhizal fungi, the fate of this C, and the N and P return to the plant. As nutrient demand shifts relative to supply, (A) plants optimally allocate C below ground to roots as well as to (B) mycorrhizas and free-living microbes to enhance nutrient access. C that is directly allocated to mycorrhiza incurs a respiratory cost before it can allocated to (C) maintaining or increasing biomass or to (D) enhancing hyphal uptake through enzymes, foraging, or organic acids. *H*, Hydrogen; *SOM*, Soil organic matter.

nutrient acquisition can be partitioned between root growth, root exudation, and mycorrhizal growth, we refer to this as the *C cascade* (Fig. 26.2). Following first principles, C allocation to mycorrhizal fungi is a function of plant nutrient demand and the size of the nutrient pool (Correa et al., 2011, 2012; Johnson et al., 2013). For example, if elevated CO_2 enhances plant growth, nutrient demand increases relative to the static nutrient pool, leading to increased C allocation to mycorrhizas (Drake et al., 2011). In contrast, N or P fertilization, which increases the nutrient pool, would decrease C allocation to mycorrhizal fungi (Lilleskov et al., 2001, 2002). However, owing to nutrient colimitation and the stoichiometric flexibility of plants and mycorrhizas, the magnitude and direction of C allocation to mycorrhizal fungi can be context dependent. As a result, scalars that predict C allocation solely as a linear function of nutrient demand and the available nutrient pool maybe applicable only under a limited set of conditions.

Predicting dynamic C allocation to mycorrhizas remains the central challenge for integrating mycorrhizal fungi into global-scale models, because many TBMs still allocate C within the plant based on fixed ratios of wood/leaves/roots. However, there are new empirical approaches for generating data that the models can use, as well as new modeling approaches that can overcome data limitations. On the empirical side, ingrowth cores and isotopic tools hold promise for populating C allocation response surfaces to limiting nutrients (Buscot et al., 2000; Phillips et al., 2012; Tedersoo et al., 2014). On the modeling side, data derived from new empirical approaches can feed directly into frameworks that optimize plant C allocation to maximize growth (Rastetter and Shaver, 1992; Brzostek et al., 2014; Dybzinski et al., 2015). The strength of optimization approaches is that they account for the dynamic interplay between strategies to compete for light, water, and nutrients. An added benefit is that integrating optimal C allocation into the models may address gaps outside of mycorrhizal fungi, including root responses to shifts in water and nutrient availability. However, there is also an added layer of complexity in these approaches. Plants have other strategies besides mycorrhizal-mediated uptake to increase the supply of nutrients to support growth (i.e., root biomass investment, root exudation of organic acids and/or enzymes, rhizosphere priming, N fixation, and N retranslocation). As a result, optimization models need to predict the optimal C allocation across these diverse strategies to maximize plant nutrition.

Once plant C is invested in mycorrhizas, models must predict how the fungi allocate C for maintenance respiration, growth (increasing biomass and surface area), reproduction, and nutrient acquisition [i.e., enzyme or organic acid production] (Fig. 26.2C and D). The first step in the C cascade of mycorrhizal allocation is controlled by CUE, which determines the respiratory costs of molecular biosynthesis. Whereas there is evidence that CUE differs between fungi and bacteria (Sinsabaugh et al., 2016), evidence that it differs between mycorrhizal types is sparse. Given differences in life history traits (e.g., longevity, biomass, pathogen resistance), CUE likely differs between mycorrhizal types. Whether it differs enough to impact ecosystem processes remains an open question, and to be conservative, models could use a common value across mycorrhizal types.

After respiratory costs are accounted for (and assuming that these costs do not differ appreciably between different pathways), the next assumption is that mycorrhizas allocate C to maximize uptake and transfer of the most limiting plant nutrient (Johnson, 2010). In this way, mycorrhizas maximize their C subsidy from the plant and thus their fitness. Although this approach does not consider the C costs of supporting mycorrhizal fungi that immobilize nutrients or act as parasites (Johnson et al., 1997; Klironomos, 2003), the lack of information about the commonness of these interactions preclude their inclusion in models. Moreover, both data and theory suggest that plants can down regulate C transfers and sanction nonbeneficial symbionts (Kiers and Denison, 2008; Bever, 2015). Thus the assumption that mycorrhizas maintain a diverse portfolio to maximize nutrient uptake is viable for models.

Similar to plant C allocation, optimization approaches may be critical for predicting dynamic C allocation to mycorrhizas. From the perspective of the fungi, the most beneficial nutrient uptake strategy should match the dominant nutrient form available in the ecosystem. Following this assumption, EcM fungi would prioritize enzyme production to mobilize N or P in systems dominated by organic nutrients (Brzostek and Finzi, 2011; Brzostek et al., 2013). In systems where competition for dissolved inorganic nutrients is a function of surface area and foraging into nutrient-rich patches, fungi would prioritize maintaining and expanding biomass production (Rillig, 2004a). Moreover, the ability to forage appears to differ between mycorrhizal types, because evidence suggests that EcM fungi but not AM fungi are able to produce more hyphae in response to nutrient hotspots (Chen et al., 2016). Finally, in systems where P release from primary minerals limits NPP, organic acid production would dominate (van Scholl et al., 2008). To minimize model complexity, the nutrient return on mycorrhizal allocation would depend on other existing model subroutines that control nutrient mobilization by enzymes or organic acids and nutrient competition between plants and microbes (Wieder et al., 2013; Sulman et al., 2014; Yang et al., 2014).

The terms of trade between mycorrhizas and plants are the last link in the symbiosis. Once mycorrhizas acquire nutrients, they must first maintain their own stoichiometric balance before transferring N or P to the plant. Despite the measurement ease on the analytical side, C/N/P ratios (and their variation) for mycorrhizas are rare, reflecting their cryptic nature. Another key area of uncertainty is the timescale at which mycorrhizal nutrient returns impact future plant C investment. Plants likely maintain a constituent level of C allocation to mycorrhizas that they up-regulate during periods of nutrient stress. This would be difficult to represent in models because they have a limited memory across time that makes it hard to represent a C cost that leads to a lagged benefit (Anderegg et al., 2015).

Mycorrhizas have impacts on soil processes after death, because their turnover is one of the largest fluxes of C, N, and P to soils in northern latitudes (Fig. 26.2C) (Clemmensen et al., 2013). Evidence that EcM fungi have greater longevity than AM fungi (Zhu and Miller, 2003; Ekblad et al., 2013) provides support for unique parameters governing turnover for each mycorrhizal type. Initially these parameters could be fixed but as models become more sophisticated, turnover rates could respond dynamically to temperature and moisture constraints. After turnover occurs, mycorrhizal necromass could feed directly into soil enzyme models. These models include representations of enzymes that can remobilize N and P in mycorrhizal tissues. Importantly, these models can also determine whether mycorrhizal necromass contributes to stable C storage by examining which fraction is respired and which fraction enters protected C pools. This level of complexity is feasible because these models are beginning to include protection processes (Sulman et al., 2014) that with minor modifications could capture the ability of fungi to aid soil aggregation (Rillig and Mummey, 2006).

One common thread across all of these critical functions is that there remain significant gaps in our empirical understanding of how they vary across spatial and temporal variation in environmental drivers. As such, these functions are not only highlighted here for the modeling community but also represent critical research needs for the empirical community. However, these data gaps do not represent an insurmountable roadblock for the integration of mycorrhizal fungi into TBMs. Numerous processes in existing TBMs were initially developed or are still based on limited datasets, including soil respiration responses to temperature (Lloyd and Taylor, 1994), drought scalars to model the response of photosynthesis to water stress (Oleson et al., 2013), and plant C allocation to shoots versus roots (Litton et al., 2007). As TBMs matured, the uncertainty in these processes has either declined as new empirical data is used to parameterize the models, or these processes have been brought to the forefront for continued refinement.

26.4 MYCORRHIZAL FUNGI AS TRAIT INTEGRATORS

Most TBMs use PFTs, which are lists of parameter values that separate plant physiology and growth forms for different ecosystem types (e.g., broadleaf deciduous trees, needleleaf evergreen trees, shrubs, grasses). Whereas these PFTs represent some diversity in plant traits, nearly all of the traits are above ground as a result of the measurement gap between above- and below-ground processes. For example, in version 4.0 of the CLM, one of the most widely used TBMs, there are over 30 parameters that govern the structure and function of canopy leaves compared with only two parameters that dictate root structure and none that control root function (Oleson et al., 2013). These parameters are also static despite the ability of plants and microbes to plastically alter traits as environmental conditions shift. Finally, within a biome, diversity in plant traits is dictated by leaf habit or physical constraints, neglecting the growing consensus that mycorrhizal fungi affect and respond to ecosystem characteristics. In defense of the models, only recently has the empirical community accelerated efforts to synthesize root data into large-scale databases that modelers can use to improve below ground representations (e.g., Fine-Root Ecology Database, roots.ornl.gov).

There is global-scale evidence that the functional abilities of EcM and AM fungi are related to characteristics of the ecosystems they inhabit. AM plants generally occur in regions where climate does not constrain nutrient release from decomposing plant detritus, in dry regions, and in highly weathered soils where P often limits NPP. On the other hand, most EcM plants occur in middle and high latitude (or elevation) ecosystems where cold, wet climate slows decomposition and creates the "need" for a fungal symbiont that can mine nutrients from SOM (Fig. 26.3) (Read and Perez-Moreno, 2003). Thus there appears to be a link between the traits and nutrient acquisition strategies of mycorrhizas and the ecosystems they inhabit. On one hand, this would indicate that explicit mycorrhizal functioning may do little to improve the performance of large-scale models. However, it is important to note that considering mycorrhizal fungi as trait integrators is a hypothesis (Phillips et al., 2013). Including explicit mycorrhizal dynamics in large-scale models would enable us to test this hypothesis at large spatial scales using the models (e.g., by turning plant allocation to mycorrhizal fungi "on" and "off"), something that is impractical to do with field experiments.

The mycorrhizal-driven nutrient acquisition strategy is correlated with key ecosystem characteristics that govern the pace and strength of plant–microbial interactions. Moving along a continuum from AM to EcM fungi, there are corresponding trends of decreasing litter quality (with increasing soil and litter C/N ratios), increasing limitation by N over P,

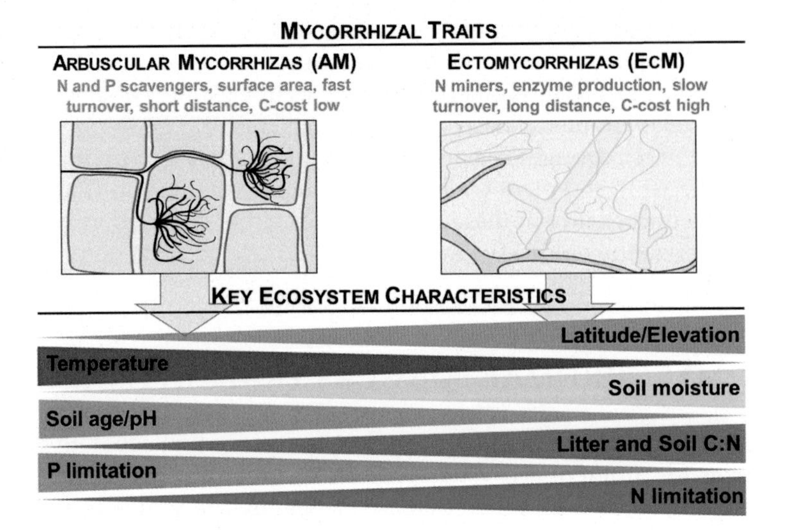

FIGURE 26.3 Mycorrhizal traits and ecosystem characteristics are intimately linked. Across gradients in ecosystem trait space, arbuscular mycorrhizal and ectomycorrhizal fungi possess traits that allow them to provide the greatest nutrient benefit to the plant for the lowest relative carbon cost. *C*, Carbon; *N*, nitrogen; *P*, phosphorus.

and increasing reliance on mycorrhizal symbionts to mobilize organic nutrients and not only scavenge inorganic nutrients from soils (Fig. 26.3). Although these trends were long viewed to hold across biomes and latitudinal gradients (Read and Perez-Moreno, 2003), there is emerging evidence that these dynamics operate within ecosystems [e.g., temperate forests (Phillips et al., 2013), tropical forests (Waring et al., 2016, Corrales et al., 2016)]. This within-PFT variation has important implication for the modsels because the relative abundance of mycorrhizal types within a plot, landscape, or region may lead to predictable and scalable effects on soil C and nutrient cycling. For example, nitrification, litter decomposition, and soil C/N ratio have all been shown to track the abundance of EcM versus AM trees at the plot scale (Phillips et al., 2013; Midgley et al., 2015).

Within a given ecosystem, the most advantageous mycorrhizal symbionts possess traits that allow them to provide nutrients to the plant at the lowest C cost (Fig. 26.3). For example, AM fungi are efficient scavengers for inorganic nutrients in ecosystems with rapid rates of decomposition (Read and Perez-Moreno, 2003; Chapman et al., 2006). Further, AM fungi possess traits that enhance their dominance in P-limited systems, including high absorptive surface area, high affinity P-transporters, and the ability to produce organic acids that can solubilize P (Bolan, 1991; Marschner and Dell, 1994; Koide and Kabir, 2000). As soil and litter become more complex, decomposition rates decline and the need for a fungal symbiont that can mine N and P from complex SOM increases (Fig. 26.3). EcM fungi possess broad enzymatic capabilities that allow them to mobilize organic forms of N and P from chemically protected SOM (Chalot and Brun, 1998). Thus there is a tight linkage between mycorrhizal traits and ecosystem characteristics that models can exploit (Fig. 26.3). While this tight linkage may not be universal (e.g., dipterocarp and podocarp-dominated ecosystems), this framework may still be viable for global models, given their scale of inference and emphasis on C and nutrient cycling.

By using mycorrhizal association as a trait integrator with the hypothesis that AM plants are more similar to one another than EcM plants in terms of multiple traits, models can take a significant first step in refining below-ground processes. Above ground the use of trait distributions for parameters such as maximum rate of carboxylation, leaf longevity, and leaf density have improved leaf-function representations over existing formulations that rely on a single constant for an entire PFT (Scheiter et al., 2013; Verheijen et al., 2013). Empirical evidence from temperate and tropical forests (Phillips et al., 2013; Waring et al., 2016; Corrales et al., 2016) suggests that this same approach could be adapted to use mycorrhizal association to govern traits integral to below-ground processes including leaf and root litter C/N/P ratios, chemical protection of SOM, and soil pH. One obstacle to this approach is the need for improved maps of mycorrhizal association at the global scale. However, most biomes are predominantly monomycorrhizal (e.g., grasslands, boreal forests), and in mixed-mycorrhizal systems like temperate forests, new remote sensing approaches coupled with national-level inventory data have the potential to produce a high resolution map that can be ingested by the models (Fisher et al., 2016). Furthermore, Soudzilovskaia et al. (2015a) have published a global map of mycorrhizal root colonization intensity by AM and EcM fungi, using their relationships with climate variables and soil chemistry.

26.5 CHALLENGES MOVING FORWARD

In order to model the critical mycorrhizal functions outlined in Fig. 26.2 at global scales, there needs to be data that models can use to parameterize and validate new mycorrhizal function. Empirically, measuring the ecosystem impacts of mycorrhizas has historically been difficult because of the cryptic nature of mycorrhizas, the lack of distinct AM or EcM biomarkers, and the difficulty in culturing mycorrhizas. Data are crucially lacking with regard to how much C plants transfer to mycorrhizas, how much N and/or P they get in return, and what mycorrhizas do with the C they get from plants (Chapter 23). On the plant side, there are measurements and estimates for how much C the plants allocate to mycorrhizas and how much nutrients they get in return but these are limited in their geographical and temporal resolution coverage (Högberg and Högberg, 2002; Hobbie, 2006; Hobbie and Hobbie, 2008). For mycorrhizal fungi, the data are even more sparse and most methods are indirect in nature; however, new molecular tools may help estimate the fate of plant C once mycorrhizas receive it (Trivedi et al., 2013). Furthermore, most empirical research focuses solely on one mycorrhizal type or the other, leading to model structures that are monomycorrhizal in nature as well. Whereas a monomycorrhizal focus works at the ecosystem scale, it hinders our ability to integrate mycorrhizas into global-scale models because of the dominance of both AM and EcM fungi both within and across biomes (Phillips et al., 2013; Soudzilovskaia et al., 2015a).

One promising avenue for parameterizing and validating new mycorrhizal functions at the global scale involves linking model development with measurement design. Models benefit the most from research networks that use standardized methodologies across time and space to generate data (e.g., Fluxnet, Nutnet, Droughtnet). For example, applying a new method to quantitatively estimate the contribution of mycorrhizas to both plant and soil C stocks across biomes and ecosystems has the potential to generate data that can be used to validate new model structures (Soudzilovskaia et al., 2015b). Beyond observations, new ecosystem-scale, global-change experiments need to prioritize the generation of data to capture shifts in plant–mycorrhizal function, spanning gradients in nutrients, temperature, and CO_2.

The distribution of EcM versus AM fungi and their partners is shifting and global change will likely intensify these shifts in the future. These shifts occur as biomes move through processes like woody encroachment or forest plantation establishment in grasslands (Richardson et al., 1994; Jackson et al., 2002). Although dynamic global vegetation models may be able to capture these binary biome shifts, it is unlikely they will be able to capture within biomes shifts in mycorrhizal association (e.g., decline in EcM–*Quercus* abundance in eastern U.S. temperate forests) (Nowacki and Abrams, 2008). Thus there is a need to build on remote sensing techniques to map mycorrhizas to determine how fast these shifts are occurring (Fisher et al., 2016), but also there is a need to develop algorithms in dynamic global vegetation models to dynamically predict mycorrhizal status (Medvigy et al., 2009).

The most important challenge to integrating mycorrhizas into TBMs is to balance the complexity of mycorrhizal function with the simplicity necessary for global model representations. Not all of the functional aspects of mycorrhizas need to be included in TBMs. In Fig. 26.2, we have highlighted the critical mycorrhizal functions that likely will impact TBM predictions of the land C sink. However, this list needs to be continually refined and updated as new empirical data emerges or model development occurs. Furthermore, most model representations of mycorrhizal fungi are too complex or specialized to be applicable at the global scale. Thus there is a need to distill the essential processes of existing model frameworks so that they can be integrated into simplified subroutines that can be inserted into TBMs.

26.6 CONCLUSION

Mycorrhizal fungi dynamically link plants with soil microbes, and their function directly impacts C and nutrient cycling at global scales. As such, the lack of mycorrhizal fungi in the current generation of TBMs is a critical omission that hinders our ability to predict the size of the future land C sink. As a first pass, existing models could leverage the growing support for distinct trait differences between EcM and AM plants to develop new mycorrhizal-based PFTs. In order to fully integrate mycorrhizas, however, these models need to dynamically represent differences between EcM and AM fungi in the amount of C they receive from plants, how they further allocate this C, and the nutrient benefit they confer to plants. Global models do not need to reinvent the wheel but could use the strengths of existing model structures highlighted here to inform new model subroutines. Whereas there are challenges to integrating mycorrhizal fungi into TBMs, empirical evidence that AM and EcM plants respond differently to global change highlight the need for mycorrhizal integration at global scales.

References

Anderegg, W.R.L., Schwalm, C., Biondi, F., Camarero, J.J., Koch, G., Litvak, M., Ogle, K., Shaw, J.D., Shevliakova, E., Williams, A.P., Wolf, A., Ziaco, E., Pacala, S., 2015. Pervasive drought legacies in forest ecosystems and their implications for carbon cycle models. Science 349 (6247), 528–532.

Arora, V.K., Boer, G.J., Friedlingstein, P., Eby, M., Jones, C.D., Christian, J.R., Bonan, G., Bopp, L., Brovkin, V., Cadule, P., 2013. Carbon–concentration and carbon–climate feedbacks in CMIP5 Earth system models. Journal of Climate 26 (15), 5289–5314.

Averill, C., Turner, B.L., Finzi, A.C., 2014. Mycorrhiza-mediated competition between plants and decomposers drives soil carbon storage. Nature 505 (7484), 543.

Bever, J.D., 2015. Preferential allocation, physio-evolutionary feedbacks, and the stability and environmental patterns of mutualism between plants and their root symbionts. New Phytologist 205 (4), 1503–1514.

Bolan, N.S., 1991. A critical-review on the role of mycorrhizal fungi in the uptake of phosphorus by plants. Plant and Soil 134 (2), 189–207.

Brzostek, E.R., Finzi, A.C., 2011. Substrate supply, fine roots, and temperature control proteolytic enzyme activity in temperate forest soils. Ecology 92 (4), 892–902.

Brzostek, E.R., Fisher, J.B., Phillips, R.P., 2014. Modeling the carbon cost of plant nitrogen acquisition: mycorrhizal trade-offs and multipath resistance uptake improve predictions of retranslocation. Journal of Geophysical Research-Biogeosciences 119 (8), 1684–1697.

Brzostek, E.R., Greco, A., Drake, J.E., Finzi, A.C., 2013. Root carbon inputs to the rhizosphere stimulate extracellular enzyme activity and increase nitrogen availability in temperate forest soils. Biogeochemistry 115 (1–3), 65–76.

Buscot, F., Munch, J.C., Charcosset, J.Y., Gardes, M., Nehls, U., Hampp, R., 2000. Recent advances in exploring physiology and biodiversity of ectomycorrhizas highlight the functioning of these symbioses in ecosystems. FEMS Microbiology Reviews 24 (5), 601–614.

Chalot, M., Brun, A., 1998. Physiology of organic nitrogen acquisition by ectomycorrhizal fungi and ectomycorrhizas. FEMS Microbiology Reviews 22 (1), 21–44.

Chapman, S.K., Langley, J.A., Hart, S.C., Koch, G.W., 2006. Plants actively control nitrogen cycling: uncorking the microbial bottleneck. New Phytologist 169 (1), 27–34.

Chen, W., Koide, R.T., Adams, T.S., DeForest, J.L., Cheng, L., Eissenstat, D.M., 2016. Root morphology and mycorrhizal symbioses together shape nutrient foraging strategies of temperate trees. Proceedings of the National Academy of Sciences. 201601006.

Cheng, W.X., Parton, W.J., Gonzalez-Meler, M.A., Phillips, R., Asao, S., McNickle, G.G., Brzostek, E., Jastrow, J.D., 2014. Synthesis and modeling perspectives of rhizosphere priming. New Phytologist 201 (1), 31–44.

Clemmensen, K.E., Bahr, A., Ovaskainen, O., Dahlberg, A., Ekblad, A., Wallander, H., Stenlid, J., Finlay, R.D., Wardle, D.A., Lindahl, B.D., 2013. Roots and associated fungi drive long-term carbon sequestration in boreal forest. Science 339 (6127), 1615–1618.

Cleveland, C.C., Houlton, B.Z., Smith, W.K., Marklein, A.R., Reed, S.C., Parton, W., Del Grosso, S.J., Running, S.W., 2013. Patterns of new versus recycled primary production in the terrestrial biosphere. Proceedings of the National Academy of Sciences of the United States of America 110 (31), 12733–12737.

Corrales, A., Mangan, S.A., Turner, B.L., Dalling, J.W., 2016. An ectomycorrhizal nitrogen economy facilitates monodominance in a neotropical forest. Ecology Letters 19 (4), 383–392.

Correa, A., Gurevitch, J., Martins-Loucao, M.A., Cruz, C., 2012. C allocation to the fungus is not a cost to the plant in ectomycorrhizae. Oikos 121 (3), 449–463.

Correa, A., Hampp, R., Magel, E., Martins-Loucao, M.A., 2011. Carbon allocation in ectomycorrhizal plants at limited and optimal N supply: an attempt at unraveling conflicting theories. Mycorrhiza 21 (1), 35–51.

Deckmyn, G., Campioli, M., Muys, B., Kraigher, H., 2011. Simulating C cycles in forest soils: including the active role of micro-organisms in the ANAFORE forest model. Ecological Modelling 222 (12), 1972–1985.

Drake, J.E., Gallet-Budynek, A., Hofmockel, K.S., Bernhardt, E.S., Billings, S.A., Jackson, R.B., Johnsen, K.S., Lichter, J., McCarthy, H.R., McCormack, M.L., Moore, D.J.P., Oren, R., Palmroth, S., Phillips, R.P., Pippen, J.S., Pritchard, S.G., Treseder, K.K., Schlesinger, W.H., DeLucia, E.H., Finzi, A.C., 2011. Increases in the flux of carbon belowground stimulate nitrogen uptake and sustain the long-term enhancement of forest productivity under elevated CO(2). Ecology Letters 14 (4), 349–357.

Dybzinski, R., Farrior, C.E., Pacala, S.W., 2015. Increased forest carbon storage with increased atmospheric CO_2 despite nitrogen limitation: a game-theoretic allocation model for trees in competition for nitrogen and light. Global Change Biology 21 (3), 1182–1196.

Ekblad, A., Wallander, H., Godbold, D.L., Cruz, C., Johnson, D., Baldrian, P., Bjork, R.G., Epron, D., Kieliszewska-Rokicka, B., Kjoller, R., Kraigher, H., Matzner, E., Neumann, J., Plassard, C., 2013. The production and turnover of extramatrical mycelium of ectomycorrhizal fungi in forest soils: role in carbon cycling. Plant and Soil 366 (1–2), 1–27.

Fernandez, C.W., Kennedy, P.G., 2016. Revisiting the 'Gadgil effect': do interguild fungal interactions control carbon cycling in forest soils? New Phytologist 209 (4), 1382–1394.

Fisher, J.B., Huntzinger, D.N., Schwalm, C.R., Sitch, S., 2014. Modeling the terrestrial biosphere. Annual Review of Environment and Resources 39, 91–123.

Fisher, J.B., Sitch, S., Malhi, Y., Fisher, R.A., Huntingford, C., Tan, S.-Y., 2010. Carbon cost of plant nitrogen acquisition: a mechanistic, globally-applicable model of plant nitrogen uptake and fixation. Global Biogeochemical Cycles 24 (GB1014). http://dx.doi.org/10.1029/2009GB003621.

Fisher, J.B., Sweeney, S., Brzostek, E.R., Evans, T.P., Johnson, D.J., Myers, J.A., Bourg, N.A., Wolf, A.T., Howe, R.W., Phillips, R.P., 2016. Tree–mycorrhizal associations detected remotely from canopy spectral properties. Global Change Biology 22 (7), 2596–2607.

Fitter, A.H., 1991. Costs and benefits of mycorrhizas - implications for functioning under natural conditions. Experientia 47 (4), 350–355.

Fransson, P.M.A., Johansson, E.M., 2010. Elevated CO_2 and nitrogen influence exudation of soluble organic compounds by ectomycorrhizal root systems. FEMS Microbiology Ecology 71 (2), 186–196.

Franklin, O., Johansson, J., Dewar, R.C., Dieckmann, U., McMurtrie, R.E., Brannstrom, A., Dybzinski, R., 2012. Modeling carbon allocation in trees: a search for principles. Tree Physiology 32 (6), 648–666.

Franklin, O., Nasholm, T., Hogberg, P., Hogberg, M.N., 2014. Forests trapped in nitrogen limitation - an ecological market perspective on ectomycorrhizal symbiosis. New Phytologist 203 (2), 657–666.

Friedlingstein, P., Meinshausen, M., Arora, V.K., Jones, C.D., Anav, A., Liddicoat, S.K., Knutti, R., 2014. Uncertainties in CMIP5 climate projections due to carbon cycle feedbacks. Journal of Climate 27 (2), 511–526.

Gashaw Deressa, T., Schenk, M.K., 2008. Contribution of roots and hyphae to phosphorus uptake of mycorrhizal onion (*Allium cepa* L.)—a mechanistic modeling approach. Journal of Plant Nutrition and Soil Science 171 (5), 810–820.

Hobbie, E.A., 2006. Carbon allocation to ectomycorrhizal fungi correlates with belowground allocation in culture studies. Ecology 87 (3), 563–569.

Hobbie, E.A., Hobbie, J.E., 2008. Natural abundance of (15)N in nitrogen-limited forests and tundra can estimate nitrogen cycling through mycorrhizal fungi: a review. Ecosystems 11 (5), 815–830.

Högberg, M.N., Högberg, P., 2002. Extramatrical ectomycorrhizal mycelium contributes one-third of microbial biomass and produces, together with associated roots, half the dissolved organic carbon in a forest soil. New Phytologist 154 (3), 791–795.

Högberg, P., Högberg, M.N., Göttlicher, S.G., Betson, N.R., Keel, S.G., Metcalfe, D.B., Campbell, C., Schindlbacher, A., Hurry, V., Lundmark, T., 2008. High temporal resolution tracing of photosynthate carbon from the tree canopy to forest soil microorganisms. New Phytologist 177 (1), 220–228.

Jackson, R.B., Banner, J.L., Jobbágy, E.G., Pockman, W.T., Wall, D.H., 2002. Ecosystem carbon loss with woody plant invasion of grasslands. Nature 418 (6898), 623–626.

Johnson, N.C., 2010. Resource stoichiometry elucidates the structure and function of arbuscular mycorrhizas across scales. New Phytologist 185 (3), 631–647.

Johnson, N.C., Angelard, C., Sanders, I.R., Kiers, E.T., 2013. Predicting community and ecosystem outcomes of mycorrhizal responses to global change. Ecology Letters 16, 140–153.

Johnson, N.C., Graham, J.H., Smith, F.A., 1997. Functioning of mycorrhizal associations along the mutualism-parasitism continuum. New Phytologist 135 (4), 575–586.

Johnson, N.C., Hoeksema, J.D., Bever, J.D., Chaudhary, V.B., Gehring, C., Klironomos, J., Koide, R., Miller, R.M., Moore, J., Moutoglis, P., 2006. From Lilliput to Brobdingnag: extending models of mycorrhizal function across scales. Bioscience 56 (11), 889–900.

Kaiser, C., Kilburn, M.R., Clode, P.L., Fuchslueger, L., Koranda, M., Cliff, J.B., Solaiman, Z.M., Murphy, D.V., 2015. Exploring the transfer of recent plant photosynthates to soil microbes: mycorrhizal pathway vs. direct root exudation. New Phytologist 205 (4), 1537–1551.

Kiers, E.T., Denison, R.F., 2008. Sanctions, Cooperation, and the Stability of Plant-Rhizosphere Mutualisms. Annual Review of Ecology, Evolution, and Systematics. Annual Reviews, Palo Alto, pp. 215–236.

Klironomos, J.N., 2003. Variation in plant response to native and exotic arbuscular mycorrhizal fungi. Ecology 84 (9), 2292–2301.

Koide, R.T., Kabir, Z., 2000. Extraradical hyphae of the mycorrhizal fungus *Glomus intraradices* can hydrolyse organic phosphate. New Phytologist 148 (3), 511–517.

Landis, F.C., Fraser, L.H., 2008. A new model of carbon and phosphorus transfers in arbuscular mycorrhizas. New Phytologist 177 (2), 466–479.

Lilleskov, E.A., Fahey, T.J., Horton, T.R., Lovett, G.M., 2002. Belowground ectomycorrhizal fungal community change over a nitrogen deposition gradient in Alaska. Ecology 83 (1), 104–115.

Lilleskov, E.A., Fahey, T.J., Lovett, G.M., 2001. Ectomycorrhizal fungal aboveground community change over an atmospheric nitrogen deposition gradient. Ecological Applications 11 (2), 397–410.

Lindahl, B.D., Ihrmark, K., Boberg, J., Trumbore, S.E., Högberg, P., Stenlid, J., Finlay, R.D., 2007. Spatial separation of litter decomposition and mycorrhizal nitrogen uptake in a boreal forest. New Phytologist 173 (3), 611–620.
Lindahl, B.D., Tunlid, A., 2015. Ectomycorrhizal fungi - potential organic matter decomposers, yet not saprotrophs. New Phytologist 205 (4), 1443–1447.
Litton, C.M., Raich, J.W., Ryan, M.G., 2007. Carbon allocation in forest ecosystems. Global Change Biology 13 (10), 2089–2109.
Lloyd, J., Taylor, J.A., 1994. On the temperature dependence of soil respiration. Functional Ecology 8 (3), 315–323.
Marschner, H., Dell, B., 1994. Nutrient-uptake in mycorrhizal symbiosis. Plant and Soil 159 (1), 89–102.
McCormack, M.L., Dickie, I.A., Eissenstat, D.M., Fahey, T.J., Fernandez, C.W., Guo, D., Helmisaari, H.S., Hobbie, E.A., Iversen, C.M., Jackson, R.B., 2015. Redefining fine roots improves understanding of below-ground contributions to terrestrial biosphere processes. New Phytologist 207 (3), 505–518.
Medvigy, D., Wofsy, S.C., Munger, J.W., Hollinger, D.Y., Moorcroft, P.R., 2009. Mechanistic scaling of ecosystem function and dynamics in space and time: Ecosystem Demography model version 2. Journal of Geophysical Research-Biogeosciences 114 (G1).
Meyer, A., Grote, R., Polle, A., Butterbach-Bahl, K., 2010. Simulating mycorrhiza contribution to forest C- and N cycling-the MYCOFON model. Plant and Soil 327 (1–2), 493–517.
Midgley, M.G., Brzostek, E., Phillips, R.P., 2015. Decay rates of leaf litters from arbuscular mycorrhizal trees are more sensitive to soil effects than litters from ectomycorrhizal trees. Journal of Ecology 103 (6), 1454–1463.
Mohan, J.E., Cowden, C.C., Baas, P., Dawadi, A., Frankson, P.T., Helmick, K., Hughes, E., Khan, S., Lang, A., Machmuller, M., 2014. Mycorrhizal fungi mediation of terrestrial ecosystem responses to global change: mini-review. Fungal Ecology 10, 3–19.
Näsholm, T., Kielland, K., Ganeteg, U., 2009. Uptake of organic nitrogen by plants. New Phytologist 182 (1), 31–48.
Norby, R.J., Zak, D.R., 2011. Ecological lessons from free-air CO_2 enrichment (FACE) experiments. Annual Review of Ecology, Evolution, and Systematics 42 (1), 181.
Nowacki, G.J., Abrams, M.D., 2008. The demise of fire and "Mesophication" of forests in the eastern United States. Bioscience 58 (2), 123–138.
Oleson, K., Lawrence, D.M., Bonan, G.B., Drewniak, B., Huang, M., Koven, C.D., Levis, S., Li, F., Riley, W.J., Subin, Z.M., Swenson, S., Thornton, P.E., Bozbiyik, A., Fisher, R., Heald, C.L., Kluzek, E., Lamarque, J.-F., Lawrence, P.J., Leung, L.R., Lipscomb, W., Muszala, S.P., Ricciuto, D.M., Sacks, W.J., Sun, Y., Tang, J., Yang, Z.-L., 2013. Technical Description of Version 4.5 of the Community Land Model (CLM). NCAR Technical Note NCAR/TN-503+STR: 420 pp.
Orwin, K.H., Kirschbaum, M.U.F., St John, M.G., Dickie, I.A., 2011. Organic nutrient uptake by mycorrhizal fungi enhances ecosystem carbon storage: a model-based assessment. Ecology Letters 14 (5), 493–502.
Peay, K.G., Kennedy, P.G., Bruns, T.D., 2008. Fungal community ecology: a hybrid beast with a molecular master. Bioscience 58 (9), 799–810.
Phillips, R.P., Brzostek, E., Midgley, M.G., 2013. The mycorrhizal-associated nutrient economy: a new framework for predicting carbon-nutrient couplings in temperate forests. New Phytologist 199 (1), 41–51.
Phillips, R.P., Meier, I.C., Bernhardt, E.S., Grandy, A.S., Wickings, K., Finzi, A.C., 2012. Roots and fungi accelerate carbon and nitrogen cycling in forests exposed to elevated CO_2. Ecology Letters 15 (9), 1042–1049.
Pozo, M.J., Azcon-Aguilar, C., 2007. Unraveling mycorrhiza-induced resistance. Current Opinion in Plant Biology 10 (4), 393–398.
Rastetter, E.B., Shaver, G.R., 1992. A model of multiple-element limitation for acclimating vegetation. Ecology 73 (4), 1157–1174.
Read, D.J., Perez-Moreno, J., 2003. Mycorrhizas and nutrient cycling in ecosystems - a journey towards relevance? New Phytologist 157 (3), 475–492.
Richardson, D.M., Williams, P.A., Hobbs, R.J., 1994. Pine invasions in the Southern Hemisphere: determinants of spread and invadability. Journal of Biogeography 511–527.
Rillig, M.C., 2004a. Arbuscular mycorrhizae and terrestrial ecosystem processes. Ecology Letters 7 (8), 740–754.
Rillig, M.C., 2004b. Arbuscular mycorrhizae, glomalin, and soil aggregation. Canadian Journal of Soil Science 84 (4), 355–363.
Rillig, M.C., Aguilar-Trigueros, C.A., Bergmann, J., Verbruggen, E., Veresoglou, S.D., Lehmann, A., 2015. Plant root and mycorrhizal fungal traits for understanding soil aggregation. New Phytologist 205 (4), 1385–1388.
Rillig, M.C., Mummey, D.L., 2006. Mycorrhizas and soil structure. New Phytologist 171 (1), 41–53.
Scheiter, S., Langan, L., Higgins, S.I., 2013. Next-generation dynamic global vegetation models: learning from community ecology. New Phytologist 198 (3), 957–969.

Schimel, J.P., Bennett, J., 2004. Nitrogen mineralization: challenges of a changing paradigm. Ecology 85 (3), 591–602.

Schnepf, A., Roose, T., 2006. Modelling the contribution of arbuscular mycorrhizal fungi to plant phosphate uptake. New Phytologist 171 (3), 669–682.

Schnepf, A., Roose, T., Schweiger, P., 2008. Impact of growth and uptake patterns of arbuscular mycorrhizal fungi on plant phosphorus uptake—a modelling study. Plant and Soil 312 (1–2), 85–99.

Shi, M., Fisher, J.B., Brzostek, E.R., Phillips, R.P., 2016. Carbon cost of plant nitrogen acquisition: global carbon cycle impact from an improved plant nitrogen cycle in the Community Land Model. Global Change Biology 22 (3), 1299–1314.

Sinsabaugh, R.L., Turner, B.L., Talbot, J.M., Waring, B.G., Powers, J.S., Kuske, C.R., Moorhead, D.L., Follstad Shah, J.J., 2016. Stoichiometry of microbial carbon use efficiency in soils. Ecological Monographs 86 (2), 172–189.

Smithwick, E.A.H., Lucash, M.S., McCormack, M.L., Sivandran, G., 2014. Improving the representation of roots in terrestrial models. Ecological Modelling 291, 193–204.

Soudzilovskaia, N.A., Douma, J.C., Akhmetzhanova, A.A., Bodegom, P.M., Cornwell, W.K., Moens, E.J., Treseder, K.K., Tibbett, M., Wang, Y.P., Cornelissen, J.H.C., 2015a. Global patterns of plant root colonization intensity by mycorrhizal fungi explained by climate and soil chemistry. Global Ecology and Biogeography 24 (3), 371–382.

Soudzilovskaia, N.A., Heijden, M.G.A., Cornelissen, J.H.C., Makarov, M.I., Onipchenko, V.G., Maslov, M.N., Akhmetzhanova, A.A., Bodegom, P.M., 2015b. Quantitative assessment of the differential impacts of arbuscular and ectomycorrhiza on soil carbon cycling. New Phytologist 208 (1), 280–293.

Sulman, B.N., Phillips, R.P., Oishi, A.C., Shevliakova, E., Pacala, S.W., 2014. Microbe-driven turnover offsets mineral-mediated storage of soil carbon under elevated CO_2. Nature Climate Change 4 (12), 1099–1102.

Talbot, J.M., Allison, S.D., Treseder, K.K., 2008. Decomposers in disguise: mycorrhizal fungi as regulators of soil C dynamics in ecosystems under global change. Functional Ecology 22 (6), 955–963.

Tedersoo, L., Bahram, M., Polme, S., Koljalg, U., Yorou, N.S., Wijesundera, R., Villarreal Ruiz, L., Vasco-Palacios, A.M., Pham Quang, T., Suija, A., Smith, M.E., Sharp, C., Saluveer, E., Saitta, A., Rosas, M., Riit, T., Ratkowsky, D., Pritsch, K., Poldmaa, K., Piepenbring, M., Phosri, C., Peterson, M., Parts, K., Paertel, K., Otsing, E., Nouhra, E., Njouonkou, A.L., Nilsson, R.H., Morgado, L.N., Mayor, J., May, T.W., Majuakim, L., Lodge, D.J., Lee, S.S., Larsson, K.-H., Kohout, P., Hosaka, K., Hiiesalu, I., Henkel, T.W., Harend, H., Guo, L.-D., Greslebin, A., Grelet, G., Geml, J., Gates, G., Dunstan, W., Dunk, C., Drenkhan, R., Dearnaley, J., De Kesel, A., Tan, D., Chen, X., Buegger, F., Brearley, F.Q., Bonito, G., Anslan, S., Abell, S., Abarenkov, K., 2014. Global diversity and geography of soil fungi. Science 346 (6213), 1256688.

Terrer, C., Vicca, S., Hungate, B.A., Phillips, R.P., Prentice, I.C., 2016. Mycorrhizal association as a primary control of the CO2 fertilization effect. Science 353 (6294), 72–74.

Thomas, R.Q., Brookshire, E.N.J., Gerber, S., 2015. Nitrogen limitation on land: how can it occur in Earth system models? Global Change Biology 21 (5), 1777–1793.

Treseder, K.K., Allen, M.F., 2000. Mycorrhizal fungi have a potential role in soil carbon storage under elevated CO_2 and nitrogen deposition. New Phytologist 147 (1), 189–200.

Treseder, K.K., Balser, T.C., Bradford, M.A., Brodie, E.L., Dubinsky, E.A., Eviner, V.T., Hofmockel, K.S., Lennon, J.T., Levine, U.Y., MacGregor, B.J., Pett-Ridge, J., Waldrop, M.P., 2012. Integrating microbial ecology into ecosystem models: challenges and priorities. Biogeochemistry 109 (1–3), 7–18.

Trivedi, P., Anderson, I.C., Singh, B.K., 2013. Microbial modulators of soil carbon storage: integrating genomic and metabolic knowledge for global prediction. Trends in Microbiology 21 (12), 641–651.

van der Heijden, M.G.A., Martin, F.M., Selosse, M.-A., Sanders, I.R., 2015. Mycorrhizal ecology and evolution: the past, the present, and the future. New Phytologist 205 (4), 1406–1423.

van Scholl, L., Kuyper, T.W., Smits, M.M., Landeweert, R., Hoffland, E., van Breemen, N., 2008. Rock-eating mycorrhizas: their role in plant nutrition and biogeochemical cycles. Plant and Soil 303 (1–2), 35–47.

Verheijen, L.M., Brovkin, V., Aerts, R., Bonisch, G., Cornelissen, J.H.C., Kattge, J., Reich, P.B., Wright, I.J., van Bodegom, P.M., 2013. Impacts of trait variation through observed trait-climate relationships on performance of an Earth system model: a conceptual analysis. Biogeosciences 10 (8), 5497–5515.

Walder, F., van der Heijden, M.G.A., 2015. Regulation of resource exchange in the arbuscular mycorrhizal symbiosis. Nature Plants 1 (11), 15159.

Walker, A.P., Zaehle, S., Medlyn, B.E., De Kauwe, M.G., Asao, S., Hickler, T., Parton, W., Ricciuto, D.M., Wang, Y.-P., Warlind, D., Norby, R.J., 2015. Predicting long-term carbon sequestration in response to CO_2 enrichment: how and why do current ecosystem models differ? Global Biogeochemical Cycles 29 (4), 476–495.

Wallander, H., Ekblad, A., Godbold, D.L., Johnson, D., Bahr, A., Baldrian, P., Björk, R.G., Kieliszewska-Rokicka, B., Kjøller, R., Kraigher, H., 2013. Evaluation of methods to estimate production, biomass and turnover of ectomycorrhizal mycelium in forests soils–a review. Soil Biology and Biochemistry 57, 1034–1047.

Wang, Y.P., Law, R.M., Pak, B., 2010. A global model of carbon, nitrogen and phosphorus cycles for the terrestrial biosphere. Biogeosciences 7 (7), 2261–2282.

Warren, J.M., Hanson, P.J., Iversen, C.M., Kumar, J., Walker, A.P., Wullschleger, S.D., 2015. Root structural and functional dynamics in terrestrial biosphere models - evaluation and recommendations. New Phytologist 205 (1), 59–78.

Waring, B.G., Adams, R., Branco, S., Powers, J.S., 2016. Scale-dependent variation in nitrogen cycling and soil fungal communities along gradients of forest composition and age in regenerating tropical dry forests. New Phytologist 209 (2), 845–854.

Wieder, W.R., Bonan, G.B., Allison, S.D., 2013. Global soil carbon projections are improved by modelling microbial processes. Nature Climate Change 3 (10), 909–912.

Yang, X., Thornton, P.E., Ricciuto, D.M., Post, W.M., 2014. The role of phosphorus dynamics in tropical forests - a modeling study using CLM-CNP. Biogeosciences 11 (6), 1667–1681.

Yin, H., Wheeler, E., Phillips, R.P., 2014. Root-induced changes in nutrient cycling in forests depend on exudation rates. Soil Biology and Biochemistry 78, 213–221.

Zaehle, S., Medlyn, B.E., De Kauwe, M.G., Walker, A.P., Dietze, M.C., Hickler, T., Luo, Y., Wang, Y.-P., El-Masri, B., Thornton, P., Jain, A., Wang, S., Warlind, D., Weng, E., Parton, W., Iversen, C.M., Gallet-Budynek, A., McCarthy, H., Finzi, A.C., Hanson, P.J., Prentice, I.C., Oren, R., Norby, R.J., 2014. Evaluation of 11 terrestrial carbon-nitrogen cycle models against observations from two temperate Free-Air CO_2 Enrichment studies. New Phytologist 202 (3), 803–822.

Zhu, Y.G., Miller, R.M., 2003. Carbon cycling by arbuscular mycorrhizal fungi in soil-plant systems. Trends in Plant Science 8 (9), 407–409.

Index

'*Note*: Page numbers followed by "f" indicate figures, "t" indicate tables, and "b" indicate boxes.'

N

T

U

W

Z